Topics in Mathematics

VECTOR ANALYSIS AND GEOMETRY

B.A./B.Sc. Ist Year

Topics in Mathematics

VECTOR ANALYSIS AND GEOMETRY

By

Dr. KULBHUSHAN PARKASH
Head, Deptt. of Mathematics
D.A.V. College (Lahore)
Ambala City

OM P. CHUG
Formerly, Head, Deptt. of Mathematics
Hindu College,
Sonepat

R.S. DAHIYA
Formerly, Head, Deptt. of Mathematics
A.I.J.H.M. College
Rohtak

UNIVERSITY SCIENCE PRESS

(An Imprint of Laxmi Publications Pvt. Ltd.)

BANGALORE • CHENNAI • COCHIN • GUWAHATI • HYDERABAD
JALANDHAR • KOLKATA • LUCKNOW • MUMBAI • RANCHI
NEW DELHI • BOSTON, USA

Published by :
UNIVERSITY SCIENCE PRESS
(An Imprint of Laxmi Publications Pvt. Ltd.)
113, Golden House, Daryaganj,
New Delhi-110002

Phone : 011-43 53 25 00
Fax : 011-43 53 25 28

www.laxmipublications.com
info@laxmipublications.com

Price : ₹ **375.00** *Only.*

Fourth Edition : 2008
Reprint : 2009, 2013

OFFICES

© **Bangalore**	080-26 75 69 30	© **Chennai**	044-24 34 47 26
© **Cochin**	0484-237 70 04, 405 13 03	© **Guwahati**	0361-254 36 69, 251 38 81
© **Hyderabad**	040-24 65 23 33	© **Jalandhar**	0181-222 12 72
© **Kolkata**	033-22 27 43 84	© **Lucknow**	0522-220 99 16
© **Mumbai**	022-24 91 54 15, 24 92 78 69	© **Ranchi**	0651-220 44 64

UTM-9721-375-TOPICS MATH-I P III-PAR

C—7316/013/09

Typeset at : Goswami Associates, Delhi.

Printed at : Mehra Offset Press, New Delhi.

PREFACE

The present book of 'Vector Analysis and Geometry' has been written as a textbook according to the latest syllabus in Mathematics issued by the U.G.C. for B.A./B.Sc. I (Paper III) of various universities and in particular by K.U., Kurukshetra and M.D.U., Rohtak. The book has been thoroughly revised keeping in view :

(*i*) the various invaluable suggestions received from teachers and students.

(*ii*) the recent question papers issued by K.U., Kurukshetra and M.D.U., Rohtak.

The text of the book has been prepared with the following salient features :

(*i*) The language of the book is simple and easy to understand.

(*ii*) Each topic has been presented in a **systematic, simple, lucid** and **exhaustive manner**.

(*iii*) A large number of important solved examples properly selected from the previous university question papers have been provided to enable the student to have a clear grasp of the subject and to equip him for attempting problems in the university examination without any difficulty.

(*iv*) A large number of 'notes' and 'remarks' have been added for a better understanding of the subject.

No effort has been left out to keep the book free from errors and mistakes but even then some errors might have crept in.

In fact no pains have been shared to make the book interesting and useful.

Suggestions for further improvement of the book will be welcomed.

—AUTHORS

SYLLABUS

BM 103—VECTOR ANALYSIS AND GEOMETRY

(For Maharishi Dayanand University, Rohtak)

	Maximum Marks	*Time*
B.A.	: 30	3 hours
B.Sc.	: 45	3 hours

Section I (2 Questions)

Scalar and vector product of three vectors. Product of four vectors. Reciprocal Vectors. Vector differentiation. Gradient, Divergence and Curl.

Section II (2 Questions)

Vector integration. Theorems of Gauss, Green, Stokes and problems based on these.

Section III (3 Questions)

General equation of second degree. Tracing of conics. System of conics. Confocal conics. Polar equation of a conic.

Section IV (3 Questions)

Central conicoids. Paraboloids. Plane Sections of Conicoids. Generating lines. Confocal Conicoids. Reduction of Second degree equations. Sphere, Cone, Cylinder.

Note. The examiner is requested to set **ten** questions in all selecting questions sectionwise as indicated in the syllabus. The candidate is required to attempt **five** questions selecting at least **one** question from each section.

SYLLABUS

BM 103—VECTOR ANALYSIS AND GEOMETRY

(For Kurukshetra University)

	Maximum Marks	*Time*
B.A.	: 30	3 hours
B.Sc.	: 45	3 hours

Section I (2 Questions)

Scalar and vector product of three vectors. Product of four vectors. Reciprocal vectors. Vector differentiation. Gradient, Divergence and Curl.

Section II (2 Questions)

Vector integration. Theorems of Gauss, Green, Stokes and problems based on these.

Section III (3 Questions)

General equation of second degree. Tracing of conics. System of conics. Confocal conics. Polar equation of a conic. Sphere, Cone, Cylinder.

Section IV (3 Questions)

Central conicoids. Paraboloids. Plane Sections of Conicoids. Generating lines. Confocal Conicoids. Reduction of Second degree equations.

Note. The examiner is requested to set **ten** questions in all selecting questions sectionwise as indicated in the syllabus. The candidate is required to attempt **five** questions selecting at least **one** question from each section.

CONTENTS

*Reviews III, IV and chapters 10, 11, 12 are in section III in K.U., Kurukshetra but are in section IV in M.D.U., Rohtak

PART A : VECTOR ANALYSIS

SECTION I

1

Algebra of Vectors

1. Scalars and Vectors

A scalar is a quantity that is characterized solely by magnitude such as mass, time or temperature.

The value of a scalar is an ordinary number.

A vector is a quantity that is characterized by both magnitude and direction, such as displacement velocity or force.

A vector is represented by a letter with an arrow over it, $\vec{a}$ and its magnitude is denoted by $|\vec{a}|$ or a. In printed books as in this book, a vector is represented by bold face type, such as **a**.

Geometrically, a vector **a** is represented by a directed line segment $\vec{PQ}$. The vector **a** has a direction from P to Q. The point P is called the initial point and the point Q is called the terminal point of **a**. The length $|\vec{PQ}|$ of the line segment is the magnitude of **a**.

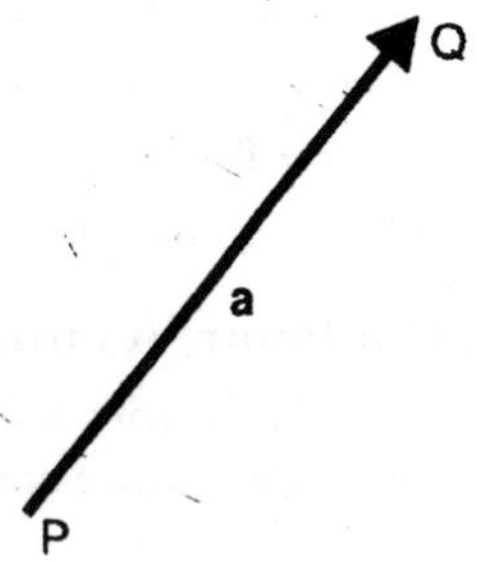

When the initial point of a vector is fixed, it is called a *bound* or *localized* or *fixed* vector, otherwise it is called a *free vector*.

2. Equality of Vectors

Two (free) vectors **a** *and* **b** *are equal and we write* **a** = **b** *if they have the same magnitude and direction.* It does not mean that two equal vectors coincide in space. This definition also does not apply to fixed vectors because only one vector can have a given magnitude, direction and initial point.

3. Multiplication of Vectors by Scalars

The product of a vector **a** *by a scalar t is a vector t***a** *with magnitude* $|t|$ *times the magnitude of* **a** *and with direction the same as or opposite to that of* **a** *according as* $t > 0$ *or* $t < 0$. *If* $t = 0$, *t***a** *is a null vector (zero vector)* **0**.

Since 0**a** = **0**, for any vector **a**, therefore the null vector has zero magnitude but any direction.

Since (– 1) **a** = – **a**, so a negative vector – **a** is a vector whose magnitude is that of **a** but whose direction is opposite to that of **a**.

A vector having unit magnitude is called a unit vector. If $\mathbf{a} \neq \mathbf{0}$, then $\dfrac{\mathbf{a}}{|\mathbf{a}|} = \dfrac{\mathbf{a}}{a}$ is a unit vector having the same direction as that of **a**. It is also denoted by $\hat{a}$.

Any vector **a** can be represented by a unit vector $\hat{a}$ multiplied by the magnitude of **a** ; *i.e.*, $\mathbf{a} = a\ \hat{a}$. *Two vectors* **a** *and* **b** *are parallel if and only if* $\mathbf{a} = m\mathbf{b}$ for some scalar m.

4. Scalar Multiplication Satisfies the following properties :

1. $|m\mathbf{a}| = |m|\,|\mathbf{a}|$
2. $m(n\mathbf{a}) = (mn)\mathbf{a}$
3. $n(m\mathbf{a}) = m(n\mathbf{a})$.

5. Addition and Subtraction of Vectors

The sum or resultant of two vectors **a** and **b** is a vector **c** represented by $\mathbf{c} = \mathbf{a} + \mathbf{b}$ and can be determined geometrically as follows :

If the initial point of **b** is placed at the terminal point of **a**, the resultant **c** is the vector whose initial point is at the initial point of **a** and whose terminal point is at the terminal point of **b**.

Given two vectors **a** and **b**, the difference $\mathbf{a} - \mathbf{b}$ is given by $\mathbf{a} + (-\mathbf{b})$.

6. Vector Addition Satisfies the following properties :

1. $\mathbf{a} + \mathbf{b} = \mathbf{b} + \mathbf{a}$ (Commutative law)
2. $(\mathbf{a} + \mathbf{b}) + \mathbf{c} = \mathbf{a} + (\mathbf{b} + \mathbf{c})$ (Associative law)
3. $m(\mathbf{a} + \mathbf{b}) = m\mathbf{a} + m\mathbf{b}$ (Distributive law)
4. $(m + n)\mathbf{a} = m\mathbf{a} + n\mathbf{a}$ (Scalar distributive law)
5. $\mathbf{a} + \mathbf{0} = \mathbf{a}$
6. $\mathbf{a} + (-\mathbf{a}) = \mathbf{0}$.

7. Coplanar Vectors

The vectors which are in one plane or parallel to the same plane are called the coplanar vectors. The plane containing these vectors or plane parallel to these vectors is called the plane of vectors.

A vector **c** *is coplanar with the vectors* **a** *and* **b** *if there exist scalars m and n such that*

$$\mathbf{c} = m\mathbf{a} + n\mathbf{b}.$$

8. Vectors in the Rectangular Coordinate System

Any vector **a** in 3 dimensional space, can be represented with initial point at the origin O of a rectangular coordinate system. If (a_1, a_2, a_3) are the coordinates of the terminal point of the vector **a** with initial point at O, then the vector **a** can be represented by $\mathbf{a} = a_1\mathbf{i} + a_2\mathbf{j} + a_3\mathbf{k}$ or $\mathbf{a} = (a_1, a_2, a_3)$. The vectors $a_1\mathbf{i}$, $a_2\mathbf{j}$ and $a_3\mathbf{k}$ are called the *rectangular component vectors* or simply *component vectors* of **a** in the *x*, *y* and *z* directions respectively.

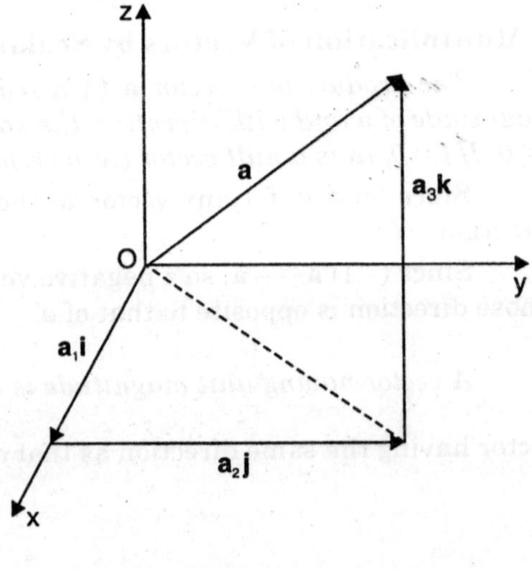

The magnitude of **a** is

$$a = |\,a\,| = \sqrt{a_1^2 + a_2^2 + a_3^2}$$

The null vector **0** is a vector whose components are all zero *i.e.*, $\mathbf{0} = (0, 0, 0)$.

Note. The same vector will have different components in different coordinate systems. But, we shall consider only the standard rectangular coordinate system.

If $\mathbf{a} = (a_1, a_2, a_3)$, $\mathbf{b} = (b_1, b_2, b_3)$, then we have

1. $\mathbf{a} = \mathbf{b}$ if and only if $a_1 = b_1$, $a_2 = b_2$, $a_3 = b_3$.

2. $m\mathbf{a} = (ma_1, ma_2, ma_3)$ for any scalar m.

3. $-\mathbf{a} = (-a_1, -a_2, -a_3)$.

4. $0 = 0\mathbf{a} = (0, 0, 0)$.

5. $\mathbf{a} + \mathbf{b} = (a_1 + b_1, a_2 + b_2, a_3 + b_3)$

6. $\mathbf{a} - \mathbf{b} = (a_1 - b_1, a_2 - b_2, a_3 - b_3)$.

9. Position Vector

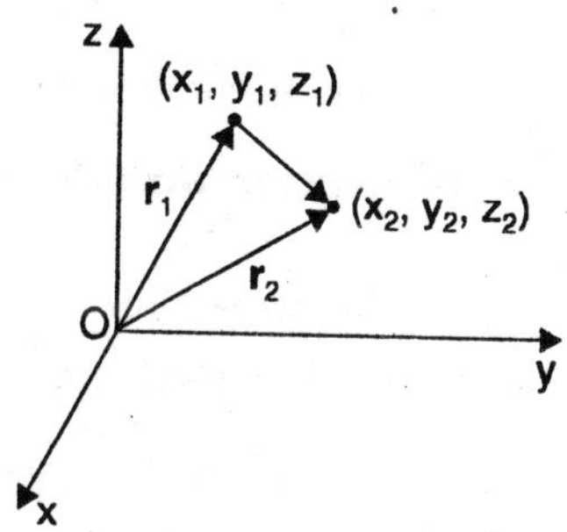

For each point (x, y, z) in three dimensional space, there exists a unique position vector $\mathbf{r}$ defined by

$$\mathbf{r} = (x, y, z) = x\mathbf{i} + y\mathbf{j} + z\mathbf{k}.$$

If $\mathbf{R}$ is a vector with its initial point at (x_1, y_1, z_1) and its terminal point at (x_2, y_2, z_2), then $\mathbf{R}$ is given by

$$\mathbf{R} = \mathbf{r}_2 - \mathbf{r}_1$$
$$= (x_2 - x_1)\mathbf{i} + (y_2 - y_1)\mathbf{j} + (z_2 - z_1)\mathbf{k}$$

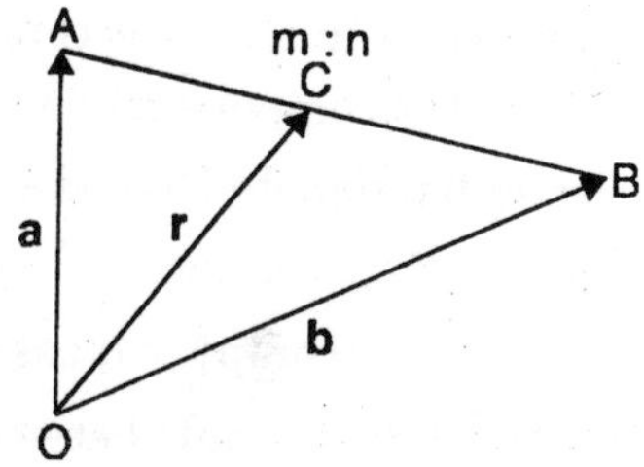

If the position vectors of two points A and B are $\mathbf{a}$ and $\mathbf{b}$ (w.r.t. some point O as origin), and a point C divides the line segment AB in the ratio $m : n$, then the position vector $\mathbf{r}$ of C is given by

$$\mathbf{r} = \frac{n\mathbf{a} + m\mathbf{b}}{n + m} \quad \text{(and is independent of origin)}$$

SCALAR AND VECTOR PRODUCTS

10. Scalar Product or Dot Product

The scalar product or dot product (sometimes also called the inner product) of two vectors **a** *and* **b** *is denoted by* **a . b** *and is a scalar defined by*

$$\mathbf{a} \cdot \mathbf{b} = |\mathbf{a}|\,|\mathbf{b}|\cos\theta = ab\cos\theta$$

where θ *is the angle between* **a** *and* **b** *and* $0 \le \theta \le \pi$.

It is clear from the definition that the angle θ between the two vectors $\mathbf{a}$ and $\mathbf{b}$ is given by

$$\cos\theta = \frac{\mathbf{a} \cdot \mathbf{b}}{|\mathbf{a}||\mathbf{b}|} = \frac{\mathbf{a} \cdot \mathbf{b}}{ab} \quad \text{or } \theta = \cos^{-1}\left(\frac{\mathbf{a} \cdot \mathbf{b}}{ab}\right)$$

provided $\mathbf{a} \neq \mathbf{0}$ and $\mathbf{b} \neq \mathbf{0}$.

Two vectors $\mathbf{a}$ and $\mathbf{b}$ are perpendicular (or orthogonal) to each other if the angle θ between them is $\frac{\pi}{2}$. Thus, $\mathbf{a}$ and $\mathbf{b}$ are perpendicular if and only if $\mathbf{a} \cdot \mathbf{b} = 0$. Since $\mathbf{0}$ has arbitrary direction, it can be considered to be perpendicular to any vector $\mathbf{a}$ and *vice-versa*.

If $\mathbf{a} = (a_1, a_2, a_3)$ *and* $\mathbf{b} = (b_1, b_2, b_3)$, *then* $\mathbf{a} \cdot \mathbf{b}$ *is given by*

$$\mathbf{a} \cdot \mathbf{b} = a_1 b_1 + a_2 b_2 + a_3 b_3.$$

Further, $$\mathbf{i} \cdot \mathbf{i} = \mathbf{j} \cdot \mathbf{j} = \mathbf{k} \cdot \mathbf{k} = 1$$

and $$\mathbf{i} \cdot \mathbf{j} = \mathbf{j} \cdot \mathbf{k} = \mathbf{k} \cdot \mathbf{i} = 0.$$

11. The Scalar Product has the following properties :

1. $\mathbf{a} \cdot \mathbf{b} = \mathbf{b} \cdot \mathbf{a}$ (Commutative law)
2. $\mathbf{a} \cdot (\mathbf{b} + \mathbf{c}) = \mathbf{a} \cdot \mathbf{b} + \mathbf{a} \cdot \mathbf{c}$ (Distributive law)
3. $(m\mathbf{a}) \cdot \mathbf{b} = \mathbf{a} \cdot (m\mathbf{b}) = m(\mathbf{a} \cdot \mathbf{b})$ for a scalar m
4. $\mathbf{a} \cdot \mathbf{a} = |\mathbf{a}|^2 = a^2$ for each $\mathbf{a}$
5. $\mathbf{a} \cdot \mathbf{a} = 0$ if and only if $\mathbf{a} = \mathbf{0}$.

12. Projection and Component of a Vector

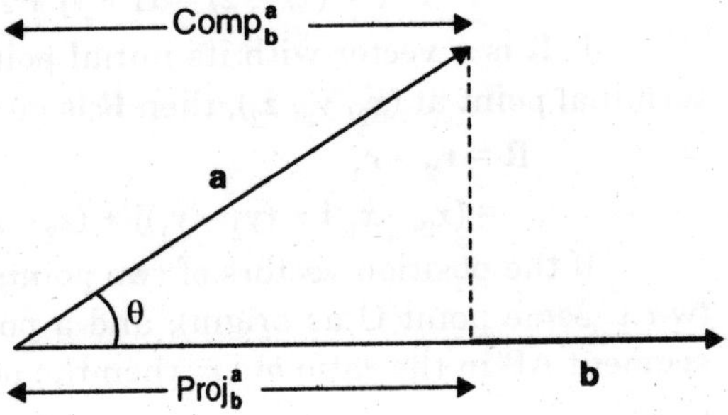

The projection of a vector $\mathbf{a}$ on a vector $\mathbf{b}$ is denoted by $\text{proj}_{\mathbf{b}}^{\mathbf{a}}$ and is a vector given by

$$\text{proj}_{\mathbf{b}}^{\mathbf{a}} = (a \cos \theta)\, \mathbf{e_b} = (\mathbf{a} \cdot \mathbf{e_b})\, \mathbf{e_b}$$

where θ is the angle between $\mathbf{a}$ and $\mathbf{b}$ and $\mathbf{e_b}$ is a unit vector in the direction of $\mathbf{b}$.

The component of a vector $\mathbf{a}$ along non-zero vector $\mathbf{b}$ is denoted by $\text{comp}_{\mathbf{b}}^{\mathbf{a}}$ and is a scalar given by

$$\text{comp}_{\mathbf{b}}^{\mathbf{a}} = a \cos \theta = \mathbf{a} \cdot \mathbf{e_b}$$

where θ is the angle between $\mathbf{a}$ and $\mathbf{b}$.

From the definition of $\mathbf{a} \cdot \mathbf{b}$, it follows that

$$\mathbf{a} \cdot \mathbf{b} = ab \cos \theta = a\,(b \cos \theta) = a\, \text{comp}_{\mathbf{a}}^{\mathbf{b}}$$

and $$\mathbf{a} \cdot \mathbf{b} = ab \cos \theta = b\,(a \cos \theta) = b\, \text{comp}_{\mathbf{a}}^{\mathbf{b}}.$$

13. Direction Cosines of a Vector

If a vector $\mathbf{a}$ makes angles α, β and γ with the positive directions of x, y and z axes respectively, then the three numbers $\cos \alpha$, $\cos \beta$ and $\cos \gamma$ are called the direction cosines (d.c.'s) of $\mathbf{a}$.

Clearly, $$\cos \alpha = \mathbf{i} \cdot \mathbf{e_a},\ \cos \beta = \mathbf{j} \cdot \mathbf{e_a},\ \cos \gamma = \mathbf{k} \cdot \mathbf{e_a}$$

where $\mathbf{e_a}$ is the unit vector along $\mathbf{a}$.

14. Vector Product or Cross Product

The vector product or cross product (sometimes also called the outer product) of two vectors $\mathbf{a}$ *and* $\mathbf{b}$ *is denoted by* $\mathbf{a} \times \mathbf{b}$ *and is a vector defined by*

$$\mathbf{a} \times \mathbf{b} = |\mathbf{a}|\,|\mathbf{b}| \sin \theta\, \mathbf{e} = ab \sin \theta\, \mathbf{e}$$

where (i) θ *is the angle between* $\mathbf{a}$ *and* $\mathbf{b}$ *such that* $0 \le \theta \le \pi$,

(ii) $\mathbf{e}$ *is a unit vector perpendicular to both* $\mathbf{a}$ *and* $\mathbf{b}$, *and* (*iii*) **a**, **b**, **e** *form a right handed triad of vectors*. (That is, if we place a skrew at the common point of intersection of **a** and **b** and rotate it with right hand from **a** to **b**, the skrew will advance in the positive direction of **e**).

If **a** = **b** or **a** is parallel to **b**, then sin θ = 0 and we define **a** × **b** = **0**.

Geometrically, | **a** × **b** | represents the area of the parallelogram with **a** and **b** forming the adjacent sides.

If $\mathbf{a} = (a_1, a_2, a_3)$ and $\mathbf{b} = (b_1, b_2, b_3)$, then their cross product **a** × **b** can be easily written with the help of the following determinant

$$\begin{vmatrix} \mathbf{i} & \mathbf{j} & \mathbf{k} \\ a_1 & a_2 & a_3 \\ b_1 & b_2 & b_3 \end{vmatrix}$$

and is given by $\quad \mathbf{a} \times \mathbf{b} = (a_2b_3 - a_3b_2, a_3b_1 - a_1b_3, a_1b_2 - a_2b_1)$

Further, $\quad \mathbf{i} \times \mathbf{i} = \mathbf{j} \times \mathbf{j} = \mathbf{k} \times \mathbf{k} = \mathbf{0},$

$$\mathbf{i} \times \mathbf{j} = \mathbf{k}, \mathbf{j} \times \mathbf{k} = \mathbf{i}, \mathbf{k} \times \mathbf{i} = \mathbf{j}$$

and $\quad \mathbf{j} \times \mathbf{i} = -\mathbf{k}, \mathbf{k} \times \mathbf{j} = -\mathbf{i}, \mathbf{i} \times \mathbf{k} = -\mathbf{j}.$

15. Vector Product has the following properties :

1. **a** × **b** = – **b** × **a** (Anticommutative law)
2. **a** × (**b** + **c**) = **a** × **b** + **a** × **c** (Distributive law)
3. (*m***a**) × **b** = **a** × (*m***b**) = *m***a** × **b**
4. **a** × **a** = **0**.

EXERCISE 1

1. If **a** = 3**i** – **j** – 4**k**, **b** = – 2**i** + 4**j** – 3**k**, **c** = **i** + 2**j** – **k**, find
   (*i*) **a** – **b** + 3**c** (*ii*) | **a** + **b** + **c** |
   (*iii*) | 3**a** – 2**b** + 4**c** | (*iv*) a unit vector parallel to 3**a** – 2**b** + 4**c**.
2. (*a*) Show that the vectors **a** = **i** + **k**, **b** = **i** + **j** + **k** and **c** = 2**i** + **k** are position vectors of vertices of a right angled triangle.
   (*b*) Find the lengths of medians of the triangle.
3. Using dot product, find the angle between the lines AB and AC where A, B, C are the points (1, 2, – 1), (2, 0, 3) and (3, – 1, 2) respectively.
4. Use dot product to show that the triangle with the vertices A(– 1, 4, – 1), B(5, 8, 5), C(11, – 4, 7) is a right angle triangle.
   [**Hint.** Show $\overrightarrow{AB} \cdot \overrightarrow{BC} = 0$.]
5. Are the points A(– 2, – 1, 4), B(2, 1, 2), C(– 4, – 2, 6) collinear ?
6. If **a** = (– 1, 0, 2), **b** = (2, 3, 6), **c** = (2, – 2, 1), find each of the following :
   (*i*) The projection of **a** on **b** (*ii*) The projection of **c** on 2**a** – **b**
   (*iii*) The projection of **a** – 2**b** + **c** on *y*-axis
   (*iv*) The projection of **a** – 2**b** + **c** on the line joining the points (0, 0, 0) and (1, 1, 1).
7. Using dot product, prove the triangle inequality
   $$|\mathbf{a} + \mathbf{b}| \le |\mathbf{a}| + |\mathbf{b}|.$$
   [**Hint.** Consider the square of each side.]

8. Using dot product, prove the *cosine formula*

$$c^2 = a^2 + b^2 - 2ab \cos C.$$

9. For any two vectors **a** and **b**, show that

$$|\,\mathbf{a}\,.\,\mathbf{b}\,| \leq |\,\mathbf{a}\,|\,|\,\mathbf{b}\,|. \qquad \text{(\textit{Cauchy-Schwarz} inequality)}$$

10. Given two vectors $\mathbf{a} = (2, 4, 6) = 2\mathbf{i} + 4\mathbf{j} + 6\mathbf{k}$, $\mathbf{b} = (1, -3, 2) = \mathbf{i} - 3\mathbf{j} + 2\mathbf{k}$,

find (*i*) $\mathbf{a}\,.\,\mathbf{b}$ (*ii*) $\mathbf{a} \times \mathbf{b}$.

11. For arbitrary vectors **a** and **b**, verify *Lagrange's identity*

$$(\mathbf{a} \times \mathbf{b})^2 = |\,\mathbf{a}\,|^2\,|\,\mathbf{b}\,|^2 - (\mathbf{a}\,.\,\mathbf{b})^2.$$

12. Find the area of the parallelogram formed by the vectors $\mathbf{a} = 2\mathbf{i} - 3\mathbf{j}$ and $\mathbf{b} = 4\mathbf{j} + 5\mathbf{k}$.

13. Find the area of a triangle having vertices A(2, 3, – 1), B(1, – 2, 3), C(– 2, 1, 4).

14. Use vector product to prove sine formula :

$$\frac{a}{\sin A} = \frac{b}{\sin B} = \frac{c}{\sin C}$$

in a triangle ABC.

15. Find a unit vector perpendicular to the plane of the vectors **a** and **b**, if

$$\mathbf{a} = \mathbf{i} + 3\mathbf{j} - 2\mathbf{k} \text{ and } \mathbf{b} = 2\mathbf{i} - \mathbf{j} - \mathbf{k}.$$

16. Show that the vector product is not associative.

[Hint. Prove by counter example, consider $\mathbf{i} \times (\mathbf{i} \times \mathbf{j})$**]**

17. Find the area of the parallelogram whose vertices are (3, 1, 1), (– 1, 1, – 2), (0, 4, – 5), (– 4, 4, – 8).

Answers

1. (*i*) $8\mathbf{i} + \mathbf{j} - 4\mathbf{k}$ (*ii*) $\sqrt{93}$ (*iii*) $\sqrt{398}$ (*iv*) $\dfrac{3\mathbf{a} - 2\mathbf{b} + 4\mathbf{c}}{\sqrt{398}}$

2. (*b*) $\dfrac{\sqrt{2}}{2}, \dfrac{\sqrt{5}}{2}, \dfrac{\sqrt{5}}{2}$ **3.** $\theta = \cos^{-1} \dfrac{20}{\sqrt{21}\,.\,\sqrt{22}}$ **5.** No. $\overrightarrow{AB}$ is not parallel to $\overrightarrow{BC}$

6. (*i*) $\left(\dfrac{20}{49}, \dfrac{30}{49}, \dfrac{60}{49}\right)$ (*ii*) $\left(\dfrac{16}{29}, \dfrac{12}{29}, \dfrac{8}{29}\right)$ (*iii*) (0, – 8, 0) (*iv*) $\left(\dfrac{-20}{3}, \dfrac{-20}{3}, \dfrac{-20}{3}\right)$

10. (*i*) 2 (*ii*) (26, 2, – 10) **12.** $\sqrt{389}$ square units

13. $\dfrac{1}{2}\sqrt{734}$ square units **15.** $\left(\dfrac{-5}{\sqrt{83}}, \dfrac{-3}{\sqrt{83}}, \dfrac{-7}{\sqrt{83}}\right)$

17. $15\sqrt{2}$ square units.

16. Triple Product of Vectors

Since the vector product of two vectors **a** and **b** is a vector $\mathbf{a} \times \mathbf{b}$, therefore, it is meaningful to combine with a third vector **c** to form the scalar product $\mathbf{c}\,.\,(\mathbf{a} \times \mathbf{b})$ or the vector product $\mathbf{c} \times (\mathbf{a} \times \mathbf{b})$. In fact, if we are given three vectors **a**, **b** and **c**, then the meaningful products of these vectors taken all at a time would be one of the following :

(*i*) $\mathbf{a}\,.\,(\mathbf{b} \times \mathbf{c})$

(*ii*) $(\mathbf{a} \times \mathbf{b})\,.\,\mathbf{c}$

(*iii*) $(\mathbf{a} \times \mathbf{b}) \times \mathbf{c}$

(*iv*) $\mathbf{a} \times (\mathbf{b} \times \mathbf{c})$.

The products (*i*) and (*ii*) are called the *scalar triple product* because their product is a scalar and the products (*iii*) and (*iv*) are called the *vector triple product* because their product is a vector quantity.

We first deal with the scalar triple product.

17. Scalar Triple Product

The scalar triple product of 3 vectors **a**, **b** *and* **c** *is a scalar* **a** . (**b** × **c**) *or simply* **a** . **b** × **c**.

The expression (**a** . **b**) × **c** *is undefined since it indicates the vector product of a scalar with a vector. So, we can omit the parenthesis in* **a** . (**b** × **c**) *and write* **a** . **b** × **c**. It is also denoted by [**abc**]. (K.U. 2006)

18. Properties of Scalar Triple Product

(1) Geometrical interpretation of a . b × c.

Consider a parallelopiped whose adjacent sides are **a**, **b**, **c**. Now, | **b** × **c** | gives the area of the parallelogram OBDC having adjacent sides represented by the vectors **b** and **c**.

If h is the altitude of the parallelopiped, then

$$h = |\mathbf{a}|\,|\cos\theta|,$$

where θ is the angle between **a** and **b** × **c**.

∴ Volume of parallelopiped $= h \times$ area of parallelogram OBDC

$$= |\mathbf{a}|\,|\mathbf{b}\times\mathbf{c}|\,|\cos\theta| = |\mathbf{a}\,.\,\mathbf{b}\times\mathbf{c}|$$

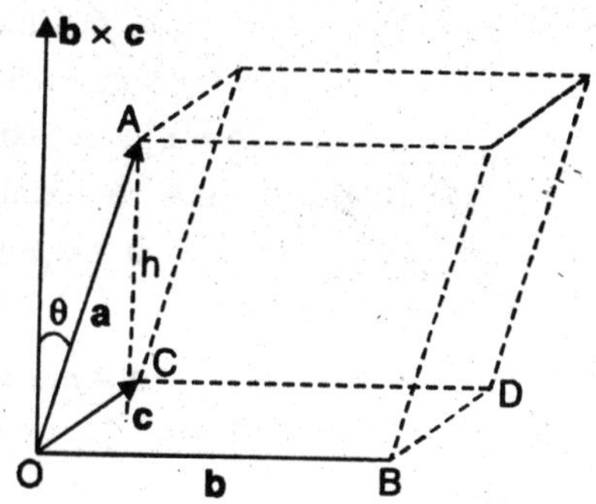

Hence, the scalar triple product **a** . **b** × **c** equals the volume of the parallelopiped having the given vectors **a**, **b**, **c** as its three coterminus edges. Further, one can easily see that the triple product **a** . **b** × **c** is positive or negative according as the vectors **a**, **b** and **c** form a right-handed triad of vectors or left-handed triad of vectors.

(2) The fact that **a** . **b** × **c** is the volume (or its negative) of the parallelopiped having **a**, **b**, **c** as its coterminus edges, shows that the value of the scalar triple product remains unchanged by a cyclic change in the order of three vectors **a**, **b**, **c**. Hence, we have

$$\mathbf{a}\,.\,\mathbf{b}\times\mathbf{c} = \mathbf{b}\,.\,\mathbf{c}\times\mathbf{a} = \mathbf{c}\,.\,\mathbf{a}\times\mathbf{b}.$$

(3) The same geometrical interpretation shows that the dot and the cross may be interchanged without altering the value of the scalar triple product. Hence,

$$\mathbf{a}\,.\,\mathbf{b}\times\mathbf{c} = \mathbf{a}\times\mathbf{b}\,.\,\mathbf{c}.$$

(4) **Condition of Coplanarity.** *The vectors* **a**, **b** *and* **c** *are coplanar if and only if*

$$\mathbf{a}\,.\,\mathbf{b}\times\mathbf{c} = 0.$$ (K.U. 2006)

For, if the vectors **a**, **b**, **c** are coplanar then the altitude of the parallelopiped that they form is zero. Hence, **a** . **b** × **c** = 0. Conversely, if **a** . **b** × **c** = 0, the altitude of the parallelopiped is zero and that means that the vectors **a**, **b** and **c** must be coplanar.

Alternative Proof

(*i*) Let **a, b, c** be three non-zero, non-collinear coplanar vectors, we shall show that

$$[\mathbf{a}\;\mathbf{b}\;\mathbf{c}] = 0.$$

Now, $\mathbf{a} \times \mathbf{b}$ is perpendicular to the plane of **a** and **b**.

$\therefore$ $\mathbf{a} \times \mathbf{b} \perp \mathbf{c}$ ($\because$ **a, b, c** are coplanar)

$\Rightarrow$ $(\mathbf{a} \times \mathbf{b}) \cdot \mathbf{c} = 0 \Rightarrow [\mathbf{a}\ \ \mathbf{b}\ \ \mathbf{c}] = 0.$

(*ii*) Conversely, let $[\mathbf{a\,b\,c}] = 0$, where **a, b, c** are non-zero non- collinear vectors. We shall show that **a, b, c** are coplanar.

Now, $[\mathbf{a}\ \ \mathbf{b}\ \ \mathbf{c}] = 0 \Rightarrow (\mathbf{a} \times \mathbf{b}) \cdot \mathbf{c} = 0$

$\Rightarrow$ $\mathbf{a} \times \mathbf{b} = \mathbf{0}$ or $\mathbf{c} = \mathbf{0}$ or $\mathbf{a} \times \mathbf{b} \perp \mathbf{c}$

$\Rightarrow$ $\mathbf{a} \times \mathbf{b} \perp \mathbf{c}$

($\because$ $\mathbf{c} \neq \mathbf{0}$ and $\mathbf{a} \times \mathbf{b} \neq \mathbf{0}$ because **a, b** are non-zero, non-collinear vectors)

But, $\mathbf{a} \times \mathbf{b}$ is a vector perpendicular to the plane of **a** and **b**.

$\therefore$ **c** lies in the plane of **a** and **b**

$\Rightarrow$ **a, b, c** are coplanar.

(5) *If any two vectors in a scalar triple product are equal, then the product is zero, i.e.*

$$\mathbf{a} \cdot \mathbf{a} \times \mathbf{c} = 0,\ \mathbf{c} \cdot \mathbf{b} \times \mathbf{c} = 0 \ \text{ and } \ \mathbf{a} \cdot \mathbf{b} \times \mathbf{b} = 0$$

For, from the definition of vector product, $\mathbf{a} \times \mathbf{c}$ is perpendicular to **a**. Hence, their dot product is zero, *i.e.*, $\mathbf{a} \cdot \mathbf{a} \times \mathbf{c} = 0$. Similarly, since $\mathbf{b} \times \mathbf{c}$ is perpendicular to **c**, their scalar product is zero, *i.e.*, $\mathbf{c} \cdot (\mathbf{b} \times \mathbf{c}) = 0$.

Also, $\mathbf{b} \times \mathbf{b} = \mathbf{0}$ and so $\mathbf{a} \cdot \mathbf{b} \times \mathbf{b} = \mathbf{a} \cdot \mathbf{0} = 0$.

(6) Scalar triple product when two of the three vectors are parallel is zero. For, let **a** and **b** of three given vectors **a**, **b** and **c** be parallel.

Then, $\mathbf{b} = m\mathbf{a}$ for some scalar m.

$\therefore$ $[\mathbf{a\,b\,c}] = \mathbf{a} \cdot (\mathbf{b} \times \mathbf{c}) = \mathbf{a} \cdot (m\mathbf{a} \times \mathbf{c}) = m\mathbf{a} \cdot (\mathbf{a} \times \mathbf{c}) = m[\mathbf{a\,a\,c}] = 0.$

(7) To calculate $\mathbf{a} \cdot \mathbf{b} \times \mathbf{c}$ in numerical examples, the following form is useful. If

$$\mathbf{a} = a_1\mathbf{i} + a_2\mathbf{j} + a_3\mathbf{k},\ \mathbf{b} = b_1\mathbf{i} + b_2\mathbf{j} + b_3\mathbf{k},$$

$$\mathbf{c} = c_1\mathbf{i} + c_2\mathbf{j} + c_3\mathbf{k},\ \text{then}\ \mathbf{a} \cdot \mathbf{b} \times \mathbf{c} = \begin{vmatrix} a_1 & a_2 & a_3 \\ b_1 & b_2 & b_3 \\ c_1 & c_2 & c_3 \end{vmatrix}$$

For, $$\mathbf{b} \times \mathbf{c} = \begin{vmatrix} \mathbf{i} & \mathbf{j} & \mathbf{k} \\ b_1 & b_2 & b_3 \\ c_1 & c_2 & c_3 \end{vmatrix} = (b_2c_3 - b_3c_2)\mathbf{i} + (b_3c_1 - b_1c_3)\mathbf{j} + (b_1c_2 - b_2c_1)\mathbf{k}$$

$$\therefore \quad \mathbf{a} \cdot \mathbf{b} \times \mathbf{c} = (a_1\mathbf{i} + a_2\mathbf{j} + a_3\mathbf{k}) \cdot [(b_2c_3 - b_3c_2)\mathbf{i} + (b_3c_1 - b_1c_3)\mathbf{j} + (b_1c_2 - b_2c_1)\mathbf{k}]$$

$$= a_1(b_2c_3 - b_3c_2) + a_2(b_3c_1 - b_1c_3) + a_3(b_1c_2 - b_2c_1)$$

$$= \begin{vmatrix} a_1 & a_2 & a_3 \\ b_1 & b_2 & b_3 \\ c_1 & c_2 & c_3 \end{vmatrix}.$$

(8) The scalar triple product is distributive :

$$\mathbf{a} \cdot (\mathbf{b} + \mathbf{c}) \times (\mathbf{d} + \mathbf{e}) = \mathbf{a} \cdot (\mathbf{b} \times \mathbf{d}) + \mathbf{a} \cdot (\mathbf{b} \times \mathbf{e}) + \mathbf{a} \cdot (\mathbf{c} \times \mathbf{d}) + \mathbf{a} \cdot (\mathbf{c} \times \mathbf{e})$$

For, using distributive property of vector product in L.H.S., we have

$$\mathbf{a} \cdot (\mathbf{b} + \mathbf{c}) \times (\mathbf{d} + \mathbf{e}) = \mathbf{a} \cdot [(\mathbf{b} \times \mathbf{d}) + (\mathbf{b} \times \mathbf{e}) + (\mathbf{c} \times \mathbf{d}) + (\mathbf{c} \times \mathbf{e})]$$

$$= \mathbf{a} \cdot \mathbf{b} \times \mathbf{d} + \mathbf{a} \cdot \mathbf{b} \times \mathbf{e} + \mathbf{a} \cdot \mathbf{c} \times \mathbf{d} + \mathbf{a} \cdot \mathbf{c} \times \mathbf{e}$$

$$= \text{R.H.S.}$$

Example 1. *Find the value of* λ *for which the vectors* $\mathbf{i} - \mathbf{j} + \mathbf{k}$, $2\mathbf{i} + \mathbf{j} - \mathbf{k}$ *and* $\lambda\mathbf{i} - \mathbf{j} + \lambda\mathbf{k}$ *are coplanar.*

Sol. The three vectors will be coplanar if their scalar triple product is zero.

i.e., if
$$\begin{vmatrix} 1 & -1 & 1 \\ 2 & 1 & -1 \\ \lambda & -1 & \lambda \end{vmatrix} = 0$$

i.e., if $1(\lambda - 1) + 1(2\lambda + \lambda) + 1(-2 - \lambda) = 0$

i.e., if $\lambda = 1$ (on simplification)

Thus, the three vectors are coplanar if $\lambda = 1$.

Example 2. *Find the volume of a parallelopiped whose three coterminus edges are represented by* $\mathbf{a} = 2\mathbf{i} - 3\mathbf{j} + 4\mathbf{k}$, $\mathbf{b} = \mathbf{i} + 2\mathbf{j} - \mathbf{k}$ *and* $\mathbf{c} = 3\mathbf{i} - \mathbf{j} + 2\mathbf{k}$.

Sol. Volume of the parallelopiped with three coterminus edges **a**, **b**, **c** is $\mathbf{a} \cdot \mathbf{b} \times \mathbf{c}$

$$= \begin{vmatrix} 2 & -3 & 4 \\ 1 & 2 & -1 \\ 3 & -1 & 2 \end{vmatrix} = 2(4 - 1) + 3(2 + 3) + 4(-1 - 6) = -7$$

The negative sign shows that **a**, **b**, **c** form a left-handed triad of vectors. Thus, volume of the parallelopiped is 7 cubic units.

Example 3. *Show that* $\mathbf{i} \cdot \mathbf{j} \times \mathbf{k} = \mathbf{j} \cdot \mathbf{k} \times \mathbf{i} = \mathbf{k} \cdot \mathbf{i} \times \mathbf{j} = 1$

and $\mathbf{i} \cdot \mathbf{k} \times \mathbf{j} = \mathbf{j} \cdot \mathbf{i} \times \mathbf{k} = \mathbf{k} \cdot \mathbf{j} \times \mathbf{i} = -1.$

Sol.
$$\mathbf{i} \cdot \mathbf{j} \times \mathbf{k} = \begin{vmatrix} 1 & 0 & 0 \\ 0 & 1 & 0 \\ 0 & 0 & 1 \end{vmatrix} = 1$$

Similarly, one can proceed with the remaining ones.

Example 4. *Show that the vectors* $\mathbf{a} = (2, 0, 1)$, $\mathbf{b} = (0, 3, 4)$ *and* $\mathbf{c} = (8, -3, 0)$ *are coplanar. Also, express* **c** *as a linear combination of* **a** *and* **b**.

Sol. a, b, c are coplanar

if $\mathbf{a} \cdot \mathbf{b} \times \mathbf{c} = 0$

i.e., if
$$\begin{vmatrix} 2 & 0 & 1 \\ 0 & 3 & 4 \\ 8 & -3 & 0 \end{vmatrix} = 0 \quad \text{which one can see is true.}$$

Hence, **a**, **b**, **c** are coplanar.

To express **c** as a linear combination of **a** and **b**, let $\mathbf{c} = \lambda_1\mathbf{a} + \lambda_2\mathbf{b}$ for some scalars λ_1, λ_2

$\Rightarrow \quad (8, -3, 0) = \lambda_1(2, 0, 1) + \lambda_2(0, 3, 4)$

$\Rightarrow \quad (8, -3, 0) = (2\lambda_1, 0, \lambda_1) + (0, 3\lambda_2, 4\lambda_2)$

$\Rightarrow \quad (8, -3, 0) = (2\lambda_1, 3\lambda_2, \lambda_1 + 4\lambda_2)$

$\Rightarrow \quad 8 = 2\lambda_1, -3 = 3\lambda_2, 0 = \lambda_1 + 4\lambda_2$

$\Rightarrow \quad \lambda_1 = 4, \lambda_2 = -1$ and $\mathbf{c} = 4\mathbf{a} - \mathbf{b}$.

Example 5. *Show that the four points with position vectors*

$$5\mathbf{i} + 6\mathbf{j} + 2\mathbf{k},\ \mathbf{i},\ 4\mathbf{i} + 10\mathbf{j} + 5\mathbf{k} \text{ and } -3\mathbf{i} + 5\mathbf{j} + 5\mathbf{k}$$

are coplanar.

Sol. If A, B, C, D denote the four given points, then

$$\vec{OA} = 5\mathbf{i} + 6\mathbf{j} + 2\mathbf{k}, \qquad \vec{OB} = \mathbf{i}, \qquad \vec{OC} = 4\mathbf{i} + 10\mathbf{j} + 5\mathbf{k}$$

and
$$\vec{OD} = -3\mathbf{i} + 5\mathbf{j} + 5\mathbf{k}.$$

$$\therefore \qquad \vec{AB} = \vec{OB} - \vec{OA} = -4\mathbf{i} - 6\mathbf{j} - 2\mathbf{k},$$

$$\vec{BC} = \vec{OC} - \vec{OB} = 3\mathbf{i} + 10\mathbf{j} + 5\mathbf{k},$$

$$\vec{CD} = \vec{OD} - \vec{OC} = -7\mathbf{i} - 5\mathbf{j}.$$

Now, the vectors $\vec{AB}$, $\vec{BC}$ and $\vec{CD}$ are coplanar if their scalar triple product is zero.

i.e., if
$$\begin{vmatrix} -4 & -6 & -2 \\ 3 & 10 & 5 \\ -7 & -5 & 0 \end{vmatrix} = 0 \quad \text{which is true (verify).}$$

Hence, the points A, B, C, D are coplanar.

Example 6. *If* **a**, **b**, **c** *are the position vectors of the points A, B, C, show that the vector* **a × b + b × c + c × a** *is perpendicular to the plane ABC.*

Sol.
$$\vec{AB} = \mathbf{b} - \mathbf{a}, \ \vec{BC} = \mathbf{c} - \mathbf{b}, \ \vec{CA} = \mathbf{a} - \mathbf{c}$$

Now, $(\mathbf{a} \times \mathbf{b} + \mathbf{b} \times \mathbf{c} + \mathbf{c} \times \mathbf{a}) \cdot \vec{AB}$

$= (\mathbf{a} \times \mathbf{b} + \mathbf{b} \times \mathbf{c} + \mathbf{c} \times \mathbf{a}) \cdot (\mathbf{b} - \mathbf{a})$

$= (\mathbf{a} \times \mathbf{b}) \cdot \mathbf{b} + (\mathbf{b} \times \mathbf{c}) \cdot \mathbf{b} + (\mathbf{c} \times \mathbf{a}) \cdot \mathbf{b} - (\mathbf{a} \times \mathbf{b}) \cdot \mathbf{a} - (\mathbf{b} \times \mathbf{c}) \cdot \mathbf{a} - (\mathbf{c} \times \mathbf{a}) \cdot \mathbf{a} = 0$

$\therefore$ **a × b + b × c + c × a** is perpendicular to AB.

Similarly, **a × b + b × c + c × a** is perpendicular to both BC and CA. Hence, it is perpendicular to the plane ABC.

Example 7. *For any three vectors* **a**, **b**, **c**, *prove that* $[\mathbf{a} + \mathbf{b}, \mathbf{b} + \mathbf{c}, \mathbf{c} + \mathbf{a}] = 2[\mathbf{a}\ \mathbf{b}\ \mathbf{c}]$.

Sol. $[\mathbf{a} + \mathbf{b}, \mathbf{b} + \mathbf{c}, \mathbf{c} + \mathbf{a}] = \{(\mathbf{a} + \mathbf{b}) \times (\mathbf{b} + \mathbf{c})\} \cdot (\mathbf{c} + \mathbf{a})$

$= (\mathbf{a} \times \mathbf{b} + \mathbf{a} \times \mathbf{c} + \mathbf{b} \times \mathbf{b} + \mathbf{b} \times \mathbf{c}) \cdot (\mathbf{c} + \mathbf{a})$

$= (\mathbf{a} \times \mathbf{b} + \mathbf{a} \times \mathbf{c} + \mathbf{b} \times \mathbf{c}) \cdot (\mathbf{c} + \mathbf{a})$

$= (\mathbf{a} \times \mathbf{b}) \cdot \mathbf{c} + (\mathbf{a} \times \mathbf{b}) \cdot \mathbf{a} + (\mathbf{a} \times \mathbf{c}) \cdot \mathbf{c} + (\mathbf{a} \times \mathbf{c}) \cdot \mathbf{a} + (\mathbf{b} \times \mathbf{c}) \cdot \mathbf{c} + (\mathbf{b} \times \mathbf{c}) \cdot \mathbf{a}$

(by distributive law)

$= [\mathbf{a}\ \ \mathbf{b}\ \ \mathbf{c}] + [\mathbf{b}\ \ \mathbf{c}\ \ \mathbf{a}]$

(As scalar triple product is zero when two vectors are equal)

$= 2[\mathbf{a}\ \ \mathbf{b}\ \ \mathbf{c}].$

Remark. From the above example, it follows that vectors a, b, c are coplanar iff [a + b, b + c, c + a] are coplanar.

Example 8. *If* **a**, **b**, **c** *are three non-zero vectors such that* $[\mathbf{a}\ \mathbf{b}\ \mathbf{c}] = 0$ *and* **b**, **c** *are non-parallel, then show that* $\mathbf{a} = \lambda\mathbf{b} + \mu\mathbf{c}$ *for some scalars* λ *and* μ.

Sol. $[\mathbf{a}\ \ \mathbf{b}\ \ \mathbf{c}] = 0 \Rightarrow \mathbf{a} \cdot (\mathbf{b} \times \mathbf{c}) = 0$

$\Rightarrow \qquad \mathbf{a} \perp \mathbf{b} \times \mathbf{c} \qquad (\because\ \mathbf{a} \neq 0, \mathbf{b} \times \mathbf{c} \neq 0)$

But **b × c** is perpendicular to the plane of **b** and **c**.

$\therefore$ **a** lies in the plane of **b** and **c**

$\Rightarrow$ there exist scalars λ and μ such that $\mathbf{a} = \lambda\mathbf{b} + \mu\mathbf{c}$.

Example 9. *The vector* $\vec{OP} = 5\mathbf{i} + 12\mathbf{j} + 13\mathbf{k}$ *turns through an angle* $\frac{\pi}{2}$ *about O passing through the positive side of the y-axis on its way. Find the vector in the new position.*

Sol. Let $\vec{OQ} = x\mathbf{i} + y\mathbf{j} + z\mathbf{k}$ be the vector in the new position. Then,

$$|\vec{OP}| = |\vec{OQ}| \Rightarrow x^2 + y^2 + z^2 = 338 \quad ...(1)$$

Since $\vec{OP} \perp \vec{OQ}$, $\vec{OP} \cdot \vec{OQ} = 0$

$$\Rightarrow \quad 5x + 12y + 13z = 0 \quad ...(2)$$

Also, $\vec{OP}$, **j** and $\vec{OQ}$ are coplanar

$$\therefore \quad [\vec{OP} \ \ \mathbf{j} \ \ \vec{OQ}] = 0$$

$$\Rightarrow \begin{vmatrix} 5 & 12 & 13 \\ 0 & 1 & 0 \\ x & y & z \end{vmatrix} = 0 \Rightarrow -13x + 5z = 0$$

$$\Rightarrow \quad z = \frac{13}{5}x \quad ...(3)$$

From (2) and (3), we get $\quad y = -\frac{97}{30}x$

$\therefore$ from (1), $\quad x = \pm 30\sqrt{\frac{2}{97}}$.

Since the vector in its new position makes an obtuse angle with x-axis, therefore

$$x = -30\sqrt{\frac{2}{97}}.$$

$$\therefore \quad \vec{OQ} = \frac{2}{\sqrt{97}}(-30\mathbf{i} + 97\mathbf{j} - 78\mathbf{k}).$$

Example 10. *If* **a, b, c** *and* **l, m, n** *are two triads of vectors, prove that*

$$[\mathbf{a} \ \ \mathbf{b} \ \ \mathbf{c}]\,[\mathbf{l} \ \ \mathbf{m} \ \ \mathbf{n}] = \begin{vmatrix} \mathbf{a}.\mathbf{l} & \mathbf{b}.\mathbf{l} & \mathbf{c}.\mathbf{l} \\ \mathbf{a}.\mathbf{m} & \mathbf{b}.\mathbf{m} & \mathbf{c}.\mathbf{m} \\ \mathbf{a}.\mathbf{n} & \mathbf{b}.\mathbf{n} & \mathbf{c}.\mathbf{n} \end{vmatrix}.$$

Sol. Let $\quad \mathbf{a} = a_1\mathbf{i} + a_2\mathbf{j} + a_3\mathbf{k}$, $\mathbf{b} = b_1\mathbf{i} + b_2\mathbf{j} + b_3\mathbf{k}$,
$\mathbf{c} = c_1\mathbf{i} + c_2\mathbf{j} + c_3\mathbf{k}$ and $\mathbf{l} = l_1\mathbf{i} + l_2\mathbf{j} + l_3\mathbf{k}$, etc.

$$\therefore \quad [\mathbf{a} \ \ \mathbf{b} \ \ \mathbf{c}]\,[\mathbf{l} \ \ \mathbf{m} \ \ \mathbf{n}] = \begin{vmatrix} a_1 & a_2 & a_3 \\ b_1 & b_2 & b_3 \\ c_1 & c_2 & c_3 \end{vmatrix} \times \begin{vmatrix} l_1 & l_2 & l_3 \\ m_1 & m_2 & m_3 \\ n_1 & n_2 & n_3 \end{vmatrix} \quad \text{(Multiplying row by row)}$$

$$= \begin{vmatrix} \Sigma a_1 l_1 & \Sigma a_1 m_1 & \Sigma a_1 n_1 \\ \Sigma b_1 l_1 & \Sigma b_1 m_1 & \Sigma b_1 n_1 \\ \Sigma c_1 l_1 & \Sigma c_1 m_1 & \Sigma c_1 n_1 \end{vmatrix} = \begin{vmatrix} \mathbf{a}.\mathbf{l} & \mathbf{a}.\mathbf{m} & \mathbf{a}.\mathbf{n} \\ \mathbf{b}.\mathbf{l} & \mathbf{b}.\mathbf{m} & \mathbf{b}.\mathbf{n} \\ \mathbf{c}.\mathbf{l} & \mathbf{c}.\mathbf{m} & \mathbf{c}.\mathbf{n} \end{vmatrix} = \begin{vmatrix} \mathbf{a}.\mathbf{l} & \mathbf{b}.\mathbf{l} & \mathbf{c}.\mathbf{l} \\ \mathbf{a}.\mathbf{m} & \mathbf{b}.\mathbf{m} & \mathbf{c}.\mathbf{m} \\ \mathbf{a}.\mathbf{n} & \mathbf{b}.\mathbf{n} & \mathbf{c}.\mathbf{n} \end{vmatrix}.$$

Example 11. *For any three vectors* **a, b, c** *prove that*

$$[\mathbf{a}\ \ \mathbf{b}\ \ \mathbf{c}]^2 = \begin{vmatrix} \mathbf{a}\cdot\mathbf{a} & \mathbf{a}\cdot\mathbf{b} & \mathbf{a}\cdot\mathbf{c} \\ \mathbf{b}\cdot\mathbf{a} & \mathbf{b}\cdot\mathbf{b} & \mathbf{b}\cdot\mathbf{c} \\ \mathbf{c}\cdot\mathbf{a} & \mathbf{c}\cdot\mathbf{b} & \mathbf{c}\cdot\mathbf{c} \end{vmatrix}.$$

Sol. In the above example, replace **l, m, n** by **a, b, c** respectively.

Example 12. *If* $\begin{vmatrix} (x-a)^2 & (x-b)^2 & (x-c)^2 \\ (y-a)^2 & (y-b)^2 & (y-c)^2 \\ (z-a)^2 & (z-b)^2 & (z-c)^2 \end{vmatrix} = 0$, *and* $\mathbf{X} = (x^2, x, 1)$, $\mathbf{Y} = (y^2, y, 1)$, $\mathbf{Z} = (z^2, z, 1)$ *are non-coplanar, then prove that* $\mathbf{a} = (a^2, a, 1)$, $\mathbf{b} = (b^2, b, 1)$, $\mathbf{c} = (c^2, c, 1)$ *are coplanar.*

Sol. Since **X, Y, Z** are non-coplanar,

$$\therefore \qquad [\mathbf{X}\ \mathbf{Y}\ \mathbf{Z}] \neq 0$$

$$\Rightarrow \qquad \begin{vmatrix} x^2 & x & 1 \\ y^2 & y & 1 \\ z^2 & z & 1 \end{vmatrix} \neq 0 \qquad \ldots(1)$$

Now, $\begin{vmatrix} (x-a)^2 & (x-b)^2 & (x-c)^2 \\ (y-a)^2 & (y-b)^2 & (y-c)^2 \\ (z-a)^2 & (z-b)^2 & (z-c)^2 \end{vmatrix} = 0 \quad \Rightarrow \quad \begin{vmatrix} x^2 & x & 1 \\ y^2 & y & 1 \\ z^2 & z & 1 \end{vmatrix} \times \begin{vmatrix} 1 & -2a & a^2 \\ 1 & -2b & b^2 \\ 1 & -2c & c^2 \end{vmatrix} = 0$

$$\Rightarrow \qquad \begin{vmatrix} a^2 & a & 1 \\ b^2 & b & 1 \\ c^2 & c & 1 \end{vmatrix} = 0 \qquad \text{[By (1)]}$$

$$\Rightarrow \qquad [\mathbf{a}\ \ \mathbf{b}\ \ \mathbf{c}] = 0$$

$\Rightarrow$ **a, b, c** are coplanar.

19. Volume of a Tetrahedron *(K.U. 2003 S)*

Let OABC be a tetrahedron whose three coterminus edges OA, OB and OC represent the vectors **a, b, c** respectively.

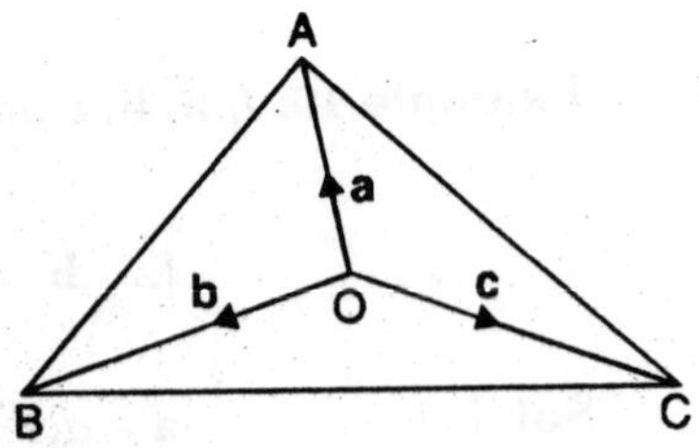

Area of ΔOBC $= \frac{1}{2}\,\mathbf{b} \times \mathbf{c}$ (which is a vector perpendicular to the plane of ΔOBC).

$\therefore$ Volume of tetrahedron OABC

$$= \tfrac{1}{3}\ \text{(area of base)} \times \text{height}$$

$$= \tfrac{1}{3}\left(\tfrac{1}{2}\,\mathbf{b} \times \mathbf{c}\right) \cdot \mathbf{a}$$

$$= \tfrac{1}{6}\,(\mathbf{b} \times \mathbf{c}) \cdot \mathbf{a} = \tfrac{1}{6}\,[\mathbf{abc}].$$

Cor. 1. If **a, b, c, d** are the position vectors of the vertices A, B, C, D of a tetrahedron ABCD, then

$$\vec{DA} = \mathbf{a} - \mathbf{d},\ \vec{DB} = \mathbf{b} - \mathbf{d},\ \vec{DC} = \mathbf{c} - \mathbf{d}$$

$\therefore$ Volume of the tetrahedron $= \frac{1}{6}\,[\mathbf{a} - \mathbf{d}\ \ \ \mathbf{b} - \mathbf{d}\ \ \ \mathbf{c} - \mathbf{d}]$.

Cor. 2. Four points A, B, C, D with position vectors **a**, **b**, **c**, **d** are coplanar if the volume of the tetrahedron ABCD is zero.

i.e., if $\frac{1}{6}\,[\mathbf{a}-\mathbf{d}\;\; \mathbf{b}-\mathbf{d}\;\; \mathbf{c}-\mathbf{d}] = 0$

i.e., if $[(\mathbf{a}-\mathbf{d})\times(\mathbf{b}-\mathbf{d})].\,(\mathbf{c}-\mathbf{d}) = 0$

i.e., if $[\mathbf{bcd}] - [\mathbf{bca}] + [\mathbf{bda}] - [\mathbf{cda}] = 0.$

Example 13. *Find the volume of the tetrahedron whose vertices are A(2, 4, 8), B(3, 6, 2), C(2, 2, 6) and D(– 1, 4, 3).*

Sol.
$$\vec{AB} = (3\mathbf{i}+6\mathbf{j}+2\mathbf{k}) - (2\mathbf{i}+4\mathbf{j}+8\mathbf{k}) = \mathbf{i}+2\mathbf{j}-6\mathbf{k}$$
$$\vec{AC} = (2\mathbf{i}+2\mathbf{j}+6\mathbf{k}) - (2\mathbf{i}+4\mathbf{j}+8\mathbf{k}) = -2\mathbf{j}-2\mathbf{k}$$
$$\vec{AD} = (-\mathbf{i}+4\mathbf{j}+3\mathbf{k}) - (2\mathbf{i}+4\mathbf{j}+8\mathbf{k}) = -3\mathbf{i}-5\mathbf{k}$$

$\therefore$ Volume of tetrahedron $= \frac{1}{6}\,[\vec{AB}\;\vec{AC}\;\vec{AD}]$

$$= \frac{1}{6}\begin{vmatrix} 1 & 2 & -6 \\ 0 & -2 & -2 \\ -3 & 0 & -5 \end{vmatrix} = \frac{58}{6} = 9\frac{2}{3} \text{ c. units.}$$

20. Vector Triple Product

The vector triple product of three vectors **a**, **b**, **c** is the vector $\mathbf{a}\times(\mathbf{b}\times\mathbf{c})$.

Unlike scalar triple product, we need parentheses because the expression $\mathbf{a}\times\mathbf{b}\times\mathbf{c}$ is ambiguous, the result depending upon whether we form $\mathbf{a}\times\mathbf{b}$ first or $\mathbf{b}\times\mathbf{c}$ first.

In general, $\mathbf{a}\times(\mathbf{b}\times\mathbf{c}) \neq (\mathbf{a}\times\mathbf{b})\times\mathbf{c}$

[For, let $\mathbf{a}=\mathbf{i}$, $\mathbf{b}=\mathbf{i}$ and $\mathbf{c}=\mathbf{j}$

Then, $\mathbf{a}\times(\mathbf{b}\times\mathbf{c}) = \mathbf{i}\times(\mathbf{i}\times\mathbf{j}) = \mathbf{i}\times\mathbf{k} = -\mathbf{j}$

and $(\mathbf{a}\times\mathbf{b})\times\mathbf{c} = (\mathbf{i}\times\mathbf{i})\times\mathbf{j} = \mathbf{0}\times\mathbf{j} = \mathbf{0}$

Thus, $\mathbf{a}\times(\mathbf{b}\times\mathbf{c}) \neq (\mathbf{a}\times\mathbf{b})\times\mathbf{c}$].

Geometrical interpretation of a × (b × c)

The vector $\mathbf{a}\times(\mathbf{b}\times\mathbf{c})$ is perpendicular to **a** as well as $\mathbf{b}\times\mathbf{c}$. But, $\mathbf{b}\times\mathbf{c}$ is perpendicular to the plane of **b** and **c**. Therefore, $\mathbf{a}\times(\mathbf{b}\times\mathbf{c})$ lies in the plane of **b** and **c**, and is perpendicular to **a**.

Similarly, $(\mathbf{a}\times\mathbf{b})\times\mathbf{c}$ lies in the plane of **a** and **b**, and is perpendicular to **c**.

Fundamental Identities for the vector triple product

Prove. $\mathbf{a}\times(\mathbf{b}\times\mathbf{c}) = (\mathbf{a}\,.\,\mathbf{c})\,\mathbf{b} - (\mathbf{a}\,.\,\mathbf{b})\,\mathbf{c}$ ...(I)

and $(\mathbf{a}\times\mathbf{b})\times\mathbf{c} = (\mathbf{a}\,.\,\mathbf{c})\,\mathbf{b} - (\mathbf{b}\,.\,\mathbf{c})\,\mathbf{a}$ ...(II)

These identities can be remembered by the middle term rule.

The vector product of three vectors is equal to the middle vector whose co-efficient is the scalar product of the remaining vectors minus the other vector in the parentheses whose co-efficient is the scalar product of the remaining vectors.

Proof. Let us prove (I).

If any one of the vectors **a**, **b** or **c** is **0**, then (I) is trivially true.

If **b** and **c** are parallel, then $\mathbf{b} = m\mathbf{c}$ or $\mathbf{c} = n\mathbf{b}$ for some scalars m and n. Hence, both sides of (I) are again **0**.

So, we assume that **a**, **b**, **c** are non-zero vectors and that **b** and **c** are not parallel.

Since **a** . **b** × **a** = 0 [∵ Scalar product with two equal vectors is always zero]

∴ Taking **a** = **b** × **c**, **b** = **a**, we have

$$(\mathbf{b} \times \mathbf{c}) \cdot \mathbf{a} \times (\mathbf{b} \times \mathbf{c}) = 0$$

By the condition of coplanarity, this implies that the vectors **b**, **c** and **a** × (**b** × **c**) are coplanar. Hence, there exist scalars m and n such that

$$\mathbf{a} \times (\mathbf{b} \times \mathbf{c}) = m\mathbf{b} - n\mathbf{c} \quad \text{...(1)}$$

Taking dot product with **a** on both sides, we have

$$\mathbf{a} \cdot [\mathbf{a} \times (\mathbf{b} \times \mathbf{c})] = m(\mathbf{a} \cdot \mathbf{b}) - n(\mathbf{a} \cdot \mathbf{c}) \quad \text{...(2)}$$

But L.H.S. is zero . (∵ a scalar triple product having two identical vectors is zero)

$$\therefore \quad m(\mathbf{a} \cdot \mathbf{b}) - n(\mathbf{a} \cdot \mathbf{c}) = 0 \quad \text{...(3)}$$

Now, if **a.b** = 0, then **a** is perpendicular to **b**. But, **a** is perpendicular to **a** × (**b** × **c**). (By L.H.S. of (2)). So, **a** is perpendicular to the plane of **b** and **c**. Hence, **a** is parallel to **b** × **c**. This implies **a.c** = 0 and **a** × (**b** × **c**) = 0

and the identity (I) is true.

Similarly, (I) is true if **a.c** = 0

So, let **a.b** ≠ 0 and **a.c** ≠ 0.

Then, from (3), $$\frac{m}{\mathbf{a} \cdot \mathbf{c}} = \frac{n}{\mathbf{a} \cdot \mathbf{b}} = \lambda \text{ (say)}$$

$$\Rightarrow \quad m = \lambda\,(\mathbf{a}.\mathbf{c}),\ n = \lambda\,(\mathbf{a}.\mathbf{b})$$

∴ (1) becomes $\mathbf{a} \times (\mathbf{b} \times \mathbf{c}) = \lambda\,[(\mathbf{a} \cdot \mathbf{c})\,\mathbf{b} - (\mathbf{a} \cdot \mathbf{b})\mathbf{c}]$ ...(4)

This is a vector identity. So it is true for each λ.

To find λ, take **a** = **i**, **b** = **i** and **c** = **j**.

Then, from (4), we find that λ = 1

∴ (4) becomes, $\mathbf{a} \times (\mathbf{b} \times \mathbf{c}) = (\mathbf{a} \cdot \mathbf{c})\,\mathbf{b} - (\mathbf{a} \cdot \mathbf{b})\,\mathbf{c}$.

Alternative proof.

Let $\mathbf{a} = a_1\mathbf{i} + a_2\mathbf{j} + a_3\mathbf{k}$, $\mathbf{b} = b_1\mathbf{i} + b_2\mathbf{j} + b_3\mathbf{k}$, $\mathbf{c} = c_1\mathbf{i} + c_2\mathbf{j} + c_3\mathbf{k}$.

$$\therefore \quad \mathbf{b} \times \mathbf{c} = \begin{vmatrix} \mathbf{i} & \mathbf{j} & \mathbf{k} \\ b_1 & b_2 & b_3 \\ c_1 & c_2 & c_3 \end{vmatrix} = \Sigma\,(b_2c_3 - b_3c_2)\mathbf{i}$$

$$\therefore \quad \mathbf{a} \times (\mathbf{b} \times \mathbf{c}) = \begin{vmatrix} \mathbf{i} & \mathbf{j} & \mathbf{k} \\ a_1 & a_2 & a_3 \\ b_2c_3 - b_3c_2 & b_3c_1 - b_1c_3 & b_1c_2 - b_2c_1 \end{vmatrix}$$

$$= \Sigma\,[a_2(b_1c_2 - b_2c_1) - a_3(b_3c_1 - b_1c_3)]\,\mathbf{i} \quad \text{...(1)}$$

Also, $$\mathbf{a} \cdot \mathbf{c} = a_1c_1 + a_2c_2 + a_3c_3$$

and $$\mathbf{a} \cdot \mathbf{b} = a_1b_1 + a_2b_2 + a_3b_3$$

$$\therefore \quad (\mathbf{a} \cdot \mathbf{c})\,\mathbf{b} - (\mathbf{a} \cdot \mathbf{b})\,\mathbf{c} = (a_1c_1 + a_2c_2 + a_3c_3)\,(b_1\mathbf{i} + b_2\mathbf{j} + b_3\mathbf{k}) - (a_1b_1 + a_2b_2 + a_3b_3)\,(c_1\mathbf{i} + c_2\mathbf{j} + c_3\mathbf{k})$$

$$= \Sigma\,(a_1b_1c_1 + a_2b_1c_2 + a_3b_1c_3 - a_1b_1c_1 - a_2b_2c_1 - a_3b_3c_1)\,\mathbf{i}$$

$$= \Sigma\,(a_2(b_1c_2 - b_2c_1) - a_3(b_3c_1 - b_1c_3)]\,\mathbf{i} \quad \text{...(2)}$$

$\therefore$ From (1) and (2), we get

$$\mathbf{a} \times (\mathbf{b} \times \mathbf{c}) = (\mathbf{a} \cdot \mathbf{c})\mathbf{b} - (\mathbf{a} \cdot \mathbf{b})\,\mathbf{c}.$$

To prove the (II) identity

Replacing **a**, **b** and **c** in (I), with **c**, **a** and **b**, we have

$$\mathbf{c} \times (\mathbf{a} \times \mathbf{b}) = (\mathbf{c} \cdot \mathbf{b})\,\mathbf{a} - (\mathbf{c} \cdot \mathbf{a})\,\mathbf{b}$$

$$\Rightarrow \quad -(\mathbf{a} \times \mathbf{b}) \times \mathbf{c} = (\mathbf{c} \cdot \mathbf{b})\,\mathbf{a} - (\mathbf{c} \cdot \mathbf{a})\,\mathbf{b}$$

$$\Rightarrow \quad (\mathbf{a} \times \mathbf{b}) \times \mathbf{c} = (\mathbf{c} \cdot \mathbf{a})\,\mathbf{b} - (\mathbf{c} \cdot \mathbf{b})\,\mathbf{a} = (\mathbf{a} \cdot \mathbf{c})\,\mathbf{b} - (\mathbf{b} \cdot \mathbf{c})\,\mathbf{a}.$$

Example 14. *Let* $\mathbf{a} = (1, 2, -3)$, $\mathbf{b} = (2, -1, 1)$, $\mathbf{c} = (-1, 1, -1)$. *Calculate* $\mathbf{a} \times (\mathbf{b} \times \mathbf{c})$ *and verify*

$$\mathbf{a} \times (\mathbf{b} \times \mathbf{c}) = (\mathbf{a} \cdot \mathbf{c})\,\mathbf{b} - (\mathbf{a} \cdot \mathbf{b})\,\mathbf{c}.$$

Sol.

$$\mathbf{b} \times \mathbf{c} = \begin{vmatrix} \mathbf{i} & \mathbf{j} & \mathbf{k} \\ 2 & -1 & 1 \\ -1 & 1 & -1 \end{vmatrix} = 0\mathbf{i} + \mathbf{j} + \mathbf{k} = (0, 1, 1)$$

$$\therefore \quad \text{L.H.S.} = \mathbf{a} \times (\mathbf{b} \times \mathbf{c}) = \begin{vmatrix} \mathbf{i} & \mathbf{j} & \mathbf{k} \\ 1 & 2 & -3 \\ 0 & 1 & 1 \end{vmatrix} = 5\mathbf{i} - \mathbf{j} + \mathbf{k} = (5, -1, 1)$$

$$(\mathbf{a} \cdot \mathbf{c})\mathbf{b} = [(1)(-1) + (2)(1) + (-3)(-1)]\,(2, -1, 1) = 4(2, -1, 1) = (8, -4, 4)$$

$$(\mathbf{a} \cdot \mathbf{b})\mathbf{c} = [(1)(2) + (2)(-1) + (-3)(1)]\,(-1, 1, -1)$$

$$= -3(-1, 1, -1) = (3, -3, 3)$$

$$\therefore \quad \text{R.H.S.} = (\mathbf{a} \cdot \mathbf{c})\mathbf{b} - (\mathbf{a} \cdot \mathbf{b})\mathbf{c} = (8, -4, 4) - (3, -3, 3) = (5, -1, 1)$$

Hence, L.H.S. = R.H.S.

Example 15. *Verify Jacobi's identity* :

$$\mathbf{a} \times (\mathbf{b} \times \mathbf{c}) + \mathbf{b} \times (\mathbf{c} \times \mathbf{a}) + \mathbf{c} \times (\mathbf{a} \times \mathbf{b}) = \mathbf{0},$$

for any vectors **a**, **b**, **c**.

Sol. From fundamental identity of vector triple product, we have

$$\mathbf{a} \times (\mathbf{b} \times \mathbf{c}) = (\mathbf{a} \cdot \mathbf{c})\,\mathbf{b} - (\mathbf{a} \cdot \mathbf{b})\,\mathbf{c}$$

$$\mathbf{b} \times (\mathbf{c} \times \mathbf{a}) = (\mathbf{b} \cdot \mathbf{a})\,\mathbf{c} - (\mathbf{b} \cdot \mathbf{c})\,\mathbf{a}$$

and

$$\mathbf{c} \times (\mathbf{a} \times \mathbf{b}) = (\mathbf{c} \cdot \mathbf{b})\,\mathbf{a} - (\mathbf{c} \cdot \mathbf{a})\,\mathbf{b}$$

Adding these, we get the Jacobi's identity.

Example 16. *For any vector* **a**, *prove that* $\mathbf{i} \times (\mathbf{a} \times \mathbf{i}) + \mathbf{j} \times (\mathbf{a} \times \mathbf{j}) + \mathbf{k} \times (\mathbf{a} \times \mathbf{k}) = 2\mathbf{a}$.

Sol. Let $\mathbf{a} = a_1\mathbf{i} + a_2\mathbf{j} + a_3\mathbf{k}$

$$\therefore \quad \mathbf{i} \times (\mathbf{a} \times \mathbf{i}) = (\mathbf{i} \cdot \mathbf{i})\mathbf{a} - (\mathbf{i} \cdot \mathbf{a})\mathbf{i} = \mathbf{a} - a_1\mathbf{i}$$

Similarly, $\mathbf{j} \times (\mathbf{a} \times \mathbf{j}) = \mathbf{a} - a_2\mathbf{j}$ and $\mathbf{k} \times (\mathbf{a} \times \mathbf{k}) = \mathbf{a} - a_3\mathbf{k}$

$$\therefore \quad \text{L.H.S.} = 3\mathbf{a} - (a_1\mathbf{i} + a_2\mathbf{j} + a_3\mathbf{k}) = 3\mathbf{a} - \mathbf{a} = 2\mathbf{a}.$$

Example 17. *Show that the vectors* $\mathbf{a} \times (\mathbf{b} \times \mathbf{c})$, $\mathbf{b} \times (\mathbf{c} \times \mathbf{a})$ *and* $\mathbf{c} \times (\mathbf{a} \times \mathbf{b})$ *are coplanar.*

Sol. Let $\mathbf{p} = \mathbf{a} \times (\mathbf{b} \times \mathbf{c})$, $\mathbf{q} = \mathbf{b} \times (\mathbf{c} \times \mathbf{a})$ and $\mathbf{r} = \mathbf{c} \times (\mathbf{a} \times \mathbf{b})$. Then,

$$\mathbf{p} + \mathbf{q} + \mathbf{r} = 0 \qquad \text{(Jacobi's identity)}$$

$$\Rightarrow \quad \mathbf{p} = (-1)\mathbf{q} + (-1)\mathbf{r}$$

$\Rightarrow$ **p**, **q**, **r** are coplanar.

21. Product of Four Vectors

1. The expression $(\mathbf{a} \times \mathbf{b}) \cdot (\mathbf{c} \times \mathbf{d})$ is called *the scalar product of four vectors* **a, b, c, d.**

$$(\mathbf{a} \times \mathbf{b}) \cdot (\mathbf{c} \times \mathbf{d}) = \begin{vmatrix} \mathbf{a.c} & \mathbf{a.d} \\ \mathbf{b.c} & \mathbf{b.d} \end{vmatrix}$$

For, $\quad (\mathbf{a} \times \mathbf{b}) \cdot (\mathbf{c} \times \mathbf{d}) = (\mathbf{a} \times \mathbf{b}) \cdot \mathbf{r}$ where $\mathbf{r} = \mathbf{c} \times \mathbf{d}$

$$= \mathbf{a} \cdot (\mathbf{b} \times \mathbf{r}) = \mathbf{a} \cdot (\mathbf{b} \times (\mathbf{c} \times \mathbf{d}))$$

$$= \mathbf{a} \cdot [(\mathbf{b} \cdot \mathbf{d})\mathbf{c} - (\mathbf{b} \cdot \mathbf{c})\mathbf{d}] = (\mathbf{a} \cdot \mathbf{c})(\mathbf{b} \cdot \mathbf{d}) - (\mathbf{a} \cdot \mathbf{d})(\mathbf{b} \cdot \mathbf{c})$$

$$= \begin{vmatrix} \mathbf{a.c} & \mathbf{a.d} \\ \mathbf{b.c} & \mathbf{b.d} \end{vmatrix}.$$

2. The expression $(\mathbf{a} \times \mathbf{b}) \times (\mathbf{c} \times \mathbf{d})$ is called *the vector product of four vectors* **a, b, c, d.**

Geometrical Interpretation

$(\mathbf{a} \times \mathbf{b}) \times (\mathbf{c} \times \mathbf{d})$ is a vector perpendicular to both $\mathbf{a} \times \mathbf{b}$ and $\mathbf{c} \times \mathbf{d}$. But, $\mathbf{a} \times \mathbf{b}$ represents a vector perpendicular to the plane of **a** and **b**.

Therefore, $(\mathbf{a} \times \mathbf{b}) \times (\mathbf{c} \times \mathbf{d})$ represents a vector coplanar with **a** and **b**. Similarly, $(\mathbf{a} \times \mathbf{b}) \times (\mathbf{c} \times \mathbf{d})$.

Hence, $(\mathbf{a} \times \mathbf{b}) \times (\mathbf{c} \times \mathbf{d})$ *is a vector parallel to the line of intersection of two planes—one being parallel to the plane of* **a** *and* **b**, *and the other parallel to the plane of* **c** *and* **d**.

Moreover, if **a, b, c, d** are coplanar, then $(\mathbf{a} \times \mathbf{b}) \times (\mathbf{c} \times \mathbf{d}) = \mathbf{0}$.

For, $\mathbf{a} \times \mathbf{b}$ is perpendicular to the plane of **a** and **b**, and $\mathbf{c} \times \mathbf{d}$ is perpendicular to **c** and **d**. Hence, $\mathbf{a} \times \mathbf{b}$ and $\mathbf{c} \times \mathbf{d}$ are parallel as they are perpendicular to the same plane of **a, b, c, d.**

Hence, $(\mathbf{a} \times \mathbf{b}) \times (\mathbf{c} \times \mathbf{d}) = \mathbf{0}$.

Example 18. *Show that* $\quad (\mathbf{a} \times \mathbf{b}) \times (\mathbf{c} \times \mathbf{d}) = [\mathbf{abd}]\,\mathbf{c} - [\mathbf{abc}]\,\mathbf{d}$

and $\quad (\mathbf{a} \times \mathbf{b}) \times (\mathbf{c} \times \mathbf{d}) = [\mathbf{cda}]\,\mathbf{b} - [\mathbf{cdb}]\,\mathbf{a}$.

Sol. Let $\quad \mathbf{a} \times \mathbf{b} = \mathbf{e}$

Using the fundamental identity (I), we have

$$(\mathbf{a} \times \mathbf{b}) \times (\mathbf{c} \times \mathbf{d})$$

$$= \mathbf{e} \times (\mathbf{c} \times \mathbf{d}) = (\mathbf{e} \cdot \mathbf{d})\,\mathbf{c} - (\mathbf{e} \cdot \mathbf{c})\,\mathbf{d}$$

$$= (\mathbf{a} \times \mathbf{b} \cdot \mathbf{d})\,\mathbf{c} - (\mathbf{a} \times \mathbf{b} \cdot \mathbf{c})\,\mathbf{d} = [\mathbf{abd}]\,\mathbf{c} - [\mathbf{abc}]\,\mathbf{d}$$

Similarly, if we take $\mathbf{c} \times \mathbf{d} = \mathbf{f}$ and use the fundamental identity (II), we get the second result.

Example 19. *Show that* $\quad (\mathbf{a} \times \mathbf{b}) \cdot (\mathbf{b} \times \mathbf{c}) \times (\mathbf{c} \times \mathbf{a}) = [\mathbf{abc}]^2$.

Sol. From above example, we have

$$(\mathbf{b} \times \mathbf{c}) \times (\mathbf{c} \times \mathbf{a}) = [\mathbf{bca}]\,\mathbf{c} - [\mathbf{bcc}]\,\mathbf{a}$$

$$= [\mathbf{abc}]\,\mathbf{c} \qquad [\because [\mathbf{bcc}] = \mathbf{0} \text{ and } [\mathbf{bca}] = [\mathbf{abc}]]$$

$$\therefore \quad (\mathbf{a} \times \mathbf{b}) \cdot (\mathbf{b} \times \mathbf{c}) \times (\mathbf{c} \times \mathbf{a})$$

$$= (\mathbf{a} \times \mathbf{b}) \cdot [\mathbf{abc}]\mathbf{c} = [\mathbf{abc}]\,(\mathbf{a} \times \mathbf{b}) \cdot \mathbf{c}$$

$$= [\mathbf{abc}][\mathbf{abc}] = [\mathbf{abc}]^2.$$

Example 20. *Prove that* $\quad (\mathbf{a} \times \mathbf{b}) \times (\mathbf{a} \times \mathbf{c}) \cdot \mathbf{d} = \mathbf{a} \cdot \mathbf{d}\,[\mathbf{abc}]$.

Sol. Let $\mathbf{a} \times \mathbf{b} = \mathbf{e}$

$$\therefore \quad (\mathbf{a} \times \mathbf{b}) \times (\mathbf{a} \times \mathbf{c}) \cdot \mathbf{d} = \{\mathbf{e} \times (\mathbf{a} \times \mathbf{c})\} \cdot \mathbf{d}$$

$$= \{(\mathbf{e} \cdot \mathbf{c})\,\mathbf{a} - (\mathbf{e} \cdot \mathbf{a})\,\mathbf{c}\} \cdot \mathbf{d} = \{(\mathbf{a} \times \mathbf{b} \cdot \mathbf{c})\,\mathbf{a} - (\mathbf{a} \times \mathbf{b} \cdot \mathbf{a})\,\mathbf{c}\} \cdot \mathbf{d}$$

$$= \{[\mathbf{abc}]\,\mathbf{a} - \mathbf{0c}\} \cdot \mathbf{d} \qquad (\because \mathbf{a} \times \mathbf{b} \cdot \mathbf{a} = 0)$$

$$= [\mathbf{abc}]\,\mathbf{a} \cdot \mathbf{d} = \mathbf{a} \cdot \mathbf{d}\,[\mathbf{abc}].$$

Example 21. *Prove that* $(\mathbf{a} \times \mathbf{b}) \times \mathbf{c} = \mathbf{a} \times (\mathbf{b} \times \mathbf{c})$ *iff* $(\mathbf{c} \times \mathbf{a}) \times \mathbf{b} = \mathbf{0}$.

Sol. $(\mathbf{a} \times \mathbf{b}) \times \mathbf{c} = \mathbf{a} \times (\mathbf{b} \times \mathbf{c})$

iff $-\mathbf{c} \times (\mathbf{a} \times \mathbf{b}) = \mathbf{a} \times (\mathbf{b} \times \mathbf{c})$

i.e., iff $-[(\mathbf{c} \cdot \mathbf{b})\mathbf{a} - (\mathbf{c} \cdot \mathbf{a})\mathbf{b}] = (\mathbf{a} \cdot \mathbf{c})\mathbf{b} - (\mathbf{a} \cdot \mathbf{b})\mathbf{c}$

i.e., iff $-(\mathbf{c} \cdot \mathbf{b})\mathbf{a} + (\mathbf{a} \cdot \mathbf{c})\mathbf{b} = (\mathbf{a} \cdot \mathbf{c})\mathbf{b} - (\mathbf{a} \cdot \mathbf{b})\mathbf{c}$

i.e., iff $-(\mathbf{c} \cdot \mathbf{b})\mathbf{a} = -(\mathbf{a} \cdot \mathbf{b})\mathbf{c}$

i.e., iff $(\mathbf{c} \cdot \mathbf{b})\mathbf{a} = (\mathbf{b} \cdot \mathbf{a})\mathbf{c}$

i.e., iff $(\mathbf{b} \cdot \mathbf{a})\mathbf{c} - (\mathbf{b} \cdot \mathbf{c})\mathbf{a} = \mathbf{0}$

i.e., iff $\mathbf{b} \times (\mathbf{c} \times \mathbf{a}) = \mathbf{0}$

i.e., iff $-(\mathbf{c} \times \mathbf{a}) \times \mathbf{b} = \mathbf{0}$

i.e., iff $(\mathbf{c} \times \mathbf{a}) \times \mathbf{b} = \mathbf{0}$.

Example 22. *Show that* $(\mathbf{a} \times \mathbf{b}) \times \mathbf{c} = \mathbf{a} \times (\mathbf{b} \times \mathbf{c})$ *iff* **a** *and* **c** *are collinear.*

Sol. Proceeding as in the above example, we get

$$(\mathbf{c} \cdot \mathbf{b})\mathbf{a} = (\mathbf{b} \cdot \mathbf{a})\mathbf{c}$$

i.e., **a** and **c** are collinear.

Example 23. *If* $\mathbf{a} = -\mathbf{i} + \mathbf{j} + \mathbf{k}$, $\mathbf{b} = 2\mathbf{j} + \mathbf{k}$, *find the vector* $\vec{X}$ *satisfying the conditions :*

(*i*) *that it is coplanar with* **a** *and* **b** (*ii*) *that it is perpendicular to* **b**

(*iii*) *that* $\mathbf{a} \cdot \vec{X} = 7$.

Sol. Since $\vec{X}$ is in the plane of **a** and **b** and is perpendicular to **b**,

$$\therefore \quad \vec{X} = \lambda[\mathbf{b} \times (\mathbf{a} \times \mathbf{b})]$$

$$\Rightarrow \quad \vec{X} = \lambda[(\mathbf{b} \cdot \mathbf{b})\mathbf{a} - (\mathbf{b} \cdot \mathbf{a})\mathbf{b}]$$

$$= \lambda[5(-\mathbf{i} + \mathbf{j} + \mathbf{k}) - (-1)(2\mathbf{i} + 0\mathbf{j} + \mathbf{k})]$$

$$= \lambda[-3\mathbf{i} + 5\mathbf{j} + 6\mathbf{k}]$$

Also, $\mathbf{a} \cdot \vec{X} = 7 \quad \Rightarrow \quad (-\mathbf{i} + \mathbf{j} + \mathbf{k}) \cdot \lambda(-3\mathbf{i} + 5\mathbf{j} + 6\mathbf{k}) = 7$

$$\Rightarrow \quad \lambda(3 + 5 + 6) = 7 \quad \Rightarrow \quad \lambda = \frac{1}{2}$$

$$\therefore \quad \vec{X} = \frac{1}{2}(-3\mathbf{i} + 5\mathbf{j} + 6\mathbf{k}).$$

Example 24. *If* **a**, **b**, **c** *are 3 unit vectors such that* $\mathbf{b} \times (\mathbf{c} \times \mathbf{a}) = \frac{1}{2}\mathbf{c}$, *find angles which* **b** *makes with* **c** *and* **a** ; **c** *and* **a** *being non-parallel.* (M.D.U. 2007)

Sol. Since $\mathbf{b} \times (\mathbf{c} \times \mathbf{a}) = \frac{1}{2}\mathbf{c}$

$$\therefore \quad (\mathbf{b} \cdot \mathbf{a})\,\mathbf{c} - (\mathbf{b} \cdot \mathbf{c})\,\mathbf{a} = \tfrac{1}{2}\mathbf{c} \quad \Rightarrow \quad (\mathbf{b} \cdot \mathbf{a} - \tfrac{1}{2})\,\mathbf{c} - (\mathbf{b} \cdot \mathbf{c})\,\mathbf{a} = 0$$

$$\Rightarrow \quad x\mathbf{c} - y\mathbf{a} = 0 \text{ when } x = \mathbf{b} \cdot \mathbf{a} - \tfrac{1}{2},\ y = \mathbf{b} \cdot \mathbf{c}$$

Since **b** and **a** are non-parallel, it follows that $x = 0, y = 0$

$$\Rightarrow \quad \mathbf{b} \cdot \mathbf{a} = \tfrac{1}{2} \quad \text{and} \quad \mathbf{b} \cdot \mathbf{c} = 0$$

$$\Rightarrow \quad 1 \cdot 1 \cos\theta = \tfrac{1}{2} \quad \text{and} \quad 1 \cdot 1 \cos\phi = 0 \qquad [\because |\mathbf{a}| = |\mathbf{b}| = |\mathbf{c}| = 1]$$

i.e., $\theta = \frac{\pi}{3}$ and $\phi = \frac{\pi}{2}$ where $\theta = \angle \mathbf{a}, \mathbf{b}$ and $\phi = \angle \mathbf{b}, \mathbf{c}$.

22. Reciprocal Sets of Vectors

If **a, b, c** *are three non-zero, non-collinear and non-coplanar vectors, then the three vectors* **a′, b′, c′** *defined by* $\mathbf{a}' = \frac{\mathbf{b}\times\mathbf{c}}{[\mathbf{a}\ \mathbf{b}\ \mathbf{c}]}$, $\mathbf{b}' = \frac{\mathbf{c}\times\mathbf{a}}{[\mathbf{a}\ \mathbf{b}\ \mathbf{c}]}$, $\mathbf{c}' = \frac{\mathbf{a}\times\mathbf{b}}{[\mathbf{a}\ \mathbf{b}\ \mathbf{c}]}$ *are called the reciprocal set of vectors for the vectors* **a, b, c.** (K.U.K. 2007)

Properties of Reciprocal Set of Vectors

If $\mathbf{a}' = \frac{\mathbf{b}\times\mathbf{c}}{[\mathbf{abc}]}$, $\mathbf{b}' = \frac{\mathbf{c}\times\mathbf{a}}{[\mathbf{abc}]}$, $\mathbf{c}' = \frac{\mathbf{a}\times\mathbf{b}}{[\mathbf{abc}]}$

where $[\mathbf{abc}] \neq 0$, *then*

(*i*) $\mathbf{a}' . \mathbf{a} = \mathbf{b}' . \mathbf{b} = \mathbf{c}' . \mathbf{c} = 1$ (K.U.K. 2007)

(*ii*) $\mathbf{a}' . \mathbf{b} = \mathbf{a}' . \mathbf{c} = 0,\ \mathbf{b}' . \mathbf{a} = \mathbf{b}' . \mathbf{c} = 0,\ \mathbf{c}' . \mathbf{a} = \mathbf{c}' . \mathbf{b} = 0$

(*iii*) $[\mathbf{a}'\ \mathbf{b}'\ \mathbf{c}'] = \frac{1}{[\mathbf{a}\ \mathbf{b}\ \mathbf{c}]}$ (K.U.K. 2007)

(*iv*) **a′, b′, c′** *are non-coplanar if* **a, b, c** *are non-coplanar*

(*v*) $\mathbf{a} = \frac{\mathbf{b}'\times\mathbf{c}'}{[\mathbf{a}'\mathbf{b}'\mathbf{c}']}$, $\mathbf{b} = \frac{\mathbf{c}'\times\mathbf{a}'}{[\mathbf{a}'\mathbf{b}'\mathbf{c}']}$, $\mathbf{c} = \frac{\mathbf{a}'\times\mathbf{b}'}{[\mathbf{a}'\mathbf{b}'\mathbf{c}']}$

(*vi*) $\mathbf{a}\times\mathbf{a}' + \mathbf{b}\times\mathbf{b}' + \mathbf{c}\times\mathbf{c}' = \mathbf{0}$ (*vii*) $\mathbf{a} . \mathbf{a}' + \mathbf{b} . \mathbf{b}' + \mathbf{c} . \mathbf{c}' = 3.$

Solution. (*i*) $\mathbf{a}' . \mathbf{a} = \mathbf{a} . \mathbf{a}' = \mathbf{a} . \frac{\mathbf{b}\times\mathbf{c}}{[\mathbf{abc}]} = \frac{\mathbf{a} . \mathbf{b}\times\mathbf{c}}{[\mathbf{abc}]} = \frac{[\mathbf{abc}]}{[\mathbf{abc}]} = 1$

Similarly, $\mathbf{b}' . \mathbf{b} = \mathbf{c}' . \mathbf{c} = 1.$

(*ii*) $$\mathbf{a}' . \mathbf{b} = \mathbf{b} . \mathbf{a}' = \mathbf{b} . \frac{\mathbf{b}\times\mathbf{c}}{[\mathbf{abc}]} = \frac{\mathbf{b} . \mathbf{b}\times\mathbf{c}}{[\mathbf{abc}]} = \frac{[\mathbf{bbc}]}{[\mathbf{abc}]} = 0$$

Similarly, the other results follow.

(*iii*) $$[\mathbf{a}'\ \mathbf{b}'\ \mathbf{c}'] = \frac{\mathbf{b}\times\mathbf{c}}{[\mathbf{abc}]} . \frac{\mathbf{c}\times\mathbf{a}}{[\mathbf{abc}]} \times \frac{\mathbf{a}\times\mathbf{b}}{[\mathbf{abc}]}$$
$$= \frac{(\mathbf{b}\times\mathbf{c}) . (\mathbf{c}\times\mathbf{a})\times(\mathbf{a}\times\mathbf{b})}{[\mathbf{abc}]^3} = \frac{(\mathbf{a}\times\mathbf{b}) . (\mathbf{b}\times\mathbf{c})\times(\mathbf{c}\times\mathbf{a})}{[\mathbf{abc}]^3}$$
$$= \frac{[\mathbf{abc}]^2}{[\mathbf{abc}]^3} \quad \text{(Example 19)}$$
$$= \frac{1}{[\mathbf{abc}]}.$$

(*iv*) If **a, b, c** are non-coplanar, then $[\mathbf{abc}] \neq 0$. Then, from part (*iii*) above, $[\mathbf{a}'\ \mathbf{b}'\ \mathbf{c}'] \neq 0$. This implies **a′, b′, c′** are non-coplanar.

(*v*) Since $\mathbf{b}' = \frac{\mathbf{c}\times\mathbf{a}}{[\mathbf{abc}]}$ and $\mathbf{c}' = \frac{\mathbf{a}\times\mathbf{b}}{[\mathbf{abc}]}$,

it follows that **a** is orthogonal to both **b′** and **c′**.

Let $$\mathbf{a} = \lambda(\mathbf{b}'\times\mathbf{c}') \quad \text{...(1)}$$

$\therefore$ $$\mathbf{a}' . \mathbf{a} = \mathbf{a} . \mathbf{a}' = \lambda(\mathbf{b}'\times\mathbf{c}') . \mathbf{a}'$$
$$= \lambda[\mathbf{b}'\ \mathbf{c}'\ \mathbf{a}'] = \lambda[\mathbf{a}'\ \mathbf{b}'\ \mathbf{c}']$$

But $\quad \mathbf{a}' . \mathbf{a} = 1$

$\therefore \quad \lambda[\mathbf{a}'\ \mathbf{b}'\ \mathbf{c}'] = 1$

$$\therefore \quad \lambda = \frac{1}{[\mathbf{a}'\mathbf{b}'\mathbf{c}']}$$

$$\therefore \text{ from (1),} \quad \mathbf{a} = \frac{\mathbf{b}' \times \mathbf{c}'}{[\mathbf{a}'\mathbf{b}'\mathbf{c}']}$$

Similarly, the other results follow.

[*Two sets* **a**, **b**, **c** *and* **a′**, **b′**, **c′** *defined above are called reciprocal sets of vectors.*]

$$(vi)\ \mathbf{a} \times \mathbf{a}' + \mathbf{b} \times \mathbf{b}' + \mathbf{c} \times \mathbf{c}' = \frac{1}{[\mathbf{abc}]} \{\mathbf{a} \times (\mathbf{b} \times \mathbf{c}) + \mathbf{b} \times (\mathbf{c} \times \mathbf{a}) + \mathbf{c} \times (\mathbf{a} \times \mathbf{b})\}$$

$$= \frac{1}{[\mathbf{abc}]} [0] = \mathbf{0} \qquad \text{[Jacobi's identity, Example 15]}$$

(*vii*) Simple.

Example 25. *Given* $\mathbf{a} = 2\mathbf{i} - \mathbf{j} + 3\mathbf{k}$, $\mathbf{b} = 2\mathbf{i} + \mathbf{j} - \mathbf{k}$ *and* $\mathbf{c} = \mathbf{i} + 3\mathbf{j} - \mathbf{k}$, *find the reciprocal triad* **a′**, **b′**, **c′** *and verify* $[\mathbf{a}\ \mathbf{b}\ \mathbf{c}]\ [\mathbf{a}'\ \mathbf{b}'\ \mathbf{c}'] = 1$. (K.U. 2003 ; M.D.U. 2007)

Sol.
$$[\mathbf{a}\ \mathbf{b}\ \mathbf{c}] = \begin{vmatrix} 2 & -1 & 3 \\ 2 & 1 & -1 \\ 1 & 3 & -1 \end{vmatrix} = 2(-1+3) + 1(-2+1) + 3(6-1) = 18 \neq 0$$

If **a′**, **b′**, **c′** is the set of vectors reciprocal to **a**, **b**, **c** then

$$\mathbf{a}' = \frac{\mathbf{b} \times \mathbf{c}}{[\mathbf{a}\ \mathbf{b}\ \mathbf{c}]}, \quad \mathbf{b}' = \frac{\mathbf{c} \times \mathbf{a}}{[\mathbf{a}\ \mathbf{b}\ \mathbf{c}]}, \quad \mathbf{c}' = \frac{\mathbf{a} \times \mathbf{b}}{[\mathbf{a}\ \mathbf{b}\ \mathbf{c}]}$$

$$\therefore \quad \mathbf{a}' = \frac{1}{18} \begin{vmatrix} \mathbf{i} & \mathbf{j} & \mathbf{k} \\ 2 & 1 & -1 \\ 1 & 3 & -1 \end{vmatrix} = \frac{1}{18} (2\mathbf{i} + \mathbf{j} + 5\mathbf{k})$$

$$\mathbf{b}' = \frac{1}{18} \begin{vmatrix} \mathbf{i} & \mathbf{j} & \mathbf{k} \\ 1 & 3 & -1 \\ 2 & -1 & 3 \end{vmatrix} = \frac{1}{18} (8\mathbf{i} - 5\mathbf{j} - 7\mathbf{k})$$

$$\mathbf{c}' = \frac{1}{18} \begin{vmatrix} \mathbf{i} & \mathbf{j} & \mathbf{k} \\ 2 & -1 & 3 \\ 2 & 1 & -1 \end{vmatrix} = \frac{1}{18} (-2\mathbf{i} + 8\mathbf{j} + 4\mathbf{k})$$

$$[\mathbf{a}'\ \mathbf{b}'\ \mathbf{c}'] = \begin{vmatrix} \frac{2}{18} & \frac{1}{18} & \frac{5}{18} \\ \frac{8}{18} & \frac{-5}{18} & \frac{-7}{18} \\ \frac{-2}{18} & \frac{8}{18} & \frac{4}{18} \end{vmatrix} = \frac{1}{(18)^3} \begin{vmatrix} 2 & 1 & 5 \\ 8 & -5 & -7 \\ -2 & 8 & 4 \end{vmatrix} = \frac{1}{(18)^3} \times 324 = \frac{1}{18}$$

Hence, $\quad [\mathbf{a}\ \mathbf{b}\ \mathbf{c}]\ [\mathbf{a}'\ \mathbf{b}'\ \mathbf{c}'] = 1.$

EXERCISE 2

1. If $\mathbf{a} = 2\mathbf{i} + \mathbf{j} + 3\mathbf{k}$, $\mathbf{b} = -\mathbf{i} + 2\mathbf{j} + \mathbf{k}$, $\mathbf{c} = 3\mathbf{i} + \mathbf{j} + 2\mathbf{k}$, find
   (*i*) $\mathbf{a} \cdot (\mathbf{b} \times \mathbf{c})$ and
   (*ii*) $(\mathbf{a} \times \mathbf{b}) \cdot \mathbf{c}$. Hence, show that $\mathbf{a} \cdot (\mathbf{b} \times \mathbf{c}) = (\mathbf{a} \times \mathbf{b}) \cdot \mathbf{c}$.
2. Find the volume of the parallelopiped whose coterminus edges are
   $$\mathbf{i} + \mathbf{j} + 2\mathbf{k},\ \mathbf{i} + 2\mathbf{j} + 4\mathbf{k},\ 2\mathbf{i} + \mathbf{j} + 3\mathbf{k}.$$
3. Find λ so that the vectors $\mathbf{a} = 2\mathbf{i} - \mathbf{j} + \mathbf{k}$, $\mathbf{b} = \mathbf{i} + 2\mathbf{j} - 3\mathbf{k}$, $\mathbf{c} = 3\mathbf{i} + \lambda\mathbf{j} + 5\mathbf{k}$ are coplanar.
4. Three concurrent edges OA, OB, OC of a parallelopiped represented by the three vectors $2\mathbf{i} + \mathbf{j} - \mathbf{k}$, $\mathbf{i} + 2\mathbf{j} + 3\mathbf{k}$, $-3\mathbf{i} - \mathbf{j} + \mathbf{k}$. Find the area of the face formed by OA and OB and also the volume of the solid.
5. Show that $[\mathbf{a} - \mathbf{b}\ \ \mathbf{b} - \mathbf{c}\ \ \mathbf{c} - \mathbf{a}] = 0$.
6. Show that $\mathbf{a} \cdot \mathbf{b} \times \mathbf{c} = \mathbf{a} \times \mathbf{b} \cdot \mathbf{c}$.
7. If $\mathbf{a} = (1, 2, 1)$ $\mathbf{b} = (2, 0, -1)$, $\mathbf{c} = (0, -1, 2)$, evaluate
   (*i*) $[\mathbf{abc}]$ (*ii*) $\mathbf{a} \times (\mathbf{b} \times \mathbf{c})$ (*iii*) $(\mathbf{a} \times \mathbf{b}) \times \mathbf{c}$
   (*iv*) $(\mathbf{a} \times \mathbf{b}) \times (\mathbf{b} \times \mathbf{c})$ and (*v*) $(\mathbf{b} \cdot \mathbf{c})(\mathbf{a} \times \mathbf{b})$.
8. For vectors **a**, **b**, **c** of question 7, verify that
   (*i*) $\mathbf{a} \times (\mathbf{b} \times \mathbf{c}) = (\mathbf{a} \cdot \mathbf{c})\,\mathbf{b} - (\mathbf{a} \cdot \mathbf{b})\,\mathbf{c}$ and (*ii*) $(\mathbf{a} \times \mathbf{b}) \times \mathbf{c} = (\mathbf{a} \cdot \mathbf{c})\,\mathbf{b} - (\mathbf{b} \cdot \mathbf{c})\,\mathbf{a}$.
9. Show that **a**, **b**, **c** are coplanar iff $\mathbf{a} \times \mathbf{b}$, $\mathbf{b} \times \mathbf{c}$, $\mathbf{c} \times \mathbf{a}$ are coplanar.
10. Show that the set of unit vectors **i**, **j**, **k** is self reciprocal.
11. If **a** and **b** lie in a plane normal to the plane containing **c** and **d**, show that $(\mathbf{a} \times \mathbf{b}) \cdot (\mathbf{c} \times \mathbf{d}) = 0$.
12. Prove that $$[\mathbf{a} \times \mathbf{b}, \mathbf{b} \times \mathbf{c}, \mathbf{c} \times \mathbf{a}] = \begin{vmatrix} \mathbf{a} \cdot \mathbf{a} & \mathbf{a} \cdot \mathbf{b} & \mathbf{a} \cdot \mathbf{c} \\ \mathbf{b} \cdot \mathbf{a} & \mathbf{b} \cdot \mathbf{b} & \mathbf{b} \cdot \mathbf{c} \\ \mathbf{c} \cdot \mathbf{a} & \mathbf{c} \cdot \mathbf{b} & \mathbf{c} \cdot \mathbf{c} \end{vmatrix}.$$
    [**Hint.** $(\mathbf{a} \times \mathbf{b}, \mathbf{b} \times \mathbf{c}, \mathbf{c} \times \mathbf{a}) = [\mathbf{a}\ \mathbf{b}\ \mathbf{c}]^2$]
13. For vectors **a**, **b**, **c**, **l**, **m** prove : $$[\mathbf{a}\ \mathbf{b}\ \mathbf{c}](\mathbf{l} \times \mathbf{m}) = \begin{vmatrix} \mathbf{a} & \mathbf{b} & \mathbf{c} \\ \mathbf{a} \cdot \mathbf{l} & \mathbf{b} \cdot \mathbf{l} & \mathbf{c} \cdot \mathbf{l} \\ \mathbf{a} \cdot \mathbf{m} & \mathbf{b} \cdot \mathbf{m} & \mathbf{c} \cdot \mathbf{m} \end{vmatrix}.$$
14. Prove : $\mathbf{a} \times \{\mathbf{b} \times (\mathbf{c} \times \mathbf{d})\} = (\mathbf{b} \cdot \mathbf{d})(\mathbf{a} \times \mathbf{c}) - (\mathbf{b} \cdot \mathbf{c})(\mathbf{a} \times \mathbf{d})$.
15. If $\mathbf{a} = 3\mathbf{i} - \mathbf{j} + \mathbf{k}$, $\mathbf{b} = \mathbf{i} + 3\mathbf{j} - \mathbf{k}$, $\mathbf{c} = -\mathbf{i} + \mathbf{j} + 3\mathbf{k}$, state which of the following is meaningful and evaluate any one of those that are meaningful $(\mathbf{a} \cdot \mathbf{b}) \times \mathbf{c}$, $\mathbf{a} \times (\mathbf{b} \times \mathbf{c})$, $(\mathbf{a} \times \mathbf{b}) \cdot \mathbf{c}$.
16. If $\mathbf{a} = 2\mathbf{i} - \mathbf{j} + 3\mathbf{k}$, $\mathbf{b} = 2\mathbf{i} + \mathbf{j} - \mathbf{k}$, $\mathbf{c} = \mathbf{i} + 3\mathbf{j} - \mathbf{k}$, find the reciprocal triad of vectors $\mathbf{a}'$, $\mathbf{b}'$, $\mathbf{c}'$. (*M.D.U. 2003*)
17. Show that the volume of the tetrahedron ABCD is $\frac{1}{6}$ [AB, AC, AD]. Hence, find the volume of the tetrahedron formed by the points (1, 1, 1), (2, 1, 3), (3, 2, 2) and (3, 3, 4). (*K.U. 2003 S*)

Answers

1. -10 2. 1 cu. unit 3. $\lambda = -4$ 4. $\sqrt{83}$ sq. units, 5 cubic units

7. (*i*) -11 (*ii*) $\mathbf{j} - 2\mathbf{k}$ (*iii*) $2\mathbf{i} + 4\mathbf{j} + 2\mathbf{k}$ (*iv*) $-22\mathbf{i} + 11\mathbf{k}$ (*v*) $4\mathbf{i} - 6\mathbf{j} + 8\mathbf{k}$

15. $-2\mathbf{i} - 2\mathbf{j} + \mathbf{k}$, 36.

2

Differentiation of Vectors

1. Scalar and Vector Functions

Scalar Function. *Let t be a scalar variable. Then, a rule to be denoted by 'f' which associates to each t in some interval $a \leq t \leq b$ (say), a unique scalar f(t), is called a scalar function of the scalar variable t. The scalar f(t) is called the value of f at t.*

Generally, the scalar variable t represents either time or a set of co-ordinates.

Vector Function. *Let t be a scalar variable. Then, a rule '**f**' which associates to each t in some interval $a \leq t \leq b$ (say), a unique vector **f**(t), is called a vector function of the scalar variable t. The vector **f**(t) is called the value of **f** at t.*

In rectangular coordinate system, the vector function **f** can be expressed in component form as :

$$\mathbf{f} = (f_1, f_2, f_3) = f_1\mathbf{i} + f_2\mathbf{j} + f_3\mathbf{k}$$

f_1, f_2 and f_3 are scalar functions of t and are called components of **f**.

Note. Sometimes we denote the vector function **f** of the scalar variable t by **r**, and write $\mathbf{r} = \mathbf{f}(t)$, then **r** may or may not be the position vector of a point.

If the components of the vector **r** along the three axes are r_1, r_2, r_3, then

$$\mathbf{r} = r_1\mathbf{i} + r_2\mathbf{j} + r_3\mathbf{k}.$$

But, when **r** is the position vector of a point (x, y, z), then

$$\mathbf{r} = x\mathbf{i} + y\mathbf{j} + z\mathbf{k}.$$

2. Limit and Continuity

*Let a vector function **f** be defined for all values of t in some neighbourhood about a point t_0, except possibly at t_0. Then, a vector **a** is called the limit of **f** as $t \to t_0$ if given any $\varepsilon > 0$, however small, there exists a real number $\delta > 0$ such that*

$$|\mathbf{f}(t) - \mathbf{a}| < \varepsilon, \text{ whenever } 0 < |t - t_0| < \delta$$

and we write

$$\lim_{t \to t_0} \mathbf{f}(t) = \mathbf{a} \quad \text{or} \quad \underset{t \to t_0}{Lt}\, \mathbf{f}(t) = \mathbf{a}.$$

The same definition holds for the limit of a scalar function if **f** is replaced by a scalar function f, and vector **a** by a scalar a.

We state below (without proof) some important theorems on limits of a vector function.

3. Theorem. *If* $\mathbf{f} = f_1\mathbf{i} + f_2\mathbf{j} + f_3\mathbf{k}$ *and* $\mathbf{a} = a_1\mathbf{i} + a_2\mathbf{j} + a_3\mathbf{k}$, *then*

$$\underset{t \to t_0}{Lt}\, \mathbf{f}(t) = \mathbf{a} \text{ if and only if } \underset{t \to t_0}{Lt}\, f_i(t) = a_i,\ i = 1, 2, 3.$$

4. Theorem. *Let* **f** *and* **g** *be vector functions of a scalar variable t. Let* ϕ *be a scalar function of t.*

Then,

(i) $\underset{t \to t_0}{\text{Lt}} [\mathbf{f}(t) + \mathbf{g}(t)] = \underset{t \to t_0}{\text{Lt}} \mathbf{f}(t) + \underset{t \to t_0}{\text{Lt}} \mathbf{g}(t)$

(ii) $\underset{t \to t_0}{\text{Lt}} [\mathbf{f}(t) - \mathbf{g}(t)] = \underset{t \to t_0}{\text{Lt}} \mathbf{f}(t) - \underset{t \to t_0}{\text{Lt}} \mathbf{g}(t)$

(iii) $\underset{t \to t_0}{\text{Lt}} [\mathbf{f}(t) \cdot \mathbf{g}(t)] = [\underset{t \to t_0}{\text{Lt}} \mathbf{f}(t)] \cdot [\underset{t \to t_0}{\text{Lt}} \mathbf{g}(t)]$

(iv) $\underset{t \to t_0}{\text{Lt}} [\mathbf{f}(t) \times \mathbf{g}(t)] = [\underset{t \to t_0}{\text{Lt}} \mathbf{f}(t)] \times [\underset{t \to t_0}{\text{Lt}} \mathbf{g}(t)]$

(v) $\underset{t \to t_0}{\text{Lt}} [\phi(t) \times \mathbf{f}(t)] = [\underset{t \to t_0}{\text{Lt}} \phi(t)] [\underset{t \to t_0}{\text{Lt}} \mathbf{f}(t)]$

A vector function **f** *is said to be continuous at* $t = t_0$ *if it is defined in some neighbourhood of* t_0 *and* $\underset{t \to t_0}{\text{Lt}} \mathbf{f}(t) = \mathbf{f}(t_0)$.

Analogous to Theorem 1, we have the following theorem (whose proof is omitted).

5. Theorem. *If* $\mathbf{f} = f_1 \mathbf{i} + f_2 \mathbf{j} + f_3 \mathbf{k}$ *is a function of t, then* **f** *is continuous at* t_0 *if and only if* f_1, f_2 *and* f_3 *are all continuous at* t_0.

6. Differentiation of Vectors

Let $\mathbf{r} = \mathbf{f}(t)$ *be a vector function of the scalar variable t. Let* δt *be a small change in t and* $\delta \mathbf{r}$ *be the corresponding change in* **r**.

Then, $\underset{\delta t \to 0}{Lt} \dfrac{\delta \mathbf{r}}{\delta t} = \underset{\delta t \to 0}{Lt} \dfrac{\mathbf{f}(t + \delta t) - \mathbf{f}(t)}{\delta t}$,

if exists is called the derivative of the vector function **r** *w.r.t. the scalar t and is denoted by*

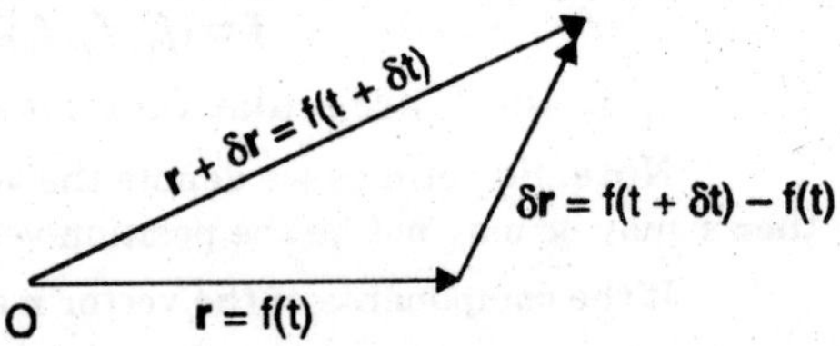

$$\frac{d\mathbf{r}}{dt} \quad or \quad \frac{d\mathbf{f}}{dt} \quad or \quad \frac{d}{dt}\mathbf{f}(t) \quad or \quad \mathbf{f}'(t)$$

We also say that the vector function **f** is differentiable.

If the vector function **f** is in component form

$$\mathbf{f} = f_1 \mathbf{i} + f_2 \mathbf{j} + f_3 \mathbf{k},$$

then $\mathbf{f}'(t)$ exists if and only if the derivatives

$$f_i'(t) = \underset{\delta t \to 0}{\text{Lt}} \frac{f_i(t + \delta t) - f_i(t)}{\delta t}$$

of the scalar functions f_i, $(i = 1, 2, 3)$ exist.

Then, we can write $\mathbf{f}'(t) = f_1'(t)\,\mathbf{i} + f_2'(t)\,\mathbf{j} + f_3'(t)\,\mathbf{k}$

or
$$\frac{d\mathbf{f}}{dt} = \frac{df_1}{dt}\mathbf{i} + \frac{df_2}{dt}\mathbf{j} + \frac{df_3}{dt}\mathbf{k}.$$

Higher derivatives of a vector function are defined in a manner similar to that for a scalar function of a single variable. Thus,

$$f^2(t) = \frac{d^2\mathbf{f}}{dt^2} = \frac{d}{dt}\left[\frac{d}{dt}\mathbf{f}(t)\right]$$

and so on.

Like elementary calculus, a differentiable vector function is necessarily continuous but not conversely.

Unless otherwise stated, here we shall deal with those functions that are differentiable to any order needed in a particular discussion.

7. Rules for differentiation of vector functions

The rules for differentiation of vector functions are similar to those for scalar functions—with one exception :

To differentiate the vector product of vector function, the order of factors must be preserved because the vector product is not commutative.

If **f**, **g** and **h** are differentiable vector functions and ϕ is a scalar function of a scalar variable t, then

1. $\frac{d}{dt}(\mathbf{f} \pm \mathbf{g}) = \frac{d\mathbf{f}}{dt} \pm \frac{d\mathbf{g}}{dt}$

2. $\frac{d}{dt}(\mathbf{f} \cdot \mathbf{g}) = \mathbf{f} \cdot \frac{d\mathbf{g}}{dt} + \mathbf{g} \cdot \frac{d\mathbf{f}}{dt}$

[Since $\mathbf{f} \cdot \mathbf{g} = \mathbf{g} \cdot \mathbf{f}$

$\therefore$ In evaluating $\frac{d}{dt}(\mathbf{f} \cdot \mathbf{g})$, we do not care about the order of **f** and **g**]

3. $\frac{d}{dt}(\mathbf{f} \times \mathbf{g}) = \mathbf{f} \times \frac{d\mathbf{g}}{dt} + \frac{d\mathbf{f}}{dt} \times \mathbf{g}$

[Since $\mathbf{f} \times \mathbf{g} = -\,\mathbf{g} \times \mathbf{f}$,

$\therefore$ In evaluating $\frac{d}{dt}(\mathbf{f} \times \mathbf{g})$, we must keep the order of **f** and **g** in tact]

4. $\frac{d}{dt}(\phi\,\mathbf{f}) = \phi\,\frac{d\mathbf{f}}{dt} + \frac{d\phi}{dt}\,\mathbf{f}$

5. $\frac{d}{dt}(\mathbf{fgh}) = \left[\frac{d\mathbf{f}}{dt}\,\mathbf{gh}\right] + \left[\mathbf{f}\,\frac{d\mathbf{g}}{dt}\,\mathbf{h}\right] + \left[\mathbf{fg}\,\frac{d\mathbf{h}}{dt}\right]$

where [**fgh**] denotes the scalar triple product of **f**, **g** and **h**.

[In evaluating $\frac{d}{dt}$ [**fgh**], we must keep the cyclic order of **f**, **g** and **h** in tact]

6. $\frac{d}{dt}(\mathbf{f} \times (\mathbf{g} \times \mathbf{h})) = \frac{d\mathbf{f}}{dt} \times (\mathbf{g} \times \mathbf{h}) + \mathbf{f} \times \left[\frac{d\mathbf{g}}{dt} \times \mathbf{h}\right] + \mathbf{f} \times \left[\mathbf{g} \times \frac{d\mathbf{h}}{dt}\right]$

8. Derivative of a Function of a Function (Chain rule)

Let **r** be a differentiable function of a scalar variable s and s be a differentiable function of another scalar variable t, then **r** is a differentiable function of t.

Let δt be a small change in t and δs, $\delta \mathbf{r}$ denote the corresponding changes in s and **r** respectively.

Now,
$$\frac{\delta \mathbf{r}}{\delta t} = \frac{\delta \mathbf{r}}{\delta s} \cdot \frac{\delta s}{\delta t}$$

Taking limits as $\delta t \to 0$ (and hence $\delta s \to 0$), we have

$$\underset{\delta t \to 0}{\text{Lt}} \frac{\delta \mathbf{r}}{\delta t} = \left(\underset{\delta t \to 0}{\text{Lt}} \frac{\delta \mathbf{r}}{\delta s}\right) \cdot \left(\underset{\delta t \to 0}{\text{Lt}} \frac{\delta s}{\delta t}\right) \quad \text{or} \quad \frac{d\mathbf{r}}{dt} = \frac{d\mathbf{r}}{ds} \cdot \frac{ds}{dt}.$$

9. Constant Vectors

A vector is said to be constant if both its magnitude and direction do not change.

10. Theorem. *The derivative of a constant vector is a zero vector.*

Proof. Let **c** be a constant vector function

$\therefore$ If $$\mathbf{r} = \mathbf{c}$$

then $$\mathbf{r} + \delta\mathbf{r} = \mathbf{c} \quad \Rightarrow \quad \delta\mathbf{r} = \mathbf{0}$$

Dividing by δt, $$\frac{\delta\mathbf{r}}{\delta t} = \mathbf{0}$$

Taking limits as $\delta t \to 0$,

$$\underset{\delta t \to 0}{\text{Lt}} \frac{\delta\mathbf{r}}{\delta t} = \mathbf{0} \quad \Rightarrow \quad \frac{d\mathbf{r}}{dt} = \mathbf{0}.$$

11. Theorem. *A vector function* **f** *of a scalar variable t is constant if and only if* $\frac{d\mathbf{f}}{dt} = \mathbf{0}$

Proof. (*i*) If **f** is a constant vector function of a scalar variable t, then proceeding as above,

$$\frac{d\mathbf{f}}{dt} = \mathbf{0}.$$

(*ii*) Conversely, let $$\frac{d\mathbf{f}}{dt} = \mathbf{0}$$

Let $$\mathbf{f} = f_1\mathbf{i} + f_2\mathbf{j} + f_3\mathbf{k}$$

where f_1, f_2, f_3 are the components of **f** and are scalar functions of t.

$$\Rightarrow \quad \frac{d\mathbf{f}}{dt} = \frac{df_1}{dt}\mathbf{i} + \frac{df_2}{dt}\mathbf{j} + \frac{df_3}{dt}\mathbf{k} \quad \Rightarrow \quad \mathbf{0} = \frac{df_1}{dt}\mathbf{i} + \frac{df_2}{dt}\mathbf{j} + \frac{df_3}{dt}\mathbf{k}$$

By definition of equality of two vectors,

$$\frac{df_1}{dt} = 0, \quad \frac{df_2}{dt} = 0, \quad \frac{df_3}{dt} = 0$$

$\Rightarrow$ f_1, f_2, f_3 are constant scalars (*i.e.*, independent of t)

$\Rightarrow$ **f** is a constant vector function.

12. Theorem. *The necessary and sufficient condition for the vector function* **f** *of a scalar variable t, to have constant magnitude is*

$$\mathbf{f} \cdot \frac{d\mathbf{f}}{dt} = 0.$$ **(M.D.U. 2003)**

Proof. (*i*) *Condition is necessary*

Let **f** be a vector function of a scalar variable t and have constant magnitude.

$$\therefore \quad \mathbf{f} \cdot \mathbf{f} = |\mathbf{f}|^2 = \text{constant} \qquad (\because |\mathbf{f}| \text{ is constant})$$

$$\Rightarrow \quad \frac{d}{dt}(\mathbf{f} \cdot \mathbf{f}) = 0 \quad \Rightarrow \quad \mathbf{f} \cdot \frac{d\mathbf{f}}{dt} + \frac{d\mathbf{f}}{dt} \cdot \mathbf{f} = 0$$

$$\Rightarrow \quad 2\mathbf{f} \cdot \frac{d\mathbf{f}}{dt} = 0 \quad \Rightarrow \quad \mathbf{f} \cdot \frac{d\mathbf{f}}{dt} = 0.$$

(*ii*) *Condition is sufficient*

We are given $$\mathbf{f} \cdot \frac{d\mathbf{f}}{dt} = 0 \quad \Rightarrow \quad \mathbf{f} \cdot \frac{d\mathbf{f}}{dt} + \frac{d\mathbf{f}}{dt} \cdot \mathbf{f} = 0$$

$$\Rightarrow \quad \frac{d}{dt}(\mathbf{f} \cdot \mathbf{f}) = 0 \quad \Rightarrow \quad \mathbf{f} \cdot \mathbf{f} = \text{constant}$$

$$\Rightarrow \quad |\mathbf{f}|^2 = \text{constant} \quad \Rightarrow \quad |\mathbf{f}| \text{ is constant.}$$

Cor. *The derivative of a vector of constant magnitude is perpendicular to the vector.*

13. Theorem. *The necessary and sufficient condition for a vector function* **f** *of a scalar variable t to have constant direction is*

$$\mathbf{f} \times \frac{d\mathbf{f}}{dt} = \mathbf{0}. \qquad \text{(K.U. 2000, 2002, 2003)}$$

Proof. Let **f** be a vector function of scalar variable t.

Let **F** be a vector function of modulus unity for all t.

If $|\mathbf{f}| = f$, then $\quad \mathbf{f} = f\mathbf{F}$

$$\Rightarrow \quad \frac{d\mathbf{f}}{dt} = f\frac{d\mathbf{F}}{dt} + \frac{df}{dt}\mathbf{F}$$

$$\Rightarrow \quad \mathbf{f} \times \frac{d\mathbf{f}}{dt} = (f\mathbf{F}) \times \left(f\frac{d\mathbf{F}}{dt} + \frac{df}{dt}\mathbf{F}\right) = f^2\,\mathbf{F} \times \frac{d\mathbf{F}}{dt} + f\frac{df}{dt}\mathbf{F} \times \mathbf{F}$$

$$= f^2\,\mathbf{F} \times \frac{d\mathbf{F}}{dt} \qquad (\because\ \mathbf{F} \times \mathbf{F} = \mathbf{0}) \qquad ...(1)$$

Condition is necessary *(K.U. 2006)*

Since $\mathbf{f} = f\mathbf{F}$, **f** and **F** have the same direction. Hence, if **f** has constant direction, then **F** is a constant vector

$$\Rightarrow \quad \frac{d\mathbf{F}}{dt} = \mathbf{0}$$

$\therefore$ From (1), $\quad \mathbf{f} \times \frac{d\mathbf{f}}{dt} = \mathbf{0}.$

Condition is sufficient

Let $\quad \mathbf{f} \times \frac{d\mathbf{f}}{dt} = \mathbf{0}.$

Then, from (1), $\quad f^2\,\mathbf{F} \times \frac{d\mathbf{F}}{dt} = \mathbf{0}$

$$\Rightarrow \quad \mathbf{F} \times \frac{d\mathbf{F}}{dt} = \mathbf{0} \qquad ...(2) \quad (\because\ f \text{ is not constantly zero})$$

Since **F** is of constant magnitude

$$\therefore \quad \mathbf{F} \cdot \frac{d\mathbf{F}}{dt} = 0 \qquad ...(3)$$

$\therefore$ From (2) and (3), $\quad \frac{d\mathbf{F}}{dt} = \mathbf{0}$

$\Rightarrow$ **F** is a constant vector.

But, magnitude of **F** is unity (constant). Hence, direction of **F** is also constant.

Since $\mathbf{f} = f\mathbf{F}$ and direction of **F** is constant.

$\therefore$ Direction of **f** is also constant.

Cor. The derivative of a vector of constant direction is collinear with it.

Conclusion. From the above theorems, we conclude :

(*i*) $\frac{d\mathbf{f}}{dt} = \mathbf{0}$, **f** is a constant vector function both in magnitude and direction.

(*ii*) $\mathbf{f} \cdot \frac{d\mathbf{f}}{dt} = 0$, **f is a vector function constant in magnitude.**

(*iii*) $\mathbf{f} \times \frac{d\mathbf{f}}{dt} = 0$, **f is a vector function of constant direction.**

Example 1. *If* $\mathbf{r} = t^2\mathbf{i} - 3t\mathbf{j} + (2t + 1)\mathbf{k}$, *find at* $t = 0$, *the value of*

(*i*) $\frac{d\mathbf{r}}{dt}$ (*ii*) $\frac{d^2\mathbf{r}}{dt^2}$ (*iii*) $\left|\frac{d\mathbf{r}}{dt}\right|$ (*iv*) $\left|\frac{d^2\mathbf{r}}{dt^2}\right|$.

Sol. $\mathbf{r} = t^2\mathbf{i} - 3t\mathbf{j} + (2t + 1)\mathbf{k}$

(*i*) $\because$
$$\frac{d\mathbf{r}}{dt} = \frac{d}{dt}(t^2)\mathbf{i} + \frac{d}{dt}(-3t)\mathbf{j} + \frac{d}{dt}(2t+1)\mathbf{k} = 2t\mathbf{i} - 3\mathbf{j} + 2\mathbf{k}$$

When $t = 0$, value of $\frac{d\mathbf{r}}{dt}$ is $-3\mathbf{j} + 2\mathbf{k}$.

(*ii*)
$$\frac{d^2\mathbf{r}}{dt^2} = \frac{d}{dt}\left(\frac{d\mathbf{r}}{dt}\right) = \frac{d}{dt}(2t\mathbf{i} - 3\mathbf{j} + 2\mathbf{k})$$
$$= \frac{d}{dt}(2t)\mathbf{i} + \frac{d}{dt}(-3)\mathbf{j} + \frac{d}{dt}(2)\mathbf{k} = 2\mathbf{i}$$

When $t = 0$, value of $\frac{d^2\mathbf{r}}{dt^2}$ is $2\mathbf{i}$

(*iii*)
$$\left|\frac{d\mathbf{r}}{dt}\right| = |\ 2t\mathbf{i} - 3\mathbf{j} + 2\mathbf{k}\ |$$
$$= \sqrt{(2t^2) + (-3)^2 + 2^2} = \sqrt{4t^2 + 9 + 4}$$

When $t = 0$, value of $\left|\frac{d\mathbf{r}}{dt}\right|$ is $\sqrt{0+9+4} = \sqrt{13}$.

(*iv*)
$$\left|\frac{d^2\mathbf{r}}{dt^2}\right| = \sqrt{2^2 + 0^2 + 0^2} = 2$$

When $t = 0$, value of $\frac{d^2\mathbf{r}}{dt^2}$ is 2.

Example 2. *If* $\mathbf{a} = 5t^2\mathbf{i} + t\mathbf{j} - t^3\mathbf{k}$ *and* $\mathbf{b} = \sin t\mathbf{i} - \cos t\mathbf{j}$, *find*

(*i*) $\frac{d}{dt}(\mathbf{a} \cdot \mathbf{b})$ (*ii*) $\frac{d}{dt}(\mathbf{a} \cdot \mathbf{a})$

(*iii*) $\frac{d}{dt}(\mathbf{a} \times \mathbf{b})$ (*iv*) *value of* $\frac{d}{dt}(\mathbf{a} \cdot \mathbf{b})$ *at* $t = \frac{\pi}{2}$.

Sol. Here, $\mathbf{a} = 5t^2\mathbf{i} + t\mathbf{j} - t^3\mathbf{k}$ and $\mathbf{b} = \sin t\mathbf{i} - \cos t\mathbf{j}$

$\therefore$
$$\frac{d\mathbf{a}}{dt} = 10t\mathbf{i} + \mathbf{j} - 3t^2\mathbf{k} \quad \text{and} \quad \frac{d\mathbf{b}}{dt} = \cos t\mathbf{i} + \sin t\mathbf{j}$$

(*i*)
$$\frac{d}{dt}(\mathbf{a} \cdot \mathbf{b}) = \mathbf{a} \cdot \frac{d\mathbf{b}}{dt} + \frac{d\mathbf{a}}{dt} \cdot \mathbf{b}$$
$$= (5t^2\mathbf{i} + t\mathbf{j} - t^3\mathbf{k}) \cdot (\cos t\mathbf{i} + \sin t\mathbf{j} + 0\mathbf{k})$$
$$+ (10t\mathbf{i} + \mathbf{j} - 3t^2\mathbf{k}) \cdot (\sin t\mathbf{i} - \cos t\mathbf{j} + 0\mathbf{k})$$

$$= (5t^2 \cos t + t \sin t) + (10t \sin t - \cos t)$$

$[\because \mathbf{i} \cdot \mathbf{i} = 1 \text{ ; etc. and } \mathbf{i} \cdot \mathbf{j} = 0, \text{ etc.}]$

$$= (5t^2 - 1) \cos t + 11t \sin t.$$

Second Method

$$\mathbf{a} \cdot \mathbf{b} = (5t^2\mathbf{i} + t\mathbf{j} - t^3\mathbf{k}) \cdot (\sin t\mathbf{i} - \cos t\mathbf{j}) = 5t^2 \sin t - t \cos t$$

$$\therefore \quad \frac{d}{dt}(\mathbf{a} \cdot \mathbf{b}) = \frac{d}{dt}(5t^2 \sin t - t \cos t)$$

$$= 5t^2 \cos t + 10t \sin t + t \sin t - \cos t$$

$$= (5t^2 - 1) \cos t + 11t \sin t.$$

(*ii*)
$$\frac{d}{dt}(\mathbf{a} \cdot \mathbf{a}) = \mathbf{a} \cdot \frac{d\mathbf{a}}{dt} + \frac{d\mathbf{a}}{dt} \cdot \mathbf{a} = 2\mathbf{a} \cdot \frac{d\mathbf{a}}{dt}$$

$$= 2(5t^2\mathbf{i} + t\mathbf{j} - t^3\mathbf{k}) \cdot (10t\mathbf{i} + \mathbf{j} - 3t^2\mathbf{k})$$

$$= 2(50t^3 + t + 3t^5) = 100t^3 + 2t + 6t^5.$$

(*iii*)
$$\frac{d}{dt}(\mathbf{a} \times \mathbf{b}) = \frac{d\mathbf{a}}{dt} \times \mathbf{b} + \mathbf{a} \times \frac{d\mathbf{b}}{dt}$$

$$= (10t\mathbf{i} + \mathbf{j} - 3t^2\mathbf{k}) \times (\sin t\mathbf{i} - \cos t\mathbf{j}) + (5t^2\mathbf{i} + t\mathbf{j} - t^3\mathbf{k}) \times (\cos t\mathbf{i} + \sin t\mathbf{j})$$

$$= \begin{vmatrix} \mathbf{i} & \mathbf{j} & \mathbf{k} \\ 10t & 1 & -3t^2 \\ \sin t & -\cos t & 0 \end{vmatrix} + \begin{vmatrix} \mathbf{i} & \mathbf{j} & \mathbf{k} \\ 5t^2 & t & -t^3 \\ \cos t & \sin t & 0 \end{vmatrix}$$

(on simplification)

$$= (t^3 \sin t - 3t^2 \cos t)\mathbf{i} - (t^3 \cos t + 3t^2 \sin t)\mathbf{j} + (5t^2 \sin t - 11t \cos t - \sin t)\mathbf{k}$$

(*iv*) From part (*i*),

$$\frac{d}{dt}(\mathbf{a} \cdot \mathbf{b}) = (5t^2 - 1) \cos t + 11t \sin t$$

When $t = \frac{\pi}{2}$, value of $\frac{d}{dt}(\mathbf{a} \cdot \mathbf{b})$ is $\left[5\left(\frac{\pi}{2}\right)^2 - 1\right] \cos \frac{\pi}{2} + 11\frac{\pi}{2} \sin \frac{\pi}{2}$ or $\frac{11}{2}\pi.$

Example 3. *If* **a**, **b** *are constant vectors,* ω *is constant and* **r** *is a vector function of a scalar t given by* **r** = *cos* ω*t* **a** + *sin* ω*t* **b**, *show that*

(*i*) $\frac{d^2\mathbf{r}}{dt^2} + \omega^2\mathbf{r} = \mathbf{0}$ (*ii*) $\mathbf{r} \times \frac{d\mathbf{r}}{dt} = \omega(\mathbf{a} \times \mathbf{b}).$

Sol. (*i*) Since $\mathbf{r} = \cos \omega t\, \mathbf{a} + \sin \omega t\, \mathbf{b}$

$$\therefore \quad \frac{d\mathbf{r}}{dt} = \frac{d}{dt}(\cos \omega t)\, \mathbf{a} + \frac{d}{dt}(\sin \omega t)\mathbf{b} \qquad (\because \mathbf{a}, \mathbf{b} \text{ are constant vectors})$$

$$= -\omega \sin \omega t\, \mathbf{a} + \omega \cos \omega t\, \mathbf{b}$$

$$\therefore \quad \frac{d^2\mathbf{r}}{dt^2} = \frac{d}{dt}(-\omega \sin \omega t)\mathbf{a} + \frac{d}{dt}(\omega \cos \omega t)\mathbf{b}$$

$$= -\omega^2 \cos \omega t\, \mathbf{a} - \omega^2 \sin \omega t\, \mathbf{b}$$

$$= -\omega^2 (\cos \omega t\, \mathbf{a} + \sin \omega t\, \mathbf{b}) = -\omega^2\mathbf{r}$$

$$\Rightarrow \quad \frac{d^2\mathbf{r}}{dt^2} + \omega^2\mathbf{r} = \mathbf{0}.$$

(*ii*) $$\mathbf{r} \times \frac{d\mathbf{r}}{dt} = (\cos \omega t\, \mathbf{a} + \sin \omega t\, \mathbf{b}) \times (-\omega \sin \omega t\, \mathbf{a} + \omega \cos \omega t\, \mathbf{b})$$

$$= \omega \cos^2 \omega t\, \mathbf{a} \times \mathbf{b} - \omega \sin^2 \omega t\, \mathbf{b} \times \mathbf{a} \qquad [\because \ \mathbf{a} \times \mathbf{a} = \mathbf{0}, \mathbf{b} \times \mathbf{b} = \mathbf{0}]$$

$$= \omega \cos^2 \omega t\, \mathbf{a} \times \mathbf{b} + \omega \sin^2 \omega t\, \mathbf{a} \times \mathbf{b} \qquad [\because \ \mathbf{a} \times \mathbf{b} = -\mathbf{b} \times \mathbf{a}]$$

$$= \omega (\cos^2 \omega t + \sin^2 \omega t)(\mathbf{a} \times \mathbf{b}) = \omega (\mathbf{a} \times \mathbf{b}).$$

Example 4. *If* $\hat{\mathbf{r}}$ *is a unit vector in the direction of* $\mathbf{r}$, *then prove that*

$$\hat{\mathbf{r}} \times \frac{d\hat{\mathbf{r}}}{dt} = \frac{1}{r^2} \mathbf{r} \times \frac{d\mathbf{r}}{dt}, \text{ where } |\mathbf{r}| = r.$$

Sol. Since $\qquad \mathbf{r} = r\hat{\mathbf{r}} \qquad \therefore \quad \hat{\mathbf{r}} = \frac{\mathbf{r}}{r}$

Differentiating w.r.t. t, we have

$$\frac{d\hat{\mathbf{r}}}{dt} = \frac{1}{r}\frac{d\mathbf{r}}{dt} - \frac{1}{r^2}\frac{dr}{dt}\mathbf{r}$$

$$\Rightarrow \quad \hat{\mathbf{r}} \times \frac{d\hat{\mathbf{r}}}{dt} = \frac{\mathbf{r}}{r} \times \left(\frac{1}{r}\frac{d\mathbf{r}}{dt} - \frac{1}{r^2}\frac{dr}{dt}\mathbf{r}\right) = \frac{1}{r^2}\mathbf{r}\frac{d\mathbf{r}}{dt} - \frac{1}{r^3}\frac{dr}{dt}\mathbf{r} \times \mathbf{r}$$

$$= \frac{1}{r^2}\mathbf{r} \times \frac{d\mathbf{r}}{dt} \qquad (\because \ \mathbf{r} \times \mathbf{r} = \mathbf{0})$$

Example 5. *If* $\mathbf{r}$ *is a vector function of a scalar* t *and modulus of* $\mathbf{r}$ *is* r, *and* $\mathbf{a}$, $\mathbf{b}$ *are constant vectors, then differentiate the following w.r.t.* t *:*

(*i*) $\mathbf{r}^2 + \dfrac{1}{\mathbf{r}^2}$ $\qquad$ (*ii*) $\dfrac{\mathbf{r} \times \mathbf{a}}{\mathbf{r} \cdot \mathbf{a}}$ $\qquad$ (*iii*) $\dfrac{\mathbf{r} + \mathbf{a}}{\mathbf{r}^2 + \mathbf{a}^2}$.

Sol. (*i*) $$\frac{d}{dt}\left(\mathbf{r}^2 + \frac{1}{\mathbf{r}^2}\right) = \frac{d}{dt}(\mathbf{r}^2) + \frac{d}{dt}\left(\frac{1}{\mathbf{r}^2}\right)$$

$$= \frac{d}{dt}(r^2) + \frac{d}{dt}\frac{1}{r^2} \qquad (\because \ \mathbf{r}^2 = r^2)$$

$$= 2r\frac{dr}{dt} - \frac{2}{r^3}\frac{dr}{dt}.$$

(*ii*) $$\frac{d}{dt}\left(\frac{\mathbf{r} \times \mathbf{a}}{\mathbf{r} \cdot \mathbf{a}}\right) = \frac{d}{dt}\left((\mathbf{r} \times \mathbf{a})\frac{1}{\mathbf{r} \cdot \mathbf{a}}\right) = \frac{d}{dt}(\mathbf{r} \times \mathbf{a})\frac{1}{\mathbf{r} \cdot \mathbf{a}} + (\mathbf{r} \times \mathbf{a})\frac{d}{dt}\left(\frac{1}{\mathbf{r} \cdot \mathbf{a}}\right)$$

$$= \left(\frac{d\mathbf{r}}{dt} \times \mathbf{a} + \mathbf{r} \times \frac{d\mathbf{a}}{dt}\right)\frac{1}{\mathbf{r} \cdot \mathbf{a}} + (\mathbf{r} \times \mathbf{a})\left\{-\frac{1}{(\mathbf{r} \cdot \mathbf{a})^2}\frac{d}{dt}(\mathbf{r} \cdot \mathbf{a})\right\}$$

$$= \left(\frac{d\mathbf{r}}{dt} \times \mathbf{a} + \mathbf{0}\right)\frac{1}{\mathbf{r} \cdot \mathbf{a}} + (\mathbf{r} \times \mathbf{a})\left\{-\frac{1}{(\mathbf{r} \cdot \mathbf{a})^2}\left(\frac{d\mathbf{r}}{dt} \cdot \mathbf{a} + \mathbf{r} \cdot \mathbf{0}\right)\right\}$$

$$\left(\because \ \frac{d\mathbf{a}}{dt} = \mathbf{0}\right)$$

$$= \frac{\dfrac{d\mathbf{r}}{dt} \times \mathbf{a}}{\mathbf{r} \cdot \mathbf{a}} - \frac{(\mathbf{r} \times \mathbf{a})\left(\dfrac{d\mathbf{r}}{dt} \cdot \mathbf{a}\right)}{(\mathbf{r} \cdot \mathbf{a})^2}$$

$$= \frac{\left(\frac{d\mathbf{r}}{dt} \times \mathbf{a}\right)(\mathbf{r} . \mathbf{a}) - (\mathbf{r} \times \mathbf{a})\left(\frac{d\mathbf{r}}{dt} . \mathbf{a}\right)}{(\mathbf{r} . \mathbf{a})^2}$$

(*iii*) $$\frac{d}{dt}\left(\frac{\mathbf{r}+\mathbf{a}}{\mathbf{r}^2+\mathbf{a}^2}\right) = \frac{(\mathbf{r}^2+\mathbf{a}^2)\frac{d}{dt}(\mathbf{r}+\mathbf{a}) - (\mathbf{r}+\mathbf{a})\frac{d}{dt}(\mathbf{r}^2+\mathbf{a}^2)}{(\mathbf{r}^2+\mathbf{a}^2)^2}$$ ($\because$ $\mathbf{r}^2 + \mathbf{a}^2$ is a scalar)

$$= \frac{(\mathbf{r}^2+\mathbf{a}^2)\left(\frac{d\mathbf{r}}{dt}+\frac{d\mathbf{a}}{dt}\right) - (\mathbf{r}+\mathbf{a})\left(\frac{d\mathbf{r}^2}{dt}+\frac{d\mathbf{a}^2}{dt}\right)}{(\mathbf{r}^2+\mathbf{a}^2)^2}$$

$$= \frac{(\mathbf{r}^2+\mathbf{a}^2)\frac{d\mathbf{r}}{dt} - (\mathbf{r}+\mathbf{a})\left(2\mathbf{r} . \frac{d\mathbf{r}}{dt}\right)}{(\mathbf{r}^2+\mathbf{a}^2)^2}$$

$$\left[\because \frac{d\mathbf{a}}{dt} = \mathbf{0}, \frac{d\mathbf{a}^2}{dt} = \mathbf{0}, \frac{d\mathbf{r}^2}{dt} = \frac{d}{dt}(\mathbf{r} . \mathbf{r}) = 2\mathbf{r} . \frac{d\mathbf{r}}{dt}\right]$$

EXERCISE 1

1. Find $\frac{d\mathbf{r}}{dt}$ and $\frac{d^2\mathbf{r}}{dt^2}$ when

   (*i*) $\mathbf{r} = (t + 4)\,\mathbf{i} + (t^2 + t + 7)\,\mathbf{j} + (t^3 + t^2 + t + 2)\,\mathbf{k}$

   (*ii*) $\mathbf{r} = (1 - \cos t)\,\mathbf{i} + (t - \sin t)\,\mathbf{j}$.

2. If $\mathbf{r} = a \cos t\,\mathbf{i} + a \sin t\,\mathbf{j} + t\mathbf{k}$, find the value of

   (*i*) $\frac{d\mathbf{r}}{dt}$ (*ii*) $\frac{d^2\mathbf{r}}{dt^2}$ (*iii*) $\left|\frac{d^2\mathbf{r}}{dt^2}\right|$.

3. If $\mathbf{a} = t^2\mathbf{i} - t\mathbf{j} + (2t + 1)\,\mathbf{k}$ and $\mathbf{b} = (2t - 3)\mathbf{i} + \mathbf{j} - t\mathbf{k}$, find

   (*i*) $\frac{d}{dt}(\mathbf{a} . \mathbf{b})$ (*ii*) $\frac{d}{dt}(\mathbf{a} \times \mathbf{b})$ (*iii*) $\frac{d}{dt}(\mathbf{a}^2)$

   when $t = 1$.

4. If f is a differentiable vector function of the scalar variable t and if $|\mathbf{f}| = f$, then show that

   (*i*) $\frac{d}{dt}(\mathbf{f}^2) = 2f\frac{df}{dt}$ (*ii*) $\mathbf{f} . \frac{d\mathbf{f}}{dt} = f\frac{df}{dt}$.

5. If $\mathbf{a} = \sin\theta\mathbf{i} + \cos\theta\mathbf{j} + \theta\mathbf{k}$, $\mathbf{b} = \cos\theta\mathbf{i} - \sin\theta\mathbf{j} - 3\mathbf{k}$ and $\mathbf{c} = 2\mathbf{i} + 3\mathbf{j} - \mathbf{k}$, find $\frac{d}{d\theta}[\mathbf{a} \times (\mathbf{b} \times \mathbf{c})]$ at

   (*i*) $\theta = 0$ (*ii*) $\theta = \frac{\pi}{2}$. *(M.D.U. 2007)*

6. Evaluate the following :

   (*i*) $\frac{d}{dt}\left(\mathbf{r} . \frac{d\mathbf{r}}{dt} \times \frac{d^2\mathbf{r}}{dt^2}\right)$ (*ii*) $\frac{d^2}{dt^2}\left[\mathbf{r}\ \frac{d\mathbf{r}}{dt}\ \frac{d^2\mathbf{r}}{dt^2}\right]$.

7. If $\mathbf{r}$ is a vector function of a scalar t and $\mathbf{a}$, $\mathbf{b}$ are constant vectors, $|\mathbf{r}| = r$, differentiate the following w.r.t. t :

(i) $\mathbf{r} \cdot \mathbf{a}$ (ii) $\mathbf{r} \times \mathbf{a}$ (iii) $\mathbf{r} \cdot \dfrac{d\mathbf{r}}{dt}$

(iv) $\mathbf{r} \times \dfrac{d\mathbf{r}}{dt}$ (v) $r^2\mathbf{r} + (\mathbf{a} \cdot \mathbf{r})\, \mathbf{b}$ (vi) $(a\mathbf{r} + r\mathbf{b})^2$.

8. Prove $\mathbf{a} \times \dfrac{d^2\mathbf{b}}{dt^2} - \dfrac{d^2\mathbf{a}}{dt^2} \times \mathbf{b} = \dfrac{d}{dt}\left(\mathbf{a} \times \dfrac{d\mathbf{b}}{dt} - \dfrac{d\mathbf{a}}{dt} \times \mathbf{b}\right)$.

9. If $\mathbf{a}$ has constant magnitude, show that $\mathbf{a}$ and $\dfrac{d\mathbf{a}}{dt}$ are perpendicular provided $\left|\dfrac{d\mathbf{a}}{dt}\right| \neq 0$.

Answers

1. (i) $\dfrac{d\mathbf{r}}{dt} = \mathbf{i} + (2t+1)\mathbf{j} + (3t^2 + 2t + 1)\mathbf{k}$; $\dfrac{d^2\mathbf{r}}{dt^2} = 2\mathbf{j} + (6t+2)\mathbf{k}$

(ii) $\dfrac{d\mathbf{r}}{dt} = (\sin t)\mathbf{i} + (1 - \cos t)\mathbf{j}$; $\dfrac{d^2\mathbf{r}}{dt^2} = \cos t\,\mathbf{i} + \sin t\,\mathbf{j}$.

2. (i) $\dfrac{d\mathbf{r}}{dt} = -a \sin t\,\mathbf{i} + a \cos t\,\mathbf{j} + \mathbf{k}$ (ii) $\dfrac{d^2\mathbf{r}}{dt^2} = -a \cos t\,\mathbf{i} - a \sin t\,\mathbf{j}$

(iii) $\left|\dfrac{d^2\mathbf{r}}{dt^2}\right| = a$

3. (i) -6 (ii) $7\mathbf{j} + 3\mathbf{k}$ (iii) 18

5. (i) $7\mathbf{i} + 6\mathbf{j} - 6\mathbf{k}$ (ii) $\left(4 + \dfrac{\pi}{2}\right)\mathbf{i} + 13\mathbf{j} + 9\mathbf{k}$

6. (i) $\left[\mathbf{r}\ \dfrac{d\mathbf{r}}{dt}\ \dfrac{d^3\mathbf{r}}{dt^3}\right]$ (ii) $\left[\mathbf{r}\ \dfrac{d^2\mathbf{r}}{dt^2}\ \dfrac{d^3\mathbf{r}}{dt^3}\right] + \left[\mathbf{r}\ \dfrac{d\mathbf{r}}{dt}\ \dfrac{d^4\mathbf{r}}{dt^4}\right]$

7. (i) $\dfrac{d\mathbf{r}}{dt} \cdot \mathbf{a}$ (ii) $\dfrac{d\mathbf{r}}{dt} \times \mathbf{a}$ (iii) $\left(\dfrac{d\mathbf{r}}{dt}\right)^2 + \mathbf{r} \cdot \dfrac{d^2\mathbf{r}}{dt^2}$

(iv) $\mathbf{r} \times \dfrac{d^2\mathbf{r}}{dt^2}$ (v) $2r\dfrac{dr}{dt}\mathbf{r} + r^2\dfrac{d\mathbf{r}}{dt} + \left(\mathbf{a} \cdot \dfrac{d\mathbf{r}}{dt}\right)\mathbf{b}$

(vi) $2(a\mathbf{r} + r\mathbf{b}) \cdot \left[a\dfrac{d\mathbf{r}}{dt} + \dfrac{dr}{dt}\mathbf{b}\right]$.

14. Partial Derivatives of Vector Functions of more than one Variable

So far, we have considered vector functions of a single scalar variable t. But, we come across with many vector functions of more than one independent scalar variable. For example, in Mechanics, we come across with vector functions involving coordinates (x, y, z) of any point in space and time t.

If $\mathbf{f}$ is a vector function of independent variables x, y and z, we write $\mathbf{f} = \mathbf{f}(x, y, z)$.

Let δx be a small change in x, while other variables remain unchanged. Let $\delta\mathbf{f}$ be the corresponding change in $\mathbf{f}$.

Then, $$\delta\mathbf{f} = \mathbf{f}(x + \delta x, y, z) - \mathbf{f}(x, y, z)$$

Dividing by δx, $$\frac{\delta\mathbf{f}}{\delta x} = \frac{\mathbf{f}(x + \delta x, y, z) - \mathbf{f}(x, y, z)}{\delta x}$$

Taking limits as $\delta x \to 0$, we have

$$\underset{\delta x \to 0}{\text{Lt}} \frac{\delta \mathbf{f}}{\delta x} = \underset{\delta x \to 0}{\text{Lt}} \frac{\mathbf{f}(x + \delta x, y, z) - \mathbf{f}(x, y, z)}{\delta x}$$

If this limit exists, we denote it by $\frac{\partial \mathbf{f}}{\partial x}$ or $\mathbf{f}_x$ and is called the partial derivative of **f** w.r.t. x.

Similarly, we define the partial derivatives

$$\frac{\delta \mathbf{f}}{\delta y} = \underset{\delta y \to 0}{\text{Lt}} \frac{\mathbf{f}(x, y + \delta y, z) - \mathbf{f}(x, y, z)}{\delta y}$$

$$\frac{\delta \mathbf{f}}{\delta z} = \underset{\delta z \to 0}{\text{Lt}} \frac{\mathbf{f}(x, y, z + \delta z) - \mathbf{f}(x, y, z)}{\delta y}$$

etc., provided the limits exist.

Higher-order partial derivatives may also be defined in the same way. For example,

$$\mathbf{f}_{xx} = \frac{\partial^2 \mathbf{f}}{\partial x^2} = \frac{\partial}{\partial x}\left(\frac{\partial \mathbf{f}}{\partial x}\right), \qquad \mathbf{f}_{yy} = \frac{\partial^2 \mathbf{f}}{\partial y^2} = \frac{\partial}{\partial y}\left(\frac{\partial \mathbf{f}}{\partial y}\right),$$

$$\mathbf{f}_{xy} = \frac{\partial}{\partial y}\left(\frac{\partial \mathbf{f}}{\partial x}\right) = \frac{\partial^2 \mathbf{f}}{\partial y \partial x}, \qquad \mathbf{f}_{yx} = \frac{\partial}{\partial x}\left(\frac{\partial \mathbf{f}}{\partial y}\right) = \frac{\partial^2 \mathbf{f}}{\partial x \partial y}$$

If **f** has continuous partial derivatives of second order, then

$$\frac{\partial^2 \mathbf{f}}{\partial x \partial y} = \frac{\partial^2 \mathbf{f}}{\partial y \partial x}.$$

15. Rules for Partial Differentiation of Vectors

Rules for partial differentiation of vectors are similar to those used in elementary calculus for scalar functions. Thus, if **f** and **g** are differentiable vector functions of x, y, z and ϕ is a differentiable scalar function of x, y, z then (using x as an example)

1. $$\frac{\partial}{\partial x}(\mathbf{f} + \mathbf{g}) = \frac{\partial \mathbf{f}}{\partial x} + \frac{\partial \mathbf{g}}{\partial x}$$

2. $$\frac{\partial}{\partial x}(\phi \mathbf{f}) = \phi \frac{\partial \mathbf{f}}{\partial x} + \frac{\partial \phi}{\partial x} \mathbf{f}$$

3. $$\frac{\partial}{\partial x}(\mathbf{f} \cdot \mathbf{g}) = \mathbf{f} \cdot \frac{\partial \mathbf{g}}{\partial x} + \frac{\partial \mathbf{f}}{\partial x} \cdot \mathbf{g}$$

4. $$\frac{\partial}{\partial x}(\mathbf{f} \times \mathbf{g}) = \mathbf{f} \times \frac{\partial \mathbf{g}}{\partial x} + \frac{\partial \mathbf{f}}{\partial x} \times \mathbf{g}$$

5. $$\frac{\partial}{\partial y \partial x}(\mathbf{f} \cdot \mathbf{g}) = \frac{\partial}{\partial y}\left\{\frac{\partial}{\partial x}(\mathbf{f} \cdot \mathbf{g})\right\} = \frac{\partial}{\partial y}\left\{\mathbf{f} \cdot \frac{\partial \mathbf{g}}{\partial x} + \frac{\partial \mathbf{f}}{\partial x} \cdot \mathbf{g}\right\}$$

$$= \frac{\partial}{\partial y}\left(\mathbf{f} \cdot \frac{\partial \mathbf{g}}{\partial x}\right) + \frac{\partial}{\partial y}\left(\frac{\partial \mathbf{f}}{\partial x} \cdot \mathbf{g}\right) = \mathbf{f} \cdot \frac{\partial^2 \mathbf{g}}{\partial y \partial x} + \frac{\partial \mathbf{f}}{\partial y} \cdot \frac{\partial \mathbf{g}}{\partial x} + \frac{\partial \mathbf{f}}{\partial x} \cdot \frac{\partial \mathbf{g}}{\partial x} + \frac{\partial^2 \mathbf{f}}{\partial y \partial x} \cdot \mathbf{g}$$

6. $$\frac{\partial^2}{\partial x \partial y}(\mathbf{f} \times \mathbf{g}) = \frac{\partial}{\partial x}\left\{\frac{\partial}{\partial y}(\mathbf{f} \times \mathbf{g})\right\} = \frac{\partial}{\partial x}\left(\mathbf{f} \times \frac{\partial \mathbf{g}}{\partial y} + \frac{\partial \mathbf{f}}{\partial y} \times \mathbf{g}\right)$$

$$= \frac{\partial}{\partial x}\left(\mathbf{f} \times \frac{\partial \mathbf{g}}{\partial y}\right) + \frac{\partial}{\partial x}\left(\frac{\partial \mathbf{f}}{\partial y} \times \mathbf{g}\right)$$

$$= \mathbf{f} \times \frac{\partial^2 \mathbf{g}}{\partial x \partial y} + \frac{\partial \mathbf{f}}{\partial x} \times \frac{\partial \mathbf{g}}{\partial y} + \frac{\partial \mathbf{f}}{\partial y} \times \frac{\partial \mathbf{g}}{\partial x} + \frac{\partial^2 \mathbf{f}}{\partial x \partial y} \times \mathbf{g}$$

If $\mathbf{f}(x, y, z) = f_1(x, y, z)\mathbf{i} + f_2(x, y, z)\mathbf{j} + f_3(x, y, z)\mathbf{k}$, then

$$\frac{\partial \mathbf{f}}{\partial x} = \frac{\partial f_1}{\partial x}\mathbf{i} + \frac{\partial f_2}{\partial x}\mathbf{j} + \frac{\partial f_3}{\partial x}\mathbf{k} \text{ etc.}$$

If a vector function **f** is a function of x, y, z and the variables x, y, z are expressed in terms of new variables s, t (say), then the vector function **f** will be a function of s and t. Then, we can write

$$\frac{\partial \mathbf{f}}{\partial t} = \frac{\partial \mathbf{f}}{\partial x}\frac{\partial x}{\partial t} + \frac{\partial \mathbf{f}}{\partial y}\frac{\partial y}{\partial t} + \frac{\partial \mathbf{f}}{\partial z}\frac{\partial z}{\partial t}, \text{ etc.}$$

16. Total Differentials

Differentials of vectors follow rules similar to those of elementary calculus. For example,

1. If $\mathbf{f} = f_1\mathbf{i} + f_2\mathbf{j} + f_3\mathbf{k}$, then $d\mathbf{f} = df_1\mathbf{i} + df_2\mathbf{j} + df_3\mathbf{k}$
2. $d(\mathbf{f}.\mathbf{g}) = \mathbf{f}\,.\,d\mathbf{g} + d\mathbf{f}.\mathbf{g}$
3. $d(\mathbf{f} \times \mathbf{g}) = \mathbf{f} \times d\mathbf{g} + d\mathbf{f} \times \mathbf{g}$
4. If $\mathbf{f} = \mathbf{f}(x, y, z)$, then $d\mathbf{f} = \frac{\partial \mathbf{f}}{\partial x}dx + \frac{\partial \mathbf{f}}{\partial y}dy + \frac{\partial \mathbf{f}}{\partial z}dz$, etc.

Example 6. *If* $\mathbf{f} = e^{xy}\mathbf{i} + (x - y)\mathbf{j} + (x \sin y)\mathbf{k}$, *find*

(*i*) $\frac{\partial \mathbf{f}}{\partial x}$ (*ii*) $\frac{\partial \mathbf{f}}{\partial y}$ (*iii*) $\frac{\partial^2 \mathbf{f}}{\partial x^2}$ (*iv*) $\frac{\partial \mathbf{f}}{\partial x} \times \frac{\partial \mathbf{f}}{\partial y}$.

Sol. We are given $\mathbf{f} = e^{xy}\mathbf{i} + (x - y)\mathbf{j} + (x \sin y)\mathbf{k}$

(*i*) $$\frac{\partial \mathbf{f}}{\partial x} = \frac{\partial}{\partial x}(e^{xy})\mathbf{i} + \frac{\partial}{\partial x}(x - y)\mathbf{j} + \frac{\partial}{\partial x}(x \sin y)\mathbf{k} = y\,e^{xy}\mathbf{i} + \mathbf{j} + \sin y\mathbf{k}.$$

(*ii*) $$\frac{\partial \mathbf{f}}{\partial y} = \frac{\partial}{\partial y}(e^{xy})\mathbf{i} + \frac{\partial}{\partial y}(x - y)\mathbf{j} + \frac{\partial}{\partial y}(x \sin y)\mathbf{k} = xe^{xy}\mathbf{i} - \mathbf{j} + x \cos y\mathbf{k}.$$

(*iii*) $$\frac{\partial^2 \mathbf{f}}{\partial x^2} = \frac{\partial}{\partial x}\left(\frac{\partial \mathbf{f}}{\partial x}\right) = \frac{\partial}{\partial x}(y\,e^{xy})\mathbf{i} + \frac{\partial}{\partial x}(1)\mathbf{j} + \frac{\partial}{\partial x}(\sin y)\mathbf{k} = y^2\,e^{xy}\mathbf{i}.$$

(*iv*) $$\frac{\partial \mathbf{f}}{\partial x} \times \frac{\partial \mathbf{f}}{\partial y} = \begin{vmatrix} \mathbf{i} & \mathbf{j} & \mathbf{k} \\ ye^{xy} & 1 & \sin y \\ xe^{xy} & -1 & x\cos y \end{vmatrix}$$

$$= (x \cos y + \sin y)\mathbf{i} + (\sin y - y \cos y)\,x\,e^{xy}\mathbf{j} - (x + y)\,e^{xy}\mathbf{k}.$$

Example 7. *If* $\mathbf{f} = (2x^2y - x^4)\,\mathbf{i} + (e^{xy} - y \sin x)\,\mathbf{j} + x^2 \cos y\,\mathbf{k}$, *verify that* $\frac{\partial^2 \mathbf{f}}{\partial y \partial x} = \frac{\partial^2 \mathbf{f}}{\partial x \partial y}$.

Sol. Here $$\mathbf{f} = (2x^2y - x^4)\,\mathbf{i} + (e^{xy} - y \sin x)\,\mathbf{j} + x^2 \cos y\,\mathbf{k}$$

$$\therefore \quad \frac{\partial \mathbf{f}}{\partial x} = (4xy - 4x^3)\,\mathbf{i} + (ye^{xy} - y \cos x)\,\mathbf{j} + 2x \cos y\,\mathbf{k}$$

$$\therefore \quad \frac{\partial^2 \mathbf{f}}{\partial y \partial x} = \frac{\partial}{\partial y}\left(\frac{\partial \mathbf{f}}{\partial x}\right) = 4x\,\mathbf{i} + (xye^{xy} + e^{xy} - \cos x)\,\mathbf{j} - 2x \sin y\,\mathbf{k} \qquad ...(1)$$

Also, $$\frac{\partial \mathbf{f}}{\partial y} = 2x^2\,\mathbf{i} + (xe^{xy} - \sin x)\,\mathbf{j} - x^2 \sin y\,\mathbf{k}$$

$$\therefore \quad \frac{\partial^2 \mathbf{f}}{\partial x \partial y} = \frac{\partial}{\partial x}\left(\frac{\partial \mathbf{f}}{\partial y}\right) = 4x\,\mathbf{i} + (xye^{xy} + e^{xy} - \cos x)\,\mathbf{j} - 2x \sin y\,\mathbf{k} \qquad ...(2)$$

$\therefore$ From (1) and (2), we have

$$\frac{\partial^2 \mathbf{f}}{\partial y \partial x} = \frac{\partial^2 \mathbf{f}}{\partial x \partial y}.$$

Example 8. *If* $\mathbf{a} = x^2yz\,\mathbf{i} - 2xz^3\,\mathbf{j} + xz^2\,\mathbf{k}$ *and* $\mathbf{b} = 2z\,\mathbf{i} + y\,\mathbf{j} - x^2\,\mathbf{k}$, *find the value of* $\frac{\partial^2}{\partial x \partial y}(\mathbf{a} \times \mathbf{b})$ *at the point (1, 0, – 2).* (M.D.U. 2007)

Sol. $$\mathbf{a} \times \mathbf{b} = \begin{vmatrix} \mathbf{i} & \mathbf{j} & \mathbf{k} \\ x^2yz & -2xz^3 & xz^2 \\ 2z & y & -x^2 \end{vmatrix}$$

$$= (2x^3z^3 - xyz^2)\,\mathbf{i} + (x^4yz + 4xz^3)\,\mathbf{j} + (x^2y^2z + 4xz^4)\,\mathbf{k}$$

$$\therefore \quad \frac{\partial}{\partial y}(\mathbf{a} \times \mathbf{b}) = -xz^2\,\mathbf{i} + x^4z\,\mathbf{j} + 2x^2yz\,\mathbf{k}$$

$$\therefore \quad \frac{\partial^2}{\partial x \partial y}(\mathbf{a} \times \mathbf{b}) = \frac{\partial}{\partial x}\left(\frac{\partial}{\partial y}(\mathbf{a} \times \mathbf{b})\right) = \frac{\partial}{\partial x}(-xz^2\,\mathbf{i} + x^4z\,\mathbf{j} + 2x^2yz\,\mathbf{k})$$

$$= -z^2\,\mathbf{i} + 4x^3z\,\mathbf{j} + 4xyz\,\mathbf{k}$$

$\therefore$ Value of $\frac{\partial^2}{\partial x \partial y}(\mathbf{a} \times \mathbf{b})$ at (1, 0, – 2) is $-4\,\mathbf{i} + 4(1)(-2) = -4\mathbf{i} - 8\mathbf{j}$.

EXERCISE 2

1. Given $\mathbf{f} = x^2\,\mathbf{i} + y^2\,\mathbf{j} + z^2\,\mathbf{k}$, show that $\mathbf{f}_x$, $\mathbf{f}_y$, $\mathbf{f}_z$ are perpendicular to each other.
2. If $\mathbf{f} = (2x^2y - x^4)\,\mathbf{i} + (e^{xy} - y \sin x)\,\mathbf{j} + (x^2 \cos y)\,\mathbf{k}$, find

   (*i*) $\frac{\partial \mathbf{f}}{\partial x}$ (*ii*) $\frac{\partial \mathbf{f}}{\partial y}$ (*iii*) $\frac{\partial^2 \mathbf{f}}{\partial x^2}$

   (*iv*) $\frac{\partial^2 \mathbf{f}}{\partial y^2}$ (*v*) $\frac{\partial^2 \mathbf{f}}{\partial x \partial y}$.
3. If $\phi(x, y, z) = xy^2z$ and $\mathbf{f} = xz\,\mathbf{i} - xy^2\,\mathbf{j} + yz^2\,\mathbf{k}$, find $\frac{\partial^3}{\partial x^2 \partial z}(\phi\,\mathbf{f})$ at the point (2, – 1, 1).
4. If $\mathbf{f} = xyz\,\mathbf{i} + xz^2\,\mathbf{j} - y^3\,\mathbf{k}$ and $\mathbf{g} = x^3\,\mathbf{i} - xyz\,\mathbf{j} + x^2z\,\mathbf{k}$, calculate

   (*i*) $\frac{\partial^2 \mathbf{f}}{\partial x \partial y}$ at the origin (*ii*) $\frac{\partial^2 \mathbf{f}}{\partial y^2} \times \frac{\partial^2 \mathbf{g}}{\partial x^2}$ at the point (1, 1, 0).

Answers

2. (*i*) $(4xy - 4x^3)\,\mathbf{i} + (ye^{xy} - y\cos x)\,\mathbf{j} + 2x\cos y\,\mathbf{k}$
   (*ii*) $2x^2\,\mathbf{i} + (xe^{xy} - \sin x)\,\mathbf{j} - x^2 \sin y\,\mathbf{k}$ (*iii*) $(4y - 12x^2)\,\mathbf{i} + (y^2e^{xy} + y\sin x)\,\mathbf{j} + 2\cos y\,\mathbf{k}$
   (*iv*) $x^2 e^{xy}\,\mathbf{j} - x^2\cos y\,\mathbf{k}$ (*v*) $4x\mathbf{i} + (xye^{xy} + e^{xy} - \cos x)\,\mathbf{j} - 2x\sin y\,\mathbf{k}$.
3. $4\mathbf{i} - 2\mathbf{j}$ 4. (*i*) 0 (*ii*) $-36\mathbf{j}$.

17. Space Curves

If $\mathbf{r} = \mathbf{f}(t)$ denotes a vector function of a scalar variable t and $\mathbf{r}$ is the position vector of a point $P(x, y, z)$, in space, then as t varies, so does $\mathbf{r}$ and therefore the end point P or $\mathbf{r}$ describes a continuous curve. Therefore, a curve C can always be represented by expressing the position vector $\mathbf{r}$ of any point on it as a function of some scalar variable t.

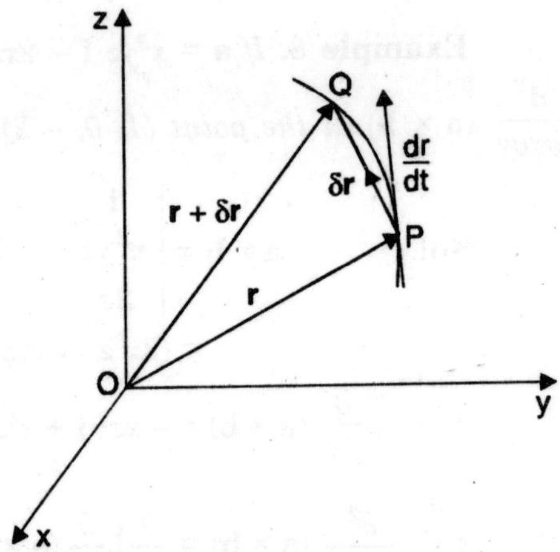

For example,

(*i*) the vector equation of a circle (having parametric equations : $x = a\cos t$, $y = a\sin t$) is $\mathbf{r} = (a\cos t)\,\mathbf{i} + (a\sin t)\,\mathbf{j} + 0\mathbf{k}$, a being a constant.

(*ii*) the vector equation of a parabola (having parametric equations ($x = at^2$, $y = 2at$) is $\mathbf{r} = at^2\,\mathbf{i} + 2at\,\mathbf{j} + 0\mathbf{k}$, a being a constant.

If $a \le t \le b$, the end point of the vector $\mathbf{r}(a)$ is called initial point of the curve C and the end point of the vector $\mathbf{r}(b)$ is called the end point of C. If $\mathbf{r}(a) = \mathbf{r}(b)$, then curve C is a closed curve. That is, a curve is closed if its initial and end points coincide. If there exist t_1 and t_2 ($t_1 \ne t_2$ and $a < t_1, t_2 < b$) such that $\mathbf{r}(t_1) = \mathbf{r}(t_2) = \mathbf{r}_0$ (say), then $\mathbf{r}_0$ is called a multiple point. Geometrically, a multiple point occurs where curve C crosses itself or is tangent to itself. A curve C is called a simple curve if it has no multiple points. A curve which is both simple and closed, is called a simple closed curve.

Let P be a fixed point on the curve having position vector $\mathbf{r}$ corresponding to some t. Let Q be the point corresponding to $t + \delta t$ and have position vector $\mathbf{r} + \delta\mathbf{r}$. Then,

$$\underset{\delta t \to 0}{\text{Lt}} \frac{\overrightarrow{PQ}}{\delta t} = \underset{\delta t \to 0}{\text{Lt}} \frac{\delta \mathbf{r}}{\delta t} = \frac{d\mathbf{r}}{dt}$$

is a vector tangent to the space curve C at P. *This vector is called the tangent vector to C at P.*

A point t on a space curve C is called singular or non-singular according as the value of $\dfrac{d\mathbf{r}}{dt}$ is zero or non-zero at t_i. The direction of the space curve C at a non-singular point P is the same as that of the tangent vector to C at P.

A vector function which has a continuous derivative and no singular points is called a smooth vector function.

If the parameter t is the time, then $\dfrac{d\mathbf{r}}{dt}$ represents the velocity $\mathbf{v}$ with which the end point of $\mathbf{r}$ describes the curve and $\dfrac{d\mathbf{v}}{dt} = \dfrac{d^2\mathbf{r}}{dt^2}$ represents its acceleration $\mathbf{a}$ along the curve.

Example 9. *A particle moves along the curve $x = 4 \cos t$, $y = 4 \sin t$, $z = 6t$*

Find the velocity and acceleration at time $t = 0$ and $t = \frac{\pi}{2}$. Find also the magnitudes of velocity and acceleration at any time t.

Sol. If **r** is the position vector of the particle at any time t, then

$$\mathbf{r} = x\mathbf{i} + y\mathbf{j} + z\mathbf{k}$$
$$= 4 \cos t\mathbf{i} + 4 \sin t\mathbf{j} + 6t\mathbf{k}$$

If **v** and **a** denote the velocity and acceleration at any time t, then

$$\mathbf{v} = \frac{d\mathbf{r}}{dt} = -4 \sin t\mathbf{i} + 4 \cos t\mathbf{j} + 6\mathbf{k}$$

and
$$\mathbf{a} = \frac{d^2\mathbf{r}}{dt^2} = -4 \cos t\mathbf{i} - 4 \sin t\mathbf{j}$$

$\therefore$ When $t = 0$, $\quad \mathbf{v} = 4\mathbf{j} + 6\mathbf{k}$ and $\mathbf{a} = -4\mathbf{i}$

When $t = \frac{\pi}{2}$, $\quad \mathbf{v} = -4\mathbf{i} + 6\mathbf{k}$ and $\mathbf{a} = -4\mathbf{j}$

Magnitude of velocity at t is $\sqrt{(-4 \sin t)^2 + (4 \cos t)^2 + 6^2} = 2\sqrt{13}$

Magnitude of acceleration at t is $\sqrt{(-4 \cos t)^2 + (-4 \sin t)^2} = 4$.

Example 10. *A particle moves along the curve $x = t^3 + 1$, $y = t^2$, $z = 2t + 5$, where t is the time. Find the component of its velocity and acceleration at time $t = 1$ in the direction of* $\mathbf{i} + \mathbf{j} + 3\mathbf{k}$.

Sol. If **r** is the position vector of the particle at any time t, then

$$\mathbf{r} = (t^3 + 1)\,\mathbf{i} + t^2\mathbf{j} + (2t + 5)\mathbf{k}$$

(*i*) Velocity $= \frac{d\mathbf{r}}{dt} = 3t^2\mathbf{i} + 2t\mathbf{j} + 2\mathbf{k}$ ...(1)

$\therefore$ Velocity at $t = 1$ is $3\mathbf{i} + 2\mathbf{j} + 2\mathbf{k}$

Unit vector in the direction of $\mathbf{i} + \mathbf{j} + 3\mathbf{k} = \frac{\mathbf{i}+\mathbf{j}+3\mathbf{k}}{\sqrt{1^2+1^2+3^2}} = \frac{\mathbf{i}+\mathbf{j}+3\mathbf{k}}{\sqrt{11}}$

$\therefore$ The required component of velocity in the direction of $\mathbf{i} + \mathbf{j} + 3\mathbf{k}$ is

$$(3\mathbf{i} + 2\mathbf{j} + 2\mathbf{k}) \cdot \frac{(\mathbf{i}+\mathbf{j}+3\mathbf{k})}{\sqrt{11}} = \frac{3+2+6}{\sqrt{11}} = \frac{11}{\sqrt{11}} = \sqrt{11}.$$

(*ii*) Acceleration $= \frac{d^2\mathbf{r}}{dt^2} = 6t\mathbf{i} + 2\mathbf{j}$ [By (1)]

$\therefore$ Acceleration at $t = 1$ is $6\mathbf{i} + 2\mathbf{j}$.

$\therefore$ The required component of acceleration $= (6\mathbf{i} + 2\mathbf{j}) \cdot \frac{(\mathbf{i}+\mathbf{j}+3\mathbf{k})}{\sqrt{11}} = \frac{6+2}{\sqrt{11}} = \frac{8}{\sqrt{11}}$.

Example 11. *Find the unit tangent vector at any point on the curve $x = a \cos t$, $y = a \sin t$, $z = bt$.*

Sol. If **r** is the position vector of a point at t, then

$$\mathbf{r} = a \cos t\,\mathbf{i} + a \sin t\,\mathbf{j} + bt\,\mathbf{k}$$

$$\Rightarrow \qquad \frac{d\mathbf{r}}{dt} = -a \sin t\mathbf{i} + a \cos t\mathbf{j} + b\mathbf{k}$$

$$\Rightarrow \qquad \left|\frac{d\mathbf{r}}{dt}\right| = \sqrt{(-a \sin t)^2 + (a \cos t)^2 + b^2} = \sqrt{a^2 + b^2}$$

$\therefore$ Unit tangent vector at any point (x, y, z) is

$$\frac{\frac{d\mathbf{r}}{dt}}{\left|\frac{d\mathbf{r}}{dt}\right|} = \frac{-a \sin t\mathbf{i} + a \cos t\mathbf{j} + b\mathbf{k}}{\sqrt{a^2 + b^2}}.$$

EXERCISE 3

1. A particle moves along the curve $x = 3t^2, y = t^2 - 2t, z = t^3$
   Find the velocity and acceleration at $t = 1$.
2. If **a**, **b**, **c** are constant vectors, show that vector $\mathbf{r} = \mathbf{a}t^2 + \mathbf{b}t + \mathbf{c}$ is the position vector of a point moving with constant acceleration.
3. A particle moves so that its position vector is given by $\mathbf{r} = \cos \omega t\, \mathbf{i} + \sin \omega t\, \mathbf{j}$, where ω is a constant. Show that (*i*) the velocity **v** of the particle is perpendicular to **r**, (*ii*) the acceleration is directed towards the origin and has magnitude proportional to the distance from the origin, (*iii*) $\mathbf{r} \times \mathbf{v}$ is a constant vector.
4. Find the unit tangent vector to the curves represented by the following vector functions at the points specified :

   (*i*) $\mathbf{r} = \left(t - \frac{t^3}{3}\right)\mathbf{i} + t^2\mathbf{j} + \left(t + \frac{t^3}{3}\right)\mathbf{k}$, at $t = 1$

   (*ii*) $\mathbf{r} = a(t - \sin t)\,\mathbf{i} + a(1 - \cos t)\,\mathbf{j}$, at any point t.
5. A particle moves along the curve $x = 3t^2, y = t^2 - 2t, z = t^3$.
   Find its velocity and acceleration at $t = 1$ in the direction of $\mathbf{i} + \mathbf{j} - \mathbf{k}$.

Answers

1. $\mathbf{v} = 6\mathbf{i} + 3\mathbf{k}$ $\quad \mathbf{a} = 6\mathbf{i} + 2\mathbf{j} + 6\mathbf{k}$, $|\mathbf{v}| = 3\sqrt{5}$

4. (*i*) $\frac{1}{\sqrt{2}}\mathbf{j} + \frac{1}{\sqrt{2}}\mathbf{k}$ $\qquad$ (*ii*) $\left(\sin \frac{t}{2}\right)\mathbf{i} + \left(\cos \frac{t}{2}\right)\mathbf{j}$ $\qquad$ 5. (*i*) $\sqrt{3}$ $\qquad$ (*ii*) $\frac{2\sqrt{3}}{3}$.

3

Gradient, Divergence and Curl

SCALAR AND VECTOR POINT FUNCTIONS

1. Scalar Point Function

If for each point P of a region R, there corresponds a scalar denoted by $\phi(P)$ or $\phi(x, y, z)$, then ϕ is called a scalar point function for the region R. For example, the temperature ϕ(P) at any point P of a body occupying a certain region is a scalar point function. The distance of a point P from a fixed point O is also a scalar point function.

2. Vector Point Function

If for each point P of a region R, there corresponds a vector denoted by $\mathbf{f}(P)$, *then* $\mathbf{f}$ *is called a vector point function for the region R.*

For example, the acceleration $\mathbf{a}$(P) of a particle at any time t occupying the position P in a certain region is a vector point function.

SCALAR AND VECTOR FIELDS

3. Scalar Field

The set of points of the region R together with the functional values $\phi(P)$ is said to be a scalar field over R. For example, the temperature distribution in a medium is a scalar field.

4. Vector Field

The set of points of the region R together with the vectors $\mathbf{f}(P)$ *is said to be a vector field over R.* For example, velocity of a moving fluid at any instant is a vector field.

Note. *A function is said to be single valued or uniform function, if it has only one value at each point. If it has more than one value at each point, it is called a multiple valued function.*

5. Level Surfaces

If ϕ is a uniform continuous point function, then through any point P(x, y, z) of the region considered, we can draw a surface such that at each point on it, the function has the same value as at P. Such a surface is called a level surface of the function. For example,

$$f(x, y, z) = c$$

where c is a constant, is a level surface.

No two level surfaces corresponding to different values of the function can intersect because otherwise, there would be two different values of the function ϕ for each point of intersection, which is contrary to the assumption that the function is uniform.

6. The Vector Differential Operator ∇

(read 'del' or 'nabla') is defined by $\nabla \equiv \mathbf{i}\dfrac{\partial}{\partial x} + \mathbf{j}\dfrac{\partial}{\partial y} + \mathbf{k}\dfrac{\partial}{\partial z}$

This vector operator possesses properties like ordinary vectors and is useful in defining three quantities—the gradient, the divergence and the curl, which arise in practical applications.

(*K.U. 2003 S*)

7. Gradient of a Scalar Function

If a scalar function ϕ is defined and differentiable at each point (x, y, z) in a certain region of space (*i.e.*, ϕ defines a differentiable scalar field), then the gradient of ϕ, written as grad ϕ or $\nabla \phi$, is a vector defined by

$$\nabla \phi = \left(\mathbf{i}\frac{\partial}{\partial x} + \mathbf{j}\frac{\partial}{\partial y} + \mathbf{k}\frac{\partial}{\partial z}\right)\phi = \mathbf{i}\frac{\partial \phi}{\partial x} + \mathbf{j}\frac{\partial \phi}{\partial y} + \mathbf{k}\frac{\partial \phi}{\partial z} = \frac{\partial \phi}{\partial x}\mathbf{i} + \frac{\partial \phi}{\partial y}\mathbf{j} + \frac{\partial \phi}{\partial z}\mathbf{k}$$

Note that (i) $\nabla\phi$ is a vector whose three components are $\dfrac{\partial \phi}{\partial x}, \dfrac{\partial \phi}{\partial y}, \dfrac{\partial \phi}{\partial z}$.

(ii) If ϕ is a scalar point function, then $\nabla \phi$ is a vector point function (*i.e.*, $\nabla\phi$ defines a vector field).

8. Properties of Gradient

If ϕ and ψ are two scalar point functions, then

1. $\nabla (\phi \pm \psi) = \nabla \phi \pm \nabla \psi$ (*i.e.*, grad $(\phi \pm \psi)$ = grad ϕ ± grad ψ)
2. $\nabla (\phi\psi) = \phi\nabla\psi + \psi\nabla\phi$ (*i.e.*, grad $(\phi\psi) = \phi$ grad $\psi + \psi$ grad ϕ)
3. $\nabla\left(\dfrac{\phi}{\psi}\right) = \dfrac{\psi\nabla\phi - \phi\nabla\psi}{\psi^2}, \ \psi \neq 0$
4. $\nabla (c\phi) = c\nabla\phi$, where c is a constant
5. ϕ is a constant if and only if $\nabla\phi = 0$.

Proof. 1.
$$\nabla(\phi \pm \psi) = \left(\mathbf{i}\frac{\partial}{\partial x} + \mathbf{j}\frac{\partial}{\partial y} + \mathbf{k}\frac{\partial}{\partial z}\right)(\phi \pm \psi)$$
$$= \mathbf{i}\frac{\partial}{\partial x}(\phi \pm \psi) + \mathbf{j}\frac{\partial}{\partial y}(\phi \pm \psi) + \mathbf{k}\frac{\partial}{\partial z}(\phi \pm \psi)$$
$$= \mathbf{i}\left(\frac{\partial \phi}{\partial x} \pm \frac{\partial \psi}{\partial x}\right) + \mathbf{j}\left(\frac{\partial \phi}{\partial y} \pm \frac{\partial \psi}{\partial y}\right) + \mathbf{k}\left(\frac{\partial \phi}{\partial z} \pm \frac{\partial \psi}{\partial z}\right)$$
$$= \left(\mathbf{i}\frac{\partial \phi}{\partial x} + \mathbf{j}\frac{\partial \phi}{\partial y} + \mathbf{k}\frac{\partial \phi}{\partial z}\right) \pm \left(\mathbf{i}\frac{\partial \psi}{\partial x} + \mathbf{j}\frac{\partial \psi}{\partial y} + \mathbf{k}\frac{\partial \psi}{\partial z}\right)$$
$$= \left(\mathbf{i}\frac{\partial}{\partial x} + \mathbf{j}\frac{\partial}{\partial y} + \mathbf{k}\frac{\partial}{\partial z}\right)\phi \pm \left(\mathbf{i}\frac{\partial}{\partial x} + \mathbf{j}\frac{\partial}{\partial y} + \mathbf{k}\frac{\partial}{\partial z}\right)\psi$$
$$= \nabla\phi \pm \nabla\psi.$$

2.
$$\nabla (\phi\psi) = \left(\mathbf{i}\frac{\partial}{\partial x} + \mathbf{j}\frac{\partial}{\partial y} + \mathbf{k}\frac{\partial}{\partial z}\right)(\phi\psi)$$
$$= \mathbf{i}\frac{\partial}{\partial x}(\phi\psi) + \mathbf{j}\frac{\partial}{\partial y}(\phi\psi) + \mathbf{k}\frac{\partial}{\partial z}(\phi\psi)$$

$$= \mathbf{i}\left(\phi\frac{\partial\psi}{\partial x}+\psi\frac{\partial\phi}{\partial x}\right)+\mathbf{j}\left(\phi\frac{\partial\psi}{\partial y}+\psi\frac{\partial\phi}{\partial y}\right)+\mathbf{k}\left(\phi\frac{\partial\psi}{\partial z}+\psi\frac{\partial\phi}{\partial z}\right)$$

$$= \phi\left(\mathbf{i}\frac{\partial\psi}{\partial x}+\mathbf{j}\frac{\partial\psi}{\partial y}+\mathbf{k}\frac{\partial\psi}{\partial z}\right)+\psi\left(\mathbf{i}\frac{\partial\phi}{\partial x}+\mathbf{j}\frac{\partial\phi}{\partial y}+\mathbf{k}\frac{\partial\phi}{\partial z}\right)$$

$$= \phi\nabla\psi + \psi\nabla\phi.$$

3. Using $\quad \dfrac{\partial}{\partial x}\left(\dfrac{\phi}{\psi}\right)=\dfrac{\psi\dfrac{\partial\phi}{\partial x}-\phi\dfrac{\partial\psi}{\partial x}}{\psi^2}$, etc.,

one may proceed as in the proof of property 2.

4. Proof is simple and left to the reader as an exercise.

5. (*i*) Let ϕ be a constant

$$\Rightarrow \qquad \frac{\partial\phi}{\partial x}=\frac{\partial\phi}{\partial y}=\frac{\partial\phi}{\partial z}=0$$

By definition of $\nabla\phi$,

$$\nabla\phi = \mathbf{i}\frac{\partial\phi}{\partial x}+\mathbf{j}\frac{\partial\phi}{\partial y}+\mathbf{k}\frac{\partial\phi}{\partial z} = \mathbf{i}\,.\,0+\mathbf{j}\,.\,0+\mathbf{k}\,.\,0 = \mathbf{0}.$$

(*ii*) Conversely, let $\nabla\phi = \mathbf{0}$

$$\Rightarrow \quad \mathbf{i}\frac{\partial\phi}{\partial x}+\mathbf{j}\frac{\partial\phi}{\partial y}+\mathbf{k}\frac{\partial\phi}{\partial z}=\mathbf{0}. \qquad \Rightarrow \quad \frac{\partial\phi}{\partial x}=0, \frac{\partial\phi}{\partial y}=0, \frac{\partial\phi}{\partial z}=0.$$

$\Rightarrow$ ϕ is independent of x, y and z

$\Rightarrow$ ϕ is a constant.

Example 1. *If $\phi(x, y, z) = 3xy^2 - y^3z^2$, find $\nabla\phi$ (or grad ϕ) at the point $(-1, 2, -1)$.*

Sol.
$$\nabla\phi = \left(\mathbf{i}\frac{\partial}{\partial x}+\mathbf{j}\frac{\partial}{\partial y}+\mathbf{k}\frac{\partial}{\partial z}\right)(3xy^2 - y^3z^2)$$

$$= \mathbf{i}\frac{\partial}{\partial x}(3xy^2 - y^3z^2)+\mathbf{j}\frac{\partial}{\partial y}(3xy^2 - y^3z^2)+\mathbf{k}\frac{\partial}{\partial z}(3xy^2 - y^3z^2)$$

$$= 3y^2\,\mathbf{i} + (6xy - 3y^2z^2)\,\mathbf{j} + (-2y^3z)\,\mathbf{k}$$

$\therefore$ at $(-1, 2, -1)$, $\nabla\phi$ is $12\,\mathbf{i} - 24\,\mathbf{j} + 16\,\mathbf{k}$.

Example 2. (*i*) *If $\mathbf{r} = x\,\mathbf{i} + y\,\mathbf{j} + z\,\mathbf{k}$, find ∇r, where $|\mathbf{r}| = r$.*

Or

If $r = (x^2 + y^2 + z^2)^{1/2}$, find grad r.

(*ii*) *Find ∇r^n and hence $\nabla\left(\dfrac{1}{r}\right)$.*

Sol. (*i*)
$$r = |\mathbf{r}| = \sqrt{x^2+y^2+z^2}$$

$$\Rightarrow \qquad r^2 = x^2+y^2+z^2 \qquad \Rightarrow \quad 2r\frac{\partial r}{\partial x}=2x,\ 2r\frac{\partial r}{\partial y}=2y,\ 2r\frac{\partial r}{\partial z}=2z$$

$$\Rightarrow \quad \frac{\partial r}{\partial x} = \frac{x}{r}, \frac{\partial r}{\partial y} = \frac{y}{r}, \frac{\partial r}{\partial z} = \frac{z}{r} \qquad ...(1)$$

$$\therefore \quad \nabla r = \left(\mathbf{i}\frac{\partial}{\partial x} + \mathbf{j}\frac{\partial}{\partial y} + \mathbf{k}\frac{\partial}{\partial r}\right) r = \mathbf{i}\frac{\partial r}{\partial x} + \mathbf{j}\frac{\partial r}{\partial y} + \mathbf{k}\frac{\partial r}{\partial z}$$

$$= \mathbf{i}\left(\frac{x}{r}\right) + \mathbf{j}\left(\frac{y}{r}\right) + \mathbf{k}\left(\frac{z}{r}\right) \qquad \text{[By (1)]}$$

$$= \frac{x\mathbf{i} + y\mathbf{j} + z\mathbf{k}}{r} = \frac{\mathbf{r}}{r} = \hat{\mathbf{r}} \qquad ...(2)$$

(*ii*)

$$\nabla r^n = \left(\mathbf{i}\frac{\partial}{\partial x} + \mathbf{j}\frac{\partial}{\partial y} + \mathbf{k}\frac{\partial}{\partial z}\right) r^n = \mathbf{i}\frac{\partial r^n}{\partial x} + \mathbf{j}\frac{\partial r^n}{\partial y} + \mathbf{k}\frac{\partial r^n}{\partial z}$$

$$= \mathbf{i}\left(nr^{n-1}\frac{\partial r}{\partial x}\right) + \mathbf{j}\left(nr^{n-1}\frac{\partial r}{\partial y}\right) + \mathbf{k}\left(nr^{n-1}\frac{\partial r}{\partial z}\right)$$

$$= nr^{n-1}\left[\mathbf{i}\left(\frac{x}{r}\right) + \mathbf{j}\left(\frac{y}{r}\right) + \mathbf{k}\left(\frac{z}{r}\right)\right] \qquad \text{[By (1)]}$$

$$= nr^{n-1}\frac{x}{r} = nr^{n-2}\,\mathbf{r} \quad \text{or} \quad nr^{n-1}\,\hat{\mathbf{r}}$$

Letting $n = -1$, we have

$$\nabla\left(\frac{1}{r}\right) = -\frac{1}{r^3}\,\mathbf{r} \quad \text{or} \quad -\frac{1}{r^2}\,\hat{\mathbf{r}}.$$

Example 3. *If $\phi = \phi(u)$ and $u = u(x, y, z)$, then show that* $\nabla \phi = \dfrac{d\phi}{du}\nabla u$. (K.U. 2006)

Sol. By definition, $\nabla \phi = \dfrac{\partial \phi}{\partial x}\mathbf{i} + \dfrac{\partial \phi}{\partial y}\mathbf{j} + \dfrac{\partial \phi}{\partial z}\mathbf{k} = \dfrac{d\phi}{du}\dfrac{\partial u}{\partial x}\mathbf{i} + \dfrac{d\phi}{du}\dfrac{\partial u}{\partial y}\mathbf{j} + \dfrac{d\phi}{du}\dfrac{\partial u}{\partial z}\mathbf{k}$

$$= \frac{d\phi}{du}\left(\frac{\partial u}{\partial x}\mathbf{i} + \frac{\partial u}{\partial y}\mathbf{j} + \frac{\partial u}{\partial z}\mathbf{k}\right) = \frac{d\phi}{du}\nabla u.$$

Example 4. *For any vector* **a**, *show that* $\nabla(\mathbf{a} \cdot \mathbf{r}) = \mathbf{a}$ *where* **r** *is the position vector of a point. Hence, show that*

$$grad\ [\mathbf{r}\ \mathbf{a}\mathbf{b}] = \mathbf{a} \times \mathbf{b}.$$

Sol. Let $\quad \mathbf{a} = a_1\,\mathbf{i} + a_2\,\mathbf{j} + a_3\,\mathbf{k}$ and $\mathbf{r} = x\,\mathbf{i} + y\,\mathbf{j} + z\,\mathbf{k}$

Then, $\quad \mathbf{a} \cdot \mathbf{r} = a_1 x + a_2 y + a_3 z$

$$\therefore \quad \nabla(\mathbf{a} \cdot \mathbf{r}) = \mathbf{i}\frac{\partial}{\partial x}(a_1 x + a_2 y + a_3 z) + \mathbf{j}\frac{\partial}{\partial y}(a_1 x + a_2 y + a_3 z) + \mathbf{k}\frac{\partial}{\partial z}(a_1 x + a_2 y + a_3 z)$$

$$= a_1\mathbf{i} + a_2\mathbf{j} + a_3\mathbf{k} = \mathbf{a}$$

Now, $\quad$ grad $[\mathbf{r}\ \mathbf{a}\ \mathbf{b}]$ = grad $[\mathbf{r} \cdot (\mathbf{a} \times \mathbf{b})] = \mathbf{a} \times \mathbf{b}$.

Example 5. *If $\phi(x, y) = \log\sqrt{x^2 + y^2}$, show that*

$$\nabla\phi = \frac{\mathbf{r} - (\mathbf{k} \cdot \mathbf{r})\mathbf{k}}{[\mathbf{r} - (\mathbf{k} \cdot \mathbf{r})\mathbf{k}] \cdot [\mathbf{r} - (\mathbf{k} \cdot \mathbf{r})\mathbf{k}]}.$$

Sol.
$$\nabla\phi = \nabla \log \sqrt{(x^2+y^2)} = \nabla \tfrac{1}{2} \log (x^2+y^2)$$
$$= \tfrac{1}{2} \nabla \log (x^2+y^2) = \frac{1}{2}\left(\mathbf{i}\frac{\partial}{\partial x} + \mathbf{j}\frac{\partial}{\partial y} + \mathbf{k}\frac{\partial}{\partial z}\right) \log (x^2+y^2)$$
$$= \frac{1}{2}\left[\mathbf{i}\frac{\partial}{\partial x}\log(x^2+y^2) + \mathbf{j}\frac{\partial}{\partial y}\log(x^2+y^2) + \mathbf{k}\frac{\partial}{\partial z}(x^2+y^2)\right]$$
$$= \frac{1}{2}\left[\frac{2x}{x^2+y^2}\mathbf{i} + \frac{2y}{x^2+y^2}\mathbf{j} + 0\mathbf{k}\right] = \frac{x\mathbf{i}+y\mathbf{j}}{x^2+y^2} \qquad ...(1)$$

Also,
$$\frac{\mathbf{r}-(\mathbf{k}\cdot\mathbf{r})\mathbf{k}}{[\mathbf{r}-(\mathbf{k}\cdot\mathbf{r})\mathbf{k}]\cdot[\mathbf{r}-(\mathbf{k}\cdot\mathbf{r})\mathbf{k}]} = \frac{\mathbf{r}-z\mathbf{k}}{[\mathbf{r}-z\mathbf{k}]\cdot[\mathbf{r}-z\mathbf{k}]}$$
$$= \frac{x\mathbf{i}+y\mathbf{j}}{(x\mathbf{i}+y\mathbf{j})\cdot(x\mathbf{i}+y\mathbf{j})} \qquad (\because \; \mathbf{r} = x\mathbf{i}+y\mathbf{j}+z\mathbf{k})$$
$$= \frac{x\mathbf{i}+y\mathbf{j}}{x^2+y^2} \qquad ...(2) \quad (\because \; \mathbf{i}\cdot\mathbf{i} = 1, \mathbf{i}\cdot\mathbf{j} = 0)$$

From (1) and (2), $\nabla\phi = \dfrac{\mathbf{r}-(\mathbf{k}\cdot\mathbf{r})\mathbf{k}}{[\mathbf{r}-(\mathbf{k}\cdot\mathbf{r})\mathbf{k}]\cdot[\mathbf{r}-(\mathbf{k}\cdot\mathbf{r})\mathbf{k}]}$.

EXERCISE 1

1. If $\phi(x, y, z) = x^2y + y^2x + z^2$, find $\nabla\phi$ at the point (1, 1, 1).
2. If $\phi = (x^2+y^2+z^2)\, e^{-\sqrt{(x^2+y^2+z^2)}}$, find $\nabla\phi$.
3. Show that $\nabla f(r) \times \mathbf{r} = \mathbf{0}$, where $\mathbf{r} = x\mathbf{i}+y\mathbf{j}+z\mathbf{k}$.
4. Show that $\dfrac{d\phi}{ds} = \nabla\phi \cdot \dfrac{d\mathbf{r}}{ds}$, where $\mathbf{r} = x\mathbf{i}+y\mathbf{j}+z\mathbf{k}$.
5. If $\mathbf{a} = 2x^2\mathbf{i} - 3yz\mathbf{j} + xz^2\mathbf{k}$ and $\phi = 2z^2 - x^3y$, find $\mathbf{a} \times \nabla\phi$ at (1, – 1, 1).
6. Prove $\nabla f(r) = f'(r)\, \nabla r$, where $\mathbf{r} = x\mathbf{i}+y\mathbf{j}+z\mathbf{k}$ and $|\mathbf{r}| = r$.
7. If $\phi = \dfrac{1}{r}$, where $r^2 = x^2+y^2+z^2$, find $\nabla\phi$.

Answers

1. $3\mathbf{i} + 3\mathbf{j} + 2\mathbf{k}$
2. $e^{-r}\,[2-r]\,\mathbf{r}$, where $r = \sqrt{x^2+y^2+z^2}$
5. $(13\mathbf{i} - 5\mathbf{j} - 11\mathbf{k})$
7. $-\dfrac{\mathbf{r}}{r^3}$.

9. $\nabla\phi$ is perpendicular to the surface S given by $\phi(x, y, z) = c$, c being a constant, at the point P(x, y, z)

(K.U. 2003 S)

Let a space curve **C** that lies on the surface S be represented by $\mathbf{r} = x\mathbf{i}+y\mathbf{j}+z\mathbf{k}$, where x, y, z are scalar functions of a scalar t, then differentiating $\phi(x, y, z) = c$ w.r.t. t, we have $\dfrac{d\phi}{dt} = 0$

$$\Rightarrow \quad \frac{\partial \phi}{\partial x}\frac{dx}{dt} + \frac{\partial \phi}{\partial y}\frac{dy}{dt} + \frac{\partial \phi}{\partial z}\frac{dz}{dt} = 0$$

$$\Rightarrow \quad \left(\frac{\partial \phi}{\partial x}\mathbf{i} + \frac{\partial \phi}{\partial y}\mathbf{j} + \frac{\partial \phi}{\partial z}\mathbf{k}\right) \cdot \left(\frac{dx}{dt}\mathbf{i} + \frac{dy}{dt}\mathbf{j} + \frac{dz}{dt}\mathbf{k}\right) = 0 \quad \Rightarrow \quad \nabla\phi \cdot \frac{d\mathbf{r}}{dt} = 0$$

But $\frac{d\mathbf{r}}{dt}$ is the vector tangent to the curve C at the point P.

$\therefore$ $\nabla\phi$ is perpendicular to the curve C at P.

The same reasoning is applicable to every smooth curve on the surface S that passes through P. Therefore, $\nabla\phi$ is perpendicular to every such curve at P. Hence, **$\nabla\phi$ is perpendicular to the surface S.**

Remark. It can be proved that $|\nabla\phi|$ gives the rate of change of ϕ along the normal to the surface at P.

TANGENT PLANE AND NORMAL TO A LEVEL SURFACE

10. Tangent Plane

Let **a** be the position vector of a given point A on the surface $\phi(x, y, z) = c$, (where c is a constant). If **r** is the position vector of any point P on the tangent plane at A, then $\vec{AP} = \mathbf{r} - \mathbf{a}$ and it lies in the tangent plane. But, grad ϕ is normal to the surface.

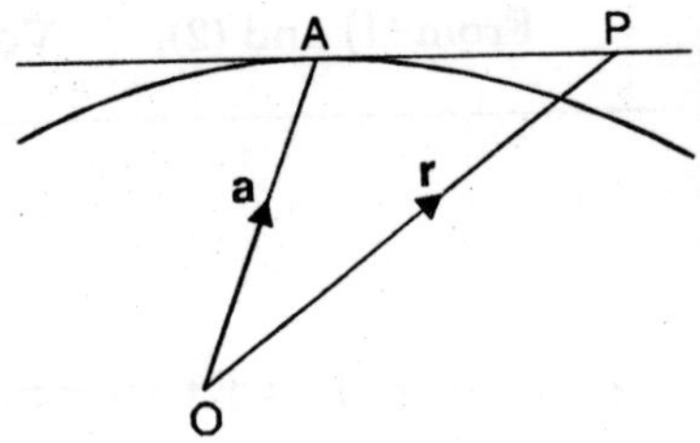

$\therefore$ $(\mathbf{r} - \mathbf{a}) \cdot \text{grad}\, \phi = 0$, which is the equation of the tangent plane. If $\mathbf{a} = (x, y, z)$ and $\mathbf{r} = (X, Y, Z)$, then equation of tangent plane can be written as

$$(X - x)\frac{\partial \phi}{\partial x} + (Y - y)\frac{\partial \phi}{\partial y} + (Z - z)\frac{\partial \phi}{\partial z} = 0.$$

11. Normal

If **r** is the position vector of any point P on the normal at A having position vector **a**, then $\vec{AP} = \mathbf{r} - \mathbf{a}$.

Also, grad ϕ is along the normal to the surface.

$\therefore$ $(\mathbf{r} - \mathbf{a}) \times \text{grad}\, \phi = \mathbf{0}$, which is the equation of the normal at the point A.

If $\mathbf{a} = (x, y, z)$ and $\mathbf{r} = (X, Y, Z)$, then

$$\mathbf{r} - \mathbf{a} = (X - x)\mathbf{i} + (Y - y)\mathbf{j} + (Z - z)\mathbf{k}$$

and

$$\text{grad}\, \phi = \frac{\partial \phi}{\partial x}\mathbf{i} + \frac{\partial \phi}{\partial y}\mathbf{j} + \frac{\partial \phi}{\partial z}\mathbf{k}.$$

$\therefore$ $\mathbf{r} - \mathbf{a}$ is parallel to grad ϕ if $\mathbf{r} - \mathbf{a} = p$ grad ϕ for some scalar p.

i.e.,

$$(X - x)\,\mathbf{i} + (Y - y)\,\mathbf{j} + (Z - z)\,\mathbf{k} = p\left(\frac{\partial \phi}{\partial x}\mathbf{i} + \frac{\partial \phi}{\partial y}\mathbf{j} + \frac{\partial \phi}{\partial z}\mathbf{k}\right)$$

Comparing coefficients of i, j, k, we get

$$X - x = p\frac{\partial \phi}{\partial x},\ Y - y = p\frac{\partial \phi}{\partial y},\ Z - z = p\frac{\partial \phi}{\partial z}$$

or
$$\frac{X-x}{\frac{\partial \phi}{\partial x}}=\frac{Y-y}{\frac{\partial \phi}{\partial y}}=\frac{Z-z}{\frac{\partial \phi}{\partial z}}$$
are equations of the normal at **A.**

Example 6. *Find the equation of tangent plane to the surface $2xz^2 - 3xy - 4x = 7$ at the point (1, – 1, 2).*

Sol. Here, $\phi(x, y, z) \equiv 2xz^2 - 3xy - 4x = 7$

$$\therefore \quad \nabla\phi = \frac{\partial \phi}{\partial x}\mathbf{i} + \frac{\partial \phi}{\partial y}\mathbf{j} + \frac{\partial \phi}{\partial z}\mathbf{k} = (2z^2 - 3y - 4)\mathbf{i} - 3x\mathbf{j} + 4xz\mathbf{k}$$

$\therefore$ $\nabla\phi$ at (1, – 1, 2) is $7\mathbf{i} - 3\mathbf{j} + 8\mathbf{k}$.

$\therefore$ The vector equation of a plane passing through a point whose position vector is $\mathbf{a} = (1, -1, 2)$ is $(\mathbf{r} - \mathbf{a}) \cdot \text{grad}\, \phi = 0$.

or $$[(x\mathbf{i} + y\mathbf{j} + z\mathbf{k}) - (\mathbf{i} - \mathbf{j} + 2\mathbf{k})] \cdot (7\mathbf{i} - 3\mathbf{j} + 8\mathbf{k}) = 0$$
or $$7(x - 1) - 3(y + 1) + 8(z - 2) = 0$$
or $$7x - 3y + 8z - 26 = 0.$$

Example 7. *Find a unit normal to the surface $x^2y + 2xz = 4$ at the point (2, – 2, 3). Find the equations of normal to the surface at (2, – 2, 3).*

Sol. Here, $\phi(x, y, z) \equiv x^2y + 2xz = 4$

$$\therefore \quad \nabla\phi = \frac{\partial \phi}{\partial x}\mathbf{i} + \frac{\partial \phi}{\partial y}\mathbf{j} + \frac{\partial \phi}{\partial z}\mathbf{k} = (2xy + 2z)\,\mathbf{i} + x^2\mathbf{j} + 2x\mathbf{k}$$

$\therefore$ $\nabla\phi$ at (2, – 2, 3) is $-2\mathbf{i} + 4\mathbf{j} + 4\mathbf{k}$

Since $\nabla\phi$ is normal to the surface $\phi(x, y, z) = 4$

$\therefore$ Unit normal to the surface is $\dfrac{\nabla\phi}{|\nabla\phi|}$

$$= \frac{-2\mathbf{i} + 4\mathbf{j} + 4\mathbf{k}}{\sqrt{(-2)^2 + 4^2 + 4^2}} = \frac{-1}{3}\mathbf{i} + \frac{2}{3}\mathbf{j} + \frac{2}{3}\mathbf{k}$$

Another unit normal is $\dfrac{1}{3}\mathbf{i} - \dfrac{2}{3}\mathbf{j} - \dfrac{2}{3}\mathbf{k}$

Equations of the normal to the surface at the point (2, – 2, 3) are

$$\frac{X-2}{\frac{\partial \phi}{\partial x}}=\frac{Y-(-2)}{\frac{\partial \phi}{\partial y}}=\frac{Z-3}{\frac{\partial \phi}{\partial z}} \quad \text{or} \quad \frac{X-2}{-2}=\frac{Y+2}{4}=\frac{Z-3}{4}.$$

Example 8. *Find the angle between the surfaces $x^2 + y^2 + z^2 = 9$ and $z = x^2 + y^2 - z - 3$ at the point (2, – 1, 2).* (M.D.U. 2007)

Sol. The angle between the surfaces at the point (2, – 1, 2) is the angle between their normals to the surfaces at the point (2, – 1, 2).

Let $\phi_1(x, y, z) = x^2 + y^2 + z^2 - 9$; $\phi_2(x, y, z) = x^2 + y^2 - z - 3$

If $\mathbf{n}_1$, $\mathbf{n}_2$ are normal vectors to the two surfaces at (2, – 1, 2), then

$$\mathbf{n}_1 = \text{grad}\, \phi_1 \text{ at } (2, -1, 2)$$

$$= \frac{\partial \phi_1}{\partial x}\mathbf{i} + \frac{\partial \phi_1}{\partial y}\mathbf{j} + \frac{\partial \phi_1}{\partial z}\mathbf{k} \text{ at } (2, -1, 2)$$

$$= 2x\mathbf{i} + 2y\mathbf{j} + 2z\mathbf{k} \text{ at } (2, -1, 2) = 4\mathbf{i} - 2\mathbf{j} + 4\mathbf{k}$$

and $$\mathbf{n}_2 = \text{grad } \phi_2 \text{ at } (2, -1, 2)$$

$$= \frac{\partial \phi_2}{\partial x}\mathbf{i} + \frac{\partial \phi_2}{\partial y}\mathbf{j} + \frac{\partial \phi_2}{\partial z}\mathbf{k} \text{ at } (2, -1, 2)$$

$$= 2x\mathbf{i} + 2y\mathbf{j} - \mathbf{k} \text{ at } (2, -1, 2) = 4\mathbf{i} - 2\mathbf{j} - \mathbf{k}.$$

If θ is the angle between the two surfaces, then

$$\cos\theta = \frac{\mathbf{n}_1 \cdot \mathbf{n}_2}{|\mathbf{n}_1||\mathbf{n}_2|} = \frac{(4\mathbf{i} - 2\mathbf{j} + 4\mathbf{k}) \cdot (4\mathbf{i} - 2\mathbf{j} - \mathbf{k})}{|4\mathbf{i} - 2\mathbf{j} + 4\mathbf{k}| \cdot |4\mathbf{i} - 2\mathbf{j} - \mathbf{k}|}$$

$$= \frac{16 + 4 - 4}{\sqrt{16 + 4 + 16}\sqrt{16 + 4 + 1}} = \frac{16}{\sqrt{36}\sqrt{21}} = \frac{8\sqrt{21}}{63}.$$

Example 9. *Find the tangent line to the curve of intersection of surfaces $x^2 + y^2 + z^2 = 1$ and $x + y + z = 1$ at the point (1, 0, 0).*

Sol. Let $$\phi_1(x, y, z) = x^2 + y^2 + z^2 - 1 \quad \text{...(1)}$$

$$\phi_2(x, y, z) = x + y + z - 1 \quad \text{...(2)}$$

If n_1 and n_2 are normals to the two surfaces at P(1, 0, 0), then

$$n_1 = \text{grad } \phi_1 \text{ at } (1, 0, 0) = 2\mathbf{i}$$

and $$n_2 = \text{grad } \phi_2 \text{ at } (1, 0, 0) = \mathbf{i} + \mathbf{j} + \mathbf{k}$$

The tangent line to the curve of intersection of (1) and (2) at (1, 0, 0) is the line of intersection of two planes—one passing through the (1, 0, 0) and perpendicular to n_1 and the other passing through (1, 0, 0) and perpendicular to n_2.

$\therefore$ Equations of tangents of tangent line are :

$$2(x - 1) + 0y + 0z = 0$$

and $$1(x - 1) + 1y + 1z = 0$$

or $$x - 1 = 0 \quad \text{and} \quad y + z = 0.$$

12. Directional Derivatives

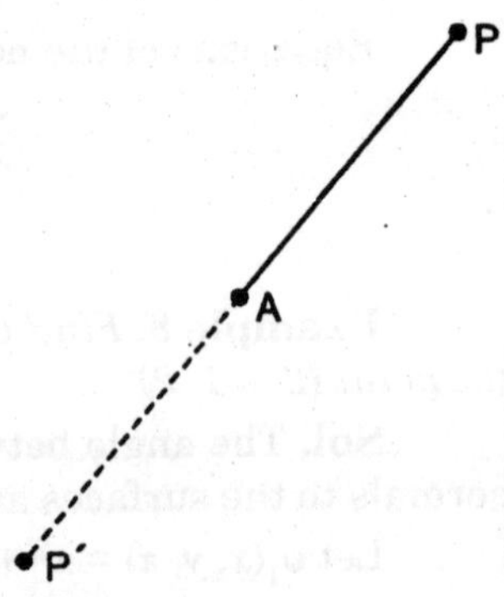

Let ϕ be a scalar point function and A any given point in its region of definition. Let AP be any line drawn on one side of A and a point P taken on it.

Then, $\displaystyle \underset{P \to A}{Lt} \frac{\phi(P) - \phi(A)}{AP}$, *if exists, is called the directional derivative of the scalar point function at A along the direction of the line AP. The length AP is to be regarded as positive.*

The directional derivative along AP′ is negative of that along AP.

Similarly, *if* **f**(*P*) *is a vector point function and if* $\displaystyle \underset{P \to A}{Lt} \frac{\mathbf{f}(P) - \mathbf{f}(A)}{AP}$ *exists, then this limit is defined as the directional derivative of the vector point function along the direction of the line AP.*

Note that the directional derivative of a scalar point function is a scalar and the directional derivative of a vector point function is a vector.

13. Directional Derivatives of a Scalar Point Function along Co-ordinate axes

Let A(x, y, z) be a given point.

Let P$(x+\delta x, y, z)$ be a point on a line through A drawn parallel to the positive direction of x-axis so that AP = δx and $\delta x > 0$.

∴ Directional derivative of the scalar point function ϕ along AP is

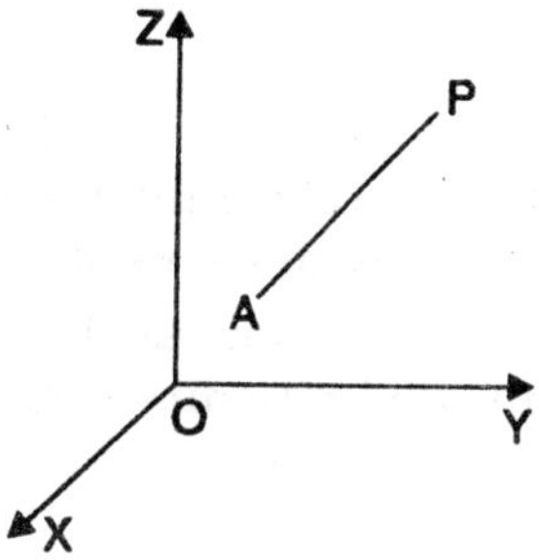

$$\operatorname*{Lt}_{P\to A} \frac{\phi(P)-\phi(A)}{AP} = \operatorname*{Lt}_{\delta x\to 0} \frac{\phi(x+\delta x, y, z)-\phi(x, y, z)}{\delta x}$$

$$= \frac{\partial \phi}{\partial x} \text{ (partial derivative of } \phi \text{ w.r.t. } x)$$

Similarly, the directional derivatives of the scalar point function ϕ along positive directions of y-axis and z-axis are $\frac{\partial \phi}{\partial y}$ and $\frac{\partial \phi}{\partial z}$ respectively.

The directional derivatives along negative directions of axes are $-\frac{\partial \phi}{\partial x}, \frac{\partial \phi}{\partial y}$ and $-\frac{\partial \phi}{\partial z}$.

14. Directional Derivatives of a Vector Point Function along Coordinate axes

Arguing as for a scalar point function, one can see that the directional derivative of a vector point function **f** along the positive directions of coordinate axes are :

$$\frac{\partial \mathbf{f}}{\partial x}, \frac{\partial \mathbf{f}}{\partial y} \text{ and } \frac{\partial \mathbf{f}}{\partial z} \text{ respectively.}$$

15. Directional Derivative of a Scalar Point Function along any line whose Direction cosines are l, m, n.

Let AP be any line drawn on one side of the point A(x, y, z).

Let l, m, n be the direction cosines of AP. If AP = r, then coordinates of the point P are $(x+lr, y+mr, z+nr)$.

Directional derivative of the scalar point function ϕ along AP is

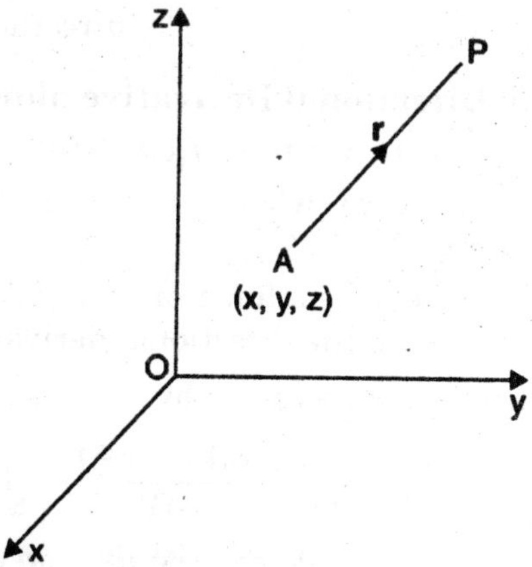

$$= \operatorname*{Lt}_{P\to A} \frac{\phi(P)-\phi(A)}{AP}$$

$$= \operatorname*{Lt}_{r\to 0} \frac{\phi(x+lr, y+mr, z+nr)-\phi(x, y, z)}{r}$$

[Suppose that ϕ possesses continuous first order partial derivatives, then by Taylor's theorem]

$$= \operatorname*{Lt}_{r\to 0} \frac{\phi(x, y, z)+\left(lr\frac{\partial \phi}{\partial x}+mr\frac{\partial \phi}{\partial y}+nr\frac{\partial \phi}{\partial z}\right)+\ldots\ldots-\phi(x, y, z)}{r} = l\frac{\partial \phi}{\partial x}+m\frac{\partial \phi}{\partial y}+n\frac{\partial \phi}{\partial z}.$$

Since $\frac{\partial \phi}{\partial x}, \frac{\partial \phi}{\partial y}, \frac{\partial \phi}{\partial z}$ are the directional derivatives along the coordinate axes.

$\therefore$ *The directional derivative along any line can be expressed in terms of the directional derivatives along the coordinate axes and the direction cosines of the line.*

Now, if we draw a line AP′ through A in the direction opposite to that of AP, then direction cosines of AP′ are $-l, -m, -n$. Hence, the directional derivative along AP′ is

$$-l\frac{\partial\phi}{\partial x}-m\frac{\partial\phi}{\partial y}-n\frac{\partial\phi}{\partial z}$$

16. Directional derivative of a vector point function f along any line with direction cosines l, m, n.

One can easily prove that the directional derivative of a vector point function **f** along any line with direction cosines l, m, n is

$$l\frac{\partial \mathbf{f}}{\partial x}+m\frac{\partial \mathbf{f}}{\partial y}+n\frac{\partial \mathbf{f}}{\partial z}$$

We have seen that if a point function (scalar or vector) possesses continuous first order partial derivatives, then it possesses directional derivatives along every direction. We say that the point function is continuously differentiable if it possesses continuous first order partial derivatives. Hereafter, we assume that every point function, scalar or vector, which we deal with is a continuously differentiable.

17. The directional derivative of a scalar point function ϕ along the direction of a unit vector $\hat{\mathbf{a}}$ can be written as grad $\phi\,.\,\hat{\mathbf{a}}$.

For, if $\hat{\mathbf{a}}=l\mathbf{i}+m\mathbf{j}+n\mathbf{k}$ is the unit vector along the line with direction cosines l, m, n and

grad $\phi=\frac{\partial\phi}{\partial x}\mathbf{i}+\frac{\partial\phi}{\partial y}\mathbf{j}+\frac{\partial\phi}{\partial z}\mathbf{k}$, then

$$\text{grad}\,\phi\,.\,\hat{\mathbf{a}}=\left(\frac{\partial\phi}{\partial x}\mathbf{i}+\frac{\partial\phi}{\partial y}\mathbf{j}+\frac{\partial\phi}{\partial z}\mathbf{k}\right).(l\mathbf{i}+m\mathbf{j}+n\mathbf{k})=l\frac{\partial\phi}{\partial x}+m\frac{\partial\phi}{\partial y}+n\frac{\partial\phi}{\partial z}$$

= directional derivative of ϕ along the direction of the unit vector $\hat{\mathbf{a}}$.

18. Directional Derivative along a Curve

Let $\mathbf{r}=\mathbf{f}(s)=f_1(s)\mathbf{i}+f_2(s)\mathbf{j}+f_3(s)\mathbf{k}$

be any curve, where $f_i (i = 1, 2, 3)$ is a scalar function of the arc length s.

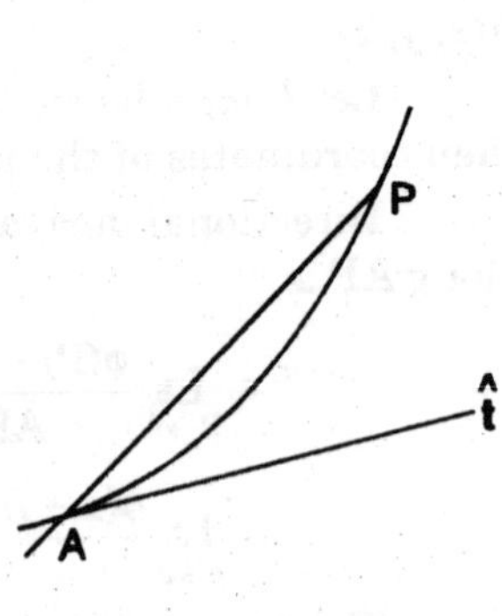

Let ϕ be a function of s. If A and P are any two points on the curve, then the directional derivative of ϕ at A along the unit tangent vector $\hat{\mathbf{t}}$ is given by

$$=\underset{P\to A}{\text{Lt}}\frac{\phi(P)-\phi(A)}{AP}=\underset{\delta s\to 0}{\text{Lt}}\frac{\phi(s+\delta s)-\phi(s)}{\delta s}=\frac{d\phi}{ds}$$

$$=\frac{\partial\phi}{\partial x}\frac{dx}{ds}+\frac{\partial\phi}{\partial y}\frac{dy}{ds}+\frac{\partial\phi}{\partial z}\frac{dz}{ds}\quad\text{(which can be written as)}$$

$$=\left(\mathbf{i}\frac{\partial\phi}{\partial x}+\mathbf{j}\frac{\partial\phi}{\partial y}+\mathbf{k}\frac{\partial\phi}{\partial z}\right).\left(\mathbf{i}\frac{dx}{ds}+\mathbf{j}\frac{dy}{ds}+\mathbf{k}\frac{dz}{ds}\right)$$

$$=\text{grad}\,\phi\,.\,\hat{\mathbf{t}}\qquad\left(\because \frac{d\mathbf{r}}{ds}=\frac{dx}{ds}\mathbf{i}+\frac{dy}{ds}\mathbf{j}+\frac{dz}{ds}\mathbf{k}=\hat{\mathbf{t}}\right)$$

19. We close the section with the following two important results (without proofs).

(*a*) If $\hat{n}$ is a unit vector normal to the level surface $\phi(x, y, z) = c$ at a point P(x, y, z) and n is the distance of P from some fixed point A in the direction of $\hat{n}$, then grad $\phi = \dfrac{d\phi}{dn}\,\hat{n}$.

(*b*) Grad ϕ is a vector in the direction of which the maximum value of the directional derivative of ϕ, occurs.

Thus, the directional derivative is maximum along the normal to the surface and maximum value is $|\text{ grad } \phi \,|$.

Example 10. *Find the directional derivative of $\phi(x, y, z) = x^2yz + 4xz^2$ at $(1, -2, -1)$ in the direction $2\mathbf{i} - \mathbf{j} - 2\mathbf{k}$.*

Sol. Here, $\phi(x, y, z) = x^2yz + 4xz^2$.

$$\therefore \quad \text{grad } \phi = \nabla\phi = \frac{\partial\phi}{\partial x}\mathbf{i} + \frac{\partial\phi}{\partial y}\mathbf{j} + \frac{\partial\phi}{\partial z}\mathbf{k} = (2xyz + 4z^2)\,\mathbf{i} + (x^2z)\,\mathbf{j} + (x^2y + 8xz)\,\mathbf{k}$$

$$= 8\mathbf{i} - \mathbf{j} - 10\mathbf{k} \text{ at } (1, -2, -1).$$

If $\hat{\mathbf{a}}$ is a unit vector in the direction of $2\mathbf{i} - \mathbf{j} - 2\mathbf{k}$, then

$$\hat{\mathbf{a}} = \frac{2\mathbf{i} - \mathbf{j} - 2\mathbf{k}}{\sqrt{4 + 1 + 4}} = \frac{2}{3}\mathbf{i} - \frac{1}{3}\mathbf{j} - \frac{2}{3}\mathbf{k}$$

$\therefore$ Required directional derivative

$$= \nabla\phi \,.\, \hat{\mathbf{a}} = (8\mathbf{i} - \mathbf{j} - 10\mathbf{k}) \,.\, \left(\frac{2}{3}\mathbf{i} - \frac{1}{3}\mathbf{j} - \frac{2}{3}\mathbf{k}\right) = \frac{16}{3} + \frac{1}{3} + \frac{20}{3} = \frac{37}{3}.$$

Example 11. *Find the directional derivative of the function $\phi = \dfrac{y}{x^2 + y^2}$ at the point (0, 1) along a line making an angle of 60° with the positive direction of x-axis.*

Sol. $$\text{grad } \phi = \nabla\phi = \frac{\partial\phi}{\partial x}\mathbf{i} + \frac{\partial\phi}{\partial y}\mathbf{j} \quad \left(\because \frac{\partial\phi}{\partial z} = 0\right)$$

$$= \frac{-2xy}{(x^2 + y^2)^2}\mathbf{i} + \frac{x^2 - y^2}{(x^2 + y^2)^2}\mathbf{j}$$

$$= -\mathbf{j} \text{ at the point } (0, 1).$$

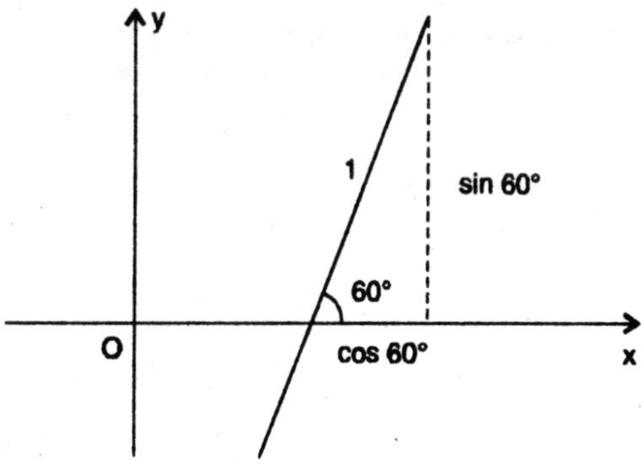

If $\hat{\mathbf{a}}$ is a unit vector along the line making an angle of 60° with x-axis, then

$$\hat{\mathbf{a}} = \cos 60°\,\mathbf{i} + \sin 60°\,\mathbf{j} = \frac{1}{2}\mathbf{i} + \frac{\sqrt{3}}{2}\mathbf{j}$$

$\therefore$ Required directional derivative

$$= \nabla\phi \,.\, \hat{\mathbf{a}} = (-\mathbf{j}) \,.\, \left(\frac{1}{2}\mathbf{i} + \frac{\sqrt{3}}{2}\mathbf{j}\right) = -\frac{\sqrt{3}}{2}.$$

Example 12. *In what direction from the point $(1, 1, -1)$ is the direction derivative of $\phi(x, y, z) = x^2 - 2y^2 + 4z^2$ a maximum ? Also find the magnitude of this maximum.*

Sol. Here, $\phi(x, y, z) = x^2 - 2y^2 + 4z^2$

$$\therefore \quad \nabla\phi = \frac{\partial\phi}{\partial x}\mathbf{i} + \frac{\partial\phi}{\partial y}\mathbf{j} + \frac{\partial\phi}{\partial z}\mathbf{k}$$

$$= 2x\mathbf{i} - 4y\mathbf{j} + 8z\mathbf{k} = 2\mathbf{i} - 4\mathbf{j} - 8\mathbf{k} \text{ at the point } (1, 1, -1).$$

The directional derivative of ϕ is maximum in the direction of $\nabla\phi = 2\mathbf{i} - 4\mathbf{j} - 8\mathbf{k}$.

The magnitude of this maximum is $|\nabla\phi|$

$$= \sqrt{(2)^2 + (-4)^2 + (-8)^2} = \sqrt{4 + 16 + 64} = \sqrt{84} = 2\sqrt{21}.$$

EXERCISE 2

1. Find a unit vector which is perpendicular to the surface $z = x^2 + y^2$ at the point (1, 2, 5).
2. Find the unit outward drawn normal to the surface $(x - 1)^2 + y^2 + (z + 2)^2 = 9$ at the point (3, 1, – 4).
3. Find the equation of tangent plane to the surface $xz^2 + x^2y = z - 1$ at the point (1, – 3, 2).
4. Find the equations for the tangent plane and the normal line to the surface $z = x^2 + y^2$ at the point (2, – 1, 5). *(M.D.U. 2003)*
5. Find the directional derivative of $\phi = 4xz^3 - 3x^2y^2z$ at (2, – 1, 2) in the direction $2\mathbf{i} - 3\mathbf{j} + 6\mathbf{k}$.
6. Find the directional derivative of $\phi = x^2 - y^2 + 2z^2$ at the point P(1, 2, 3) in a direction towards the point Q(5, 0, 4) *i.e.*, in the direction of the line PQ. *(K.U. 2000 S)*

   [**Hint.** $\vec{PQ} = (5\mathbf{i} + 0\mathbf{j} + 4\mathbf{k}) - (\mathbf{i} + 2\mathbf{j} + 3\mathbf{k}) = 4\mathbf{i} - 2\mathbf{j} + \mathbf{k}$. Now proceed as in example 10.]
7. In what direction from the point (1, 3, 2) is the directional derivative of $\phi(x, y, z) = 2xz - y^2$ a maximum ? What is the magnitude of this maximum ?
8. Find the constants a, b, c so that the directional derivative of $\phi(x, y, z) = axy^2 + byz + cz^2x^3$ at (1, 2, – 1) has a maximum of magnitude 64 in a direction parallel to z-axis.
9. Find the acute angle between the surfaces $xy^2z - 3x = z^2$ and $3x^2 - y^2 + 2z = 1$ at the point (1, – 2, 1).
10. Find the constants a and b so that the surface $ax^2 - byz = (a + 2)x$ will be orthogonal to the surface $4x^2y + z^2 = 4$ at the point (1, – 1, 2). *(M.D.U. 2000)*

Answers

1. $\dfrac{2\mathbf{i} + 4\mathbf{j} - \mathbf{k}}{\pm\sqrt{21}}$ 2. $(2\mathbf{i} + \mathbf{j} - 2\mathbf{k})/3$ 3. $2x - y - 3z + 1 = 0$

4. $4x - 2y - z = 5, \dfrac{x-2}{4} = \dfrac{y+1}{-2} = \dfrac{z-5}{-1}$ 5. $\dfrac{376}{7}$

6. $\dfrac{4}{3}\sqrt{21}$ 7. In the direction of the vector $4\mathbf{i} - 6\mathbf{j} + 2\mathbf{k}$, $2\sqrt{14}$

8. $a = 6, b = 24, c = -8$ 9. $\cos^{-1}\dfrac{\sqrt{6}}{14}$ 10. $a = \dfrac{5}{2}, b = 1.$

---

20. Divergence of a Vector Function

If a vector function $\mathbf{f} = f_1\mathbf{i} + f_2\mathbf{j} + f_3\mathbf{k}$, *where* f_1, f_2, f_3 *are scalar functions, is defined and differentiable at each point (x, y, z) in a certain region of space (i.e.,* **f** *defines a differentiable vector field), then divergence of* **f**, *written as* $\nabla \cdot \mathbf{f}$ *or div.* **f** *is defined by*

$$\nabla \cdot \mathbf{f} = \left(\mathbf{i}\frac{\partial}{\partial x} + \mathbf{j}\frac{\partial}{\partial y} + \mathbf{k}\frac{\partial}{\partial z}\right) \cdot (f_1\mathbf{i} + f_2\mathbf{j} + f_3\mathbf{k}) = \frac{\partial f_1}{\partial x} + \frac{\partial f_2}{\partial y} + \frac{\partial f_3}{\partial z}.$$

Note that the divergence of a vector **f** is a scalar. Thus, *divergence of a vector point function is a scalar point function.*

Further, a vector **f** is said to be solenoidal, if $\nabla \cdot \mathbf{f} = 0$.

21. Properties of Divergence

If **f** *and* **g** *are two vector point functions and* ϕ *is a scalar point function, then*

1. $\nabla \cdot (\mathbf{f}+\mathbf{g}) = \nabla \cdot \mathbf{f} + \nabla \cdot \mathbf{g}$ [*i.e.*, div $(\mathbf{f}+\mathbf{g})$ = div **f** + div **g**]

2. $\nabla \cdot (\phi \mathbf{f}) = (\nabla \phi) \cdot \mathbf{f} + \phi (\nabla \cdot \mathbf{f})$ [*i.e.*, div $(\phi\mathbf{f})$ = (grad ϕ). **f** + ϕ (div **f**)]

3. $\nabla \cdot \mathbf{f} = 0$ if **f** is a constant vector.

Proof. 1. Let
$$\mathbf{f} = f_1\mathbf{i} + f_2\mathbf{j} + f_3\mathbf{k},$$
$$\mathbf{g} = g_1\mathbf{i} + g_2\mathbf{j} + g_3\mathbf{k}$$

$\therefore$
$$\mathbf{f}+\mathbf{g} = (f_1+g_1)\,\mathbf{i} + (f_2+g_2)\,\mathbf{j} + (f_3+g_3)\,\mathbf{k}$$

$\therefore$
$$\nabla \cdot (\mathbf{f}+\mathbf{g}) = \left(\mathbf{i}\frac{\partial}{\partial x} + \mathbf{j}\frac{\partial}{\partial y} + \mathbf{k}\frac{\partial}{\partial z}\right) \cdot [(f_1+g_1)\mathbf{i} + (f_2+g_2)\mathbf{j} + (f_3+g_3)\mathbf{k}]$$
$$= \frac{\partial}{\partial x}(f_1+g_1) + \frac{\partial}{\partial y}(f_2+g_2) + \frac{\partial}{\partial z}(f_3+g_3)$$
$$= \left(\frac{\partial f_1}{\partial x} + \frac{\partial f_2}{\partial y} + \frac{\partial f_3}{\partial z}\right) + \left(\frac{\partial g_1}{\partial x} + \frac{\partial g_2}{\partial y} + \frac{\partial g_3}{\partial z}\right)$$
$$= \nabla \cdot \mathbf{f} + \nabla \cdot \mathbf{g}.$$

2.
$$\phi\mathbf{f} = \phi f_1\mathbf{i} + \phi f_2\mathbf{j} + \phi f_3\mathbf{k}$$

$\therefore$
$$\nabla \cdot (\phi\mathbf{f}) = \left(\mathbf{i}\frac{\partial}{\partial x} + \mathbf{j}\frac{\partial}{\partial y} + \mathbf{k}\frac{\partial}{\partial z}\right) \cdot (\phi f_1\mathbf{i} + \phi f_2\mathbf{j} + \phi f_3\mathbf{k})$$
$$= \frac{\partial}{\partial x}(\phi f_1) + \frac{\partial}{\partial y}(\phi f_2) + \frac{\partial}{\partial z}(\phi f_3)$$
$$= \left(\frac{\partial \phi}{\partial x} f_1 + \phi\frac{\partial f_1}{\partial x}\right) + \left(\frac{\partial \phi}{\partial y} f_2 + \phi\frac{\partial f_2}{\partial y}\right) + \left(\frac{\partial \phi}{\partial z} f_3 + \phi\frac{\partial f_3}{\partial z}\right)$$
$$= \left(\frac{\partial \phi}{\partial x} f_1 + \frac{\partial \phi}{\partial y} f_2 + \frac{\partial \phi}{\partial z} f_3\right) + \left(\phi\frac{\partial f_1}{\partial x} + \phi\frac{\partial f_2}{\partial y} + \phi\frac{\partial f_3}{\partial z}\right)$$
$$= \left(\frac{\partial \phi}{\partial x}\mathbf{i} + \frac{\partial \phi}{\partial y}\mathbf{j} + \frac{\partial \phi}{\partial z}\mathbf{k}\right) \cdot (f_1\mathbf{i} + f_2\mathbf{j} + f_3\mathbf{k}) + \phi\left(\frac{\partial f_1}{\partial x} + \frac{\partial f_2}{\partial y} + \frac{\partial f_3}{\partial z}\right)$$
$$= (\nabla\phi) \cdot \mathbf{f} + \phi(\nabla \cdot \mathbf{f}).$$

3. By definition,
$$\nabla \cdot \mathbf{f} = \left(\mathbf{i}\frac{\partial}{\partial x} + \mathbf{j}\frac{\partial}{\partial y} + \mathbf{k}\frac{\partial}{\partial z}\right) \cdot \mathbf{f} = \mathbf{i} \cdot \frac{\partial \mathbf{f}}{\partial x} + \mathbf{j} \cdot \frac{\partial \mathbf{f}}{\partial y} + \mathbf{k} \cdot \frac{\partial \mathbf{f}}{\partial z}$$
$$= \mathbf{i} \cdot \mathbf{0} + \mathbf{j} \cdot \mathbf{0} + \mathbf{k} \cdot \mathbf{0} = 0 \qquad \left(\because \ \mathbf{f}\text{, being a constant vector, } \frac{\partial \mathbf{f}}{\partial x} = 0 \text{ etc.}\right)$$

Example 13. *If* $\mathbf{f} = x^2z\mathbf{i} - 2y^3z^2\mathbf{j} + xy^2\mathbf{k}$, *find* $\nabla \cdot \mathbf{f}$ *(or div* $\mathbf{f}$*) at the point* $(-1, 1, 2)$.

Sol.

$$\nabla \cdot \mathbf{f} = \left(\mathbf{i}\frac{\partial}{\partial x} + \mathbf{j}\frac{\partial}{\partial y} + \mathbf{k}\frac{\partial}{\partial z}\right) \cdot (x^2z\mathbf{i} - 2y^3z^2\mathbf{j} + xy^2\mathbf{k})$$

$$= \frac{\partial}{\partial x}(x^2z) + \frac{\partial}{\partial y}(-2y^3z^2) + \frac{\partial}{\partial z}(xy^2)$$

$$= 2xz - 6y^2z^2 = 2(-1)(2) - 6(1)^2(2)^2$$

$$= -28 \text{ at the point } (-1, 1, 2).$$

Example 14. *Show that* div $(r^n\mathbf{r}) = (n + 3)\, r^n$, *where* $r = |\,\mathbf{r}\,|$ *and* $\mathbf{r} = x\mathbf{i} + y\mathbf{j} + z\mathbf{k}$. *If* $r^n\mathbf{r}$ *is solenoidal then* $n + 3 = 0$. (K.U. 2000)

Sol. We know, $r^2 = x^2 + y^2 + z^2$

$$\therefore \quad 2r\frac{\partial r}{\partial x} = 2x \quad \text{or} \quad \frac{\partial r}{\partial x} = \frac{x}{r} \qquad \text{...(1)}$$

Similarly, $\dfrac{\partial r}{\partial y} = \dfrac{y}{r}$ and $\dfrac{\partial r}{\partial z} = \dfrac{z}{r}$ ...(2)

Now,

$$\text{div}\,(r^n\mathbf{r}) = \text{div}\,[r^n(x\mathbf{i} + y\mathbf{j} + z\mathbf{k})]$$

$$= \text{div}\,[(r^nx)\mathbf{i} + (r^ny)\mathbf{j} + (r^nz)\mathbf{k}]$$

$$= \frac{\partial}{\partial x}(r^nx) + \frac{\partial}{\partial y}(r^ny) + \frac{\partial}{\partial z}(r^nz) \qquad \text{...(3)}$$

But,

$$\frac{\partial}{\partial x}(r^nx) = r^n + xnr^{n-1}\frac{\partial r}{\partial x}$$

$$= r^n + nr^{n-1}x\frac{x}{r} \qquad \text{[By (1)]}$$

$$= r^n + nx^2r^{n-2}$$

Similarly, using (2), we have

$$\frac{\partial}{\partial y}(r^ny) = r^n + ny^2r^{n-2} \text{ and } \frac{\partial}{\partial z}(r^nz) = r^n + nz^2r^{n-2}$$

$\therefore$ From (3), we have

$$\text{div}\,(r^n\mathbf{r}) = (r^n + nx^2r^{n-2}) + (r^n + ny^2r^{n-2}) + (r^n + nz^2r^{n-2})$$

$$= 3r^n + nr^{n-2}(x^2 + y^2 + z^2)$$

$$= 3r^n + nr^{n-2}r^2 = (3 + n)r^n.$$

If $r^n\mathbf{r}$ is solenoidal, then div $(r^n\mathbf{r}) = 0$ and hence $n + 3 = 0$ $(\because \ r \neq 0)$

Example 15. *Evaluate* div $\left(\dfrac{\mathbf{r}}{r}\right)$ *when* $\mathbf{r}$ *and* r *have their usual meanings. Hence, show that* grad div $\left(\dfrac{\mathbf{r}}{r}\right) = -\dfrac{2}{r^3}\,\mathbf{r}$.

Or

If $\mathbf{r} = x\mathbf{i} + y\mathbf{j} + z\mathbf{k}$ *and* $(\mathbf{r})$ r, *show that* $\nabla \cdot \left(\dfrac{\mathbf{r}}{r}\right) = \dfrac{2}{r}$ (K.U. 2006)

Sol.

$$\text{div}\left(\frac{\mathbf{r}}{r}\right) = \nabla \cdot \left(\frac{\mathbf{r}}{r}\right) = \nabla \cdot \left[\left(\frac{1}{r}\right)\mathbf{r}\right]$$

$$= \nabla \cdot \left(\frac{1}{r}\right)\mathbf{r} + \left(\frac{1}{r}\right)\nabla \cdot \mathbf{r} \qquad \text{...(1)} \quad [\because \ \nabla \cdot (\phi\mathbf{f}) = (\nabla\phi) \cdot \mathbf{f} + \phi(\nabla \cdot \mathbf{f})]$$

Now, $$\nabla\left(\frac{1}{r}\right)=\left(\mathbf{i}\frac{\partial}{\partial x}+\mathbf{j}\frac{\partial}{\partial y}+\mathbf{k}\frac{\partial}{\partial z}\right)\left(\frac{1}{r}\right)$$

$$=\mathbf{i}\frac{\partial}{\partial x}\left(\frac{1}{r}\right)+\mathbf{j}\frac{\partial}{\partial y}\left(\frac{1}{r}\right)+\mathbf{k}\frac{\partial}{\partial z}\left(\frac{1}{r}\right)$$

$$=\mathbf{i}\left(-\frac{1}{r^2}\frac{\partial r}{\partial x}\right)+\mathbf{j}\left(\frac{1}{r^2}\frac{\partial r}{\partial y}\right)+\mathbf{k}\left(-\frac{1}{r^3}\frac{\partial r}{\partial z}\right)$$

$$=\mathbf{i}\left(-\frac{1}{r^2}\frac{x}{r}\right)+\mathbf{j}\left(-\frac{1}{r^2}\frac{y}{r}\right)+\mathbf{k}\left(-\frac{1}{r^3}\frac{z}{r}\right) \qquad \left(\because \frac{\partial r}{\partial x}=\frac{x}{r}, \text{etc.}\right)$$

$$=-\frac{1}{r^3}(x\mathbf{i}+y\mathbf{j}+z\mathbf{k})=-\frac{\mathbf{r}}{r^3} \qquad ...(2)$$

Also, $$\nabla\cdot\mathbf{r}=\left(\mathbf{i}\frac{\partial}{\partial x}+\mathbf{j}\frac{\partial}{\partial y}+\mathbf{k}\frac{\partial}{\partial z}\right)\cdot(x\mathbf{i}+y\mathbf{j}+z\mathbf{k})$$

$$=\frac{\partial x}{\partial x}+\frac{\partial y}{\partial y}+\frac{\partial z}{\partial z}=1+1+1=3 \qquad ...(3)$$

Using (2) and (3) in (1), we have

$$\text{div}\left(\frac{\mathbf{r}}{r}\right)=-\frac{\mathbf{r}^2}{r^3}+\left(\frac{1}{r}\right)(3)=-\frac{1}{r}+\frac{3}{r}=\frac{2}{r} \qquad (\because \mathbf{r}^2=r^2)$$

$\therefore$ $$\text{grad div}\left(\frac{\mathbf{r}}{r}\right)=\text{grad}\left(\frac{2}{r}\right)$$

$$=\nabla\left(\frac{2}{r}\right)=2\nabla\left(\frac{1}{r}\right)=2\left(-\frac{\mathbf{r}}{r^3}\right) \qquad \text{[from (2)]}$$

$$=-\frac{2}{r^3}\mathbf{r}.$$

Example 16. (*i*) *Interpret the symbol* $\mathbf{f}\cdot\nabla$

(*ii*) *Give a possible meaning to* $(\mathbf{f}\cdot\nabla)\,\mathbf{g}$.

Sol. (*i*) If $\mathbf{f}=f_1\mathbf{i}+f_2\mathbf{j}+f_3\mathbf{k}$, then

$$\mathbf{f}\cdot\nabla\equiv(f_1\mathbf{i}+f_2\mathbf{j}+f_3\mathbf{k})\cdot\left(\mathbf{i}\frac{\partial}{\partial x}+\mathbf{j}\frac{\partial}{\partial y}+\mathbf{k}\frac{\partial}{\partial z}\right)$$

$$\equiv f_1\frac{\partial}{\partial x}+f_2\frac{\partial}{\partial y}+f_3\frac{\partial}{\partial z} \text{ is an operator.}$$

Thus, $$(\mathbf{f}\cdot\nabla)\phi=\left(f_1\frac{\partial}{\partial x}+f_2\frac{\partial}{\partial y}+f_3\frac{\partial}{\partial z}\right)\phi \qquad ...(1)$$

$$=f_1\frac{\partial\phi}{\partial x}+f_2\frac{\partial\phi}{\partial y}+f_3\frac{\partial\phi}{\partial z}$$

This infact, is the same as $\mathbf{f}\cdot(\nabla\phi)$.

(*ii*) Replacing ϕ by $\mathbf{g}=g_1\mathbf{i}+g_2\mathbf{j}+g_3\mathbf{k}$ in (1), we have

$$(\mathbf{f}\cdot\nabla)\,\mathbf{g}=\left(f_1\frac{\partial}{\partial x}+f_2\frac{\partial}{\partial y}+f_3\frac{\partial}{\partial z}\right)\mathbf{g}=f_1\frac{\partial\mathbf{g}}{\partial x}+f_2\frac{\partial\mathbf{g}}{\partial y}+f_3\frac{\partial\mathbf{g}}{\partial z}$$

$$= \left(f_1 \frac{\partial g_1}{\partial x} + f_2 \frac{\partial g_1}{\partial y} + f_3 \frac{\partial g_1}{\partial z}\right) \mathbf{i} + \left(f_1 \frac{\partial g_2}{\partial x} + f_2 \frac{\partial g_2}{\partial y} + f_3 \frac{\partial g_2}{\partial z}\right) \mathbf{j}$$

$$+ \left(f_1 \frac{\partial g_3}{\partial x} + f_2 \frac{\partial g_3}{\partial y} + f_3 \frac{\partial g_3}{\partial z}\right) \mathbf{k}.$$

EXERCISE 3

1. Prove : div $\mathbf{r} = 3$, where $\mathbf{r} = x\mathbf{i} + y\mathbf{j} + z\mathbf{k}$.
2. Determine the constant a so that the vector $\mathbf{f} = (x + 3y)\,\mathbf{i} + (y - 2z)\,\mathbf{j} + (x + az)\,\mathbf{k}$ is solenoidal.
   [**Hint.** Vector is solenoidal if div $\mathbf{f} = 0$, *i.e.*, if $2 + a = 0$.]
3. If $\mathbf{a}$ is a constant vector, show that div $(\mathbf{r} \times \mathbf{a}) = 0$.
   [**Hint.** If $\mathbf{r} = x\mathbf{i} + y\mathbf{j} + z\mathbf{k}$ and $\mathbf{a} = a_1\mathbf{i} + a_2\mathbf{j} + a_3\mathbf{k}$,
   then $\mathbf{r} \times \mathbf{a} = (a_3 y - a_2 z)\,\mathbf{i} + (a_1 z - a_3 x)\,\mathbf{j} + (a_2 x - a_1 y)\mathbf{k}$.]
4. Evaluate div $\mathbf{f}$ (or $\nabla \cdot \mathbf{f}$), where $\mathbf{f} = 2x^2z\mathbf{i} - xy^2z\mathbf{j} + 3y^2x\mathbf{k}$ at (1, 1, 1).
5. Find div $\mathbf{f}$, where $\mathbf{f} = \text{grad}\,(x^3 + y^3 + z^3 - 3xyz)$.
6. If $\phi = 3x^2y$ and $\psi = xz^2 - 2y$, find grad [(grad ϕ) . (grad ψ)].
7. If $\mathbf{r} = x\mathbf{i} + y\mathbf{j} + z\mathbf{k}$ and $r = |\,\mathbf{r}\,| = \sqrt{x^2 + y^2 + z^2}$, then prove

   (*i*) div $(\hat{\mathbf{r}}) = \dfrac{2}{r}$ (*K.U. 1993, 96*) (*ii*) div $\left(\dfrac{\mathbf{r}}{r^3}\right) = 0$.

   [**Hint.** Use $\nabla \cdot (\phi\mathbf{f}) = \nabla\phi \cdot \mathbf{f} + \phi(\nabla \cdot \mathbf{f})$, where ϕ is a scalar and $\mathbf{f}$ is a vector.]
8. If $\mathbf{w}$ is a constant vector and $\mathbf{v} = \mathbf{w} \times \mathbf{r}$, prove that div $\mathbf{v} = 0$.
9. If $\mathbf{f} = 2yz\mathbf{i} - x^2y\mathbf{j} + xz^2\mathbf{k}$, $\mathbf{g} = x^2\mathbf{i} + yz\mathbf{j} - xy\mathbf{k}$ and $\phi = 2x^2yz^3$, find
   (*i*) $(\mathbf{f} \cdot \nabla)\phi$ (*ii*) $\mathbf{f} \cdot \nabla\phi$ (*iii*) $(\mathbf{f} \cdot \nabla)\mathbf{g}$.
10. If $\mathbf{f} = \dfrac{x\mathbf{i} + y\mathbf{j} + z\mathbf{k}}{\sqrt{x^2 + y^2 + z^2}}$, show that $\nabla \cdot \mathbf{f} = \dfrac{2}{\sqrt{x^2 + y^2 + z^2}}$. (*M.D.U. 2007*)
11. Show that div $\left[\dfrac{f(r)\mathbf{r}}{r}\right] = \dfrac{1}{r^2}\dfrac{d}{dr}(r^2 f(r))$, where r and $\mathbf{r}$ have their usual meanings.
    (*K.U. 2003, 2007*)

Answers

4. $+2$ 5. $6(x + y + z)$ 6. $(6yz^2 - 12x)\mathbf{i} + 6xz^2\mathbf{j} + 12xyz\mathbf{k}$
9. (*i*) $8xy^2z^4 - 2x^4yz^3 + 6x^3yz^4$ (*ii*) $8xy^2z^4 - 2x^4yz^3 + 6x^3yz^4$
   (*iii*) $4xyz\,\mathbf{i} - (x^2yz - xyz^2)\,\mathbf{j} - (-2y^2z - x^3y)\,\mathbf{k}$.

---

22. Curl of a Vector Function

If a vector function $\mathbf{f} = f_1\mathbf{i} + f_2\mathbf{j} + f_3\mathbf{k}$ *where* f_1, f_2, f_3 *are scalar functions, is defined and differentiable at each point (x, y, z) in a certain region of space, then the curl or rotation of* **f**, *written as* $\nabla \times \mathbf{f}$ *or curl* **f** *or rot* **f** *is defined by*

$$\nabla \times \mathbf{f} = \left(\mathbf{i}\frac{\partial}{\partial x} + \mathbf{j}\frac{\partial}{\partial y} + \mathbf{k}\frac{\partial}{\partial z}\right) \times (f_1\mathbf{i} + f_2\mathbf{j} + f_3\mathbf{k})$$

$$= \left(\frac{\partial f_3}{\partial y} - \frac{\partial f_2}{\partial z}\right)\mathbf{i} + \left(\frac{\partial f_1}{\partial z} - \frac{\partial f_3}{\partial x}\right)\mathbf{j} + \left(\frac{\partial f_2}{\partial x} - \frac{\partial f_1}{\partial y}\right)\mathbf{k}$$

which can be easily written as $\begin{vmatrix} \mathbf{i} & \mathbf{j} & \mathbf{k} \\ \dfrac{\partial}{\partial x} & \dfrac{\partial}{\partial y} & \dfrac{\partial}{\partial z} \\ f_1 & f_2 & f_3 \end{vmatrix}$

Note that $\nabla \times \mathbf{f}$ is a vector point function and in the expansion of the determinant, the operators $\frac{\partial}{\partial x}, \frac{\partial}{\partial y}, \frac{\partial}{\partial z}$ must preceed f_1, f_2, f_3.

A vector **f** *for which* $\nabla \times \mathbf{f} = \mathbf{0}$, *is called an irrotational vector.*

23. Properties of Curl

If **f** and **g** are two vector point functions and ϕ a scalar point function, all having continuous second order partial derivatives, then

1. $\nabla \times (\mathbf{f} + \mathbf{g}) = \nabla \times \mathbf{f} + \nabla \times \mathbf{g}$
   [*i.e.*, curl $(\mathbf{f} + \mathbf{g})$ = curl **f** + curl **g**]
2. $\nabla \times (\phi\mathbf{f}) = (\nabla\phi) \times \mathbf{f} + \phi(\nabla \times \mathbf{f})$
   [*i.e.*, curl $(\phi\mathbf{f})$ = (grad ϕ) $\times$ **f** + ϕ (curl **f**)] (*K.U. 2003 ; M.D.U. 2003*)
3. $\nabla \times \mathbf{f} = \mathbf{0}$ if **f** is a constant vector
4. $\nabla \times (\mathbf{f} \times \mathbf{g}) = (\mathbf{g} \cdot \nabla)\,\mathbf{f} - (\mathbf{f} \cdot \nabla)\,\mathbf{g} + \mathbf{f}\,(\nabla \cdot \mathbf{g}) - (\nabla \cdot \mathbf{f})\mathbf{g}$
5. $\nabla \cdot (\mathbf{f} \times \mathbf{g}) = \mathbf{g} \cdot (\nabla \times \mathbf{f}) - \mathbf{f} \cdot (\nabla \times \mathbf{g})$
   [*i.e.*, div $(\mathbf{f} \times \mathbf{g}) = \mathbf{g}$. (curl **f**) $-$ **f** . (curl **g**)] (*K.U. 2003 S*)

Proof. 1. Let $\mathbf{f} = f_1\mathbf{i} + f_2\mathbf{j} + f_3\mathbf{k}$ and $\mathbf{g} = g_1\mathbf{i} + g_2\mathbf{j} + g_3\mathbf{k}$

Then $\mathbf{f} + \mathbf{g} = (f_1 + g_1)\mathbf{i} + (f_2 + g_2)\,\mathbf{j} + (f_3 + g_3)\mathbf{k}$

$$\therefore \qquad \nabla \times (\mathbf{f} \times \mathbf{g}) = \begin{vmatrix} \mathbf{i} & \mathbf{j} & \mathbf{k} \\ \dfrac{\partial}{\partial x} & \dfrac{\partial}{\partial y} & \dfrac{\partial}{\partial z} \\ f_1 + g_1 & f_2 + g_2 & f_3 + g_3 \end{vmatrix}$$

$$= \left[\frac{\partial}{\partial y}(f_3 + g_3) - \frac{\partial}{\partial z}(f_2 + g_2)\right]\mathbf{i} + \left[\frac{\partial}{\partial z}(f_1 + g_1) - \frac{\partial}{\partial x}(f_3 + g_3)\right]\mathbf{j}$$

$$+ \left[\frac{\partial}{\partial x}(f_2 + g_2) - \frac{\partial}{\partial y}(f_1 + g_1)\right]\mathbf{k}$$

$$= \left[\left(\frac{\partial f_3}{\partial y} - \frac{\partial f_2}{\partial z}\right)\mathbf{i} + \left(\frac{\partial f_1}{\partial z} - \frac{\partial f_3}{\partial x}\right)\mathbf{j} + \left(\frac{\partial f_2}{\partial x} - \frac{\partial f_1}{\partial y}\right)\mathbf{k}\right]$$

$$+ \left[\left(\frac{\partial g_3}{\partial y} - \frac{\partial g_2}{\partial z}\right)\mathbf{i} + \left(\frac{\partial g_1}{\partial z} - \frac{\partial g_3}{\partial x}\right)\mathbf{j} + \left(\frac{\partial g_2}{\partial x} - \frac{\partial g_1}{\partial y}\right)\mathbf{k}\right]$$

$$= \nabla \times \mathbf{f} + \nabla \times \mathbf{g}.$$

2. $$\nabla \times (\phi \mathbf{f}) = \nabla \times (\phi f_1 \mathbf{i} + \phi f_2 \mathbf{j} + \phi f_3 \mathbf{k})$$

$$= \begin{vmatrix} \mathbf{i} & \mathbf{j} & \mathbf{k} \\ \dfrac{\partial}{\partial x} & \dfrac{\partial}{\partial y} & \dfrac{\partial}{\partial z} \\ \phi f_1 & \phi f_2 & \phi f_3 \end{vmatrix}$$

$$= \left[\frac{\partial}{\partial y}(\phi f_3) - \frac{\partial}{\partial z}(\phi f_2)\right]\mathbf{i} + \left[\frac{\partial}{\partial z}(\phi f_1) - \frac{\partial}{\partial x}(\phi f_3)\right]\mathbf{j} + \left[\frac{\partial}{\partial x}(\phi f_2) - \frac{\partial}{\partial y}(\phi f_1)\right]\mathbf{k}$$

$$= \left[\phi\frac{\partial f_3}{\partial y} + \frac{\partial \phi}{\partial y} f_3 - \phi\frac{\partial f_2}{\partial z} - \frac{\partial \phi}{\partial z} f_2\right]\mathbf{i} + \left[\phi\frac{\partial f_1}{\partial z} + \frac{\partial \phi}{\partial z} f_1 - \phi\frac{\partial f_3}{\partial x} - \frac{\partial \phi}{\partial x} f_3\right]\mathbf{j}$$

$$+ \left[\phi\frac{\partial f_2}{\partial x} + \frac{\partial \phi}{\partial x} f_2 - \phi\frac{\partial f_1}{\partial y} - \frac{\partial \phi}{\partial y} f_1\right]\mathbf{k}$$

$$= \phi\left[\left(\frac{\partial f_3}{\partial y} - \frac{\partial f_2}{\partial z}\right)\mathbf{i} + \left(\frac{\partial f_1}{\partial z} - \frac{\partial f_3}{\partial x}\right)\mathbf{j} + \left(\frac{\partial f_2}{\partial x} - \frac{\partial f_1}{\partial y}\right)\mathbf{k}\right]$$

$$+ \left[\left(\frac{\partial \phi}{\partial y} f_3 - \frac{\partial \phi}{\partial z} f_2\right)\mathbf{i} + \left(\frac{\partial \phi}{\partial z} f_1 - \frac{\partial \phi}{\partial x} f_3\right)\mathbf{j} + \left(\frac{\partial \phi}{\partial x} f_2 - \frac{\partial \phi}{\partial y} f_1\right)\mathbf{k}\right]$$

$$= \phi(\nabla \times \mathbf{f}) + \begin{vmatrix} \mathbf{i} & \mathbf{j} & \mathbf{k} \\ \dfrac{\partial \phi}{\partial x} & \dfrac{\partial \phi}{\partial y} & \dfrac{\partial \phi}{\partial z} \\ f_1 & f_2 & f_3 \end{vmatrix}$$

$$= \phi(\nabla \times \mathbf{f}) + (\nabla \phi) \times \mathbf{f}.$$

3. $$\nabla \times \mathbf{f} = \mathbf{i} \times \frac{\partial \mathbf{f}}{\partial x} + \mathbf{j} \times \frac{\partial \mathbf{f}}{\partial y} + \mathbf{k} \times \frac{\partial \mathbf{f}}{\partial z} \qquad ...(1)$$

But, **f** is a constant vector,

$\therefore$ $\dfrac{\partial \mathbf{f}}{\partial x}, \dfrac{\partial \mathbf{f}}{\partial y}, \dfrac{\partial \mathbf{f}}{\partial z}$ are all zero (null vectors)

$\therefore$ From (1), $\nabla \times \mathbf{f} = \mathbf{0}$.

4. $$\nabla \times (\mathbf{f} \times \mathbf{g}) = \left(\mathbf{i}\frac{\partial}{\partial x} + \mathbf{j}\frac{\partial}{\partial y} + \mathbf{k}\frac{\partial}{\partial z}\right) \times (\mathbf{f} \times \mathbf{g})$$

$$= \mathbf{i} \times \frac{\partial}{\partial x}(\mathbf{f} \times \mathbf{g}) + \mathbf{j} \times \frac{\partial}{\partial y}(\mathbf{f} \times \mathbf{g}) + \mathbf{k} \times \frac{\partial}{\partial z}(\mathbf{f} \times \mathbf{g})$$

$$= \mathbf{i} \times \left(\frac{\partial \mathbf{f}}{\partial x} \times \mathbf{g} + \mathbf{f} \times \frac{\partial \mathbf{g}}{\partial x}\right) + \mathbf{j} \times \left(\frac{\partial \mathbf{f}}{\partial y} \times \mathbf{g} + \mathbf{f} \times \frac{\partial \mathbf{g}}{\partial y}\right) + \mathbf{k} \times \left(\frac{\partial \mathbf{f}}{\partial z} \times \mathbf{g} + \mathbf{f} \times \frac{\partial \mathbf{g}}{\partial z}\right)$$

$$= \Sigma\left(\mathbf{i} \times \left(\frac{\partial \mathbf{f}}{\partial x} \times \mathbf{g}\right)\right) + \Sigma\left(\mathbf{i} \times \left(\mathbf{f} \times \frac{\partial \mathbf{g}}{\partial x}\right)\right)$$

[But, $\mathbf{a} \times (\mathbf{b} \times \mathbf{c}) = (\mathbf{a} \cdot \mathbf{c})\,\mathbf{b} - (\mathbf{a} \cdot \mathbf{b})\mathbf{c}$]

$$= \Sigma\left((\mathbf{i} \cdot \mathbf{g})\frac{\partial \mathbf{f}}{\partial x} - \left(\mathbf{i} \cdot \frac{\partial \mathbf{f}}{\partial x}\right)\mathbf{g}\right) + \Sigma\left(\mathbf{i} \cdot \frac{\partial \mathbf{g}}{\partial x}\mathbf{f} - (\mathbf{i} \cdot \mathbf{f})\frac{\partial \mathbf{g}}{\partial x}\right)$$

$$= \Sigma\left((\mathbf{i} \cdot \mathbf{g})\frac{\partial \mathbf{f}}{\partial x}\right) - \Sigma\left(\mathbf{i} \cdot \frac{\partial \mathbf{f}}{\partial x}\right)\mathbf{g} + \Sigma\left(\left(\mathbf{i} \cdot \frac{\partial \mathbf{g}}{\partial x}\right)\mathbf{f}\right) - \Sigma\left((\mathbf{i} \cdot \mathbf{f})\frac{\partial \mathbf{g}}{\partial x}\right)$$

$$= \Sigma g_1 \frac{\partial \mathbf{f}}{\partial x} - \left(\Sigma \frac{\partial f_1}{\partial x}\right)\mathbf{g} + \left(\Sigma \frac{\partial g_1}{\partial x}\right)\mathbf{f} - \Sigma f_1 \frac{\partial \mathbf{g}}{\partial x}$$

[Letting $\mathbf{f} = f_1\mathbf{i} + f_2\mathbf{j} + f_3\mathbf{k}$ and $\mathbf{g} = g_1\mathbf{i} + g_2\mathbf{j} + g_3\mathbf{k}$]

$$= (\mathbf{g} \cdot \nabla)\mathbf{f} - (\nabla \cdot \mathbf{f})\mathbf{g} + (\nabla \cdot \mathbf{g})\mathbf{f} - (\mathbf{f} \cdot \nabla)\mathbf{g}$$

$$= (\mathbf{g} \cdot \nabla)\mathbf{f} - (\nabla \cdot \mathbf{f})\mathbf{g} + \mathbf{f}(\nabla \cdot \mathbf{g}) - (\mathbf{f} \cdot \nabla)\mathbf{g}.$$

5. $$\nabla \cdot (\mathbf{f} \times \mathbf{g}) = \Sigma \mathbf{i} \cdot \frac{\partial}{\partial x} (\mathbf{f} \times \mathbf{g})$$

$$= \Sigma \mathbf{i} \cdot \left(\frac{\partial \mathbf{f}}{\partial x} \times \mathbf{g} + \mathbf{f} \times \frac{\partial \mathbf{g}}{\partial x}\right) = \Sigma\left(\mathbf{i} \cdot \left(\frac{\partial \mathbf{f}}{\partial x} \times \mathbf{g}\right)\right) + \Sigma\left(\mathbf{i} \cdot \left(\mathbf{f} \times \frac{\partial \mathbf{g}}{\partial x}\right)\right)$$

[But in a scalar triple product **[a b c]**, **a . (b × c) = (a × b) . c**
i.e., the position of dot and cross are interchangeable, provided we maintain the cyclic order. Further, for every change of cyclic order, there is a change of sign]

$$= \Sigma\left[\left(\mathbf{i} \times \frac{\partial \mathbf{f}}{\partial x} \cdot \mathbf{g}\right]\right] - \Sigma\left[\left(\mathbf{i} \times \frac{\partial \mathbf{g}}{\partial x}\right) \cdot \mathbf{f}\right]$$

$$= \left[\Sigma\left(\mathbf{i} \times \frac{\partial \mathbf{g}}{\partial x}\right)\right] \cdot \mathbf{g} - \left[\Sigma\left(\mathbf{i} \times \frac{\partial \mathbf{g}}{\partial x}\right)\right] \cdot \mathbf{f}$$

$$= (\nabla \times \mathbf{f}) \cdot \mathbf{g} - (\nabla \times \mathbf{g}) \cdot \mathbf{f} = \mathbf{g} \cdot (\nabla \times \mathbf{f}) - \mathbf{f} \cdot (\nabla \times \mathbf{g}).$$

Example 17. *If* $\mathbf{f} = x^2z\mathbf{i} - 2y^3z^2\mathbf{j} + xy^2z\mathbf{k}$,
find $\nabla \times \mathbf{f}$ *(or curl* **f***) at the point (1, – 1, 1).*

Sol. $$\text{curl } \mathbf{f} = \nabla \times \mathbf{f}$$

$$= \left(\mathbf{i}\frac{\partial}{\partial x} + \mathbf{j}\frac{\partial}{\partial y} + \mathbf{k}\frac{\partial}{\partial z}\right) \times (x^2z\mathbf{i} - 2y^3z^2\mathbf{j} + xy^2z\mathbf{k}]$$

$$= \begin{vmatrix} \mathbf{i} & \mathbf{j} & \mathbf{k} \\ \frac{\partial}{\partial x} & \frac{\partial}{\partial y} & \frac{\partial}{\partial z} \\ x^2z & -2y^3z^2 & xy^2z \end{vmatrix}$$

$$= \left[\frac{\partial}{\partial y}(xy^2z) - \frac{\partial}{\partial z}(-2y^3z^2)\right]\mathbf{i} + \left[\frac{\partial}{\partial z}(x^2z) - \frac{\partial}{\partial x}(xy^2z)\right]\mathbf{j} + \left[\frac{\partial}{\partial x}(-2y^3z^2) - \frac{\partial}{\partial y}(x^2z)\right]\mathbf{k}$$

$$= (2xyz + 4y^3z)\mathbf{i} + (x^2 - y^2z)\mathbf{j} + 0\mathbf{k}$$

$= -6\mathbf{i}$ **at the point** (1, – 1, 1).

Example 18. *Prove that curl* $\mathbf{r} = \mathbf{0}$, *where* $\mathbf{r} = x\mathbf{i} + y\mathbf{j} + z\mathbf{k}$.

Sol. $$\text{Curl } \mathbf{r} = \nabla \times \mathbf{r}$$

$$= \begin{vmatrix} \mathbf{i} & \mathbf{j} & \mathbf{k} \\ \frac{\partial}{\partial x} & \frac{\partial}{\partial y} & \frac{\partial}{\partial z} \\ x & y & z \end{vmatrix}$$

$$= \left(\frac{\partial z}{\partial y} - \frac{\partial y}{\partial z}\right)\mathbf{i} + \left(\frac{\partial x}{\partial z} - \frac{\partial z}{\partial x}\right)\mathbf{j} + \left(\frac{\partial y}{\partial x} - \frac{\partial x}{\partial y}\right)\mathbf{k}$$

$$= 0\mathbf{i} + 0\mathbf{j} + 0\mathbf{k} = \mathbf{0}.$$

Example 19. *If* $\mathbf{f} = \mathbf{w} \times \mathbf{r}$, *prove* $\mathbf{w} = \frac{1}{2}$ *curl* $\mathbf{f}$, *where* $\mathbf{w}$ *is a constant.* (K.U. 2006)

Sol. Let $\mathbf{w} = w_1\mathbf{i} + w_2\mathbf{j} + w_3\mathbf{k}$

$$\therefore \quad \mathbf{w} \times \mathbf{r} = \begin{vmatrix} \mathbf{i} & \mathbf{j} & \mathbf{k} \\ w_1 & w_2 & w_3 \\ x & y & z \end{vmatrix} = (w_2 z - w_3 y)\mathbf{i} + (w_3 x - w_1 z)\mathbf{j} + (w_1 y - w_2 x)\mathbf{k}$$

$$\therefore \quad \text{curl } \mathbf{f} = \nabla \times \mathbf{f} = \nabla \times (\mathbf{w} \times \mathbf{r})$$

$$= \begin{vmatrix} \mathbf{i} & \mathbf{j} & \mathbf{k} \\ \frac{\partial}{\partial x} & \frac{\partial}{\partial y} & \frac{\partial}{\partial z} \\ w_2 z - w_3 y & w_3 x - w_1 z & w_1 y - w_2 x \end{vmatrix}$$

$$= 2(w_1\mathbf{i} + w_2\mathbf{j} + w_3\mathbf{k}) = 2\mathbf{w}$$

Hence, $\mathbf{w} = \frac{1}{2}$ curl $\mathbf{f}$.

Example 20. *Find the constants a, b, c so that*

$$\mathbf{f} = (x + 2y + az)\mathbf{i} + (bx - 3y - z)\mathbf{j} + (4x + cy + 2z)\mathbf{k}$$

is irrotational.

Sol.

$$\nabla \times \mathbf{f} = \begin{vmatrix} \mathbf{i} & \mathbf{j} & \mathbf{k} \\ \frac{\partial}{\partial x} & \frac{\partial}{\partial y} & \frac{\partial}{\partial z} \\ x + 2y + az & bx - 3y - z & 4x + cy + 2z \end{vmatrix} \quad \text{(On simplification)}$$

$$= (c + 1)\mathbf{i} + (a - 4)\mathbf{j} + (b - 2)\mathbf{k}$$

Now, $\mathbf{f}$ is irrotational if $\nabla \times \mathbf{f} = 0$

i.e., if $c + 1 = 0, a - 4 = 0, b - 2 = 0$

i.e., if $c = -1, a = 4, b = 2.$

Example 21. *Prove curl* $\frac{\mathbf{a} \times \mathbf{r}}{r^3} = -\frac{\mathbf{a}}{r^3} + \frac{3\mathbf{r}}{r^5}(\mathbf{a} \cdot \mathbf{r})$ *where* $\mathbf{a}$ *is a constant vector.*

(K.U. 2007)

Sol.

$$\text{curl } \frac{\mathbf{a} \times \mathbf{r}}{r^3} = \nabla \times \frac{\mathbf{a} \times \mathbf{r}}{r^3} = \Sigma \mathbf{i} \times \frac{\partial}{\partial x}\left(\frac{\mathbf{a} \times \mathbf{r}}{r^3}\right) \qquad \text{...(1)}$$

Now,

$$\frac{\partial}{\partial x}\left(\frac{\mathbf{a} \times \mathbf{r}}{r^3}\right) = -\frac{3}{r^4}\frac{\partial r}{\partial x}(\mathbf{a} \times \mathbf{r}) + \frac{1}{r^3}\left(\mathbf{a} \times \frac{\partial \mathbf{r}}{\partial x}\right)$$

$$= -\frac{3x}{r^5}(\mathbf{a} \times \mathbf{r}) + \frac{1}{r^3}\mathbf{a} \times \mathbf{i} \quad \left[\because \frac{\partial r}{\partial x} = \frac{x}{r} \text{ and } \frac{\partial \mathbf{r}}{\partial x} = \frac{\partial}{\partial x}(x\mathbf{i} + y\mathbf{j} + z\mathbf{k}) = \mathbf{i}\right]$$

$$\therefore \quad \mathbf{i} \times \frac{\partial}{\partial x}\left(\frac{\mathbf{a} \times \mathbf{r}}{r^3}\right) = -\frac{3x}{r^5}\mathbf{i} \times (\mathbf{a} \times \mathbf{r}) + \frac{1}{r^3}\mathbf{i} \times (\mathbf{a} \times \mathbf{i})$$

$$= -\frac{3x}{r^5}[(\mathbf{i}\,.\,\mathbf{r})\mathbf{a} - (\mathbf{i}\,.\,\mathbf{a})\mathbf{r}] + \frac{1}{r^3}[(\mathbf{i}\,.\,\mathbf{i})\mathbf{a} - (\mathbf{i}\,.\,\mathbf{a})\mathbf{i}]$$

$$= -\frac{3x^2}{r^5}\mathbf{a} + \frac{3\mathbf{r}}{r^3}(x\mathbf{i}\,.\,\mathbf{a}) + \frac{1}{r^3}\mathbf{a} - \frac{1}{r^3}(\mathbf{i}\,.\,\mathbf{a})\mathbf{i}$$

∴ From (1),

$$\text{curl}\left(\frac{\mathbf{a}\times\mathbf{r}}{r^3}\right) = -\frac{3}{r^5}(\Sigma x^2)\mathbf{a} + \frac{3}{r^5}(\Sigma x\mathbf{i})\mathbf{r}\,.\,\mathbf{a} - \frac{3}{r^3}\mathbf{a} - \frac{\mathbf{a}}{r^3}$$

$[\because \;\; \Sigma x\mathbf{i} = \mathbf{r},\ \Sigma(\mathbf{i}\,.\,\mathbf{a})\mathbf{i} = \Sigma a_1\mathbf{i} = \mathbf{a}]$

$$= -\frac{\mathbf{a}}{r^3} + \frac{3\mathbf{r}}{r^5}(\mathbf{a}\,.\,\mathbf{r}).$$

EXERCISE 4

1. If $\mathbf{f} = xy^2\mathbf{i} + 2x^2yz\mathbf{j} - 3yz^2\mathbf{k}$, find curl **f** at the point (1, – 1, 1).
2. Find the curl of the vector function :
   (*a*) $\mathbf{f} = y(x + z)\mathbf{i} + z(x + y)\mathbf{j} + x(y + z)\mathbf{k}$ (*b*) $\mathbf{f} = e^{xyz}(\mathbf{i} + \mathbf{j} + \mathbf{k})$.
3. If $\mathbf{f} = (x + y + 1)\mathbf{i} + \mathbf{j} + (-x - y)\mathbf{k}$, prove that $\mathbf{f}$. curl $\mathbf{f} = 0$.
4. Find 'a' so that the vector $\mathbf{f} = (axy - z^3)\mathbf{i} + (a - 2)x^2\mathbf{j} + (1 - a)xz^2\mathbf{k}$ is irrotational.
   [**Hint.** curl $\mathbf{f} = \mathbf{0}$.]
5. Show that the vector $\mathbf{f} = (\sin y + z)\mathbf{i} + (x\cos y - z)\mathbf{j} + (x - y)\mathbf{k}$ is irrotational.
6. If **a** is a constant vector, find curl $(\mathbf{r}\times\mathbf{a})$.
7. Show that $r^n\mathbf{r}$ is irrotational.
8. If **f** and **g** are irrotational, prove that $\mathbf{f}\times\mathbf{g}$ is solenoidal.
   [**Hint.** div $(\mathbf{f}\times\mathbf{g}) = \mathbf{g}$. curl $\mathbf{f} - \mathbf{f}$ curl $\mathbf{g}$.]
9. Evaluate λ, μ, ν so that the vector **f** given by $\mathbf{f} = (2x + 3y + \lambda z)\mathbf{i} + (\mu x + 2y + 3z)\mathbf{i} + (2x + \nu y + 3z)\mathbf{k}$ is irrotational.

Answers

1. $-\mathbf{i} - 2\mathbf{k}$
2. (*a*) $-(y\mathbf{i} + z\mathbf{j} + x\mathbf{k})$ (*b*) $e^{xyz}[x(z - y)\mathbf{i} + y(x - z)\mathbf{j} + z(y - z)\mathbf{k}]$
4. $a = 4$
6. $-2\mathbf{a}$
9. $\lambda = 2$, $\mu = 3$, $\nu = 3$.

---

24. Operations with ∇ and Second Order Differential Functions

The operator $\nabla \equiv \mathbf{i}\frac{\partial}{\partial x} + \mathbf{j}\frac{\partial}{\partial y} + \mathbf{k}\frac{\partial}{\partial z}$ is not a vector, but an operator and may be considered a **symbolic** vector. Thus, if ϕ is a scalar field, then $\phi\nabla$ is an operator but $\nabla\phi$ gives a vector function known as *gradient* of ϕ. Similarly, if **f** is a differentiable vector function, then $\mathbf{f}\,.\,\nabla$ and $\mathbf{f}\times\nabla$ are operations but $\nabla\,.\,\mathbf{f}$ and $\nabla\times\mathbf{f}$ give scalar and vector functions respectively known as divergence of **f** and curl of **f** respectively.

In other words, ∇ operates only on what follows it. The order of symbols which appear in expressions using ∇ is important. For example, $(\mathbf{f}\,.\,\nabla)\mathbf{g} \neq \mathbf{g}(\mathbf{f}\,.\,\nabla)$, because the left hand side is a vector field, while the right hand side is an operator.

We have seen that if ϕ is a scalar point function and **f** is a vector point function, then div **f** is a scalar function whereas grad ϕ and curl **f** are vector functions. Therefore, grad ϕ, curl **f**, being vector functions, must possess divergence as well as, curl. The div **f**, being a scalar function, must possess gradient. Hence, we can evaluate div (grad ϕ) and curl (grad ϕ), div (curl **f**) and curl (curl **f**) ; grad (div **f**). These are called second order differential functions.

25. Laplacian Operator

If ϕ is a scalar point function, then gradient of ϕ is

$$\nabla\phi = \frac{\partial\phi}{\partial x}\mathbf{i} + \frac{\partial\phi}{\partial y}\mathbf{j} + \frac{\partial\phi}{\partial z}\mathbf{k}$$

$$\therefore \quad \text{div (grad } \phi) = \nabla \cdot (\nabla\phi)$$

$$= \left(\mathbf{i}\frac{\partial}{\partial x} + \mathbf{j}\frac{\partial}{\partial y} + \mathbf{k}\frac{\partial}{\partial z}\right) \cdot \left(\frac{\partial\phi}{\partial x}\mathbf{i} + \frac{\partial\phi}{\partial y}\mathbf{j} + \frac{\partial\phi}{\partial z}\mathbf{k}\right)$$

$$= \frac{\partial}{\partial x}\left(\frac{\partial\phi}{\partial x}\right) + \frac{\partial}{\partial y}\left(\frac{\partial\phi}{\partial y}\right) + \frac{\partial}{\partial z}\left(\frac{\partial\phi}{\partial z}\right) = \frac{\partial^2\phi}{\partial x^2} + \frac{\partial^2\phi}{\partial y^2} + \frac{\partial^2\phi}{\partial z^2}$$

Since divergence of gradient $\nabla \cdot \nabla$ is written as ∇^2, therefore, $\nabla \cdot (\nabla\phi)$ is written as

$$\nabla \cdot \nabla\phi = \nabla^2\phi$$

The operator $$\nabla^2 \equiv \frac{\partial^2}{\partial x^2} + \frac{\partial^2}{\partial y^2} + \frac{\partial^2}{\partial z^2}$$

is called the **Laplacian operator** and $\nabla^2\phi = \dfrac{\partial^2\phi}{\partial x^2} + \dfrac{\partial^2\phi}{\partial y^2} + \dfrac{\partial^2\phi}{\partial z^2}$.

Further, a scalar, point function ϕ is said to be harmonic if it is continuous, has continuous second order partial derivatives and satisfies the **Laplace's equation** : $\nabla^2\phi = 0$.

Example 22. *If* **f** *is a vector point function and ϕ is a scalar point function, then show that*

(*i*) $\nabla \times \nabla\phi = \mathbf{0}$ (*i.e., curl (grad ϕ) =* **0**)

(*ii*) $\nabla \cdot (\nabla \times \mathbf{f}) = 0$ (*i.e., div. (curl* **f**) *= 0*)

(*iii*) $\nabla \times (\nabla \times \mathbf{f}) = \nabla(\nabla \cdot \mathbf{f}) - \nabla^2\mathbf{f}$ (*i.e., curl (curl* **f**) *= grad (div* **f**) $-$ ∇^2 **f**).

Sol. (*i*) $$\nabla \times (\nabla\phi) = \nabla \times \left(\frac{\partial\phi}{\partial x}\mathbf{i} + \frac{\partial\phi}{\partial y}\mathbf{j} + \frac{\partial\phi}{\partial z}\mathbf{k}\right)$$

$$= \begin{vmatrix} \mathbf{i} & \mathbf{j} & \mathbf{k} \\ \dfrac{\partial}{\partial x} & \dfrac{\partial}{\partial y} & \dfrac{\partial}{\partial z} \\ \dfrac{\partial\phi}{\partial x} & \dfrac{\partial\phi}{\partial y} & \dfrac{\partial\phi}{\partial z} \end{vmatrix}$$

$$= \left[\frac{\partial}{\partial y}\left(\frac{\partial\phi}{\partial z}\right) - \frac{\partial}{\partial z}\left(\frac{\partial\phi}{\partial y}\right)\right]\mathbf{i} + \left[\frac{\partial}{\partial z}\left(\frac{\partial\phi}{\partial x}\right) - \frac{\partial}{\partial x}\left(\frac{\partial\phi}{\partial z}\right)\right]\mathbf{j}$$

$$+ \left[\frac{\partial}{\partial x}\left(\frac{\partial\phi}{\partial y}\right) - \frac{\partial}{\partial y}\left(\frac{\partial\phi}{\partial x}\right)\right]\mathbf{k}$$

$$= \left(\frac{\partial^2\phi}{\partial y\partial z} - \frac{\partial^2\phi}{\partial z\partial y}\right)\mathbf{i} + \left(\frac{\partial^2\phi}{\partial z\partial x} - \frac{\partial^2\phi}{\partial x\partial z}\right)\mathbf{j} + \left(\frac{\partial^2\phi}{\partial x\partial y} - \frac{\partial^2\phi}{\partial y\partial x}\right)\mathbf{k}$$

$$= 0\mathbf{i} + 0\mathbf{j} + 0\mathbf{k} = \mathbf{0} \qquad \left[\because \frac{\partial^2\phi}{\partial y\partial z} = \frac{\partial^2\phi}{\partial z\partial y}, \text{ etc.}\right]$$

[Here, we assume that ϕ has continuous second order partial derivatives].

(ii)
$$\nabla \cdot (\nabla \times \mathbf{f}) = \nabla \cdot \begin{vmatrix} \mathbf{i} & \mathbf{j} & \mathbf{k} \\ \frac{\partial}{\partial x} & \frac{\partial}{\partial y} & \frac{\partial}{\partial z} \\ f_1 & f_2 & f_3 \end{vmatrix}$$

$$= \nabla \cdot \left[\left(\frac{\partial f_3}{\partial y} - \frac{\partial f_2}{\partial z}\right)\mathbf{i} + \left(\frac{\partial f_1}{\partial z} - \frac{\partial f_3}{\partial x}\right)\mathbf{j} + \left(\frac{\partial f_2}{\partial x} - \frac{\partial f_1}{\partial y}\right)\mathbf{k}\right]$$

$$= \frac{\partial}{\partial x}\left(\frac{\partial f_3}{\partial y} - \frac{\partial f_2}{\partial z}\right) + \frac{\partial}{\partial y}\left(\frac{\partial f_1}{\partial z} - \frac{\partial f_3}{\partial x}\right) + \frac{\partial}{\partial z}\left(\frac{\partial f_2}{\partial x} - \frac{\partial f_1}{\partial y}\right)$$

$$= \frac{\partial^2 f_3}{\partial x\partial y} - \frac{\partial^2 f_2}{\partial x\partial z} + \frac{\partial^2 f_1}{\partial y\partial z} - \frac{\partial^2 f_3}{\partial y\partial x} + \frac{\partial^2 f_2}{\partial z\partial x} - \frac{\partial^2 f_1}{\partial z\partial y} = 0$$

$$\left(\because \frac{\partial^2 f_3}{\partial x\partial y} = \frac{\partial^2 f_3}{\partial y\partial x}, \text{ etc.}\right)$$

[Here, we assume that **f** has continuous second order partial derivatives].

(iii)
$$\nabla \times (\nabla \times \mathbf{f}) = \nabla \times \begin{vmatrix} \mathbf{i} & \mathbf{j} & \mathbf{k} \\ \frac{\partial}{\partial x} & \frac{\partial}{\partial y} & \frac{\partial}{\partial z} \\ f_1 & f_2 & f_3 \end{vmatrix}$$

$$= \nabla \times \left[\left(\frac{\partial f_3}{\partial y} - \frac{\partial f_2}{\partial z}\right)\mathbf{i} + \left(\frac{\partial f_1}{\partial z} - \frac{\partial f_3}{\partial x}\right)\mathbf{j} + \left(\frac{\partial f_2}{\partial x} - \frac{\partial f_1}{\partial y}\right)\mathbf{k}\right]$$

$$= \begin{vmatrix} \mathbf{i} & \mathbf{j} & \mathbf{k} \\ \frac{\partial}{\partial x} & \frac{\partial}{\partial y} & \frac{\partial}{\partial z} \\ \frac{\partial f_3}{\partial y} - \frac{\partial f_2}{\partial z} & \frac{\partial f_1}{\partial z} - \frac{\partial f_3}{\partial x} & \frac{\partial f_2}{\partial x} - \frac{\partial f_1}{\partial y} \end{vmatrix}$$

$$= \Sigma\left[\frac{\partial}{\partial y}\left(\frac{\partial f_2}{\partial x} - \frac{\partial f_1}{\partial y}\right) - \frac{\partial}{\partial z}\left(\frac{\partial f_1}{\partial z} - \frac{\partial f_3}{\partial x}\right)\right]\mathbf{i}$$

$$= \Sigma\left[\frac{\partial^2 f_2}{\partial y\partial x} - \frac{\partial^2 f_1}{\partial y^2} - \frac{\partial^2 f_1}{\partial z^2} + \frac{\partial^2 f_3}{\partial z\partial x}\right]\mathbf{i}$$

$$= \Sigma\left[\left(\frac{\partial^2 f_2}{\partial y\partial x} + \frac{\partial^2 f_3}{\partial z\partial x}\right) - \left(\frac{\partial^2 f_1}{\partial y^2} + \frac{\partial^2 f_1}{\partial z^2}\right)\right]\mathbf{i}$$

$$\left(\text{Adding and subtracting } \frac{\partial^2 f_1}{\partial x^2}\right)$$

$$= \Sigma \left[\left(\frac{\partial^2 f_1}{\partial x^2} + \frac{\partial^2 f_1}{\partial x \partial y} + \frac{\partial^2 f_3}{\partial x \partial z} \right) - \left(\frac{\partial^2 f_1}{\partial x^2} + \frac{\partial^2 f_1}{\partial y^2} + \frac{\partial^2 f_1}{\partial z^2} \right) \right] \mathbf{i}$$

$$= \Sigma \left[\left(\frac{\partial^2 f_1}{\partial x^2} + \frac{\partial^2 f_2}{\partial x \partial y} + \frac{\partial^2 f_3}{\partial x \partial z} \right) \mathbf{i} - \Sigma \left(\frac{\partial^2 f_1}{\partial x^2} + \frac{\partial^2 f_1}{\partial y^2} + \frac{\partial^2 f_1}{\partial z^2} \right) \right] \mathbf{i}$$

$$= \Sigma \frac{\partial}{\partial x} \left(\frac{\partial f_1}{\partial x} + \frac{\partial f_2}{\partial y} + \frac{\partial f_3}{\partial z} \right) \mathbf{i} - \Sigma \left(\frac{\partial^2}{\partial x^2} + \frac{\partial^2}{\partial y^2} + \frac{\partial^2}{\partial z^2} \right) f_1 \mathbf{i}$$

$$= \nabla \left(\frac{\partial f_1}{\partial x} + \frac{\partial f_2}{\partial y} + \frac{\partial f_3}{\partial z} \right) - \nabla^2 \mathbf{f}$$

$$= \nabla(\nabla \cdot \mathbf{f}) - \nabla^2 \mathbf{f}.$$

Note. Treating ∇ as a vector, the results of repeated application of ∇ can be easily written down.

Thus, $\nabla \cdot \nabla\phi = \nabla^2\phi$ $\qquad (\because \mathbf{a} \cdot \mathbf{a} = a^2)$

$\nabla \times \nabla\phi = \mathbf{0}$ $\qquad (\because \mathbf{a} \times \mathbf{a} = \mathbf{0})$

$\nabla \cdot (\nabla \times \mathbf{f}) = 0$ $\qquad (\because [\mathbf{a}\ \mathbf{a}\ \mathbf{b}] = 0)$

$\nabla \times (\nabla \times \mathbf{f}) = \nabla(\nabla \cdot \mathbf{f}) - \nabla^2\mathbf{f}$

(by expanding as a vector triple product)

Example 23. *If* $\mathbf{f} = x^2y\mathbf{i} + xz\mathbf{j} + 2yz\mathbf{k}$, *find the div (curl* $\mathbf{f}$*).*

Sol. curl $\mathbf{f} = \nabla \times \mathbf{f}$

$$= \begin{vmatrix} \mathbf{i} & \mathbf{j} & \mathbf{k} \\ \frac{\partial}{\partial x} & \frac{\partial}{\partial y} & \frac{\partial}{\partial z} \\ x^2 y & xz & 2yz \end{vmatrix}$$

$$= \left[\frac{\partial}{\partial y}(2yz) - \frac{\partial}{\partial z}(xz) \right] \mathbf{i} + \left[\frac{\partial}{\partial z}(x^2 y) - \frac{\partial}{\partial x}(2yz) \right] \mathbf{j} + \left[\frac{\partial}{\partial x}(xz) - \frac{\partial}{\partial y}(x^2 y) \right] \mathbf{k}$$

$$= (2z - x)\mathbf{i} + 0 \cdot \mathbf{j} + (z - x^2)\mathbf{k}$$

$\therefore$ $\quad$ div (curl $\mathbf{f}$) $= \nabla \cdot [(2z - x)\mathbf{i} + 0\mathbf{j} + (z - x^2)\mathbf{k}]$

$$= \frac{\partial}{\partial x}(2z - x) + \frac{\partial}{\partial y}(0) + \frac{\partial}{\partial z}(z - x^2)$$

$$= -1 + 0 + 1 = 0.$$

Example 24. *If* $\phi = 2x^3y^2z^4$, *find* $\nabla \cdot \nabla\phi$ *(or div grad* ϕ*).*

Sol.
$$\nabla\phi = \mathbf{i}\frac{\partial}{\partial x}(2x^3y^2z^4) + \mathbf{j}\frac{\partial}{\partial y}(2x^3y^2z^4) + \mathbf{k}\frac{\partial}{\partial z}(2x^3y^2z^4)$$

$$= 6x^2y^2z^4\mathbf{i} + 4x^3yz^4\mathbf{j} + 8x^3y^2z^3\mathbf{k}$$

$\because$
$$\nabla \cdot \nabla\phi = \left(\mathbf{i}\frac{\partial}{\partial x} + \mathbf{j}\frac{\partial}{\partial y} + \mathbf{k}\frac{\partial}{\partial z} \right) \cdot (6x^2y^2z^4\,\mathbf{i} + 4x^3yz^4\,\mathbf{j} + 8x^3y^2z^3\,\mathbf{k})$$

$$= \frac{\partial}{\partial x}(6x^2y^2z^4) + \frac{\partial}{\partial y}(4x^3yz^4) + \frac{\partial}{\partial z}(8x^3y^2z^3)$$

$$= 12xy^2z^4 + 4x^3z^4 + 24x^3y^2z^2.$$

Example 25. *Show that the function* $\frac{1}{r}$, *where* $r = |\mathbf{r}| = \sqrt{x^2 + y^2 + z^2}$ *is a harmonic function, provided* $r \neq 0$.

Or

Show that $\nabla^2\left(\frac{1}{r}\right) = 0$, *where* $r = \sqrt{x^2 + y^2 + z^2}$. (K.U. 2000 S)

Sol.
$$\nabla^2\left(\frac{1}{r}\right) = \left(\frac{\partial^2}{\partial x^2} + \frac{\partial^2}{\partial y^2} + \frac{\partial^2}{\partial z^2}\right)\left(\frac{1}{\sqrt{x^2 + y^2 + z^2}}\right)$$
$$= \frac{\partial^2}{\partial x^2}(x^2 + y^2 + z^2)^{-1/2} + \frac{\partial^2}{\partial y^2}(x^2 + y^2 + z^2)^{-1/2} + \frac{\partial^2}{\partial z^2}(x^2 + y^2 + z^2)^{-1/2} \quad ...(1)$$

But $\frac{\partial}{\partial x}(x^2 + y^2 + z^2)^{-1/2} = -x(x^2 + y^2 + z^2)^{-3/2}$

$$\therefore \quad \frac{\partial^2}{\partial x^2}(x^2 + y^2 + z^2)^{-1/2} = \frac{\partial}{\partial x}[-x(x^2 + y^2 + z^2)^{-3/2}]$$
$$= 3x^2(x^2 + y^2 + z)^{-5/2} - (x^2 + y^2 + z^2)^{-3/2}$$

which can be written as
$$= \frac{2x^2 - y^2 - z^2}{(x^2 + y^2 + z^2)^{5/2}}$$

Similarly,
$$\frac{\partial^2}{\partial y^2}(x^2 + y^2 + z^2)^{-1/2} = \frac{2y^2 - x^2 - z^2}{(x^2 + y^2 + z^2)^{5/2}}$$

and
$$\frac{\partial^2}{\partial z^2}(x^2 + y^2 + z^2)^{-1/2} = \frac{2z^2 - x^2 - y^2}{(x^2 + y^2 + z^2)^{5/2}}$$

Substituting the values in (1), we have
$$\nabla^2\left(\frac{1}{r}\right) = 0$$

Since $\frac{1}{r}$ *satisfies* $\nabla^2\phi = 0$ *therefore* $\frac{1}{r}$ *is a harmonic function.*

Example 26. *If* $\mathbf{f} = 2yz\,\mathbf{i} - x^2y\,\mathbf{j} + xz^2\mathbf{k}$ *and* $\phi = 2x^2yz^3$. *Find*

(*i*) $(\mathbf{f} \times \nabla)\phi$ (*ii*) $\mathbf{f} \times \nabla\phi$.

Sol. (*i*)
$$(\mathbf{f} \times \nabla)\phi = \left[(2yz\,\mathbf{i} - x^2y\,\mathbf{j} + xz^2\,\mathbf{k}) \times \left(\mathbf{i}\frac{\partial}{\partial x} + \mathbf{j}\frac{\partial}{\partial y} + \mathbf{k}\frac{\partial}{\partial z}\right)\right]\phi$$
$$= \begin{vmatrix} \mathbf{i} & \mathbf{j} & \mathbf{k} \\ 2yz & -x^2y & xz^2 \\ \frac{\partial}{\partial x} & \frac{\partial}{\partial y} & \frac{\partial}{\partial z} \end{vmatrix}\phi.$$
$$= \left[\mathbf{i}\left(-x^2y\frac{\partial}{\partial z} - xz^2\frac{\partial}{\partial y}\right) + \mathbf{j}\left(xz^2\frac{\partial}{\partial x} - 2yz\frac{\partial}{\partial z}\right) + \mathbf{k}\left(2yz\frac{\partial}{\partial y} + x^2y\frac{\partial}{\partial x}\right)\right]\phi$$

$$= -\left(x^2 y \frac{\partial \phi}{\partial z} + xz^2 \frac{\partial \phi}{\partial y}\right)\mathbf{i} + \left(xz^2 \frac{\partial \phi}{\partial x} - 2yz \frac{\partial \phi}{\partial z}\right)\mathbf{j} + \left(2yz \frac{\partial \phi}{\partial y} + x^2 y \frac{\partial \phi}{\partial x}\right)\mathbf{k}$$

$$= -(6x^4y^2z^2 + 2x^3z^5)\mathbf{i} + (4x^2yz^5 - 12x^2y^2z^3)\mathbf{j} + (4x^2yz^4 + 4x^3y^2z^3)\mathbf{k}.$$

(On simplification)

(*ii*) $$\mathbf{f} \times \nabla\phi = (2yz\mathbf{i} - x^2y\mathbf{j} + xz^2\mathbf{k}) \times \left(\frac{\partial \phi}{\partial x}\mathbf{i} + \frac{\partial \phi}{\partial y}\mathbf{j} + \frac{\partial \phi}{\partial z}\mathbf{k}\right)$$

$$= \begin{vmatrix} \mathbf{i} & \mathbf{j} & \mathbf{k} \\ 2yz & -x^2y & xz^2 \\ \frac{\partial \phi}{\partial x} & \frac{\partial \phi}{\partial y} & \frac{\partial \phi}{\partial z} \end{vmatrix}$$

$$= -\left(x^2 y \frac{\partial \phi}{\partial z} + xz^2 \frac{\partial \phi}{\partial y}\right)\mathbf{i} + \left(xz^2 \frac{\partial \phi}{\partial x} - 2yz \frac{\partial \phi}{\partial z}\right)\mathbf{j} + \left(2yz \frac{\partial \phi}{\partial y} + x^2 y \frac{\partial \phi}{\partial x}\right)\mathbf{k}$$

(On simplification)

$$= -(6x^4y^2z^2 + 2x^3z^5)\mathbf{i} + (4x^2yz^5 - 12x^2y^2z^3)\mathbf{j} + (4x^2yz^4 + 4x^3y^2z^3)\mathbf{k}.$$

Example 27. *Show that* $\nabla^2 f(r) = \frac{2}{r} f'(r) + f''(r)$.

Sol. $$\nabla^2 f(r) = \frac{\partial^2}{\partial x^2} f(r) + \frac{\partial^2}{\partial y^2} f(r) + \frac{\partial^2}{\partial z^2} f(r)$$

$$= \sum \frac{\partial^2}{\partial x^2} f(r) = \sum \frac{\partial}{\partial x}\left(\frac{\partial}{\partial x} f(r)\right)$$

$$= \sum \frac{\partial}{\partial x}\left[f'(r)\frac{\partial r}{\partial x}\right] = \Sigma \frac{\partial}{\partial x}\left[f'(r)\frac{x}{r}\right] \qquad \left(\because \frac{\partial r}{\partial x} = \frac{x}{r}\right)$$

$$= \sum \left[\frac{f'(r)}{r} + f''(r)\frac{x}{r}\frac{\partial r}{\partial x} + xf'(r)\left(-\frac{1}{r^2}\right)\frac{\partial r}{\partial x}\right]$$

$$= \sum \left[\frac{f'(r)}{r} + \frac{x^2}{r^2} f''(r) - \frac{x^2}{r^3} f'(r)\right]$$

$$= \frac{3}{r} f'(r) + \frac{f''(r)}{r^2}(x^2 + y^2 + z^2) - \frac{f'(r)}{r^3}(x^2 + y^2 + z^2)$$

$$= \frac{3}{r} f'(r) + \frac{f''(r)}{r^2} r^2 - \frac{f'(r)}{r^3} r^2 \qquad (\because x^2 + y^2 + z^2 = r^2)$$

$$= \frac{2}{r} f'(r) + f''(r).$$

Example 28. *Show that div. (grad)* $r^n = \nabla^2 r^n = n(n+1)r^{n-2}$. *(K.U. 2003 S)*

Sol. $$\text{grad } r^n = \left(\mathbf{i}\frac{\partial}{\partial x} + \mathbf{j}\frac{\partial}{\partial y} + \mathbf{k}\frac{\partial}{\partial z}\right) r^n$$

$$= \mathbf{i} nr^{n-1}\frac{\partial r}{\partial x} + \mathbf{j} nr^{n-1}\frac{\partial r}{\partial y} + \mathbf{k} nr^{n-1}\frac{\partial r}{\partial z} = nr^{n-1}\left[\mathbf{i}\frac{\partial r}{\partial x} + \mathbf{j}\frac{\partial r}{\partial y} + \mathbf{k}\frac{\partial r}{\partial z}\right]$$

$$= nr^{n-1}\left[\mathbf{i}\left(\frac{x}{r}\right)+\mathbf{j}\left(\frac{y}{r}\right)+\mathbf{k}\left(\frac{z}{r}\right)\right] \qquad \left(\because \frac{\partial r}{\partial x}=\frac{x}{r}, \text{etc.}\right)$$

$$= nr^{n-2}\,[x\,\mathbf{i}+y\,\mathbf{j}+z\,\mathbf{k}]$$

$\therefore$ $$\text{div}\,(\text{grad}\, r^n) = \nabla\,.\,[nr^{n-2}\,.\,(x\,\mathbf{i}+y\,\mathbf{j}+z\,\mathbf{k})]$$

$$= \left(\mathbf{i}\frac{\partial}{\partial x}+\mathbf{j}\frac{\partial}{\partial y}+\mathbf{k}\frac{\partial}{\partial z}\right).\,[nr^{n-2}\,(x\,\mathbf{i}+y\,\mathbf{j}+z\,\mathbf{k})]$$

$$= n\left[\frac{\partial}{\partial x}(xr^{n-2})+\frac{\partial}{\partial y}(yr^{n-2})+\frac{\partial}{\partial z}(zr^{n-2})\right]$$

$$= n\left[\left\{r^{n-2}+(n-2)xr^{n-3}\frac{\partial r}{\partial x}\right\}+\left\{r^{n-2}+(n-2)yr^{n-3}\frac{\partial r}{\partial y}\right\}+\left\{r^{n-2}+(n-2)zr^{n-3}\frac{\partial r}{\partial z}\right\}\right]$$

$$= 3nr^{n-2}+n(n-2)r^{n-3}\left[x\frac{\partial r}{\partial x}+y\frac{\partial r}{\partial y}+z\frac{\partial r}{\partial z}\right]$$

$$= 3nr^{n-2}+n(n-2)r^{n-3}\left[x\left(\frac{x}{r}\right)+y\left(\frac{y}{r}\right)+z\left(\frac{z}{r}\right)\right] \qquad \left[\because \frac{\partial r}{\partial x}=\frac{x}{r}, \text{etc.}\right]$$

$$= 3nr^{n-2}+n(n-2)r^{n-4}\,(x^2+y^2+z^2)$$

$$= 3nr^{n-2}+n(n-2)r^{n-2}$$

$$= [3n+n(n-2)]r^{n-2} = n(n+1)r^{n-2}.$$

Example 29. *Prove that* $\nabla^2\,(r\mathbf{r}) = \left(\frac{4}{r}\right)\mathbf{r}$. (M.D.U. 2007 ; K.U. 2007)

Sol. $$\nabla^2(r\mathbf{r}) = \frac{\partial^2}{\partial x^2}(r\mathbf{r})+\frac{\partial^2}{\partial y^2}(r\mathbf{r})+\frac{\partial^2}{\partial z^2}(r\mathbf{r})$$

$$= \Sigma\frac{\partial^2}{\partial x^2}(r\mathbf{r}) = \Sigma\frac{\partial}{\partial x}\left[\frac{\partial}{\partial x}(r\,\mathbf{r})\right] = \frac{\partial}{\partial x}\left[\frac{\partial r}{\partial x}\mathbf{r}+r\frac{\partial \mathbf{r}}{\partial x}\right]$$

$$= \Sigma\frac{\partial}{\partial x}\left[\frac{x}{r}\mathbf{r}+r\mathbf{i}\right] \qquad \left[\because \frac{\partial \mathbf{r}}{\partial x}=\frac{\partial}{\partial x}(x\mathbf{i}+y\mathbf{j}+z\mathbf{k})=\mathbf{i}\right]$$

$$= \Sigma\left[\left\{\frac{\mathbf{r}}{r}+\frac{x}{r}\frac{\partial \mathbf{r}}{\partial x}-\frac{x}{r^2}\frac{\partial r}{\partial x}\mathbf{r}\right\}+\mathbf{i}\frac{\partial r}{\partial x}\right]$$

$$= \Sigma\left[\left\{\frac{\mathbf{r}}{r}+\frac{x}{r}(i)-\frac{x}{r^2}\left(\frac{x}{r}\right)\mathbf{r}\right\}+\mathbf{i}\frac{x}{r}\right] \qquad \left[\because \frac{\partial r}{\partial x}=\frac{x}{r}, \frac{\partial \mathbf{r}}{\partial x}=\mathbf{i}\ \text{etc.}\right]$$

$$= \frac{3}{r}\mathbf{r}+\frac{2}{r}(x\mathbf{i}+y\mathbf{j}+z\mathbf{k})-\frac{1}{r^3}\mathbf{r}\,(x^2+y^2+z^2) = \frac{4}{r}\mathbf{r}.$$

EXERCISE 5

1. Find curl (curl **f**) of the function $\mathbf{f} = y(x + z)\mathbf{i} + z(x + y)\mathbf{j} + x(y + z)\mathbf{k}$.
2. If $\mathbf{f} = 2xz^2\mathbf{i} - yz\mathbf{j} + 3xz^3\mathbf{k}$ and $\phi = x^2yz$, find
   (*i*) $\nabla \times \mathbf{f}$ (*ii*) curl $(\phi\mathbf{f})$
   (*iii*) $\nabla \times (\nabla \times \mathbf{f})$ (*iv*) $\nabla[\mathbf{f} \cdot \text{curl } \mathbf{f}]$ at (1, 1, 1).
3. If $\mathbf{f} = yz^2\,\mathbf{i} - 3xz^2\,\mathbf{j} + 2xyz\,\mathbf{k}$, $\mathbf{g} = 3x\,\mathbf{i} + 4z\,\mathbf{j} - xy\,\mathbf{k}$ and $\phi = xyz$, find
   (*i*) $\mathbf{f} \times (\nabla\phi)$ (*ii*) $(\mathbf{f} \times \nabla)\phi$
   (*iii*) $(\nabla \times \mathbf{f}) \times \mathbf{g}$ (*iv*) $\mathbf{g} \cdot \nabla \times \mathbf{f}$.
4. Find $\mathbf{f} \times (\nabla \times \mathbf{g})$ and $(\mathbf{f} \times \nabla) \times \mathbf{g}$ at the point (1, – 1, 2) if
   $$\mathbf{f} = xz^2\,\mathbf{i} + 2y\,\mathbf{j} - 3xz\,\mathbf{k} \text{ and } \mathbf{g} = 3xz\,\mathbf{i} + 2yz\,\mathbf{j} - z^2\,\mathbf{k}.$$
5. Show that $\nabla^2\left(\dfrac{x}{r^3}\right) = 0$, where $r = \sqrt{x^2 + y^2 + z^2}$. (*K.U. 2001 S*)
   $\left[\textbf{Hint.} \text{ Use } \dfrac{\partial r}{\partial x} = \dfrac{x}{r}, \text{ etc.}\right]$
6. Show that $\nabla^2\left[\nabla \cdot \left(\dfrac{\mathbf{r}}{r^2}\right)\right] = 2r^{-4}$.
7. Prove that curl $(\phi \text{ grad } \phi) = \mathbf{0}$.
8. Show that curl (grad r^n) $= \nabla \times \nabla r^n = \mathbf{0}$.
9. Show that curl $(\mathbf{r} \sin r) = \mathbf{0}$.
10. Find $\nabla^2 r^n$ and show that $\nabla^2\left(\dfrac{1}{r}\right) = 0$.
11. Prove that $\nabla^2\left(\dfrac{x}{r^2}\right) = -\dfrac{2x}{r^4}$.

Answers

1. $\mathbf{i} + \mathbf{j} + \mathbf{k}$ 2. (*i*) $\mathbf{i} + \mathbf{j}$ (*ii*) $5\mathbf{i} - 3\mathbf{j} - 4\mathbf{k}$ (*iii*) $5\mathbf{i} + 3\mathbf{k}$
(*iv*) $-2\mathbf{i} + \mathbf{j} + 8\mathbf{k}$ 3. (*i*) $-5x^2yz^2\,\mathbf{i} + xy^2z^2\,\mathbf{j} + 4xyz^3\,\mathbf{k}$
(*ii*) $-5x^2yz^2\,\mathbf{i} + xy^2z^2\,\mathbf{j} + 4xyz^3\,\mathbf{k}$ (*iii*) $16z^3\,\mathbf{i} + (8x^2yz - 12xz^2)\,\mathbf{j} + 32xz^2\,\mathbf{k}$
(*iv*) $24x^2z + 4xyz^2$ 4. $18\mathbf{i} - 12\mathbf{j} + 16\mathbf{k}$, $4\mathbf{j} + 76\mathbf{k}$

10. $n(n + 1)r^{n-2}$.

SECTION II

4

Vector Integration—Gauss's, Green's and Stokes' Theorems

1. Integration of Vector Functions

We define integration as the reverse process of differentiation.

If **f** and **F** are two vector functions of the scalar variable t such that

$$\frac{d\mathbf{F}}{dt} = \mathbf{f},$$

then **F** is called the indefinite integral of **f** w.r.t. t and we write

$$\int \mathbf{f}\, dt = \mathbf{F}.$$

The function **f**, which is to be integrated is called the integrand.

If **c** is an arbitrary constant vector independent of t, then

$$\frac{d}{dt}(\mathbf{F} + \mathbf{c}) = \frac{d\mathbf{F}}{dt} = \mathbf{f}$$

$$\therefore \qquad \int \mathbf{f}\, dt = \mathbf{F} + \mathbf{c}$$

Thus, the integral **F** is indefinite to the extent of an additive arbitrary constant vector **c**. That is why we call **F** the indefinite integral. **c** is called the constant of integration.

If $\dfrac{d\mathbf{F}}{dt} = \mathbf{f}$ for all values of t in the interval $[a, b]$, the definite integral between the limits $t = a$ and $t = b$ is written as

$$\int_a^b \mathbf{f}\, dt = \Big[\mathbf{F}\Big]_a^b = \mathbf{F}(b) - \mathbf{F}(a).$$

We state without proof, the following theorem :

2. Theorem

If $\mathbf{f} = f_1\,\mathbf{i} + f_2\,\mathbf{j} + f_3\,\mathbf{k}$, where f_1, f_2, f_3 are scalar functions of the scalar variable t, then

$$\int \mathbf{f}\, dt = \mathbf{i}\int f_1\, dt + \mathbf{j}\int f_2\, dt + \mathbf{k}\int d_3\, dt.$$

Thus, to integrate a vector function, we should integrate its components.

3. Some Standard Results

Since the process of integration is the reverse of differentiation, we can obtain some standard results for integration from the corresponding results of differentiation.

Let **f** and **g** be two vector functions of a scalar t.

1. Since $$\frac{d}{dt}(\mathbf{f}\,.\,\mathbf{g}) = \frac{d\mathbf{f}}{dt}\,.\,\mathbf{g} + \mathbf{f}\,.\,\frac{d\mathbf{f}}{dt}$$

$\therefore$ $$\int \left(\frac{d\mathbf{f}}{dt}.\mathbf{g} + \mathbf{f}.\frac{d\mathbf{g}}{dt}\right) dt = \mathbf{f}\,.\,\mathbf{g} + c$$

where c is the constant of integration and is a scalar because the integrand is a scalar.

2. Since $$\frac{d}{dt}\,\mathbf{f}^2 = 2\mathbf{f}\,.\,\frac{d\mathbf{f}}{dt}$$

$\therefore$ $$\int \left(2\mathbf{f}\,.\frac{d\mathbf{f}}{dt}\right) dt = \mathbf{f}^2 + c$$

where c (scalar) is the constant of integration.

3. Since $$\frac{d}{dt}\left(\frac{d\mathbf{f}}{dt}\right)^2 = 2\frac{d\mathbf{f}}{dt}\,.\,\frac{d^2\mathbf{f}}{dt^2}$$

$\therefore$ $$\int \left(2\frac{d\mathbf{f}}{dt}\,.\,\frac{d^2\mathbf{f}}{dt^2}\right) dt = \left(\frac{d\mathbf{f}}{dt}\right)^2 + c$$

where c (scalar) is the constant of integration.

4. Since $$\frac{d}{dt}(\mathbf{f} \times \mathbf{g}) = \frac{d\mathbf{f}}{dt} \times \mathbf{g} + \mathbf{f} \times \frac{d\mathbf{g}}{dt}$$

$\therefore$ $$\int \left(\frac{d\mathbf{f}}{dt} \times \mathbf{g} + \mathbf{f} \times \frac{d\mathbf{g}}{dt}\right) dt = \mathbf{f} \times \mathbf{g} + \mathbf{c}$$

where **c** is the constant of integration.

Here **c** is a vector because the integrand is a vector.

5. Since $\frac{d}{dt}(\mathbf{a} \times \mathbf{f}) = \mathbf{a} \times \frac{d\mathbf{f}}{dt}$, where **a** is a constant vector.

$\therefore$ $$\int \left(\mathbf{a} \times \frac{d\mathbf{f}}{dt}\right) dt = \mathbf{a} \times \mathbf{f} + \mathbf{c},$$

where **c** (vector) is the constant of integration.

6. If **c** is a constant scalar,
$$\int c\,\mathbf{f}dt = c \int \mathbf{f}\,dt.$$

7. $$\int (\mathbf{f} \pm \mathbf{g})\,dt = \int \mathbf{f}\,dt \pm \int \mathbf{g}\,dt.$$

Remark. *The constant of integration is of the same nature as the integrand.*

Example 1. *If* $\mathbf{f} = (t + t^2)\,\mathbf{i} + 2t^2\,\mathbf{j} - 3t\,\mathbf{k}$ *find*

(*i*) $\int \mathbf{f}\,dt$ (*ii*) $\int_1^2 \mathbf{f}\,dt.$

Sol. (*i*) $$\int \mathbf{f}dt = \int \{(t + t^2)\,\mathbf{i} + 2t^2\,\mathbf{j} - 3t\,\mathbf{k}\}\,dt$$
$$= \mathbf{i}\int (t + t^2)dt + \mathbf{j}\int 2t^2\,dt + \mathbf{k}\int -3t\,dt$$

$$= \mathbf{i}\left(\frac{t^2}{2}+\frac{t^3}{3}\right)+\mathbf{j}\left(2\frac{t^3}{3}\right)+\mathbf{k}\left(-3\frac{t^2}{2}\right)+\mathbf{c}$$

where **c** is a constant vector

$$= \left(\frac{t^2}{2}+\frac{t^3}{3}\right)\mathbf{i}+\frac{2}{3}t^3\,\mathbf{j}-\frac{3}{2}\,t^2\mathbf{k}+\mathbf{c}.$$

(ii)
$$\int_1^2 \mathbf{f}\,dt = \int_1^2 \{(t+t^2)\,\mathbf{i}+2t^2\,\mathbf{j}-3t\,\mathbf{k}\}dt$$

$$= \mathbf{i}\int_1^2 t+t^2\,dt+\mathbf{j}\int_1^2 2t^2\,dt+\mathbf{k}\int_1^2 -3t\,dt$$

$$= \mathbf{i}\left[\frac{t^2}{2}+\frac{t^3}{3}\right]_1^2+\mathbf{j}\left[\frac{2}{3}t^3\right]_1^2+\mathbf{k}\left[\frac{-3t^2}{2}\right]_1^2 = \frac{23}{6}\,\mathbf{i}+\frac{14}{3}\,\mathbf{j}-\frac{9}{2}\,\mathbf{k}.$$

Example 2. *If* $\mathbf{r} = 5t^2\,\mathbf{i}+t\,\mathbf{j}-t^3\,\mathbf{k}$, *prove that*

$$\int_1^2 \mathbf{r}\times\frac{d^2\mathbf{r}}{dt^2}\,dt = -14\,\mathbf{i}+75\,\mathbf{j}-15\,\mathbf{k}. \qquad \text{(K.U. 2006)}$$

Sol. Since
$$\frac{d}{dt}\left[\mathbf{r}\times\frac{d\mathbf{r}}{dt}\right] = \mathbf{r}\times\frac{d^2\mathbf{r}}{dt^2}+\frac{d\mathbf{r}}{dt}\times\frac{d\mathbf{r}}{dt}$$

$$= \mathbf{r}\times\frac{d^2\mathbf{r}}{dt^2} \qquad \left\{\because\ \frac{d\mathbf{r}}{dt}\times\frac{d\mathbf{r}}{dt}=\mathbf{0}\right\}$$

∴ Integrating both sides w.r.t. t, we have

$$\int\left(\mathbf{r}\times\frac{d^2\mathbf{r}}{dt^2}\right)dt = \mathbf{r}\times\frac{d\mathbf{r}}{dt}+c$$

where c is constant of integration.

$$\therefore \quad \int_1^2\left(\mathbf{r}\times\frac{d^2\mathbf{r}}{dt^2}\right)dt = \left[\mathbf{r}\times\frac{d\mathbf{r}}{dt}\right]_1^2 \qquad \text{...(1)}$$

Let us find $\mathbf{r}\times\dfrac{d\mathbf{r}}{dt}$

Since
$$\mathbf{r} = 5t^2\,\mathbf{i}+t\,\mathbf{j}-t^3\,\mathbf{k}$$

$$\therefore \quad \frac{d\mathbf{r}}{dt} = 10t\,\mathbf{i}+\mathbf{j}-3t^2\,\mathbf{k}$$

$$\therefore \quad \mathbf{r}\times\frac{d\mathbf{r}}{dt} = \begin{vmatrix}\mathbf{i} & \mathbf{j} & \mathbf{k}\\ 5t^2 & t & -t^3\\ 10t & 1 & -3t^2\end{vmatrix}$$

$$= \mathbf{i}\,(-3t^3+t^3)-\mathbf{j}\,(-15t^4+10t^4)+\mathbf{k}\,(5t^2-10t^2)$$

$$= -2t^3\,\mathbf{i}+5t^4\,\mathbf{j}-5t^2\,\mathbf{k} \qquad \text{...(2)}$$

∴ Using (2) in (1), we have

$$\int_1^2\left(\mathbf{r}\times\frac{d^2\mathbf{r}}{dt^2}\right)dt = \left[-2t^3\,\mathbf{i}+5t^4\,\mathbf{j}-5t^2\,\mathbf{k}\right]_1^2$$

$$= (-16+2)\,\mathbf{i}+(80-5)\,\mathbf{j}+(-20+5)\,\mathbf{k} = -14\,\mathbf{i}+75\,\mathbf{j}-15\,\mathbf{k}.$$

Example 3. *Given that* $\mathbf{r} = 2\mathbf{i} - \mathbf{j} + 2\mathbf{k}$ *when* $t = 2$

$= 4\mathbf{i} - 2\mathbf{j} + 3\mathbf{k}$ *when* $t = 3$,

show that $$\int_2^3 \left(\mathbf{r} \cdot \frac{d\mathbf{r}}{dt}\right) dt = 10.$$

Sol. Since $$\int \left(\mathbf{r} \cdot \frac{d\mathbf{r}}{dt}\right) dt = \frac{1}{2}\,\mathbf{r}^2 + \mathbf{c}$$

$$\therefore \quad \int_2^3 \left(\mathbf{r} \cdot \frac{d\mathbf{r}}{dt}\right) dt = \left[\frac{1}{2}\,\mathbf{r}^2\right]_2^3 \qquad ...(1)$$

when $t = 3$, $\mathbf{r} = 4\mathbf{i} - 2\mathbf{j} + 3\mathbf{k}$

$\therefore$ $\mathbf{r}^2 = \mathbf{r}.\mathbf{r} = 4.4 + (-2)(-2) + 3.3 = 29$

when $t = 2$, $\mathbf{r} = 2\mathbf{i} - \mathbf{j} + 2\mathbf{k}$

$\therefore$ $\mathbf{r}^2 = \mathbf{r}.\mathbf{r} = 2.2 + (-1)(-1) + 2.2 = 9$

$\therefore$ From (1), $$\int_2^3 \left(\mathbf{r} \cdot \frac{d\mathbf{r}}{dt}\right) dt = \frac{1}{2}\,(29) - \frac{1}{2}\,(9) = 10.$$

Example 4. *The acceleration of a particle at any time $t \geq 0$ is given by*

$$\mathbf{a} = \frac{d\mathbf{v}}{dt} = 12 \cos 2t\,\mathbf{i} - 8 \sin 2t\,\mathbf{j} + 16t\,\mathbf{k}.$$

If the velocity **v** *and displacement* **r** *are zero at $t = 0$, find* **v** *and r at any time.*

Sol. $$\frac{d\mathbf{v}}{dt} = 12 \cos 2t\,\mathbf{i} - 8 \sin 2t\,\mathbf{j} + 16t\,\mathbf{k}.$$

Integrating, we have

$$\mathbf{v} = \mathbf{i} \int 12 \cos 2t\, dt + \mathbf{j} \int -8 \sin 2t\, dt + \mathbf{k} \int 16t\, dt$$
$$= 6 \sin 2t\,\mathbf{i} + 4 \cos 2t\,\mathbf{j} + 8t^2\,\mathbf{k} + c \qquad ...(1)$$

where **c** (vector) is constant of integration.

When $t = 0$, $\mathbf{v} = \mathbf{0}$

$\therefore$ $0 = 0\,\mathbf{i} + 4\,\mathbf{j} + 0\,\mathbf{k} + \mathbf{c} \quad \Rightarrow \quad \mathbf{c} = -4\mathbf{j}$

$\therefore$ From (1), $\mathbf{v} = 6 \sin 2t\,\mathbf{i} + (4 \cos 2t - 4)\,\mathbf{j} + 8t^2\,\mathbf{k}$

$$\Rightarrow \quad \frac{d\mathbf{r}}{dt} = 6 \sin 2t\,\mathbf{i} + (4 \cos 2t - 4)\,\mathbf{j} + 8t^2\,\mathbf{k}$$

Integrating, we have

$$\mathbf{r} = \mathbf{i} \int 6 \sin 2t\, dt + \mathbf{j} \int (4 \cos 2t - 4)\, dt + \mathbf{k} \int 8t^2\, dt$$
$$= -3 \cos 2t\,\mathbf{i} + (2 \sin 2t - 4t)\mathbf{j} + \frac{8}{3}\,t^3\,\mathbf{k} + \mathbf{c}_1 \qquad ...(2)$$

where $\mathbf{c}_1$ (vector) is constant of integration.

But, when $t = 0$, $\mathbf{r} = \mathbf{0}$,

$\therefore$ $\mathbf{0} = -3\,\mathbf{i} + 0\,\mathbf{j} + 0\,\mathbf{k} + \mathbf{c}_1 \quad \Rightarrow \quad \mathbf{c}_1 = 3\mathbf{i}$

$\therefore$ From (2), $$\mathbf{r} = (3 - 3 \cos 2t)\,\mathbf{i} + (2 \sin 2t - 4t)\,\mathbf{j} + \frac{8}{3}\,t^3\,\mathbf{k}.$$

EXERCISE 1

1. If $\mathbf{r} = (t - t^2)\,\mathbf{i} + 2t^3\,\mathbf{j} - 3\mathbf{k}$, find

   (i) $\int \mathbf{r}\,dt$ (ii) $\int_1^2 \mathbf{r}\,dt.$

2. Find the value of **r** satisfying the equation $\dfrac{d^2\mathbf{r}}{dt^2} = \mathbf{a}$

   where **a** is a constant vector. Also, it is given that when $t = 0$, $\mathbf{r} = \mathbf{0}$ and $\dfrac{d\mathbf{r}}{dt} = \mathbf{u}$.

3. Find the value of **r** satisfying the equation $\dfrac{d^2\mathbf{r}}{dt^2} = t\mathbf{a} + \mathbf{b}$

   where **a** and **b** are constant vectors.

4. The acceleration of a particle at any time t is given by $e^t\,\mathbf{i} + e^{2t}\,\mathbf{j} + \mathbf{k}$
   Find **v**, given that $\mathbf{v} = \mathbf{i} + \mathbf{j}$ at $t = 0$.

5. Evaluate $\int_1^2 (\mathbf{a} \,.\, \mathbf{b} \times \mathbf{c})\,dt$, when $\mathbf{a} = t\mathbf{i} - 3\mathbf{j} + 2t\mathbf{k}$, $\mathbf{b} = \mathbf{i} - 2\mathbf{j} + 2\mathbf{k}$, $\mathbf{c} = 3\mathbf{i} + t\mathbf{j} + \mathbf{k}$.

6. If $\mathbf{r} = 2t\mathbf{i} + 3t^2\,\mathbf{j} - t^3\mathbf{k}$, evaluate $\int_1^2 (\dot{\mathbf{r}} \times \ddot{\mathbf{r}})\,dt$

   where dot denotes differentiation w.r.t. t.

Answers

1. (i) $\left(\dfrac{t^2}{2} - \dfrac{t^3}{3}\right)\mathbf{i} + \dfrac{t^4}{2}\,\mathbf{j} - 3t\mathbf{k} + \mathbf{c}$ (ii) $-\dfrac{5}{6}\,\mathbf{i} + \dfrac{15}{2}\,\mathbf{j} - 3\mathbf{k}$

2. $\mathbf{r} = \dfrac{1}{2}\,t^2\mathbf{a} + t\mathbf{u}$ 3. $\mathbf{r} = \dfrac{1}{6}\,t^3\mathbf{a} + \dfrac{1}{2}\,t^2\mathbf{b} + t\mathbf{c} + \mathbf{d}$

4. $\mathbf{v} = e^t\,\mathbf{i} + \dfrac{1}{2}\,(e^{2t} + 1)\,\mathbf{j} + t\mathbf{k}$ 5. $-\dfrac{23}{3}$ 6. $(-42\mathbf{i} + 18\mathbf{j} + 12\mathbf{k})$.

4. Line Integrals

An integral which is to be evaluated along a curve is called a line integral.

Let $\mathbf{r} = x\mathbf{i} + y\mathbf{j} + z\mathbf{k}$ be the position of a point $P(x, y, z)$ on a curve C, where x, y, z are functions of a scalar parameter t, $a \le t \le b$.

Let $t = a$ at A and $t = b$ at B.

Let $\mathbf{f} = f_1\mathbf{i} + f_2\mathbf{j} + f_3\mathbf{k}$ be a vector point function defined and continuous along C. If s denotes the arc length of the curve C, then $\dfrac{d\mathbf{r}}{ds} = \mathbf{t}$ is a unit vector along the tangent to the curve C at P. The component of the vector **f** along the tangent at P is

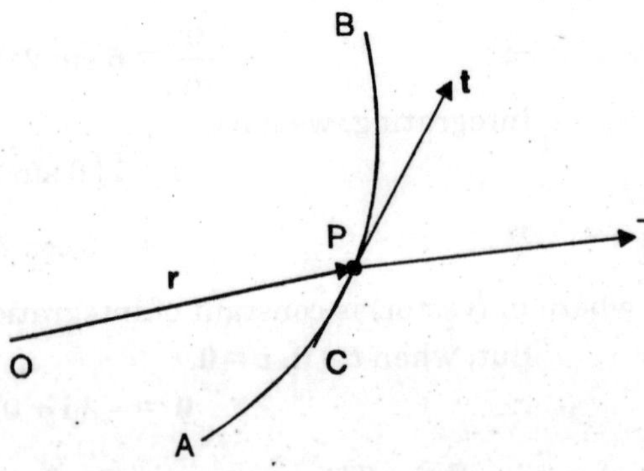

$$f_t = \mathbf{f} \,.\, \mathbf{t} = \mathbf{f} \,.\, \frac{d\mathbf{r}}{ds}.$$

For points on C, f_t is a function of s and $\int_A^B f_t\, ds$ is known as the tangent line integral or just line integral of **f** along C.

But,
$$f_t\, ds = \mathbf{f} \cdot \mathbf{t}\, ds = \mathbf{f} \cdot \frac{d\mathbf{r}}{ds}\, ds = \mathbf{f} \cdot d\mathbf{r}.$$

$\therefore$
$$\int_A^B f_t\, ds = \int_A^B \mathbf{f} \cdot \mathbf{t}\, ds = \int_A^B \mathbf{f} \cdot d\mathbf{r} \text{ and is also written as } \int_C \mathbf{f} \cdot d\mathbf{r}.$$

Note that
$$\int_A^B \mathbf{f} \cdot d\mathbf{r} = -\int_B^A \mathbf{f} \cdot d\mathbf{r}.$$

The differential displacement vector $d\mathbf{r}$ along the curve C is defined by
$$d\mathbf{r} = dx\mathbf{i} + dy\mathbf{j} + dz\mathbf{k}.$$

$\therefore$
$$\mathbf{f} \cdot d\mathbf{r} = (f_1\mathbf{i} + f_2\mathbf{j} + f_3\mathbf{k}) \cdot (dx\mathbf{i} + dy\mathbf{j} + dz\mathbf{k})$$
$$= f_1 dx + f_2 dy + f_3 dz.$$

Therefore, *in component form*, the above line integral can be written as
$$\int_C \mathbf{f} \cdot d\mathbf{r} = \int_C (f_1\, dx + f_2\, dy + f_3\, dz).$$

If the curve C is given by the parametric equations :

$x = x(t),\ y = y(t),\ z = z(t)$, then we can write
$$\int_C \mathbf{f} \cdot d\mathbf{r} = \int_a^b \left[f_1 \frac{dx}{dt} + f_2 \frac{dy}{dt} + f_3 \frac{dz}{dt} \right] dt,$$
where a, b are values of the parameter t at A and B.

5. Circulation

If C is a simple closed curve (i.e., a curve which does not intersect itself anywhere), then the tangent line integral of **f** *around C is called the circulation of* **f** *about C and is often denoted by*
$$\oint_C \mathbf{f} . d\mathbf{r} = \oint_C (f_1\, dx + f_2\, dy + f_3\, dz).$$

Example 5. *If* $\phi = xy$, *evaluate* $\int_C \phi\, d\mathbf{r}$ *from (0, 0, 0) to (1, 1, 0) along (i) the curve C represented by* $y = x^2$, $z = 0$ *and (ii) the straight line C that joins (0, 0, 0) and (1, 1, 0).*

Sol. (*i*) Let $x = t$ in $y = x^2$. Then, $y = t^2$ and the parametric equations of the curve C are :
$$x = t,\ y = t^2,\ z = 0,\ 0 \le t \le 1$$

$\therefore$
$$dx = dt,\ dy = 2t dt \text{ and } dz = 0$$

$\therefore$
$$d\mathbf{r} = dx\mathbf{i} + dy\mathbf{j} + dz\mathbf{k} = dt\mathbf{i} + 2t\, dt\, \mathbf{j}$$

and the function $\phi = xy = t(t^2) = t^3$.

$\therefore$
$$\int_{(0,0,0)}^{(1,1,0)} \phi\, d\mathbf{r} = \int_0^1 t^3 (dt\mathbf{i} + 2t\, dt\mathbf{j})$$
$$= \mathbf{i}\int_0^1 t^3\, dt + \mathbf{j}\int_0^1 2t^4\, dt = \mathbf{i}\left[\frac{t^4}{4}\right]_0^1 + \mathbf{j}\left[\frac{2t^5}{5}\right]_0^1 = \frac{1}{4}\mathbf{i} + \frac{2}{5}\mathbf{j}.$$

(*ii*) Equations of the straight line C are

$$\frac{x-0}{1-0}=\frac{y-0}{1-0}=\frac{z-0}{0-0}=t \text{ (say)}$$

$$\Rightarrow \qquad x=t,\ y=t,\ z=0,\ 0\le t\le 1$$

as the parametric equations of the straight line C.

$\therefore$ Along this path $\quad d\mathbf{r}=dx\mathbf{i}+dy\mathbf{j}+dz\mathbf{k}=dt\mathbf{i}+dt\mathbf{j}$

and $\quad \phi = xy = t\,.\,t = t^2,$

We have $\quad \displaystyle\int_{(0,0,0)}^{(1,1,0)} \phi\, d\mathbf{r} = \int_0^1 t^2(\mathbf{i}+\mathbf{j})\,dt$

$$=\mathbf{i}\int_0^1 t^2\,dt+\mathbf{j}\int_0^1 t^2\,dt=\mathbf{i}\left[\frac{t^3}{3}\right]_0^1+\mathbf{j}\left[\frac{t^3}{3}\right]_0^1=\frac{1}{3}\mathbf{i}+\frac{1}{3}\mathbf{j}.$$

Remark. *The results of the same integral over different paths are different.*

Example 6. *Evaluate* $\oint_C d\mathbf{r}$ *around the circle represented by* $x^2+y^2=a^2$, $z=0$, *where a is a constant.*

Sol. Parametric equations of the circle $x^2+y^2=a^2$, $z=0$ are :

$$x=a\cos t,\ y=a\sin t,\ z=0,\ 0\le t\le 2\pi.$$

$$\Rightarrow \qquad dx=-a\sin t\,dt,\ dy=a\cos t\,dt \text{ and } dz=0$$

$$\therefore \qquad d\mathbf{r}=dx\mathbf{i}+dy\mathbf{j}+dz\mathbf{k}=-a\sin t\,dt\,\mathbf{i}+a\cos t\,dt\,\mathbf{j}$$

$$\therefore \qquad \oint_C d\mathbf{r}=\int_0^{2\pi}(-a\sin t\,\mathbf{i}+a\cos t\,\mathbf{j})\,dt$$

$$=\mathbf{i}\int_0^{2\pi}-a\sin t\,dt+\mathbf{j}\int_0^{2\pi}a\cos t\,dt$$

$$=\mathbf{i}\Big[a\cos t\Big]_0^{2\pi}+\mathbf{j}\Big[a\sin t\Big]_0^{2\pi}$$

$$=\mathbf{i}\,[a\cos 2\pi-a\cos 0]+\mathbf{j}[a\sin 2\pi-a\sin 0]$$

$$=\mathbf{i}(a-a)+\mathbf{j}(0-0)=\mathbf{0}.$$

Example 7. *If* $\mathbf{f}=3xy\,\mathbf{i}-y^2\mathbf{j}$, *evaluate* $\int_C \mathbf{f}\,.\,d\mathbf{r}$, *where C is the curve in the xy-plane,* $y=2x^2$, *from (0, 0) to (1, 2).*

Sol. Since the integration is performed in the *xy*-plane ($z=0$), we can take $\mathbf{r}=x\mathbf{i}+y\mathbf{j}$.

$$\therefore \qquad \int_C \mathbf{f}\,.\,d\mathbf{r}=\int_C(3xy\mathbf{i}-y^2\mathbf{j})\,.\,(dx\mathbf{i}+dy\mathbf{j}) \qquad \left[\begin{array}{l}\because\ \mathbf{r}=x\mathbf{i}+y\mathbf{j}\\ \therefore\ d\mathbf{r}=dx\mathbf{i}+dy\mathbf{j}\end{array}\right]$$

$$=\int_C 3xy\,dx-y^2\,dy \qquad \ldots(1)$$

First Method. Let $x=t$ in $y=2x^2$. Then, parametric equations of C are $x=t$, $y=2t^2$. At the point (0, 0), $x=0$ and so $t=0$. Again at the point (1, 2), $x=1$ and so $t=1$.

$$\therefore \quad \text{From (1),} \quad \int_C \mathbf{f}\,.\,d\mathbf{r}=\int_{t=0}^1\left(3xy\frac{dx}{dt}-y^2\frac{dy}{dt}\right)dt$$

$$= \int_{t=0}^{1} (3 \,.\, t \,.\, 2t^2 \,.\, 1 - 4t^4 \,.\, 4t)\, dt \qquad \left[\because\ x = t, y = 2t^2,\ \therefore\ \frac{dx}{dt} = 1, \frac{dy}{dt} = 4t\right]$$

$$= \int_{t=0}^{1} (6t^3 - 16t^5)\, dt = \left[6\frac{t^4}{4} - 16\frac{t^6}{6}\right]_0^1 = \frac{6}{4} - \frac{16}{6} = -\frac{7}{6}.$$

Second Method. Substitute $y = 2x^2$ directly in (1), where x varies from 0 to 1 and y varies from 0 to 2.

$$\therefore \qquad \int_C \mathbf{f} \,.\, d\mathbf{r} = \int_{x=0}^{1} 3x \,.\, 2x^2\, dx - \int_{y=0}^{2} y^2\, dy$$

$$= \left[6\frac{x^4}{4}\right]_0^1 - \left[\frac{y^3}{3}\right]_0^2 = \frac{6}{4} - \frac{8}{3} = -\frac{7}{6}$$

Note that if the curve is traversed in the opposite sense, *i.e.*, from (1, 2) to (0, 0), the value of the integral would be $\frac{7}{6}$ instead of $-\frac{7}{6}$.

Example 8. *Find the total work done in moving a particle in a force field given by* $\mathbf{f} = 3xy\mathbf{i} - 5z\mathbf{j} + 10x\mathbf{k}$ *along the curve* $x = t^2 + 1$, $y = 2t^2$, $z = t^3$ *from* $t = 1$ *to* $t = 2$.

Sol. [**Work done by a force.** Let a force **f** act on a particle. Let the particle be **displaced** along a path C in space. If **r** denotes the position vector of a point P on C, then $\frac{d\mathbf{r}}{ds}$ is a unit vector tangent to C at P in the direction of increasing s. The component of force **f** along tangent to C is $\mathbf{f} \,.\, \frac{d\mathbf{r}}{ds}$.

$\therefore$ Work done by **f** during a small displacement ds of the particle along

$$C = \left(\mathbf{f} \,.\, \frac{d\mathbf{r}}{ds}\right) ds = \mathbf{f} \,.\, d\mathbf{r}.$$

$\therefore$ Total work done by **f** in its displacement along $C = \int_C \mathbf{f} \,.\, d\mathbf{r}$

$$\text{Total work done } = \int_C \mathbf{f} \,.\, d\mathbf{r} = \int_C \mathbf{f} \,.\, \frac{d\mathbf{r}}{dt}\, dt$$

$$= \int_C (3xy\mathbf{i} - 5z\mathbf{j} + 10x\mathbf{k}) \,.\, \left(\frac{dx}{dt}\mathbf{i} + \frac{dy}{dt}\mathbf{j} + \frac{dz}{dt}\mathbf{k}\right) dt$$

$$= \int_C \left(3xy\frac{dx}{dt} - 5z\frac{dy}{dt} + 10x\frac{dz}{dt}\right) dt$$

$$= \int_{t=1}^{2} [3(t^2 + 1)(2t^2)(2t) - (5t^3)(4t) + 10(t^2 + 1)(3t^2)]\, dt$$

$$= \int_1^2 (12t^5 + 10t^4 + 12t^3 + 30t^2)\, dt$$

$$= \left[\frac{12t^6}{6} + \frac{10t^5}{5} + \frac{12t^4}{4} + \frac{30t^3}{3}\right]_1^2 = 303.$$

Example 9. *Find work done in moving a particle in a field of force given by* $\mathbf{F} = 3x^2\,\mathbf{i} + (2xz - y)\mathbf{j} + z\mathbf{k}$ *along (i) the line joining the points (0, 0, 0) and (2, 1, 3).* (M.D.U. 2003)
(ii) the curve defined by $x^2 = 4y$, $3x^3 = 8z$ *from* $x = 0$ *to* $x = 2$.

Sol. (*i*) The equations of the line joining the points A(0, 0, 0) and B(2, 1, 3) are

$$\frac{x-0}{2-0} = \frac{y-0}{1-0} = \frac{z-0}{3-0} = t \text{ (say)}$$

$$\Rightarrow \qquad x = 2t,\ y = t,\ z = 3t.$$

Point A(0, 0, 0) corresponds to $t = 0$ and point B(2, 1, 3) corresponds to $t = 1$.

Also $\qquad \mathbf{r} = x\mathbf{i} + y\mathbf{j} + z\mathbf{k} \Rightarrow d\mathbf{r} = dx\mathbf{i} + dy\mathbf{j} + dz\mathbf{k}$

$$\therefore \qquad \mathbf{F} \cdot d\mathbf{r} = (3x^2\mathbf{i} + (2xz - y)\mathbf{j} + z\mathbf{k}) \cdot (dx\mathbf{i} + dy\mathbf{j} + dz\mathbf{k})$$
$$= 3x^2\,dx + (2xz - y)dy + zdz$$

$\therefore$ Total work done by **F** in its displacement from A to B

$$= \int_{AB} \mathbf{F} \cdot d\mathbf{r} = \int_{AB} [3x^2dx + (2xz - y)dy + zdz]$$

$$= \int_0^1 [3(2t)^2 \cdot 2dt + (2 \cdot 2t \cdot 3t - t)dt + 3t \cdot 3dt]$$

$$[\because x = 2t,\ y = t,\ z = 3t \Rightarrow dx = 2dt,\ dy = dt,\ dz = 3dt]$$

$$= \int_0^1 (36t^2 + 8t)dt = \left[\frac{36t^3}{3} + \frac{8t^2}{2}\right]_0^1 = 16.$$

(*ii*) Letting $x = t$ in $x^2 = 4y$, $3x^3 = 8z$, the parametric equations of the curve are

$$x = t,\ y = \frac{t^2}{4},\ z = \frac{3t^3}{8}$$

and t varies from 0 to 2. (as x varies from 0 to 2)

$$\therefore \text{ Required work done } = \int_0^2 \left[3t^2\,dt + \left(2t\,\frac{3t^3}{8} - \frac{t^2}{4}\right)\frac{2t}{4}\,dt + \frac{3t^3}{8} \cdot \frac{9t^2}{8}\,dt\right]$$

$$= \int_0^2 \left[\frac{5}{64}t^5 + 3t^2 - \frac{t^3}{8}\right] dt = 16 \text{ (on calculation).}$$

Example 10. *Find the work done by the force* $\mathbf{f} = x\mathbf{i} - z\mathbf{j} + 2y\mathbf{k}$ *in the displacement along the closed path C consisting of segments* C_1, C_2 *and* C_3, *where*

On C_1, $\qquad 0 \le x \le 1,\ y = x,\ z = 0$
On C_2, $\qquad 0 \le z \le 1,\ x = 1,\ y = 1$
On C_3, $\qquad 0 \le x \le 1,\ y = z = x.$

Sol. Total work done $= \int_C \mathbf{f} \cdot d\mathbf{r}$

$$= \int_C (x\mathbf{i} - z\mathbf{j} + 2y\mathbf{k}) \cdot (dx\mathbf{i} + dy\mathbf{j} + dz\mathbf{k})$$

$$= \int_C x\,dx - z\,dy + 2y\,dz$$

$$= \int_{C_1} (x\,dx - z\,dy + 2y\,dz) + \int_{C_2} (x\,dx - z\,dy + 2y\,dz)$$
$$+ \int_{C_3} x\,dx - z\,dy + 2y\,dz$$

$$= \int_{C_1} x\,dx + \int_{C_2} 2\,dz + \int_{C_3} 2x\,dx$$

$$= \int_0^1 x\,dx + 2\int_0^1 dz + 2\int_1^0 x\,dx = \frac{1}{2} + 2 - 1 = \frac{3}{2}.$$

Example 11. *Evaluate* $\int_C \mathbf{f} \cdot d\mathbf{r}$, *where* $\mathbf{f} = (x^2 + y^2)\mathbf{i} - 2xy\mathbf{j}$, *the curve C is the rectangle in the xy plane bounded by* $y = 0$, $x = a$, $y = b$, $x = 0$.

Sol. In the xy-plane $z = 0$

$\therefore$ $\mathbf{r} = x\mathbf{i} + y\mathbf{j}$ and $d\mathbf{r} = dx\mathbf{i} + dy\mathbf{j}$

We have, $\int_C \mathbf{f} \cdot d\mathbf{r} = \int_C [(x^2 + y^2)\mathbf{i} - 2xy\mathbf{j}] \cdot (dx\mathbf{i} + dy\mathbf{j})$

$$= \int_C [(x^2 + y^2)dx - 2xy\,dy] \qquad ...(1)$$

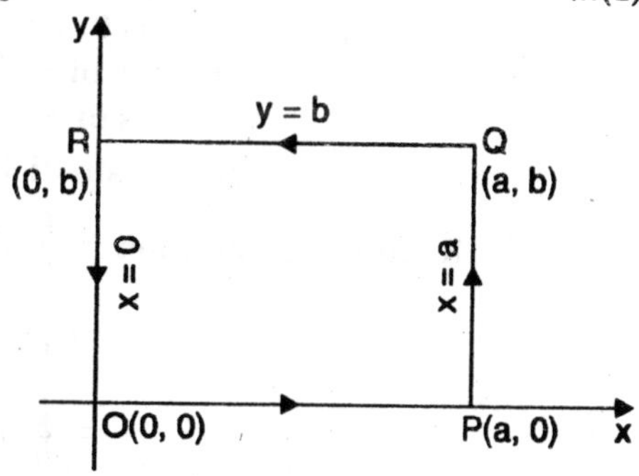

The curve C is the rectangle, OPQR, as shown

On OP, $y = 0$, $dy = 0$ and x varies from 0 to a

On PQ, $x = a$, $dx = 0$ and y varies from 0 to b

On QR, $y = b$, $dy = 0$ and x varies from a to 0

On RO, $x = 0$, $dx = 0$ and y varies from b to 0

$\therefore$ From (1), we have

$$\int_C \mathbf{f} \cdot d\mathbf{r} = \int_{OP} x^2\,dx + \int_{PQ} (-2ay)\,dy + \int_{QR} (x^2 + b^2)\,dx + \int_{RO} 0\,dy$$

$$= \int_0^a x^2 dx - 2a\int_0^b y dy + \int_a^0 (x^2 + b^2)\,dx + \int_b^0 0 dy$$

$$= \left[\frac{x^3}{3}\right]_0^a - 2a\left[\frac{y^2}{2}\right]_0^b + \left[\frac{x^3}{3} + b^2 x\right]_a^0 + 0$$

$$= \frac{a^3}{3} - ab^2 - \frac{a^3}{3} - ab^2 = -2ab^2.$$

Example 12. *Evaluate* $\int_C \mathbf{f} \cdot d\mathbf{r}$, *where* $\mathbf{f} = yz\mathbf{i} + zx\mathbf{j} + xy\mathbf{k}$ *and C is the portion of the curve* $\mathbf{r} = a \cos t\,\mathbf{i} + b \sin t\,\mathbf{j} + ct\,\mathbf{k}$, *from* $t = 0$ *to* $t = \frac{\pi}{2}$.

Sol. Along the curve C,

$$\mathbf{r} = x\mathbf{i} + y\mathbf{j} + z\mathbf{k} = a \cos t\mathbf{i} + b \sin t\mathbf{j} + ct\mathbf{k}$$

$\therefore$ Parametric equations of the curve are

$$x = a \cos t,\ y = b \sin t,\ z = ct.$$

$\therefore$ $\int_C \mathbf{f} \cdot d\mathbf{r} = \int_C (yz\mathbf{i} + zx\mathbf{j} + xy\mathbf{k}) \cdot (dx\mathbf{i} + dy\mathbf{j} + dz\mathbf{k})$

$$= \int_C yz\,dx + zx\,dy + xy\,dz = \int_C d(xyz)$$

$$= \Big[xyz\Big]_{t=0}^{t=\pi/2} = \Big[(a\cos t).(b\sin t).(ct)\Big]_0^{\pi/2}$$

$$= abc\Big[t\cos t\sin t\Big]_0^{\pi/2} = abc(0-0) = 0.$$

Example 13. *Find the circulation of* **f** *around the curve C, where* $\mathbf{f} = y\mathbf{i} + z\mathbf{j} + x\mathbf{k}$ *and C is the circle* $x^2 + y^2 = 1$, $z = 0$. (K.U. 2003)

Sol. By definition, the circulation of **f** along the curve C is $\oint_C \mathbf{f}\,.\,d\mathbf{r}$, where $\mathbf{r} = x\mathbf{i} + y\mathbf{j} + z\mathbf{k}$.

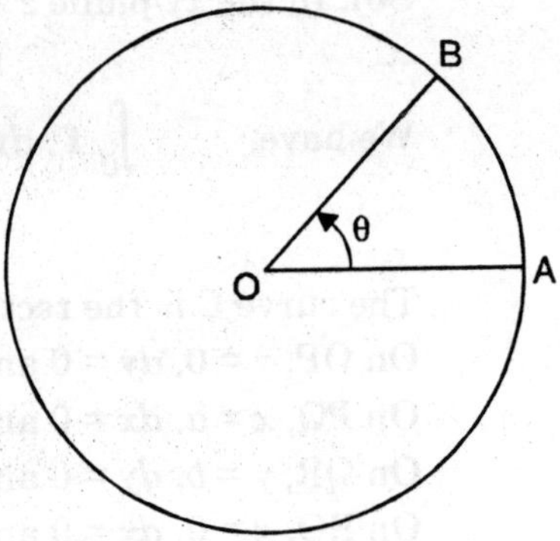

Now, the circle is $x^2 + y^2 = 1$, $z = 0$

Its parametric equations are

$$x = \cos\theta,\ y = \sin\theta,\ z = 0$$

$$\therefore \quad \mathbf{r} = \cos\theta\mathbf{i} + \sin\theta\mathbf{j} + 0\mathbf{k} = \cos\theta\mathbf{i} + \sin\theta\mathbf{j}$$

$$\Rightarrow \quad d\mathbf{r} = (-\sin\theta\mathbf{i} + \cos\theta\mathbf{j})\,d\theta$$

$$\therefore \text{ Required circulation } = \oint_C \mathbf{f}\,.\,d\mathbf{r}.$$

$$= \oint_C (y\mathbf{i} + z\mathbf{j} + x\mathbf{k})\,.\,(-\sin\theta\mathbf{i} + \cos\theta\mathbf{j})\,d\theta$$

$$= \oint_C (-y\sin\theta + z\cos\theta)\,d\theta = \oint_0^{2\pi} (-\sin^2\theta + 0)\,d\theta$$

$$= -\oint_C^{2\pi} \frac{1-\cos 2\theta}{2}\,d\theta = -\frac{1}{2}\left[\theta - \frac{\sin 2\theta}{2}\right]_0^{2\pi} = -\pi.$$

EXERCISE 2

1. Evaluate $\int_C \phi\,d\mathbf{r}$ for $\phi = x^3y + 2y$ from (1, 1, 0) to (2, 4, 0) along
   (*i*) the parabola $y = x^2$, $z = 0$; and
   (*ii*) the straight line joining (1, 1, 0) and (2, 4, 0).
2. Evaluate $\int_C \mathbf{f}\,.\,d\mathbf{r}$ for $\mathbf{f} = y\mathbf{i} + (x+z)^2\,\mathbf{j} + (x-z)^2\,\mathbf{k}$ from (0, 0, 0) to (2, 4, 0) along (*i*) the parabola $y = x^2$, $z = 0$; and (*ii*) the straight line $y = 2x$ in xy plane.
3. Find the work done when a force $\mathbf{f} = (x^2 - y^2 + x)\,\mathbf{i} - (2xy + y)\,\mathbf{j}$ moves a particle in xy-plane from (0, 0) to (1, 1) along the parabola $y^2 = x$. (*K.U. 1996*)
4. If $\mathbf{f} = (3x^2 + 6y)\,\mathbf{i} - 14yz\mathbf{j} + 20xz^2\mathbf{k}$, evaluate $\int_C \mathbf{f}\,.\,d\mathbf{r}$, where
   (*i*) C is the straight line joining the points (0, 0, 0) to (1, 1, 1)
   (*ii*) C is given by $x = t$, $y = t^2$, $z = t^3$ joining the points (0, 0, 0) to (1, 1, 1). (*K.U. 2007*)

5. Evaluate $\int_C \mathbf{f}\,.\,d\mathbf{r}$, where $\mathbf{f} = xy\mathbf{i} + yz\mathbf{j} + zx\mathbf{k}$ and the curve C is $\mathbf{r} = t\mathbf{i} + t^2\mathbf{j} + t^3\mathbf{k}$, t varying from -1 to 1.

6. If $\mathbf{f} = (2x + y)\mathbf{i} + (3y - x)\,\mathbf{j}$, evaluate $\int_C \mathbf{f}\,.\,d\mathbf{r}$, where C is the curve in the xy-plane consisting of the straight lines from (0, 0) to (2, 0) and then to (3, 2). *(K.U. 2003 S ; M.D.U. 2007)*

7. If $\phi = 2xyz^2$, $\mathbf{f} = xy\mathbf{i} - z\mathbf{j} + x^2\mathbf{k}$ and C is the curve $x = t^2$, $y = 2t$, $z = t^3$ from $t = 0$ to $t = 1$, evaluate the line integrals

   (*i*) $\int_C \phi d\mathbf{r}$ (*ii*) $\int_C \mathbf{f} \times d\mathbf{r}$.

8. Find the work done in moving a particle once around a circle C in xy-plane, if the circle has centre at the origin and radius 3 and if the force field is given by

$$\mathbf{f} = (2x - y + z)\,\mathbf{i} + (x + y - z^2)\,\mathbf{j} + (3x - 2y + 4z)\,\mathbf{k}.$$

   [**Hint.** Parametric equations of the circle are $x = 3 \cos t$, $y = 3 \sin t$, where t varies from 0 to 2π]

9. If $\mathbf{f} = x\mathbf{i} + 2y\mathbf{j} + z\mathbf{k}$, show that the value of $\int_C \mathbf{f}\,.\,d\mathbf{r}$ along any path joining the points (0, 0, 0) to (1, 1, 1) is 2.

$$\left[\textbf{Hint. } \int_C \mathbf{f}\,.\,d\mathbf{r} = \int_C x dx + 2y\,dy + z dz = \int_C d\left(\frac{x^2}{2} + y^2 + \frac{z^2}{2}\right)\right]$$

10. Show that $\oint_C \mathbf{r}\,.\,d\mathbf{r}$ along the circle C represented by $x^2 + y^2 = a^2$, $z = 0$ is zero.

11. Evaluate $\oint_C \mathbf{r} \times d\mathbf{r}$ along the circle C represented by $x^2 + y^2 = a^2$, $z = 0$.

12. Given $\mathbf{f} = (x^2 - y^2)\mathbf{i} + 2xy\mathbf{j}$ and $\mathbf{r} = x\mathbf{i} + y\mathbf{j}$. Find the value of $\int \mathbf{f}\,.\,d\mathbf{r}$ around the rectangular boundary $x = 0$, $y = 0$, $x = a$, $y = b$.

Answers

1. (*i*) $\frac{91}{6}\mathbf{i} + \frac{359}{7}\mathbf{j}$ (*ii*) $\frac{161}{10}\mathbf{i} + \frac{483}{10}\mathbf{j}$ **2.** (*i*) $\frac{32}{3}$ (*ii*) $\frac{28}{3}$

3. $-\frac{2}{3}$ **4.** (*i*) $\frac{13}{3}$ (*ii*) 5 **5.** $\frac{10}{7}$

6. 11 **7.** (*i*) $\frac{8}{11}\mathbf{i} + \frac{4}{5}\mathbf{j} + \mathbf{k}$ (*ii*) $-\frac{9}{10}\mathbf{i} - \frac{2}{3}\mathbf{j} + \frac{7}{5}\mathbf{k}$ **8.** 18π

11. $2\pi a^2 k$ **12.** $2ab^2$.

6. Surface Integral

Any integral which is to be evaluated over a surface is called a surface integral. Consider a surface S of finite area. Let ϕ be a scalar point function defined over S. Divide the area S into n sub-areas $\delta S_1, \delta S_2, \ldots\ldots, \delta S_n$. In each area δS_i, choose an arbitrary point $P_i\ (x_i, y_i, z_i)$ and form the sum

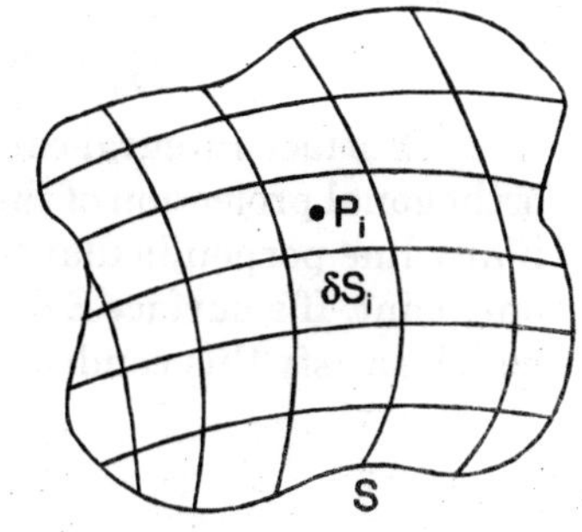

$$\sum_{i=1}^{n} \phi\,(P_i)\delta S_i, \text{ where } \phi(P_i) \equiv \phi(x_i, y_i, z_i).$$

If as $n \to \infty$ in such a way that each sub-area δS_i reduces to a point, the sum tends to a finite limit and is independent to the mode of sub-division, then the value of this limit is called the surface integral of ϕ over S and is denoted by $\iint \phi \, dS$.

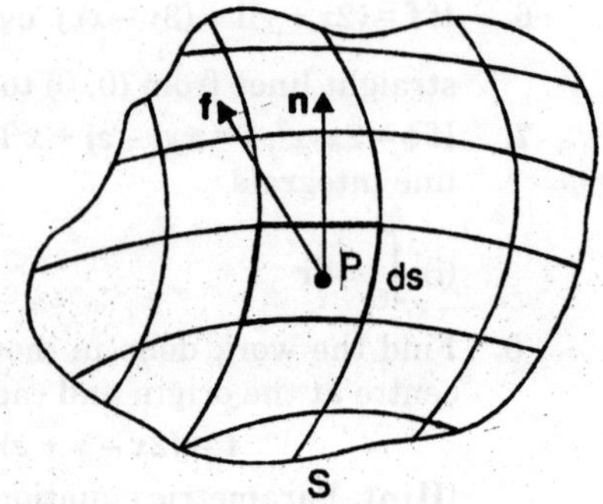

Let **f** be a vector point function defined and continuous over a surface S. Let P be any point on the surface S and let **n** be the unit vector at P in the direction of the outward drawn normal to the surface S at P.

Then, **f** . **n** is the normal component of **f** at P. The integral of **f** . **n** over S is called the (normal) surface integral of **f** over S and is denoted by

$$\iint_S \mathbf{f} \cdot \mathbf{n} \, dS$$

It is also called the flux of **f** over S.

If we associate with the differential of surface area dS a vector $d\mathbf{S}$ whose magnitude is dS and whose direction is that of **n**, then

$$d\mathbf{S} = \mathbf{n} dS$$

and
$$\iint_S \mathbf{f} \cdot \mathbf{n} \, dS = \iint_S \mathbf{f} \cdot d\mathbf{S} \text{ or } \iint f_n dS,$$

where $f_n = \mathbf{f} \cdot \mathbf{n}$ denotes the component of **f** in the direction of the outward normal at **P**.

If the outward drawn normal to the surface S at P makes angles α, β, γ with the positive direction of x, y, z axes respectively, and l, m, n are the direction cosines of this outward normal, then

$$l = \cos \alpha, \; m = \cos \beta, \; n = \cos \gamma.$$

Also,
$$\mathbf{n} = \cos \alpha \mathbf{i} + \cos \beta \mathbf{j} + \cos \gamma \mathbf{k} = l\mathbf{i} + m\mathbf{j} + n\mathbf{k}.$$

If $\mathbf{f} = f_1\mathbf{i} + f_2\mathbf{j} + f_3\mathbf{k}$, then $\mathbf{f} \cdot \mathbf{n} = (f_1\mathbf{i} + f_2\mathbf{j} + f_3\mathbf{k}) \cdot (l\mathbf{i} + m\mathbf{j} + n\mathbf{k})$

$$= f_1 l + f_2 m + f_3 n = f_1 \cos \alpha + f_2 \cos \beta + f_3 \cos \gamma$$

$$\therefore \quad \iint_S \mathbf{f} \cdot \mathbf{n} \, dS = \iint_S (f_1 \cos \alpha + f_2 \cos \beta + f_3 \cos \gamma) \, dS.$$

Now, $dS \cos \alpha$ = orthogonal projection of the area dS on the yz-plane $= dy\, dz$,

$dS \cos \beta$ = orthogonal projection of the area dS on the zx-plane $= dz dx$; and

$dS \cos \gamma$ = orthogonal projection of the area dS on the xy-plane $= dx\, dy$.

$$\therefore \quad \iint_S \mathbf{f} \cdot \mathbf{n} \, dS = \iint_S (f_1 \, dy \, dz + f_2 \, dz \, dx + f_3 \, dx \, dy).$$

Surface integral can be easily evaluated by expressing it as double integral taken over orthogonal projection of the surface S on one of the coordinate planes. But, this is possible only if any line perpendicular to the coordinate plane chosen meets the surface S is not more than one point. If a surface S does not satisfy this condition, then it can be subdivided into surfaces which satisfy this condition.

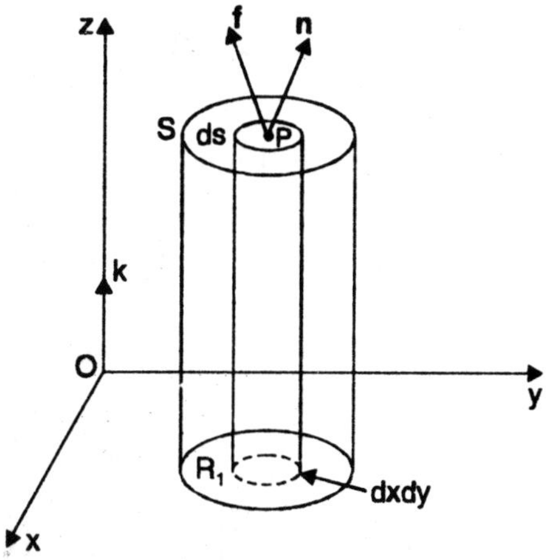

Let S be a surface such that any line perpendicular to xy-plane meets S in not more than one point. Then, the equation of the surface S can be written as $z = h(x, y)$.

Let the projection of S on xy-plane be R_1.

The projection of dS on the xy-plane = $dS \cos \gamma$, where γ is the acute angle which the normal **n** at P to the surface S makes with z-axis.

$\therefore \quad dS \cos \gamma = dx\, dy.$

Also, $\quad \cos \gamma = |\mathbf{n}\,.\,\mathbf{k}|$,

where **k** is the unit vector along z-axis.

$$\therefore \quad dS = \frac{dx\, dy}{|\mathbf{n}\,.\,\mathbf{k}|}.$$

Hence, $$\iint_S \mathbf{f}\,.\,\mathbf{n}\, dS = \iint_{R_1} \mathbf{f}\,.\,\mathbf{n} \frac{dx\, dy}{|\mathbf{n}\,.\,\mathbf{k}|}.$$

Similarly, $$\iint_S \mathbf{f}\,.\,\mathbf{n}\, dS = \iint_{R_2} \mathbf{f}\,.\,\mathbf{n} \frac{dx\, dy}{|\mathbf{n}\,.\,\mathbf{i}|} = \iint_{R_3} \mathbf{f}\,.\,\mathbf{n} \frac{dx\, dy}{|\mathbf{n}\,.\,\mathbf{j}|}$$

where R_2, R_3 are the projections of S on yz-plane and zx-plane respectively.

Example 14. *Evaluate* $\iint_S \mathbf{f}\,.\,\mathbf{n}\, dS$, *where* $\mathbf{f} = yz\mathbf{i} + zx\mathbf{j} + xy\mathbf{k}$ *and S is that part of the surface of the sphere* $x^2 + y^2 + z^2 = 1$ *which lies in the first octant.* (M.D.U. 2007)

Sol. A vector normal to surface S is given by

$$\nabla(x^2 + y^2 + z^2 - 1) = 2x\mathbf{i} + 2y\mathbf{j} + 2z\mathbf{k}$$

$\therefore \quad \mathbf{n}$ = a unit normal vector at any point (x, y, z) of S

$$= \frac{2x\mathbf{i} + 2y\mathbf{j} + 2z\mathbf{k}}{|2x\mathbf{i} + 2y\mathbf{j} + 2z\mathbf{k}|}$$

$$= \frac{2x\mathbf{i} + 2y\mathbf{j} + 2z\mathbf{k}}{\sqrt{4x^2 + 4y^2 + 4z^2}} = x\mathbf{i} + y\mathbf{j} + z\mathbf{k} \qquad (\because \ x^2 + y^2 + z^2 = 1 \text{ on S})$$

Now, $\iint_S \mathbf{f}\,.\,\mathbf{n}\, dS = \iint_R \mathbf{f}\,.\,\mathbf{n} \dfrac{dx\, dy}{|\mathbf{n}\,.\,\mathbf{k}|}$, where R is the region of projection of S on xy-plane.

R is bounded by x-axis, y-axis and the circle $x^2 + y^2 = 1$, $z = 0$.

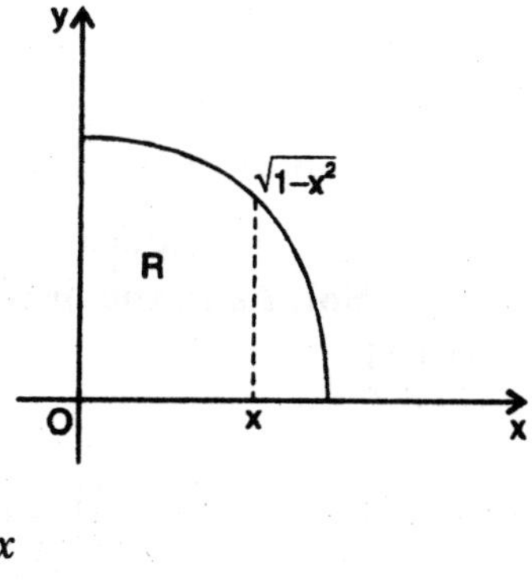

But, $\quad \mathbf{f}\,.\,\mathbf{n} = (yz\mathbf{i} + zx\mathbf{j} + xy\mathbf{k})\,.\,(x\mathbf{i} + y\mathbf{j} + z\mathbf{k})$

$= 3xyz$

and $\quad \mathbf{n}\,.\,\mathbf{k} = (x\mathbf{i} + y\mathbf{j} + z\mathbf{k})\,.\,\mathbf{k} = z$

$\therefore \quad |\mathbf{n}\,.\,\mathbf{k}| = z$

Hence, $\iint_S \mathbf{f}\,.\,\mathbf{n}\, dS$

$$= \iint_R \frac{3xyz}{z}\, dx\, dy = 3 \iint_R xy\, dx\, dy$$

$$= 3 \int_0^1 \left(\int_0^{\sqrt{1-x^2}} xy\, dy \right) dx = 3 \int_0^1 x \left[\frac{y^2}{2} \right]_0^{\sqrt{1-x^2}} dx$$

$$= \frac{3}{2}\int_0^1 x(1-x^2)dx = \frac{3}{2}\left[\frac{x^2}{2}-\frac{x^4}{4}\right]_0^1 = \frac{3}{8}.$$

Example 15. *Evaluate* $\iint_S \mathbf{f}\cdot\mathbf{n}\,dS$, *where* $\mathbf{f} = z\mathbf{i} + x\mathbf{j} - 3y^2z\mathbf{k}$ *and S is the surface of the cylinder* $x^2 + y^2 = 16$, *included in the first octant between* $z = 0$ *and* $z = 5$. **(M.D.U. 2003)**

Sol. A vector normal to the surface S is given by

$$\nabla(x^2 + y^2 - 16) = 2x\mathbf{i} + 2y\mathbf{j}$$

$\therefore$ $\mathbf{n}$ = unit normal at any point of S

$$= \frac{2x\mathbf{i} + 2y\mathbf{j}}{\sqrt{4x^2+4y^2}} = \frac{x\mathbf{i}+y\mathbf{j}}{4} \qquad (\because x^2 + y^2 = 16 \text{ on S})$$

Now, $$\iint \mathbf{f}\cdot\mathbf{n}\,dS = \iint_R \mathbf{f}\cdot\mathbf{n}\,\frac{dx\,dz}{|\mathbf{n}\cdot\mathbf{j}|},$$

where R is the region of projection of S on xz-plane.

Note that we have not taken projection of S on xy-plane because the surface S is perpendicular to the xy-plane. So we get a curve and not area.

But, $$\mathbf{f}\cdot\mathbf{n} = (z\mathbf{i} + x\mathbf{j} - 3y^2z\mathbf{k})\cdot\frac{(x\mathbf{i}+y\mathbf{j})}{4} = \frac{1}{4}(xz + xy)$$

and $$\mathbf{n}\cdot\mathbf{j} = \left(\frac{x\mathbf{i}+y\mathbf{j}}{4}\right)\cdot\mathbf{j} = \frac{y}{4}$$

$\therefore$ Required surface integral $$= \iint_R \frac{xz+xy}{4}\,\frac{dx\,dz}{y/4}$$

$$= \int_{z=0}^{5}\int_{x=0}^{4}\left(\frac{xz}{\sqrt{16-x^2}} + x\right)dx\,dz \qquad (\because y = \sqrt{16-x^2} \text{ on S})$$

$$= \int_{z=0}^{5}\left[\int_{x=0}^{4}\left(\left(-\frac{1}{2}\right)(-2x)(16-x^2)^{-1/2}\,z + x\right)dx\right]dz$$

$$= \int_{z=0}^{5}\left[-\frac{1}{2}\,\frac{(16-x^2)^{1/2}}{1/2}\,z + \frac{x^2}{2}\right]_0^4 dz$$

$$= \int_0^5 (4z+8)\,dz = \left[\frac{4z^2}{2} + 8z\right]_0^5 = 90.$$

Example 16. *Evaluate* $\iint_S \phi\mathbf{n}\,ds$, *where* $\phi = \frac{3}{8}xyz$ *and S is the surface of the cylinder* $x^2 + y^2 = 16$ *included in the first octant between* $z = 0$ *and* $z = 5$.

Sol. As in the previous example, a unit normal vector $\mathbf{n}$ at any point (x, y, z) of S is $\frac{x\mathbf{i}+y\mathbf{j}}{4}$.

$$\therefore \quad \iint_S \phi\mathbf{n}\,dS = \iint_R \phi\mathbf{n}\,\frac{dx\,dz}{|\mathbf{n}\cdot\mathbf{j}|} \qquad \ldots(1)$$

where R is the projection of S on xz-plane.

Note that we have not taken projection of S on xy-plane because the surface S is perpendicular to the xy-plane.

Now, $$\mathbf{n}\cdot\mathbf{j}=\left(\frac{x\mathbf{i}+y\mathbf{j}}{4}\right)\cdot\mathbf{j}=\frac{y}{4}$$

$\therefore$ From (1), the required surface integral is

$$\iint_R \frac{3}{8}xyz\,\frac{(x\mathbf{i}+y\mathbf{j})}{4}\,\frac{dx\,dz}{\frac{y}{4}}=\frac{3}{8}\iint_R xz(x\mathbf{i}+y\mathbf{j})\,dx\,dz$$

$$=\frac{3}{8}\int_{z=0}^{5}\int_{x=0}^{4}(x^2z\mathbf{i}+xz\sqrt{16-x^2}\,\mathbf{j})\,dx\,dz$$

$$=\frac{3}{8}\int_{z=0}^{5}\left[\left[\int_{x=0}^{4}x^2\,z\mathbf{i}+\left(-\frac{1}{2}\right)z(-2x)(16-x^2)^{1/2}\,\mathbf{j}\right]dx\right]dz$$

$$=\frac{3}{8}\int_{z=0}^{5}\left[\left[\frac{x^3}{3}z\mathbf{i}-\frac{1}{2}z\frac{(16-x^2)^{3/2}}{3/2}\mathbf{j}\right]_{x=0}^{4}\right]dz$$

$$=\frac{3}{8}\int_{z=0}^{5}\left(\frac{64}{3}z\mathbf{i}+\frac{64}{3}z\mathbf{j}\right)dz$$

$$=8\int_{z=0}^{5}(z\mathbf{i}+z\mathbf{j})\,dz=100\mathbf{i}+100\mathbf{j}.$$

Example 17. *Evaluate* $\iint_S \mathbf{f}\cdot\mathbf{n}\,dS$, *where* $\mathbf{f}=(x+y^2)\mathbf{i}-2x\mathbf{j}+2yz\mathbf{k}$ *and S is the surface of the plane* $2x+y+2z=6$ *in the first octant.* (K.U. 2000 S)

Sol. A vector normal to the surface S is given by

$$\nabla(2x+y+2z-6)=2\mathbf{i}+\mathbf{j}+2\mathbf{k}$$

$\therefore$ $\mathbf{n}$ = unit vector normal to the surface S

$$=\frac{2\mathbf{i}+\mathbf{j}+2\mathbf{k}}{\sqrt{4+1+4}}=\frac{2}{3}\mathbf{i}+\frac{1}{3}\mathbf{j}+\frac{2}{3}\mathbf{k}$$

$$\therefore\quad \mathbf{k}\cdot\mathbf{n}=\mathbf{k}\cdot\left(\frac{2}{3}\mathbf{i}+\frac{1}{3}\mathbf{j}+\frac{2}{3}\mathbf{k}\right)=\frac{2}{3}$$

$$\therefore\quad \iint_S \mathbf{f}\cdot\mathbf{n}\,dS=\iint_R \mathbf{f}\cdot\mathbf{n}\,\frac{dxdy}{|\mathbf{k}\cdot\mathbf{n}|}$$

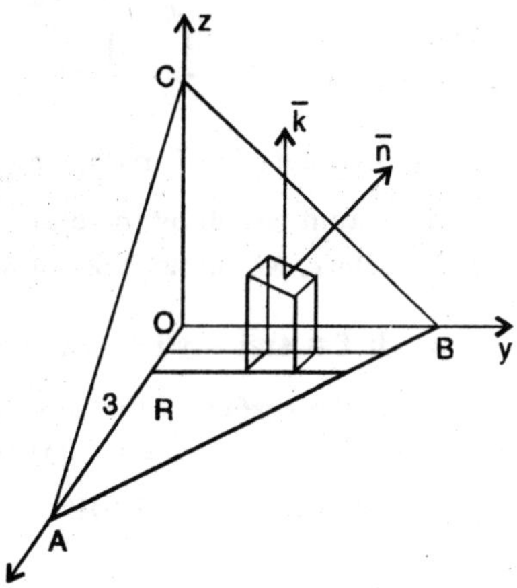

where R is the region of projection of S, *i.e.*, of triangle ABC on the xy-plane. Thus the region R is the triangle OAB bounded by x-axis, y-axis and the line $2x+y=6$, $z=0$.

Now, $$\mathbf{f}\cdot\mathbf{n}=[(x+y^2)\mathbf{i}-2x\mathbf{j}+2yz\mathbf{k}]\cdot\left(\frac{2}{3}\mathbf{i}+\frac{1}{3}\mathbf{j}+\frac{2}{3}\mathbf{k}\right)$$

$$=\frac{2}{3}(x+y^2)-\frac{2}{3}x+\frac{4}{3}yz=\frac{2}{3}y^2+\frac{4}{3}yz$$

$$=\frac{2}{3}y^2+\frac{4}{3}y\left(\frac{6-2x-y}{2}\right)\qquad\left(\because\ \text{on the plane } 2x+y+2z=6,\ z=\frac{6-2x-y}{2}\right)$$

$$=\frac{4}{3}y(3-x).$$

Hence, $\iint_S \mathbf{f} \cdot \mathbf{n}\, dS = \iint_R \mathbf{f} \cdot \mathbf{n} \dfrac{dxdy}{|\mathbf{k} \cdot \mathbf{n}|}$

$$= \iint_R \frac{4}{3} y(3-x) \cdot \frac{3}{2}\, dx\, dy = \int_{x=0}^{3} \int_{y=0}^{6-2x} 2y(3-x)\, dy\, dx$$

$$= \int_0^3 2(3-x) \left[\frac{y^2}{2}\right]_0^{6-2x} dx = \int_0^3 (3-x)(6-2x)^2\, dx,$$

$$= 4\int_0^3 (3-x)^3\, dx = 4\left[\frac{(3-x)^4}{-4}\right]_0^3 = -(0-81) = 81.$$

EXERCISE 3

1. Evaluate $\iint_S \mathbf{f} \cdot \mathbf{n}\, dS$, where $\mathbf{f} = 18z\mathbf{i} - 12\mathbf{j} + 3y\mathbf{k}$ and S is that part of the plane $2x + 3y + 6z = 12$ which lies in the first quadrant.

   [**Hint.** $\iint_S \mathbf{f} \cdot \mathbf{n}\, ds = \iint_R \mathbf{f} \cdot \mathbf{n} \dfrac{dx \cdot dy}{|\mathbf{n} \cdot \mathbf{k}|} = \iint_R (6-2x)\, dx\, dy$

   To evaluate the double integral over R, keep x fixed and integrate w.r.t. y from $y = 0$ to $y = \dfrac{12-2x}{3}$; then integrate w.r.t. x from $x = 0$ to $x = 6$. In this way, R is completely covered and the surface integral becomes

$$\int_{x=0}^{6} \int_{y=0}^{\frac{12-2x}{3}} (6-2x)\, dy\, dx = \int_{x=0}^{6} \left(24 - 12x + \frac{4x^2}{3}\right) dx = 24.]$$

2. Evaluate $\iint_S \mathbf{f} \cdot \mathbf{n}\, dS$, where $\mathbf{f} = y\mathbf{i} + 2x\mathbf{j} - z\mathbf{k}$ and S is the surface of the plane $2x + y = 6$ in the first octant cut off by the plane $z = 4$. *(M.D.U. 2007)*

   [Here, we cannot take projection xy-plane, because the surface S is perpendicular to xy-plane].

3. If $\mathbf{f} = 4xz\mathbf{i} - y^2\mathbf{j} + yz\mathbf{k}$, evaluate $\iint_S \mathbf{f} \cdot \mathbf{n}\, dS$, where S is the surface of the cube bounded by $x = 0$, $x = 1$, $y = 0$, $y = 1$, $z = 0$, $z = 1$. *(M.D.U. 2000 ; K.U. 2003, 2003 S)*

   [**Hint.** For face OADB, $\mathbf{n} = \mathbf{i}, x = 1$

   For face OBEC, $\mathbf{n} = -\mathbf{i}, x = 0$

   For face BEGD, $\mathbf{n} = \mathbf{j}, y = 1$

   For face OCFA, $\mathbf{n} = -\mathbf{j}, y = 0$, etc.]

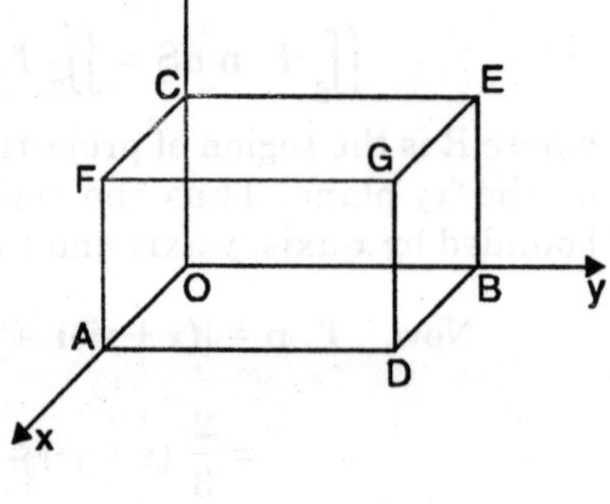

Answers

1. 24 2. 108 3. 3/2.

7. Volume Integral

Let ϕ be a scalar point function defined throughout a given region of volume V. Divide the given region into n sub-regions of volumes $\delta V_1, \delta V_2, \ldots\ldots, \delta V_n$. Let $P(x_i, y_i, z_i)$ be any point inside or on the boundary of the sub-region of volume δV_i. Form the sum $\sum_{i=1}^{n} \phi(P_i)\delta V_i$, where $\phi(P_i) = \phi(x_i, y_i, z_i)$. If as $n \to \infty$ in such a way that each δV_i tends to zero, the sum tends to a finite limit and is independent of the mode of sub-division, then this limit is called the volume integral of ϕ over the volume V and is denoted by $\iiint_V \phi\, dV$.

If we sub-divide the region of volume V into small cuboids by drawing lines parallel to the coordinate axes, then $dV = dx\, dy\, dz$ and we can write the above volume integral as $\iiint_V \phi\, dx\, dy\, dz$.

Likewise if **f** is a vector point function defined in the given region of volume V, we define and denote the vector volume integral of **f** over V by $\iiint_V \mathbf{f}\, dV$.

Example 18. *Evaluate $\iiint_V \phi\, dV$, where $\phi = 45x^2y$ and V is the enclosed region bounded by the plane $4x + 2y + z = 8$, $x = 0$, $y = 0$, $z = 0$.*

Sol. The equation of the plane $4x + 2y + z = 8$ can be written as

$$\frac{x}{2} + \frac{y}{4} + \frac{z}{8} = 1$$

It meets x-axis in A(2, 0, 0). Consider a thin column PQ parallel to z-axis. This cuts the boundary at $z = 0$ (base of column PQ) and $z = 8 - 4x - 2y$ (top of column PQ). Also, projection of the plane $4x + 2y + z = 8$ on xy-plane is bounded by $x = 0$, $y = 0$ and $2x + y = 4$.

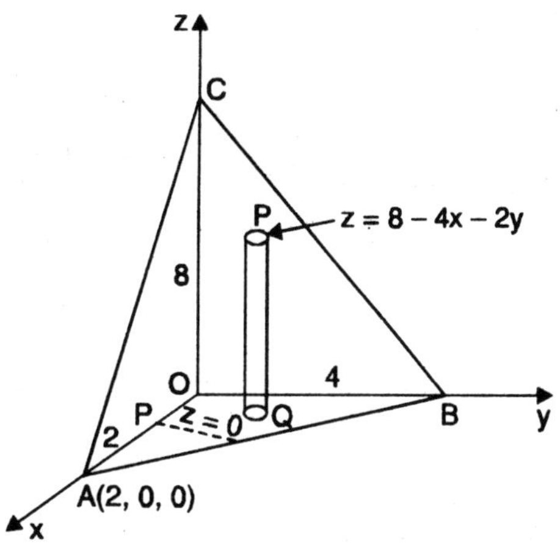

Hence, the limits for z are from 0 to $8 - 4x - 2y$; for y are from 0 to $4 - 2x$ and for x are from 0 to 2.

$$\therefore \quad \iiint_V = \int_{x=0}^{2} \int_{y=0}^{4-2x} \int_{z=0}^{8-4x-2y} 45x^2y\, dz\, dy\, dx \quad [\because \; dV = dz\, dy\, dx]$$

$$= 45 \int_{x=0}^{2} \int_{y=0}^{4-2x} x^2y \Big[z \Big]_{z=0}^{8-4x-2y} dy\, dx$$

$$= 45 \int_{x=0}^{2} \int_{y=0}^{4-2x} x^2y(8 - 4x - 2y)\, dy\, dx$$

$$= 45 \int_{x=0}^{2} \int_{y=0}^{4-2x} [x^2(8 - 4x)y - 2x^2y^2]\, dy\, dx$$

$$= 45 \int_{x=0}^{2} \left[x^2(8 - 4x)\frac{y^2}{2} - 2x^2 \frac{y^3}{3} \right]_{y=0}^{4-2x} dx$$

point $(x, y, 0)$ of R, meets the boundary of S in two points. Let the z-coordinates of these points be

$$z = \phi_1(x, y),\ z = \phi_2(x, y) \qquad \text{where } \phi_2(x, y) \geq \phi_1(x, y).$$

Now, $$\iiint_V \frac{\partial f_3}{\partial z}\, dV = \iiint_V \frac{\partial f_3}{\partial z}\, dz\, dy\, dx \qquad (\because\ dV = dz\, dy\, dx)$$

$$= \iint_R \left(\int_{z=\phi_1(x,y)}^{\phi_2(x,y)} \frac{\partial f_3}{\partial z}\, dz \right) dy\, dx = \iint_R \Big[f_3(x, y, z) \Big]_{z=\phi_1}^{\phi_2} dy\, dx$$

$$= \iint_R [f_3(x, y, \phi_2) - f_3(x, y, \phi_1)]\, dy\, dx \qquad \text{...(1)}$$

Let us denote the portion of the surface S corresponding to $z = \phi_1(x, y)$ and $z = \phi_2(x, y)$ by S_1 and S_2 respectively.

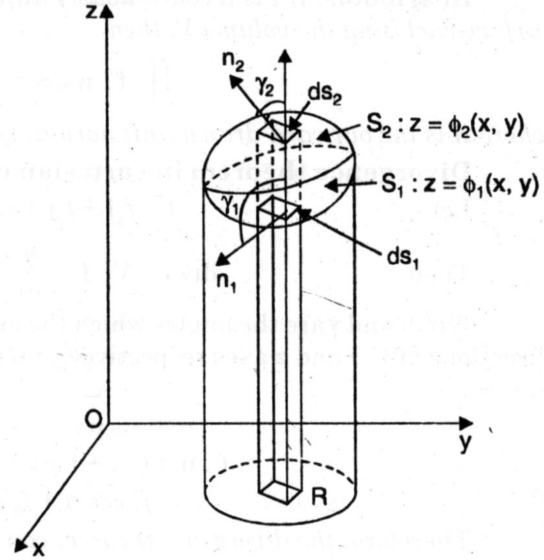

For the upper portion S_2, let $\mathbf{n}_2$ be the unit outward drawn normal vector making an angle γ_2 with $\mathbf{k}$ as shown in the figure. Here γ_2 is an acute angle.

If dS_2 is projected on xy-plane, then this projection $dydx$ of dS_2 is given by

$$dydx = \cos\gamma_2\, dS_2 = \mathbf{k}\,.\,\mathbf{n}\, dS_2$$
$$(\because\ \mathbf{k}\,.\,\mathbf{n}_2 = \cos\gamma_2)$$

Similarly, if $\mathbf{n}_1$ is the unit outward drawn normal vector to the lower surface S_1, making an angle γ_1 with $\mathbf{k}$, as shown in the figure, then γ_1 is an obtuse angle and $dydx$

$$= \cos(\pi - \gamma_1)\, dS_1$$
$$= -\cos\gamma_1\, dS_1$$
$$= -\mathbf{k}\,.\,\mathbf{n}_1\, dS_1$$

$$\therefore \qquad \iint_R f_3(x, y, \phi_2)\, dy\, dx = \iint_{S_2} f_3\, \mathbf{k}\,.\,\mathbf{n}_2\, dS_2 \qquad \text{...(2)}$$

and $$\iint_R f_3(x, y, \phi_1)\, dy\, dx = -\iint_{S_1} f_3\, \mathbf{k}\,.\,\mathbf{n}_1\, dS_1 \qquad \text{...(3)}$$

$\therefore$ Using (2) and (3) in (1), we have

$$\iiint_V \frac{\partial f_3}{\partial z}\, dV = \iint_{S_2} f_3\mathbf{k}\,.\,\mathbf{n}_2\,.\, dS_2 + \iint_{S_1} f_3\mathbf{k}\,.\,\mathbf{n}_1\, dS_1$$

$$= \iint_S f_3\mathbf{k}.\,\mathbf{n}\, dS \qquad \text{...(4)}$$

Similarly, by projecting S on other coordinate planes, we can show that

$$\iiint_V \frac{\partial f_1}{\partial x}\, dV = \iint_S f_1\mathbf{i}\,.\,\mathbf{n}\, dS \qquad \text{...(5)}$$

and $$\iiint_V \frac{\partial f_2}{\partial y}\, dV = \iint_S f_2\mathbf{j}\,.\,\mathbf{n}\, dS \qquad \text{...(6)}$$

Adding (4), (5) and (6), we have

$$\iiint_V \left(\frac{\partial f_1}{\partial x}+\frac{\partial f_2}{\partial y}+\frac{\partial f_3}{\partial z}\right) dV = \iint_S (f_1\mathbf{i}+f_2\mathbf{j}+f_3\mathbf{k}) \cdot \mathbf{n}\, dS = \iint_S \mathbf{f} \cdot \mathbf{n}\, dS$$

The theorem can be extended to surfaces which are such that lines parallel to the coordinate axes meet them in more than two points. In this case, we sub-divide the region bounded by S into sub-regions whose surfaces satisfy this condition. We apply the theorem to each sub-region and then add the results.

Remark. *The significance of divergence theorem lies in the fact that a surface integral may be expressed as a volume integral and vice-versa.*

Deductions from Gauss's Theorem

1. *Prove that* $\iiint_V \mathbf{f} \cdot \nabla\phi\, dV = \iint_S \phi\, \mathbf{f} \cdot \mathbf{n}\, dS - \iiint_V \phi\, div\, \mathbf{f}\, dV$

where **n** *is the outward drawn unit normal vector to S.*

Sol. By Gauss's theorem,

$$\iint_S \mathbf{F} \cdot \mathbf{n}\, dS = \iiint_V \text{div}\, \mathbf{F}\, dV$$

Putting $\mathbf{F} = \phi\mathbf{f}$, we have

$$\iint_S \phi\mathbf{f} \cdot \mathbf{n}\, dS = \iiint_V \text{div}\,(\phi\mathbf{f})\, dV$$

$$\Rightarrow \quad \iint_S \phi\mathbf{f} \cdot \mathbf{n}\, dS = \iiint_V (\phi\, \text{div}\, \mathbf{f} + \mathbf{f} \cdot \nabla\phi)\, dV$$

$$= \iiint_V \phi\, \text{div}\, \mathbf{f}\, dV + \iiint_V \mathbf{f} \cdot \nabla\phi\, dV$$

$$\Rightarrow \quad \iiint_V \mathbf{f} \cdot \nabla\phi\, dV = \iint_S \phi\mathbf{f} \cdot \mathbf{n}\, dS - \iiint_V \phi\, \text{div}\, \mathbf{f}\, dV.$$

2. *Prove that* $\iint_S \mathbf{f} \times \mathbf{n}\, dS = -\iiint_V curl\, \mathbf{f}\, dV.$ (K.U. 2006)

Sol. By Gauss's theorem,

$$\iint_S \mathbf{F} \cdot \mathbf{n}\, dS = \iiint_V \text{div}\, \mathbf{F}\, dV$$

Putting $\mathbf{F} = \mathbf{a} \times \mathbf{f}$, where **a** is an arbitrary constant vector, we have

$$\iint_S (\mathbf{a} \times \mathbf{f}) \cdot \mathbf{n}\, dS = \iiint_V \nabla \cdot (\mathbf{a} \times \mathbf{f})\, dV$$

$$\Rightarrow \quad \iint_S \mathbf{a} \cdot (\mathbf{f} \times \mathbf{n})\, dS = -\iiint_V \mathbf{a} \cdot (\nabla \times \mathbf{f})\, dV$$

($\because$ $\mathbf{a} \cdot (\mathbf{b} \times \mathbf{c}) = (\mathbf{a} \times \mathbf{b}) \cdot \mathbf{c}$ and $\mathbf{a} \cdot \mathbf{b} \times \mathbf{c} = -\mathbf{b} \cdot \mathbf{a} \times \mathbf{c}$)

$$\Rightarrow \quad \mathbf{a} \cdot \iint_S \mathbf{f} \times \mathbf{n}\, dS = -\mathbf{a} \cdot \iiint_V \nabla \times \mathbf{f}\, dV$$

($\because$ **a** is a constant vector)

$$\Rightarrow \quad \mathbf{a} \cdot \left[\iint_S \mathbf{f} \times \mathbf{n}\, dS + \iiint_V \nabla \times \mathbf{f}\, dV\right] = 0$$

$$\Rightarrow \quad \iint_S \mathbf{f} \times \mathbf{n}\, dS + \iiint_V \nabla \times \mathbf{f}\, dV = 0$$

($\because$ **a** is an arbitrary constant vector)

$$\Rightarrow \quad \iint_S \mathbf{n} \times \mathbf{f}\, dS = \iiint_V \nabla \times \mathbf{f}\, dV.$$

3. *Show that* $\iint_S \phi \mathbf{n}\, dS = \iiint_V \text{grad}\, \phi\, dV.$

Sol. By Gauss's theorem,

$$\iint_S \mathbf{f} \cdot \mathbf{n}\, dS = \iiint_V \text{div}\, \mathbf{f}\, dV$$

Putting $\mathbf{f} = \mathbf{a}\phi$, where $\mathbf{a}$ is an arbitrary constant vector, we have

$$\iint_S (\mathbf{a}\phi) \cdot \mathbf{n}\, dS = \iiint_V \text{div}\, (\mathbf{a}\phi)\, dV$$

$$\Rightarrow \quad \iint_S (\mathbf{a}\phi) \cdot \mathbf{n}\, dS = \iiint_V \nabla \cdot (\mathbf{a}\phi)\, dV$$

$$\Rightarrow \quad \mathbf{a} \cdot \iint_S \phi \mathbf{n}\, dS = \mathbf{a} \cdot \iiint_V \nabla\phi\, dV \qquad (\because \text{ a is a constant vector})$$

$$\Rightarrow \quad \mathbf{a} \cdot \left[\iint_S \phi \mathbf{n}\, dS - \iiint_V \nabla\phi\, dV\right] = 0$$

$$\Rightarrow \quad \iint_S \phi \mathbf{n}\, dS - \iiint_V \nabla\phi\, dV = 0 \qquad (\because \text{ a is an arbitrary vector})$$

$$\Rightarrow \quad \iint_S \phi \mathbf{n}\, dS = \iiint_V \nabla\phi\, dV.$$

Example 20. *Evaluate* $\iint \mathbf{r} \cdot \mathbf{n}\, dS$*, where S is a closed surface.*

Sol. By the divergence theorem,

$$\iint_S \mathbf{r} \cdot \mathbf{n}\, dS = \iiint_V \nabla \cdot \mathbf{r}\, dV$$

$$= \iiint_V \left(\mathbf{i}\frac{\partial}{\partial x} + \mathbf{j}\frac{\partial}{\partial y} + \mathbf{k}\frac{\partial}{\partial z}\right) \cdot (x\mathbf{i} + y\mathbf{j} + z\mathbf{k})\, dV$$

$$= \iiint_V \left(\frac{\partial x}{\partial x} + \frac{\partial y}{\partial y} + \frac{\partial z}{\partial z}\right) dV = 3 \iiint_V dV = 3V$$

where V is the volume enclosed by S.

Example 21. *Show that* $\iint_S \mathbf{n}\, dS = \mathbf{0}$ *for any closed surface S.*

Sol. Let $\mathbf{c}$ be any constant vector.

$$\therefore \quad \mathbf{c} \cdot \iint_S \mathbf{n}\, dS = \iint_S \mathbf{c} \cdot \mathbf{n}\, dS$$

$$= \iiint_V \textit{div}\, \mathbf{c}\, dV \qquad \text{(By divergence theorem)}$$

$$= 0 \qquad (\because \ \textit{div}\, \mathbf{c} = 0, \text{ as } \mathbf{c} \text{ is a constant vector})$$

$$\Rightarrow \quad \iint_S \mathbf{n}\, dS = \mathbf{0}.$$

Example 22. *If* $\mathbf{n}$ *is the outward drawn unit normal to any closed surface S, show that*

$$\iiint_V \textit{div}\, \mathbf{n}\, dV = S.$$

Sol. By divergence theorem,

$$\iiint_V \text{div}\, \mathbf{n}\, dS = \iint_S \mathbf{n} \cdot \mathbf{n}\, dS = \iint_S dS = S.$$

Example 23. *Prove* $\iiint_V \frac{1}{r^2}\, dV = \iint_S \frac{\mathbf{r}\,.\,\mathbf{n}}{r^2}\, dS.$

(K.U. 1999 S, 2000, 2007 ; M.D.U. 2007)

Sol. $\iint_S \frac{\mathbf{r}\,.\,\mathbf{n}}{r^2}\, dS = \iint_S \left(\frac{\mathbf{r}}{r^2}\right).\,\mathbf{n}\, dS$

$$= \iiint_V \nabla\,.\left(\frac{\mathbf{r}}{r^2}\right) dV \qquad \text{...(1)}$$

(By divergence theorem)

But,

$$\nabla\,.\left(\frac{\mathbf{r}}{r^2}\right) = \frac{1}{r^2}(\nabla\,.\,\mathbf{r}) + \mathbf{r}\,.\,\nabla\left(\frac{1}{r^2}\right)$$

$$= \frac{1}{r^2}(3) + \mathbf{r}\,.\left(\frac{-2}{r^3}\nabla r\right)$$

$$= \frac{3}{r^2} - \frac{2}{r^3}\left(\mathbf{r}\,.\,\frac{\mathbf{r}}{r}\right) \qquad (\because\ \text{div}\, \mathbf{r} = 3\,;\text{etc.}) \left(\because\ \nabla r = \frac{\mathbf{r}}{r}\right)$$

$$= \frac{3}{r^2} - \frac{2}{r^4}(\mathbf{r}\,.\,\mathbf{r}) = \frac{3}{r^2} - \frac{2}{r^4}r^2 = \frac{1}{r^2}$$

$\therefore$ from (1), $\iint_S \frac{\mathbf{r}\,.\,\mathbf{n}}{r^2}\, dS = \iiint_V \frac{1}{r^2}\, dV.$

Example 24. *Evaluate* $\iint_S x^2\, dydz + y^2\, dzdx + 2z(xy - x - y)\, dxdy$ *where S is the surface of the cube* $0 \le x \le 1,\ 0 \le y \le 1,\ 0 \le z \le 1.$

Sol. Since, by the divergence theorem,

$$\iiint_V \left(\frac{\partial f_1}{\partial x} + \frac{\partial f_2}{\partial y} + \frac{\partial f_3}{\partial z}\right) dx\, dy\, dz$$

$$= \iint_S f_1\, dy\, dz + f_2\, dz\, dx + f_3\, dx\, dy$$

$\therefore \quad \iint_S x^2\, dy\, dz + y^2\, dz\, dx + 2z(xy - x - y)\, dx\, dy$

$$= \iiint_V \left[\frac{\partial}{\partial x}(x^2) + \frac{\partial}{\partial y}(y^2) + \frac{\partial}{\partial z}\{2z\,(xy - x - y)\}\right] dz\, dy\, dx$$

$$= \int_{x=0}^{1}\int_{y=0}^{1}\int_{z=0}^{1} (2x + 2y + 2xy - 2x - 2y)\, dz\, dy\, dx$$

$$= 2\int_{x=0}^{1}\int_{y=0}^{1}\int_{z=0}^{1} xy\, dz\, dy\, dx$$

$$= 2\int_{x=0}^{1}\int_{y=0}^{1} xy\Big[z\Big]_0^1 dy\, dx = 2\int_{x=0}^{1}\int_{y=0}^{1} xy\, dy\, dx$$

$$= 2\int_{x=0}^{1} x\left[\frac{y^2}{2}\right]_0^1 dx = 2\,.\,\frac{1}{2}\int_{x=0}^{1} x\, dx = \left[\frac{x^2}{2}\right]_0^1 = \frac{1}{2}.$$

Example 25. *If S is any closed surface enclosing a volume V and* $\mathbf{f} = x\mathbf{i} + 2y\mathbf{j} + 3z\mathbf{k}$, *show that*

$$\iint_S \mathbf{f}\,.\,\mathbf{n}\, dS = 6V.$$

Sol. By the divergence theorem,

$$\iint_S \mathbf{f}\,.\,\mathbf{n}\,dS = \iiint_V \nabla\,.\,\mathbf{f}dV$$

$$= \iiint_V \left(\mathbf{i}\frac{\partial}{\partial x} + \mathbf{j}\frac{\partial}{\partial y} + \mathbf{k}\frac{\partial}{\partial z}\right).\,(x\mathbf{i} + 2y\mathbf{j} + 3z\mathbf{k})\,dV$$

$$= \iiint_V \left(\frac{\partial}{\partial x}(x) + \frac{\partial}{\partial y}(2y) + \frac{\partial}{\partial z}(3z)\right) dV$$

$$= \iiint_V (1 + 2 + 3)\,dV = 6\iiint_V dV = 6V.$$

Example 26. *Verify Gauss's divergence theorem for the function* $\mathbf{f} = 4xz\mathbf{i} - y^2\mathbf{j} + yz\mathbf{k}$ *over the surface S of the cube bounded by* $x = 0, x = 1, y = 0, y = 1, z = 0, z = 1$. (M.D.U. 2007)

Sol. By the divergence theorem,

$$\iint_S \mathbf{f}\,.\,\mathbf{n}\,dS = \iiint_V \nabla\,.\,\mathbf{f}\,dV$$

$$= \iiint_V \left[\frac{\partial}{\partial x}(4xz) + \frac{\partial}{\partial y}(-y^2) + \frac{\partial}{\partial z}(yz)\right] dV$$

$$= \iiint_V (4z - 2y + y)\,dV = \iiint_V (4z - y)\,dV$$

$$= \int_{x=0}^{1}\int_{y=0}^{1}\int_{z=0}^{1} (4z - y)\,dz\,dy\,dx$$

$$= \int_{x=0}^{1}\int_{y=0}^{1}\left[2z^2 - yz\right]_{z=0}^{1} dy\,dx = \int_{x=0}^{1}\int_{y=0}^{1}(2 - y)\,dy\,dx$$

$$= \frac{3}{2}\text{ (on evaluation)} \qquad ...(1)$$

We shall now, calculate $\iint_S \mathbf{f}\,.\,\mathbf{n}\,dS$ *over the six faces of the cube separately and then add them to get the value of* $\iint_S \mathbf{f}\,.\,\mathbf{n}\,dS$.

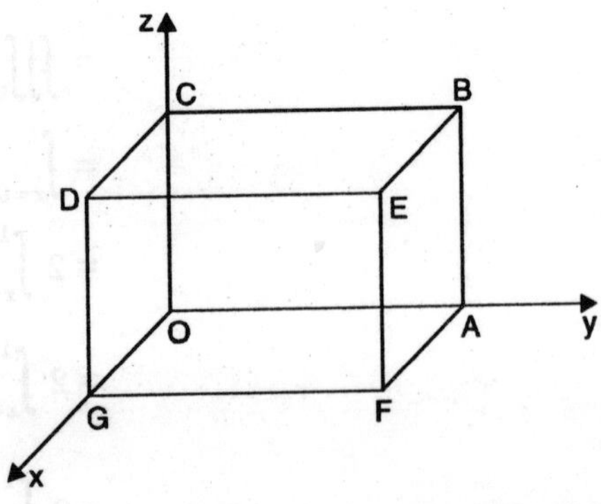

For face DEFG, $\mathbf{n} = \mathbf{i}, x = 1$

$$\therefore \iint_{DEFG} \mathbf{f}\,.\,\mathbf{n} = \int_0^1\int_0^1 (4z\mathbf{i} - y^2\mathbf{j} + yz\mathbf{k})\,.\,\mathbf{i}\,dy\,dz$$

$$= \int_0^1\int_0^1 4z\,dy\,dz = 2 \text{ (verify)} \quad ...(i)$$

For face ABCO, $\mathbf{n} = -\mathbf{i}, x = 0$

$$\therefore \iint_{ABCO} \mathbf{f}\,.\,\mathbf{n}\,dS$$

$$= \int_0^1\int_0^1 (-y^2\mathbf{j} + yz\mathbf{k})\,.\,(-\mathbf{i})\,dy\,dz$$

$$= 0 \qquad ...(ii) \text{ (Verify)}$$

For face ABEF, $\mathbf{n} = \mathbf{j}, y = 1$

$$\therefore \iint_{ABEF} \mathbf{f}\,.\,\mathbf{n}\,dS = \int_0^1\int_0^1 (4xz\,\mathbf{i} - \mathbf{j} + z\mathbf{k})\,.\,\mathbf{j}\,dx\,dz$$

$$= \int_0^1 \int_0^1 - dx\, dz = -1 \qquad ...(iii)$$

For face OGDC, $\mathbf{n} = -\mathbf{j}$, $y = 0$

$$\therefore \qquad \iint_{OGDC} \mathbf{f}\,.\,\mathbf{n}\, dS = \int_0^1 \int_0^1 (4xz\,\mathbf{i})\,.\,\mathbf{j}\, dx\, dz = 0 \qquad ...(iv)$$

For face BCDE, $\mathbf{n} = \mathbf{k}$, $z = 1$

$$\therefore \qquad \iint_{BCDE} \mathbf{f}\,.\,dS = \int_0^1 \int_0^1 (4x\,\mathbf{i} - y^2\mathbf{j} + y\mathbf{k})\,.\,\mathbf{k}\, dx\, dy$$

$$= \int_0^1 \int_0^1 y\, dx\, dy = \frac{1}{2} \text{ (Verify)} \qquad ...(v)$$

For face AFGO, $\mathbf{n} = -\mathbf{k}$, $z = 0$

$$\therefore \qquad \iint_{ABGO} \mathbf{f}\,.\,\mathbf{n}\, dS = \int_0^1 \int_0^1 (-y^2\,\mathbf{j})\,.\,(-\mathbf{k})\, dx\, dy = 0 \qquad ...(vi)$$

Adding (*i*) to (*vi*), we have

$$\iint_S \mathbf{f}\,.\,\mathbf{n}\, dS = 2 + 0 + (-1) + 0 + \frac{1}{2} + 0 = \frac{3}{2} \qquad ...(2)$$

From (1) and (2), we find that

$$\iint_S \mathbf{f}\,.\,\mathbf{n}\, dS = \iiint_V \nabla\,.\,\mathbf{f}\, dV$$

Hence, the divergence theorem is verified.

Example 27. *Verify divergence theorem for* $\mathbf{f} = 4x\mathbf{i} - 2y^2\mathbf{j} + z^2\mathbf{k}$ *taken over the region bounded by the cylinder* $x^2 + y^2 = 4$, $z = 0$, $z = 3$. (M.D.U. 2003)

Sol. By the divergence theorem,

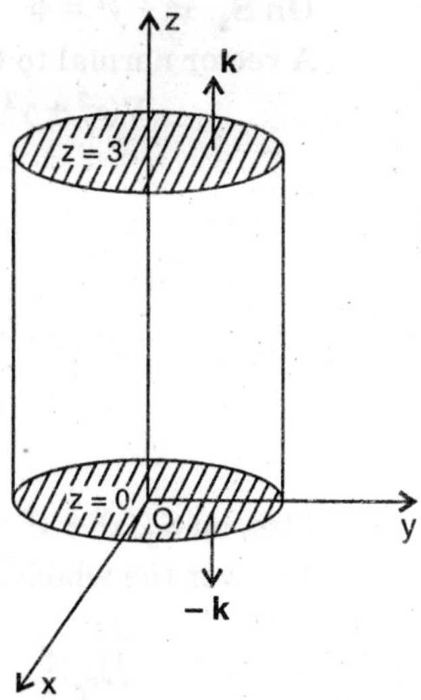

$$\iint_S \mathbf{f}\,.\,\mathbf{n}\, dS = \iiint_V \nabla\,.\,\mathbf{f}\, dV$$

$$= \iiint_V \left[\frac{\partial}{\partial x}(4x) + \frac{\partial}{\partial y}(-2y^2) + \frac{\partial}{\partial z}(z^2)\right] dV$$

$$= \iiint_V (4 - 4y + 2z)\, dx\, dy\, dz$$

$$= \int_{x=-2}^{2} \int_{y=-\sqrt{4-x^2}}^{\sqrt{4-x^2}} \int_{z=0}^{3} (4 - 4y + 2z)\, dz\, dy\, dx$$

$$= \int_{x=-2}^{2} \int_{y=-\sqrt{4-x^2}}^{\sqrt{4-x^2}} \left[4z - 4yz + z^2\right]_0^3 dy\, dx$$

$$= \int_{x=-2}^{2} \left(\int_{y=-\sqrt{4-x^2}}^{\sqrt{4-x^2}} (12 - 12y + 9)\, dy\right) dx$$

$$= \int_{x=-2}^{2} \int_{y=-\sqrt{4-x^2}}^{\sqrt{4-x^2}} 21\, dy\, dz \qquad \left[\because\ 12y \text{ is an odd function, } \int_{-a}^{a} 12y\, dy = 0\right]$$

$$= \int_{-2}^{2} 42\sqrt{4 - x^2}\, dx = 84 \int_0^2 \sqrt{4 - x^2}\, dx$$

$$= 84\left[\frac{x\sqrt{4-x^2}}{2} + \frac{4}{2}\sin^{-1}\frac{x}{2}\right]_0^2 = 84\,[2\sin^{-1} 1]$$

$$= 84\left[2\times\frac{\pi}{2}\right] = 84\pi \qquad ...(1)$$

To evaluate the surface integral, we divide the closed surface S of the cylinder into three parts :

S_1 : the circular base in the plane $z = 0$

S_2 : the circular top in the plane $z = 3$

S_3 : the curved surface of the cylinder, given by the equation

$$x^2 + y^2 = 4.$$

Also, $$\iint_S \mathbf{f}\,.\,\mathbf{n}\,dS = \iint_{S_1} \mathbf{f}\,.\,\mathbf{n}\,dS + \iint_{S_2} \mathbf{f}\,.\,\mathbf{n}\,dS + \iint_{S_3} \mathbf{f}\,.\,\mathbf{n}\,dS. \qquad ...(2)$$

On S_1 ($z = 0$), $\mathbf{n} = -\mathbf{k}$, $\mathbf{f} = 4x\mathbf{i} - 2y^2\mathbf{j}$

$$\therefore \quad \iint_{S_1} \mathbf{f}\,.\,\mathbf{n}\,dS = \iint_{S_1} 0\,dS = 0 \qquad ...(i)$$

On S_2 ($z = 3$), $\mathbf{n} = \mathbf{k}$, $\mathbf{f} = 4x\mathbf{i} - 2y^2\mathbf{j} + 9\mathbf{k}$

$$\therefore \quad \mathbf{f}\,.\,\mathbf{n} = (4x)(0) + (-2y^2)(0) + 9(1) = 9$$

$$\therefore \quad \iint_{S_2} \mathbf{f}\,.\,\mathbf{n}\,dS = \iint_{S_2} 9\,dx\,dy = 9\iint_{S_2} dx\,dy$$

$$= 9 \times \text{area of } S_2 = 9(\pi(2)^2) = 36\pi \qquad ...(ii)$$

On S_3, $x^2 + y^2 = 4$

A vector normal to the surface S_3, is given by

$$\nabla(x^2 + y^2 - 4) = 2x\mathbf{i} + 2y\mathbf{j}$$

$$\therefore \quad \mathbf{n} = \text{unit vector normal to } S_3 = \frac{2x\mathbf{i} + 2y\mathbf{j}}{\sqrt{4x^2 + 4y^2}}$$

$$= \frac{x\mathbf{i} + y\mathbf{j}}{2} \qquad (\because\ x^2 + y^2 = 4)$$

$$\therefore \quad \mathbf{f}\,.\,\mathbf{n} = (4x\mathbf{i} - 2y^2\mathbf{j} + z^2\mathbf{k})\,.\left(\frac{x\mathbf{i} + y\mathbf{j}}{2}\right)$$

$$= 2x^2 - y^3.$$

Also, on S_3, $x^2 + y^2 = 4$, $x = 2\cos\theta$, $y = 2\sin\theta$, and $dS = 2d\theta\,dz$.

To cover the whole surface S_3, z varies from 0 to 3 and θ varies from 0 to 2π.

$$\therefore \quad \iint_{S_3} \mathbf{f}\,.\,\mathbf{n}\,dS = \int_{\theta=0}^{2\pi}\int_{z=0}^{3} [2(2\cos\theta)^2 - (2\sin\theta)^3]\,2\,dz\,d\theta$$

$$= \int_0^{2\pi} 16(\cos^2\theta - \sin^3\theta) \times 3\,d\theta$$

$$= 48\int_0^{2\pi} (\cos^2\theta - \sin^3\theta)\,d\theta = 48\pi \qquad ...(iii)$$

$$\left[\because \int_0^{2\pi} \cos^2\theta\, d\theta = 2\int_0^{\pi} \cos^2\theta\, d\theta = 4\int_0^{\pi/2} \cos^2\theta\, d\theta = 4\times\frac{1}{2}\times\frac{\pi}{2} = \pi \text{ and } \int_0^{2\pi} \sin^3\theta\, d\theta = 0\right]$$

Using (*i*), (*ii*) and (*iii*) in (2), we get

$$\iint_S \mathbf{f}\,.\,\mathbf{n}\, dS = 0 + 36\pi + 48\pi = 84\pi \qquad ...(3)$$

Equality of (1) and (3) verifies the divergence theorem.

Example 28. *Evaluate* $\iint_S \mathbf{f}\,.\,\mathbf{n}\, dS$, *where* $\mathbf{f} = ax\,\mathbf{i} + by\,\mathbf{j} + cz\,\mathbf{k}$ *and S is the surface of the sphere* $x^2 + y^2 + z^2 = 1$.

Sol. By the divergence theorem,

$$\iint_S \mathbf{f}\,.\,\mathbf{n}\, dS = \iiint_V \nabla\,.\,\mathbf{f}\, dV$$

$$= \iiint_V \left(\mathbf{i}\frac{\partial}{\partial x} + \mathbf{j}\frac{\partial}{\partial y} + \mathbf{k}\frac{\partial}{\partial z}\right).\,(ax\,\mathbf{i} + by\,\mathbf{j} + cz\,\mathbf{k})\, dV$$

$$= \iiint_V (a + b + c)\, dV = (a + b + c)\iiint_V dV$$

$$= (a + b + c)\,V = (a + b + c)\,\frac{4}{3}\pi$$

$$\left[\because V = \text{volume of the sphere is } \frac{4}{3}\pi(1)^3\right]$$

Example 29. *Evaluate* $\iint_S (y^2z^2\,\mathbf{i} + z^2x^2\,\mathbf{j} + z^2y^2\,\mathbf{k})\,.\,\mathbf{n}\, dS$

where S is the part of the sphere $x^2 + y^2 + z^2 = 1$ *above the xy-plane and bounded by xy-plane.*

Sol. Let $\mathbf{f} = y^2z^2\,\mathbf{i} + z^2x^2\,\mathbf{j} + z^2y^2\,\mathbf{k}$

$$\therefore \quad \text{div}\,\mathbf{f} = \nabla\,.\,\mathbf{f} = \left(\mathbf{i}\frac{\partial}{\partial x} + \mathbf{j}\frac{\partial}{\partial y} + \mathbf{k}\frac{\partial}{\partial z}\right).\,(y^2z^2\,\mathbf{i} + z^2x^2\,\mathbf{j} + z^2y^2\,\mathbf{k})$$

$$= \frac{\partial}{\partial x}(y^2z^2) + \frac{\partial}{\partial y}(z^2x^2) + \frac{\partial}{\partial z}(z^2y^2) = 2zy^2 \qquad ...(1)$$

By Gauss's Theorem,

$$\iint_S \mathbf{f}\,.\,\mathbf{n}\, dS = \iiint_V \text{div}\,\mathbf{f}\, dV = \iiint_V 2zy^2\, dz\, dy\, dx$$

(By (1) and $dV = dz\, dy\, dx$)

where V is the volume enclosed by the surface S, *i.e.*, V is the volume of the hemi-sphere $x^2 + y^2 + z^2 = 1$ above the $z = 0$ plane.

Here, z varies from 0 to $\sqrt{1 - x^2 - y^2}$, y varies from $-\sqrt{1 - x^2}$ to $\sqrt{1 - x^2}$ and x varies from -1 to 1.

$$\therefore \quad \iint_S \mathbf{f}\,.\,\mathbf{n}\, dS = \iiint_V \nabla\,.\,\mathbf{f}\, dV = \int_{x=-1}^{1}\int_{y=-\sqrt{1-x^2}}^{\sqrt{1-x^2}}\int_{z=0}^{\sqrt{1-x^2-y^2}} 2zy^2\, dz\, dy\, dx$$

$$= \int_{-1}^{1}\int_{-\sqrt{1-x^2}}^{\sqrt{1-x^2}} y^2\left[z^2\right]_0^{\sqrt{1-x^2-y^2}} dy\, dx$$

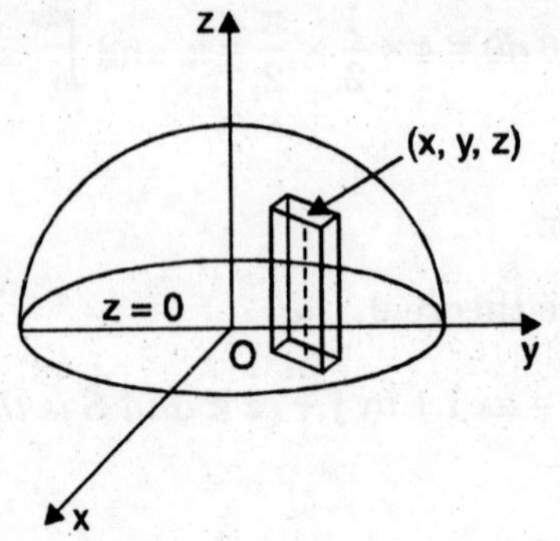

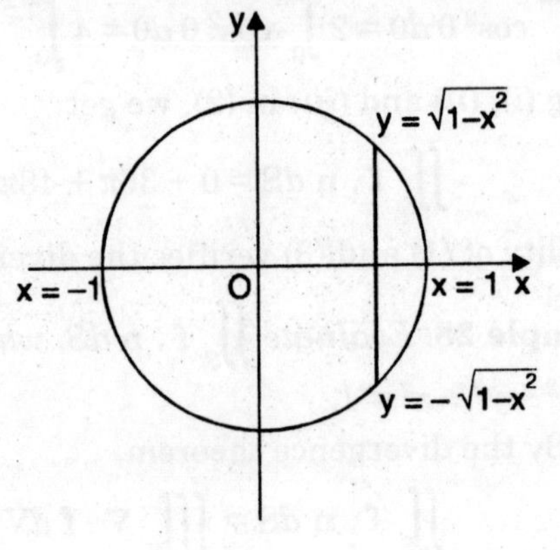

$$= \int_{-1}^{1}\left(\int_{-\sqrt{1-x^2}}^{\sqrt{1-x^2}} y^2(1-x^2-y^2)\,dy\right)dx = \int_{-1}^{1}\left[\frac{y^3}{3}(1-x^2)-\frac{y^5}{5}\right]_{-\sqrt{1-x^2}}^{\sqrt{1-x^2}} dx$$

$$= 2\int_{-1}^{1}\left[\frac{1}{3}(1-x^2)^{5/2}-\frac{1}{5}(1-x^2)^{5/2}\right]dx$$

$$= \frac{4}{15}\int_{-1}^{1}(1-x^2)^{5/2}\,dx = \frac{8}{15}\int_{0}^{1}(1-x^2)^{5/2}\,dx$$

$$\left[\because \int_{-a}^{a} f(x)\,dx = 2\int_{0}^{a} f(x)\,dx, \text{ if } f(x)=f(-x)\right]$$

$$= \frac{8}{15}\int_{0}^{\pi/2}(1-\sin^2\theta)^{5/2}\cos\theta\,d\theta, \text{ where } x=\sin\theta$$

$$= \frac{8}{15}\int_{0}^{\pi/2}\cos^6\theta\,d\theta = \frac{8}{15}\cdot\frac{5.3.1}{6.4.2}\frac{\pi}{2} = \frac{\pi}{12}.$$

Example 30. *Evaluate* $\iint_S \mathbf{f}\cdot\mathbf{n}\,dS$ *over the entire surface of the region above the xy-plane bounded by the cone* $z^2 = x^2 + y^2$ *and the plane* $z = 4$, *if* $\mathbf{f} = 4xz\,\mathbf{i} + xyz^2\,\mathbf{j} + 3z\,\mathbf{k}$.

Sol. If V is the volume enclosed by S, then V is bounded by the surfaces $z = 0$, $z = 4$, $z^2 = x^2 + y^2$.

By divergence theorem, $\iint_S \mathbf{f}\cdot\mathbf{n}\,dS = \iint_V \nabla\cdot\mathbf{f}\,dV$

$$= \iiint_V\left[\frac{\partial}{\partial x}(4xz)+\frac{\partial}{\partial y}(xyz^2)+\frac{\partial}{\partial z}(3z)\right]dV$$

$$= \iiint_V (4z + xz^2 + 3)\,dV$$

$$= \int_{z=0}^{4}\int_{y=-z}^{z}\int_{x=-\sqrt{z^2-y^2}}^{\sqrt{z^2-y^2}} (4z + xz^2 + 3)\,dx\,dy\,dz$$

$$= 2\int_{z=0}^{4}\int_{y=-z}^{z}\int_{0}^{\sqrt{z^2-y^2}} (4z+3)\,dx\,dy\,dz \qquad \left[\because \int_{-\sqrt{z^2-y^2}}^{\sqrt{z^2-y^2}} x\,dx = 0\right]$$

$$= 2\int_{z=0}^{4}\int_{y=-z}^{z} (4z+3)\sqrt{z^2-y^2}\, dy\, dz$$

$$= 4\int_0^4 \left(\int_0^z (4z+3)\sqrt{z^2-y^2}\, dy\right) dz$$

$$= 4\int_0^4 (4z+3)\left[\frac{y\sqrt{z^2-y^2}}{2} + \frac{z^2}{2}\sin^{-1}\frac{y}{z}\right]_0^z dz$$

$$= 4\int_0^4 (4z+3)\left[\frac{z^2}{2}\sin^{-1} 1\right] dz$$

$$= 4\times\frac{\pi}{4}\int_0^4 (4z^3+3z^2)\, dz \qquad \left[\because \sin^{-1} 1 = \frac{\pi}{2}\right]$$

$$= \pi\left[z^4 + z^3\right]_0^4 = \pi(256+64) = 320\pi.$$

Example 31. *Evaluate* $\iint_S (\nabla\times \mathbf{f})\,.\,\mathbf{n}\, dS$, *where S is the surface of the cone* $z = 2 - \sqrt{x^2+y^2}$ *above xy plane and* $\mathbf{f} = (x-z)\mathbf{i} + (x^3+yz)\mathbf{j} - 3xy^2\mathbf{k}$. (M.D.U. 2007)

Sol. Here S is not a closed surface.

The surface $z = 2 - \sqrt{x^2+y^2}$ meets the xy-plane in a circle C given by $x^2+y^2 = 4$, $z = 0$. Let S_1 be the plane region bounded by the circle C. If S′ is the surface consisting of the surfaces S and S_1 ; then **S′ is a closed** surface.

∴ By divergence theorem, we have

$$\iint_{S'} (\nabla\times \mathbf{f})\,.\,\mathbf{n}\, dS = \iiint_V \nabla\,.\,(\nabla\times\mathbf{f})\, dV = 0 \qquad [\because \nabla\,.\,(\nabla\times\mathbf{f}) = 0]$$

$$\Rightarrow \iint_{S} (\nabla\times \mathbf{f})\,.\,\mathbf{n}\, dS + \iint_{S_1} (\nabla\times \mathbf{f})\,.\,\mathbf{n}\, dS = 0 \qquad [\because \text{S}' \text{ consists of S and } S_1]$$

$$\Rightarrow \iint_{S} (\nabla\times \mathbf{f})\,.\,\mathbf{n}\, dS - \iint_{S_1} (\nabla\times \mathbf{f})\,.\,\mathbf{k}\, dS = 0 \qquad [\because \text{On } S_1,\ \mathbf{n} = -\mathbf{k}]$$

$$\Rightarrow \iint_{S} (\nabla\times \mathbf{f})\,.\,\mathbf{n}\, dS = \iint_{S_1} (\nabla\times \mathbf{f})\,.\,\mathbf{k}\, dS \qquad ...(1)$$

But,
$$\nabla\times\mathbf{f} = \begin{vmatrix} \mathbf{i} & \mathbf{j} & \mathbf{k} \\ \dfrac{\partial}{\partial x} & \dfrac{\partial}{\partial y} & \dfrac{\partial}{\partial z} \\ x-z & x^3+yz & -3xy^2 \end{vmatrix}$$

$$= (-6xy - y)\mathbf{i} + (-1 + 3y^2)\mathbf{j} + (3x^2 - 0)\mathbf{k}$$

∴ $(\nabla\times\mathbf{f})\,.\,\mathbf{k} = 3x^2$

∴ From (1),

$$\iint_{S} (\nabla\times \mathbf{f})\,.\,\mathbf{n}\, dS = \iint_{S_1} 3x^2\, dS = \int_0^{2\pi}\int_0^2 3(r^2\cos^2\theta)\, r\, dr\, d\theta$$

(Changing to polar coordinates)

$$= 3\int_0^{2\pi}\int_0^{2}(r^3\cos^2\theta)\,dr\,d\theta = 3\int_0^{2\pi}\left[\frac{r^4}{4}\right]_0^2\cos^2\theta\,d\theta$$

$$= 12\int_0^{2\pi}\cos^2\theta\,d\theta = 12\pi.$$

EXERCISE 5

1. Evaluate $\iint_S \mathbf{f}\,.\,\mathbf{n}\,dS$, where $\mathbf{f} = 4xy\,\mathbf{i} + yz\,\mathbf{j} - xy\,\mathbf{k}$ and S is the surface of the cube bounded by the planes $x = 0, x = 2, y = 0, y = 2, z = 0$ and $z = 2$. Verify Gauss's divergence theorem for f. *(M.D.U. 2000)*
2. Verify Gauss's theorem to show that
$$\iint_S [(x^3 - yz)\,\mathbf{i} - 2x^2y\,\mathbf{j} + 2\mathbf{k}]\,.\,\mathbf{n}\,dS = \frac{1}{3}a^5$$
where S denotes the surface of the cube bounded by the planes $x = 0, x = a, y = 0, y = a, z = 0$ and $z = a$.
3. Show that $V = \iint_S x\,dy\,dz = \iint_S y\,dz\,dx = \iint_S z\,dx\,dy$
$$= \frac{1}{3}\iint_S (x\,dy\,dz + y\,dz\,dx + z\,dxdy)$$
where V is volume of a region bounded by a surface S.
4. Prove that $\frac{1}{3}\iint_S \mathbf{r}\,.\,\mathbf{n}\,dS$ is the volume of the space enclosed by the surface S.
5. For a closed surface S and a vector point function **f**, show that $\iint_S \mathbf{n}\,.\,\text{curl}\,\mathbf{f}\,dS = 0$.
6. Use the divergence theorem in Cartesian form to evaluate
$$\iint_S (x^3 - yz)\,dy\,dz - 2x^2y\,dz\,dx + zdx\,dy$$
over the surface of a cube bounded by the coordinate planes and the planes $x = y = z = a$.
7. If S is any closed surface enclosing a volume V and $\mathbf{f} = x\mathbf{i} + 2y\mathbf{j} + 7z\mathbf{k}$, show that
$$\iint \mathbf{f}\,.\,\mathbf{n}\,dS = 10\,V.$$
8. Evaluate $\iint_S \mathbf{f}\,.\,\mathbf{n}\,dS$, where $\mathbf{f} = (2x + 3z)\mathbf{i} - (xz + y)\mathbf{j} + (y^2 + 2z)\mathbf{k}$ and S is the surface of the sphere having centre at $(3, -1, 2)$ and radius 3.
[**Hint.** Volume of sphere $= \frac{4}{3}\pi(3)^3 = 36\pi$
Proceed as in example 25.]
9. State and prove Gauss's divergence theorem. *(K.U. 2001 S)*

Answers

1. 40 6. $\frac{1}{3}a^5 + a^3$ 8. 108π.

9. Green's Theorem*

If ϕ and ψ are two scalar point functions having continuous second order derivatives in a region V bounded by a closed surface S, then

$$\iiint_V (\phi\nabla^2 \psi - \psi\nabla^2\phi)\, dV = \iint_S (\phi\nabla\psi - \psi\nabla\phi) \,.\, \mathbf{n}\, dS \qquad \text{(M.D.U. 2003)}$$

Proof. By Gauss's divergence theorem, we have

$$\iiint_V \nabla \,.\, \mathbf{f}\, dV = \iint_S \mathbf{f} \,.\, \mathbf{n}\, dS \qquad \text{...(1)}$$

Putting $\mathbf{f} = \phi\nabla\psi$, we get

$$\nabla \,.\, \mathbf{f} = \nabla \,.\, (\phi\nabla\psi)$$
$$= \phi(\nabla \,.\, \nabla\, \psi) + (\nabla\phi) \,.\, (\nabla\psi) = \phi\nabla^2\psi + (\nabla\phi) \,.\, (\nabla\psi)$$

Also, $\mathbf{f} \,.\, \mathbf{n} = (\phi\nabla\psi) \,.\, \mathbf{n}$

$\therefore$ (1) becomes,

$$\iiint_V [\phi\nabla^2 \psi + (\nabla\phi) \,.\, (\nabla\psi)]\, dV = \iint_S (\phi\nabla\psi) \,.\, \mathbf{n}\, dS \qquad \text{...(2)}$$

This is called Green's first theorem (or identity).

Interchanging ϕ and ψ in (2), we have

$$\iiint_V [\psi\nabla^2\phi + (\nabla\psi) \,.\, (\nabla\phi)]\, dV = \iint_S (\psi\nabla\phi) \,.\, \mathbf{n}\, dS \qquad \text{...(3)}$$

(2) – (3) gives,

$$\iiint_V (\phi\nabla^2\psi - \psi\nabla^2\phi)\, dV = \iint_S (\phi\nabla\psi - \psi\nabla\phi) \,.\, \mathbf{n}\, dS \qquad \text{...(4)}$$

This is also called Green's second theorem (or identity).

Hence, the theorem.

Green's theorem can also be stated as

$$\iiint_V (\phi\, \nabla^2 \psi - \psi\nabla^2\phi)\, dV = \iint_S \left(\phi \frac{\partial\psi}{\partial n} - \psi \frac{\partial\phi}{\partial n}\right) dS$$

where $\frac{\partial\psi}{\partial n}$ and $\frac{\partial\phi}{\partial n}$ denote the directional derivatives of ϕ and ψ respectively along the outward drawn normal at any point of S.

$\left[\text{For, } \nabla\psi = \frac{\partial\psi}{\partial n}\mathbf{n} \text{ and } \nabla\phi = \frac{\partial\phi}{\partial n}\mathbf{n}\right.$

$$\therefore \quad (\phi\nabla\psi - \psi\nabla\phi) \,.\, \mathbf{n} = \left(\phi \frac{\partial\psi}{\partial n}\mathbf{n} - \psi \frac{\partial\phi}{\partial n}\mathbf{n}\right) = \phi \frac{\partial\psi}{\partial n} - \psi \frac{\partial\phi}{\partial n}$$

$\therefore$ (4) becomes

$$\left.\iiint_V (\phi\nabla^2\psi - \psi\nabla^2\phi)\, dV = \iint_S \left(\phi \frac{\partial\psi}{\partial n} - \psi \frac{\partial\phi}{\partial n}\right) dS\right].$$

*After the name of an English mathematician G. Green (1793—1841).

REDUCTION OF SURFACE INTEGRAL TO LINE INTEGRAL

10. Stokes' Theorem*

If **f** *is any continuously differentiable vector point function and S is a surface bounded by a simple closed curve C, then*

$$\oint_C \mathbf{f} \cdot d\mathbf{r} = \iint_S \text{curl } \mathbf{f} \cdot \mathbf{n} \, dS$$

where **n** is the unit normal vector at any point of S and is drawn in the sense in which a right-handed skrew would move when rotated in the sense of description of C. *(K.U. 2001)*

In other words : *The line integral of the tangential component of a vector* **f** *taken around a simple closed curve C is equal to the surface integral of the normal component of the curl of* **f** *taken over any surface S having C as its boundary.*

Cartesian form of Stokes' Theorem for Plane** *(K.U. 2001)*

Let us take a system of Cartesian rectangular coordinate axes such that the plane of the given surface S is the *xy*-plane and *z*-axis lies along the direction of the normal vector **n**.

Here, the normal vector is constant.

Let $$\mathbf{f} = f_1 \mathbf{i} + f_2 \mathbf{j} + f_3 \mathbf{k}$$

$\therefore$ $$\mathbf{f} \cdot d\vec{\mathbf{r}} = f_1 \, dx + f_2 \, dy$$

Now, $$\iint_S \text{curl } \mathbf{f} \cdot \mathbf{n} \, dS = \iint_S \text{curl } \mathbf{f} \cdot \mathbf{k} \, dS = \iint_S \left(\frac{\partial f_2}{\partial x} - \frac{\partial f_1}{\partial y} \right) dx \, dy$$

$\therefore$ Stokes' theorem for plane becomes

$$\oint_C (f_1 \, dx + f_2 \, dy) = \iint_S \left(\frac{\partial f_2}{\partial x} - \frac{\partial f_1}{\partial y} \right) dx \, dy$$

Let us, now prove it.

Let the region S be such that any line parallel to either axis meet the boundary C in at the most two points.

Let the region be $x = a$, $x = b$, $y = \phi_1(x)$, $y = \phi_2(x)$, where $\phi_2(x) \geq \phi_1(x)$.

The boundary C is split up into two arcs C_1 and C_2 as shown.

$$\therefore \quad \iint_S \frac{\partial f_1}{\partial y} \, dx \, dy = \int_a^b \left[\int_{\phi_1(x)}^{\phi_2(x)} \frac{\partial f_1}{\partial y} \, dy \right] dx = \int_a^b [f_1(x, \phi_2(x)) - f_1(x, \phi_1(x))] \, dx$$

$$= - \iint_{C_2} f_1(x, y) \, dx - \iint_{C_1} f_1(x, y) \, dx$$

$$= - \oint_C f_1(x, y) \, dx \qquad \ldots(i)$$

---

*After the name of an English mathematician George Gabriel Stokes (1819—1903) and provides us a relation between a surface integral taken over a surface to a line integral around the boundary curve of the surface.

***Stokes' theorem for a plane is also known as Green's theorem* and provides a relation connecting a line integral along a closed curve and the surface integral over the surface bounded by the closed curve. It has important applications in Physics and Mathematics.

Similarly,

$$\iint_S \frac{\partial f_2}{\partial x}\, dx\, dy = \oint_C f_2(x, y)\, dy \qquad ...(ii)$$

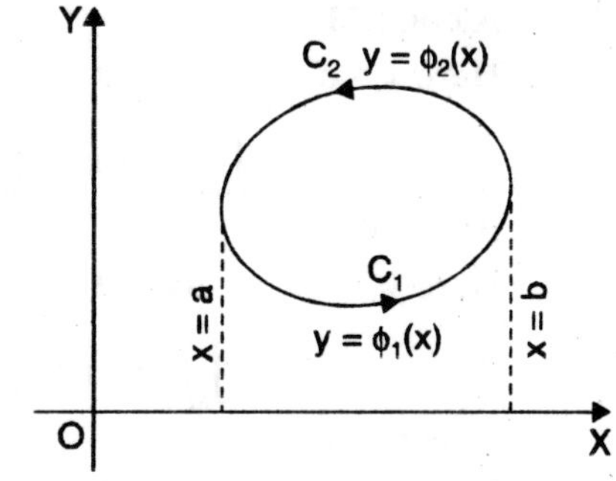

∴ From (*i*) and (*ii*),

$$\oint_C (f_2\, dy + f_1\, dx) = \iint_S \left(\frac{\partial f_2}{\partial x} - \frac{\partial f_1}{\partial y}\right) dx\, dy$$

Hence the result.

If the region S is such that a line parallel to either axes meets the boundary C in more than two points, we divide S into a finite number of sub-regions such that the boundary of each is met in at the most two points by any line parallel to either axis. By applying the above result to each sub-region and then adding the results, we find that the theorem is true for S. This is because the line integrals along the boundary curves will cancel in pairs.

Cartesian form of Stokes' theorem in space. Let $\mathbf{f} = f_1\,\mathbf{i} + f_2\,\mathbf{j} + f_3\,\mathbf{k}$. Let the outward drawn normal vector $\mathbf{n}$ make angles α, β, γ with positive directions of *x*, *y*, *z* axes respectively so that

$$\mathbf{n} = \cos\alpha\,\mathbf{i} + \cos\beta\,\mathbf{j} + \cos\gamma\,\mathbf{k}$$

Then,
$$\nabla \times \mathbf{f} = \begin{vmatrix} \mathbf{i} & \mathbf{j} & \mathbf{k} \\ \frac{\partial}{\partial x} & \frac{\partial}{\partial y} & \frac{\partial}{\partial z} \\ f_1 & f_2 & f_3 \end{vmatrix} = \left(\frac{\partial f_3}{\partial y} - \frac{\partial f_2}{\partial z}\right)\mathbf{i} + \left(\frac{\partial f_1}{\partial z} - \frac{\partial f_3}{\partial x}\right)\mathbf{j} + \left(\frac{\partial f_2}{\partial x} - \frac{\partial f_1}{\partial y}\right)\mathbf{k}$$

∴
$$(\nabla \times \mathbf{f}) \cdot \mathbf{n} = \left(\frac{\partial f_3}{\partial y} - \frac{\partial f_2}{\partial z}\right)\cos\alpha + \left(\frac{\partial f_1}{\partial z} - \frac{\partial f_3}{\partial x}\right)\cos\beta + \left(\frac{\partial f_2}{\partial x} - \frac{\partial f_1}{\partial y}\right)\cos\gamma$$

Also,
$$\mathbf{f} \cdot d\mathbf{r} = (f_1\,\mathbf{i} + f_2\,\mathbf{j} + f_3\mathbf{k}) \cdot (dx\,\mathbf{i} + dy\,\mathbf{j} + dz\,\mathbf{k})$$
$$= f_1\, dx + f_2\, dy + f_3\, dz$$

∴ Stokes' theorem becomes

$$\int_C f_1\, dx + f_2\, dy + f_3\, dz = \iint_S \left[\left(\frac{\partial f_3}{\partial y} - \frac{\partial f_2}{\partial z}\right)\cos\alpha + \left(\frac{\partial f_1}{\partial z} - \frac{\partial f_3}{\partial x}\right)\cos\beta + \left(\frac{\partial f_2}{\partial x} - \frac{\partial f_1}{\partial y}\right)\cos\gamma\right] dS$$

Proof. Proof is beyond the scope of this book.

Example 32. *Verify Stokes' theorem for the function* $\mathbf{f} = x^2\mathbf{i} + xy\mathbf{j}$, *integrated round the square in the plane z = 0 whose sides are along the lines x = 0, x = a, y = 0, y = a.*

(K.U. 2000, 2000 S)

Sol. Here,
$$\mathbf{f} = x^2\mathbf{i} + xy\mathbf{j}$$

∴
$$\mathbf{f} \cdot d\mathbf{r} = (x^2\mathbf{i} + xy\mathbf{j}) \cdot (dx\mathbf{i} + dy\mathbf{j})$$
$$= x^2 dx + xy dy$$

∴
$$\oint_C \mathbf{f} \cdot d\mathbf{r} = \oint_C (x^2\, dx + xy\, dy)$$

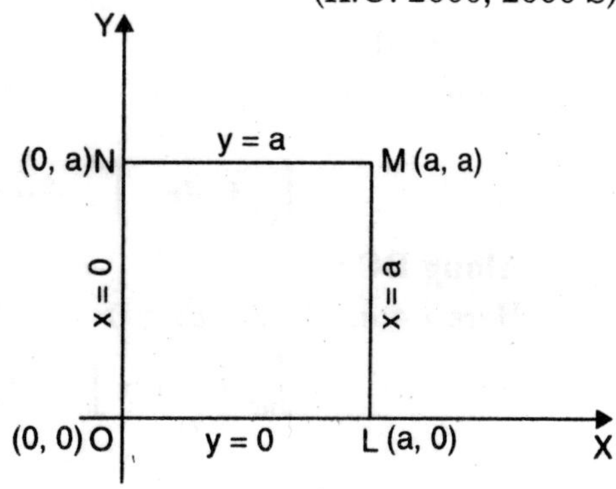

Along OL : Here $y = 0$

∴ $dy = 0$

∴
$$\int_{OL} \mathbf{f} \cdot d\mathbf{r} = \int_0^a x^2\, dx = \frac{a^3}{3}$$

Along LM :

Here $x = a \quad \therefore \quad dx = 0$

$$\therefore \qquad \int_{LM} \mathbf{f} \cdot d\mathbf{r} = \int_0^a ay\, dy = \frac{a^3}{2}$$

Along MN :

Here $y = a \quad \therefore \quad dy = 0$

$$\therefore \qquad \int_{MN} \mathbf{f} \cdot \mathbf{r} = \int_a^0 x^2\, dx = -\frac{a^3}{3}$$

Along NO :

Here $x = 0 \quad \therefore \quad dx = 0$

$$\therefore \qquad \int_{NO} \mathbf{f} \cdot d\mathbf{r} = 0$$

$$\therefore \qquad \oint_C \mathbf{f} \cdot d\mathbf{r} = \int_{OL} \mathbf{f} \cdot d\mathbf{r} + \int_{LM} \mathbf{f} \cdot d\mathbf{r} + \int_{MN} \mathbf{f} \cdot d\mathbf{r} + \int_{NO} \mathbf{f} \cdot d\mathbf{r}$$

$$= \frac{a^3}{3} + \frac{a^3}{2} - \frac{a^3}{3} + 0 = \frac{a^3}{2} \qquad \ldots(1)$$

Now, $$\text{curl } \mathbf{f} = \begin{vmatrix} \mathbf{i} & \mathbf{j} & \mathbf{k} \\ \frac{\partial}{\partial x} & \frac{\partial}{\partial y} & \frac{\partial}{\partial z} \\ x^2 & xy & 0 \end{vmatrix} = y\mathbf{k}$$

Since the surface lies in the xy-plane, $\mathbf{n} = \mathbf{k}$.

$$\therefore \qquad \text{curl } \mathbf{f} \cdot \mathbf{n} = y\mathbf{k} \cdot \mathbf{k} = y$$

$$\therefore \qquad \iint_S \text{curl } \mathbf{f} \cdot \mathbf{n}\, dS = \int_{x=0}^a \int_{y=0}^a y\, dx\, dy = \int_{x=0}^a \left[\frac{y^2}{2}\right]_0^a dx$$

$$= \int_0^a \frac{a^2}{2}\, dx = \frac{a^2}{2}\Big[x\Big]_0^a = \frac{a^3}{2} \qquad \ldots(2)$$

Equality of (1) and (2) verifies Stokes' theorem.

Example 33. *Verify Stokes' theorem for the function* $\mathbf{f} = (x^2 + y^2)\mathbf{i} - 2xy\mathbf{j}$, *taken round the rectangle bounded by* $x = \pm a$, $y = 0$, $y = b$. (K.U. 2007)

Sol. Here, $$\mathbf{f} = (x^2 + y^2)\,\mathbf{i} - 2xy\,\mathbf{j}$$

$$\therefore \qquad \mathbf{f} \cdot d\mathbf{r} = [(x^2 + y^2)\,\mathbf{i} - 2xy\,\mathbf{j}] \cdot [dx\mathbf{i} + dy\mathbf{j}] = (x^2 + y^2)dx - 2xy\, dy$$

$$\Rightarrow \qquad \oint_C \mathbf{f} \cdot d\mathbf{r} = \oint_C [(x^2 + y^2)\, dx - 2xy\, dy]$$

Along AB :

Here $y = 0. \quad \therefore \quad dy = 0$

$$\therefore \qquad \int_{AB} \mathbf{f} \cdot d\mathbf{r} = \int_{-a}^a x^2\, dx = \left[\frac{x^3}{3}\right]_{-a}^a = \frac{2a^3}{3}$$

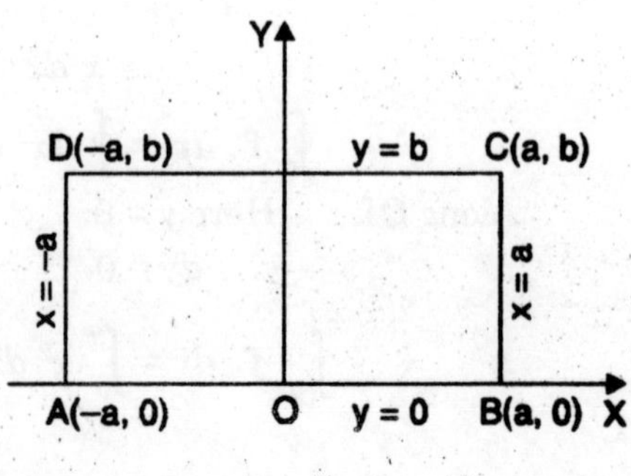

Along BC :

Here $x = a, \quad \therefore \quad dx = 0$

$$\therefore \qquad \int_{BC} \mathbf{f} \cdot d\mathbf{r} = \int_0^b -2ay\, dy = -ab^2$$

Along CD :

Here $y = b$. $\therefore$ $dy = 0$

$$\therefore \quad \int_{CD} \mathbf{f} \cdot d\mathbf{r} = \int_a^{-a} (x^2 + b^2)\, dx = \left[\frac{x^3}{3} + b^2 x\right]_a^{-a} = -2\left(\frac{a^3}{3} + ab^2\right)$$

Along DA :

Here $x = -a$ $\therefore$ $dx = 0$

$$\therefore \quad \int_{DA} \mathbf{f} \cdot d\mathbf{r} = \int_b^0 (2ay\, dy) = -ab^2$$

Adding the integrals over the rectangle ABCD, we get

$$\int_C \mathbf{f}\, d\mathbf{r} = \frac{2a^3}{3} - ab^2 - \frac{2a^3}{3} - 2ab^2 - ab^2 = -4ab^2 \quad \text{...(1)}$$

Now,
$$\text{curl } \mathbf{f} = \begin{vmatrix} \mathbf{i} & \mathbf{j} & \mathbf{k} \\ \frac{\partial}{\partial x} & \frac{\partial}{\partial y} & \frac{\partial}{\partial z} \\ x^2 + y^2 & -2xy & 0 \end{vmatrix} = (-2y - 2y)\mathbf{k} = -4y\mathbf{k}$$

For the rectangle ABCD, $\mathbf{n} = \mathbf{k}$

$$\therefore \quad \text{curl } \mathbf{f} \cdot \mathbf{n} = -4y\, \mathbf{k} \cdot \mathbf{k} = -4y$$

$$\therefore \quad \iint_S \text{curl } \mathbf{f} \cdot \mathbf{n}\, dS = \int_{x=-a}^{a} \int_{y=0}^{b} -4y\, dx\, dy$$

$$= -4\left[x\right]_{-a}^{a} \cdot \left[\frac{y^2}{2}\right]_0^b = -4\,[a + a]\left[\frac{b^2}{2} - 0\right] = -4ab^2 \quad \text{...(2)}$$

(1) and (2) verify Stokes' theorem.

Example 34. *Verify Stoke's theorem for the function* $\mathbf{f} = (2x - y)\,\mathbf{i} - yz^2\mathbf{j} - y^2z\mathbf{k}$, *where S is the upper half surface of the sphere* $x^2 + y^2 + z^2 = 1$ *and C is its boundary.*

Sol. Clearly, C the boundary of the upper half of the sphere S is the circle $x^2 + y^2 = 1$ in the xy-plane. The parametric equations of C can be taken as $x = \cos t$, $y = \sin t$, $z = 0$, $0 \le t < 2\pi$.

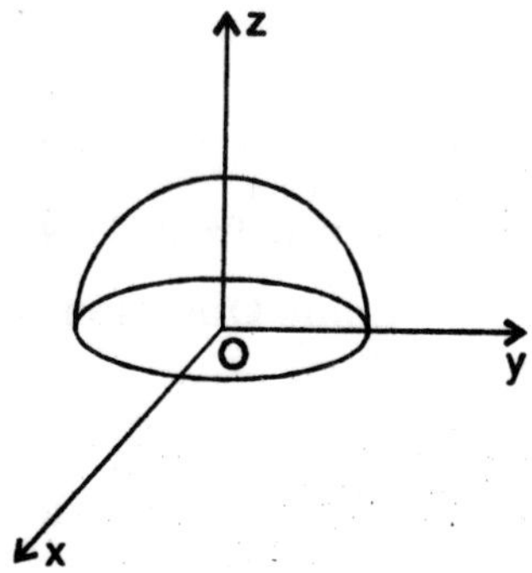

$$\therefore \quad \oint_C f \cdot dr = \oint_C (f_1\, dx + f_2\, dy + f_3\, dz)$$

$$= \oint_C ((2x - y)\, dx - yz^2\, dy - y^2z\, dz)$$

$$= \int_0^{2\pi} (2\cos t - \sin t)(-\sin t)\, dt$$

$$= \int_0^{2\pi} (-2\cos t \sin t + \sin^2 t)\, dt$$

$$= \left[\cos^2 t\right]_0^{2\pi} + 4\int_0^{\pi/2} \sin^2 t\, dt = 0 + 4 \cdot \frac{1}{2} \cdot \frac{\pi}{2} = \pi \quad \text{...(1)}$$

Also,
$$\text{curl } \mathbf{f} = \begin{vmatrix} \mathbf{i} & \mathbf{j} & \mathbf{k} \\ \frac{\partial}{\partial x} & \frac{\partial}{\partial y} & \frac{\partial}{\partial z} \\ 2x - y & -yz^2 & -y^2z \end{vmatrix} = \mathbf{i}(-2yz + 2yz) - \mathbf{j}(0 - 0) + \mathbf{k}(0 + 1) = \mathbf{k}$$

$$\therefore \quad \iint_S \text{curl } \mathbf{f} \cdot \mathbf{n} \, dS = \iint_S \mathbf{k} \cdot \mathbf{n} \, dS = \iint_R \mathbf{k} \cdot \mathbf{n} \frac{dxdy}{\mathbf{n} \cdot \mathbf{k}}$$

where R is the projection of S on xy-plane

$$= \iint_R dxdy = \text{area of region R}$$

$$= \pi(1)^2 = \pi \qquad \text{...(2)}$$

Equality of (1) and (2) verifies Stokes' theorem.

Remark. *Gauss' divergence theorem is a generalization of Stokes' theorem in plane (also known as Green's theorem) where the (plane) region R and its closed boundary (curve) C are replaced by a (space) region V and its closed boundary (surface) S. For this reason, the divergence theorem is often called Green's theorem in space.*

Example 35. *Evaluate by Stokes' theorem* $\oint_C (e^x \, dx + 2y \, dy - dz)$.

where C is the curve $x^2 + y^2 = 4$, $z = 2$. (K.U. 2002)

Sol. Here, $\mathbf{f} = e^x \, \mathbf{i} + 2y\mathbf{j} - \mathbf{k}$

$$\therefore \quad \oint_C \mathbf{f} \cdot d\mathbf{r} = \oint_C (f_1 \, dx + f_2 \, dy + f_3 \, dz)$$

$$= \int_C (e^x \, dx + 2y \, dy - dz)$$

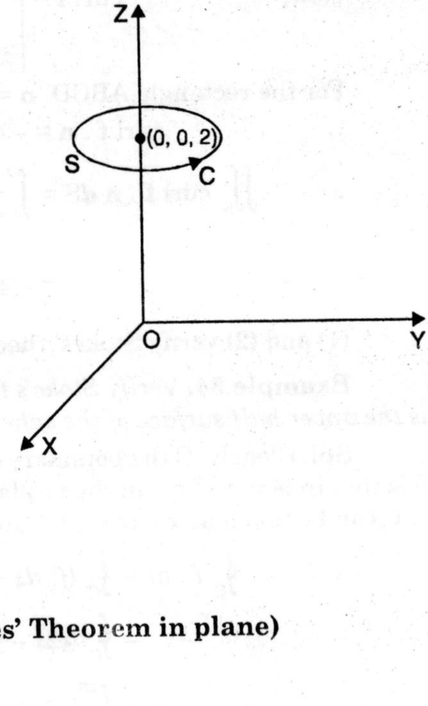

By Stokes' theorem,

$$\oint_C \mathbf{f} \cdot d\mathbf{r} = \iint_S \text{curl } \mathbf{f} \cdot \mathbf{n} \, dS \qquad \text{...(1)}$$

where S is the surface whose boundary C is given by the circle $x^2 + y^2 = 4$, $z = 2$. This circle has centre (0, 0, 2) and radius 2. Also, $\mathbf{n} = \mathbf{k}$.

Now,
$$\text{curl } \mathbf{f} = \begin{vmatrix} \mathbf{i} & \mathbf{j} & \mathbf{k} \\ \frac{\partial}{\partial x} & \frac{\partial}{\partial y} & \frac{\partial}{\partial z} \\ e^x & 2y & -1 \end{vmatrix} = \mathbf{0}$$

$\Rightarrow$ $\quad$ curl $\mathbf{f} \cdot \mathbf{n} = 0$.

Hence, from (1), $\oint_C \mathbf{f} \cdot d\mathbf{r} = 0$.

Some Questions of Green's theorem (or Stokes' Theorem in plane)

$$\oint_C (f_1 dx + f_2 dy) = \iint_S \left(\frac{\partial f_2}{\partial x} - \frac{\partial f_1}{\partial y} \right) dx \, dy$$

where $f = f_1 \mathbf{i} + f_2 \mathbf{j}$, $\mathbf{n} = \mathbf{k}$ and $dS = dx \, dy$.

Example 36. *Verify Green's theorem in the plane for* $\oint_C (xy + y^2) \, dx + x^2 \, dy$

where C is the closed curve of the region bounded by $y = x$ *and* $y = x^2$. (K.U. 2006, 2003, 2003 S)

Sol. $y = x^2$ is a parabola and $y = x$ is a straight line through the origin both intersecting at the points (0, 0) and (1, 1).

The positive direction in traversing C is as shown in the diagram.

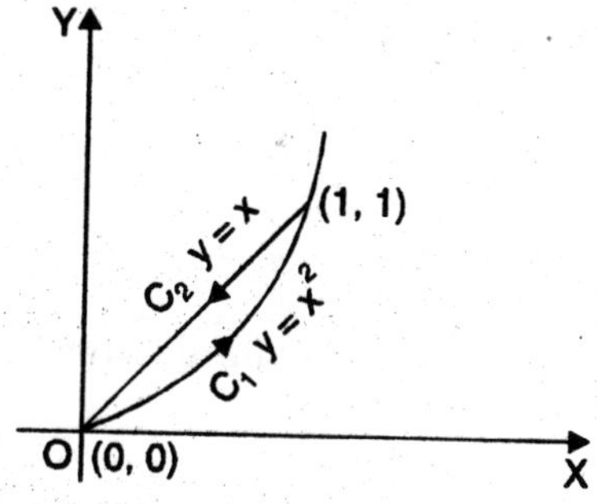

Line integral along $C_1 : y = x^2$

Here $dy = 2x\, dx$ and x varies from 0 to 1.

$$\therefore \int_{C_1} (xy + y^2)\, dx + x^2\, dy$$

$$= \int_0^1 (x \,.\, x^2 + x^4)\, dx + x^2 \,.\, 2x\, dx$$

$$= \int_0^1 (3x^3 + x^4)\, dx$$

$$= \left[\frac{3x^3}{4} + \frac{x^5}{5}\right]_0^1 = \frac{19}{20} \qquad \ldots(i)$$

Line integral along $C_2 : y = x$

Here $dy = dx$ and x varies from 1 to 0

$$\therefore \int_{C_2} (xy + y^2) dx + x^2\, dy = \int_1^0 (x \,.\, x + x^2)\, dx + x^2\, dx$$

$$= \int_1^0 3x^2\, dx = \left[x^3\right]_1^0 = -1 \qquad \ldots(ii)$$

Adding (*i*) and (*ii*), we have

$$\oint_C (xy + y^2)\, dx + x^2\, dy = \int_{C_1} + \int_{C_2} = \frac{19}{20} - 1 = -\frac{1}{20} \qquad \ldots(1)$$

Again, $\iint_S \left(\frac{\partial f_2}{\partial x} - \frac{\partial f_1}{\partial y}\right) dx\, dy$

$$= \iint_S \left(\frac{\partial}{\partial x}(x^2) - \frac{\partial}{\partial y}(xy + y^2)\right) dx\, dy$$

$$= \int_{x=0}^1 \int_{y=x^2}^x [2x - (x + 2y)]\, dy\, dx$$

$$= \int_{x=0}^1 \left(\int_{y=x^2}^x (x - 2y)\, dy\right) dx$$

$$= \int_0^1 \left[xy - y^2\right]_{y=x^2}^x . \, dx = \int_0^1 [x(x - x^2) + (x^2 - x^4)]\, dx$$

$$= \int_0^1 (x^4 - x^3)\, dx = -\frac{1}{20} \qquad \ldots(2)$$

Hence, from (1) and (2), Green's theorem is verified.

Example 37. *Show that the area bounded by a simple closed curve C is given by*

$$\frac{1}{2}\left(\oint xdy - ydx\right).$$

Hence, find the area of the ellipse

$$x = a \cos \theta, y = b \sin \theta \left(or\ \frac{x^2}{a^2} + \frac{y^2}{b^2} = 1\right). \qquad \text{(K.U. 2003)}$$

Sol. By Green's theorem,

$$\oint_C f_1 dx + f_2 dy = \iint_S \left(\frac{\partial f_2}{\partial x} - \frac{\partial f_1}{\partial y}\right) dx\, dy$$

Putting $f_1 = -y,\ f_2 = x$, we have

$$\oint_C x\, dy - y\, dx = \iint_S \left(\frac{\partial}{\partial x}(x) - \frac{\partial}{\partial y}(-y)\right) dx\, dy = 2\iint_S dx\, dy = 2A$$

where A is the required area.

Thus, $$A = \frac{1}{2}\left(\oint_C xdy - ydx\right)$$

Area of ellipse $$= \frac{1}{2}\left(\oint_C xdy - ydx\right)$$

$$= \frac{1}{2}\int_0^{2\pi} (a\cos\theta)(b\cos\theta)\, d\theta - (b\sin\theta)(-a\sin\theta)\, d\theta$$

$$= \frac{1}{2}\int_0^{2\pi} ab(\cos^2\theta + \sin^2\theta)\, d\theta = \frac{1}{2}\int_0^{2\pi} ab\, d\theta = \pi ab.$$

Example 38. *Evaluate* $\oint_C \mathbf{f}\cdot d\mathbf{r}$ *by Stokes' theorem where* $\mathbf{f} = y^2\mathbf{i} + x^2\mathbf{j} - (x+z)\mathbf{k}$ *and C is the boundary of the triangle with vertices at (0, 0, 0), (1, 0, 0), (1, 1, 0).*

Sol. We have $$\text{curl } \mathbf{f} = \begin{vmatrix} \mathbf{i} & \mathbf{j} & \mathbf{k} \\ \frac{\partial}{\partial x} & \frac{\partial}{\partial y} & \frac{\partial}{\partial z} \\ y^2 & x^2 & -(x+z) \end{vmatrix} = 0\mathbf{i} + \mathbf{j} + 2(x-y)\,\mathbf{k}$$

We see that z-coordinate of each vertex of triangle is zero. Therefore, the triangle lies in xy-plane.

$\therefore$ $$\mathbf{n} = \mathbf{k}$$

$\therefore$ $$\text{curl } \mathbf{f}\cdot\mathbf{n} = [\mathbf{j} + 2(x-y)\mathbf{k}]\cdot\mathbf{k} = 2(x-y)$$

In the figure, only xy-plane is considered.

Equation of line OB is $y = x$.

$\therefore$ By Stokes' theorem,

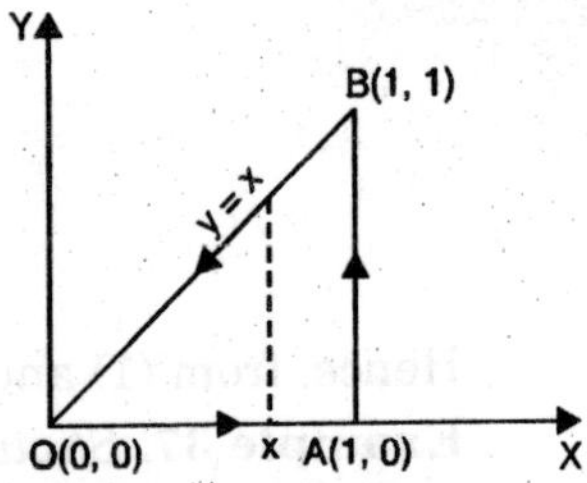

$$\oint_C \mathbf{f}\cdot d\mathbf{r} = \iint_S (\text{curl } \mathbf{f}\cdot\mathbf{n})\, ds$$

$$= \int_{x=0}^{1}\int_{y=0}^{x} 2(x-y)\, dx\, dy$$

$$= 2\int_{x=0}^{1}\left[xy - \frac{y^2}{2}\right]_{y=0}^{x} dx$$

$$= 2\int_{x=0}^{1}\left[x^2 - \frac{x^2}{2}\right] dx = 2\int_0^1 \frac{x^2}{2}\, dx = \frac{1}{3}.$$

Example 39. *Verify Green's theorem in the plane for*

$$\oint_C (3x^2 - 8y^2)\, dx + (4y - 6xy)\, dy$$

where C is the boundary of the region bounded by the parabolas :

$$y = \sqrt{x} \text{ and } y = x^2. \qquad \text{(M.D.U. 2007)}$$

Sol. The parabolas $y = \sqrt{x}$ *i.e.*, $y^2 = x$ and $y = x^2$ intersect at O(0, 0) and A(1, 1).

Here, $f_1 = 3x^2 - 8y^2,\ f_2 = 4y - 6xy$

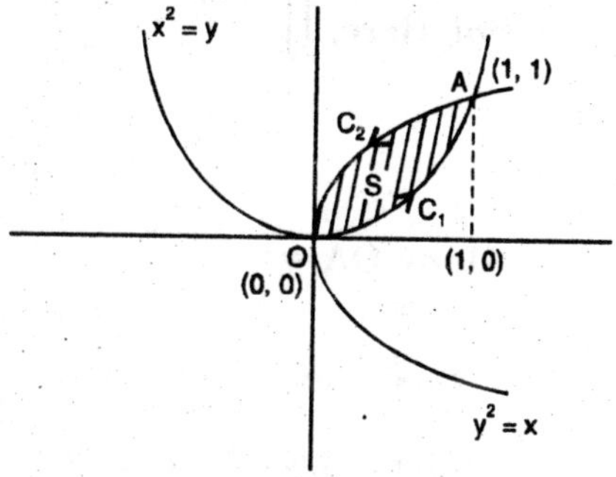

$$\therefore \quad \frac{\partial f_1}{\partial y} = -16y,\ \frac{\partial f_2}{\partial x} = -6y$$

$$\therefore \quad \frac{\partial f_2}{\partial x} - \frac{\partial f_1}{\partial y} = 10y$$

By Green's theorem,

$$\oint_C (f_1\, dx + f_2\, dy) = \iint_S \left(\frac{\partial f_2}{\partial x} - \frac{\partial f_1}{\partial y}\right) dx\, dy$$

where S is the region bounded by the curve C.

$$\text{R.H.S.} = \iint_S \left(\frac{\partial f_2}{\partial x} - \frac{\partial f_1}{\partial y}\right) dx\, dy$$

$$= \int_{x=0}^{1} \left(\int_{y=x^2}^{\sqrt{x}} 10y\, dy\right) dx = \int_0^1 5\left[y^2\right]_{x^2}^{\sqrt{x}} dx$$

$$= 5\int_0^1 (x - x^4)\, dx = 5\left[\frac{x^2}{2} - \frac{x^5}{5}\right]_0^1$$

$$= 5\left(\frac{1}{2} - \frac{1}{5}\right) = 5\left(\frac{3}{10}\right) = \frac{3}{2} \qquad \text{...(1)}$$

$$\text{L.H.S.} = \oint_C (f_1\, dx + f_2\, dy) = \int_{C_1} (f_1\, dx + f_2\, dy) + \int_{C_2} (f_1\, dx + f_2\, dy)$$

Consider $\int_{C_1} (f_1\, dx + f_2\, dy)$ ...(2)

Along C_1, $x^2 = y \Rightarrow 2x\, dx = dy$. x varies from 0 to 1.

$$\therefore \quad \int_{C_1} (f_1\, dx + f_2\, dy) = \int_0^1 (3x^2 - 8x^4)\, dx + (4x^2 - 6x \,.\, x^2) \,.\, 2x\, dx$$

$$= \int_0^1 (3x^2 + 8x^3 - 20x^4)\, dx = \left[x^3 + 2x^4 - 4x^5\right]_0^1 = -1 \qquad \text{...}(i)$$

Along C_2, $y^2 = x \Rightarrow 2y\, dy = dx$ and y varies from 1 to 0.

$$\therefore \quad \int_{C_2} (f_1\, dx + f_2\, dy) = \int_1^0 (3y^4 - 8y^2)\, 2y\, dy + (4y - 6y^2 \,.\, y)\, dy$$

$$= \int_1^0 (4y - 22y^3 + 6y^5)\, dy = \left[2y^2 - \frac{11}{2}y^4 + y^6\right]_1^0$$

$$= \frac{5}{2} \qquad \text{...}(ii)$$

Using (*i*) and (*ii*) in (2), L.H.S. $= -1 + \dfrac{5}{2} = \dfrac{3}{2}$ ...(3)

The equality of (1) and (3) verifies Green's theorem.

Example 40. *Verify Green's theorem in plane for* $\oint_C (3x^2 - 8y^2)\, dx + (4y - 6xy)\, dy$ *where C is the boundary of the region bounded by* $x = 0, y = 0, x + y = 1$.

Sol. Here, $\iint_S \left(\dfrac{\partial f_2}{\partial x} - \dfrac{\partial f_1}{\partial y}\right) dx\, dy = \int_0^1 \int_0^{1-x} 10y\, dy\, dx$

$$= \int_0^1 5\left[y^2\right]_0^{1-x} dx = 5\int_0^1 (1-x)^2\, dx = 5\left[\frac{(1-x)^3}{-3}\right]_0^1 = \frac{5}{3} \qquad ...(1)$$

Along OA, $y = 0 \Rightarrow$ $dy = 0$ and x varies from 0 to 1

$\therefore$ Line integral along OA $= \int_0^1 3x^2\, dx = \left[x^3\right]_0^1 = 1$

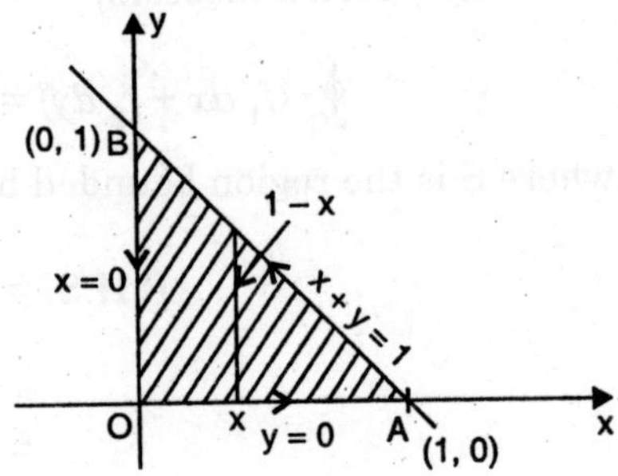

Along AB, $y = 1 - x \Rightarrow dy = -dx$. Also, x varies from 1 to 0

$\therefore$ Line integral along AB

$$= \int_1^0 [3x^2 - 8(1-x)^2]\, dx + [4(1-x) - 6x(1-x)](-dx)$$

$$= \int_1^0 (3x^2 - 8 + 16x - 8x^2 - 4 + 4x + 6x - 6x^2)\, dx$$

$$= \int_1^0 (-12 + 26x - 11x^2)\, dx = \left[-12x + 13x^2 - 11\frac{x^3}{3}\right]_1^0 = \frac{8}{3}.$$

Along BO, $x = 0 \Rightarrow$ $dx = 0$ and y varies from 1 to 0.

$\therefore$ Line integral along BO $= \int_1^0 4y\, dy = \left[2y^2\right]_1^0 = -2$

$\therefore$ Line integral along the curve C (*i.e.*, along OABO)

$$= 1 + \frac{8}{3} - 2 = \frac{5}{3}$$

i.e., $$\oint_C (f_2\, dx + f_1\, dy) = \frac{5}{3} \qquad ...(2)$$

The equality of (1) and (2) verifies Green's theorem.

Example 41. *Using Green's theorem in plane, evaluate*

$$\oint_C [(2x^2 - y^2)dx + (x^2 + y^2)dy],$$

where C is the boundary in the xy-plane of the area enclosed by the x-axis and the semi-circle $x^2 + y^2 = 1$ *in the upper half of xy-plane.*

Sol. By Green's theorem,

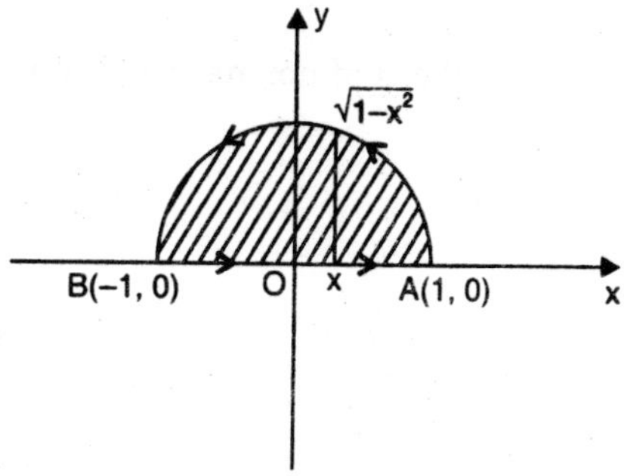

$$\oint_C (f_1\, dx + f_2\, dy) = \iint_S \left(\frac{\partial f_2}{\partial x} - \frac{\partial f_1}{\partial y}\right) dx\, dy \qquad ...(1)$$

Here, $f_1 = 2x^2 - y^2,\ f_2 = x^2 + y^2$

$\therefore$ $$\frac{\partial f_2}{\partial x} - \frac{\partial f_1}{\partial y} = 2x - (-2y) = 2(x + y)$$

Region S is bounded by :

$$x = -1,\ x = 1,\ y = 0,\ y = \sqrt{1 - x^2}$$

$\therefore$ From (1),

$$= \iint_S 2(x + y)\, dx\, dy = \int_{-1}^{1} \int_0^{\sqrt{1-x^2}} 2(x + y)\, dy\, dx$$

$$= 2\int_{-1}^{1} \left[xy + \frac{y^2}{2}\right]_0^{\sqrt{1-x^2}} dx$$

$$= 2\int_{-1}^{1} \left[x\sqrt{1-x^2} + \frac{1}{2}(1 - x^2)\right] dx$$

$$= \int_{-1}^{1} (1 - x^2)\, dx \qquad \left[\because \int_{-1}^{1} x\sqrt{1-x^2}\, dx = 0\right]$$

$$= \frac{4}{3}.$$

Example 42. *Apply Stokes' Theorem to evaluate* $\int_C (ydx + zdy + xdz)$, *where C is the curve of intersection of the sphere* $x^2 + y^2 + z^2 = a^2$ *and the plane* $x + z = a$.

Sol. The curve C in fact is a circle in the plane $x + z = a$ having $A(a, 0, 0)$, $B(0, 0, a)$ as the extremities of the diameter.

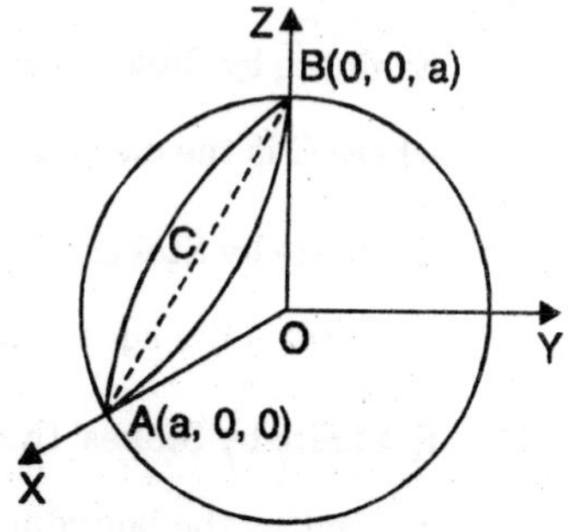

$$\therefore \int_C (ydx + zdy + xdz) = \int_C (y\mathbf{i} + z\mathbf{j} + x\mathbf{k}) \cdot d\mathbf{r}$$

$$= \int_C \mathbf{f} \cdot d\mathbf{r}, \text{ where } \mathbf{f} = y\mathbf{i} + z\mathbf{j} + x\mathbf{k}$$

$$= \iint_S \text{curl}\, \mathbf{f} \cdot \mathbf{n}\, ds$$

where S is the circle on AB as diameter

$$\left[\text{Now, curl } \mathbf{f} = \begin{vmatrix} \mathbf{i} & \mathbf{j} & \mathbf{k} \\ \frac{\partial}{\partial x} & \frac{\partial}{\partial y} & \frac{\partial}{\partial z} \\ y & z & x \end{vmatrix} = -\mathbf{i} - \mathbf{j} - \mathbf{k}\right.$$

Direction ratios of normal to the plane $x + z = a$ are $< 1, 0, 1 >$.

$\therefore$ d.c.'s of normal to plane are $< \frac{1}{\sqrt{2}}, 0, \frac{1}{\sqrt{2}} >$

$$\therefore \qquad \mathbf{n} = \frac{1}{\sqrt{2}}\mathbf{i} + \frac{1}{\sqrt{2}}\mathbf{k}\Bigg]$$

$$= \iint_S -(\mathbf{i}+\mathbf{j}+\mathbf{k}) \cdot \left(\frac{\mathbf{i}+\mathbf{k}}{\sqrt{2}}\right) ds$$

$$= -\frac{2}{\sqrt{2}}\iint_S ds = -\sqrt{2} \cdot \pi\left(\frac{2}{\sqrt{2}}\right)^2$$

$$\left[\because \text{ area of surface} = \text{area of circle with radius } \frac{a}{\sqrt{2}} = \frac{AB}{2}\right]$$

$$= -\frac{\pi a^2}{\sqrt{2}}.$$

EXERCISE 6

1. Verify Stokes' theorem for the function $\mathbf{f} = z\mathbf{i} + x\mathbf{j} + y\mathbf{k}$ where the curve is the unit circle in the *xy*-plane bounding the hemi-sphere $z = \sqrt{1 - x^2 - y^2}$.
2. Verify Stokes' theorem for $\mathbf{f} = y\mathbf{i} + 2\mathbf{j} + x\mathbf{k}$, where S is the upper half surface of the sphere $x^2 + y^2 + z^2 = 1$ and C is its boundary. *(M.D.U. 2007)*
3. Using Stokes' theorem, evaluate $\oint_C (xy\,dx + xy^2\,dy)$ taken round the positively oriented square C with vertices (1, 0), (0, – 1), (– 1, 0) and (0, 1).
4. Using Stokes' theorem, prove that $\oint_C \mathbf{r} \cdot d\mathbf{r} = 0$.
   [**Hint.** Curl $\mathbf{r} = \mathbf{0}$.]
5. Evaluate by Stokes' theorem $\oint_C (yz\,dx + xz\,dy + xy\,dz)$ where C is the curve $x^2 + y^2 = 1$, $x = y^2$.
6. Evaluate by Stokes' theorem $\oint_C xy\,dx + xy^2\,dy$ where C is the square in the *xy*-plane with vertices (1, 0), (– 1, 0), (0, 1), (0, – 1). *(K.U. 2001 S)*
7. Evaluate by Stokes' theorem $\oint_C (\sin z\,dx - \cos x\,dy + \sin y\,dz)$ where C is the boundary of the rectangle $0 \le x \le \pi$, $0 \le y \le 1$, $z = 3$.
8. Verify Green's theorem is the *xy*-plane for $\oint_C (xy^2 - 2xy)\,dx + (x^2y + 3)\,dy$ around the boundary C of the region enclosed by $y^2 = 8x$ and $x = 2$.

9. Evaluate by Green's theorem $\oint_C (x^2 - \cosh y)\, dx + (y + \sin x)\, dy$

where C is the rectangle with vertices (0, 0), (π, 0), (π, 1), (0, 1).

10. Evaluate $\oint_C (y - \sin x)\, dx + \cos x\, dy$, where C is the triangle having vertices (0, 0), $\left(\frac{\pi}{2}, 0\right)$ and $\left(\frac{\pi}{2}, 1\right)$

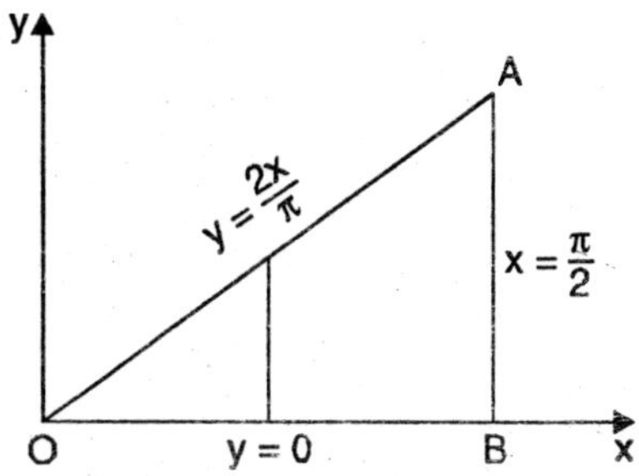

(*i*) directly

(*ii*) by using Green's theorem in the plane.

$$\left[\textbf{Hint. } \oint_C (f_1\, dx + f_2\, dy) = \iint_S \left(\frac{\partial f_2}{\partial x} - \frac{\partial f_1}{\partial y}\right) dy\, dx = -\int_0^{\pi/2} \int_0^{2x/\pi} (1 + \sin x)\, dy\, dx.\right]$$

11. Evaluate by the Green's theorem in space (*i.e.*, divergence theorem)

$$\iint_S 4xzdydz - y^2 dz\, dx + yz\, dx\, dy$$

where S is the surface of the cube bounded by the planes :

$$x = 0,\ y = 0,\ z = 0,\ x = 1,\ y = 1,\ z = 1.$$

Answers

1. $-\pi$	**2.** $-\pi$	**5.** 0	**6.** $\frac{4}{3}$
7. 2	**8.** $\frac{128}{5}$	**9.** $\pi(\cosh 1 - 1)$	**10.** $-\frac{2}{\pi} - \frac{\pi}{4}$.

PART B : GEOMETRY

SECTION III

(Students should note that Reviews III, IV and Chapters 10, 11, 12, are in Section III in K.U. Syllabus but are in Section IV in M.D.U. Syllabus)

Review I

Change of Origin

1. Change of Origin

To change the origin without changing the direction of axes.*

Let P(x, y) be any point referred to OX, OY called as the old axes. Let O′ be the new origin and O′X′, O′Y′ the new axes parallel to the old axes.

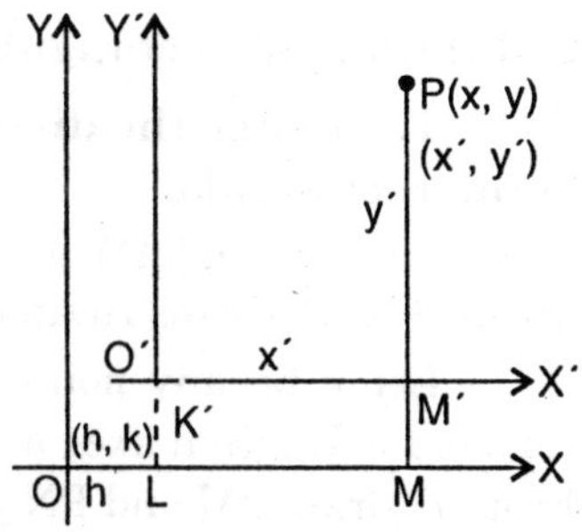

Let the co-ordinates of O′ referred to the old axes be (h, k), so that

$$OL = h, LO' = k$$

Let (x', y') be the co-ordinates of P referred to the new axes. Draw PM ⊥ on OX meeting O′X′ in M′

Then $OM = x, MP = y\ ;\ O'M' = x', M'P = y'$.

Now $x = OM = OL + LM = OL + O'M'$

$$= h + x', \qquad \ldots(1)$$

$$y = MP = MM' + M'P = LO' + M'P$$

$$= k + y'. \qquad \ldots(2)$$

Hence $\left.\begin{array}{l}\mathbf{x = h + x'}\\ \mathbf{y = k + y'}\end{array}\right]$ are the old co-ordinates in terms of the new co-ordinates x' and y'.

Cor. *To express the new co-ordinates in terms of old co-ordinates.*

From (1) and (2), we have $x = h + x', y = k + y'$

$$\mathbf{x' = x - h, y' = y - k.}$$

1. Rule to find the equation of a curve referred to the new axes when its equation is given referred to old axes.

(*i*) *Write $x' + h$ for x and $y' + k$ for y in the given equation, where (h, k) are the co-ordinates of the new origin referred to old axes.*

(*ii*) *In the resulting equation drop the dashes i.e., change x', y' to x, y.*

2. Rule to find the equation of a curve referred to the old axes when its equation is given referred to new axes.

(*i*) *In the given equation change x, y to x', y',*

(*ii*) *Now put $x' = x - h$, and $y' = y - k$.*

The resulting equation is the required equation.

---

*Such a change of axes is also known as Translation of axes.

Example 1. *On shifting the origin to the point (1, – 1), the axes remaining parallel to the original axes, the equation of a curve becomes* $4x^2 + y^2 + 3x - 4y + 2 = 0$.

Find the original equation.

Sol. The given equation is $4x^2 + y^2 + 3x - 4y + 2 = 0$ ...(1)

The new origin is (1, – 1)

Changing x, y to x', y' in (1), we have

$$4x'^2 + y'^2 + 3x' - 4y' + 2 = 0. \quad ...(2)$$

Putting $x' = x - 1$, $y' = y - (-1) = y + 1$ in (2), we have

$$4(x-1)^2 + (y+1)^2 + 3(x-1) - 4(y+1) + 2 = 0$$

i.e.,

$$4x^2 + y^2 - 5x - 2y = 0$$

which is the required equation.

2. Rotation of Rectangular Axes

To change the direction of axes without changing the origin, both the systems being rectangular

Let OX and OY be the original axes and OX′ and OY′ be the new axes rotated through an angle 'θ'.

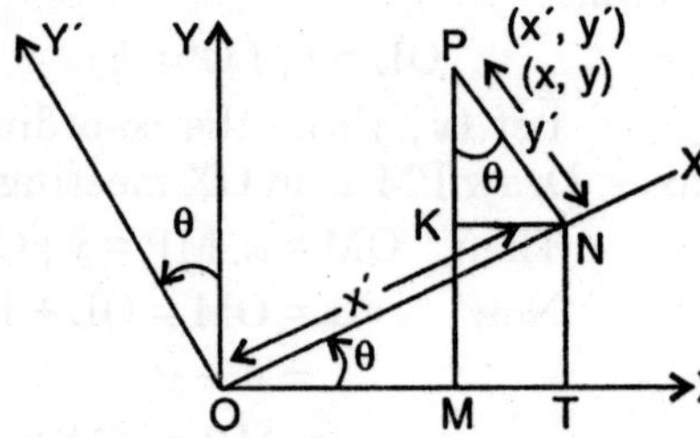

Let P be any point whose co-ordinates are (x, y) referred to original axes and (x', y') referred to new axes. From P, draw PM and PN perpendiculars on OX and OX′ respectively and from N, draw NT and NK perpendiculars on OX and MP respectively, then ∠KPN = θ. From the figure, we have

$$x = OM = OT - MT = OT - KN$$

$$= ON \cos\theta - NP \sin\theta$$

∴ $$\mathbf{x = x' \cos\theta - y' \sin\theta} \quad ...(1)$$

Again $$y = MP = MK + KP = TN + KP = ON \sin\theta + NP \cos\theta$$

∴ $$\mathbf{y = x' \sin\theta + y' \cos\theta} \quad ...(2)$$

Equations (1) and (2) are the equations of transformation from old to new system.

Cor. *Solving equations (1) and (2) for x' and y', we have*

$$x' = x \cos\theta + y \sin\theta \quad ...(3)$$

$$y' = -x \sin\theta + y \cos\theta \quad ...(4)$$

Equations (3) and (4) are the equations of transformation from the new to old system.

Remember. *The following table helps us to remember the above formulae :*

(*i*) *To get* x, multiply elements in the x-row by the corresponding elements at the top and add the products so formed. Similarly, for y.

(*ii*) *To get* x', multiply the elements in its column by the corresponding elements of the first column and add the products so formed. Similarly, for y'.

NEW → / OLD ↓	x′	y′
x	cos θ	– sin θ
y	sin θ	cos θ

Note. *When the origin is shifted to the point (h, k) and the axes are rotated through an angle θ, then we can perform the change of axes either one by one or simultaneously by using the equations*

$$x = h + x' \cos\theta - y' \sin\theta$$
$$y = k + x' \sin\theta + y' \cos\theta.$$

Example 1. *Find the transformed equation of $17x^2 - 16xy + 17y^2 - 225 = 0$ when the axes are rotated through 45°.*

Sol. The given equation is $\quad 17x^2 - 16xy - 17y^2 - 225 = 0 \quad$ ...(1)

Equations of transformation are

$$x = x' \cos 45° - y' \sin 45° = \frac{x' - y'}{\sqrt{2}}$$
$$y = x' \sin 45° + y' \cos 45° = \frac{x' + y'}{\sqrt{2}}$$

Putting these values of x and y, we have

$$17\left(\frac{x'-y'}{\sqrt{2}}\right)^2 - 16\left(\frac{x'-y'}{\sqrt{2}}\right)\left(\frac{x'+y'}{\sqrt{2}}\right) + 17\left(\frac{x'+y'}{\sqrt{2}}\right)^2 - 225 = 0$$

or $$\frac{17}{2}(x'-y')^2 - \frac{16}{2}(x'^2 - y'^2) + \frac{17}{2}(x'+y')^2 - 225 = 0$$

or $$9x'^2 + 25y'^2 = 225.$$

or suppressing the dashes, we get the required transformed equation as

$$9x^2 + 25y^2 = 225.$$

Example 2. *Show that it is always possible to rotate the axes through a suitable angle so as to remove the term of xy from the equation $ax^2 + 2hxy + by^2 = 1$.*

Sol. The given equation is $ax^2 + 2hxy + by^2 = 1$. ...(1)

Let us rotate the axes through an angle θ the equations of transformation are

$$x = x' \cos\theta - y' \sin\theta \quad ...(2)$$
$$y = x' \sin\theta + y' \cos\theta \quad ...(3)$$

Using (2) and (3), equation (1) becomes

$$a(x' \cos\theta - y' \sin\theta)^2 + 2h(x' \cos\theta - y' \sin\theta)(x' \sin\theta + y' \cos\theta) + b(x' \sin\theta + y' \cos\theta)^2 = 1 \quad ...(4)$$

To remove the term of xy, (or of $x'y'$),

Put the co-efficient of $x'y'$ from (4) = 0

i.e., $$2(b - a)\sin\theta\cos\theta + 2h(\cos^2\theta - \sin^2\theta) = 0$$

or $$(b - a)\sin 2\theta + 2h\cos 2\theta = 0$$

or $$\tan 2\theta = \frac{2h}{a-b} \quad \text{or} \quad \theta = \frac{1}{2}\tan^{-1}\left(\frac{2h}{a-b}\right). \quad ...(5)$$

Now whatever the real values of a, b and h may be, it is always possible to find a value of θ lying between $-\pi/4$ and $\pi/4$ so as to satisfy equation (5).

Substituting this value of θ in (4), the resulting equation will be free from the term of $x'y'$ and therefore, after suppressing the dashes, the equation will not contain any term of xy.

Example 3. *Transform the equation $14x^2 - 4xy + 11y^2 - 44x - 58y + 71 = 0$ to rectangular axes through the point (2, 3) and inclined at an angle $\tan^{-1}(2)$.*

Sol. The given equation is

$$14x^2 - 4xy + 11y^2 - 44x - 58y + 71 = 0 \quad ...(1)$$

Shift the origin to (2, 3) by replacing x by $x + 2$ and y by $y + 3$.

$\therefore$ (1) becomes, $14(x + 2)^2 - 4(x + 2)(y + 3) + 11(y + 3)^2 - 44(x + 2) - 58(y + 3) + 71 = 0$

or
$$14x^2 - 4xy + 11y^2 - 60 = 0 \quad ...(2)$$

Rotate the axes through an angle $\theta = \tan^{-1}(2)$.

Since, $\tan\theta = 2 \quad \therefore \quad \sin\theta = \frac{2}{\sqrt{5}}$ and $\cos\theta = \frac{1}{\sqrt{5}}$.

$\therefore$ Equations of transformation are

$$x = x'\cos\theta - y'\sin\theta = \frac{x' - 2y'}{\sqrt{5}}$$

$$y = x'\sin\theta + y'\cos\theta = \frac{2x' + y'}{\sqrt{5}}$$

Substituting in (2), we have

$$14\left(\frac{x' - 2y'}{\sqrt{5}}\right)^2 - 4\left(\frac{x' - 2y'}{\sqrt{5}}\right)\left(\frac{2x' + y'}{\sqrt{5}}\right) + 11\left(\frac{2x' + y'}{\sqrt{5}}\right)^2 - 60 = 0$$

or
$$14(x'^2 - 4x'y' + 4y'^2) - 4(2x'^2 - 3x'y' - 2y'^2) + 11(4x'^2 + 4x'y' + y'^2) - 300 = 0$$

or
$$50x'^2 + 75y'^2 = 300 \quad \text{or} \quad 2x'^2 + 3y'^2 = 12$$

$\therefore$ Suppressing the dashes, the required equation is

$$2x^2 + 3y^2 = 12.$$

Review II

1. Standard Results of The Parabola

1. Conic

Def. *The locus of a point which moves in a plane so that its distance from a fixed point bears a constant ratio to its distance from a fixed straight line is called a* **conic.**

The fixed point is called the **focus** *and is generally denoted by S.*

The fixed straight line is called the **directrix.** The constant ratio is called the **eccentricity** *and is denoted by the letter* **e.**

The straight line passing through the focus and perpendicular to the directrix is called **the axis of the conic.**

The point where the axis meets the conic is called the **vertex** of the conic.

The conic is called **parabola** if $e = 1$

The conic is called **ellipse** if $e < 1$

The conic is called **hyperbola** if $e > 1$.

Note. Conic is a brief name of conic section.

2. Parabola

Def. *A parabola is the locus of a point which moves in a plane so that its distance from a fixed point is equal to its distance from a fixed straight line.*

The fixed point is called the focus and the fixed straight line is called the directrix of the parabola.

EQUATION OF A PARABOLA IN THE STANDARD FORM

3. To find the equation of a parabola in the form $y^2 = 4ax$

Let S be the focus and ZM the directrix of the parabola.

Draw SZ ⊥ on ZM and bisect SZ at A. Let ZS = $2a$ so that

$$ZA = AS = a.$$

By definition, A lies on the parabola.

Take ZS produced as the x-axis, A as origin and AY, a line through A ⊥ to ZX as the y-axis.

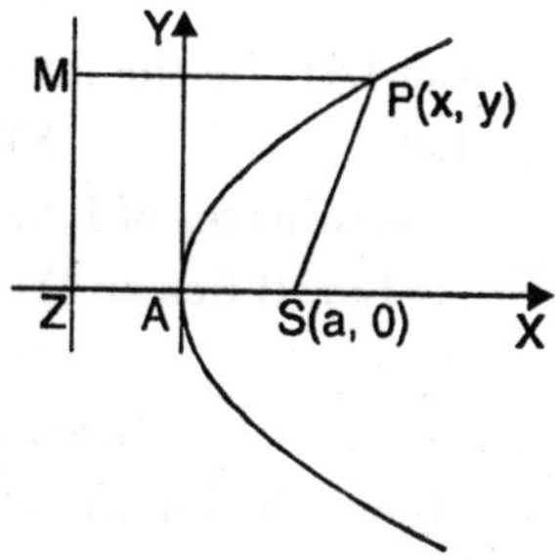

∴ Co-ordinates of S are $(a, 0)$. ZM is a line || to y-axis at a distance a to its left.

∴ It equation is $x = -a$ or $x + a = 0$ ...(1)

Let P(x, y) be any point on the parabola.

Join SP. Draw PM ⊥ on ZM.

Then by definition, SP = PM ...(2)

Now $\quad SP = \sqrt{(x-a)^2 + (y-0)^2} = \sqrt{(x-a)^2 + y^2}$

and $\quad$ PM = length of the perpendicular from (x, y) on (1)

$$= \frac{x+a}{1} = x + a$$

Hence from (2), $\sqrt{(x-a)^2 + y^2} = (x + a)$

$$x^2 - 2ax + a^2 + y^2 = x^2 + 2ax + a^2 \qquad \text{(Squaring both sides)}$$

or $$\mathbf{y^2 = 4ax},$$

which is the required equation of the parabola.

4. Geometrical Interpretation of the Equation $y^2 = 4ax$

Let P(x, y) be any point on the parabola $y^2 = 4ax$.

Draw PN ⊥ from P on the axis of the parabola. Let S$(a, 0)$ be the focus of the parabola.

Then geometrically the equation

$$y^2 = 4ax \text{ is}$$

$$(PN)^2 = 4 \,.\, AS \,.\, AN$$

where A is the vertex of the parabola.

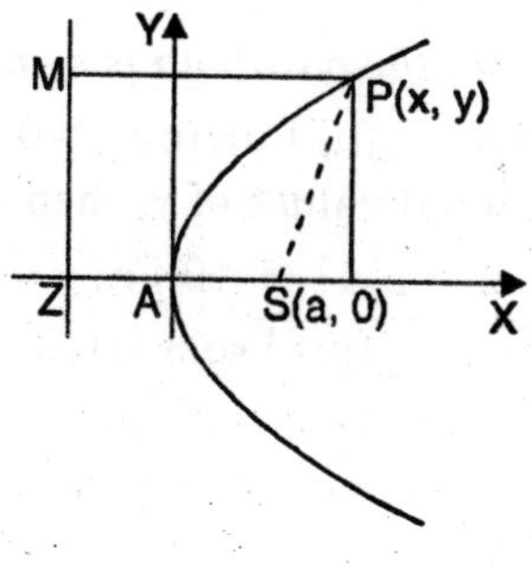

5. Focal Distance

Def. *The distance of any point on the parabola from the focus is called the* **focal distance** *of the point.*

6. To find the focal distance of any point P(x, y) on the parabola $y^2 = 4ax$.

Proof. The focal distance of the point $P(x, y) = SP$

$$= PM = ZN = ZA + AN = a + x$$

Hence focal distance of $P(x, y)$ *is* $= x + a$.

7. Latus Rectum

Def. *The latus rectum of a parabola is the chord drawn through the focus perpendicular to the axis of the parabola.*

8. To find the length of the latus rectum of the parabola

$$y^2 = 4ax$$

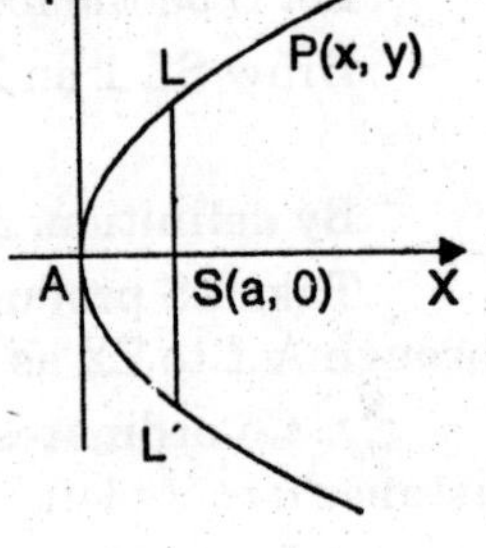

Let LSL′ be the latus rectum of the parabola $y^2 = 4ax$

Let $SL = l$

Co-ordinates of L are (a, l).

As $L(a, l)$ lies on (1)

$\therefore$ $l^2 = 4a^2$ or $l = \pm 2a$

$\therefore$ $LSL' = 2SL = 2(2a) = 4a$.

Hence the length of the latus rectum = **4a**.

Cor. 1. As the latus rectum is a straight line parallel to the *y*-axis at a distance '*a*' from it.

$\therefore$ Its equation is **x = a**.

Cor. 2. Ends of the latus rectum are $L(a, 2a)$ and $L'(a, -2a)$.

Note. Two parabolas are said to be equal when the length of latus rectum of each parabola is same.

9. Tracing the Parabola $y^2 = 4ax$

(*i*) The equation of curve is $y^2 = 4ax$ ...(1)

From this equation, $y = \pm 2\sqrt{a}\sqrt{x}$. ...(2)

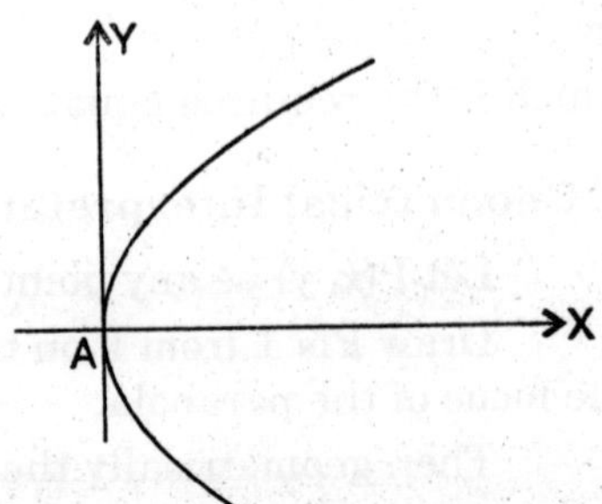

From equation (2), it is clear that for every positive value of *x* there are two equal and opposite values of *y*. Therefore, if the point (x, y) be on the curve then the point $(x, -y)$ also lies on the curve.

i.e., the parabola is symmetrical about its axis.

(*ii*) Putting $x = 0$ in (1), we get $y^2 = 0$, *i.e.*, there are two equal values of *y*, each equal to zero.

$\therefore$ $x = 0$ *i.e., y-axis touches the parabola at the origin.*

(*iii*) From (2), if *x* is negative, *y* is imaginary.

Hence no part of parabola (1) lies to the left of *y*-axis and hence the whole curve lies to the right of the *y*-axis.

(*iv*) When x is positive and increases, y also increases : when $x \to \infty$, $y \to \pm \infty$.
Hence the shape of the curve is as shown in the figure.

10. Four Standard Forms of the parabola

The figures of the four standard forms of the parabola are given below. The useful information about these parabolas are given below :

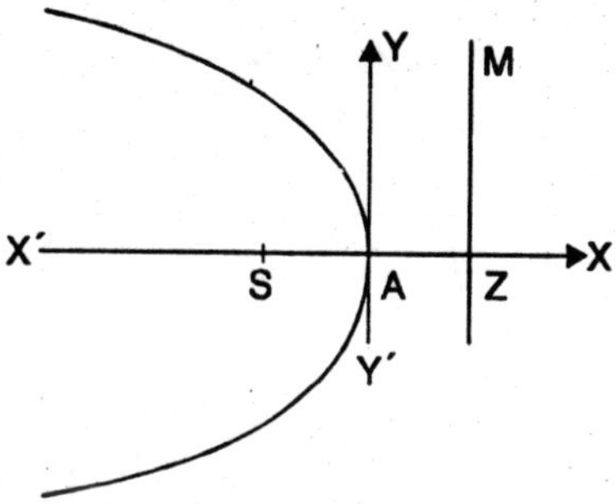

$y^2 = -4ax$
(Left-handed parabola)

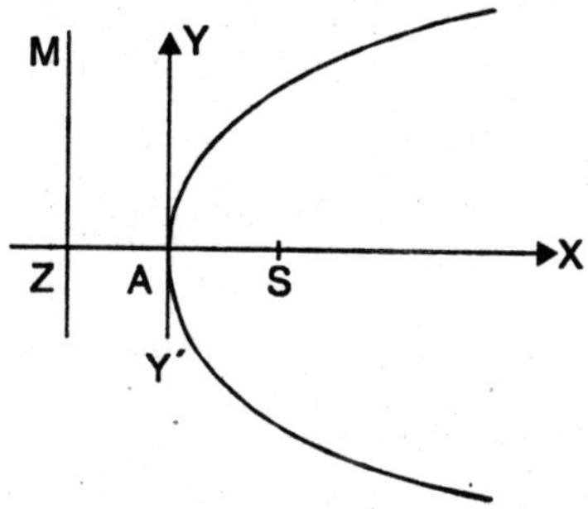

$y^2 = 4ax$
(Right-handed parabola)

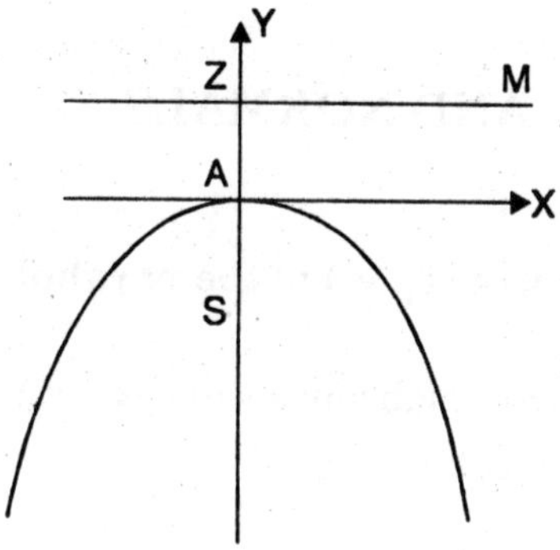

$x^2 = -4ay$
(Downward parabola)

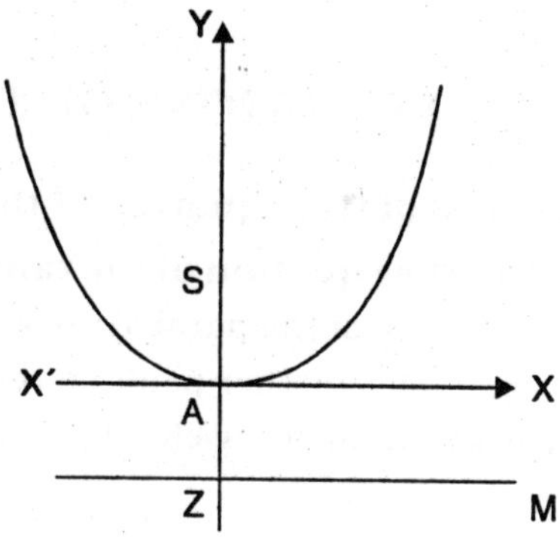

$x^2 = 4ay$
(Upward parabola)

	(*i*)	(*ii*)	(*iii*)	(*iv*)
Form	$y^2 = 4ax$	$y^2 = -4ax$	$x^2 = 4ay$	$x^2 = -4ay$
Axis	$y = 0$	$y = 0$	$x = 0$	$x = 0$
Vertex	(0, 0)	(0, 0)	(0, 0)	(0, 0)
Focus	$(a, 0)$	$(-a, 0)$	$(0, a)$	$(0, -a)$
Directrix	$x = -a$	$x = a$	$y = -a$	$y = a$.

Example 1. *Find the vertex, focus, latus rectum, axis and directrix of the parabola*

$$x^2 + 4x + 2y = 0.$$

Sol. The equation of the parabola is

$$x^2 + 4x + 2y = 0 \quad \text{or} \quad x^2 + 4x = -2y$$

or $$x^2 + 4x + 4 = 4 - 2y \quad \text{or} \quad (x + 2)^2 = -2(y - 2) \qquad ...(1)$$

Transferring the origin to $(-2, 2)$ if (x, y) transforms to (x', y') then by putting

$$x = x' - 2, y = y' + 2, \quad ...(2)$$

equation (1), becomes $x'^2 = -2y'$

which is a downward parabola.

Here $4a = 2$, $\therefore$ $a = \frac{1}{2}$

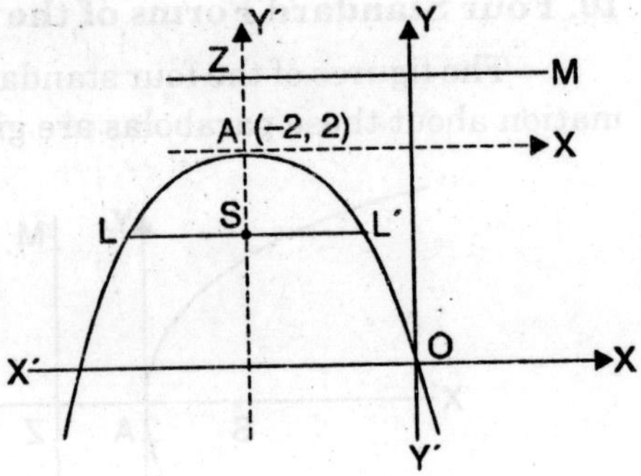

$\therefore$ With respect to the new axes

(*i*) vertex is $(0, 0)$ (*ii*) focus is $(0, -\frac{1}{2})$.

(*iii*) equation of the axis is $x' = 0$

(*iv*) equation of the directrix is $y' = \frac{1}{2}$.

Hence with respect to the old axes, by (2), we have

(*i*) vertex is $(-2, 2)$ [putting $x' = 0, y' = 0$ in (2)]

(*ii*) focus is $(-2, \frac{3}{2})$ [putting $x' = 0, y' = -\frac{1}{2}$ in (2)]

(*iii*) equation of the axis is $x = -2$ [putting $x' = 0$ in (2)]

(*iv*) equation of the directrix is $y = \frac{1}{2} + 2 = 5/2$, or $2y - 5 = 0$

(*v*) the length of the latus rectum $= 4a = 4.\frac{1}{2} = 2$.

EQUATIONS OF TANGENTS AND NORMALS

11. Point Form of the Equation of the Tangent

To find the equation of the tangent at the point (x_1, y_1) of the parabola $y^2 = 4ax$.

The equation of the parabola is $y^2 = 4ax$.

Let (x_2, y_2) be another point on the parabola in the neighbourhood of (x_1, y_1).

$\therefore$ The equation of the chord joining (x_1, y_1) and (x_2, y_2) is

$$y - y_1 = \frac{y_2 - y_1}{x_2 - x_1}(x - x_1) \quad ...(1)$$

$\because$ (x_1, y_1) and (x_2, y_2) lie on the parabola

$$\therefore \quad y_1^2 = 4ax_1 \quad ...(2)$$

$$y_2^2 = 4ax_2 \quad ...(3)$$

Subtracting (2) from (3), we have

$$y_2^2 - y_1^2 = 4a(x_2 - x_1) \quad \text{or} \quad (y_2 - y_1)(y_2 + y_1) = 4a(x_2 - x_1)$$

or
$$\frac{y_2 - y_1}{x_2 - x_1} = \frac{4a}{y_2 + y_1}$$

Substituting it in (1), the equation of the chord is

$$y - y_1 = \frac{4a}{y_2 + y_1}(x - x_1) \quad ...(4)$$

The chord joining (x_1, y_1) and (x_2, y_2) becomes tangent at (x_1, y_1) when $x_2 \to x_1$ and $y_2 \to y_1$.

$\therefore$ Putting $x_2 = x_1, y_2 = y_1$ in (4) the equation of the tangent at (x_1, y_1) is

$$y - y_1 = \frac{4a}{2y_1}(x - x_1) \qquad \text{or} \qquad yy_1 - y_1^2 = 2ax - 2ax_1$$

or $$yy_1 = 2ax + y_1^2 - 2ax_1 = 2ax + 4ax_1 - 2ax_1 \qquad \text{[From (2)]}$$
$$= 2a(x + x_1)$$

or $$\mathbf{yy_1 = 2a\,(x + x_1)},$$ which is the required equation of the tangent.

Note. Rule to write down the equation of the tangent at the point (x_1, y_1).

Change y^2 into yy_1 and x^2 into $\frac{1}{2}(x + x_1)$.

12. Equation of the Tangent in the Slope Form

To find the equation of the tangent to the parabola $y^2 = 4ax$ in the form

$$\mathbf{y = mx + \frac{a}{m}}.$$

Equation of the parabola is $y^2 = 4ax$ ...(1)

The equation of the tangent at (x_1, y_1) to (1) is $yy_1 = 2a\,(x + x_1)$ ...(2)

If m be the slope of the tangent, then $m = \frac{2a}{y_1} \qquad \therefore \quad y_1 = \frac{2a}{m}$

$\because$ (x_1, y_1) lies on (1), $\therefore$ $y_1^2 = 4ax_1$

or $$\left(\frac{2a}{m}\right)^2 = 4ax_1 \quad \text{or} \quad \frac{4a^2}{m^2} = 4ax_1 \quad \therefore \quad x_1 = \frac{a}{m^2}$$

Substituting these values of (x_1, y_1) in (2), we have

$$y\left(\frac{2a}{m}\right) = 2a\left(x + \frac{a}{m}\right) \qquad \text{or} \qquad y = mx + \frac{a}{m},$$

which is the required equation of the tangent in the slope form.

Cor. To find the point of contact. The point of contact of the tangents is (x_1, y_1) *i.e.*, $\left(\frac{\mathbf{a}}{\mathbf{m^2}}, \frac{\mathbf{2a}}{\mathbf{m}}\right)$.

13. Equation of the Normal : Point Form

To find the equation of the normal at the point (x_1, y_1) of the parabola $y^2 = 4ax$.

The equation of the parabola is $\quad y^2 = 4ax$ ...(1)

The equation of the tangent at (x_1, y_1) to (1) is $yy_1 = 2a(x + x_1)$

Slope of the tangent $= \frac{2a}{y_1}$

$\therefore$ Slope of the normal $= -\frac{y_1}{2a}$

$\therefore$ Equation of the normal at (x_1, y_1) is $y - y_1 = -\frac{y_1}{2a}(x - x_1)$.

14. Slope Form of the Normal

To find the equation of the normal to the parabola in the form

$$\mathbf{y = mx - 2am - am^3.}$$

The equation of the normal to the parabola $y^2 = 4ax$ at (x_1, y_1) is

$$y - y_1 = -\frac{y_1}{2a}(x - x_1) \qquad ...(1)$$

Let m be the slope of the normal

$$\therefore \qquad m = -\frac{y_1}{2a} \qquad \therefore \quad y_1 = -2am$$

$\because$ (x_1, y_1) lies on $\quad y^2 = 4ax \qquad \therefore \quad y_1{}^2 = 4ax_1$

or $\quad (-2am)^2 = 4ax_1 \quad$ or $\quad 4a^2m^2 = 4ax_1$

$\therefore \quad x_1 = am^2$

Substituting these values of (x_1, y_1) in (1), we have

$$y - (-2am) = m(x - am^2) \quad \text{or} \quad y + 2am = mx - am^3$$

or $\quad y = mx - 2am - am^3 \quad$ which is the required equation of the normal.

Cor. Foot of the Normal. If m is the slope of the normal, the foot of the normal is (x_1, y_1) *i.e.*, $(\mathbf{am^2, -2am})$.

A LINE AND A PARABOLA

15. To find the points of intersection of the line y = mx + c with parabola y² = 4ax

The equation of the parabola is $y^2 = 4ax$ ...(1)

and that of the line is $\quad y = mx + c$ ...(2)

To find the points of intersection, we solve (1) and (2) simultaneously.

Substituting the value of y from (2) in (1), we get

$$(mx + c)^2 = 4ax \quad \text{or} \quad m^2x^2 + 2mcx + c^2 = 4ax$$

or $\quad m^2x^2 + 2(mc - 2a)x + c^2 = 0$ ...(3)

which being quadratic in x, gives two values of x which are the abscissae of the points of intersection of (1) and (2).

Substituting the values of x thus obtained in (2), we get the corresponding values of y.

Thus a straight line meets a parabola in two points real and distinct, coincident or imaginary according as the roots of (3) are real and distinct, equal or imaginary.

16. Condition of Tangency

To find the condition that the line y = mx + c may touch the parabola y² = 4ax

The equation of the line is $\quad y = mx + c$ ...(1)

and that of the parabola is $\quad y^2 = 4ax$ ...(2)

Eliminating y from (1) and (2), we have

$$(mx + c)^2 = 4ax \text{ or } m^2x^2 + 2mcx + c^2 = 4ax$$

or $\quad m^2x^2 + 2(mc - 2a)x + c^2 = 0$ ...(3)

which is quadratic in x.

If the line (1) touches the parabola (2), the quadratic (3) has equal roots and condition for which is

$$4(mc - 2a)^2 - 4m^2c^2 = 0 \quad \text{or} \quad m^2c^2 + 4a^2 - 4mca - m^2c^2 = 0 \text{ or } a - mc = 0$$

$$\therefore \qquad \mathbf{c = \frac{a}{m}}$$

which is the required condition for the line to be tangent.

Cor. 1. Substituting the value of c in (1), we get

$$y = mx + \frac{a}{m} \qquad \text{...(4)}$$

as the equation of tangent in the slope form.

Cor. 2. To find the point of contact. Putting the value $c = \frac{a}{m}$ in (3), we get

$$m^2x^2 - 2ax + \frac{a^2}{m^2} = 0 \quad \text{or} \quad \left(mx - \frac{a}{m}\right)^2 = 0$$

which gives $x = \frac{a}{m^2}$ and therefore from (4), $y = \frac{2a}{m}$.

Hence the co-ordinates of the point of contact are $\left(\frac{a}{m^2}, \frac{2a}{m}\right)$.

17. To show that from any point three normal can be drawn to a parabola, and

(i) at least one of the three normals is real.

(ii) the algebraic sum of their slopes is zero.

(iii) the algebraic sum of the ordinates of their feet is zero.

Let the equation of the parabola be $y^2 = 4ax$ ...(1)

Let (x_1, y_1) be the given point,

Equation of any normal in the slope form is

$$y = mx - 2am - am^3 \qquad \text{...(2)}$$

If this normal passes through (x_1, y_1), then

$$y_1 = mx_1 - 2am - am^3$$

or

$$am^3 + m(2a - x_1) + y_1 = 0 \qquad \text{...(3)}$$

which being a cubic equation in m has three values of m, real or imaginary, which are the slopes of the three normals that pass through the point (x_1, y_1).

Substituting these values of m in the equation $y - y_1 = m(x - x_1)$, we get the equations of the three normals.

(i) Equation (3) is a cubic in m with all its co-efficients real.

$\therefore$ It has atleast one root real and corresponding to this real root, it has at least one real normal.

(ii) Let m_1, m_2, m_3 be the roots of the equation (3),

then

$$m_1 + m_2 + m_3 = -\frac{\text{co-eff. of } m^2}{\text{co-eff. of } m^3} = -\frac{0}{a} = 0 \qquad \text{...(4)}$$

Hence the algebraic sum of the slopes of the normal is zero.

(*iii*) The three feet of the normals are

$$(am_1^2, -2am_1), (am_2^2, -2am_2) \text{ and } (am_3^2, -2am_3)$$

$\therefore$ Sum of the ordinates of the feet

$$= (-2am_1) + (-2am_2) + (-2am_3)$$
$$= -2a[m_1 + m_2 + m_3] = 2a(0) = 0. \qquad | \text{ From (4)}$$

Note. Three points on the parabola, normal at which intersect in a point are called **Co-Normal Points.**

PARAMETRIC CO-ORDINATES

18. To express the co-ordinates of any point on the parabola $y^2 = 4ax$ in terms of one variable

The equation of the parabola is $y^2 = 4ax$ ...(1)

Equation (1) can be written as

$$\frac{y}{2a} = \frac{2x}{y} = t \text{ (say)}$$

From first and third members, $y = 2at$ ...(2)

From last two members, $x = \frac{y}{2} . t = \frac{2at^2}{2} = at^2$ | From (2)

$$\therefore \quad \mathbf{x = at^2, y = 2at,}$$

are the co-ordinates of any point on the parabola, t being the parameter.

Cor. *Parametric co-ordinates of any point on the parabola.* Since it is evident that the point $(at^2, 2at)$, t being a parameter, satisfies the equation of the parabola $y^2 = 4ax$, therefore, any point on the parabola is $\mathbf{(at^2, 2at)}$. It is generally called the point 't' on the parabola $y^2 = 4ax$.

19. To find the equation of the chord joining the points 't_1' and 't_2' on the parabola $y^2 = 4ax$. Hence to deduce the equation of the tangent at 't_1'

The equation of the parabola is $y^2 = 4ax$

The given points 't_1', 't_2', on the parabola are $(at_1^2, 2at_1)$ and $(at_2^2, 2at_2)$

The equation of the chord joining these points is

$$y - 2at_1 = \frac{2at_2 - 2at_1}{at_2^2 - at_1^2}(x - at_1^2)$$

$$= \frac{2(t_2 - t_1)}{(t_2 + t_1)(t_2 - t_1)}(x - at_1^2) = \frac{2}{t_2 + t_1}(x - at_1^2)$$

or $$(t_1 + t_2)y - 2at_1t_2 - 2at_1^2 = 2x - 2at_1^2$$

or $$\mathbf{(t_1 + t_2)y = 2x + 2at_1t_2} \qquad ...(1)$$

which is the required equation.

The chord becomes tangent at the point t_1 when $t_2 \to t_1$ *i.e.*, $t_2 = t_1$

$\therefore$ From (1), equation of the tangent at the point t_1, is

$$2t_1y = 2x + 2at_1^2 \quad \text{or} \quad \mathbf{t_1y = x + at_1^2.}$$

20. To find the equation of the normal at the point 't' to the parabola $y^2 = 4ax$

The equation of the parabola is $y^2 = 4ax$

The equation of the tangent at 't' *i.e.*, at $(at^2, 2at)$ to the parabola is $ty = x + at^2$

Its slope $= \frac{1}{t}$

$\therefore$ Slope of the normal $= -t$

$\therefore$ Equation of the normal at $(at^2, 2at)$ is $y - 2at = -t(x - at^2)$ or $y + tx = 2at + at^3$.

Example. *If the chord joining the points $(at_1^2, 2at_1)$ and $(at_2^2, 2at_2)$ on the parabola $y^2 = 4ax$ passes through the focus, show that $t_1t_2 = -1$.*

Sol. The equation of the parabola is $y^2 = 4ax$.

The equation of the chord joining $(at_1^2, 2at_1)$ and $(at_2^2, 2at_2)$ on the parabola is $(t_1 + t_2)y = 2x + 2at_1t_2$.

If it passes through the focus $(a, 0)$, then $0 = 2a + 2at_1t_2$ or $t_1t_2 = -1$ which is the required condition.

POSITION OF A POINT WITH REGARD TO A PARABOLA

21. Test for a Point

To show that the quantity $y_1^2 - 4ax_1$ is positive, zero or negative according as the point (x_1, y_1) is outside, on or inside the parabola

$$y^2 = 4ax.$$

The equation of the parabola is $y^2 = 4ax$ ...(1)

Let P be the given point (x_1, y_1). Draw PM $\perp$ X′ OX meeting the parabola in $Q(x_1, y')$ (say).

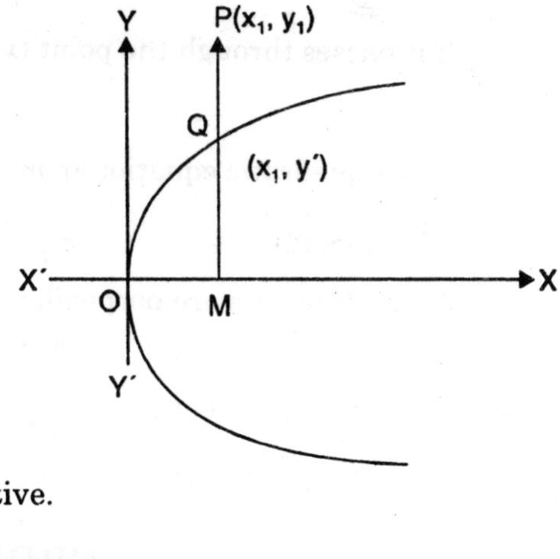

$\therefore$ The point $Q(x_1, y')$ lies on the parabola (1)

$\therefore$ $y'^2 = 4ax_1$ ...(2)

The point P lies outside, on or inside the parabola according as

MP $>$ = or $<$ MQ numerically.

i.e., according as $MP^2 > = \text{ or } < MQ^2$

or according as $MP^2 - MQ^2 > = \text{ or } < 0$

or according as $y_1^2 - y'^2 > = \text{ or } < 0$

or according as $y_1^2 - 4ax_1 > = \text{ or } < 0$ | From (2)

i.e., according as $y_1^2 - 4ax_1$ is positive, zero or negative.

22. Two Tangents from a Point to a Parabola

To show that from any point two tangents can be drawn to a parabola which are real and distinct coincident or imaginary according as the point is outside, on or inside the parabola.

Let the equation of the parabola be

$$y^2 = 4ax \quad ...(1)$$

Let the given point be (x_1, y_1).

The equation of any tangent to the parabola (1) is

$$y = mx + \frac{a}{m} \text{ for all values of } m.$$

If it passes through (x_1, y_1), then $y_1 = mx_1 + \frac{a}{m}$

or $$x_1m^2 - y_1m + a = 0 \qquad ...(2)$$

which being a quadratic equation in m gives two values of m which are the slopes of the two tangents from (x_1, y_1).

Therefore, the equations of the two tangents are obtained by substituting the values of m one by one in $y - y_1 = m(x - x_1)$.

Hence in general two tangents can be drawn from (x_1, y_1) to the parabola $y^2 = 4ax$.

These tangents are real and distinct, coincident or imaginary according as the roots of (2) are real and distinct, equal or imaginary.

i.e., according as discriminant of (2) is + ve, zero or – ve

i.e., according as $y_1^2 - 4ax_1$ is + ve, zero or – ve

i.e., according as the point (x_1, y_1) lies outside, on or inside the parabola.

23. To find the locus of the point of intersection of two perpendicular tangents to a parabola

Let the equation of the parabola be $y^2 = 4ax$ ...(1)

Let (x_1, y_1) be the point of intersection of two tangents to the parabola (1).

The equation of any tangent to parabola (1) is

$$y = mx + \frac{a}{m}$$

If it passes through the point (x_1, y_1), then

$$y_1 = mx_1 + \frac{a}{m} \quad \text{or} \quad x_1m^2 - y_1m + a = 0 \qquad ...(2)$$

It is a quadratic equation in m. Let m_1, m_2 be its roots.

$\therefore$ From (2), $$m_1m_2 = \frac{a}{x_1} \qquad ...(3)$$

As the tangents are perpendicular to each other,

$$m_1m_2 = -1 \qquad ...(4)$$

$\therefore$ From (3) and (4), $$\frac{a}{x_1} = -1 \quad \text{or} \quad x_1 = -a$$

$\therefore$ Locus of (x_1, y_1) is $x = -a$, which is the equation of the directrix.

CHORD OF CONTACT

24. Equation of the Chord of Contact

To find the equation of the chord of contact of tangents drawn to the parabola $y^2 = 4ax$ from (x_1, y_1).

The equation of the parabola is $y^2 = 4ax$. ...(1)

Let the points of contacts of two tangents from the given point $P(x_1, y_1)$ to the parabola (1) be $T(x', y')$ and $T'(x'', y'')$.

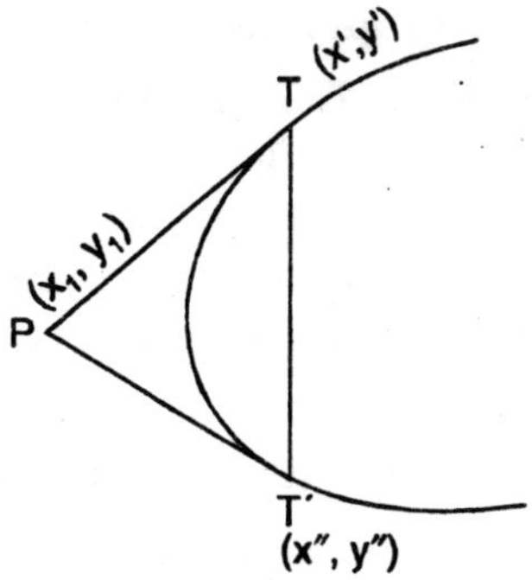

Then the equations of the tangents at T, T′ are

$$yy' = 2a(x + x') \qquad ...(2)$$

and

$$yy'' = 2a(x + x'') \qquad ...(3)$$

As these tangents pass through the point $P(x_1, y_1)$

$\therefore$
$$y_1y' = 2a(x_1 + x') \qquad ...(4)$$
$$y_1y'' = 2a(x_1 + x'') \qquad ...(5)$$

Equations (4) and (5) show that $T(x', y')$, $T'(x'', y'')$ lie on the line

$$y_1y = 2a(x_1 + x) \quad i.e., \quad yy_1 = 2a(x + x_1)$$

which is the required equation of the chord of contact.

POLE AND POLAR

25. Equation of the Polar

To find the equation of the polar of the point (x_1, y_1) with respect to the parabola $y^2 = 4ax$.

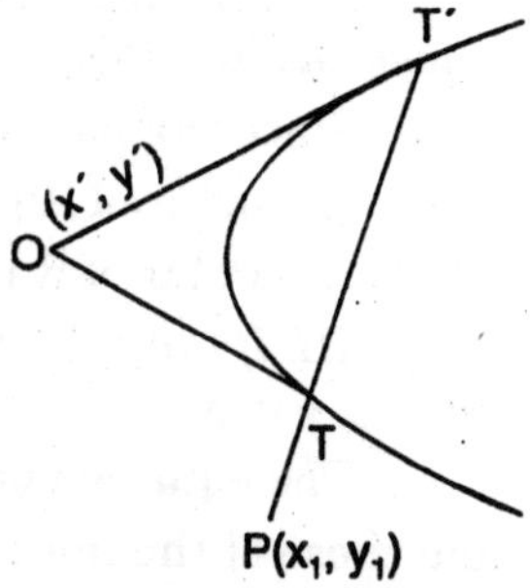

The equation of the parabola is $y^2 = 4ax$ ...(1)

Let PTT′ be any straight line through $P(x_1, y_1)$ meeting the parabola (1) in T, T′.

Let the tangents at T and T′ meet in $Q(x', y')$.

Now TT′ is the chord of contact of tangents from (x', y') to the parabola (1).

$\therefore$ Its equation is $yy' = 2a(x + x')$. As it passes through $P(x_1, y_1)$

$\therefore$
$$y_1y' = 2a(x_1 + x')$$

$\therefore$ The locus of (x', y') is

$$y_1y = 2a(x_1 + x) \quad i.e., \quad \mathbf{yy_1 = 2a(x + x_1)}$$

which is the required equation of the polar of the point $P(x_1, y_1)$.

26. Pole of a Line

To find the pole of the line $Ax + By + C = 0$ with respect to the parabola $y^2 = 4ax$.

Let (x_1, y_1) be the pole of the line $Ax + By + C = 0$ ...(1)

with respect to the parabola $y^2 = 4ax$ ...(2)

The equation of the polar of (x_1, y_1) w.r.t. (2) is

$$yy_1 = 2a(x + x_1) \quad \text{or} \quad 2ax - yy_1 + 2ax_1 = 0 \qquad ...(3)$$

The lines (1) and (3) are the same.

$\therefore$ Comparing the co-effs. of like terms, we have

$$\frac{2a}{A} = \frac{-y_1}{B} = \frac{2ax_1}{C}$$

$\therefore$
$$x_1 = \frac{C}{A} \quad \text{and} \quad y_1 = -\frac{2aB}{A}$$

Hence the required pole is $\left(\frac{C}{A}, -\frac{2aB}{A}\right)$.

27. Properties of Pole and Polar

I. To show that if the polar of a point P with respect to a parabola passes through a point Q, the polar of Q with respect to the same parabola passes through P.

Let the equation of the parabola be $y^2 = 4ax$ ...(1)

Let the given points P, Q be (x_1, y_1) and (x_2, y_2) respectively.

The equation of the polar of $P(x_1, y_1)$ w.r.t. the parabola (1) is

$$yy_1 = 2a(x + x_1)$$

$\because$ It passes through Q (x_2, y_2)

$$\therefore \quad y_2y_1 = 2a(x_2 + x_1) \quad ...(2)$$

The result (2) shows that the line

$$y_2y = 2a(x_2 + x)$$

i.e., $yy_2 = 2a(x + x_2)$ passes through $P(x_1, y_1)$

$\therefore$ The polar of Q also passes through P.

Hence, the result.

Note. Two such points are called conjugate points w.r.t. the parabola.

II. If the polars of two points P and Q with respect to a parabola meet in R, then R is the pole of the line PQ.

Since the polars of the points P and Q pass through R

$\therefore$ polar of R passes through P as well as through Q.

$\therefore$ Polar of R is the PQ *i.e.*, R is the pole of the line PQ.

III. If the pole of the line $l_1x + m_1y + n_1 = 0$ with respect to the parabola $y^2 = 4ax$ lies on the line $l_2x + m_2y + n_2 = 0$ then the pole of the line $l_2x + m_2y + n_2 = 0$ lies on the line $l_1x + m_1y + n_1 = 0$.

The equation of the parabola is $y^2 = 4ax$ ...(1)

and those of the lines are $l_1x + m_1y + n_1 = 0$ ...(2)

and $l_2x + m_2y + n_2 = 0$ ...(3)

Let (x_1, y_1) be the pole of the line (2)

$\therefore$ The equation of the polar of (x_1, y_1) w.r.t. (1) is

$$yy_1 = 2a(x + x_1) \quad \text{or} \quad 2ax - yy_1 + 2ax_1 = 0 \quad ...(4)$$

The lines (2) and (4) are the same.

$\therefore$ Comparing the co-effs. of like terms, we have

$$\frac{2a}{l_1} = \frac{-y_1}{m_1} = \frac{2ax_1}{n_1}$$

$$\therefore \quad x_1 = \frac{n_1}{l_1} \quad \text{and} \quad y_1 = -\frac{2am_1}{l_1}$$

Pole of (2) is $\left(\frac{n_1}{l_1}, -\frac{2am_1}{l_1}\right)$

If this lies on (3), then

$$l_2\left(\frac{n_1}{l_1}\right) + m_2\left(-\frac{2am_1}{l_1}\right) + n_2 = 0$$

i.e., $$l_2n_1 + l_1n_2 - 2am_1m_2 = 0 \quad ...(5)$$

The symmetry of the result (5) shows that it is also the condition that the pole of (3) also lies on (2).

Note. Two lines such that the pole of one lies on the other are called conjugate lines and the result (5) is the condition of conjugacy.

CHORD IN TERMS OF ITS MID-POINT

28. To find the equation of the chord of the parabola $y^2 = 4ax$ in terms of its mid-point (x_1, y_1).

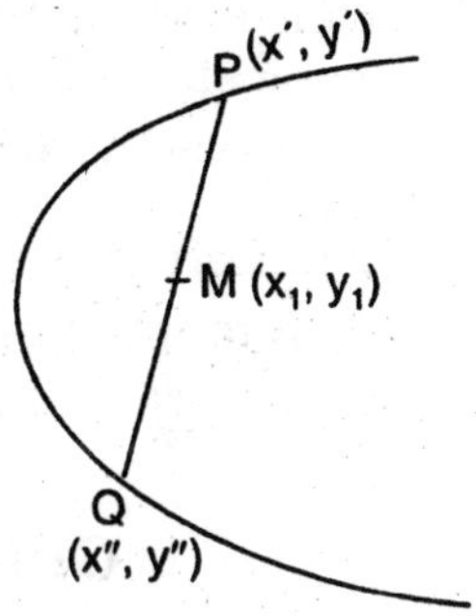

The equation of the parabola is $y^2 = 4ax$ ...(1)

Let PQ be any chord whose mid-point is $M(x_1, y_1)$.

Let (x', y') and (x'', y'') be the co-ordinates of P and Q respectively.

$$\therefore \quad x_1 = \frac{x' + x''}{2}$$

or $$x' + x'' = 2x_1 \qquad ...(2)$$

and $$y_1 = \frac{y' + y''}{2}$$

or $$y' + y'' = 2y_1 \qquad ...(3)$$

Slope of $$PQ = \frac{y'' - y'}{x'' - x'}$$

Since $P(x', y')$ and $Q(x'', y'')$ lie on the parabola (1)

$$\therefore \quad y'^2 = 4ax' \qquad ...(4)$$

and $$y''^2 = 4ax'' \qquad ...(5)$$

Subtracting (4) from (5), we have

$$y''^2 - y'^2 = 4a(x'' - x')$$

or $$\frac{y'' - y'}{x'' - x'} = \frac{4a}{y'' + y'} = \frac{4a}{2y_1} = \frac{2a}{y_1} \qquad \text{[From (3)]}$$

$\therefore$ Slope of $$PQ = \frac{2a}{y_1}$$

$\therefore$ The equation of the chord PQ through (x_1, y_1) is

$$y - y_1 = \frac{2a}{y_1}(x - x_1) \quad i.e., \quad yy_1 - y_1^2 = 2a(x - x_1)$$

i.e., $$yy_1 - 2a(x + x_1) = y_1^2 - 4ax_1 \quad i.e., \quad T = S_1$$

where $T = yy_1 - 2a(x + x_1)$ and $S_1 = y_1^2 - 4ax_1$.

29. To prove that the locus of the middle points of a system of parallel chords of a parabola is a straight line which is parallel to the axis of the parabola.

Let the equation of the parabola be $y^2 = 4ax$ ...(1)

Let m be the slope of the system of parallel chords.

Let $M(x_1, y_1)$ be the mid-point of any chord PQ of the system.

Then the equation of the chord having (x_1, y_1) as its mid-point is

$$yy_1 - 2a(x + x_1) = y_1^2 - 4ax_1 \qquad | \; T = S_1$$

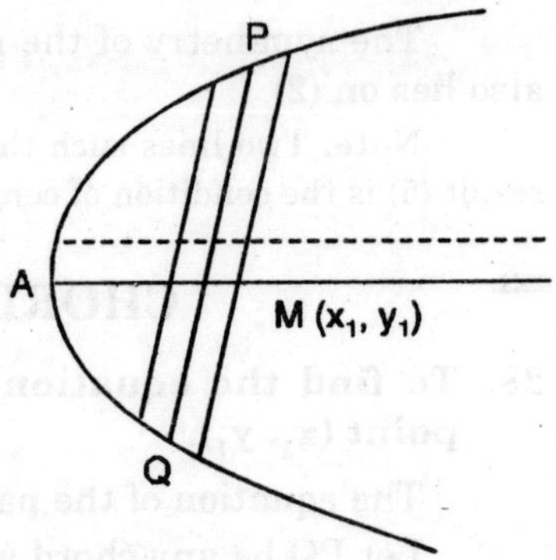

i.e., $$yy_1 - 2ax = y_1^2 - 2ax_1$$

Its slope $= \dfrac{2a}{y_1}$

Also slope of this chord $= m$

$$\therefore \quad \frac{2a}{y_1} = m \quad \text{or} \quad y_1 = \frac{2a}{m}$$

$\therefore$ Locus of (x_1, y_1) is $y = \dfrac{2a}{m}$

which is a straight line parallel to the x-axis, *i.e.*, the axis of the parabola, called the diameter of the parabola.

Hence the definition.

30. Diameter

Def. *The locus of mid-points of a system of parallel chords of a parabola is called a diameter of the parabola.*

GEOMETRICAL PROPERTIES OF PARABOLA

31. Def. Tangent, Normal, Sub-tangent and Sub-normal

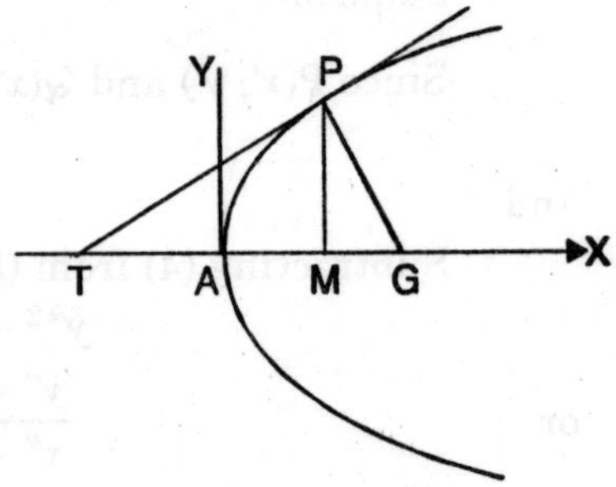

Let PT and PG be the tangent and normal at any point of a parabola $y^2 = 4ax$ meeting the axis is T and G respectively.

Let PM be the perpendicular from P on the axis of the parabola, then

(*i*) *the portion PT of the tangent is the length of the* **tangent** *at P,*

(*ii*) *the portion PG of the normal is the length of the* **Normal** *at P,*

(*iii*) *the projection TM of PT is called the* **sub-tangent** *at P,*

(*iv*) *the projection MG of PG is called the* **sub-normal** *at P.*

32. (1) The Sub-tangent at any point of a parabola is bisected at the vertex

Let P$(at^2, 2at)$ be any point on the parabola $y^2 = 4ax$.

Let the tangent at P to the parabola meet the axis of the parabola in T.

The equation of the tangent at P$(at^2, 2at)$ is

$$ty = x + at^2 \qquad \text{...(1)}$$

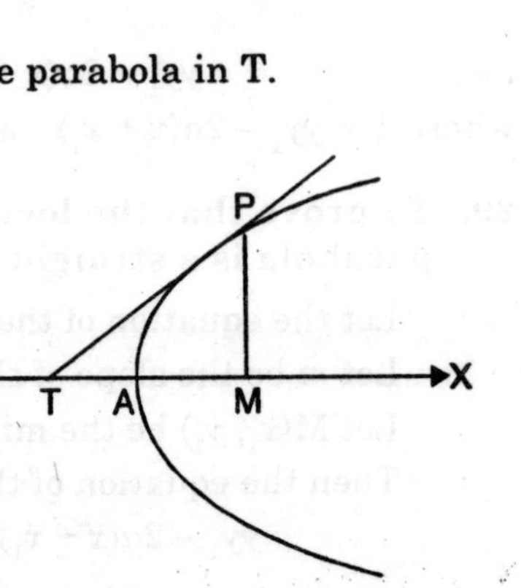

It meets the axis, $y = 0$, where $x = -at^2$

[Putting $y = 0$ in (1)]

$\therefore$ Co-ordinates of T are $(-at^2, 0)$.

Also the co-ordinates of M are $(at^2, 0)$.

Clearly the mid-point of TM is (0, 0) which is the vertex of the parabola.

Hence the sub-tangent is bisected at the vertex.

(2) The Sub-normal at any point of a parabola is constant and is equal to half the latus rectum

Let $P(at^2, 2at)$ be any point on the parabola $y^2 = 4ax$.

Let the normal at P meet the axis in G. Draw PM ⊥ to the axis, so that MG is the sub-normal.

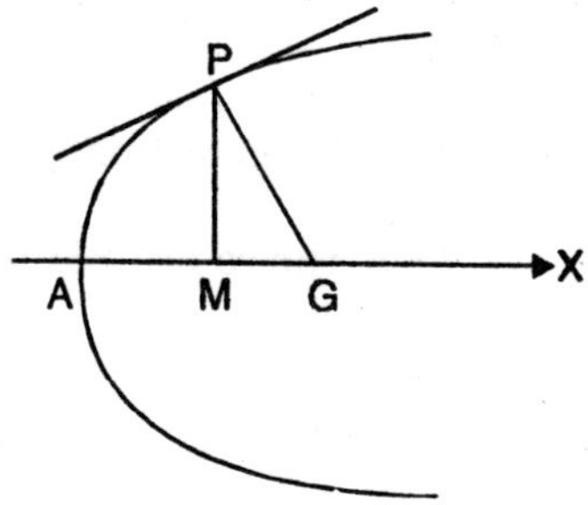

The equation of the normal at $P(at^2, 2at)$ is

$$y + tx = 2at + at^3 \qquad ...(1)$$

It meets the x-axis, $y = 0$ where

$$tx = 2at + at^3 \quad i.e., \quad x = 2a + at^2$$

[Putting $y = 0$ in (1)]

$\therefore \qquad AG = 2a + at^2$

Also $\qquad AM = at^2$

$\therefore \qquad MG = AG - AM = 2a + at^2 - at^2 = 2a$

which is constant and is equal to half the latus rectum.

(3) The locus of the foot of the perpendicular from the focus on any tangent to a parabola is the tangent at the vertex

Let the equation of the parabola be $y^2 = 4ax$.

Let $P(x, y)$ be any point on the parabola.

The equation of any tangent at $P(x, y)$ is

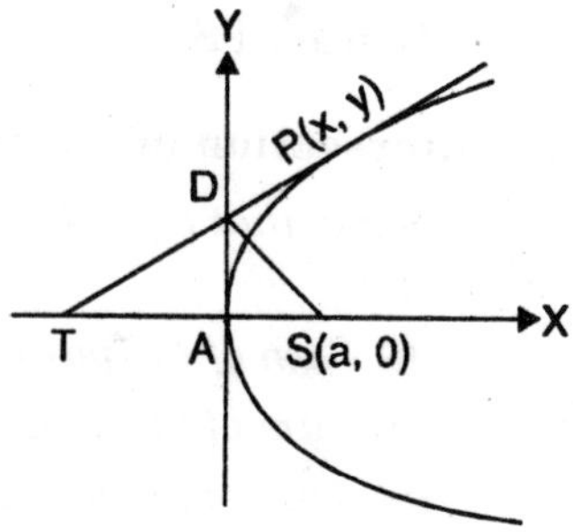

$$y = mx + \frac{a}{m} \qquad ...(1)$$

Slope of tangent = m

$\therefore$ Slope of SD which is ⊥ to

$$PT = -\frac{1}{m}$$

$\therefore$ The equation of SD through the focus $(a, 0)$ is

$$y - 0 = -\frac{1}{m}(x - a)$$

i.e.,
$$y = -\frac{1}{m}x + \frac{a}{m} \qquad ...(2)$$

Eliminating m from (1) and (2) by subtracting (2) from (1), we have

$$0 = \left(m + \frac{1}{m}\right)x \quad \text{or} \quad x = 0,$$

which is the required locus, and it is a tangent at the vertex.

(4) The portion of the tangent intercepted between the point of contact and the directrix subtends a right angle at the focus of a parabola

Let the equation of the parabola be

$$y^2 = 4ax \qquad ...(1)$$

Let $P(at^2, 2at)$ be any point on (1).

The equation on the tangent at P is

$$yt = x + at^2 \qquad ...(2)$$

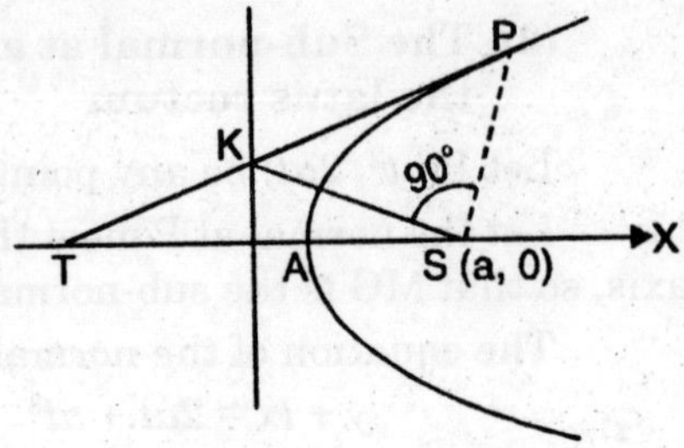

It meets the directrix, $x = -a$ where

$$yt = -a + at^2$$

i.e.,
$$y = a\left(t - \frac{1}{t}\right)$$

$\therefore$ The co-ordinates of K are $\left[-a, a\left(t - \frac{1}{t}\right)\right]$

Also, the co-ordinates of S are $(a, 0)$

$\therefore$ Slope of
$$SK = \frac{a\left(t - \frac{1}{t}\right) - 0}{-a - a} = -\frac{t^2 - 1}{2t} \quad \text{...(3)}$$

Also slope of
$$SP = \frac{2at - 0}{at^2 - a} = \frac{2t}{t^2 - 1} \quad \text{...(4)}$$

From (3) and (4), product of the slopes of SK and SP $= -\frac{t^2 - 1}{2t} \cdot \frac{2t}{t^2 - 1} = -1$

$\therefore$ SP and SK are perpendicular to each other.

Hence $\angle PSK = 1$ rt. angle.

33. Intersection of a Circle and a Parabola

Show that a circle and a parabola intersect in four points and also prove the following results :

(i) *Sum of the parameters of the four points of intersection is zero.*

(ii) *Sum of the ordinates of the four points of intersection is zero.*

(iii) *The common chords of a circle and a parabola are equally inclined to the axis.*

(iv) *If out of the four common points of a circle and a parabola, three are co-normal point, the fourth point must be the vertex of the parabola.*

Let the equations of the given parabola and circle be

$$y^2 = 4ax \quad \text{...(1)}$$

and
$$x^2 + y^2 + 2gx + 2fy + c = 0 \quad \text{...(2)}$$

Any point on the parabola (1) is $(at^2, 2at)$, it will become the point of intersection of (1) and (2), if it satisfies (2) also.

$\therefore$
$$a^2t^4 + 4a^2t^2 + 2gat^2 + 4aft + c = 0$$

or
$$a^2t^4 + (4a^2 + 2ga)t^2 + 4aft + c = 0 \quad \text{...(3)}$$

which being a fourth degree equation in t, has four roots ; say, t_1, t_2, t_3 and t_4.

Hence a circle and a parabola intersect in four points ; say P, Q, R and T and their co-ordinates are respectively $(at_1^2, 2at_1)$ $(at_2^2, 2at_2)$, $(at_3^2, 2at_3)$ and $(at_4^2, 2at_4)$.

(i) Now in equation (3), the co-efficient of $t^3 = 0$.

$\therefore$ The sum of the roots of (3) *i.e.*, $t_1 + t_2 + t_3 + t_4 = 0$ which proves the result (*i*).

(ii) Sum of the ordinates of the points of intersection

$= 2a(t_1 + t_2 + t_3 + t_4) = 0$, which proves the result. | *Using (i)* (*iii*) Slope of chord $PQ = \frac{2}{t_1 + t_2}$

Slope of chord $RS = \frac{2}{t_3 + t_4} = -\frac{2}{t_1 + t_2}$ $\quad [\because \ t_3 + t_4 = -(t_1 + t_2)]$

As the slopes of PQ and RS are equal and opposite,

$\therefore$ The common chords of the circle and parabola are equally inclined to the axis. This establishes result (*iii*).

(*iv*) If the normals at the points t_1, t_2, t_3 are concurrent, then

$$t_1 + t_2 + t_3 = 0 \qquad [\textit{Art. 2 Property (i)}]$$

and

$$t_1 + t_2 + t_3 + t_4 = 0 \qquad [\textit{Proved in Art. 34 (i)}]$$

$$\therefore \quad t_4 = 0.$$

$\therefore$ Co-ordinates of the fourth point $(at_4^2, 2at_4)$ become (0, 0) *i.e.*, the fourth point is the vertex of the parabola. Hence the result.

Example 1. *If the normals at three points P, Q, R on the parabola intersect in the point O(h, k), prove that the circle PQR passes through the vertex of the parabola.*

Also determine the equation of circle PQR.

Sol. Let the parabola be $y^2 = 4ax$

The equation of any normal to the parabola is

$$y = -tx + 2at + at^3$$

As it passes through the point (h, k).

$$\therefore \quad k = -th + 2at + at^3 \quad \text{or} \quad at^3 + t(2a - h) - k = 0 \qquad ...(1)$$

It has three roots 't_1', 't_2' and 't_3' which are the parameters of the points P, Q and R respectively.

Let the equation of the circle through the points P, Q and R be

$$x^2 + y^2 + 2gx + 2fy + c = 0 \qquad ...(2)$$

The parameters of the points of intersection of the parabola and the circle are given by substituting the co-ordinates of any point $(at^2, 2at)$ of the parabola in the equation (2) of the circle.

i.e., by

$$a^2t^4 + 4a^2t^2 + 2gat^2 + 4aft + c = 0$$

or

$$a^2t^4 + (4a^2 + 2ga)t^2 + 4aft + c = 0 \qquad ...(3)$$

Three of the roots of this equation are t_1, t_2, t_3 ; let the fourth root be t_4.

$$\therefore \quad t_1 + t_2 + t_3 + t_4 = 0$$

Also from (1),

$$t_1 + t_2 + t_3 = 0$$

$$\therefore \quad t_4 = 0$$

i.e., the fourth point is the vertex of the parabola.

$\therefore$ The circle PQR passes through the vertex of the parabola.

$$\therefore \quad c = 0.$$

Now as one root of the equation (3) is zero.

Putting $c = 0$ in (3) and dividing by at, we get

$$at^3 + 2(2a + g)t + 4f = 0 \quad \text{...(4)}$$

Now roots of (4) and (1) are the same.

$\therefore$ Comparing co-efficients in (1) and (4), we get

$$\frac{a}{a} = \frac{2(2a+g)}{2a-h} = \frac{4f}{-k}$$

or $$4a + 2g = 2a - h \quad \text{and} \quad 4f = -k$$

which give $$2g = -(2a + h) \quad \text{and} \quad 2f = -k/2.$$

$\therefore$ Substituting the values of c, $2g$ and $2f$ in (2), the required equation of the circle is $x^2 + y^2 - (2a + h)\,x - \frac{1}{2}ky = 0$ which is known as **Harvey's Circle** and the result proved is **Harvey's Theorem.**

34. Joint Equation of the Pair of Tangents

To find the joint equation of the pair of tangents from $P(x_1, y_1)$ to the parabola $y^2 = 4ax$.

Let AB be any line through $A(x_1, y_1)$ meeting the parabola

$$y^2 = 4ax \quad \text{...(1)}$$

in the points P and Q.

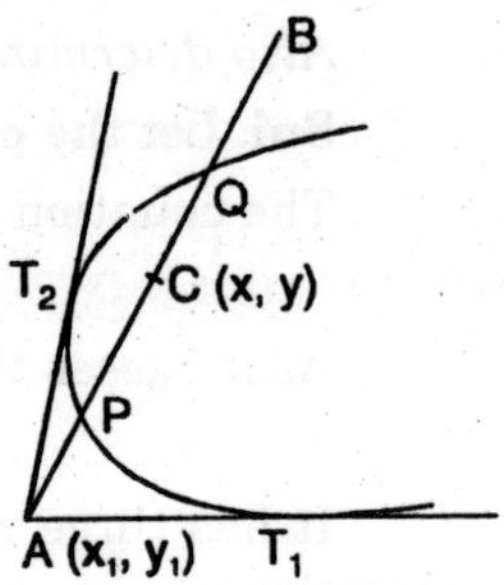

Let $C(x, y)$ be any point on the line AB.

$\therefore$ The co-ordinates of any point which divide AC in the ratio $\lambda : 1$ are $\left(\frac{\lambda x + x_1}{\lambda + 1}, \frac{\lambda y + y_1}{\lambda + 1}\right)$

Let this point represent either of the points P and Q. Then its co-ordinates must satisfy (1).

$$\therefore \quad \left(\frac{\lambda y + y_1}{\lambda + 1}\right)^2 = 4a\left(\frac{\lambda x + x_1}{\lambda + 1}\right)$$

or $$(\lambda y + y_1)^2 = 4a(\lambda x + x_1)(\lambda + 1)$$

or $$\lambda^2(y^2 - 4ax) + 2\lambda\,[yy_1 - 2a\,(x + x_1)] + (y_1^2 - 4ax_1) = 0 \quad \text{...(2)}$$

It is a quadratic equation in λ and hence gives two values of λ, one corresponding to P and the other to Q.

The line AB becomes tangent if the points P and Q coincide *i.e.*, the two values of λ in (2) are equal.

$\therefore$ For equal roots, discriminant of (2) = 0

or $$4[yy_1 - 2a(x + x_1)]^2 - 4(y^2 - 4ax)\,(y_1^2 - 4ax_1) = 0$$

or $$[yy_1 - 2a(x + x_1)]^2 = (y^2 - 4ax)\,(y_1^2 - 4ax_1)$$

which is the required equation of the tangents PT_1 and PT_2.

This gives, $$SS_1 = T^2$$

where $S = y^2 - 4ax$, $S_1 = y_1^2 - 4ax_1$ and $T = yy_1 - 2a(x + x_1)$.

2. Standard Results of The Ellipse

EQUATION OF AN ELLIPSE

1. Ellipse

Def. *An ellipse is the locus of a point which moves in a plane so that its distance from a fixed point bears a constant ratio, less than one, to its distance from a fixed straight line.*

The fixed point is called the **focus** and is generally denoted by S.

The fixed straight line is called the **directrix** of the ellipse and the constant ratio is called the **eccentricity** of the ellipse and is denoted by **e**.

Thus, if P be any point on the ellipse whose focus is S and eccentricity e, and PM is drawn perpendicular on the directrix from P, then

$$\frac{SP}{PM} = e \ (e < 1)$$

or

$$\mathbf{SP = e \cdot PM.}$$

Equation of an ellipse in the Standard Form

2. To find the equation of an ellipse in the form $\frac{x^2}{a^2} + \frac{y^2}{b^2} = 1.$

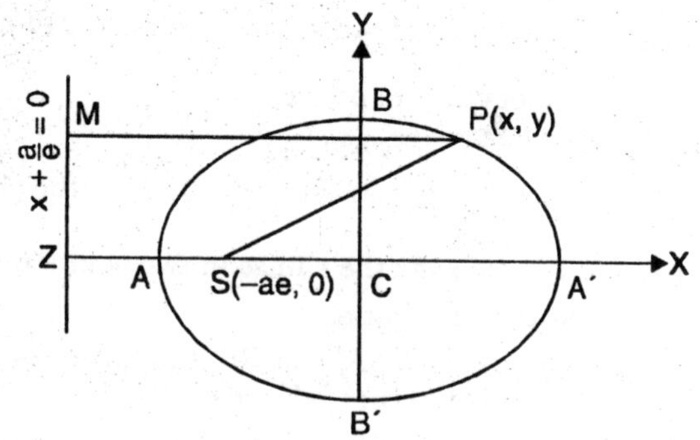

Let S be the focus, ZM the directrix and e (< 1) the eccentricity of the ellipse.

From S draw SZ perpendicular to the directrix, ZM. Divide ZS *internally* at A and *externally* at A′ in the ratio 1 : e so that

$$\frac{ZA}{AS} = \frac{1}{e} \quad \text{or} \quad AS = eZA \qquad ...(1)$$

and

$$\frac{ZA'}{A'S} = \frac{1}{e} \quad \text{or } A'S = eZA' \qquad ...(2)$$

Thus, by the definition, A and A′ lie on the ellipse. Let C be the middle point of AA′ and let AA′ = $2a$, then CA = CA′ = a

Adding (1) and (2), we have

$$AS + A'S = e[ZA + ZA']$$

or $$AA' = e(CZ - CA + CA' + CZ)$$

or $$2a = e \cdot 2CZ \qquad (\because CA = CA', AA' = 2a)$$

$$\therefore \quad \mathbf{CZ = \frac{a}{e}}. \qquad ...(3)$$

Again subtracting (1) from (2), A′S – AS = e (ZA′ – ZA)

or $$(A'C + CS) - (AC - CS) = e \cdot AA' \quad \text{or} \quad 2CS = e \cdot 2a$$

$$\therefore \quad \mathbf{CS = ae.} \qquad ...(4)$$

Take C as origin, AA′ as the x-axis and CY, a line through C and ⊥ to AA′ as the y-axis. From (3) and (4), the co-ordinates of S are $(-ae, 0)$ and the equation of ZM is

$$x = -\frac{a}{e} \quad \text{or} \quad x + \frac{a}{e} = 0. \qquad \text{...(5)}$$

Let $P(x, y)$ be any point on the ellipse. Join SP. Draw PM ⊥ on ZM. Then by definition,

$$\frac{SP}{PM} = e \quad \text{or} \quad SP = ePM \qquad \text{...(6)}$$

But $SP = \sqrt{(x+ae)^2 + (y-0)^2} = \sqrt{(x+ae)^2 + y^2}$

and PM = perpendicular from $P(x, y)$ on (5)

$$= \frac{x + \frac{a}{e}}{1} = x + \frac{a}{e}$$

∴ From (6), $\sqrt{(x+ae)^2 + y^2} = e\left(x + \frac{a}{e}\right) = ex + a$

Squaring both sides, $x^2 + 2aex + a^2e^2 + y^2 = e^2x^2 + aex + a^2$

or $x^2(1-e^2) + y^2 = a^2(1-e^2)$,

Dividing both sides by $a^2(1-e^2)$,

$$\frac{x^2}{a^2} + \frac{y^2}{a^2(1-e^2)} = 1 \qquad \because a^2(1-e^2) > 0.$$

Putting $a^2(1-e^2) = b^2$, the required equation is $\frac{x^2}{a^2} + \frac{y^2}{b^2} = 1.$

Observation : $b^2 < a^2$.

Note 1. The y-axis *i.e.*, $x = 0$ meets the ellipse $\frac{x^2}{a^2} + \frac{y^2}{b^2} = 1$, where $\frac{y^2}{b^2} = 1$ or $y = \pm b$

i.e., the y-axis meets the ellipse in two points B and B′, such that CB = CB′ = b.

Note 2. Definitions,

(*i*) The lines AA′ and BB′ are called the major and minor axes respectively. Both together are called **principal axes.**

(*ii*) The points A, A′, the extremities of the major axis are called the vertices of the ellipse.

(*iii*) C is the centre of the ellipse. The centre C bisects every chord of the ellipse which passes through it.

3. Tracing the ellipse $\frac{x^2}{a^2} + \frac{y^2}{b^2} = 1$

(*i*) **Symmetry.** Since the equation $\frac{x^2}{a^2} + \frac{y^2}{b^2} = 1$ contains even powers of x, therefore if (x, y) be any point on the ellipse, then $(-x, y)$ is also a point on the curve so that it is symmetrical about the y-axis.

Similarly, the curve is symmetrical about the x-axis because it contains even-powers of y.

(*ii*) **Intersection with axes.**

The curve meets the x-axis, *i.e.*, $y = 0$, where $\frac{x^2}{a^2} = 1$ $\therefore$ $x = \pm a$.

$\therefore$ It cuts the x-axis in point A $(-a, 0)$ and A′ $(a, 0)$.

Similarly, it cuts the y-axis, $x = 0$ in the point B$(0, b)$ and B′$(0, -b)$.

(*iii*) **Range.** From $\frac{x^2}{a^2} + \frac{y^2}{b^2} = 1$, we have

$y = \pm \frac{b}{a}\sqrt{a^2 - x^2}$, which shows that if x is numerically greater than a, y is imaginary.

$\therefore$ No portion of the curve lies to the right of A′ or to the left of A.

Again, solving for x, we have $x = \pm \frac{a}{b}\sqrt{b^2 - y^2}$. If y is greater than b numerically, x is imaginary.

$\therefore$ No portion of the curve lies above B or below B′.

(*iv*) **Shape.** From the equations

$$y = \pm \frac{b}{a}\sqrt{a^2 - x^2}$$

and

$$x = \pm \frac{a}{b}\sqrt{b^2 - y^2}$$

We conclude that as x increases, y decreases, and *vice versa*. Therefore, the curve is closed.

Hence the shape of the curve is as shown in the figure.

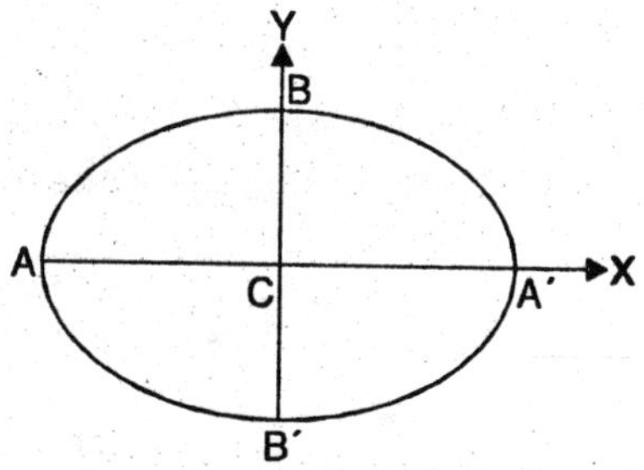

4. To prove that the ellipse has a second focus and a second directrix

Since the ellipse is symmetrical about the y-axis, therefore, if we take point S′ and Z′ on the x-axis such that CS′ = CS and CZ′ = CZ, then the point S′ will also be a focus of the ellipse and Z′M′, the line through Z′ and perpendicular to CZ′ will be the corresponding directrix.

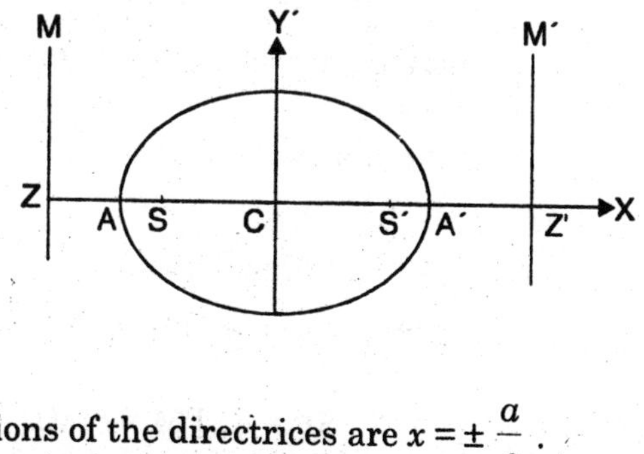

The co-ordinates of the focus S′ are $(ae, 0)$ and the equation of the second directrix is $x = \frac{a}{e}$. Therefore, the foci of the ellipse are $(\pm ae, 0)$ and the equations of the directrices are $x = \pm \frac{a}{e}$.

5. Latus Rectum

Def. A line through the focus and parallel to the directrix is called latus rectum.

To find the length of the latus rectum of the ellipse $\frac{x^2}{a^2} + \frac{y^2}{b^2} = 1$.

Let LSL′ be the latus rectum of the ellipse

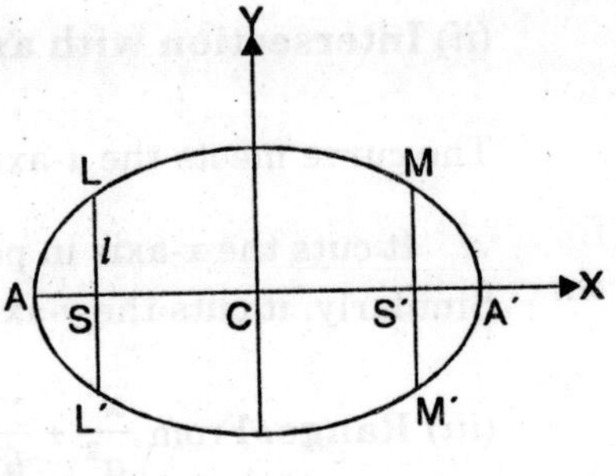

$$\frac{x^2}{a^2}+\frac{y^2}{b^2}=1 \quad \text{...(1)}$$

Let $SL = l$

$\therefore$ Co-ordinates of L are $(-ae, l)$.

Since L $(-ae, l)$ lies on the ellipse (1).

$$\therefore \quad \frac{a^2e^2}{a^2}+\frac{l^2}{b^2}=1 \quad \text{or} \quad \frac{l^2}{b^2}=1-e^2$$

or $$l^2 = b^2(1-e^2) = b^2 \cdot \frac{b^2}{a^2} \qquad \left(\because 1-e^2=\frac{b^2}{a^2}\right)$$

$$=\frac{b^4}{a^2}$$

$$\therefore \quad l = \pm\frac{b^2}{a} \qquad \therefore \quad LSL' = 2SL = 2\frac{b^2}{a}.$$

(The – ve sign corresponds to L′).

Cor. Co-ordinates of the ends of the latera recta.

The co-ordinates of the ends of latera recta are

$$L\left(-ae,\frac{b^2}{a}\right);\ L'\left(-ae,-\frac{b^2}{a}\right) \text{ and } M\left(ae,\frac{b^2}{a}\right);\ M'\left(ae,-\frac{b^2}{a}\right).$$

6. Focal Property

To show that the sum of the focal distances of a point on an ellipse is constant and equal to the major axis. *(K.U. 1995 S)*

Let P(x, y) be any point on the ellipse.

Let S and S′ be the two foci and ZM, Z′M′ the two directrices.

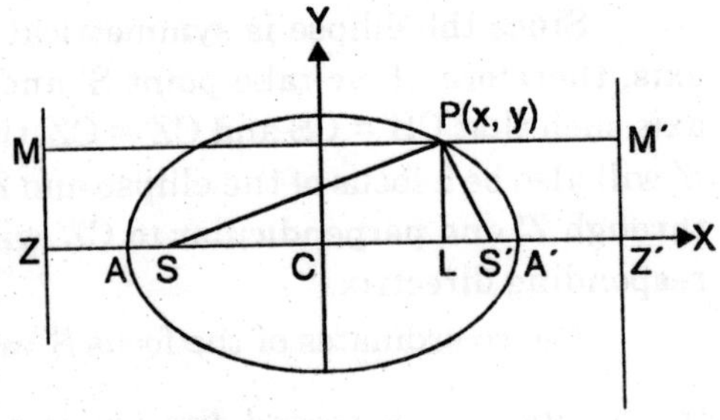

Join PS and PS′. From P draw PM and PM′ ⊥s on ZM and Z′M′ respectively.

Draw PL ⊥ on the x-axis.

By definition of the ellipse, we have

$$SP = e\,.\,PM = e\,.\,ZL = e(CL+CZ) = e\left(x+\frac{a}{e}\right) = ex + a \quad \text{...(1)}$$

Also, $$S'P = e\,.\,PM' = e\,.\,LZ' = e(CZ'-CL) = e\left(\frac{a}{e}-x\right) = a - ex \quad \text{...(2)}$$

$\therefore$ From (1) and (2), $SP + S'P = ex + a + a - ex = 2a$ (constant) = Major axis.

Cor. 1. Focal distances of any point P(x, y) on the ellipse are $a \pm ex$.

Cor. 2. Second definition of an ellipse.

The ellipse is the locus of a point which moves in a plane so that the sum of its distances from two fixed points in the plane is constant. The fixed points are the foci of the ellipse and the constant is the length of its major axis.

Note 1. The following table gives useful informations about the ellipse :

(*i*) $\frac{x^2}{a^2} + \frac{y^2}{b^2} = 1$ (*ii*) $\frac{x^2}{b^2} + \frac{y^2}{a^2} = 1.$

	$\frac{x^2}{a^2} + \frac{y^2}{b^2} = 1 \ (a > b)$	$\frac{x^2}{b^2} + \frac{y^2}{a^2} = 1 \ (a > b)$
1. Co-ordinates of foci	$(\pm ae, 0)$	$(0, \pm ae)$
2. Equation of major axis	$y = 0$	$x = 0$
3. Equation of minor axis	$x = 0$	$y = 0$
4. Length of major axis	$2a$	$2a$
5. Length of minor axis	$2b$	$2b$
6. Equation of directrices	$x = \pm \frac{a}{e}$	$\mathrm{y} = \pm \frac{a}{e}$
7. Length of latus rectum	$\frac{2b^2}{a}$	$\frac{2b^2}{a}$

Note 2. The figures of two ellipses are given below :

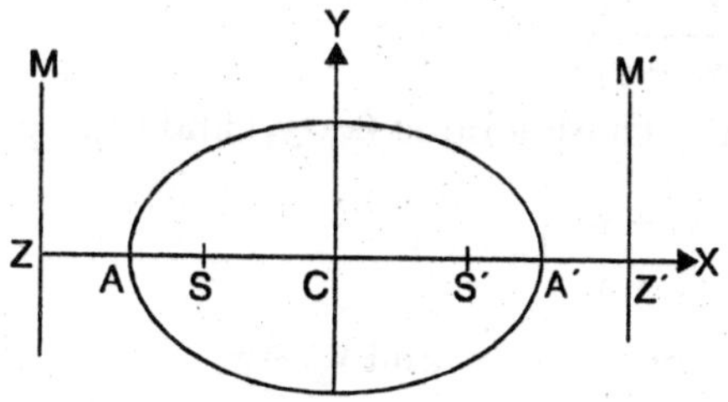

(i)

$$\frac{x^2}{a^2} + \frac{y^2}{b^2} = 1 \quad (a > b)$$

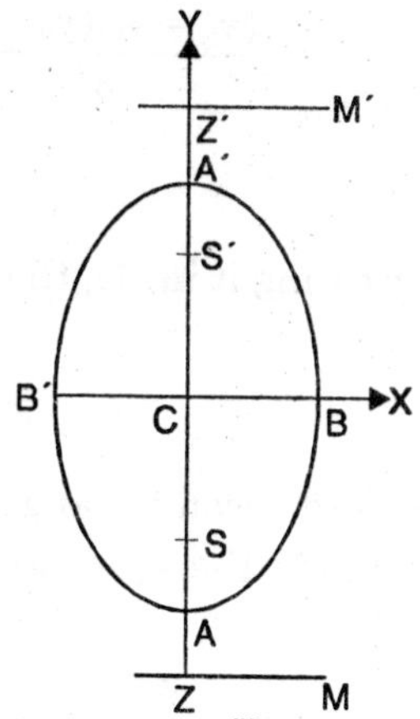

(ii)

$$\frac{x^2}{b^2} + \frac{y^2}{b^2} = 1 \quad (a > b)$$

EQUATIONS OF TANGENTS AND NORMALS

7. Point form of the Equation of the Tangent

To find the equation of the tangent at the point (x_1, y_1) of the ellipse

$$\frac{x^2}{a^2} + \frac{y^2}{b^2} = 1.$$

The equation of the ellipse is $\frac{x^2}{a^2} + \frac{y^2}{b^2} = 1.$

Let (x_2, y_2) be another point on the ellipse in the neighbourhood of (x_1, y_1).

$\therefore$ The equation of the chord joining (x_1, y_1) and (x_2, y_2) is

$$y - y_1 = \frac{y_2 - y_1}{x_2 - x_1}(x - x_1) \quad ...(1)$$

$\because$ (x_1, y_1) and (x_2, y_2) lie on the ellipse

$$\therefore \quad \frac{x_1^2}{a^2} + \frac{y_1^2}{b^2} = 1 \quad ...(2)$$

$$\frac{x_2^2}{a^2} + \frac{y_2^2}{b^2} = 1 \quad ...(3)$$

Subtracting (2) from (3), we have

$$\frac{x_2^2 - x_1^2}{a^2} + \frac{y_2^2 - y_1^2}{b^2} = 0$$

or
$$\frac{(x_2 - x_1)(x_2 + x_1)}{a^2} + \frac{(y_2 - y_1)(y_2 + y_1)}{b^2} = 0$$

or
$$\frac{(y_2 - y_1)(y_2 + y_1)}{b^2} = -\frac{(x_2 - x_1)(x_2 + x_1)}{a^2}$$

or
$$\frac{y_2 - y_1}{x_2 - x_1} = -\frac{b^2}{a^2} \cdot \frac{x_2 + x_1}{y_2 + y_1}$$

Substituting it in (1), the equation of the chord joining (x_1, y_1) and (x_2, y_2) is

$$y - y_1 = -\frac{b^2}{a^2} \cdot \frac{x_2 + x_1}{y_2 + y_1}(x - x_1) \quad ...(4)$$

The chord becomes tangent at (x_1, y_1) when $x_2 \to x_1$ and $y_2 \to y_1$

$\therefore$ From (4), the equation of the tangent at (x_1, y_1) is

$$y - y_1 = -\frac{b^2}{a^2} \cdot \frac{2x_1}{2y_1}(x - x_1) \quad \text{or} \quad \frac{yy_1}{b^2} - \frac{y_1^2}{b^2} = -\frac{xx_1}{a^2} + \frac{x_1^2}{a^2}$$

or
$$\frac{xx_1}{a^2} + \frac{yy_1}{b^2} = \frac{x_1^2}{a^2} + \frac{y_1^2}{b^2} = 1 \quad \text{[From (2)]}$$

or
$$\mathbf{\frac{xx_1}{a^2} + \frac{yy_1}{b^2} = 1}$$

which is the required equation of the tangent.

8. Equation of the Normal. Point Form

To find the equation of the normal at the point (x_1, y_1) of the ellipse

$$\frac{x^2}{a^2}+\frac{y^2}{b^2}=1.$$

The equation of the ellipse is $\frac{x^2}{a^2}+\frac{y^2}{b^2}=1$ ...(1)

The equation of the tangent at (x_1, y_1) to (1) is $\frac{xx_1}{a^2}+\frac{yy_1}{b^2}=1$

Slope of the tangent $= -\frac{\frac{x_1}{a^2}}{\frac{y_1}{b^2}} = -\frac{b^2}{a^2}\cdot\frac{x_1}{y_1}$

$\therefore$ Slope of the normal $= \frac{a^2}{b^2}\frac{y_1}{x_1}$

$\therefore$ The equation of the normal at (x_1, y_1) is

$$y-y_1=\frac{a^2}{b^2}\cdot\frac{y_1}{x_1}(x-x_1) \quad i.e., \quad \frac{yb^2}{y_1}-b^2=\frac{a^2x}{x_1}-a^2$$

or $\frac{a^2x}{x_1}-\frac{b^2y}{y_1}=a^2-b^2$ which is the required equation of the normal.

9. Equation of the Tangent in the Slope Form

To find the equation of the tangent to the ellipse $\frac{x^2}{a^2}+\frac{y^2}{b^2}=1$, in the form

$$y = mx \pm \sqrt{a^2m^2+b^2}$$

The equation of the ellipse is $\frac{x^2}{a^2}+\frac{y^2}{b^2}=1$ ...(1)

The equation of the tangent at (x_1, y_1) to (1) is

$$\frac{xx_1}{a^2}+\frac{yy_1}{b^2}=1 \quad \text{...(2)}$$

If m be the slope of the tangent, then $m=-\frac{b^2x_1}{a^2y_1}$

$$\therefore \quad y_1=-\frac{b^2x_1}{a^2m} \quad \text{...(3)}$$

$\because$ (x_1, y_1) lies on (1),

$$\therefore \quad \frac{x_1^2}{a^2}+\frac{y_1^2}{b^2}=1 \quad \text{or} \quad \frac{x_1^2}{a^2}+\frac{b^2x_1^2}{a^4m^2}=1 \quad \text{[From (3)]}$$

or $$x_1^2=\frac{a^4m^2}{a^2m^2+b^2} \qquad \therefore \quad x_1=\pm\frac{a^2m}{\sqrt{a^2m^2+b^2}}$$

$\therefore$ From (3), $\qquad y_1 = \mp \dfrac{b^2}{a^2 m} \cdot \dfrac{a^2 m}{\sqrt{a^2 m^2 + b^2}} = \mp \dfrac{b^2}{\sqrt{a^2 m^2 + b^2}}$

Substituting these values of x_1, y_1 in (2), the equation of the tangent is

$$\frac{x}{a^2}\left(\pm \frac{a^2 m}{\sqrt{a^2 m^2 + b^2}}\right) + \frac{y}{b^2}\left(\mp \frac{b^2}{\sqrt{a^2 m^2 + b^2}}\right) = 1$$

or $$\mathbf{y = mx \pm \sqrt{a^2 m^2 + b^2}}.$$

Cor. To find the point of contact. The point of contact of the tangent is

$$\left(\mp \frac{\mathbf{a^2 m}}{\sqrt{\mathbf{a^2 m^2 + b^2}}}, \pm \frac{\mathbf{b^2}}{\sqrt{\mathbf{a^2 m^2 + b^2}}}\right).$$

CONDITION OF TANGENCY

10. To find the condition that the line y = mx + c may touch the ellipse $\dfrac{x^2}{a^2} + \dfrac{y^2}{b^2} = 1$ and hence to find the equation of the tangent.

The equation of the line is $\qquad y = mx + c \qquad$ ...(1)

and the equation of the ellipse is $\qquad \dfrac{x^2}{a^2} + \dfrac{y^2}{b^2} = 1 \qquad$ ...(2)

Eliminating y from (1) and (2), we have

$$\frac{x^2}{a^2} + \frac{(mx + c)^2}{b^2} = 1$$

or $$x^2(a^2m^2 + b^2) + 2a^2mcx + a^2(c^2 - b^2) = 0 \qquad \text{...(3)}$$

If the line (1) touches the ellipse (2), the equation (3) has equal roots and the condition for which is

$$4a^4m^2c^2 = 4(a^2m^2 + b^2) \,.\, a^2(c^2 - b^2)$$

or $$a^2m^2c^2 = a^2m^2c^2 + b^2c^2 - a^2m^2b^2 - b^4 \quad \text{or} \quad c^2 = a^2m^2 + b^2$$

$\therefore$ $\qquad \mathbf{c = \pm \sqrt{a^2 m^2 + b^2}} \qquad$ which is the required condition.

Substituting the value of c in (2), the equation of the tangent is

$$\mathbf{y = mx \pm \sqrt{a^2 m^2 + b^2}}.$$

Cor. The equation of any tangent to the ellipse $\dfrac{x^2}{a^2} + \dfrac{y^2}{b^2} = 1$ in the slope form is

$$\mathbf{y = mx + \sqrt{a^2 m^2 + b^2}}.$$

11. Two Tangents from a Point to an Ellipse

To show that from any point two tangents can be drawn to an ellipse which are real and distinct, coincident or imaginary according as the point is outside, on or inside the ellipse.

Let the equation of the ellipse be $\frac{x^2}{a^2} + \frac{y^2}{b^2} = 1$ ...(1)

Let $P(x_1, y_1)$ be any point in the plane of the ellipse.

The equation of any tangent to the ellipse (1) is $y = mx + \sqrt{a^2m^2 + b^2}$ for all values of m.

If it passes through $P(x_1, y_1)$, then we have

$$y_1 = mx_1 + \sqrt{a^2m^2 + b^2} \quad \text{or} \quad y_1 - mx_1 = \sqrt{a^2m^2 + b^2}$$

Squaring both sides, $y_1^2 + m^2x_1^2 - 2mx_1y_1 = a^2m^2 + b^2$

or $$(x_1^2 - a^2)m^2 - 2x_1y_1m + (y_1^2 - b^2) = 0 \quad ...(2)$$

which being a quadratic equation in m gives two values of m which are the slopes of the two tangents from (x_1, y_1).

Therefore, the equations of the two tangents are obtained by substituting the values of m one by one in $y - y_1 = m(x - x_1)$.

Hence in general two tangents can be drawn from (x_1, y_1) to the ellipse

$$\frac{x^2}{a^2} + \frac{y^2}{b^2} = 1.$$

These tangents are real and distinct, coincident or imaginary according as the roots of (2) are real and distinct, equal or imaginary.

i.e., according as discriminant of (2) + ve, zero or – ve

i.e., according as $4x_1^2y_1^2 - 4(x_1^2 - a^2)(y_1^2 - b^2)$ is + ve, zero or – ve

i.e., according as $x_1^2y_1^2 - (x_1^2y_1^2 - b^2x_1^2 - a^2y_1^2 + a^2b^2)$ is + ve zero or – ve

or according as $b^2x_1^2 + a^2y_1^2 - a^2b^2$ is + ve, zero or – ve

or according as $\frac{x_1^2}{a^2} + \frac{y_1^2}{b^2} - 1$ is + ve, zero or – ve

i.e., according as the point (x_1, y_1) lies outside, on, or inside the ellipse.

12. Director Circle

Def. It is the locus of the point of intersection of perpendicular tangents to an ellipse.

CHORD OF CONTACT

13. Equation of the Chord of Contact

To find the equation of the chord of contact of tangents drawn to the ellipse $\frac{x^2}{a^2} + \frac{y^2}{b^2} = 1$ from (x_1, y_1).

The equation of the ellipse is

$$\frac{x^2}{a^2} + \frac{y^2}{b^2} = 1 \quad ...(1)$$

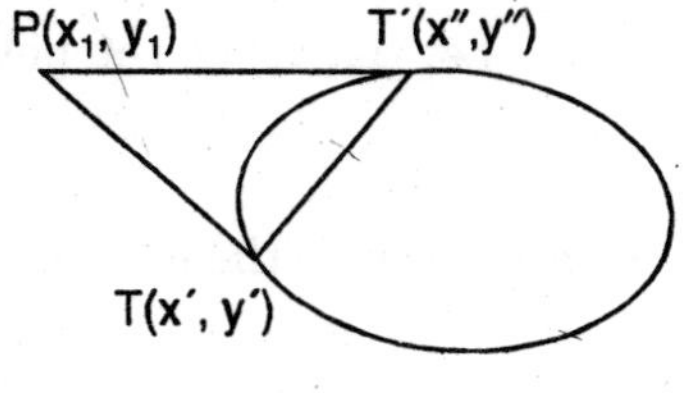

Let the points of contact of two tangents from the given point $P(x_1, y_1)$ be $T(x', y')$ and $T'(x'', y'')$.

$\therefore$ The equations of the tangents at T. T′ to (1) are

$$\frac{xx'}{a^2}+\frac{yy'}{b^2}=1 \quad ...(2) \quad \text{and} \quad \frac{xx''}{a^2}+\frac{yy''}{b^2}=1 \quad ...(3)$$

As these tangents pass through the point $P(x_1, y_1)$

$$\therefore \quad \frac{x_1x'}{a^2}+\frac{y_1y'}{b^2}=1 \quad ...(4) \quad \text{and} \quad \frac{x_1x''}{a^2}+\frac{y_1y''}{b^2}=1 \quad ...(5)$$

Equations (4) and (5) show that the points (x', y'), (x'', y'') lie on the line

$$\frac{x_1x}{a^2}+\frac{y_1y}{b^2}=1 \quad \text{or} \quad \frac{\mathbf{xx_1}}{\mathbf{a^2}}+\frac{\mathbf{yy_1}}{\mathbf{b^2}}=1$$

which is the required equation of the chord of contact.

POLE AND POLAR

14. Equation of the Polar

To find the equation of the polar of the point (x_1, y_1) with respect to the ellipse

$$\frac{x^2}{a^2}+\frac{y^2}{b^2}=1.$$

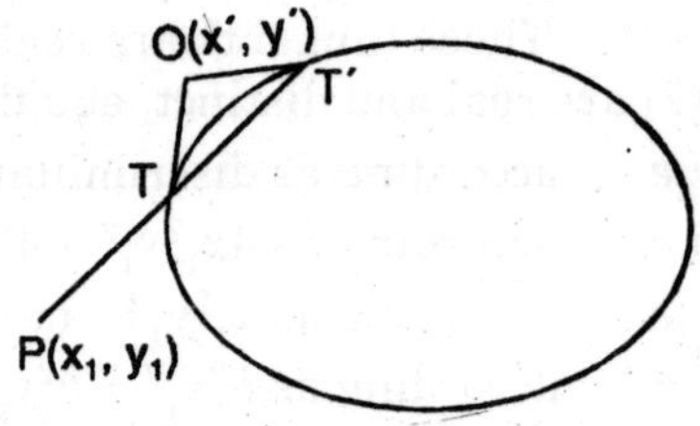

The equation of the ellipse is $\frac{x^2}{a^2}+\frac{y^2}{b^2}=1$ ...(1)

Let PTT′ be any straight line through $P(x_1, y_1)$ meeting the ellipse (1) in T, T′.

Let the tangents at T, T′ meet in $Q(x', y')$.

Now TT′ is the chord of contact of tangents from (x', y') to the ellipse (1),

$\therefore$ Its equation is $\frac{xx'}{a^2}+\frac{yy'}{b^2}=1$

As it passes through $P(x_1, y_1)$

$$\therefore \quad \frac{x_1x'}{a^2}+\frac{y_1y'}{b^2}=1$$

$\therefore$ Locus (x', y') is $\frac{x_1x}{a^2}+\frac{y_1y}{b^2}=1$ or $\frac{\mathbf{xx_1}}{\mathbf{a^2}}+\frac{\mathbf{yy_1}}{\mathbf{b^2}}=1,$

which is the required equation of the polar of the point $P(x_1, y_1)$.

15. Pole of a line

To find the pole of a line $Ax + By + C = 0$ with respect to the ellipse

$$\frac{x^2}{a^2}+\frac{y^2}{b^2}=1.$$

Let (x_1, y_1) be the pole of the line

$$Ax + By + C = 0 \quad ...(1)$$

with respect to the ellipse $\frac{x^2}{a^2}+\frac{y^2}{b^2}=1$ ...(2)

The equation of the polar of (x_1, y_1) w.r.t. (2) is

$$\frac{xx_1}{a^2} + \frac{yy_1}{b^2} = 1 \quad \text{or} \quad \frac{xx_1}{a^2} + \frac{yy_1}{b^2} - 1 = 0 \qquad \text{...(3)}$$

The lines (1) and (3) are the same.

∴ Comparing the co-efficients of like terms in (3) and (1) we have

$$\frac{\frac{x_1}{a^2}}{A} = \frac{\frac{y_1}{b^2}}{B} = \frac{-1}{C}$$

$$\therefore \qquad x_1 = -\frac{A}{B}a^2 \quad \text{and} \quad y_1 = -\frac{B}{C}b^2$$

Hence the required pole is $\left(-\frac{A}{C}a^2, -\frac{B}{C}b^2\right)$.

16. Reciprocal Properties of Pole and Polar

I. To show that if the polar of a point P with respect to an ellipse passes through a point Q, the polar of Q with respect to the same ellipse passes through P.

Let the equation of the ellipse be $\frac{x^2}{a^2} + \frac{y^2}{b^2} = 1$ ...(1)

Let the given points P, Q be (x_1, y_1) and (x_2, y_2) respectively.

The equation of the polar of $P(x_1, y_1)$ w.r.t. the ellipse (1) is

$$\frac{xx_1}{a^2} + \frac{yy_1}{b^2} = 1$$

Since it passes through $Q(x_2, y_2)$

$$\therefore \qquad \frac{x_2x_1}{a^2} + \frac{y_2y_1}{b^2} = 1 \qquad \text{...(2)}$$

The result (2) shows that the line

$$\frac{x_2x}{a^2} + \frac{y_2y}{b^2} = 1 \quad i.e., \quad \frac{xx_2}{a^2} + \frac{yy_2}{b^2} = 1 \quad \text{passes through } P(x_1, y_1).$$

∴ The polar of Q passes through P.

Hence the result.

Note. Such points are called conjugate points.

II. If the pole of a straight line $lx + my + n = 0$ w.r.t. an ellipse $\frac{x^2}{a^2} + \frac{y^2}{b^2} = 1$, lies on the straight line $l'x + m'y + n' = 0$, show that the pole of the latter with respect to the same ellipse lies on the former.

The equation of the ellipse is $\frac{x^2}{a^2} + \frac{y^2}{b^2} = 1$ ...(1)

The equation of the lines are $lx + my + n = 0$ ...(2)

and $l'x + m'y + n = 0$ ...(3)

Let (x_1, y_1) be the pole of the line (2)

$\therefore$ The equation of the polar of (x_1, y_1) w.r.t. (1) is

$$\frac{xx_1}{a^2} + \frac{yy_1}{b^2} = 1 \qquad ...(4)$$

The lines (2) and (4) are the same.

$\therefore$ Comparing the co-efficients of like terms, we have

$$\frac{\frac{x_1}{a^2}}{l} = \frac{\frac{y_1}{b^2}}{m} = \frac{-1}{n}$$

$$\therefore \qquad x_1 = -\frac{a^2 l}{n} \text{ and } y_1 = -\frac{b^2 m}{n}$$

$\therefore$ Pole of (2) is $\left(-\frac{a^2 l}{n}, -\frac{b^2 m}{n}\right)$

If this lies on (3), then $l'\left(-\frac{a^2 l}{n}\right) + m'\left(-\frac{b^2 m}{n}\right) + n' = 0$

i.e., $$a^2 ll' + b^2 mm' = nn'. \qquad ...(5)$$

The symmetry of the result (5) shows that it is also the condition that the pole of (3) lies on (2).

Note. Such lines are called conjugate lines.

PARAMETRIC CO-ORDINATES

17. Auxiliary Circle

Def. *The circle described on the major axis of an ellipse as diameter is called the* **auxiliary circle** *of the ellipse and its equation is* $x^2 + y^2 = a^2$.

18. To find a relation between the ordinates of the corresponding points of an ellipse and its auxiliary circle

Let P be any point on the ellipse $\frac{x^2}{a^2} + \frac{y^2}{b^2} = 1 \qquad ...(1)$

From P draw PN ⊥ on the major axis. Let NP produced meet the auxiliary circle in p ; join Cp.

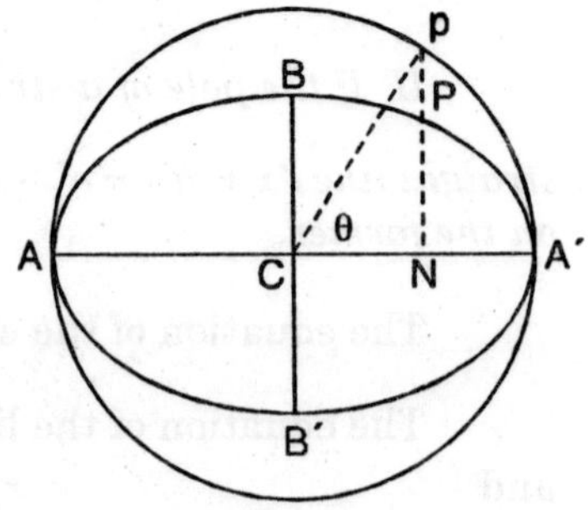

Let the co-ordinates of P and p be (x, NP) and $(x, \text{N}p)$ respectively.

Since P lies on the ellipse (1)

$$\therefore \qquad \frac{x^2}{a^2} + \frac{\text{NP}^2}{b^2} = 1 \qquad ...(2)$$

$\because$ p lies on the auxiliary circle $x^2 + y^2 = a^2$

$$\therefore \qquad x^2 + \text{N}p^2 = a^2 \quad \text{or} \quad \frac{x^2}{a^2} + \frac{\text{N}p^2}{a^2} = 1 \qquad ...(3)$$

Subtracting (3) from (2), we have

$$\frac{\text{NP}^2}{b^2} - \frac{\text{N}p^2}{a^2} = 0 \quad \text{or} \quad \frac{\text{NP}^2}{b^2} = \frac{\text{N}p^2}{a^2} \quad i.e., \quad \frac{\text{NP}^2}{\text{N}p^2} = \frac{b^2}{a^2} \quad \text{or} \quad \frac{\text{NP}}{\text{N}p} = \frac{b}{a}.$$

Therefore, the corresponding ordinates of the points on the ellipse and the auxiliary circle are in a constant ratio.

Definition. The point p in which the ordinate of P meets the auxiliary circle is called the corresponding point to P.

19. Eccentric Angle

Definition. *The angle θ which the radius from the centre to the corresponding point on the auxiliary circle makes with the major axis is called the* **eccentric angle** *of the corresponding point on the ellipse.*

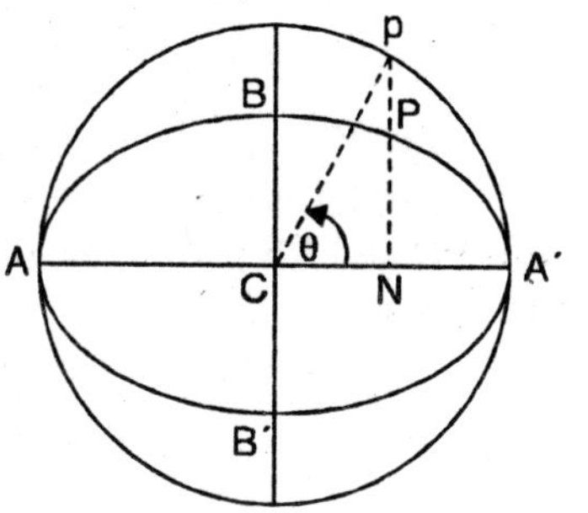

Let PN be the ordinate of any point P on the ellipse whose major axis is AA′. Produce NP to meet the auxiliary circle in p, and join Cp. Let $\angle NCp = \theta$. Then angle θ is called the **eccentric angle** of point P.

20. To express the co-ordinates of any point on the ellipse $\frac{x^2}{a^2} + \frac{y^2}{b^2} = 1$ in terms of parameter.

Let P(x, y) be any point on the ellipse $\frac{x^2}{a^2} + \frac{y^2}{b^2} = 1$ ...(1)

From P draw PN perpendicular to major axis AA′. Produce NP to meet the auxiliary circle in p.

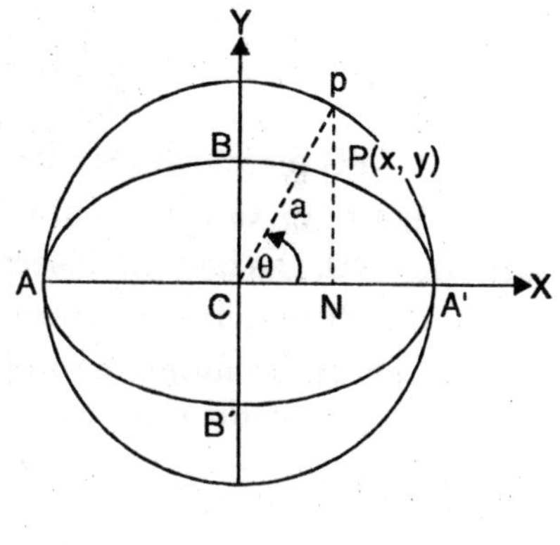

Join Cp and let $\angle NCp = \theta$.

From the right angled triangle CNp.

$$\frac{CN}{Cp} = \cos\theta$$

$$\therefore \quad CN = Cp\cos\theta$$

or $$x = a\cos\theta$$

∴ Co-ordinates of P are ($a\cos\theta$, y)

As it lies on the ellipse (1)

$$\therefore \quad \frac{a^2\cos^2\theta}{a^2} + \frac{y^2}{b^2} = 1 \quad \text{or} \quad \frac{y^2}{b^2} = 1 - \cos^2\theta = \sin^2\theta$$

or $$y^2 = b^2\sin^2\theta \qquad \therefore \quad y = b\sin\theta$$

Hence the co-ordinates of the point P are (**a cos θ, b sin θ**) and the equations **x = a cos θ, y = b sin θ** taken together are called the parametric equations of the ellipse, where θ is the parameter.

Note. The co-ordinates of any point on the ellipse are ($a\cos\theta$, $b\sin\theta$) θ being the parameter. It is called the point 'θ' of the ellipse.

GEOMETRICAL PROPERTIES OF ELLIPSE

21. Prove that the tangent and normal at any point of an ellipse bisect respectively the external and internal angles between the focal distances of the point.

Let the equation of the ellipse be $\frac{x^2}{a^2}+\frac{y^2}{b^2}=1$ ...(1)

Let $P(x_1, y_1)$ be any point on the ellipse.

Let PT be the tangent and PG be the normal. Join PS and PS′, where S and S′ are the two foci of the ellipse.

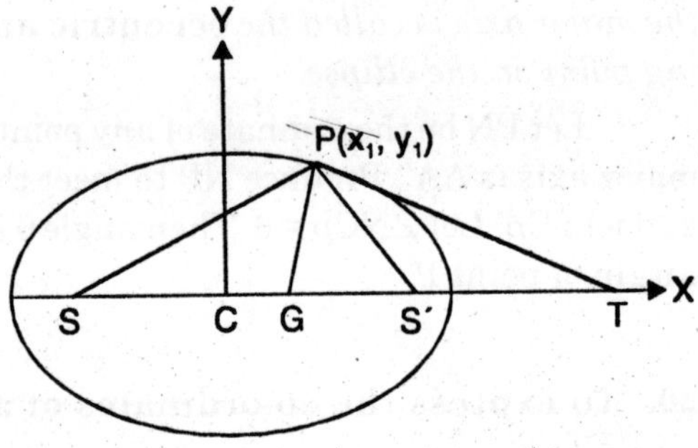

The equation of the normal at $P(x_1, y_1)$ is

$$\frac{a^2x}{x_1}-\frac{b^2y}{y_1}=a^2-b^2.$$

It meets the major axis where, putting $y = 0$

$$\therefore \quad \frac{a^2x}{x_1}=a^2-b^2$$

or
$$x=\frac{a^2-b^2}{a^2}x_1=\frac{a^2-a^2(1-e^2)}{a^2}x_1=e^2x_1$$

$$\therefore \quad CG=e^2x_1$$

Now
$$\frac{SG}{GS'}=\frac{CS+CG}{CS'-CG}=\frac{ae+e^2x_1}{ae-e^2x_1}=\frac{a+ex_1}{a-ex_1}=\frac{SP}{S'P}$$

∴ PG bisects angle SPS′

Again ∵ tangent is always ⊥ to the normal *i.e.*, PT is ⊥ to PG and since PG, the normal, bisects the interior angle between SP and S′P.

∴ PT bisects the exterior angle between SP and S′P.

Cor. *If the normal at any point P of an ellipse with centre C meets the major axis in G, and PN is perpendicular on the major axis then*

$$CG = e^2 . GN.$$ | See figure with Art. 22

22. Sub-tangent and Sub-normal

To find the lengths of the sub-tangent and sub-normal at any point of an ellipse.

Let the equation of the ellipse be

$$\frac{x^2}{a^2}+\frac{y^2}{b^2}=1.$$

Let the co-ordinates of the given point P be (x_1, y_1).

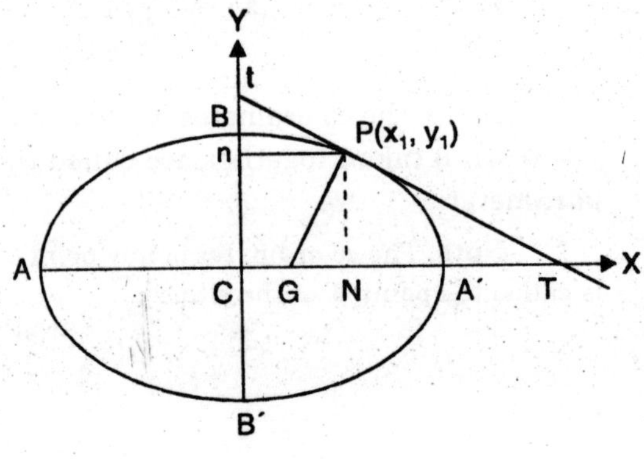

Let PT and PG be the tangent and normal at P. Draw PN ⊥ AA.

The equation of the tangent at $P(x_1, y_1)$ is

$$\frac{xx_1}{a^2}+\frac{yy_1}{b^2}=1$$

It meets x-axis, where (putting $y = 0$) in T

$$\therefore \quad \frac{xx_1}{a^2} = 1 \qquad i.e., \quad x = \frac{a^2}{x_1}$$

or
$$CT = \frac{a^2}{x_1} \qquad i.e., \quad CT.CN = a^2$$

$$\therefore \text{ Sub-tangent at } \quad P = NT = CT - CN = \frac{a^2}{x_1} - x_1 = \frac{a^2 - x_1^2}{x_1}$$

Again, the equation of the normal at $P(x_1, y_1)$ is

$$\frac{a^2 x}{x_1} - \frac{b^2 y}{y_1} = a^2 - b^2$$

It meets the x-axis, where putting $y = 0$

$$\frac{a^2 x}{x_1} = a^2 - b^2 \quad \text{or} \quad x = \frac{x_1}{a^2}(a^2 - b^2) = \frac{x_1}{a^2} a^2 e^2 \qquad | \because \; a^2 - b^2 = a^2 e^2$$

$$\therefore \quad x = e^2 x_1$$

i.e.,
$$CG = e^2.CN$$

$$\therefore \text{ Sub-normal at } P = GN = CN - CG = x_1 - e^2 x_1 = (1 - e^2)x_1 = \frac{b^2}{a^2} x_1.$$

Cor. 1. *If the tangent at any point P of an ellipse with centre C meets the major axis AA′ in T and PN is perpendicular on the major axis, then CT.CN = CA².*

Cor. 2. *If the tangent at any point P of an ellipse with centre C meets the minor axis BB′ in t and Pn is perpendicular on the minor axis, then*

$$Ct.Cn = CB^2.$$

23. To prove that the feet of the perpendiculars from the foci on any tangent to an ellipse lie on the auxiliary circle and the product of the perpendiculars is equal to the square on the semi-minor axis.

Let the equation of the ellipse be $\frac{x^2}{a^2} + \frac{y^2}{b^2} = 1$

(*a*) The equation of the any tangent to the ellipse is

$$y = mx + \sqrt{a^2 m^2 + b^2} \qquad \text{...(1)}$$

Let SM and S′M′ be the ⊥s from the foci upon the tangent (1).

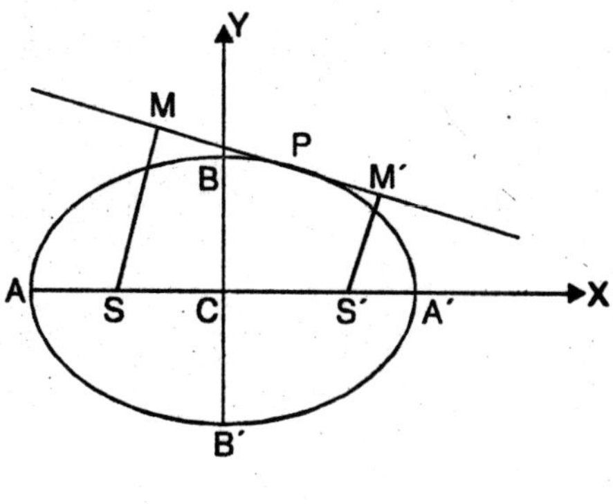

The equation of SM *i.e.,* the line from S(– *ae*, 0) and ⊥ to (1) is

$$y - 0 = -\frac{1}{m}(x + ae) \qquad \text{...(2)}$$

From (1) and (2), we have

$$y - mx = \sqrt{a^2 m^2 + b^2} \qquad \text{...(3)}$$

and
$$x + my = -ae \qquad \text{...(4)}$$

The locus of the foot of the perpendicular is obtained by eliminating m between (3) and (4),

Squaring (3) and (4) and adding

$$y^2 - 2mxy + m^2x^2 + x^2 + 2mxy + m^2y^2 = a^2m^2 + b^2 + a^2e^2$$

or $$(1 + m^2)x^2 + (1 + m^2)y^2 = a^2m^2 + a^2(1 - e^2) + a^2e^2 \qquad [\because \ b^2 = a^2(1 - e^2)]$$

$$= (1 + m^2)a^2$$

or $$x^2 + y^2 = a^2$$

which is the required locus and is the auxiliary circle of the ellipse.

Similarly, the locus of M′, the foot of the perpendicular from focus S′ is also the auxiliary circle.

(*b*) The equation of the tangent (1) can be written as

$$mx - y + \sqrt{a^2m^2 + b^2} = 0 \qquad ...(5)$$

$\therefore$ SM = length of the ⊥ from S($-ae$, 0) on (5)

$$= \frac{-aem + \sqrt{a^2m^2 + b^2}}{\sqrt{m^2 + 1}}$$

S′M′ = length of the ⊥ from S′(ae, 0) on (5)

$$= \frac{aem + \sqrt{a^2m^2 + b^2}}{\sqrt{m^2 + 1}}$$

$$\therefore \quad \text{SM.S}'\text{M}' = \frac{(a^2m^2 + b^2) - a^2e^2m^2}{m^2 + 1} = \frac{a^2m^2(1 - e^2) + b^2}{m^2 + 1}$$

$$= \frac{m^2b^2 + b^2}{m^2 + 1} = \frac{b^2(m^2 + 1)}{(m^2 + 1)} = b^2 \qquad [\because \ b^2 = a^2(1 - e^2)]$$

$$= b^2 = \text{CB}^2.$$

Hence semi-minor axis is the mean proportional between the lengths of the ⊥s from the foci on any tangent to the ellipse.

24. To find the locus of the middle points of a system of parallel chords

The equation of the ellipse is $\frac{x^2}{a^2} + \frac{y^2}{b^2} = 1$.

Let the slope of the system of parallel chords be m.

If M(x_1, y_1) be the middle point of any one chord PQ belonging to the system, then the equation of PQ is

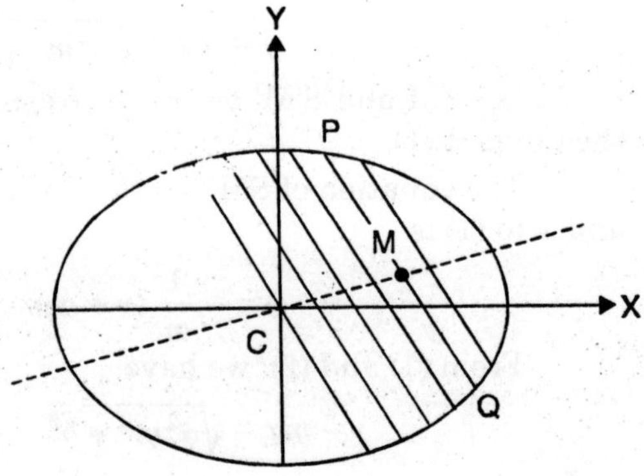

$$\frac{xx_1}{a^2} + \frac{yy_1}{b^2} - 1 = \frac{x_1^2}{a^2} + \frac{y_1^2}{b^2} - 1 \quad [\text{T} = \text{S}_1]$$

or $$\frac{xx_1}{a^2} + \frac{yy_1}{b^2} = \frac{x_1^2}{a^2} + \frac{y_1^2}{b^2}$$

Slope of PQ $= -\frac{b^2x_1}{a^2y_1} = m$ [*Given*]

$\therefore$ Locus of (x_1, y_1) is $\dfrac{b^2x}{a^2y} = m$ or $y = -\dfrac{b^2}{a^2m}x$ ...(1)

which is the required locus of the middle points of chords of slope m. It is a straight line passing through the centre of the ellipse and is called a **diameter** of the ellipse.

Def. Diameter *is the locus of the middle points of a system of parallel chords.*

25. Conjugate Diameters

The equation (1) in Art. 24 becomes

$$y = m'x \qquad ...(2)$$

if
$$m' = -\frac{b^2}{a^2m} \quad \text{or} \quad mm' = -\frac{b^2}{a^2}$$

Now the parallel chords of slope m are parallel to the diameter $y = mx$.

This shows that the diameter $y = m'x$ bisects all chords parallel to the diameter

$$y = mx \quad \text{if} \quad mm' = -\frac{b^2}{a^2} \qquad ...(3)$$

$\because$ (3) remains unchanged when we interchange m and m'.

$\therefore$ All chords parallel to the diameter $y = m'x$ will also be bisected by the diameter $y = mx$.

Hence we have the property : *Of two diameters of an ellipse, if one bisects chords parallel to the second, then second also bisects chords parallel to the first.*

Two such diameters are called conjugate diameters.

Conjugate Diameters. Def. *Two diameters such that each bisects chords parallel to the other are called conjugate diameters.*

Note. Condition for two conjugate diameters. If $y = mx$ and $y = m'x$ are conjugate diameters, then the product of slopes *i.e.*,

$$mm' = -\frac{b^2}{a^2}.$$

26. Properties of Conjugate Diameters

(*a*) *The eccentric angles of the extremities of two conjugate semi-diameters of an ellipse differ by a right angle.*

Let CP and CD be two conjugate semi-diameters of the ellipse

$$\frac{x^2}{a^2} + \frac{y^2}{b^2} = 1$$

Let the eccentric angles of P, D be ϕ, ϕ' respectively.

Then, co-ordinates of the points P and D are P($a\cos\phi$, $b\sin\phi$) and D($a\cos\phi'$, $b\sin\phi'$)

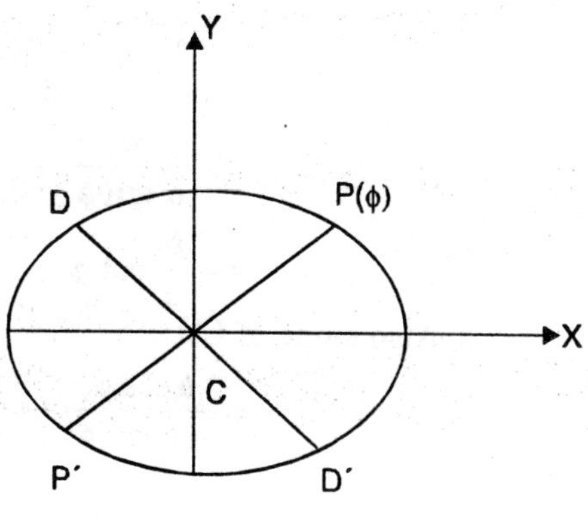

$\therefore$ Slope of CP $= m_1 = \dfrac{b\sin\phi}{a\cos\phi}$

and slope of CD $= m_2 = \dfrac{b\sin\phi'}{a\cos\phi'}$

But CP and CD are conjugate diameters,

$$\therefore \quad m_1 m_2 = -\frac{b^2}{a^2} \quad \text{or} \quad \frac{b^2 \sin\phi \sin\phi'}{a^2 \cos\phi \cos\phi'} = -\frac{b^2}{a^2}$$

or $$\sin\phi \sin\phi' + \cos\phi \cos\phi' = 0$$

or $$\cos(\phi - \phi') = 0 \quad i.e., \quad \phi \sim \phi' = \frac{\pi}{2}$$

$\therefore$ Eccentric angles differ by a right angle.

Note. If PC and DC be produced to meet the ellipse in P′ and D′ and if the eccentric angle of P be ϕ, then the eccentric angles of D, P′ and D′ are $\phi + \frac{\pi}{2}, \pi + \phi, \frac{3\pi}{2} + \phi$.

(b) *The sum of the squares of conjugate semi-diameters of an ellipse is constant and is equal to the sum of the squares of the semi-axes of the ellipse.*

Or

If CP and CD be two conjugate semi-diameters of an ellipse, then

$$CP^2 + CD^2 = a^2 + b^2.$$

Let ϕ be the eccentric angle of P.

Then, co-ordinates of P are (a, cos ϕ, b sin ϕ)

$\therefore$ The co-ordinates of D are

$$\left[a\cos\left(\frac{\pi}{2}+\phi\right), b\sin\left(\frac{\pi}{2}+\phi\right)\right] \quad \text{or} \quad (-a\sin\phi, b\cos\phi)$$

$$\therefore \quad CP^2 = a^2\cos^2\phi + b^2\sin^2\phi$$

and $$CD^2 = a^2\sin^2\phi + b^2\cos^2\phi$$

$$\therefore \quad CP^2 + CD^2 = a^2(\cos^2\phi + \sin^2\phi) + b^2(\sin^2\phi + \cos^2\phi) = a^2 + b^2$$

$\therefore$ $CP^2 + CD^2$ is constant.

(c) *To prove that the tangents at the extremities of a diameter of an ellipse are **parallel** to the conjugate diameters.*

Let the equation of the ellipse be $\frac{x^2}{a^2} + \frac{y^2}{b^2} = 1$ and $\phi, \phi + \frac{\pi}{2}$ be the eccentric angles of the extremities P, D of two conjugate diameters PCP′ and DCD′.

$\therefore$ Co-ordinates of P, D are (a cos ϕ, b sin ϕ) and ($-a$ sin ϕ, b cos ϕ).

$\therefore$ Equation of the tangent at P(a cos ϕ, b sin ϕ) is

$$\frac{x\cos\phi}{a} + \frac{y\sin\phi}{b} = 1.$$

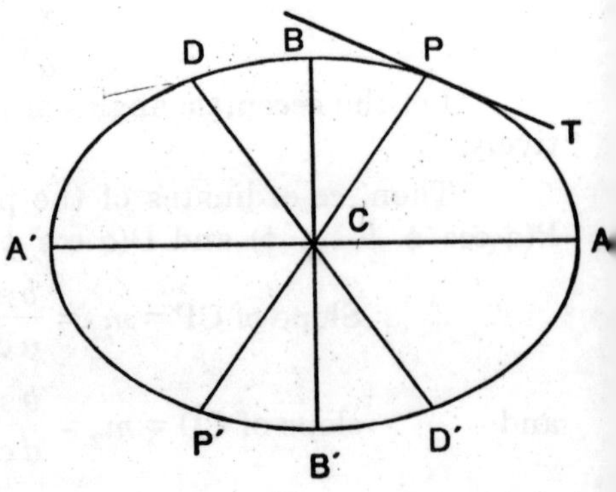

Its slope $= -\frac{b\cos\phi}{a\sin\phi}$

$$= -\frac{b}{a}\cot\phi \qquad \text{...(1)}$$

Also slope of CD

$$= \frac{b\cos\phi}{-a\sin\phi} = -\frac{b}{a}\cot\phi \qquad \text{...(2)}$$

From (1) and (2), because slopes are equal, therefore, the tangent at P is parallel to the conjugate diameter DCD′.

Similarly, we can prove that the tangent at P′ is parallel to CD, the tangent at D is parallel to CP and tangent at D′ is parallel to CP.

(d) *The area of the parallelogram formed by the tangents at the extremities of two conjugate diameters of an ellipse is constant and is equal to the product of the axes.*

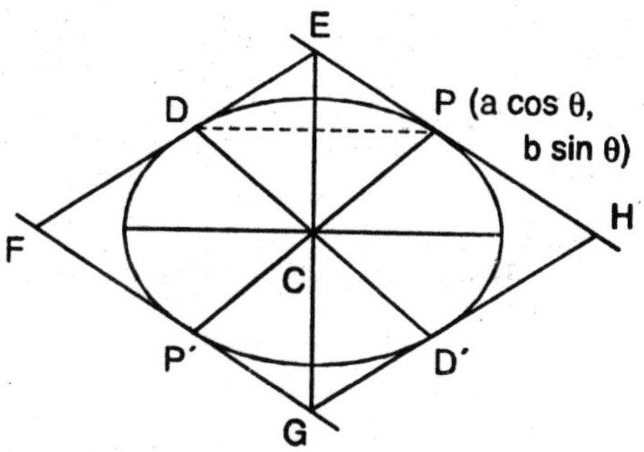

Let the equation of the ellipse be $\frac{x^2}{a^2}+\frac{y^2}{b^2}=1$ and ϕ, $\phi+\frac{\pi}{2}$ be the eccentric angles of the extremities P, D of two conjugate diameters PCP′ and DCD′.

Let EFGH be the parallelogram formed by the tangents at P, P′, D and D′. Join PD.

∴ Co-ordinates of P and D are $(a\cos\phi, b\sin\phi)$ and $(-a\sin\phi, b\cos\phi)$

∴ Area of ‖ gm EFGH = 4 times the area of ‖ gm EDCP

= 8 times the area of triangle CPD

$$= 8.\frac{1}{2}\begin{vmatrix} 0 & 0 & 1 \\ a\cos\phi & b\sin\phi & 1 \\ -a\sin\phi & b\cos\phi & 1 \end{vmatrix} = 4[ab\cos^2\phi + ab\sin^2\phi]$$

$$= 4ab = \text{constant} = 2a \times 2b = \text{Product of axes.}$$

27. Equi-Conjugate Diameters

Def. *When two conjugate diameters of an ellipse are equal in length, they are called* **equi-conjugate diameters.**

28. To find the equation of equi-conjugate diameters of the ellipse

$$\frac{x^2}{a^2}+\frac{y^2}{b^2}=1.$$

Let ϕ, $\phi+\frac{\pi}{2}$ be the eccentric angles of the extremities P, D of two conjugate diameters (Fig. Art. 26) PCP′ and DCD′.

∴ $$CP^2 = a^2\cos^2\phi + b^2\sin\phi \quad \text{...(1)}$$

$$CD^2 = a^2\sin^2\phi + b^2\cos^2\phi \quad \text{...(2)}$$

∵ PCP′ and DCD′ are equi-conjugate diameters.

∴ $$PCP' = DCD' \text{ or } CP = CD$$

i.e., $$CP^2 = CD^2$$

∴ From (1) and (2),

$$a^2\cos^2\phi + b^2\sin^2\phi = a^2\sin^2\phi + b^2\cos^2\phi$$

or $$a^2(\cos^2\phi - \sin^2\phi) = b^2(\cos^2\phi - \sin^2\phi)$$

or $$(a^2 - b^2)\cos 2\phi = 0 \quad \text{or} \quad \cos 2\phi = 0 \qquad [\because\ a^2 \neq b^2]$$

$\therefore \qquad 2\phi = \frac{\pi}{2}$ or $\frac{-\pi}{2}$ or $\phi = \frac{\pi}{4}$ or $\frac{-\pi}{4}$

Taking $\phi = \frac{\pi}{4}$, the equation of CD is $y = \frac{b}{a} \cdot \frac{\sin \pi/4}{\cos \pi/4} x = \frac{b}{a} x$.

Taking $\phi = -\frac{\pi}{4}$, the equation of CD is $y = \frac{b}{a} \frac{\sin\left(-\frac{\pi}{4}\right)}{\cos\left(-\frac{\pi}{4}\right)} x = -\frac{b}{a} x.$

Hence the equations of equi-conjugate diameters are $y = \pm \frac{b}{a} x$.

Note. The equations of equi-conjugate diameters show that these are equally inclined to the major axis of the ellipse.

Also, the tangents at the extremities of the major and minor axes are $x = \pm a$, $y = \pm b$.

The diagonals of the rectangle formed by these tangents are also given by $y = \pm \frac{b}{a} x$.

Hence the equi-conjugate diameters of an ellipse lie along the diagonals of the rectangle formed by the tangents at the ends of the major and minor axes.

29. Joint Equation of the Pair of Tangents

To find the joint equation of the pair of tangents drawn from $A(x_1, y_1)$ to the ellipse $\frac{x^2}{a^2} + \frac{y^2}{b^2} = 1$.

Let AB be any line through A (x_1, y_1) meeting the ellipse,

$$\frac{x^2}{a^2} + \frac{y^2}{b^2} = 1 \qquad \ldots(1)$$

in the points P and Q. Let C (x, y) be any point on the line AB.

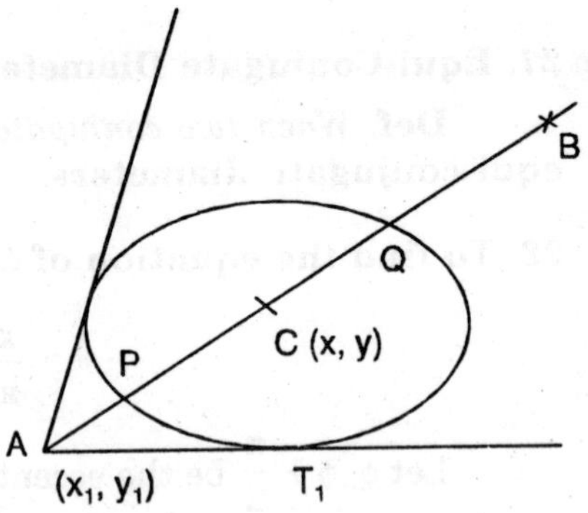

$\therefore$ The co-ordinates of any point which divide AC in the ratio $\lambda : 1$ are $\left(\frac{\lambda x + x_1}{\lambda + 1}, \frac{\lambda y + y_1}{\lambda + 1}\right)$

Let this point represent either of the points P and Q. Then its co-ordinates must satisfy (1).

$$\therefore \qquad \frac{(\lambda x + x_1)^2}{a^2} + \frac{(\lambda y + y_1)^2}{b^2} - (\lambda + 1)^2 = 0$$

or
$$\lambda^2 \left(\frac{x^2}{a^2} + \frac{y^2}{b^2} - 1\right) + 2\lambda \left(\frac{x x_1}{a^2} + \frac{y y_1}{b^2} - 1\right) + \left(\frac{x_1^2}{a^2} + \frac{y_1^2}{b^2} - 1\right) = 0 \qquad \ldots(2)$$

It is a quadratic equation in λ and hence gives two values of λ, one corresponding to P and the other to Q.

The line AB becomes tangent if the points P and Q coincide, *i.e.*, the two values of λ in (2) are equal.

$\therefore$ For equal roots, discriminant of (2) = 0

$$\therefore \qquad 4\left(\frac{xx_1}{a^2}+\frac{yy_1}{b^2}-1\right)^2 = 4\left(\frac{x^2}{a^2}+\frac{y^2}{b^2}-1\right)\left(\frac{x_1^2}{a^2}+\frac{y_1^2}{b^2}-1\right)$$

or
$$\left(\frac{x^2}{a^2}+\frac{y^2}{b^2}-1\right)\left(\frac{x_1^2}{a^2}+\frac{y_1^2}{b^2}-1\right) = \left(\frac{xx_1}{a^2}+\frac{yy_1}{b^2}-1\right)^2$$

which is the required equation of the tangents.

This give $\qquad SS_1 = T^2$

where $\qquad S \equiv \frac{x^2}{a^2}+\frac{y^2}{b^2}-1, \quad S_1 \equiv \frac{x_1^2}{a^2}+\frac{y_1^2}{b^2}-1$

and $\qquad T = \frac{xx_1}{a^2}+\frac{yy_1}{b^2}-1.$

3. Standard Results of The Hyperbola

EQUATION OF A HYPERBOLA

1. Hyperbola

Def. *A hyperbola is the locus of a point which moves in a plane so that its distance from a fixed point bears a constant ratio, greater than one, to its distance from a fixed straight line.*

The fixed point is called the **focus** and is generally denoted by S.

The fixed straight line is called the **directrix** of the hyperbola and the constant ratio is called the eccentricity of the hyperbola and is denoted by **e**.

Thus, if P be any point on the hyperbola whose focus is S and eccentricity e, and PM is drawn perpendicular on the directrix from P, then

$$\frac{SP}{PM} = e \quad (e > 1) \quad \text{or} \quad \mathbf{SP = e.PM}.$$

EQUATION OF A HYPERBOLA IN THE STANDARD FORM

2. To find the equation of a hyperbola in the form $\frac{x^2}{a^2} - \frac{y^2}{b^2} = 1$

Let S be the focus, ZM the directrix and ($e > 1$) the eccentricity of the hyperbola.

From S draw SZ perpendicular to the directrix, ZM. Divide SZ internally at A and externally at A′ in the ratio $e : 1$, so that

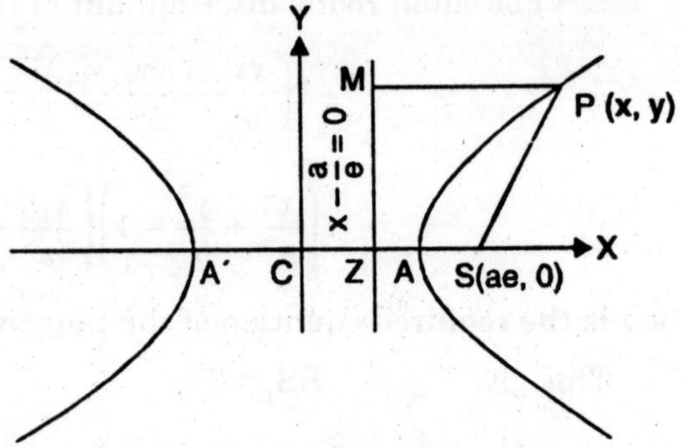

$$\frac{SA}{AZ} = \frac{e}{1} \quad \text{or} \quad SA = e.AZ \qquad ...(1)$$

or $$\frac{SA'}{A'Z} = \frac{e}{1} \quad \text{or} \quad SA' = e.A'Z \qquad ...(2)$$

Thus, by definition, A and A′ lie on the hyperbola. Let C be the middle point of AA′ and let AA′ = $2a$, then

$$CA = CA' = a$$

Adding (1) and (2), we have

$$SA + SA' = e[AZ + A'Z]$$

or $$(CS - CA)\,(CS + CA') = e\,.\,AA'$$

or $$2CS = e\,.\,2a \qquad (\because\ CA = CA',\ AA' = 2a)$$

or $$\mathbf{CS = ae.} \qquad ...(3)$$

Again, subtracting (1) from (2), we have

$$SA' - SA = e(A'Z - AZ)$$

or $$AA' = e[(CA' + CZ) - (CA - CZ)]$$

or $$2a = e\,.\,2CZ \qquad (\because\ CA = CA',\ AA' = 2a)$$

or $$CZ = \frac{a}{e}. \qquad ...(4)$$

Take C as origin, AA′ as the x-axis and CY, a line through C and ⊥ to AA′ as the y-axis.

∴ From (3) and (4), the co-ordinates of S are $(ae, 0)$ and the equation of ZM is

$$x = \frac{a}{e} \quad \text{or} \quad x - \frac{a}{e} = 0 \qquad ...(5)$$

Let P(x, y) be any point on the hyperbola. Join SP. Draw PM ⊥ on ZM.

Then by definition,

$$\frac{SP}{PM} = e \quad \text{or} \quad SP = e.PM \qquad ...(6)$$

But $$SP = \sqrt{(x - ae)^2 + (y - 0)^2} = \sqrt{(x - ae)^2 + y^2}$$

and PM = perpendicular from P(x, y) on (5)

$$= \frac{x - \frac{a}{e}}{1} = x - \frac{a}{e}$$

Substituting the values of SP and PM in (6)

$$\sqrt{(x - ae)^2 + y^2} = e.\left(x - \frac{a}{e}\right) = ex - a$$

or Squaring both sides

$$x^2 + a^2e^2 - 2aex + y^2 = e^2x^2 - 2aex + a^2$$

or $$x^2(e^2 - 1) - y^2 = a^2(e^2 - 1) \quad \text{or} \quad \frac{x^2}{a^2} - \frac{y^2}{a^2(e^2 - 1)} = 1.$$

As $$a^2(e^2 - 1) > 0.$$

Putting $a^2(e^2-1)=b^2$, the required equation is $\frac{\mathbf{x}^2}{\mathbf{a}^2}-\frac{\mathbf{y}^2}{\mathbf{b}^2}=\mathbf{1}$.

Note 1. The y-axis *i.e.*, $x=0$ meets the hyperbola $\frac{x^2}{a^2}-\frac{y^2}{b^2}=1$, where $y^2=-b^2$ or $y=\pm ib$

∴ The points of intersection are imaginary.

∴ The hyperbola does not meet the y-axis in real points.

Note 2. Definition :

(*i*) If B and B′ be two points on y-axis so that CB = CB′ = b, then BB′ is called the conjugate axis of the hyperbola. The line AA′ is called the transverse axis. Both AA′ and BB′ together are called the axes of the hyperbola.

(*ii*) The points A, A′, the extremities of the transverse axis are called the vertices of the hyperbola.

(*iii*) C is called the centre of the hyperbola. The centre C bisects every chord of the hyperbola which passes through it.

3. Rectangular Hyperbola

Def. A hyperbola is said to be rectangular when its two axes are equal, *i.e.*, $b=a$.

Its equation is $\mathbf{x^2-y^2=a^2}$.

In the case of a rectangular hyperbola, as

$$b=a \text{ and } b^2=a^2(e^2-1)$$

$$\therefore \quad a^2=a^2(e^2-1) \quad i.e., \quad 1=e^2-1 \quad \text{or} \quad e^2=2 \quad i.e., \quad e=\sqrt{2}.$$

PARAMETRIC CO-ORDINATES

4. Auxiliary Circle

The circle described on the transverse axis of a hyperbola as diameter is called the auxiliary circle of the hyperbola.

5. To express the co-ordinates of any point on the hyperbola $\frac{x^2}{a^2}-\frac{y^2}{b^2}=1$ in terms of a variable parameter

Let P(x, y) be any point on the hyperbola $\frac{x^2}{a^2}-\frac{y^2}{b^2}=1$ ...(1)

From P draw **PN** ⊥ to the transverse axis *i.e.*, the x-axis. From N draw NQ tangent to the auxiliary circle to touch it at Q. Join CQ and let ∠ACQ = θ.

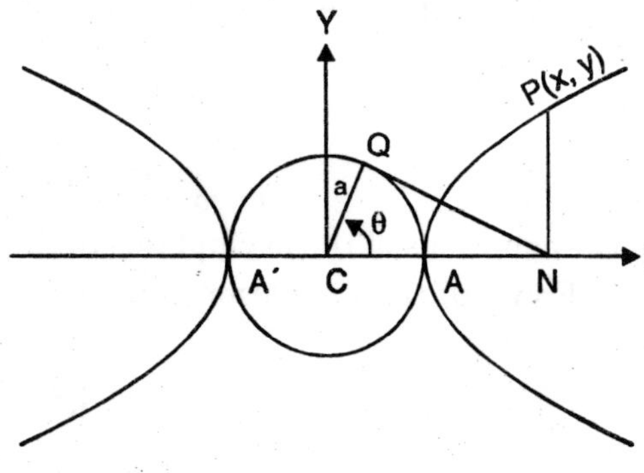

From the right angled triangle CQN,

$$\frac{\text{CN}}{\text{CQ}}=\sec\theta$$

or $$\text{CN}=\text{CQ}\sec\theta$$

or $$x=a\sec\theta$$

$\therefore$ Co-ordinates of P are $(a \sec \theta, y)$.

As it lies on the hyperbola (1)

$$\therefore \quad \frac{a^2 \sec^2 \theta}{a^2} - \frac{y^2}{b^2} = 1 \quad \text{or} \quad \frac{y^2}{b^2} = \sec^2 \theta - 1 = \tan^2 \theta$$

or
$$y^2 = b^2 \tan^2 \theta \quad \therefore \quad y = b \tan \theta$$

Hence the co-ordinates of the point P are **($\mathbf{a \sec \theta, b \tan \theta}$)** and the equations **$\mathbf{x = a \sec \theta}$, $\mathbf{y = b \tan \theta}$ taken together are called the parametric equations of the hyperbola, where θ is the parameter.**

Note. The co-ordinates of any point P on the hyperbola are $(a \sec \theta, b \tan \theta)$, θ being the parameter, P is called the point 'θ' on the hyperbola.

6. Asymptote

Def. *A straight line which meets the infinite branch of the curve in two coincident points at infinity* **but itself does not lie wholly at infinity is called asymptote to the curve.**

7. To find the equation of the asymptotes to the hyperbola

$$\frac{\mathbf{x}^2}{\mathbf{a}^2} - \frac{\mathbf{y}^2}{\mathbf{b}^2} = \mathbf{1}$$

The straight line
$$y = mx + c \quad \text{...(1)}$$

intersects the hyperbola
$$\frac{x^2}{a^2} - \frac{y^2}{b^2} = 1, \quad \text{...(2)}$$

in the points whose abscissae are given by

$$\frac{x^2}{a^2} - \frac{(mx + c)^2}{b^2} = 1$$

or
$$x^2\left(\frac{1}{a^2} - \frac{m^2}{b^2}\right) - \frac{2mc}{b^2}x - \frac{c^2}{b^2} - 1 = 0 \quad \text{...(3)}$$

For the line (1) to be asymptote, both the roots of (3) should be infinite.

$\therefore$ Co-efficients of x^2 and x are both zero separately.

i.e.,
$$\frac{1}{a^2} - \frac{m^2}{b^2} = 0 \quad \text{i.e.,} \quad m = \pm \frac{b}{a}$$

and
$$-\frac{2mc}{b^2} = 0 \quad \text{i.e.,} \quad c = 0 \qquad [\because \ m \neq 0]$$

Substituting these values of m and c in (1), we get

$$y = \pm \frac{b}{a} x \quad \text{...(4)}$$

or
$$\frac{x}{a} + \frac{y}{b} = 0 \quad \text{and} \quad \frac{x}{a} - \frac{y}{b} = 0 \quad \text{...(5)}$$

are the equations of the asymptotes to the hyperbola.

From (5), joint equation of the asymptotes is

$$\frac{\mathbf{x}^2}{\mathbf{a}^2} - \frac{\mathbf{y}^2}{\mathbf{b}^2} = \mathbf{0}.$$

Note 1. Joint equation of the asymptotes differs from that of the hyperbola only in the constant term.

2. The asymptotes are inclined at an angle $\tan^{-1}\dfrac{b}{a}$ to the axis.

3. Both these asymptotes pass through the centre (0, 0) of the hyperbola and touch the curve at the points at infinity.

Hence the asymptotes to the hyperbola may also be defined as tangents to the curve from its centre.

By using this definition of asymptotes, we can find their equation as follows :

8. Equations of Asymptotes

Equation of the hyperbola is $\dfrac{x^2}{a^2}-\dfrac{y^2}{b^2}=1.$

Equation of pair of tangents from the centre (0, 0) is

$$SS_1 = T^2$$

i.e.,
$$\left(\frac{x^2}{a^2}-\frac{y^2}{b^2}-1\right)(-1)=(-1)$$

or
$$-\frac{x^2}{a^2}+\frac{y^2}{b^2}+1=1 \quad i.e., \quad \frac{x^2}{a^2}-\frac{y^2}{b^2}=0$$

which is the same as in Art. 7.

9. Rectangular Hyperbola or Equilateral Hyperbola

A hyperbola whose asymptotes are at right angles to each other is called a rectangular hyperbola.

Now the asymptotes (a pair of lines) represented by $\dfrac{x^2}{a^2}-\dfrac{y^2}{b^2}=0$ are perpendicular when

$$\text{co-eff. of } x^2 + \text{co-eff. of } y^2 = 0$$

i.e.,
$$\frac{1}{a^2}-\frac{1}{b^2}=0 \text{ as } a^2=b^2 \text{ or } a=b.$$

$\therefore$ In the case of a rectangular hyperbola, the lengths of transverse and conjugate axes become equal and for this reason, a rectangular hyperbola is also called an **Equilateral hyperbola.**

Cor. In the case of a rectangular hyperbola $x^2 - y^2 = a^2$, $b = a$

$\therefore$ Equations of asymptotes are $y = \pm x$.

This shows that in case of a rectangular hyperbola, asymptotes are $\perp$ to each other and are inclined at an angle 45° to the axes.

10. Geometrical Construction for the Asymptotes

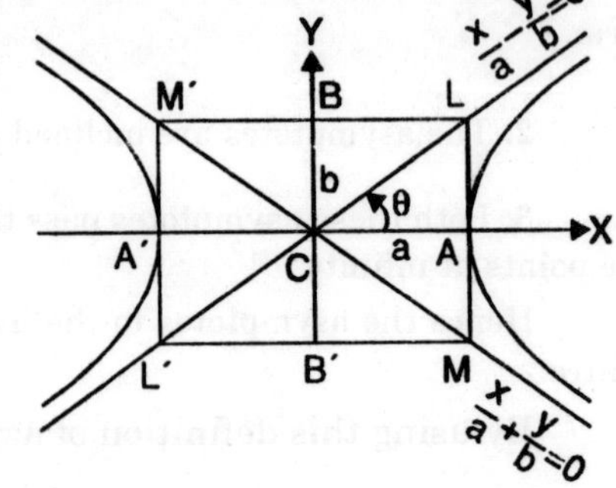

Let AA′ be the transverse axis and along the conjugate axis measure off CB and CB′, each equal to b. Through B and B′ draw parallels to the transverse axis and through A and A′ draw parallels to the conjugate axis and let these meet in L, M, L′ and M′. Thus the rectangle LML′M′ is formed. The equations of its diagonals are

$$y=\frac{b}{a}x \quad \text{and} \quad y=-\frac{b}{a}x$$

and these are, therefore, asymptotes.

11. Conjugate Hyperbola

Two hyperbolas such that the transverse and conjugate axes of either are respectively the conjugate transverse and axes of the other are called **conjugate hyperbolas.**

Let $\frac{x^2}{a^2}-\frac{y^2}{b^2}=1$ and $\frac{x^2}{b^2}-\frac{y^2}{a^2}=1$, which may be written as $\frac{x^2}{a^2}-\frac{y^2}{b^2}=-1$, **are** the equations of two conjugate hyperbolas. Clearly these two hyperbolas have the **same asymptotes.**

i.e.,
$$\frac{x^2}{a^2}-\frac{y^2}{b^2}=0.$$

These three equations can be written in the following systematic order

Hyperbola is $\frac{x^2}{a^2}-\frac{y^2}{b^2}-1=0$ ...(1)

Asymptotes are given by $\frac{x^2}{a^2}-\frac{y^2}{b^2}=0$ ...(2)

Conjugate hyperbola is $\frac{x^2}{a^2}-\frac{y^2}{b^2}+1=0$ ...(3)

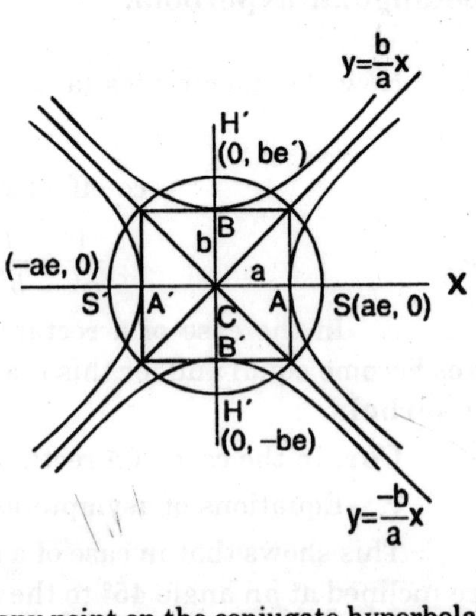

Thus we have the following **proposition :**

The equation of the asymptotes differs from that of the hyperbola by a constant and the equation of the conjugate hyperbola differs from that of the asymptotes by the same constant.

Symbolically. If the left hand sides of these equations (1), (2) and (3) be represented by H, A and H′, we have the relation

$$H+H'=2A.$$

i.e., **A is the Arithmetic Mean (A.M.) between H and H′.**

Note. ($a\tan\theta$, $b\sec\theta$) are the parametric co-ordinates of any point on the conjugate hyperbola.

Example 1. *Find the equation to the hyperbola whose asymptotes are the lines $x+2y+1=0$ and $2x+y+3=0$ and which pass through the point (1, 2).*

Sol. The joint equation of the asymptotes is

$$(x + 2y + 1)(2x + y + 3) = 0$$

Since the equation of the hyperbola differs from that of the asymptotes only by a constant, therefore,

Let $(x + 2y + 1)(2x + y + 3) + k = 0$ be the equation of the hyperbola.

As it passes through (1, 2),

$$\therefore \qquad (1 + 4 + 1)(2 + 2 + 3) + k = 0 \quad \text{or} \quad k = -42.$$

Hence the required equation of the hyperbola is

$$(x + 2y + 1)(2x + y + 3) - 42 = 0 \quad \text{or} \quad (x + 2y + 1)(2x + y + 3) = 42.$$

5

General Equation of the Second Degree Tracing of Conics

1. Introduction

We have already studied in the previous chapters that the equation of a conic section *i.e.*, a pair of straight lines, circle, parabola, ellipse and hyperbola, is of second degree in x and y. In the following article, we shall prove the converse problem that every second degree equation in x and y represents a conic.

2. To prove that the general equation of the second degree, i.e.,

$$ax^2 + 2hxy + by^2 + 2gx + 2fy + c = 0$$

always represents a conic section *(K.U. 1998 ; M.D.U. 2004)*

Let the most general equation of second degree in x and y referred to rectangular axes be

$$ax^2 + 2hxy + by^2 + 2gx + 2fy + c = 0 \qquad ...(1)$$

In order to remove the term containing xy, we rotate the axes through an angle θ. If (x', y') be the changed co-ordinates of the point $P(x, y)$, then the equations of transformations are

$$x = x' \cos\theta - y' \sin\theta$$
$$y = x' \sin\theta + y' \cos\theta$$

Substituting the values of x and y in (1), we have

$$a(x' \cos\theta - y' \sin\theta)^2 + 2h(x' \cos\theta - y' \sin\theta)(x' \sin\theta + y' \cos\theta)$$
$$+ b(x' \sin\theta + y' \cos\theta)^2 + 2g(x' \cos\theta - y' \sin\theta)$$
$$+ 2f(x' \sin\theta + y' \cos\theta) + c = 0 \qquad ...(2)$$

To remove the term containing $x'y'$, put coefficient of $x'y' = 0$

i.e., $$-2a\cos\theta\sin\theta + 2h(\cos^2\theta - \sin^2\theta) + 2b\sin\theta\cos\theta = 0$$

or $$(a - b)\sin 2\theta = 2h\cos 2\theta$$

i.e., $$\tan 2\theta = \frac{2h}{a - b}. \qquad ...(3)$$

Putting this value if θ in (2), the equation reduces to the form

$$Ax'^2 + By'^2 + 2Gx' + 2Fy' + C = 0 \qquad ...(4)$$

Now the following different cases arise :

Case I. When $A \neq 0$, $B \neq 0$

Equation can be written as

$$A\left(x'^2 + \frac{2G}{A}x'\right) + B\left(y'^2 + \frac{2F}{B}y'\right) = -C$$

or $$A\left(x'^2 + \frac{2G}{A}x' + \frac{G^2}{A^2}\right) + B\left(y'^2 + \frac{2F}{B}y' + \frac{F^2}{B^2}\right) = \frac{G^2}{A} + \frac{F^2}{B} - C$$

i.e., $$A\left(x' + \frac{G}{A}\right)^2 + B\left(y' + \frac{F}{B}\right)^2 = K$$

where $K = \frac{G^2}{A} + \frac{F^2}{B} - C$

Shifting the origin to $\left(\frac{-G}{A}, \frac{-F}{B}\right)$ by putting

$$x' = X - \frac{G}{A},\ y' = Y - \frac{F}{B}$$

The equation becomes $AX^2 + BY^2 = K$ ...(5)

Sub-case (*i*). If $K = 0$, then $AX^2 + BY^2 = 0$, which represents a *pair of straight lines* real or imaginary according as A and B are of opposite signs or of same sign.

Sub-case (*ii*). If $K \neq 0$, the equation may be written as

$$\frac{X^2}{K/A} + \frac{Y^2}{K/B} = 1$$

which represents :

(*a*) **an ellipse** if $\frac{K}{A}$ and $\frac{K}{B}$ are both positive *i.e.*, if A, B, K are of same sign.

(*b*) **a hyperbola** if $\frac{K}{A}$ and $\frac{K}{B}$ are of opposite signs, *i.e.*, if A and B are of opposite signs.

(*c*) **an imaginary ellipse** if $\frac{K}{A}$ and $\frac{K}{B}$ are both negative, *i.e.*, when A and B have same sign opposite to the sign of K.

(*d*) **a circle** if $A = B$.

Case II. Either A or $B = 0$.

Let us suppose $A = 0$, equation (4) now becomes

$$By'^2 + 2Fy' = -2Gx' - C$$

or $$B\left(y'^2 + \frac{2F}{B}y' + \frac{F^2}{B^2}\right) = -2Gx' - C + \frac{F^2}{B}$$

or $$\left(y'^2 + \frac{F}{B}\right)^2 = -\frac{2G}{B}\left\{x' - \left(\frac{F^2}{2BG} - \frac{C}{2G}\right)\right\}$$

Shifting the origin to $\left(\frac{F^2}{2BG} - \frac{C}{2G}, \frac{-F}{B}\right)$, the new equation becomes $Y^2 = -\frac{2G}{B}X$, which represents a parabola.

Note. (*i*) If $B = 0$ instead of A, the discussion is similar.

(*ii*) If $A = 0$ and also $G = 0$, equation (4) becomes

$$By'^2 + 2Fy' + C = 0$$

$$\therefore \quad y' = \frac{-2F \pm \sqrt{4F^2 - 4BC}}{2B} \quad \text{or} \quad y' = \frac{-F}{B} \pm \frac{\sqrt{F^2 - BC}}{B}$$

which represents a pair of parallel straight lines.

Thus in every case the *second degree equation represents a conic.*

3. Centre of a Conic

Definition. The centre of a conic is the point which bisects all chords passing through it.

The chords which pass through the centre of a conic are called the diameters of the conic.

4. General Equation of a Conic whose Centre lies at the Origin

Let the equation of the conic whose centre is at the origin be

$$ax^2 + 2hxy + by^2 + 2gx + 2fy + c = 0 \quad \text{...(1)}$$

As centre of the conic is at the origin (0, 0), it follows that if $P(x_1, y_1)$ is any point on the conic, then $Q(-x_1, -y_1)$ is also a point on (1).

Therefore, co-ordinates (x_1, y_1) and $(-x_1, -y_1)$ will satisfy the equation (1).

$$\therefore \quad ax_1^2 + 2hx_1y_1 + by_1^2 + 2gx_1 + 2fy_1 + c = 0 \quad \text{...(2)}$$

and

$$ax_1^2 + 2hx_1y_1 + by_1^2 - 2gx_1 - 2fy_1 + c = 0 \quad \text{...(3)}$$

Subtracting (3) from (2), we have

$$4(gx_1 + fy_1) = 0 \quad \text{or} \quad gx_1 + fy_1 = 0 \quad \text{...(4)}$$

Since equation (4) is satisfied for all values of x_1, y_1 it is an identity.

$$\therefore \quad g = 0 \quad \text{and} \quad f = 0$$

$\therefore$ equation (1) becomes

$$ax^2 + 2hxy + by^2 + c = 0$$

or

$$ax^2 + 2hxy + by^2 = -c$$

or

$$\left(-\frac{a}{c}\right)x^2 + \left(\frac{2h}{-c}\right)xy + \left(\frac{b}{-c}\right)y^2 = 1$$

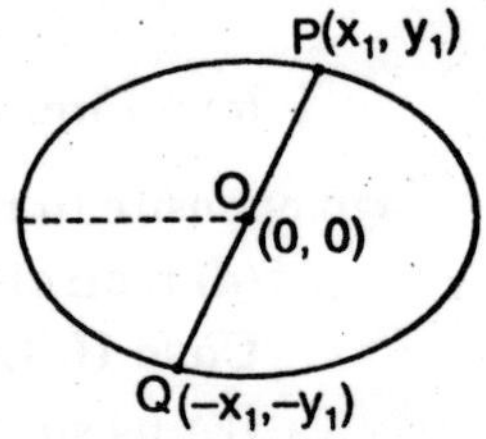

which can be put in the form

$$Ax^2 + 2Hxy + By^2 = 1.$$

It is the **STANDARD FORM** of the equation of the conic whose centre lies at the origin.

Note. The equation of the conic with centre at the origin, is free from first degree terms in x and y.

5. Co-ordinates of the Centre of the Conic

To find the co-ordinates of the centre of the conic

$$\mathbf{ax^2 + 2hxy + by^2 + 2gx + 2fy + c = 0} \quad \textit{(K.U. 1995 ; M.D.U. 2007)}$$

Let (x_1, y_1) be the centre of the conic

$$ax^2 + 2hxy + by^2 + 2gx + 2fy + c = 0 \quad \text{...(1)}$$

Shifting the origin to (x_1, y_1), axes remaining parallel to the original axes by putting $x = X + x_1$, $y = Y + y_1$ in (1), where (X, Y) are the co-ordinates in the new set of axes.

The equation of the conic referred to the new axes is

$$a(X + x_1)^2 + 2h(X + x_1)(Y + y_1) + b(Y + y_1)^2 + 2g(X + x_1) + 2f(Y + y_1) + c = 0$$

i.e.,

$$aX^2 + 2hXY + bY^2 + 2(ax_1 + hy_1 + g)X + 2(hx_1 + by_1 + f)Y + ax_1^2 + 2hx_1y_1 + by_1^2 + 2gx_1 + 2fy_1 + c = 0 \quad ...(2)$$

Since the centre of the conic (2) lies at the origin.

∴ Co-efficient of $X = 0$ and coefficient of $Y = 0$

i.e.,

$$ax_1 + hy_1 + g = 0 \quad ...(3)$$

$$hx_1 + by_1 + f = 0 \quad ...(4)$$

Solving for x_1 and y_1, we have

$$\frac{x_1}{hf - bg} = \frac{y_1}{gh - af} = \frac{1}{ab - h^2}$$

$$\therefore \quad x_1 = \frac{hf - bg}{ab - h^2}, \; y_1 = \frac{gh - af}{ab - h^2}$$

∴ Co-ordinates of the centre are $\left(\frac{hf - bg}{ab - h^2}, \frac{gh - af}{ab - h^2}\right)$ ...(5)

and the equation (2) reduces to

$$aX^2 + 2hXY + bY^2 + C_1 = 0$$

where

$$\begin{aligned} C_1 &= ax_1^2 + 2hx_1y_1 + by_1^2 + 2gx_1 + 2fy_1 + c = 0 \\ &= x_1(ax_1 + hy_1 + g) + y_1(hx_1 + by_1 + f) + gx_1 + fy_1 + c \\ &= gx_1 + fy_1 + c \qquad \text{[From (3) and (4)]} \\ &= g\left(\frac{hf - bg}{ab - h^2}\right) + f\left(\frac{gh - af}{ab - h^2}\right) + c \qquad \text{[From (5)]} \\ &= \frac{abc + 2fgh - af^2 - bg^2 - ch^2}{ab - h^2} \\ &= \frac{\Delta}{ab - h^2}, \text{ where } \Delta = abc + 2fgh - af^2 - bg^2 - ch^2 \end{aligned}$$

Hence the equation of conic, referred to parallel axes through the centre is

$$\mathbf{aX^2 + 2hXY + bY^2 + C_1 = 0}$$

where $C_1 = gx_1 + fy_1 + c$, (x_1, y_1) being the centre.

Note 1. Equation of conic referred to parallel axes, through the centre is

$$aX^2 + 2hXY + bY^2 = -C_1 \qquad \left[C_1 = \frac{\Delta}{ab - h^2}\right]$$

or dividing by $-C_1$, it can be written as

$$AX^2 + 2HXY + BY^2 = 1.$$

Note 2. If $f(x, y) = ax^2 + 2hyx + by^2 + 2gx + 2fy + c = 0$ be the equation of a conic, the centre of the conic is given by equations $\frac{\partial f}{\partial x} = 0$ and $\frac{\partial f}{\partial y} = 0$.

6. Summary of the Results

To know the nature of the general conic

$$ax^2 + 2hyx + by^2 + 2gx + 2fy + c = 0$$

We denote $\Delta = \begin{vmatrix} a & h & g \\ h & b & f \\ g & f & c \end{vmatrix} = abc + 2fgh - af^2 - bg^2 - ch^2$

Remember the following results (without proof).

Condition(s)			Nature of conic
If→			
1.	$\Delta = 0,$	$h^2 - ab \neq 0$	A pair of intersecting lines
2.	$\Delta = 0,$	$h^2 - ab = 0$	A pair of parallel lines
3.	$\Delta = 0,$	$a + b = 0$	A pair of perpendicular lines
4.	$\Delta \neq 0,$	$h^2 - ab = 0$	A parabola
5.	$\Delta \neq 0,$	$h^2 - ab < 0$	An ellipse
6.	$\Delta \neq 0,$	$h^2 - ab > 0$	A hyperbola
7.	$\Delta \neq 0,$	$h^2 - ab > 0$ $a + h = 0$	A rectangular hyperbola
8.	$\Delta \neq 0,$	$a = b,\ h = 0$	A circle
9.	If e is the eccentricity of the conic, then For parabola, $e = 1$; For ellipse, $e < 1$; For hyperbola, $e > 1$; For rectangular hyperbola $e = \sqrt{2}$.		

7. To Find the Asymptotes of the Conic

$$\mathbf{ax^2 + 2hxy + by^2 + 2gx + 2fy + c = 0}$$

Since the equation of the asymptotes differs that from the equation of the conic only in the constant term, hence let the equation of the asymptotes be

$$ax^2 + 2hxy + by^2 + 2gx + 2fy + c + \lambda = 0 \qquad \text{...(1)}$$

where λ is the arbitrary constant to be determined.

Since (1) represents a pair of straight lines, therefore, we have

$$ab(c + \lambda) + 2fgh - af^2 - bg^2 - (c + \lambda)h^2 = 0$$

or
$$abc + 2fgh - af^2 - bg^2 - ch^2 + \lambda(ab - h^2) = 0$$

or
$$\Delta + \lambda(ab - h^2) = 0 \quad \text{or} \quad \lambda = \frac{-\Delta}{ab - h^2}$$

Substituting this value in (1).

Equation of asymptotes is

$$ax^2 + 2hxy + by^2 + 2gx + 2fy + c - \frac{\Delta}{ab - h^2} = 0.$$

Note. Since the equation of the conjugate hyperbola differs from that of the asymptotes by the same constant, therefore the equation of the conjugate hyperbola is

$$ax^2 + 2hxy + by^2 + 2gx + 2fyk - \frac{2\Delta}{ab - h^2}.$$

ILLUSTRATIVE EXAMPLES

Example 1. *What conic does $13x^2 - 18xy + 37y^2 + 2x + 14y - 2 = 0$ represent ? Find its centre and the equation referred to the centre.*

Sol. The given equation is

$$13x^2 - 18xy + 37y^2 + 2x + 14y - 2 = 0 \quad \text{...(1)}$$

Comparing it with $ax^2 + 2hxy + by^2 + 2gx + 2fy + c = 0$

We have $a = 13, h = -9, b = 37, g = 1, f = 7$ and $c = -2$

$\therefore \quad \Delta = abc + 2fgh - af^2 - bg^2 - ch^2$

$= (13)(37)(-2) + 2(7)(1)(-9) - 13(49) - 37(1) + 2(81) = -1600 \neq 0$

and $\quad h^2 - ab = 81 - 13 \times 37 = 81 - 481$

$= -400 < 0$ and $\Delta \neq 0$

Therefore, the given equation represents an ellipse.

Let (x_1, y_1) be the centre of (1).

$\therefore$ Centre giving equations are

$$ax_1 + hy_1 + g = 0, \qquad hx_1 + by_1 + f = 0$$

$\Rightarrow \quad 13x_1 - 9y_1 + 1 = 0 \quad \text{...(2)} \qquad -9x_1 + 37y_1 + 7 = 0 \quad \text{...(3)}$

Solving (2) and (3), we have

$$\frac{x_1}{-63-37} = \frac{y_1}{-9-91} = \frac{1}{481-81}$$

or

$$\frac{x_1}{100} = \frac{y_1}{100} = \frac{1}{-400} \quad \therefore \quad x_1 = y_1 = -\frac{1}{4}$$

$\therefore$ Centre is $\left(-\frac{1}{4}, -\frac{1}{4}\right)$.

Equation of conic referred to centre as origin is

$$13X^2 - 18XY + 37Y^2 + C_1 = 0$$

where $C_1 = gx_1 + fy_1 + c = 1\left(-\frac{1}{4}\right) + 7\left(-\frac{1}{4}\right) - 2 = -4$

Hence required equation is $13X^2 - 18XY + 37Y^2 - 4 = 0$.

Example 2. *Find the centre of the conic $2x^2 - 72xy + 23y^2 - 4x - 28y - 48 = 0$ and show that it represents a hyperbola. Find its equations referred to the centre as origin.*

Sol. The given equation is

$$2x^2 - 72xy + 23y^2 - 4x - 28y - 48 = 0 \quad \text{...(1)}$$

Comparing it with $ax^2 + 2hxy + by^2 + 2gx + 2fy + c = 0$

We have $a = 2, h = -36, b = 23, g = -2, f = -14, c = -48$

$\therefore \quad \Delta = abc + 2fgh - af^2 - hg^2 - ch^2$

$= (2)(23)(-48) + 2(-14)(-2)(-36) - 2(196) - 23(4) + 48(-36)^2$

$= -2208 - 2016 - 392 - 92 + 62208$

$= -4708 + 62208 = 57500 \neq 0$

and $\quad h^2 - ab = 1296 - (2)(23) = 1250 > 0$ and $\Delta \neq 0$

Therefore, the given equation represents a hyperbola.

Let (x_1, y_1) be the centre of (1).

$\therefore$ Centre giving equation are

$$ax_1 + hy_1 + g = 0, \qquad hx_1 + by_1 + f = 0$$

$$\Rightarrow \quad 2x_1 - 36y_1 - 2 = 0 \qquad - 36x_1 + 23y_1 - 14 = 0$$

i.e., $x_1 - 18y_1 - 1 = 0$ ...(1) and $36x_1 - 23y_1 + 14 = 0$ ...(2)

Solving (1) and (2), we get

$$\frac{x_1}{-252-23} = \frac{y_1}{-36-14} = \frac{1}{-23+608}$$

or $$\frac{x_1}{-275} = \frac{y_1}{-50} = \frac{1}{625} \quad \Rightarrow \quad x_1 = \frac{-11}{25}, y_1 = \frac{-2}{25}$$

$\therefore$ Centre is $\left(\frac{-11}{25}, \frac{-2}{25}\right)$

The equation of the conic referred to centre as origin is

$$2X^2 - 27XY + 23Y^2 + c_1 = 0 \qquad ...(3)$$

where $c_1 = gx_1 + fy_1 + c$

$$= (-2)\left(\frac{-11}{25}\right) + (-14)\left(\frac{-2}{25}\right) - 48 = \frac{22}{25} + \frac{28}{25} - 48 = -46$$

$\therefore$ From (3), required equation is

$$2X^2 - 27XY + 23Y^2 = 46.$$

Example 3. *Find the co-ordinates of the centre of the conic and also the equation of conic referred to centre as origin of the conic*

$$14x^2 - 4xy + 11y^2 - 44x - 58y + 71 = 0. \qquad \text{(M.D.U. 1995, 2001)}$$

Sol. Comparing the given equation with

$$ax^2 + 2hxy + by^2 + 2gx + 2fy + c = 0$$

Here $a = 14, b = 11, c = 71, h = -2, g = -22, f = -29$

Let (x_1, y_1) be the centre.

$\therefore$ Centre giving equations are

$$ax_1 + hy_1 + g = 0, \qquad hx_1 + by_1 + f = 0$$

$$\Rightarrow \quad 14x_1 - 2y_1 - 22 = 0 \quad ...(1) \qquad -2x_1 + 11y_1 - 29 = 0 \quad ...(2)$$

Solving for x_1 and y_1,

or $$\frac{x_1}{58+242} = \frac{y_1}{44+406} = \frac{1}{154-4}$$

i.e., $$\frac{x_1}{300} = \frac{y_1}{450} = \frac{1}{150} \qquad \therefore \quad x_1 = 2, y_1 = 3$$

$\therefore$ Centre is (2, 3).

Equation of conic referred to centre as origin is

$$14X^2 - 4XY + 11Y^2 + C_1 = 0$$

where $C_1 = gx_1 + fy_1 + c = -22x_1 - 29y_1 + 71 = -22(2) - 29(3) + 71 = -60$

Hence required equation is

$$14X^2 - 4XY + 11Y^2 - 60 = 0.$$

Example 4. *What curves are represented by the equation*

$$x^2 + 2xy + (1 + \lambda)y^2 + 2\lambda y - 1 = 0$$

for different values of constant λ ? (K.U. 1995 S ; M.D.U. 2007)

Sol. The given equation is

$$x^2 + 2xy + (1 + \lambda)y^2 + 2\lambda y - 1 = 0 \qquad ...(1)$$

Here $h = 1, a = 1, b = 1 + \lambda$

$\therefore \quad h^2 - ab = (1)^2 - (1)(1 + \lambda) = -\lambda$

(i) If $h^2 - ab < 0$, i.e., if $-\lambda < 0 \Rightarrow \lambda > 0$, the curve represents an ellipse.

(ii) If $h^2 - ab > 0$, i.e., if $\lambda < 0$, the curve represents a hyperbola.

(iii) If $h^2 - ab = 0$, i.e., if $\lambda = 0$,

the curve is either a parabola or a pair of parallel straight lines.

$\therefore$ From (1), when $\lambda = 0$, the curve reduces to

$$x^2 + 2xy + y^2 - 1 = 0 \quad \text{or} \quad (x + y)^2 - 1 = 0$$

i.e., $$(x + y - 1)(x + y + 1) = 0$$

which represent a pair of parallel straight lines.

Example 5. *Find the asymptotes of the hyperbola*

$$x^2 - 4xy - 5y^2 + 6x + 42y - 63 = 0.$$

Also find the equation of the conjugate hyperbola.

Sol. Equation of the hyperbola is

$$x^2 - 4xy - 5y^2 + 6x + 42y - 63 = 0 \qquad ...(1)$$

Comparing it with $ax^2 + 2hxy + by^2 + 2gx + 2fy + c = 0$

Here $a = 1, \; b = -5, \; c = -63, \; f = 21, \; g = 3, h = -2$

$$\therefore \quad \Delta = abc + 2fgh - af^2 - bg^2 - ch^2$$
$$= 1(-5)(-63) + 2(21)(3)(-2) - 1 . (21)^2 + 5(3)^2 + 63(-2)^2$$
$$= 315 - 252 - 441 + 45 + 252 = -81$$

$$\therefore \quad \frac{\Delta}{ab - h^2} = \frac{-81}{-5-4} = 9$$

$\therefore$ Equation of asymptotes is

$$x^2 - 4xy - 5y^2 + 6x + 42y - 63 - 9 = 0$$

or $$x^2 - 4xy - 5y^2 + 6x + 42y - 72 = 0$$

Equation of conjugate hyperbola is

$$x^2 - 4xy - 5y^2 + 6x + 42y - 63 - \frac{2\Delta}{ab - h^2} = 0$$

or $$x^2 - 4xy - 5y^2 + 6x + 42y - 81 = 0.$$

EXERCISE 1

Find the co-ordinates of the centre and the equation of conic referred to centre as origin of the conics :

1. (i) $2x^2 + 4xy + 3y^2 - 8x - 14y + 9 = 0$ (ii) $2x^2 - 3xy + y^2 - 5x + 4y + 6 = 0$
   (iii) $x^2 - 3xy + y^2 + 10x - 10y + 21 = 0$.
2. What conic is represented by the equation $17x^2 - 12xy + 8y^2 + 46x - 28y + 17 = 0$. Also find its centre and the equation referred to the centre.
3. Find the centre of the conic $5x^2 - 6xy + 5y^2 + 22x - 26y + 29 = 0$ and show that it represents an ellipse. Find also its equation referred to centre as origin.
4. What conic does the equation $36x^2 + 24xy + 29y^2 - 72x + 126y + 81 = 0$ represents ? Find its centre and the equation referred to the centre as origin.

5. Find the asymptotes of the hyperbola $xy - 3x - 2y = 0$. What is the equation of the conjugate hyperbola ?
6. Find the asymptotes of the hyperbola $3x^2 - 5xy - 2y^2 + 17x + y + 14 = 0$. What is the equation of the conjugate hyperbola ?

Answers

1. (*i*) Centre $(-1, 3)$; $2X^2 + 4XY + 3Y^2 - 8 = 0$
   (*ii*) Centre $(2, 1)$; $2X^2 - 3XY + Y^2 + 3 = 0$
   (*iii*) Centre $(-2, 2)$; $X^2 - 3XY + Y^2 + 1 = 0$
2. Centre $(-1, 1)$; $17X^2 - 12XY + 8Y^2 = 20$
3. Centre $(-1, 2)$; $5X^2 - 6XY + 5Y^2 = 8$
4. Centre $(2, -3)$; $36X^2 + 24XY + 29Y^2 = 180.$
5. $xy - 3x - 2y + 6 = 0.$ Conjugate hyperbola $xy - 3x - 2y + 12 = 0.$
6. $3x^2 - 5xy - 2y^2 + 17x + y + 10 = 0.$ Conjugate hyperbola $3x^2 - 5xy - 2y^2 + 17x + y + 6 = 0.$

LENGTHS AND EQUATION OF AXES OF A STANDARD CONIC

8. To find the lengths and the equations of the axes of the conic $Ax^2 + 2Hxy + By^2 = 1$, $(A \neq 0, B \neq 0)$. Also find the eccentricity of the conic

Equation of the conic is

$$Ax^2 + 2Hxy + By^2 = 1 \quad \text{...(1)}$$

Consider the circle of radius r whose equation is

$$x^2 + y^2 = r^2 \quad \text{...(2)}$$

which is concentric with the conic (1).

The joint equation of the lines through the origin to the points of intersection of (1) and (2) is the homogeneous equation of second order given by the association of (1) and (2).

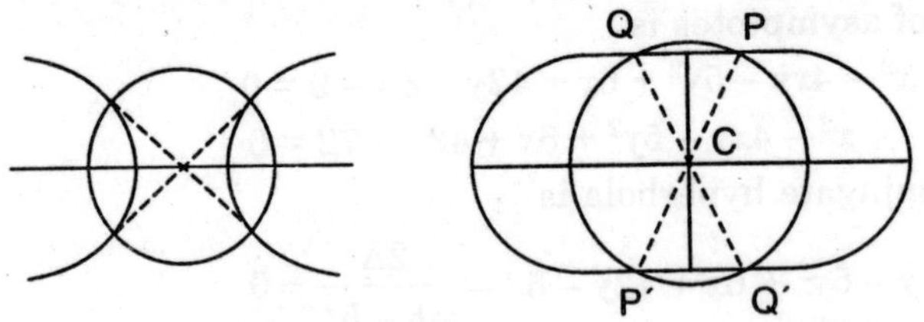

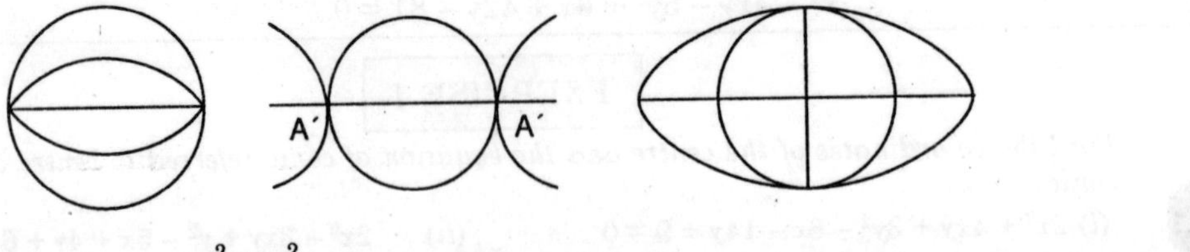

From (2), $\dfrac{x^2 + y^2}{r^2} = 1$, Putting it in (1),

$$\therefore \quad Ax^2 + 2Hxy + By^2 = \frac{x^2 + y^2}{r^2}$$

or

$$\left(A - \frac{1}{r^2}\right) x^2 + 2Hxy + \left(B - \frac{1}{r^2}\right) y^2 = 0 \quad \text{...(3)}$$

It represents two lines PCP′ and QCQ′.

As is clear from the following figures that if a concentric circle cuts the conic (1), the diameters through the points of intersection are *equally inclined to the axes* of the conic and if these two diameters coincide, they coincide with the axes of the conic and then the *radius of the circle is equal to either of the semi-axes of the conic.*

Now the two lines represented by (3) will coincide if

$$\left(A-\frac{1}{r^2}\right)\left(B-\frac{1}{r^2}\right)-H^2=0 \qquad [ab-h^2=0]$$

or
$$\frac{1}{r^4}-(A+B)\frac{1}{r^2}+(AB-H^2)=0$$

or
$$(AB-H^2)r^4-(A+B)r^2+1=0 \qquad \ldots(4)$$

It is a quadratic in r^2 and hence gives two values of r^2 say r_1^2 and r_2^2 from which the lengths of axes are obtained as under.

Case I. *When both the values of r^2 are positive.*

If both r_1^2 and r_2^2 are positive then the conic is an ellipse. Let

$$r_1^2 > r_2^2 > 0.$$

$\therefore$ Length of major axis $= 2r_1$,

Length of minor axis $= 2r_2$.

Case II. *When one value of r^2 is positive and the other value is negative.*

Let $\quad r_1^2 > 0$ and $r_2^2 < 0$.

Here conic is a hyperbola

$\therefore$ Length of transverse axis $= 2r_1$,

Length of conjugate axis $= 2\sqrt{-r_2^2}$.

To find the eccentricity of the conic $AX^2 + 2Hxy + By^2 = 1$

(*a*) *In case of ellipse*

The eccentricity '*e*' is given by

$$r_2^2 = r_1^2(1-e^2) \qquad (b^2 = a^2(1-e^2))$$

i.e.,
$$\frac{r_2^2}{r_1^2} = 1-e^2 \quad \text{or} \quad e^2 = \frac{r_1^2-r_2^2}{r_1^2} \quad \therefore \quad e = \frac{\sqrt{r_1^2-r_2^2}}{r_1} \qquad (e<1)$$

(*b*) *In case of hyperbola*

The eccentricity ($e > 1$) is given by

$$-r_2^2 = r_1^2(e^2-1) \qquad (b^2 = a^2(e^2-1))$$

or
$$\frac{-r_2^2}{r_1^2} = e^2-1, \quad i.e., \quad e^2 = \frac{r_1^2-r_2^2}{r_1^2} \quad \therefore \quad e = \frac{\sqrt{r_1^2-r_2^2}}{e}.$$

To find the equations of the axes

Multiplying equation (3) by $\left(A-\frac{1}{r^2}\right)$, we have

$$\left(A-\frac{1}{r^2}\right)^2 x^2 + 2H\left(A-\frac{1}{r^2}\right)xy + \left(A-\frac{1}{r^2}\right)\left(B-\frac{1}{r^2}\right)y^2 = 0$$

i.e., $$\left(A-\frac{1}{r^2}\right)^2 x^2 + 2H\left(A-\frac{1}{r^2}\right) xy + H^2y^2 = 0 \qquad \text{[From (3)]}$$

or $$\left(A-\frac{1}{r^2}\right) x + Hy = 0 \qquad ...(5)$$

It gives the equations of the axes, corresponding to the two values r_1^2 and r_2^2 of r^2. Thus the equations of the axes are

$$\left(A-\frac{1}{r_1^2}\right) x + Hy = 0 \quad \text{and} \quad \left(A-\frac{1}{r_2^2}\right) x + Hy = 0.$$

9. To find the foci and directrices of the conic $ax^2 + 2hxy + by^2 = 1$

Case I. *In case of ellipse*

(*i*) The lengths of the semi-major axis and semi-minor axes are respectively r_1 and r_2.

The equation of the major axis is

$$\left(a-\frac{1}{r_1^2}\right) x + hy = 0$$

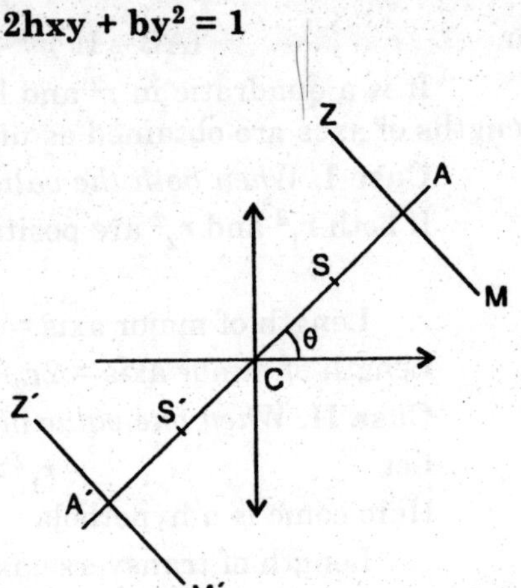

Let C be the centre, S and S′ be the foci and MZ, M′Z′ be the directrices of the ellipse. Let the major axis make angle θ with the x-axis. Let (x', y') be the co-ordinates of S.

Since S and S′ are at a distance er_1 from C

$$\therefore \quad \frac{x'}{\cos\theta} = \frac{y'}{\sin\theta} = er_1$$

or $$x' = er_1 \cos\theta\ ;\ y' = er_1 \sin\theta$$

$\therefore$ Co-ordinates of foci are $[\pm\, er_1 \cos\theta, \pm\, er_1 \sin\theta]$

But $$er_1 = \frac{\sqrt{r_1^2 - r_2^2}}{r_1} \times r_1 = \sqrt{r_1^2 - r_2^2}$$

$\therefore$ foci are $[\pm \sqrt{r_1^2 - r_2^2} \cos\theta, \pm \sqrt{r_1^2 - r_2^2} \sin\theta]$

(*ii*) To find the equation of the directrices.

Let the major axis meet the directrices at A and A′.

Since CA is perpendicular to MZ, therefore

$$CA = \frac{r_1}{e} = \frac{r_1^2}{\sqrt{r_1^2 - r_2^2}} \qquad \left|\ \because\ e = \frac{\sqrt{r_1^2 - r_2^2}}{r_1}\right.$$

$\therefore$ Equation of the directrix MZ is

$$x\cos\theta + y\sin\theta = CA$$

(line in the perpendicular form is $x\cos\alpha + y\sin\alpha = p$)

or $$x\cos\theta + y\sin\theta = \frac{r_1^2}{\sqrt{r_1^2 - r_2^2}}$$

Similarly equation of the other directrix M'Z' is

$$x \cos \theta + y \sin \theta = \frac{-r_1^2}{\sqrt{r_1^2 - r_2^2}}.$$

Case II. *In case of hyperbola*

Here the length of the semi transverse axis and conjugate axis are respectively r_1 and $\sqrt{-r_2^2}$, where $r_1^2 > 0$ and $r_2^2 < 0$.

As in case I, the co-ordinates of foci are

$$[\pm \sqrt{r_1^2 - r_2^2} \cos \theta, \pm \sqrt{r_1^2 - r_2^2} \sin \theta]$$

Equation of the directrices are also

$$x \cos \theta + y \sin \theta = \frac{\pm r_1^2}{\sqrt{r_1^2 - r_2^2}}, \text{ where } r_2^2 < 0.$$

(*iii*) Length of latus rectum $= \dfrac{2r_2^2}{r_1}$.

Note. The equations of **directrices** of the conic $ax^2 + 2hxy + by^2 + 2gx + 2fy + c = 0$ are

(*i*) Ellipse : $(x - x_1) \cos \theta + (y - y_1) \sin \theta = \pm \dfrac{r_1^2}{\sqrt{r_1^2 - r_2^2}}$

(*ii*) Hyperbola : $(x - x_1) \cos \theta + (y - y_1) \sin \theta = \pm \dfrac{r_1^2}{\sqrt{r_1^2 - r_2^2}}$, $r_2^2 < 0$

where (x_1, y_1) are the co-ordinates of the centre of the conic.

Example 1. *Find the lengths of semi-axes and their equations for the conic*

$$5x^2 - 24xy - 5y^2 + 14x + 8y - 16 = 0.$$

Also find the eccentricity. (M.D.U. 2000 ; K.U. 1998 S)

Sol. The given equation is

$$5x^2 - 24xy - 5y^2 + 14x + 8y - 16 = 0 \qquad ...(1)$$

Comparing it with

$$ax^2 + 2hxy + by^2 + 2gx + 2fy + c = 0, \text{ we get}$$

Here $a = 5$, $h = -12$, $b = -5$, $g = 7$, $f = 4$, $c = -16$

Centre giving equations are

$$ax_1 + hy_1 + g = 0 \quad \text{and} \quad hx_1 + by_1 + f = 0$$

i.e.,

$$5x_1 - 12y_1 + 7 = 0 \quad \text{and} \quad -12x_1 - 5y_1 + 4 = 0,$$

Solving

$$\frac{x_1}{-48+35} = \frac{y_1}{-84-20} = \frac{1}{-25-144}$$

$$\therefore \quad x_1 = \frac{-13}{-169} = \frac{1}{13}, \quad y_1 = \frac{-104}{-169} = \frac{8}{13}$$

$\therefore$ Centre is $\left(\dfrac{1}{13}, \dfrac{8}{13}\right)$.

Shifting the origin to $\left(\frac{1}{13}, \frac{8}{13}\right)$, equation of the conic centre as origin is

$$5X^2 - 24XY - 5Y^2 + C_1 = 0$$

where $C_1 = gx_1 + fy_1 + c = 7\left(\frac{1}{13}\right) + 4\left(\frac{8}{13}\right) - 16 = -13$

$\therefore$ Equation of the conic is

$$5X^2 - 24XY - 5Y^2 - 13 = 0 \quad \text{or} \quad \frac{5}{13}X^2 - \frac{24}{13}XY - \frac{5Y^2}{13} = 1$$

Comparing it with $\quad AX^2 + 2HXY + BY^2 = 1$

$$\therefore \quad A = \frac{5}{13},\ H = \frac{-12}{13},\ B = -\frac{5}{13}$$

Now lengths of semi-axes are given by

$$(AB - H^2)r^4 - (A + B)r^2 + 1 = 0$$

i.e.,
$$\left(\frac{-25}{169} - \frac{144}{169}\right) r^4 - (0)\, r^2 + 1 = 0$$

or
$$r^4 - 1 = 0 \quad \therefore \quad r^2 = \pm 1$$

Here one value of r^2 is positive and other negative.

$\therefore$ Conic is a hyperbola.

Let $\quad r_1^{\,2} = 1 > 0,\ r_2^{\,2} = -1 < 0$

$\therefore$ Lengths of semi-axes are r_1 and $\sqrt{-r_2^{\,2}}$ *i.e.,* 1 and 1

Equations of axes are given by $\left(A - \frac{1}{r^2}\right)X + HY = 0$

Put $r^2 = 1$

$$\therefore \quad \left(\frac{5}{13} - 1\right)X - \frac{12}{13}Y = 0 \quad \text{or} \quad 2X + 3Y = 0 \qquad ...(5)$$

Referred to old axes, the equation of transverse axis is, put

$$X = x - \frac{1}{13}\ ;\ Y = y - \frac{8}{13} \text{ in (5),}$$

$$2\left(x - \frac{1}{13}\right) + 3\left(y - \frac{8}{13}\right) = 0 \quad \text{or} \quad 2x + 3y - 2 = 0.$$

Again, put $r^2 = -1$

Equation of conjugate axes referred to new axes is

$$\left(\frac{5}{13} + 1\right)X - \frac{12}{13}Y = 0 \quad \text{or} \quad 3X - 2Y = 0$$

Referred to old axes, the equation is

$$3\left(x - \frac{1}{13}\right) - 2\left(y - \frac{8}{13}\right) = 0 \quad i.e., \quad 3x - 2y + 1 = 0$$

Eccentricity is given by $\quad -r_2^{\,2} = r_1^{\,2}(e^2 - 1)$

i.e.,
$$1 = 1(e^2 - 1) \Rightarrow e^2 = 2$$

$$\therefore \quad e = \sqrt{2}.$$

Note. As the lengths of axes of the hyperbola are equal, it is a rectangular hyperbola.

Example 2. *Show that the conic*

$$8x^2 - 24xy + 15y^2 + 48x - 48y = 0$$

represents a hyperbola and hence find its eccentricity, lengths and equations of the axes.

(M.D.U. 1998 ; K.U. 1998)

Sol. Here $a = 8, b = 15, c = 0, h = -12, g = 24, f = -24$

$\therefore \quad h^2 - ab = 144 - 120 = 24.$

$\therefore$ Conic is a hyperbola.

Let (x_1, y_1) be the centre of the conic ; centre giving equations are

$$ax_1 + hy_1 + g = 0 \quad \text{and} \quad hx_1 + by_1 + f = 0$$

i.e., $\quad 8x_1 - 12y_1 + 24 = 0 \quad \text{or} \quad 2x_1 - 3y_1 + 6 = 0 \quad$ ...(1)

and $\quad -12x_1 + 15y_1 - 24 = 0 \quad \text{or} \quad 4x_1 - 5y_1 + 8 = 0 \quad$ ...(2)

Solving for x_1 and y_1

$$\frac{x_1}{-24+30} = \frac{y_1}{24-16} = \frac{1}{-10+12} = \frac{1}{2}$$

$\therefore \quad x_1 = 3, y_1 = 4 \quad \therefore$ Centre is (3, 4)

Equation of conic with centre as origin is

$$8X^2 - 24XY + 15Y^2 + C_1 = 0$$

where $C_1 = gx_1 + fy_1 + c = 24.(3) + (-24)(4) = -24$

$\therefore$ Equation of conic is

$$8X^2 - 24XY + 15Y^2 = 24 \quad \text{or} \quad \frac{X^2}{3} - XY + \frac{5}{8}Y^2 = 1$$

Here $\quad A = \frac{1}{3}, H = -\frac{1}{2}, B = \frac{5}{8}$

$\therefore$ Length of semi-axes is given by

$$(AB - H^2)r^4 - (A + B)r^2 + 1 = 0$$

$$\left(\frac{5}{24} - \frac{1}{4}\right)r^4 - \left(\frac{1}{3} + \frac{5}{8}\right)r^2 + 1 = 0$$

or $\quad r^4 + 23r^2 - 24 = 0 \quad \text{or} \quad (r^2 + 24)(r^2 - 1) = 0.$

$\therefore \quad r^2 = 1, -24.$

Let $\quad r_1^2 = 1 > 0, r_2^2 = -24 < 0$

$\therefore$ Lengths of semi-axes are r_1 and $\sqrt{-r_2^2}$ i.e., $1, 2\sqrt{6}$

$\therefore$ Transverse axis = 2, Conjugate axis = $4\sqrt{6}$.

Equations of axes are given by $\left(A - \frac{1}{r^2}\right)X + HY = 0$

Put $r^2 = 1$,

i.e., $\quad \left(\frac{1}{3} - 1\right)X - \frac{1}{2}Y = 0 \quad \text{or} \quad 2X + 3Y = 0$

Equation of transverse axis referred to old axes is obtained by putting

$$X = x - 3, Y = y - 4$$

$$2(x - 3) + 3(y - 4) = 0$$

i.e., $\quad 2x + 3y - 18 = 0$

Again put $r^2 = -24$

Equation of conjugate axis is

$$\left(\frac{1}{3}+\frac{1}{24}\right)X-\frac{1}{2}\ Y=0 \quad \text{or} \quad 3X-4Y=0$$

Referred to old axes, equation of conjugate axis is

$$3(x-3)-4(y-4)=0 \quad \text{or} \quad 3x-4y+7=0$$

For eccentricity, use $\quad -r_2^{\,2} = r_1^{\,2}(e^2-1)$

$$24 = 1(e^2-1) \Rightarrow e^2 = 25$$

$\therefore \quad e = 5.$

Example 3. *Find the lengths of the axes, the eccentricity and the equations of the axes of the conic*

$$5x^2+6xy+5y^2+4x+12y-4=0.$$

Sol. Here $a = 5$, $h = 3$, $b = 5$, $g = 2$, $f = 6$, $c = -4$ centre (x_1, y_1) is given by

$$ax_1 + hy_1 + g = 0 \quad \text{and} \quad hx_1 + by_1 + f = 0$$

i.e., $\quad 5x_1 + 3y_1 + 2 = 0 \quad$ ...(1)

and $\quad 3x_1 + 5y_1 + 6 = 0 \quad$ ...(2)

Solving (3) and (4),

$$\frac{x_1}{8}=\frac{y_1}{-24}=\frac{1}{16} \qquad \therefore \quad x_1=\frac{1}{2},\ y_1=-\frac{3}{2}$$

$\therefore$ Centre is $\left(\frac{1}{2},-\frac{3}{2}\right)$

Equation of conic referred to parallel axes through the centre is

$$5X^2+6XY+5Y^2+C_1=0$$

where $\quad C_1 = gx_1 + fy_1 + c = 2\left(\frac{1}{2}\right)+6\left(\frac{-3}{2}\right)-4 = 1-9-4 = -12$

Equation of the conic is $\ 5X^2+6XY+5Y^2=12$

or $$\frac{5}{12}X^2+\frac{6}{12}XY+\frac{5}{12}Y^2=1$$

Comparing it with $\quad AX^2+2HXY+BY^2=1$

$$A=\frac{5}{12},\ H=\frac{3}{12},\ B=\frac{5}{12}$$

Lengths of axes giving equation is

$$(AB-H^2)r^4-(A+B)r^2+1=0$$

i.e., $$\left(\frac{25}{144}-\frac{9}{144}\right)r^4-\left(\frac{10}{12}\right)r^2+1=0 \qquad \text{or} \quad 2r^4-15r^2+18=0$$

i.e., $$2r^4-12r^2-3r^2+18=0 \qquad \text{or} \quad 2r^2(r^2-6)-3(r^2-6)=0$$

or $$(r^2-6)(2r^2-3)=0 \qquad \text{or} \quad r^2=6,\ 3/2$$

Both values of r^2 are positive.

$\therefore$ Conic is an ellipse.

Let $\quad r_1^{\,2}=6,\ r_2^{\,2}=3/2.$

Lengths of the axes are $2r_1$ and $2r_2$ *i.e.,* $\ 2\sqrt{6}$ and $\sqrt{6}$.

Eccentricity (e) is given by $r_2^2 = r_1^2 (1 - e^2)$

$$\frac{3}{2} = 6(1 - e^2) \quad i.e., \quad 1 - e^2 = \frac{1}{4}$$

$$\therefore \qquad e^2 = \frac{3}{4} \Rightarrow e = \frac{\sqrt{3}}{2}$$

Equations of the axes are given by $\left(A - \frac{1}{r^2}\right)X + HY = 0$

Put $r^2 = 6$, the equation of major axis is

$$\left(\frac{5}{12} - \frac{1}{6}\right)X + \frac{3}{12}Y = 0 \quad \text{or} \quad X + Y = 0$$

Equation of major axis referred to old axes is obtained by putting

$$X = x - \frac{1}{2}, \; Y = y + \frac{3}{2}$$

i.e.,
$$\left(x - \frac{1}{2}\right) + \left(y + \frac{3}{2}\right) = 0 \quad \text{or } x + y + 1 = 0$$

Again put $r^2 = \frac{3}{2}$

Equation of the minor axis referred to new axes is

$$\left(\frac{5}{12} - \frac{2}{3}\right)X + \frac{3}{12}Y = 0, \quad i.e., \quad X - Y = 0$$

$$\therefore \qquad \left(x - \frac{1}{2}\right) - \left(y + \frac{3}{2}\right) = 0 \quad i.e., \quad x - y - 2 = 0$$

referred to old axes.

EXERCISE 2

1. Find the lengths and equations of axes of the conics
   (*i*) $8x^2 - 4xy + 5y^2 - 16x - 14y + 17 = 0$ (*M.D.U. 2005*) (*ii*) $x^2 - 5xy + y^2 + 8x - 20y + 15 = 0$
   (*iii*) $36x^2 + 24xy + 29y^2 - 72x + 126y + 81 = 0$.
2. Find the centre, eccentricity, lengths and equations of the conic
   $x^2 + 12xy - 4y^2 - 6x + 4y + 9 = 0$.
3. Find the lengths of the axes, the eccentricity and the equations of the axes of the conic
   $5x^2 - 6xy + 5y^2 + 22x - 26y + 29 = 0$.
4. Find the lengths of the axes, the eccentricity and equations of axes of the conic
   $14x^2 - 4xy + 11y^2 - 44x - 58y + 71 = 0$. (*M.D.U. 2003*)
5. Prove that the lengths of the semi-axes of the conic $ax^2 + 2hxy + ay^2 = d$ are $\sqrt{\frac{d}{a \pm h}}$ and that their equation is $x^2 - y^2 = 0$.

Answers

1. (*i*) Length of major axis = 3, minor axis = 2 ;
   $2x - y - 1 = 0$, $2x + 4y - 11 = 0$
   (*ii*) $2\sqrt{\frac{2}{3}}$, $2\sqrt{\frac{2}{7}}$; $x + y + 4 = 0$, $x - y + 4 = 0$
   (*iii*) 6, 4 ; $4x + 3y + 1 = 0$, $3x - 4y - 18 = 0$.

2. Centre $\left(0, \frac{1}{2}\right)$; $\sqrt{5}$, $2\sqrt{2}$; $3x + 2y - 1 = 0$; $4x - 6y + 3 = 0$; $e = \sqrt{\frac{13}{5}}$.

3. Length of axes 4 ; 2 ; axes are $x - y + 3 = 0$, $x + y - 1 = 0$; $e = \frac{\sqrt{3}}{2}$.

4. $e = \frac{1}{\sqrt{3}}$; $2\sqrt{6}$; 4 ; $2x - y - 1 = 0$; $x + 2y - 8 = 0$.

---

10. Tracing of Conics

Method for tracing a conic. The following method is adopted for tracing of an ellipse or hyperbola.

Let the equation of a conic be $ax^2 + 2hxy + by^2 + 2gx + 2fy + c = 0$. ...(1)

1. First find the nature of the conic by finding

$$\Delta = abc + 2fgh - af^2 - bg^2 - ch^2$$

If $\Delta \neq 0$, and $h^2 - ab \neq 0$, then (1) is an ellipse or hyperbola.

If $h^2 - ab < 0$, then (1) is an ellipse and if $h^2 - ab > 0$, then (1) is a hyperbola.

2. Find the co-ordinates of its centre (x_1, y_1) by solving the equations

$$ax_1 + hy_1 + g = 0$$
$$hx_1 + by_1 + f = 0.$$

3. Shift the origin to the centre. The new equation of the conic can be found as

$$aX^2 + 2hXY + bY^2 + C_1 = 0 \quad ...(2)$$

where $C_1 = gx_1 + fy_1 + c$

which reduces to the form $AX^2 + 2HXY + BY^2 = 1$

where $X = x - x_1$, $Y = y - y_1$.

4. *Length of the axes of the conic*

The squares of the lengths of the semi axes of the conic are given by the equation

$$\left(A - \frac{1}{r^2}\right)\left(B - \frac{1}{r^2}\right) = H^2$$

or

$$(AB - H^2)\, r^4 - (A + B)\, r^2 + 1 = 0.$$

It is a quadratic in r^2 and hence gives two values of r^2 say r_1^2 and r_2^2 from which the lengths of the axes are obtained as under :

Case I. ***When both the values of r^2 are positive***

If both r_1^2 and r_2^2 are positive then the conic is an ellipse. Let $r_1^2 > r_2^2 > 0$, length of major axis $= 2r_1$, length of minor axis $= 2r_2$.

Case II. ***When one value of r^2 is positive and the other value is negative***

Let $r_1^2 > 0$ and $r_2^2 < 0$, then the conic is a hyperbola

Length of transverse axis $= 2r_1$. Length of conjugate axis $= 2\sqrt{-r_2^2}$.

5. *Equations of the axes.* The equations of the axes, corresponding to the two values r_1^2 and r_2^2 of r^2 referred to the centre as origin are given by

$$\left(A - \frac{1}{r_1^2}\right)X + HY = 0 \quad \text{and} \quad \left(A - \frac{1}{r_2^2}\right)X + HY = 0$$

Equations of the axes referred to the original axes are

$$\left(A - \frac{1}{r_1^2}\right)(x - x_1) + H(y - y_1) = 0 \quad \text{and} \quad \left(A - \frac{1}{r_2^2}\right)(x - x_1) + H(y - y_1) = 0$$

where (x_1, y_1) is the centre of the conic.

6. *Eccentricity.* If the eccentricity of the conic is required it is given by $e = \sqrt{\dfrac{1 - r_2^2}{r_1^2}}$.

7. *Foci.* The co-ordinates of the foci with respect to the original axes are given by

$$(x_1 \pm \sqrt{r_1^2 - r_2^2}\cos\theta,\ y_1 \pm \sqrt{r_1^2 - r_2^2}\sin\theta)$$

where θ is the inclination of the major or transverse axis to the x-axis.

8. *Latus Rectum.* The latus-rectum of conic is $\dfrac{2r_2^2}{r_1}$.

9. *Directrix.* The equations of directrices w.r.t. original axes are

$$(x - x_1)\cos\theta + (y - y_1)\sin\theta = \pm\frac{r_1^2}{\sqrt{r_1^2 - r_2^2}}.$$

10. To draw the figure of the conic more accurately, find the points, where the conic meets the original x and y axes.

Example 1. *Trace the conic $8x^2 - 4xy + 5y^2 - 16x - 14y + 17 = 0$*

Sol. The given conic is $F(x, y) \equiv 8x^2 - 4xy + 5y^2 - 16x - 14y + 17 = 0$. Here, the second degree terms do not form a perfect square, therefore the given conic represents a central conic.

Let (x_1, y_1) be the centre of the conic.

The centre giving equations are

$$\frac{\partial F}{\partial x} = 16x - 4y - 16 = 0 \quad \text{and} \quad \frac{\partial F}{\partial y} = -4x + 10y - 14 = 0$$

On solving these equations the co-ordinate of the centre are $\left(\frac{3}{2}, 2\right)$ *i.e.,* $x_1 = \frac{3}{2}, y_1 = 2$.

Equation of conic referred to parallel axes through the centre is

$$8X^2 - 4XY + 5Y^2 + C_1 = 0$$

where $C_1 = gx_1 + fy_1 + c = -8 \times \frac{3}{2} + (-7) \times 2 + 17 = -9$

$\therefore$ Equation of the conic is

$$8X^2 - 4XY + 5Y^2 = 9 \quad \text{or} \quad \frac{8}{9}X^2 - \frac{4}{9}XY + \frac{5}{9}Y^2 = 1$$

Comparing it with $AX^2 + 2HXY + BY^2 = 1$

$$A = \frac{8}{9},\ H = -\frac{2}{9},\ B = \frac{5}{9}$$

Lengths of axes giving equation is

$$(AB - H^2)r^4 - (A + B)r^2 + 1 = 0$$

i.e., $$\left(\frac{40}{81}-\frac{4}{81}\right)r^4 - \frac{13}{9}r^2 + 1 = 0 \quad \text{or} \quad 36r^4 - 117r^2 + 81 = 0$$

or $$4r^4 - 13r^2 + 9 = 0 \quad \text{or} \quad (4r^2 - 9)(r^2 - 1) = 0$$

$\therefore$ The squares of the semi-axes are given by

$$r_1^2 = \frac{9}{4}, \quad r_2^2 = 1$$

Since r_1^2 and r_2^2 are both positive, the conic is an ellipse.

Lengths of the major and minor axes are $2r_1$ and $2r_2$ *i.e.*, 3 and 2 respectively.

Eccentricity (e) is given by $r_2^2 = r_1^2(1 - e^2)$

$$1 = \frac{9}{4}(1 - e^2) \quad i.e., \quad 1 - e^2 = \frac{4}{9}$$

$$\therefore \quad e^2 = \frac{5}{9} \quad \text{or} \quad e = \frac{\sqrt{5}}{3}$$

Equations of the axes are given by

$$\left(A - \frac{1}{r^2}\right)X + HY = 0$$

Putting $r^2 = \frac{9}{4}$, the equation of the major axis is

$$\left(\frac{8}{9}-\frac{4}{9}\right)X + \left(-\frac{2}{9}\right)Y = 0 \quad \text{or} \quad 2X - Y = 0$$

Equation of major axis referred to old axes is obtained by putting

$$X = x - \frac{3}{2}, \; Y = y - 2$$

i.e., $$2\left(x - \frac{3}{2}\right) - (y - 2) = 0 \quad \text{or} \quad 2x - y - 1 = 0 \qquad ...(1)$$

Again putting $r^2 = 1$,

The equation of the minor axis referred to new axes is

$$\left(\frac{8}{9} - 1\right)X + \left(-\frac{2}{9}\right)Y = 0 \quad \text{or} \quad X + 2Y = 0$$

i.e., $$\left(x - \frac{3}{2}\right) + 2(y - 2) = 0 \quad \text{or} \quad 2x + 4y - 11 = 0 \qquad ...(2)$$

referred to old axes.

Foci. Slope of major axis = 2

Let the major axis make angle θ with the x-axis

$\therefore$ $\tan\theta = 2$. Hence $\sin\theta = \frac{2}{\sqrt{5}}$, $\cos\theta = \frac{1}{\sqrt{5}}$

Co-ordinates of the foci are

$(x_1 + lr_1 \cos\theta, y_1 + lr_1 \sin\theta)$ and $(x_1 - lr_1 \cos\theta, y_1 - lr_1 \sin\theta)$

i.e., $$\left(\frac{3}{2} + \frac{\sqrt{5}}{3}\cdot\frac{3}{2}\cdot\frac{1}{\sqrt{5}}, 2 + \frac{\sqrt{5}}{3}\cdot\frac{3}{2}\cdot\frac{2}{\sqrt{5}}\right) = (2, 3)$$

and $$\left(\frac{3}{2}-\frac{\sqrt{5}}{3}\cdot\frac{3}{2}\cdot\frac{1}{\sqrt{5}}, 2-\frac{\sqrt{5}}{3}\cdot\frac{3}{2}\cdot\frac{2}{\sqrt{5}}\right) = (1, 1)$$

Again, the conic meets the x-axis in the points, where $y = 0$

i.e., in the points given by $8x^2 - 16x + 17 = 0$

which are imaginary. $\mid \because \quad b^2 - 4ac = -\text{ve}$

Hence the conic does not meet the x-axis. Further, the conic meets the y-axis, where $x = 0$ *i.e.*, in points given by

$$5y^2 - 14y + 17 = 0.$$

This also gives imaginary values.

Hence the conic does not meet the y-axis also.

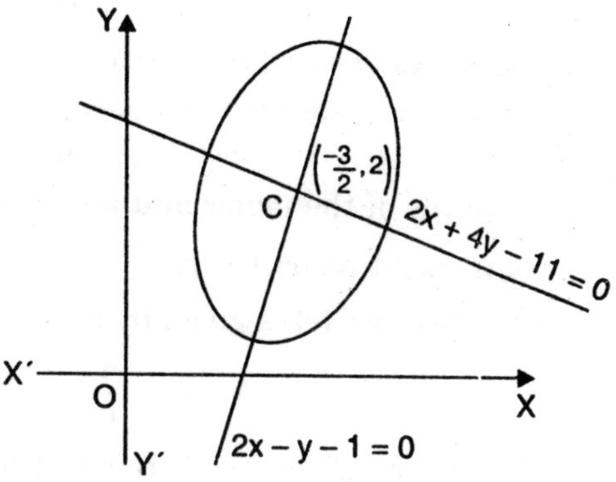

The shape of the curve is as shown in the figure.

Example 2. *Trace the curve*

$$14x^2 - 4xy + 11y^2 - 44x - 58y + 71 = 0. \qquad \text{(M.D.U. 1995, 2003)}$$

Sol. The given conic is

$$14x^2 - 4xy + 11y^2 - 44x - 58y + 71 = 0 \qquad ...(1)$$

Since the second degree terms do not form a perfect square, the given conic represents a central conic. Let (x_1, y_1) be its centre.

The centre giving equation are

$$ax_1 + hy_1 + g = 0 \quad \text{and} \quad hx_1 + by_1 + f = 0$$

i.e., $$14x_1 - 2y_1 - 22 = 0 \qquad ...(2)$$

and $$-2x_1 + 11y_1 - 29 = 0 \qquad ...(3)$$

On solving (2) and (3) simultaneously, we get

$$x_1 = 2, y_1 = 3 \quad \therefore \quad \text{centre is } (2, 3).$$

Equation of conic referred to parallel axes through the centre is

$$14X^2 - 4XY + 11Y^2 + C_1 = 0$$

where $$C_1 = gx_1 + fy_1 + c = -22(2) - 29(3) + 71 = -60$$

$\therefore$ the transformed equation is

$$14X^2 - 4XY + 11Y^2 = 60 \quad \text{or} \quad \frac{14}{60}X^2 - \frac{4}{60}XY + \frac{11}{60}Y^2 = 1$$

Comparing it with

$$AX^2 + 2HXY + BY^2 = 1$$

$$A = \frac{14}{60}, B = \frac{11}{60}, H = -\frac{2}{60}$$

Lengths of axes giving equation is

$$(AB - H^2)r^4 - (A + B)r^2 + 1 = 0$$

i.e., $$\left(\frac{154}{60^2} - \frac{4}{60^2}\right)r^4 - \left(\frac{14}{60} + \frac{11}{60}\right)r^2 + 1 = 0$$

or $$\frac{150}{(60)^2}r^4 - \frac{25}{60}r^2 + 1 = 0 \quad \text{or} \quad r^4 - 10r^2 + 24 = 0$$

or $$(r^2-6)(r^2-4) = 0 \quad \text{or} \quad r^2 = 6, 4$$

Both values of r^2 are positive. $\therefore$ Conic is an ellipse.

The squares of the semi-axes of the conic are given by

$$r_1^2 = 6 \text{ and } r_2^2 = 4$$

Lengths of the major and minor axes are $2r_1$ and $2r_2$

i.e., $2\sqrt{6}$ and 4 respectively.

Eccentricity (e) is given by $r_2^2 = r_1^2(1-e^2)$

or $$1 - e^2 = \frac{4}{6} \quad \therefore \quad e^2 = 1 - \frac{2}{3} = \frac{1}{3} \quad \therefore \quad e = \frac{1}{\sqrt{3}}$$

Equations of the axes are given by

$$\left(A - \frac{1}{r^2}\right)X + HY = 0$$

Put $r^2 = 6$, the equation of the major axis referred to new axes is

$$\left(\frac{14}{60} - \frac{1}{6}\right)X - \frac{2}{60}Y = 0 \quad \text{or} \quad 2X - Y = 0$$

Equation of major axis referred to old axes is obtained by putting

$$X = x - 2, \; Y = y - 3$$

i.e., $$2(x-2) - (y-3) = 0 \quad \text{or} \quad 2x - y - 1 = 0$$

Again put $r^2 = 4$

Equation of the minor axis referred to new axes is

$$\left(\frac{14}{60} - \frac{1}{4}\right)X - \frac{2}{60}Y = 0 \quad \text{or} \quad X + 2Y = 0$$

Equation of minor axis referred to old axes is obtained by putting

$$X = x - 2, \; Y = y - 3.$$

i.e., $$x - 2 + 2(y-3) = 0$$

or $$x + 2y - 8 = 0$$

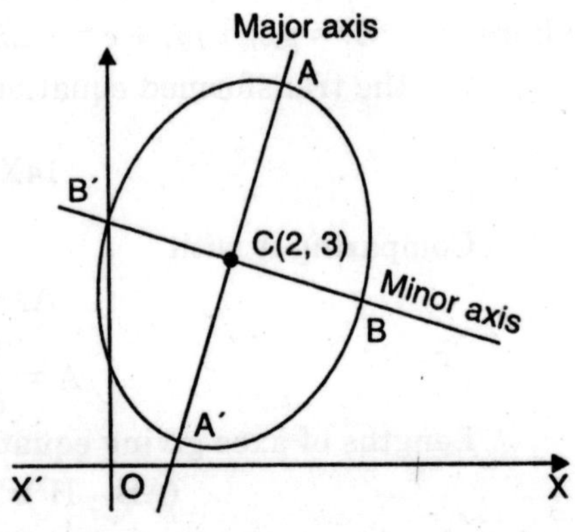

Length of Latus Rectum $= \dfrac{2r_2^2}{r_1} = \dfrac{8}{\sqrt{6}}$

The curve meets the x-axis in points, where $y = 0$, *i.e.*, by

$$14x^2 - 44x + 71 = 0$$

which gives imaginary value.

Hence the curve does not meet the x-axis .

Further the y-axis meets the curve, where $x = 0$, *i.e.*, in points given by

$$11y^2 - 58y + 71 = 0$$

$$\therefore \quad y = \frac{58 \pm \sqrt{58^2 - 44 \times 71}}{22}$$

Simplifying, $1 < y < 2$ and $3 < y < 4$.

Hence the curve is as shown.

Example 3. *Trace the conic*

$$17x^2 - 12xy + 8y^2 + 46x - 28y + 17 = 0$$

Also find its eccentricity, the equations of its axes, the co-ordinates of its foci and the equations of its directrices.

Sol. The given conic is

$$17x^2 - 12xy + 8y^2 + 46x - 28y + 17 = 0 \qquad ...(1)$$

Since the second degree terms do not form a perfect square, the given conic represents a central conic. Let (x_1, y_1) be its centre

The centre giving equations are

$$ax_1 + hy_1 + g = 0 \quad \text{and} \quad hx_1 + by_1 + f = 0$$

i.e., $$17x_1 - 6y_1 + 23 = 0 \qquad ...(2)$$

and $$-3x_1 + 4y_1 - 7 = 0 \qquad ...(3)$$

On solving (2) and (3) simultaneously, we get

$$x_1 = -1, \quad y_1 = 1 \quad \therefore \quad \text{centre is } (-1, 1).$$

Equation of conic referred to parallel axes through the centre is

$$17X^2 - 12XY + 8Y^2 + C_1 = 0$$

where $C_1 = gx_1 + fy_1 + c = 23(-1) + (-14) . 1 + 17 = -20$

$\therefore$ The transformed equation is

$$17X^2 - 12XY + 8Y^2 = 20 \quad \text{or} \quad \frac{17}{20}X^2 - \frac{12}{20}XY + \frac{8}{20}Y^2 = 1.$$

Comparing it with,

$$AX^2 + 2HXY + BY^2 = 1$$

$$A = \frac{17}{20}, \quad B = \frac{8}{20}, \quad H = -\frac{6}{20}.$$

Lengths of axes giving equation is

$$(AB - H^2)\, r^4 - (A + B)\, r^2 + 1 = 0$$

i.e., $$\left(\frac{17}{20} \cdot \frac{8}{20} - \frac{36}{20^2}\right) r^4 - \left(\frac{17}{20} + \frac{8}{20}\right) r^2 + 1 = 0$$

or $$100r^4 - 500r^2 + 400 = 0$$

or $$r^4 - 5r^2 + 4 = 0 \quad \text{or} \quad (r^2 - 1)(r^2 - 4) = 0 \quad \text{or} \quad r^2 = 4, 1$$

Both values of r^2 are positive. $\therefore$ Conic is an ellipse.

The squares of the semi-axes of the conic are given by

$$r_1^2 = 4 \quad \text{and} \quad r_2^2 = 1$$

Lengths of the major and minor axes are $2r_1$ and $2r_2$

i.e., 4 and 2 respectively.

Eccentricity (e) is given by $r_2^2 = r_1^2(1 - e^2)$

or $$e = \sqrt{1 - \frac{r_2^2}{r_1^2}} = \sqrt{1 - \frac{1}{4}} = \frac{\sqrt{3}}{2}.$$

Equations if the major axis referred to old origin is given by

$$\left(A - \frac{1}{r_1^2}\right)(x - x_1) + H(y - y_1) = 0$$

or $$\left(\frac{17}{20} - \frac{1}{4}\right)\{x - (-1)\} + \frac{-6}{20}(y - 1) = 0$$

or $$\frac{12}{20}(x + 1) - \frac{6}{20}(y - 1) = 0, \quad \text{or} \quad 2x - y + 3 = 0 \qquad \text{...(4)}$$

The equation of the minor axis, or equation of the line perpendicular to major axis (4) and passing through the centre (– 1, 1) is

$$y - 1 = -\frac{1}{2}(x + 1) \quad \text{or} \quad x + 2y - 1 = 0 \qquad \text{...(5)}$$

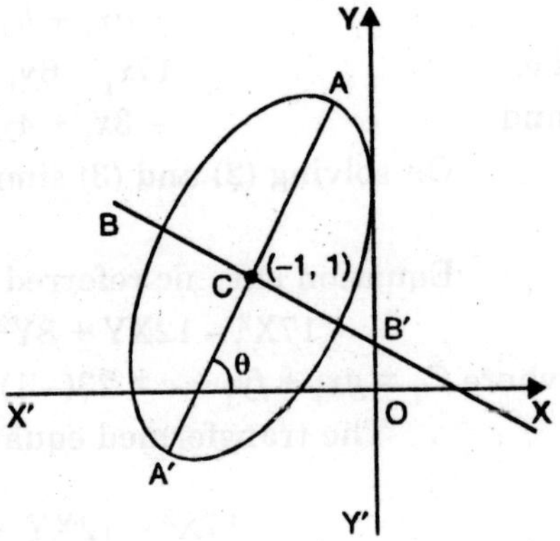

Co-ordinates of the foci. Let θ be the angle which the major axis (4) makes with the axes.

∴ $\tan\theta = 2$ and hence

$$\sin\theta = \frac{2}{\sqrt{5}} \quad \text{and} \quad \cos\theta = \frac{1}{\sqrt{5}}.$$

The co-ordinates of the foci are

$$\left(x_1 \pm \sqrt{r_1^2 - r_2^2}\cos\theta,\; y_1 \pm \sqrt{r_1^2 - r_2^2}\sin\theta\right)$$

i.e., $$\left(-1 \pm \sqrt{4-1}\,.\,\frac{1}{\sqrt{5}},\; 1 \pm \sqrt{4-1}\,.\,\frac{2}{\sqrt{5}}\right)$$

i.e., $$\left(-1 \pm \frac{\sqrt{3}}{5},\; 1 \pm 2\sqrt{\frac{3}{5}}\right)$$

Equations of the directrices. The equations of directrices w.r.t. original axes are given by

$$(x - x_1)\cos\theta + (y - y_1)\sin\theta = \pm\frac{r_1^2}{\sqrt{r_1^2 - r_2^2}}$$

i.e., $$(x + 1)\frac{1}{\sqrt{5}} + (y - 1)\frac{2}{\sqrt{5}} = \pm\frac{4}{\sqrt{3}}$$

or $$(x + 1) + 2(y - 1) = \pm 4\sqrt{\frac{5}{3}}.$$

Hence the shape of the given ellipse is as shown in the figure :

Example 4. *Trace the curve* $x^2 - 3xy + y^2 + 10x - 10y + 21 = 0.$ (M.D.U. 2005)

Sol. Here $a = 1, b = 1, c = 21, h = -\frac{3}{2}, f = -5, g = 5.$

∴ $$h^2 - ab = \frac{9}{4} - 1 = \frac{5}{4} > 0$$

∴ Conic is a hyperbola.

Let (x_1, y_1) be the centre of the conic. Centre giving equations are

$$ax_1 + hy_1 + g = 0 \quad \text{and} \quad hx_1 + by_1 + f = 0$$

i.e., $$2x_1 - 3y_1 + 10 = 0 \quad \text{and} \quad -3x_1 + 2y_1 - 10 = 0$$

On solving these equations, we have $x_1 = -2$, $y_1 = 2$.

$\therefore$ Centre is $(-2, 2)$.

Equation of the conic referred to parallel axes through the centre is

$$X^2 - 3XY + Y^2 + C_1 = 0$$

where $C_1 = gx_1 + fy_1 + c = 5(-2) + (-5) \,.\, 2 + 21 = 1$

$\therefore$ Equation of the conic is

$$X^2 - 3XY + Y^2 = -1 \quad \text{or} \quad -X^2 + 3XY - Y^2 = 1$$

Comparing it with $AX^2 + 2HXY + BY^2 = 1$.

Here $A = -1$, $H = \dfrac{3}{2}$, $B = -1$.

Lengths of axes giving equation is

$$(AB - H^2)r^4 - (A + B)r^2 + 1 = 0 \quad \text{or} \quad \left(1 - \frac{9}{4}\right) r^4 + 2r^2 + 1 = 0$$

or
$$5r^4 - 8r^2 - 4 = 0 \quad \text{or} \quad (5r^2 + 2)(r^2 - 2) = 0$$

$$\therefore \qquad r_1^{\,2} = 2 \quad \text{and} \quad r_2^{\,2} = -\frac{2}{5}$$

One value of r^2 is positive and the other is negative.

$\therefore$ Conic is a hyperbola.

$\therefore$ Lengths of semi-axes are r_1 and $\sqrt{-r_2^{\,2}}$ *i.e.*, $\sqrt{2}, \sqrt{\dfrac{2}{5}}$

$\therefore$ Transverse axis $= 2\sqrt{2}$, conjugate axis $= 2\sqrt{\dfrac{2}{5}}$

Equations of the axes are given by

$$\left(A - \frac{1}{r^2}\right) X + HY = 0$$

Put $r^2 = 2$, the equation of the transverse axis is

$$\left(-1 - \frac{1}{2}\right) X + \frac{3}{2} Y = 0 \quad \text{or} \quad X - Y = 0$$

Equation referred to old axes is obtained by putting

$$X = x + 2,\; Y = y - 2$$

i.e., $\qquad x + 2 - y + 2 = 0 \quad \text{or} \quad x - y + 4 = 0.$

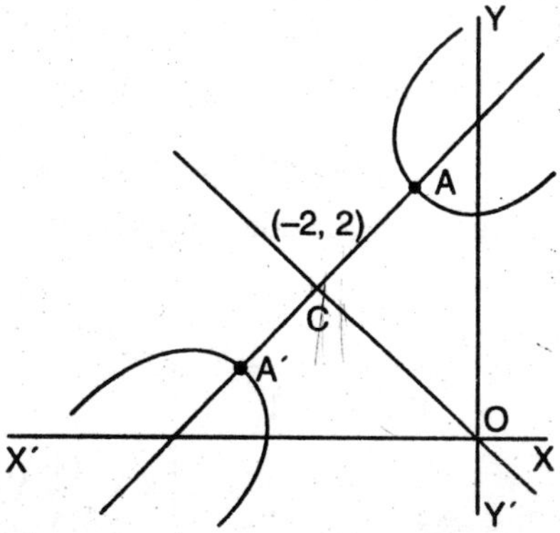

Again put $r^2 = -\dfrac{2}{5}$

Equation of conjugate axis is

$$\left(-1 + \frac{5}{2}\right) X + \frac{3}{2} Y = 0 \quad \text{or} \quad X + Y = 0$$

Put $X = x + 2$ and $Y = y - 2$.

The equation of the conjugate axis referred to old axis is

$$x + 2 + y - 2 = 0 \quad \text{or} \quad x + y = 0$$

The shape of the curve is as shown in the figure.

Example 5. *Find the centre, lengths and directions of the axes of the conic*

$$7x^2 - 48xy - 7y^2 + 110x - 20y + 100 = 0$$

and trace it.

Sol. Here $a = 7, b = -7, h = -24, g = 55, f = -10, c = 100$

$\therefore$ $$h^2 - ab = (-24)^2 + 49 = 625 > 0$$

Therefore the curve is a hyperbola.

Also '$a + b$' $= 7 - 7 = 0$

Hence the curve is a rectangular hyperbola.

Let (x_1, y_1) be the centre of the conic.

The centre giving equation are

$$ax_1 + hy_1 + g = 0 \text{ and } hx_1 + by_1 + f = 0$$

i.e., $$7x_1 - 24y_1 + 55 = 0 \qquad ...(1)$$

and $$-24x_1 - 7y_1 - 10 = 0 \qquad ...(2)$$

On solving (1) and (2) simultaneously, we get

$$x_1 = -1 \quad \text{and} \quad y_1 = 2 \quad \therefore \quad \text{centre is } (-1, 2) \qquad ...(3)$$

Also $$C_1 = gx_1 + fy_1 + c = 55(-1) + (-10)2 + 100 = 25$$

$\therefore$ The transformed equation, where origin is shifted to the centre $(-1, 2)$ is

$$7X^2 - 48XY - 7Y^2 + 25 = 0$$

or $$-\frac{7}{25}X^2 + \frac{48}{25}XY + \frac{7}{25}Y^2 = 1$$

Comparing it with $AX^2 + 2HXY + BY^2 = 1$, we have

$$A = -\frac{7}{25}, B = \frac{7}{25}, H = \frac{24}{25}$$

Now lengths of semi-axes are given by

$$(AB - H^2)r^4 - (A + B)r^2 + 1 = 0$$

i.e., $$\left(-\frac{49}{625} - \frac{576}{625}\right) r^4 + 1 = 0$$

or $$r^4 - 1 = 0 \quad \therefore \quad r^2 = \pm 1$$

Here one value of r^2 is positive and other negative and the two values are numerically equal, the curve is a rectangular hyperbola.

Let $r_1^2 = 1 > 0$ and $r_2^2 = -1 < 0$

$\therefore$ Lengths of semi-axes are r_1 and $\sqrt{-r_2^2}$ *i.e.,* 1 and 1. Equations of axes are given by

$$\left(A - \frac{1}{r^2}\right) X + HY = 0.$$

Put $r^2 = 1$,

$$\therefore \quad \left(-\frac{7}{25} - 1\right)X + \frac{24}{25}Y = 0$$

or $$4X - 3Y = 0 \qquad ...(4)$$

Referred to old axes, the equation of the transverse axis is obtained by putting $X = x + 1$, $Y = y - 2$ in (4).

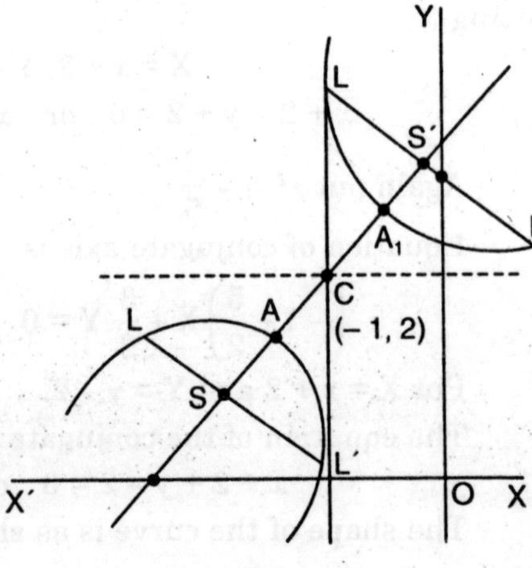

$$\therefore \qquad 4(x + 1) - 3(y - 2) = 0 \quad \text{or} \quad 4x - 3y + 10 = 0 \qquad ...(5)$$

Equation of the conjugate axis, which is perpendicular to (5) and passes through C(– 1, 2) is

$$y - 2 = \frac{3}{4}(x + 1) \quad \text{or} \quad 3x + 4y - 5 = 0 \qquad ...(6)$$

Length of latus rectum $= -\dfrac{2r_2^{\,2}}{r_1} = 2$

The shape of the curve is as shown in the figure.

EXERCISE 3

1. Trace the following conics
   (*i*) $5x^2 - 6xy + 5y^2 + 22x - 26y + 29 = 0$
   (*ii*) $36x^2 + 24xy + 29y^2 - 72x + 126y + 81 = 0$
   (*iii*) $11x^2 + 4xy + 14y^2 - 26x - 32y + 23 = 0$
   (*iv*) $x^2 + 4xy + y^2 - 2x + 2y - 6 = 0$.
2. Trace the conic : $x^2 + 12xy - 4y^2 - 6x + 4y + 9 = 0$.
   Also find the eccentricity, co-ordinates of foci and directrices of the conic. (*M.D.U. 1997*)
3. Trace the conic : $x^2 - 4xy - 2y^2 + 10x + 4y = 0$. (*M.D.U. 2007*)
4. Trace the curve $32x^2 + 52xy - 7y^2 - 64x - 52y - 148 = 0$.

Answers

1. (*i*) 4 ; 2 ; $x - y + 3 = 0$, $x + y - 1 = 0$
   (*ii*) 6 ; 4 ; $4x + 3y + 1 = 0$, $3x - 4y - 18 = 0$
   (*iii*) Centre (1, 1) ; axes $\dfrac{2\sqrt{3}}{5}, \dfrac{2\sqrt{2}}{5}$; $x + 2y - 3 = 0$, $2x - y - 1 = 0$
   (*iv*) Hyperbola centre (– 1, 1), transverse axis $x - y + 2 = 0$, length $4/\sqrt{3}$, conjugate axis $x + y = 0$, length = 4.
2. Centre $\left(0, \dfrac{1}{2}\right)$; $\sqrt{5}$, $2\sqrt{2}$; $3x + 2y - 1 = 0$; $4x - 6y + 3 = 0$; $e = \sqrt{\dfrac{13}{5}}$; foci (1, – 1), (– 1, 2) ; directrices $4x - 6y + 8 = 0$; $2x - 3y + 4 = 0$.
4. Hyperbola, centre (1, 0), transverse axis $x - 2y = 0$, length 4, conjugate axis $2x + y - 2 = 0$, length 6.

11. The Parabola in General

The standard equation of the parabola is $y^2 = 4ax$. Geometrically, it can be represented as

$$(\text{MP})^2 = 4a(\text{NP})$$

where MP is the perpendicular from P on x-axis and NP is perpendicular from P on y-axis.

Thus, in general a parabola may also be defined as the locus of a point which moves such that

$$\begin{pmatrix}\text{distance of the}\\ \text{point from } l_1\end{pmatrix}^2 = 4a\begin{pmatrix}\text{distance of the}\\ \text{point from } l_2\end{pmatrix},$$

where l_1 and l_2 are two perpendicular lines. Here l_1 is the axis of the parabola and l_2 is the tangent at the vertex.

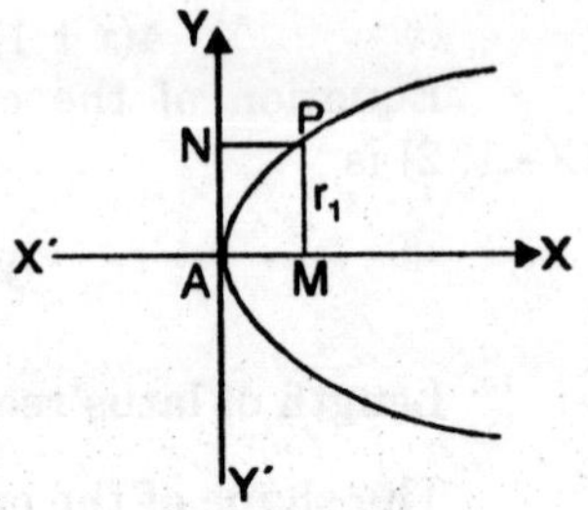

$\therefore$ If $\quad ax + by + c = 0$ and $bx - ay + c' = 0$

are two perpendicular lines, then the general equation of the parabola is

$$\left(\frac{ax+by+c}{\sqrt{a^2+b^2}}\right)^2 = K\left(\frac{bx-ay+c'}{\sqrt{a^2+b^2}}\right)$$

where (*i*) **Length of Latus Rectum** = | K |.

(*ii*) Equation of axis of parabola is $ax + by + c = 0$.

(*iii*) Equation of tangent at vertex is $bx - ay + c' = 0$.

Note. In the equation of parabola, second degree terms form a perfect square.

12. Find the axis, the latus rectum and tangent at the vertex of the parabola given by the equation

$$\mathbf{ax^2 + 2hxy + by^2 + 2gx + 2fy + c = 0.} \qquad \textit{(M.D.U. 1999)}$$

The equation $\quad ax^2 + 2hxy + by^2 + 2gx + 2fy + c = 0 \qquad ...(1)$

will represent a parabola, if $h^2 - ab = 0$,

i.e., if the second degree terms form a perfect square.

Put $a = \alpha^2$ and $b = \beta^2$, therefore $h = \alpha\beta$. Then equation (1) takes the form

$$\alpha^2x^2 + 2\alpha\beta\, xy + \beta^2y^2 + 2gx + 2fy + c = 0$$

i.e.,
$$(\alpha x + \beta y)^2 + 2gx + 2fy + c = 0$$

or
$$(\alpha x + \beta y)^2 = -2gx - 2fy - c \qquad ...(2)$$

We choose λ, so that equation (2) may be written as

$$(\alpha x + \beta y + \lambda)^2 = -2gx - 2fy - c + 2\lambda(\alpha x + \beta y) + \lambda^2$$

i.e.,
$$(\alpha x + \beta y + \lambda)^2 = 2x(\lambda\alpha - g) + 2y(\lambda\beta - f) + \lambda^2 - c \qquad ...(3)$$

The two lines $\quad \alpha x + \beta y + \lambda = 0$

and $\quad 2x(\lambda\alpha - g) + 2y(\lambda\beta - f) + \lambda^2 - c = 0$

are at right angles.

This is so if
$$\left(-\frac{\alpha}{\beta}\right)\left(-\frac{\lambda\alpha - g}{\lambda\beta - f}\right) = -1 \qquad (\because\ m_1m_2 = -1)$$

i.e.,
$$\alpha(\lambda\alpha - g) = -\beta(\lambda\beta - f)$$

i.e.,
$$\lambda = \frac{\alpha g + \beta f}{\alpha^2 + \beta^2}.$$

Equation (3) can be written as

$$\left(\frac{\alpha x + \beta y + \lambda}{\sqrt{\alpha^2+\beta^2}}\right)^2(\alpha^2+\beta^2) = 2\left\{\frac{(\lambda\alpha - g)\,x + (\lambda\beta - f)\,y + \frac{1}{2}(\lambda^2 - c)}{\sqrt{(\lambda\alpha - g)^2 + (\lambda\beta - f)^2}}\right\} \times \sqrt{(\lambda\alpha - g)^2 + (\lambda\beta - f)^2}$$

i.e.,
$$\left(\frac{\alpha x + \beta y + \lambda}{\sqrt{\alpha^2+\beta^2}}\right)^2 = \frac{2\sqrt{(\lambda\alpha - g)^2 + (\lambda\beta - f)^2}}{\alpha^2+\beta^2}\left\{\frac{(\lambda\alpha - g)\,x + (\lambda\beta - f)\,y + \frac{1}{2}(\lambda^2 - c)}{\sqrt{(\lambda\alpha - g)^2 + (\lambda\beta - f)^2}}\right\} \qquad ...(4)$$

which is of the form $p_1^{\,2} = 4ap_2$,

where p_1 and p_2 are perpendicular distances of (x, y) from the above two perpendicular lines. As such it is a parabola.

Substituting the value of $\lambda = \dfrac{\alpha g + \beta f}{\alpha^2 + \beta^2}$ in the expression for the latus rectum

$\dfrac{2\sqrt{(\lambda\alpha - g)^2 + (\lambda\beta - f)^2}}{\alpha^2 + \beta^2}$, we get the length of latus rectum as

$$\frac{2(\alpha f - \beta g)}{(\alpha^2 + \beta^2)^{3/2}}.$$

Hence (*i*) Latus rectum is $\dfrac{2(\alpha f - \beta g)}{(\alpha^2 + \beta^2)^{3/2}}$

(*ii*) The equation of axis is $\alpha x + \beta y + \lambda = 0$

(*iii*) The tangent at the vertex is

$$x(\lambda\alpha - g) + y(\lambda\beta - f) + \frac{1}{2}(\lambda^2 - c) = 0.$$

Example 1. *Find the latus rectum, equation of axis, and tangent at the vertex, and vertex of the parabola*

$$4x^2 - 4xy + y^2 - 10y - 19 = 0. \qquad \text{(M.D.U. 2003)}$$

Sol. Here $a = 4, b = 1, h = -2$. So $h^2 - ab = 0$.

Hence the conic is a parabola. The second degree terms form the perfect square $(2x - y)^2$. Writing the given equation in the form

$$(2x - y + \lambda)^2 - \mu(x + 2y + c) = 0$$

Comparing it with the given equation

Co-eff. of x : $4\lambda - \mu = 0$ ...(1)

Co-eff. of y : $-2\lambda - 2\mu = -10$ *i.e.*, $\lambda + \mu = 5$ ...(2)

Const. term : $\lambda^2 - \mu c = -19$ ...(3)

Solving (1) and (2), $\lambda = 1, \mu = 4$

Thus from (3), $1 - 4c = -19 \quad \therefore \quad c = 5$

The equation then becomes $(2x - y + 1)^2 = 4(x + 2y + 5)$

Dividing both sides by $2^2 + 1^2$

$$\left(\frac{2x - y + 1}{\sqrt{2^2 + 1^2}}\right)^2 = \frac{4}{\sqrt{2^2 + 1^2}}\left(\frac{x + 2y + 5}{\sqrt{2^2 + 1^2}}\right)$$

i.e.,

$$\left(\frac{2x - y + 1}{\sqrt{5}}\right)^2 = \frac{4}{\sqrt{5}}\left(\frac{x + 2y + 5}{\sqrt{5}}\right)$$

Thus, we have the following results :

(*i*) Latus rectum : $\dfrac{4}{\sqrt{5}}$

(*ii*) Axis of the parabola : $2x - y + 1 = 0$ ...(4)

(*iii*) Tangent at the vertex : $x + 2y + 5 = 0$ ...(5)

Its vertex A is the point of intersection of (4) and (5).

Thus, solving (4) and (5), vertex is $\left(-\dfrac{7}{5}, -\dfrac{9}{5}\right)$.

Second method. Writing the given equation in the form

$$(2x - y)^2 - 10y - 19 = 0 \qquad ...(1)$$

Select λ so that, when writing equation (1) in the form,

$$(2x - y + \lambda)^2 = 4\lambda x + (10 - 2\lambda)\, y + \lambda^2 + 19 = 0 \qquad ...(2)$$

the two lines $\quad 2x - y + \lambda = 0 \quad$ and $\quad 4\lambda x + (10 - 2\lambda)\, y + \lambda^2 + 19 = 0$

are perpendicular. This is so if

$$(2)\left(\frac{-4\lambda}{10 - 2\lambda}\right) = -1 \quad i.e., \quad \lambda = 1$$

The equation then becomes

$$(2x - y + 1)^2 = 4(x + 2y + 5)$$

Then proceed as in 1st Method.

Example 2. *Trace the conic*

$$25x^2 - 120xy + 144y^2 - 2x - 29y - 1 = 0.$$

Sol. As the second degree terms form a perfect square, it is a parabola. Writing it as,

$$(5x - 12y)^2 - 2x - 29y - 1 = 0 \qquad ...(1)$$

It again can be written in the form

$$(5x - 12y + \lambda)^2 - \mu(12x + 5y + c) = 0$$

Comparing it with the given equation

Co-eff. of x : $\quad 10\lambda - 12\mu = -2 \qquad ...(2)$

Co-eff. of y : $\quad -24\lambda - 5\mu = -29 \qquad ...(3)$

Const. term : $\quad \lambda^2 - \mu c = -1 \qquad ...(4)$

Solving (2) and (3), $\quad \lambda = \mu = 1.$

∴ From (4), $\quad c = 2.$

Hence, $\quad (5x - 12y + 1)^2 = 12x + 5y + 2$

or dividing both sides by $5^2 + 12^2$,

$$\left(\frac{5x - 12y + 1}{\sqrt{5^2 + 12^2}}\right)^2 = \frac{1}{\sqrt{5^2 + 12^2}}\left(\frac{12x + 5y + 2}{\sqrt{5^2 + 12^2}}\right)$$

i.e.,
$$\left(\frac{5x - 12y + 1}{13}\right)^2 = \frac{1}{13}\cdot\left(\frac{12x + 5y + 2}{13}\right)$$

Thus, the latus rectum $= \dfrac{1}{13}$

Axis of the parabola : $\quad 5x - 12y + 1 = 0 \qquad ...(5)$

And, tangent at the vertex : $\quad 12x + 5y + 2 = 0 \qquad ...(6)$

Solving (5) and (6),

$$\text{vertex} : \left(\frac{-29}{169}, \frac{2}{169}\right).$$

Example 3. *Find the nature of the conic*

$$x^2 + 2xy + y^2 - 2x - 1 = 0.$$

Sol. Equation of the curve is

$$x^2 + 2xy + y^2 - 2x - 1 = 0 \qquad ...(1)$$

Here $h^2 - ab = 1 - 1 = 0$

and the second degree terms form a perfect square *i.e.*, $(x + y)^2$, therefore equations (1) represents a parabola.

Equation (1) can be written as

$$(x + y)^2 = 2x + 1$$

Now consider $(x + y + \lambda)^2 = (x + y)^2 + 2\lambda(x + y) + \lambda^2$

$$= 2x + 1 + 2\lambda x + 2\lambda y + \lambda^2$$

$$= (2\lambda + 2)x + 2\lambda y + \lambda^2 + 1 \quad \text{...(2)}$$

Choose λ such that the lines $x + y + \lambda = 0$ and $(2\lambda + 2)x + 2\lambda y + \lambda^2 + 1 = 0$ are perpendicular.

$\therefore$ Product of slopes $= -1$

i.e., $$1 \cdot \frac{(2\lambda + 2)}{2\lambda} = -1 \quad \text{or} \quad 2\lambda + 2 + 2\lambda = 0 \quad \therefore \quad \lambda = -\frac{1}{2}$$

$\therefore$ Equation (2) becomes

$$\left(x + y - \frac{1}{2}\right)^2 = x - y + \frac{5}{4} \quad \text{...(3)}$$

or $$\left(\frac{x + y - \frac{1}{2}}{\sqrt{2}}\right)^2 = \frac{1}{\sqrt{2}} \frac{x - y + \frac{5}{4}}{\sqrt{2}}$$

$\therefore$ Axis of parabola is $x + y - \frac{1}{2} = 0$ ...(4)

Tangent at the vertex is $x - y + \frac{5}{4} = 0$ ...(5)

Length of Latus rectum $= \frac{1}{\sqrt{2}} = 0.7$ ...(6)

Solving the equations (4) and (5), $x = -\frac{3}{8}, y = \frac{7}{8}$

Vertex is $\left(-\frac{3}{8}, \frac{7}{8}\right)$ ...(7)

Now the latus rectum and the directrix both are parallel to the tangent at the vertex at a distance $\frac{1}{4\sqrt{2}}\left(\frac{1}{4} \text{ of L.R.}\right)$.

Any line parallel to (5) is $x - y + c = 0$

$$\therefore \quad \text{Distance from (5)} = \pm \left|\frac{c - \frac{5}{4}}{\sqrt{2}}\right| = \frac{1}{4\sqrt{2}}.$$

$$\therefore \quad c = 1 \quad \text{or} \quad \frac{3}{2}$$

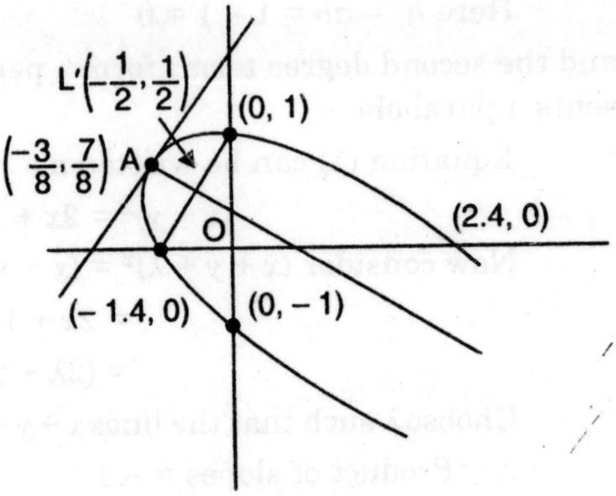

If $c = 1$, the equation of latus rectum is

$$x - y + 1 = 0 \qquad ...(8)$$

If $c = \frac{3}{2}$, directrix is $x - y + \frac{3}{2} = 0$ ...(9)

Again solving the equation (4) and (8), we have

$$x = -\frac{1}{4}, y = \frac{3}{4}$$

$\therefore$ Focus is $S\left(-\frac{1}{4}, \frac{3}{4}\right)$. ...(10)

The latus rectum meets the curve (3) in L and L′.

Solving (3) and (8), we get

ends of latus rectum $L(0, 1), L'\left(-\frac{1}{2}, \frac{1}{2}\right)$ ...(11)

The curve meets the x-axis *i.e.*, $y = 0$ at $\left(\frac{12}{5}, 0\right)$ and $\left(-\frac{7}{5}, 0\right)$ *i.e.*, (2.4, 0) and (– 1.4, 0).

The curve meets the y-axis, where $x = 0$ and $y^2 - 1 = 0$

i.e., $x = 0, y = \pm 1$

or at the points (0, 1) and (0, – 1).

The curve lies on the origin side of the tangent at the vertex. The shape of the curve is as given in the figure.

Example 4. *Trace the conic*

$$9x^2 + 24xy + 16y^2 - 2x + 14y + 1 = 0. \qquad ...(1) \text{ (K.U. 2003, 2007)}$$

Sol. Here $a = 9, h = 12, b = 16$. $\therefore$ $h^2 - ab = 144 - 144 = 0$.

Hence the conic is a parabola. The second degree terms form the perfect square $(3x + 4y)^2$

We write the given equation in the form

$$(3x + 4y)^2 = 2x - 14y - 1$$

Now consider the equation

$$(3x + 4y + \lambda)^2 = (3x + 4y)^2 + 2\lambda(3x + 4y) + \lambda^2 = 2x - 14y - 1 + 2\lambda(3x + 4y) + \lambda^2$$
$$= 2(1 + 3\lambda)x + 2(4\lambda - 7)y + \lambda^2 - 1$$

Choose λ such that the lines

$$3x + 4y + \lambda = 0 \quad \text{and} \quad 2(1 + 3\lambda)x + 2(4\lambda - 7)y + \lambda^2 - 1 = 0 \qquad ...(2)$$

are at right angles

$\therefore$ Product of their slopes = – 1

$$\Rightarrow \quad \left(-\frac{3}{4}\right)\left(-\frac{1+3\lambda}{4\lambda-7}\right) = -1$$

$$\Rightarrow \quad 3(1 + 3\lambda) + 4(4\lambda - 7) = 0 \quad \therefore \quad \lambda = 1$$

The equation (2) becomes

$$(3x + 4y + 1)^2 = 8x - 6y$$

$$25\left(\frac{3x+4y+1}{\sqrt{3^2+4^2}}\right)^2 = 10 \cdot \frac{8x-6y}{\sqrt{8^2+6^2}}$$

i.e.,
$$\left(\frac{3x+4y+1}{5}\right)^2 = \frac{2}{5} \cdot \frac{(4x-3y)}{5}$$

This is of the form $Y^2 = 4AX$, where $Y = \frac{3x+4y+1}{5}$, $X = \frac{4x-3y}{5}$, $4A = \frac{2}{5}$

Thus we have following results :

(*i*) Length of latus rectum $= \frac{2}{5}$

(*ii*) Axis of the parabola is $Y = 0$ *i.e.*, $3x + 4y + 1 = 0$

(*iii*) Tangent at the vertex is $X = 0$ *i.e.*, $4X - 3y = 0$

Its vertex A is the point of intersection of (*ii*) and (*iii*).

Thus, solving (*ii*) and (*iii*) simultaneously the co-ordinate of the vertex are given by $\left(-\frac{3}{25}, -\frac{4}{25}\right)$.

The parabola meets the x-axis, where $y = 0$ *i.e.*, in points given by

$$\therefore \quad 9x^2 - 2x + 1 = 0$$

$$\therefore \quad x = \frac{2 \pm \sqrt{4-36}}{18}, \text{ which are imaginary.}$$

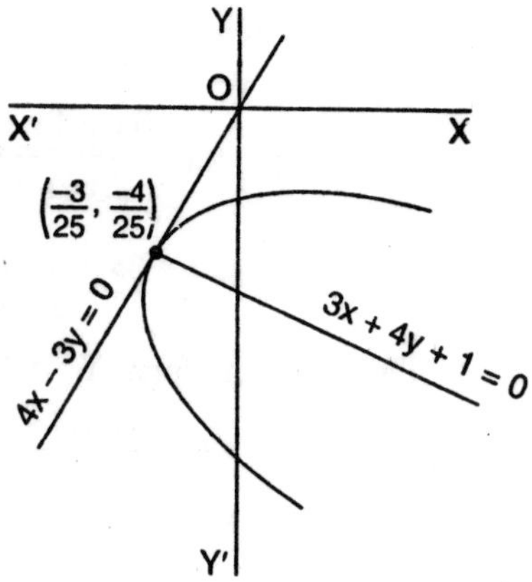

Again, the curve meets y-axis where $x = 0$ *i.e.*, in the points given by

$$16y^2 + 14y + 1 = 0$$

or
$$y = \frac{-14 \pm \sqrt{196-64}}{32} = \frac{-7 \pm \sqrt{33}}{16}$$

Thus the parabola meets the y-axis in the points $\left(0, \frac{-7 \pm \sqrt{33}}{16}\right)$.

The curve does not meet the x-axis.

(*iv*) The equation of the latus rectum is $X = A$

i.e.,
$$\frac{4x-3y}{5} = \frac{1}{10} \quad \text{or} \quad 4x - 3y = \frac{1}{2}.$$

(*v*) **Focus.** Solving the equations (*ii*) and (*iv*), the co-ordinates of focus are given as $\left(-\frac{1}{25}, -\frac{11}{50}\right)$.

Hence the shape of the curve is as shown in the figure.

Example 5. *Trace the curve*

$$9x^2 + 24xy + 16y^2 - 44x + 108y - 124 = 0.$$

Sol. Here $a = 9, b = 16, h = 12 \quad \therefore \quad h^2 = ab$

Hence the conic is a parabola. The second degree terms form the perfect square $(3x + 4y)^2$.

The equation can be written as

$$(3x + 4y)^2 = 44x - 108y + 124 \qquad ...(1)$$

Consider the equation

$$(3x + 4y + \lambda)^2 = (3x + 4y)^2 + 2\lambda(3x + 4y) + \lambda^2$$
$$= 44x - 108y + 124 + 6\lambda x + 8\lambda y + \lambda^2$$
$$= 2x(22 + 3\lambda) + 4y(-27 + 2\lambda) + \lambda^2 + 124 \qquad ...(2)$$

Choose λ such that

the two lines $3x + 4y + \lambda = 0$

and $2x(22 + 3\lambda) + 4y(-27 + 2\lambda) + \lambda^2 + 124 = 0$

are perpendicular. This is so if

$$\left(-\frac{3}{4}\right)\left(-\frac{22 + 3\lambda}{2(-27 + 2\lambda)}\right) = -1$$

or $$3(22 + 3\lambda) + 8(-27 + 2\lambda) = 0 \quad \therefore \quad \lambda = 6$$

The equation then becomes

$$(3x + 4y + 6)^2 = 20(4x - 3y + 8)$$

which can be written as

$$25\left(\frac{3x + 4y + 6}{\sqrt{3^2 + 4^2}}\right)^2 = 20\left(\frac{4x - 3y + 8}{\sqrt{3^2 + 4^2}}\right)5$$

i.e., $$\left(\frac{3x + 4y + 6}{5}\right)^2 = 4\left(\frac{4x - 3y + 8}{5}\right)$$

This is of the form $Y^2 = 4AX$, where

$$Y = \frac{3x + 4y + 6}{5}, \ X = \frac{4x - 3y + 8}{5}, \ 4A = 4$$

Hence the axis of the parabola is $Y = 0$ *i.e.,*

$$3x + 4y + 6 = 0 \qquad ...(3)$$

and tangent at the vertex is $X = 0$, *i.e.,* $4x - 3y + 8 = 0$ ...(4)

On solving the equations (3) and (4) simultaneously the co-ordinates of the vertex are given by $(-2, 0)$.

Length of latus rectum = 4.

The parabola meets x-axis, where $y = 0$ *i.e.,* in the points given by

$$9x^2 - 44x - 124 = 0$$

i.e., $$(9x - 62)(x + 2) = 0$$

i.e., $$x = \frac{62}{9} \quad \text{or} \quad -2$$

Further, the parabola meets y-axis, where $x = 0$ *i.e.,* in the points given by

$$16y^2 + 108y - 124 = 0$$

or $$4y^2 + 27y - 31 = 0$$

i.e., $$(y - 1)(4y + 31) = 0$$

$\therefore$ $$y = -\frac{31}{4} \quad \text{or} \quad 1$$

$\therefore$ The parabola meets the y-axis in the points $\left(0, -\frac{31}{4}\right)$ and (0, 1)

and the curve meets x-axis at $\left(\frac{62}{9}, 0\right)$ and $(-2, 0)$.

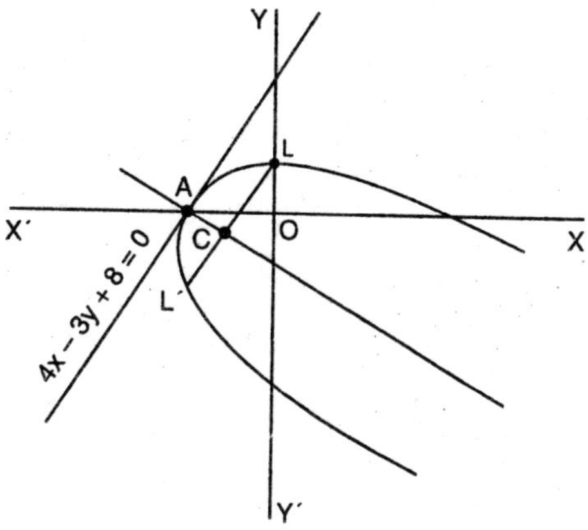

Now the equation of the latus rectum is

$$X = A$$

i.e., $$\frac{4x - 3y + 8}{5} = 1 \Rightarrow 4x - 3y + 3 = 0 \quad ...(5)$$

On solving (3) and (5), we get, the co-ordinates of focus as $\left(-\frac{6}{5}, -\frac{3}{5}\right)$.

The shape of the parabola is as shown in the figure.

EXERCISE 4

1. Find the length of latus rectum, equations of axis and the tangent at the vertex, and vertex of the parabola

$$16x^2 - 24xy + 9y^2 - 104x - 172y + 44 = 0.$$ (*M.D.U. 1999*)

2. Show that $9x^2 - 24xy + 16y^2 - 50x - 100y + 225 = 0$ represents a parabola whose axis is $3x - 4y + 5 = 0$, tangent at the vertex is $4x + 3y - 10 = 0$ and latus rectum is 4.

3. Trace the curve

(*i*) $9x^2 - 24xy + 16y^2 - 18x - 101y + 19 = 0$

(*ii*) $x^2 + 2xy + y^2 - 2x - 1 = 0$.

4. Show that the latus rectum of the parabola

$$(a^2 + b^2)(x^2 + y^2) = (bx + ay - ab)^2 \text{ is } \frac{2ab}{\sqrt{a^2 + b^2}}.$$

5. Trace the following curves :

(*i*) $16x^2 - 24xy + 9y^2 + 77x - 64y + 95 = 0$

(*ii*) $25x^2 - 120xy + 144y^2 - 2x - 29y - 1 = 0$

(*iii*) $4x^2 - 4xy + y^2 - 8x - 6y + 5 = 0$

(*iv*) $9x^2 + 24xy + 16y^2 - 2x + 14y + 1 = 0$.

Answers

1. 8 ; $4x - 3y + 2 = 0$; $3x + 4y - 1 = 0$; $\left(\frac{-1}{5}, \frac{2}{5}\right)$

3. (*i*) Parabola, Axis : $3x - 4y + 7 = 0$, tangent at vertex $4x + 3y + 2 = 0$,

latus rectum 3, vertex $\left(\frac{-29}{25}, \frac{22}{25}\right)$

(*ii*) Parabola, Axis : $2x + 2y - 1 = 0$, tangent at vertex : $4x - 4y + 5 = 0$,

vertex : $\left(\frac{-3}{8}, \frac{7}{8}\right)$, latus rectum $= \frac{1}{\sqrt{2}}$.

5. (*i*) Parabola, Axis : $4x - 3y + 10 = 0$, tangent at the vertex $3x + 4y + 5 = 0$, latus rectum $= \frac{1}{5}$.

(*ii*) Parabola, Axis : $5x - 12y + 1 = 0$, tangent at vertex $12x + 5y + 2 = 0$

Latus rectum $= \frac{1}{13}$, Vertex $\left(-\frac{29}{169}, \frac{2}{169}\right)$.

(*iii*) Parabola, Axis : $2x - y - 1 = 0$, tangent, at vertex $x + 2y - 1 = 0$

Latus rectum $= \frac{4}{\sqrt{5}}$, Vertex $\left(\frac{3}{5}, \frac{1}{5}\right)$, Focus $\left(\frac{4}{5}, \frac{3}{5}\right)$.

(*iv*) Parabola, Axis : $3x + 4y + 1 = 0$, tangent at vertex $4x - 3y = 0$,

Vertex $\left(-\frac{3}{25}, -\frac{4}{25}\right)$, Focus $\left(-\frac{1}{25}, -\frac{11}{50}\right)$.

EQUATION OF THE TANGENT AT ANY POINT TO THE CONIC

13. To find the equation of the tangent at the point (x_1, y_1) to the conic

$$\mathbf{ax^2 + 2hxy + by^2 + 2gx + 2fy + c = 0.}$$ *(M.D.U. 1998)*

Equation of the conic is

$$ax^2 + 2hxy + by^2 + 2gx + 2fy + c = 0 \quad ...(1)$$

Let $P(x_1, y_1)$ be the given point on the conic (1). Take a neighbouring point $Q(x_2, y_2)$ on the conic.

The equation of the chord PQ is

$$y - y_1 = \frac{y_2 - y_1}{x_2 - x_1}(x - x_1) \quad ...(2)$$

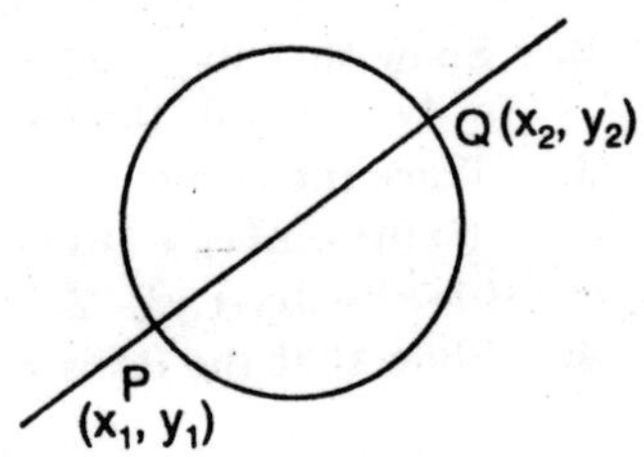

As $P(x_1, y_1)$ and $Q(x_2, y_2)$ lie on (1),

$$\therefore \quad ax_1^2 + 2hx_1y_1 + by_1^2 + 2gx_1 + 2fy_1 + c = 0 \quad ...(3)$$

and $$ax_2^2 + 2hx_2y_2 + by_2^2 + 2gx_2 + 2fy_2 + c = 0 \quad ...(4)$$

Subtracting (3) from (4), we have

$$a(x_2^2 - x_1^2) + 2h(x_2y_2 - x_1y_1) + b(y_2^2 - y_1^2) + 2g(x_2 - x_1) + 2f(y_2 - y_1) = 0$$

[Use $2(x_2y_2 - x_1y_1) = (x_2 + x_1)(y_2 - y_1) + (x_2 - x_1)(y_2 + y_1)$]

or $$a(x_2^2 - x_1^2) + h[(x_2 + x_1)(y_2 - y_1) + (x_2 - x_1)(y_2 + y_1)] + b(y_2^2 - y_1^2) + 2g(x_2 - x_1) + 2f(y_2 - y_1) = 0$$

i.e., $$(x_2 - x_1)[a(x_2 + x_1) + h(y_2 + y_1) + 2g] = -(y_2 - y_1)[b(y_2 + y_1) + h(x_2 + x_1) + 2f]$$

$$\therefore \quad \frac{y_2 - y_1}{x_2 - x_1} = -\frac{a(x_2 + x_1) + h(y_2 + y_1) + 2g}{b(y_2 + y_1) + h(x_2 + x_1) + 2f}$$

Substituting it in (1), the equation of chord PQ is

$$y - y_1 = -\frac{a(x_2 + x_1) + h(y_2 + y_1) + 2g}{b(y_2 + y_1) + h(x_2 + x_1) + 2f}(x - x_1) \quad ...(5)$$

If equation (5) is a tangent at (x_1, y_1), then $Q \to P$ *i.e.,* $x_2 \to x_1$ and $y_2 \to y_1$.

$\therefore$ Putting $x_2 = x_1$, $y_2 = y_1$ in (5), the equation of the tangent at $P(x_1, y_1)$ is

$$y - y_1 = -\frac{2(ax_1 + hy_1 + g)}{2(hx_1 + by_1 + f)}(x - x_1)$$

or $\quad (hx_1 + by_1 + f)(y - y_1) = -(ax_1 + hy_1 + g)(x - x_1)$

or $\quad (ax_1 + hy_1 + g)x + (hx_1 + by_1 + f)y = ax_1^2 + hx_1y_1 + gx_1 + hx_1y_1 + by_1^2 + fy_1$

or $\quad (ax_1 + hy_1 + g)x + (hx_1 + by_1 + f)y = ax_1^2 + 2hx_1y_1 + by_1^2 + gx_1 + fy_1$

(adding $gx_1 + fy_1 + c$ both sides)

or $\quad (ax_1 + hy_1 + g)x + (hx_1 + by_1 + f)y + gx_1 + fy_1 + c = 0$

[$\because$ (x_1, y_1) lies on (1)]

Hence, $axx_1 + h(x_1y + y_1x) + byy_1 + g(x + x_1) + f(y + y_1) + c = 0$ is the required equation of the tangent.

Note. Rule to write the equation of tangent at (x_1, y_1)

Change $x^2 \to xx_1$, $y^2 \to yy_1$, $xy \to \frac{1}{2}(xy_1 + x_1y)$,

$$x \to \frac{1}{2}(x + x_1),\ y \to \frac{1}{2}(y + y_1)\ ;\ c = c$$

or $\quad x(ax_1 + hy_1 + g) + y(hx_1 + by_1 + f) + gx_1 + fy_1 + c = 0$

For this remember the determinants $\begin{vmatrix} a & h & g \\ h & b & f \\ g & f & c \end{vmatrix}$

which plays an important role in two or three dimensional co-ordinate geometry.

A LINE AND A CONIC

14. The intersection of the general conic and the line

$$\frac{\mathbf{x} - \mathbf{x_1}}{\cos\theta} = \frac{\mathbf{y} - \mathbf{y_1}}{\sin\theta} = \mathbf{r}.$$

Let the equation of the general conic be

$$ax^2 + 2hxy + by^2 + 2gx + 2fy + c = 0 \qquad ...(1)$$

Any point on the given line

$$\frac{x - x_1}{\cos\theta} = \frac{y - y_1}{\sin\theta} = r \qquad ...(2)$$

is $(x_1 + r\cos\theta, y_1 + r\sin\theta)$.

Let this be a point, where (1) and (2) meet.

It will satisfy the equation of the conic. Hence we have

$$a(x_1 + r\cos\theta)^2 + 2h(x_1 + r\cos\theta)(y_1 + r\sin\theta)$$
$$+ b(y_1 + r\sin\theta)^2 + 2g(x_1 + r\cos\theta) + 2f(y_1 + r\sin\theta) + c = 0$$

or $\quad r^2(a\cos^2\theta + 2h\sin\theta\cos\theta + b\sin^2\theta)$

$$+ 2r\{(ax_1 + hy_1 + g)\cos\theta + (hx_1 + by_1 + f)\sin\theta\} + S_1 = 0 \qquad ...(3)$$

where $S_1 = ax_1^2 + 2hx_1y_1 + by_1^2 + 2gx_1 + 2fy_1 + c$.

The equation (3) is a quadratic equation in r, therefore gives two values of r. This shows that (1) and (2) meet in two points. As such in general, every straight line meets a conic in two points.

Deduction of the equation of the tangent at (x_1, y_1) *(M.D.U. 1998)*

To find the equation of the tangent at (x_1, y_1)

If the line (2) is a tangent to (1) at $P(x_1, y_1)$, then both the roots of equation (3) *i.e.*,

$$r^2(a \cos^2 \theta + 2h \sin \theta \cos \theta + b \sin^2 \theta) + 2r(ax_1 + by_1 + g) \cos \theta + (hx_1 + by_1 + f) \sin \theta + S_1 = 0 \quad ...(3)$$

are equal. Also the point $P(x_1, y_1)$ lies on the conic.

$\therefore$ $S_1 = 0$ *i.e.*, constant term = 0

$\therefore$ One root of (3) = 0

$\therefore$ 2nd root is also zero.

$\therefore$ Co-efficient of $r = 0$ in equation (3)

i.e.,
$$(ax_1 + hy_1 + g) \cos \theta + (hx_1 + by_1 + f) \sin \theta = 0 \quad ...(4)$$

Substituting the values of $\cos \theta$ and $\sin \theta$ from (2) in (4), we have

$$(ax_1 + hy_1 + g)(x - x_1) + (hx_1 + by_1 + f)(y - y_1) = 0$$

or
$$x(ax_1 + hy_1 + g) + y(hx_1 + by_1 + f) + (gx_1 + fy_1 + c) = S_1$$

As $S_1 = ax_1^2 + 2hx_1y_1 + by_1^2 + 2gx_1 + 2fy_1 + c = 0$, it becomes

$$\mathbf{x(ax_1 + hy_1 + g) + y(hx_1 + by_1 + f) + (gx_1 + fy_1 + c) = 0}$$

which is the required equation of the tangent.

15. Joint Equation of Pair of Tangents

To find the joint equation of the pair of tangents from a given point (x_1, y_1) to the conic

$$\mathbf{ax^2 + 2hxy + by^2 + 2gx + 2fy + c = 0}$$

Let $P(x_1, y_1)$ be the given point. Draw a straight line PAB to meet the conic in A and B.

The equation of the line PAB is of the form

$$\frac{x - x_1}{\cos \theta} = \frac{y - y_1}{\sin \theta} = r \quad ...(1)$$

Any point on it is $(x_1 + r \cos \theta, y_1 + r \sin \theta)$

Let it be either A or B. As it lies on the conic, we have

$$a(x_1 + r \cos \theta)^2 + 2h(x_1 + r \cos \theta)(y_1 + r \sin \theta) + b(y_1 + r \sin \theta)^2 + 2g(x_1 + r \cos \theta) + 2f(y_1 + r \sin \theta) + c = 0$$

or
$$r^2(a \cos^2 \theta + 2h \sin \theta \cos \theta + b \sin^2 \theta) + 2r\{(ax_1 + hy_1 + g) \cos \theta + (hx_1 + by_1 + f) \sin \theta\} + S_1 = 0 \quad ...(2)$$

where $S_1 = ax_1^2 + 2hx_1y_1 + by_1^2 + 2gx_1 + 2fy_1 + c$

which is quadratic in r and gives two values of r corresponding to the two points A and B.

If the line (1) is a tangent line from $P(x_1, y_1)$ to the conic, the two roots of equation (2) must be equal.

$\therefore$ its discriminant is zero

i.e.,
$$4[(ax_1 + hy_1 + g) \cos \theta + (hx_1 + by_1 + f) \sin \theta]^2 - 4(a \cos^2 \theta + 2h \sin \theta \cos \theta + b \sin^2 \theta) S_1 = 0$$

Eliminating θ between this equation and equation (1), we have

$$[(ax_1 + hy_1 + g)(x - x_1) + (hx_1 + by_1 + f)(y - y_1)]^2 = [a(x - x_1)^2 + 2h(x - x_1)(y - y_1) + b(y - y_1)^2]S_1$$

We take $\quad T = x(ax_1 + hy_1 + g) + y(hx_1 + by_1 + f) + (gx_1 + fy_1 + c)$

$$S = f(x, y) = ax^2 + 2hxy + by^2 + 2gx + 2fy + c$$

and $\quad S_1 = f(x_1, y_1)$

the above equation can be written as

$$[T - S_1]^2 = [S + S_1 - 2T]S_1$$

or
$$T^2 + S_1^2 - 2TS_1 = SS_1 + S_1^2 - 2TS_1$$

i.e.,
$$T^2 = SS_1$$

which is the required equation of the pair of tangents from the point (x_1, y_1).

16. Chord with a given Middle Point

To find the equation of the chord of the conic

$$\mathbf{ax^2 + 2hxy + by^2 + 2gx + 2fy + c = 0}$$

having (x_1, y_1) as its middle point. *(M.D.U. 1997, 2005)*

Equation of the conic is

$$ax^2 + 2hxy + by^2 + 2gx + 2fy + c = 0 \quad ...(1)$$

Let $M(x_1, y_1)$ be the middle point of the chord PQ of the conic (1).

Let the equation of the chord PQ through $M(x_1, y_1)$ be

$$\frac{x - x_1}{\cos\theta} = \frac{y - y_1}{\sin\theta} = r \quad ...(2)$$

Any point on (2) is $(x_1 + r\cos\theta, y_1 + r\sin\theta)$

If it is either P or Q, it must satisfy the equation (1)

$\therefore \quad a(x_1 + r\cos\theta)^2 + 2h(x_1 + r\cos\theta)(y_1 + r\sin\theta) + b(y_1 + r\sin\theta)^2$
$$+ 2g(x_1 + r\cos\theta) + 2f(y_1 + r\sin\theta) + c = 0$$

i.e., $\quad r^2(a\cos^2\theta + 2h\sin\theta\cos\theta + b\sin^2\theta)$
$$+ 2r\{(ax_1 + hy_1 + g)\cos\theta + (hx_1 + by_1 + f)\sin\theta\} + S_1 = 0 \quad ...(3)$$

where $S_1 = ax_1^2 + 2hx_1y_1 + by_1^2 + 2gx_1 + 2fy_1 + c$.

Since $M(x_1, y_1)$ is the middle point of chord PQ, the two values of r given by equation (3) are equal in magnitude but opposite in sign.

$\therefore$ Sum of roots of equation (3) is zero.

$$\therefore \quad (ax_1 + hy_1 + g)\cos\theta + (hx_1 + by_1 + f)\sin\theta = 0$$

Substituting values of $\cos\theta$ and $\sin\theta$ from (2), we have

$$(ax_1 + hy_1 + g)(x - x_1) + (hx_1 + by_1 + f)(y - y_1) = 0$$

or
$$x(ax_1 + hy_1 + g) + y(hx_1 + by_1 + f) + (gx_1 + fy_1 + c) = S_1$$

i.e.,
$$T = S_1$$

which is the required equation.

17. Locus of Middle Points of a System of Parallel Chord

To find the locus of the middle points of a system of parallel chords of the conic $\mathbf{ax^2 + 2hxy + by^2 + 2gx + 2fy + c = 0}$. *(M.D.U. 2007)*

Let the chords be parallel to $y = mx$.

Consider one chord PQ of the system of parallel chords of the conic. Let $M(x_1, y_1)$ be its middle point.

The equation of the chord PQ, from above is given by

$$T = S_1$$

i.e.,
$$x(ax_1 + hy_1 + g) + y(hx_1 + by_1 + f) + (gx_1 + fy_1 + c) = S_1$$

This is parallel to $y = mx$, if

$$-\frac{ax_1 + hy_1 + g}{hx_1 + by_1 + f} = m$$

or
$$(ax_1 + hy_1 + g) + m(hx_1 + by_1 + f) = 0$$

Therefore, replacing (x_1, y_1) by the general co-ordinates (x, y), locus of (x_1, y_1) is **(ax + hy + g) + m(hx + by + f) = 0.**

This passes through the point of intersection of the lines

$$ax + hy + g = 0 \quad \text{and} \quad hx + by + f = 0$$

which gives the centre of the conic.

Hence the locus of the middle points of parallel chords is a diameter.

Note. It is to be noted that the equation of the diameter bisecting all the chords parallel to the straight line $y = mx$ is

$$(a + hm)x + (h + bm)y + (g + fm) = 0.$$

18. Conjugate Diameters

Def. *Two diameters of a conic such that each bisects chords parallel to the other are called conjugate diameters.*

19. To find the condition that the two straight lines y = mx and y = m′ x may be parallel to the conjugate diameters of the conic

From above we have that the locus of the middle points of chords parallel to $y = mx$ is the diameter

$$(ax + hy + g) + m(hx + by + f) = 0 \quad \text{...(1)}$$

Its slope $= -\dfrac{a + mh}{h + mb}$

Since this diameter is parallel to the line $y = m'x$

$$\therefore \quad m' = -\frac{a + mh}{h + mb} \quad \text{or} \quad a + h(m + m') + b\,mm' = 0$$

which is the required condition.

It is clear from above that if **a + h(m + m′) + bmm′ = 0** *the lines* $y = mx$ *and* $y = m'x$ *is a pair of conjugate diameters.*

20. Condition of Tangency

To find the condition that the straight line lx + my + n = 0 may be a tangent to the conic

$$\mathbf{ax^2 + 2hxy + by^2 + 2gx + 2fy + c = 0.}$$

Let the line
$$lx + my + n = 0 \quad \text{...(1)}$$

touch the conic

$$ax^2 + 2hxy + by^2 + 2gx + 2fy + c = 0 \quad \text{...(2)}$$

at (x_1, y_1).

∴ Equation of the tangent to the conic (2) at (x_1, y_1) is given by

$$x(ax_1 + hy_1 + g) + y(hx_1 + by_1 + f) + (gx_1 + fy_1 + c) = 0 \quad ...(3)$$

Since it is the same as the equation of the given line (1).

∴ Comparing co-effs. of like terms, we have

$$\frac{ax_1 + hy_1 + g}{l} = \frac{hx_1 + by_1 + f}{m} = \frac{gx_1 + fy_1 + c}{n} = k \text{ (say)}$$

∴
$$ax_1 + hy_1 + g - kl = 0$$
$$hx_1 + by_1 + f - km = 0$$
$$gx_1 + fy_1 + c - kn = 0$$

and as (x_1, y_1) lies on (1)

∴
$$lx_1 + my_1 + n = 0$$

Eliminating x_1, y_1, k from these four equations, we have

$$\begin{vmatrix} a & h & g & l \\ h & b & f & m \\ g & f & c & n \\ l & m & n & o \end{vmatrix} = 0$$

which on expansion and simplification gives

$$(bc - f^2)\, l^2 + (ca - g^2)m^2 + (ab - h^2)n^2 + 2(gh - af)mn + 2(hf - bg)nl + 2(fg - ch)lm = 0$$

or
$$Al^2 + Bm^2 + Cn^2 + 2Fmn + 2Gnl + 2Hlm = 0.$$

This is the required condition, where capital letters are cofactors of corresponding small letters in the above determinant.

21. Chord of Contact

To find the equation of the chord of contact of the point (x_1, y_1) with respect to the conic

$$\mathbf{ax^2 + 2hxy + by^2 + 2gx + 2fy + c = 0}$$

Let P be the point (x_1, y_1) and $A(x', y')$, $B(x'', y'')$, the points of contact of tangents from (x_1, y_1) to the conic

$$ax^2 + 2hxy + by^2 + 2gx + 2fy + c = 0 \quad ...(1)$$

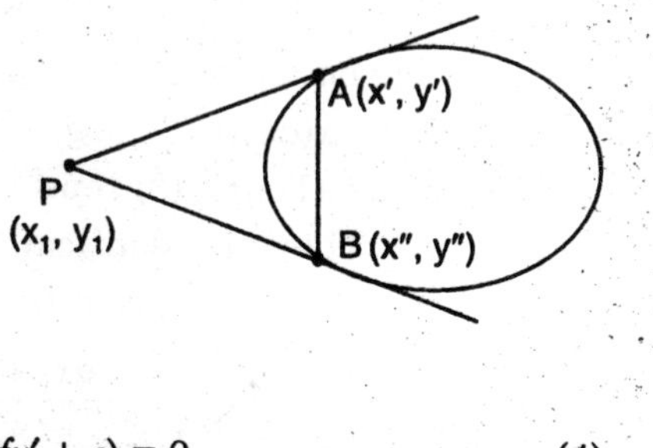

∴ Equation of tangents at A and B are

$$x(ax' + hy' + g) + y(hx' + by' + f) + (gx' + fy' + c) = 0 \quad ...(2)$$

and
$$x(ax'' + hy'' + g) + y(hx'' + by'' + f) + (gx'' + fy'' + c) = 0 \quad ...(3)$$

Since these tangents pass through the point $P(x_1, y_1)$

∴
$$x_1(ax' + hy' + g) + y_1(hx' + by' + f) + (gx' + fy' + c) = 0 \quad ...(4)$$

and
$$x_1(ax'' + hy'' + g) + y_1(hx'' + by'' + f) + (gx'' + fy'' + c) = 0 \quad ...(5)$$

From (4) and (5) it follows that the points $A(x', y')$ and $B(x'', y'')$ lie on

$$x_1(ax + hy + g) + y_1(hx + by + f) + (gx + fy + c) = 0$$

i.e.,
$$x(ax_1 + hy_1 + g) + y(hx_1 + by_1 + f) + (gx_1 + fy_1 + c) = 0$$

which being a first degree equation is a straight line and since it passes through A and B, it is the equation of chord of contact of the point (x_1, y_1).

Note. Method to write equation of chord of contact of (x_1, y_1) is same as that of writing the equation of tangent at (x_1, y_1).

22. Pole and Polar

To find the polar of the point (x_1, y_1) with respect to the conic

$$ax^2 + 2hxy + by^2 + 2gx + 2fy + c = 0$$

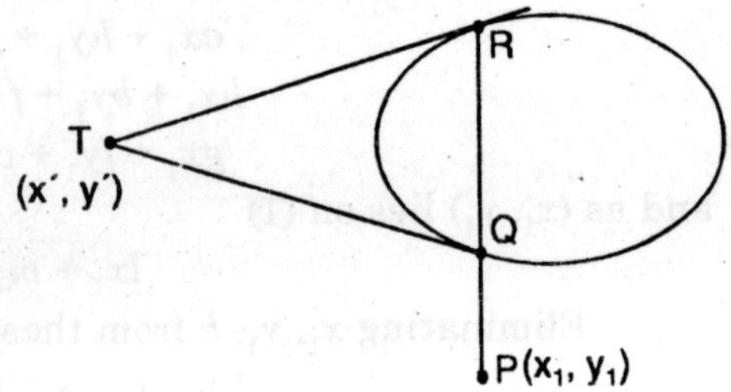

The equation of the conic is

$$ax^2 + 2hyx + by^2 + 2gx + 2fy + c = 0$$

Let PQR be any straight line through $P(x_1, y_1)$ meeting the conic in Q and R. Let the tangents at R and Q meet in $T(x', y')$.

Now QR is the chord of contact of tangents from (x', y') to the conic (1).

$\therefore$ Its equation is

$$x(ax' + hy' + g) + y(hx' + by' + f) + (gx' + fy' + c) = 0 \quad ...(2)$$

As it passes through the point $P(x_1, y_1)$

$\therefore$ $$x_1(ax' + hy' + g) + y_1(hx' + by' + f) + (gx' + fy' + c) = 0$$

$\therefore$ Locus of $T(x', y')$ is got by changing $x' \to x, y' \to y$

i.e., $$x_1(ax + hy + g) + y_1(hx + by + f) + (gx + fy + c) = 0$$

or $$x(ax_1 + hy_1 + g) + y(hx_1 + by_1 + f) + gx_1 + fy_1 + c = 0$$

which is the required equation of the polar of P.

Note. Method to write equation of polar of $P(x_1, y_1)$ is same as that of writing the equation of tangent at (x_1, y_1).

23. Pole of a Given Line

To find the pole of the line $lx + my + n = 0$ with respect to the conic

$$ax^2 + 2hxy + by^2 + 2gx + 2fy + c = 0$$

Let (x_1, y_1) be the pole of the line $lx + my + n = 0$ ...(1)

with respect to the conic

$$ax^2 + 2hxy + by^2 + 2gx + 2fy + c = 0 \quad ...(2)$$

The equation of the polar of (x_1, y_1) w.r.t. (2) is

$$x(ax_1 + hy_1 + g) + y(hx_1 + by_1 + f) + (gx_1 + fy_1 + c) = 0 \quad ...(3)$$

Since (1) and (3) represent the polar of (x_1, y_1).

$\therefore$ They are identical.

$\therefore$ Comparing co-effs. of each term, we have

$$\frac{ax_1 + hy_1 + g}{l} = \frac{hx_1 + by_1 + f}{m} = \frac{gx_1 + fy_1 + c}{n}$$

From 1st two members, we have

$$(am - hl)x_1 + (hm - bl)y_1 + (gm - fl) = 0 \quad ...(4)$$

From last two members, we get

$$(hn - gm)x_1 + (bn - fm)y_1 + (fn - cm) = 0 \quad ...(5)$$

Solving for (x_1, y_1) by cross-multiplication

$$\frac{x_1}{(hm-bl)(fn-cm)-(bn-fm)(gm-fl)}$$

$$=\frac{y_1}{(gm-fl)(hn-gm)-(am-hl)(fn-cm)}$$

$$=\frac{1}{(am-hl)(bn-fm)-(hn-gm)(hm-bl)}$$

i.e.,

$$=\frac{x_1}{(bc-f^2)\,lm+(fg-ch)\,m^2+(hf-bg)mn}$$

$$=\frac{y_1}{(fg-ch)lm+(ca-g^2)m^2+(gh-af)mn}$$

$$=\frac{1}{(hf-bg)lm+(gh-af)m^2+(ab-h^2)mn}$$

or

$$\frac{x_1}{\text{A}l+\text{H}m+\text{G}n}=\frac{y_1}{\text{H}l+\text{B}m+\text{F}n}=\frac{1}{\text{G}l+\text{F}m+\text{C}n}$$

$$\therefore \qquad x_1=\frac{\text{A}l+\text{H}m+\text{G}n}{\text{G}l+\text{F}m+\text{C}n}, \quad y_1=\frac{\text{H}l+\text{B}m+\text{F}n}{\text{G}l+\text{F}m+\text{C}n}$$

where capital letters are cofactors of corresponding small letters in Δ.

24. Director Circle

Definition. The locus of the point of intersection of two perpendicular tangents to the conic is called a director circle.

To find the equation of the director circle of the conic

$$\mathbf{ax^2 + 2hxy + by^2 + 2gx + 2fy + c = 0.}$$

The equation of the pair of tangents drawn from the point $P(x_1, y_1)$ to the conic

$$ax^2 + 2hxy + by^2 + 2gx + 2fy + c = 0 \text{ is}$$

$$SS_1 = T^2$$

i.e.,

$$(ax^2 + 2hxy + by^2 + 2gx + 2fy + c) \times (ax_1^2 + 2hx_1y_1 + by_1^2 + 2gx_1 + 2fy_1 + c)$$

$$= [axx_1 + h(xy_1 + x_1y) + byy_1 + g(x + x_1) + f(y + y_1) + c]^2$$

If these tangents are at right angles, then

Co-effs. x^2 + co-eff. of $y^2 = 0$

i.e.,

$$(a + b)(ax_1^2 + 2hx_1y_1 + by_1^2 + 2gx_1 + 2fy_1 + c) - (ax_1 + hy_1 + g)^2 - (hx_1 + by_1 + f)^2 = 0$$

i.e.,

$$(ab - h^2)(x_1^2 + y_1^2) - 2(fh - bg)x_1 - 2(gh - af)y_1 + (bc - f^2) + (ca - g^2) = 0$$

$\therefore$ Locus of (x_1, y_1) is

$$(ab - h^2)(x^2 + y^2) - 2(hf - bg)x - 2(gh - af)y + bc - f^2 + ca - g^2 = 0$$

which is the director circle and its centre is $\left(\dfrac{hf-bg}{ab-h^2}, \dfrac{gh-af}{ab-h^2}\right)$

which is the centre of the conic.

Note. If the conic is a parabola, then $ab - h^2 = 0$.

In that case the director circle degenates into the directrix of the parabola.

Its equation is $2(hf - bg)x + 2(gh - af)y - (bc - f^2) - (ca - g^2) = 0$.

Example 1. *Find the condition that the two lines given by the equation* $a'x^2 + 2h'xy + b'y^2 = 0$ *are conjugate diameters with respect to the conic*

$$ax^2 + 2hxy + by^2 = 1.$$

Sol. The given conic is

$$ax^2 + 2hxy + by^2 = 1 \quad ...(1)$$

Let the two lines given by

$$a'x^2 + 2h'xy + b'y^2 = 0, \text{ be}$$

$$y = m_1x \quad \text{and} \quad y = m_2x$$

Then,

$$m_1 + m_2 = -2h'/b' \text{ and } m_1m_2 = a'/b' \quad ...(2)$$

These lines are conjugate diameters with respect to the conic (1), if

$$a + h(m_1 + m_2) + bm_1m_2 = 0$$

i.e., if

$$a - \frac{2hh'}{b'} + \frac{ba'}{b'} = 0 \quad \text{or} \quad ab' + ba' = 2hh'$$

which is the required condition.

Example 2. *Show that all conics, which pass through the intersection of two rectangular hyperbolas are themselves rectangular hyperbolas.*

Sol. Let the two rectangular hyperbolas be

$$S \equiv ax^2 + 2hxy + by^2 + 2gx + 2fy + c = 0 \quad ...(1)$$

and

$$S' = a_1x^2 + 2h_1xy + b_1y^2 + 2g_1x + 2f_1y + c_1 = 0 \quad ...(2)$$

so that

$$a + b = 0 \quad \text{and} \quad a_1 + b_1 = 0 \quad ...(3)$$

Any conic through the intersection of (1) and (2) is

$$S + \lambda S' = 0$$

It will be a rectangular hyperbola, if

$$(a + \lambda a_1) + (b + \lambda b_1) = 0$$

i.e., if

$$(a + b) \pm \lambda(a_1 + b_1) = 0$$

which is true from (3).

Hence the result.

EXERCISE 5

1. Show that the line $lx + my = 1$, touches the conic

$$ax^2 + 2hxy + by^2 = 1$$

if

$$bl^2 - 2hlm + am^2 = ab - h^2.$$

2. Show that $4x + 5y = 14$ touches $x^2 + 4xy + y^2 - 2x + 2y - 15 = 0$ at the point (1, 2). *(M.D.U. 1998)*

3. Prove that the chord of contact of tangents drawn from an external point to the conic $ax^2 + 2hxy + by^2 = 1$, subtends a right angle at the centre if the point lies on the conic

$$x^2(a^2 + b^2) + 2h(a + b)xy + y^2(h^2 + b^2) = a + b.$$

4. Find the condition that the pair of lines, whose equation is $AX^2 + 2Hxy + By^2 = 0$ may be parallel to the conjugate diameter of the conic $ax^2 + 2hxy + by^2 = 1$.

5. Pairs of tangents are drawn to the conic $\alpha x^2 + \beta y^2 = 1$ so as to be always parallel to conjugate diameters of the conic $ax^2 + 2hxy + by^2 = 1$. Show that the locus of their points of intersection is the conic $ax^2 + 2hxy + by^2 = \frac{a}{\alpha} + \frac{b}{\beta}$.

6. If the lines, $y = mx$, $y = m_1x$ are parallel to conjugate diameters of the conic $ax^2 + 2hxy + by^2 + 2gx + 2fy + c = 0$, then $a + h(m + m_1) + bmm_1 = 0$.

7. Find the equation of the director circle of the conic

$$11x^2 + 24xy + 4y^2 - 2x + 16y + 11 = 0.$$ *(M.D.U. 1996)*

8. Find the equations to the common conjugate diameters of the conics

$$x^2 + 4xy + 6y^2 - 1 = 0 \quad \text{and} \quad 2x^2 + 6xy + 9y^2 - 1 = 0.$$

9. Find the equations of the tangents to the cone $x^2 + 4xy + 3y^2 - 5x - 6y + 3 = 0$ which are parabola to the straight line $x + 4y = 0$. *(M.D.U. 2007)*

Answers

4. $aB + bA = 2hH$

7. $x^2 + y^2 + 2x - 2y - 1 = 0$

8. $x(x + 3y) = 0$

9. $x + 4y = 5$ and $x + 4y = 8$.

6

System of Conics

1. Conic Through Five Points

One Conic and in general only one conic be drawn through five given points.

Let the general equation of a conic be

$$f(x, y) = ax^2 + 2hxy + by^2 + 2gx + 2fy + c = 0.$$

It contains six constants a, b, c, f, g, h. If we divide the equation by any of these constants say by c, we see that the number of constants is really five which are a/c, h/c, b/c, g/c and f/c. Let the equation pass through five given points. Then we get five equations from which we can determine the five constants. Hence in general we shall have only one conic passing through five points.

2. Equation of a Conic Through the Intersection of two given Conics

If S and S_1 denote the equations of two conics

$$ax^2 + 2hxy + by^2 + 2gx + 2fy + c = 0$$

and

$$a_1x^2 + 2h_1xy + b_1y^2 + 2g_1x + 2f_1y + c_1 = 0$$

then $S + \lambda S_1 = 0$ is the equation of a conic which passes through the points common to the two conics $S = 0$ and $S_1 = 0$. For, the equation $S + \lambda S_1 = 0$ is of the second degree, and therefore represents some conic and is satisfied by points common to both $S = 0$ and $S_1 = 0$.

Thus $S + \lambda S_1 = 0$ is the general equation of a conic through the points of intersection of the conics $S = 0$ and $S_1 = 0$.

3. Contacts of Conics

It is obvious that in general, two conics intersect in four points, real or imaginary. Here, we discuss the contacts of different orders when two conics intersect in four real points.

Let the four real points be A, B, C and D in which the two conics intersect.

(*i*) Contact of the zeroth order

The contact is said to be of the zeroth order if the two conics intersect in four real and distinct points [Fig. (*i*)].

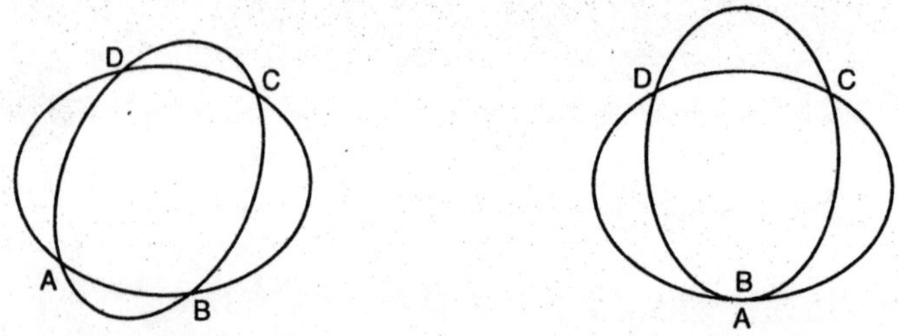

(*i*) (*ii*)

(*ii*) Contact of the first order

When two of the points of intersection of any two curves coincide, that is, when two curves touch, they are said to have **contact of the first order** at the point [Fig. (*ii*)]

Illustration. *Let two conics $S = 0$ and $S_1 = 0$ intersect at the points A, B, C, D.* Suppose two points coincide *i.e.*, B coinsides with A (as shown in the figure (*ii*). Then the conics touch at A. They have contact of first order.

(*iii*) **Double Contact**

If out of four intersecting points of the two conics, suppose that the point B coincides with A and D with C (as shown in Fig. (*iii*)), then the conics are said to have double contact at A and C.

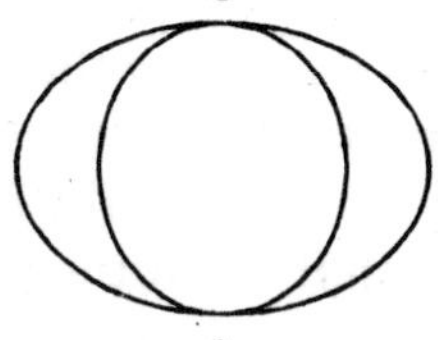

(*iii*)

(*iv*) **Contact of the second order**

If out of four intersecting points of the two conics suppose B and C coincide with A and the fourth point D is distinct from them, then the conics have three point contact at A, then the conics are said to have a contact of second order.

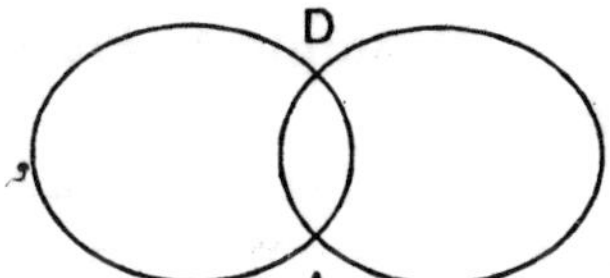

(*iv*)

(*v*) **Contact of the third order**

If all the four points *i.e.*, B, C and D coincide with A, then the conics have four point contact, then the conics are said to have a contact of third order as shown in figure (*v*).

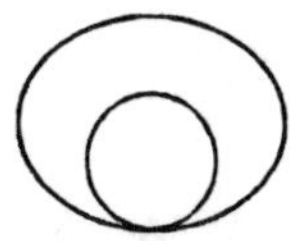

(*v*)

Thus we have,

Order of contact	*No. of coincident, points*
Zeroth	None
1st	2
2nd	3
3rd	4

(*vi*) **Oscillating Curve**

A curve which has with a given curve a contact of the highest order is called an *oscillating curve.* A circle which can be made to pass through three points has contact of the second order with a given curve at a given point and is called the *circle of curvature* at that point and its radius is called the radius of curvature.

4. To find the equation of a conic which passes through the intersection of a given conic and two given straight lines.

Let $S = ax^2 + 2hxy + by^2 + 2gx + 2fy + c = 0$ ...(1)

be the equation of the given conic

and $u = lx + my + n = 0$ ...(2)

$v = l_1x + m_1y + n_1 = 0$ ...(3)

be the equations of two given straight lines.

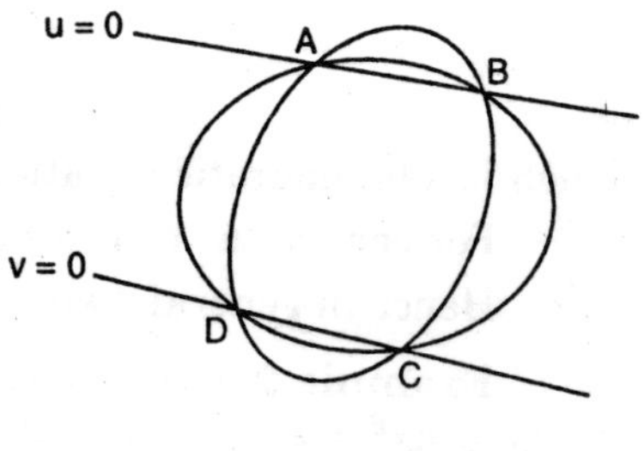

Let the lines (2) and (3) meet the curve u in points A and B and in C and D respectively.

Now consider the equation $S + \lambda uv = 0$. ...(4)

where λ is an arbitrary constant.

Equation (4) is a second degree equation in x and y and hence represents a conic passing through the points, where $S = 0$ is cut by the lines $u = 0$ and $v = 0$ *i.e.*, it passes through the points A, B, C and D.

If now the line $v = 0$ coincides with the line $u = 0$, then the equation (4) becomes $S + \lambda u^2 = 0$ and will represent a conic which cuts the conic $S = 0$ in two pairs of coincident points, where $S = 0$ is cut by the line $u = 0$. The conic $S + \lambda u^2 = 0$ touches $S = 0$ at the two points, where $u = 0$ meets the conic $S = 0$.

Hence, the equation $S + \lambda u^2 = 0$ is the general equation of conics having double contact with $S = 0$ at the points, where $u = 0$ meets it.

If the two lines $u = 0$ and $v = 0$ coincide and are tangents to the conic $S = 0$ at one point, then $S + \lambda u^2 = 0$ represents a conic passing through four coincident points and hence has a contact of third order.

Example 1. *Prove that two conics will in general intersect in four points real, or imaginary.*

Sol. Let $$S_1 = a_1x^2 + 2h_1xy + b_1y^2 + 2g_1x + 2f_1y + c_1 = 0 \quad ...(1)$$

and $$S_2 = a_2x^2 + 2h_2xy + b_2y^2 + 2g_2x + 2f_2y + c_2 = 0 \quad ...(2)$$

by the equations of two conics.

The equations (1) and (2) may be written as

$$a_1x^2 + 2(h_1y + g)x + b_1y^2 + 2f_1y + c_1 = 0$$

and $$a_2x^2 + 2(h_2y + g)x + b_2y^2 + 2f_2y + c_2 = 0$$

or as $$A_0x^2 + A_1x + A_2 = 0 \quad ...(3)$$

and $$B_0x^2 + B_1x + B_2 = 0 \quad ...(4)$$

where A_0, A_1, A_2 and B_0, B_1 and B_2 are the functions of y of degree denoted by their suffixes.

Solving (2) and (3) by cross multiplication, we have

$$\frac{x^2}{A_1B_2 - A_2B_1} = \frac{x}{A_2B_0 - A_0B_2} = \frac{1}{A_0B_1 - B_0A_1}$$

From first and 2nd members

$$x = \frac{A_1B_2 - A_2B_1}{A_2B_0 - A_0B_2} \quad ...(4)$$

From 2nd and 3rd members

$$x = \frac{A_2B_0 - A_0B_2}{A_0B_1 - B_0A_1} \quad ...(5)$$

From (4) and (5), we have

$$\frac{A_2B_0 - A_0B_2}{A_0B_1 - B_0A_1} = \frac{A_1B_2 - A_2B_1}{A_2B_0 - A_0B_2}$$

or $$(A_2B_0 - A_0B_2)^2 = (A_1B_2 - A_2B_1)(A_0B_1 - B_0A_1)$$

which is a biquadratic equation in y and hence has four roots.

For each value of y, we get a value of x.

Hence in general, two conics intersect in four points real or imaginary.

Example 2. *Find the equation of the circle which has contact of the second order with the conic $ax^2 + 2\,bxy + cy^2 + 2dx = 0$ at the origin.*

Sol. The equation of the given conic is

$$ax^2 + 2bxy + cy^2 + 2dx = 0 \qquad ...(1)$$

we know, in general, the equation to the tangent at the origin to an algebraic curve is obtained by equating to zero the terms of lowest degree.

Here, the tangent at the origin is $x = 0$.

Also the equation of any line through origin is $y = mx$.

Therefore, the equation of any conic having double contact, with the conic (1) at the origin is given by

$$ax^2 + 2bxy + cy^2 + 2dx - \lambda x (y - mx) = 0 \qquad ...(2)$$

$[S - \lambda uv = 0]$

If the equation (2) represents a circle, then the coefficient of xy is zero and coefficients of x^2 and y^2 are equal.

$$\therefore \qquad 2b - \lambda = 0 \quad i.e., \lambda = 2b \qquad ...(3)$$

and $$a + \lambda m = c \quad i.e., \quad m = \frac{c-a}{\lambda} = \frac{c-a}{2b} \qquad \text{[From (3)]}$$

substituting the values of λ and m in (2), we have

$$ax^2 + 2bxy + cy^2 + 2dx - 2bx\left[y - \left(\frac{c-a}{2b}\right)x\right] = 0$$

or $$ax^2 + 2bxy + cy^2 + 2dx - 2bxy + (c-a)x^2 = 0$$

or $$cx^2 + cy^2 + 2dx = 0$$

which is the required equation of the circle.

Example 3. *Two conics have double contact. Prove that the polar of the centre of one conic with respect to the other is parallel to the common chord.* (M.D.U. 2007)

Sol. Let the equation of a conic with centre (0, 0) be

$$ax^2 + 2hxy + by^2 = 1 \qquad ...(1)$$

Let the equation of the common chord be

$$lx + my + n = 0 \qquad ...(2)$$

Now, the equation of any conic having double contact with the conic (1) at the points of intersection of (1) and (2) is given by

$$(ax^2 + 2hxy + by^2 - 1) + \lambda(lx + my + n)^2 = 0 \qquad ...(3)$$

The equation of the polar of the centre (0, 0, 0) of conic (1) w.r.t. the conic (3) is

$$a(0 \,.\, x) + h(0 \,.\, y + x \,.\, 0) + b(0 \,.\, y) - 1 + \lambda[l^2(0) + 2lm(0)$$
$$+ m^2(0) + ln(0 + x) + mn(0 + y) + n^2] = 0$$

or $$-1 + \lambda(lnx + mny + n^2) = 0$$

or equation of the polar is

or $$lx + my = \frac{1}{\lambda n} - n$$

which is parallel to the common chord (2).

Hence the result.

Example 4. *A conic has a double contact with the parabola $y^2 = 4ax$. If the chord of contact passes through the vertex, and the conic passes through the focus of the parabola, prove that the locus of the centre of the conic is the parabola $y^2 = a(2x - a)$.*

Sol. The equation of the given parabola is

$$S = y^2 - 4ax = 0 \quad \ldots(1)$$

And, the equation of chord of contact through the vertex (0, 0) of (1) is

$$u = y - mx = 0 \quad \ldots(2)$$

Now the general equation of a conic which have double contact with the parabola (1) at extremities of the chord is of the form $S + \lambda u^2 = 0$

i.e.,
$$y^2 - 4ax + \lambda (y - mx)^2 = 0 \quad \ldots(3)$$

Since the conic (3) passes through the focus $(a, 0)$ of the parabola (1), therefore

$$0 - 4a^2 + \lambda (0 - ma)^2 = 0$$

or
$$\lambda m^2 a^2 = 4a^2 \quad \therefore \quad \lambda = \frac{4}{m^2}$$

Substituting the value of λ in (3), the equation of the conic is

$$y^2 - 4ax + \frac{4}{m^2}(y - mx)^2 = 0$$

i.e.,
$$F(x, y) = m^2 (y^2 - 4ax) + 4 (y - mx)^2 = 0 \quad \ldots(4)$$

To find the locus of the centre of the conic $F(x, y) = 0$

We eliminate m from centre giving equations

$$\frac{\partial F}{\partial x} = 0 \quad \text{and} \quad \frac{\partial F}{\partial y} = 0$$

Now
$$\frac{\partial F}{\partial x} = -4am^2 - 8m (y - mx) = 0 \quad \ldots(5)$$

and
$$\frac{\partial F}{\partial y} = 2m^2 y + 8 (y - mx) = 0 \quad \ldots(6)$$

Multiplying (6) by m and adding to (5), we get $-4am^2 + 2m^3 y = 0$

or
$$m = \frac{2a}{y} \quad \ldots(7)$$

Again, from (5), $am + 2y - 2mx = 0$ or $m = \dfrac{2y}{2x - a}$ ...(8)

From (7) and (8) eliminating m, we get

$$\frac{2y}{2x - a} = \frac{2a}{y} \quad \text{or} \quad y^2 = a(2x - a)$$

which is the **required** locus and is a parabola.

Example 5. *Find the equation of a conic passing through (1, 1) and also through the intersection of the conic*

$x^2 + 2xy + 5y^2 - 7x - 8y + 6 = 0$ with the straight lines $2x - y - 5 = 0$ and $3x + y - 11 = 0$.

Sol. Equation of the given conic is

$$S = x^2 + 2xy + 5y^2 - 7x - 8y + 6 = 0 \quad \ldots(1)$$

and the two lines are

$$u = 2x - y - 5 = 0, \quad v = 3x + y - 11 = 0 \quad \ldots(2)$$

The equation of a conic through the intersection of the conic (1) and the lines (2) is

$$x^2 + 2xy + 5y^2 - 7x - 8y + 6 + \lambda (2x - y - 5)(3x + y - 11) = 0 \qquad [S + \lambda uv = 0]$$

or
$$x^2 + 2xy + 5y^2 - 7x - 8y + 6 + \lambda (6x^2 - xy - y^2 - 37x + 6y + 55) = 0 \qquad ...(3)$$

As it passes through the point (1, 1)

$$\therefore \quad 1 + 2 + 5 - 7 - 8 + 6 + \lambda (6 - 1 - 1 - 37 + 6 + 55) = 0$$

$$\therefore \quad \lambda = \frac{1}{28}.$$

Substituting the value of λ in (3), we get

$$x^2 + 2xy + 5y^2 - 7x - 8y + 6 + \frac{1}{28}(6x^2 - xy - y^2 - 37x + 6y + 55) = 0$$

or $\quad 34x^2 + 55xy + 139y^2 - 233\,x - 218y + 223 = 0.$

which is the required equation of the conic.

Example 6. *Prove that the conics $x^2 + 2y^2 - 1 = 0$ and $3x^2 + 8xy + 10y^2 - 4x - 8y + 1 = 0$ have double contact with each other.*

Sol. The given conic are
$$x^2 + 2y^2 - 1 = 0 \qquad ...(1)$$

and
$$3x^2 + 8xy + 10y^2 - 4x - 8y + 1 = 0 \qquad ...(2)$$

From (1),
$$y^2 = \frac{1 - x^2}{2} \quad \Rightarrow \quad y = \sqrt{\frac{1 - x^2}{2}} \qquad ...(3)$$

Now (2) can be written as

$$3x^2 + 10y^2 - 4x + 1 = 8y(1 - x)$$

or
$$3x^2 + 10\left(\frac{1 - x^2}{2}\right) - 4x + 1 = \pm 8\sqrt{\frac{1 - x^2}{2}}\,(1 - x) \qquad \text{[From (3)]}$$

or
$$3x^2 + 5 - 5x^2 - 4x + 1 = \pm 8\sqrt{\frac{1 - x^2}{2}}\,(1 - x)$$

or
$$6 - 4x - 2x^2 = \pm 8\sqrt{\frac{1 - x^2}{2}}\,(1 - x)$$

or
$$3 - 2x - x^2 = \pm 2\sqrt{2}\,\sqrt{1 - x^2}\,(1 - x)$$

or
$$(1 - x)(3 + x) = \pm 2\sqrt{2}\,\sqrt{1 - x^2}\,(1 - x)$$

Squaring, $(1 - x)^2 (3 + x)^2 = 8(1 - x^2)(1 - x)^2$

or
$$(1 - x)^2 [(3 + x)^2 - 8(1 - x^2)] = 0$$

or
$$(1 - x)^2 (1 + 6x + 9x^2) = 0$$

or
$$(1 - x)^2 (1 + 3x)^2 = 0$$

$$\therefore \quad x = 1, 1, -\frac{1}{3}, -\frac{1}{3} \quad \Rightarrow \quad x = 1, -\frac{1}{3}$$

When $x = 1$, from (3), $y = 0$

When $x = -\dfrac{1}{3}$, $y = \dfrac{2}{3}$

The four points of intersection of the two conic are

$$(1, 0), (1, 0), \left(-\frac{1}{3}, \frac{2}{3}\right), \left(-\frac{1}{3}, \frac{2}{3}\right)$$

$\Rightarrow$ The two conics touch each other at two points (1, 0) and $\left(-\frac{1}{3}, \frac{2}{3}\right)$.

Hence the two conics have double contact with each other.

EXERCISE 1

1. (*a*) Find the equation of the conic passing through the five points (2, 1), (1, 0), (3, – 1), (– 1, 0) and (3, – 2).
   (*b*) Find the equation of the conic passing through the origin and the points (1, 1), (– 1, 1), (2, 0) and (3, – 2). Determine its species.
2. Any two parabolas which have a common focus and their axes in opposite directions intersect at right angles.
3. Find the equation of the parabola which touches the conic $x^2 + xy + y^2 - 2x - 2y + 1 = 0$ at the point where it is cut by the line $x + y + 1 = 0$.
4. If a circle has double contact with a conic $ax^2 + by^2 = 1$, show that the chord of contact is parallel to one or other of the axes.
5. If the conic which passes through the intersection of two given conics represents straight lines then show that at least one of these straight line is real.
6. If two conics have their axes parallel, prove that a circle will pass through their points of intersection.
7. Prove that the conics $x^2 + 3y^2 - 1 = 0$ and $2x^2 + 12xy + 39y^2 - 2x - 12y = 0$ have double contact with each other. Find the co-ordinates of the points of intersection of the tangents at the two points of contact.
8. Prove that the equation to the circle having double contact with the ellipse $\frac{x^2}{a^2} + \frac{y^2}{b^2} = 1$ at the ends of a latus rectum is $x^2 + y^2 - 2\,ae^3x = a^2(1 - e^2 - e^4)$.

   Hint. Equation of any conic having double contact will be $\left(\frac{x^2}{a^2} + \frac{y^2}{b^2} - 1\right) + \lambda(x - ae)^2 = 0$.

   It will be circle if $\frac{1}{a^2} + \lambda = \frac{1}{b^2}$. Find $\lambda = \frac{e^2}{a^2(1-e)^2}$ etc.
9. Prove that two parabolas and a rectangular hyperbola will pass through four given points.
   Hint. Conic through four points given by S = 0 and S′ = 0 is S + λS′ = 0

   i.e., $(a + \lambda a_1)x^2 + 2(h + \lambda h_1)xy + (b + \lambda b_1)y^2 = 0$ ...(1)

   where $S = ax^2 + 2hxy + by^2 = 0$

   (1) is parabola if $(h + \lambda h_1)^2 = (a + \lambda a_1)(b + \lambda b_2)$

   which is quadratic in λ and hence two parabola through four given points etc.

Answers

1. (*a*) $x^2 + 19xy + 4y^2 - 45y - 1 = 0$; (*b*) $6x^2 + 12xy + 7y^2 - 12x - 13y = 0$ (Ellipse)
3. $x^2 + y^2 - 2xy - 14x - 14y + 1 = 0$
7. (1, 2).

7

Confocal Conics

1. Definition

All those conics which have the same, two points as their foci are called confocal conic. If two conics are confocal they must have, the same axes.

The equation $$\frac{x^2}{a^2+\lambda}+\frac{y^2}{b^2+\lambda}=1$$

will, for different values of λ (λ is a parameter) represent a family of confocals. If $a < b$, the distance of a focus from the centre is

$$\sqrt{\{(a^2+\lambda)-(b^2+\lambda)\}} \quad \text{or} \quad \sqrt{(a^2-b^2)}$$

The foci are the points $[\pm\sqrt{(a^2-b^2)}, 0]$ which remain the same for all λ.

The equation of a system of confocal conics is

$$\frac{x^2}{a^2+\lambda}+\frac{y^2}{b^2+\lambda}=1$$

If λ is positive, the curve is an ellipse and if $b^2+\lambda$ is negative, the curve is a hyperbola.

2. Some Propositions on Confocal Conic

(A) *Two conics of a confocal system pass through any given point. One of these conics is an ellipse and the other a hyperbola.*

Let the equation of the given conic be $\dfrac{x^2}{a^2}+\dfrac{y^2}{b^2}=1$ ...(1)

Let (h, k) be the given point.

The equation of any confocal conic to (1) is

$$\frac{x^2}{a^2+\lambda}+\frac{y^2}{b^2+\lambda}=1 \qquad \text{...(2)}$$

This will pass through the given point (h, k) if

$$\frac{h^2}{a^2+\lambda}+\frac{k^2}{b^2+\lambda}=1 \qquad \text{...(3)}$$

or $(a^2+\lambda)(b^2+\lambda)-h^2(b^2+\lambda)-k^2(a^2+\lambda)=0$

It is a quadratic equation in λ, therefore it gives two values of λ, showing thereby that through a given point two confocal can be drawn.

Put $b^2 + \lambda = \lambda'$ in (3)

Then
$$\frac{h^2}{a^2+\lambda} + \frac{k^2}{\lambda'} = 1$$

or
$$h^2\lambda' + k^2(a^2+\lambda) = \lambda'(a^2+\lambda)$$

or
$$h^2\lambda' + k^2(\lambda' + a^2e^2) - \lambda'(\lambda' + a^2e^2) = 0$$
$$\mid \because \quad b^2 + \lambda = \lambda', \text{ then } a^2 + \lambda = a^2 + \lambda' - b^2 = a^2e^2 + \lambda'$$

or
$$\lambda'^2 - \lambda'(h^2 + k^2 - a^2e^2) - a^2e^2k^2 = 0 \qquad ...(4)$$

Since the constant in equation (4) is negative, the roots of this quadratic in λ' are both real and are of different sign. Hence there are two conics and $b^2 + \lambda$ is positive for one and negative for other, so that one conic is an ellipse and the other a hyperbola.

(B) *One and only member of confocals touches a given line.*

Let the equation of the given line be
$$lx + my + n = 0 \qquad ...(1)$$

and the equation of a system of confocals be
$$\frac{x^2}{a^2+\lambda} + \frac{y^2}{b^2+\lambda} = 1 \qquad ...(2)$$

The line (1) will touch the conic (2) if
$$(a^2+\lambda)l^2 + (b^2+\lambda)m^2 = n^2 \qquad \text{(Condition of tangency)}$$

which is a linear equation in λ and gives one, and only one, value of λ. Hence one confocal will touch the given straight line.

(C) *Two confocal conics cut one another at right angles at all their common points.*

Let the equations of two confocals be
$$\frac{x^2}{a^2+\lambda_1} + \frac{y^2}{b^2+\lambda_1} = 1 \qquad ...(1)$$

and
$$\frac{x^2}{a^2+\lambda_2} + \frac{y^2}{b^2+\lambda_2} = 1 \qquad ...(2)$$

Let (x_1, y_1) be a common point, then the co-ordinates (x_1, y_1) will satisfy the equations (1) and (2).

$\therefore$
$$\frac{x_1^2}{a^2+\lambda_1} + \frac{y_1^2}{b^2+\lambda_1} = 1 \qquad ...(3)$$

and
$$\frac{x_1^2}{a^2+\lambda_2} + \frac{y_1^2}{b^2+\lambda_2} = 1 \qquad ...(4)$$

Subtracting (4) from (3), we have
$$x_1^2\left(\frac{1}{a^2+\lambda_1} - \frac{1}{a^2+\lambda_2}\right) + y_1^2\left(\frac{1}{b^2+\lambda_1} - \frac{1}{b^2+\lambda_2}\right) = 0$$

or
$$\frac{x_1^2(\lambda_2-\lambda_1)}{(a^2+\lambda_1)(a^2+\lambda_2)} + \frac{y_1^2(\lambda_2-\lambda_1)}{(b^2+\lambda_1)(b^2+\lambda_2)} = 0$$

or
$$\frac{x_1^2}{(a^2+\lambda_1)(a^2+\lambda_2)} + \frac{y_1^2}{(b^2+\lambda_1)(b^2+\lambda_2)} = 0 \qquad ...(5)$$

[Since $\lambda_1 \neq \lambda_2$]

Now, the equations of the tangents to conics (1) and (2) at (x_1, y_1) are

$$\frac{xx_1}{a^2+\lambda_1}+\frac{yy_1}{b^2+\lambda_1}=1 \qquad ...(6)$$

and

$$\frac{xx_1}{a^2+\lambda_2}+\frac{yy_1}{b^2+\lambda_2}=1 \qquad ...(7)$$

Slope of tangent (6), $m_1=-\frac{x_1}{a^2+\lambda_1}\cdot\frac{b^2+\lambda_1}{y_1}$

Slope of tangent (7), $m_2=-\frac{x_1}{a^2+\lambda_2}\cdot\frac{b^2+\lambda_2}{y_1}$

The tangents (6) and (7) will be at right angles to each other if

$$m_1m_2=-1$$

or if

$$\left(-\frac{x_1}{a^2+\lambda_1}\cdot\frac{b^2+\lambda_1}{y_1}\right)\left(-\frac{x_1}{a^2+\lambda_2}\cdot\frac{b^2+\lambda_2}{y_1}\right)=-1$$

or if

$$\frac{x_1^2}{(a^2+\lambda_1)(a^2+\lambda_2)}+\frac{y_1^2}{(b^2+\lambda_1)(b^2+\lambda_2)}=0 \qquad ...(8)$$

which is true by virtue of (5).

Hence the tangents are at right angles.

(D) *The difference of the squares of the perpendiculars drawn from the centre on any two parallel tangents to two given confocal conics is constant.*

Let the equations of the two confocals be

$$\frac{x^2}{a^2+\lambda_1}+\frac{y^2}{b^2+\lambda_1}=1 \qquad ...(1)$$

and

$$\frac{x^2}{a^2+\lambda_2}+\frac{y^2}{b^2+\lambda_2}=1 \qquad ...(2)$$

Let the two straight lines

$$x\cos\alpha+y\sin\alpha=p_1 \qquad ...(3)$$

and

$$x\cos\alpha+y\sin\alpha=p_2 \qquad ...(4)$$

be the two parallel tangents to the confocals (1) and (2) respectively, where p_1 and p_2 are the lengths of the perpendiculars from origin to the tangents.

Since the straight lines (3) and (4) are respectively tangents to (1) and (2), so we have

$$p_1^2=(a^2+\lambda_1)\cos^2\alpha+(b^2+\lambda_1)\sin^2\alpha \qquad \text{(Condition of tangency)} \quad ...(5)$$

and

$$p_2^2=(a^2+\lambda_2)\cos^2\alpha+(b^2+\lambda_2)\sin^2\alpha \qquad ...(6)$$

Subtracting (6) from (5), we have

$$p_1^2-p_2^2=(\lambda_1-\lambda_2)(\cos^2\alpha+\sin^2\alpha)=\lambda_1-\lambda_2=\text{constant}$$

Hence the result.

(E) *The locus of the pole of a given straight line with respect to a series of confocal conics is a straight line.*

Let the equations of the confocals be

$$\frac{x^2}{a^2+\lambda}+\frac{y^2}{b^2+\lambda}=1 \qquad ...(1)$$

Let the equation of the given straight line be

$$lx + my = 1 \quad \text{...(2)}$$

Let (x', y') be the pole of the line (2) with respect to the confocals (2).

The equation of the polar of the point (x', y') w.r.t. (1) is

$$\frac{xx'}{a^2 + \lambda} + \frac{yy'}{b^2 + \lambda} = 1 \quad \text{...(3)}$$

Equations (2) and (3) are identical.

Therefore comparing the coefficients of like terms, we have

$$\frac{x'}{a^2 + \lambda} = l, \qquad \frac{y'}{b^2 + \lambda} = m$$

or

$$\frac{x'}{l} = a^2 + \lambda, \qquad \frac{y'}{m} = b^2 + \lambda$$

or

$$\frac{x'}{l} - a^2 = \frac{y'}{m} - b^2 = \lambda$$

Hence the locus of the poles is the straight line

$$\frac{x}{l} - \frac{y}{m} = a^2 - b^2$$

which is the equation of a straight line perpendicular to the line (2). One confocal of the system will touch the line (2). The point of contact will be the pole of the line w.r.t. that confocal.

Hence the locus of the poles is a straight line.

(F) ***If a tangent to one of two confocal conics be perpendicular to a tangent to the other, the locus of their point of intersection is a circle.***

Let the equations of the two confocals be

$$\frac{x^2}{a^2 + \lambda_1} + \frac{y^2}{b^2 + \lambda_1} = 1 \quad \text{...(1)}$$

and

$$\frac{x^2}{a^2 + \lambda_2} + \frac{y^2}{b^2 + \lambda_2} = 1 \quad \text{...(2)}$$

Let

$$x \cos \alpha + y \sin \alpha = p_1 \quad \text{...(3)}$$

be the equation of the tangent to (1)

Then, $(a^2 + \lambda_1) \cos^2 \alpha + (b^2 + \lambda_1) \sin^2 \alpha = p_1^2$ ...(4)

(Condition of tangency)

Again, let a straight line perpendicular to (3) given by

$$x \sin \alpha - y \cos \alpha = p_2 \quad \text{...(5)}$$

be a tangent to confocal (2).

$\therefore$ $(a^2 + \lambda_2) \sin^2 \alpha + (b^2 + \lambda_2) \cos^2 \alpha = p_2^2$ ...(6)

To find the locus of point of intersection of (3) and (5), we square (3) and (5) and add and get the required as

$$x^2(\cos^2 \alpha + \sin^2 \alpha) + y^2(\sin^2 \alpha + \cos^2 \alpha) = p_1^2 + p_2^2$$

i.e.,

$$x^2 + y^2 = p_1^2 + p_2^2$$

or

$$x^2 + y^2 = a^2 + b^2 + \lambda_1 + \lambda_2 \quad \text{[From results (4) and (6)]}$$

which is a circle.

3. To express the co-ordinates of any point in the plane of the ellipse $\frac{x^2}{a^2}+\frac{y^2}{b^2}=1$ in terms of the axes of the conics through it with the ellipse

Let (x_1, y_1) be a point in the plane of the ellipse

$$\frac{x^2}{a^2}+\frac{y^2}{b^2}=1$$

Let the confocals through it be

$$\frac{x^2}{a^2+\lambda_1}+\frac{y^2}{b^2+\lambda_1}=1 \quad \text{and} \quad \frac{x_1^2}{b^2+\lambda_2}+\frac{y_1^2}{a^2+\lambda_2}=1$$

where λ_1 and λ_2 are the parameters of the confocals.

Therefore

$$\frac{x_1^2}{a^2+\lambda_1}+\frac{y_1^2}{b^2+\lambda_1}=1 \quad \text{and} \quad \frac{x_1^2}{a^2+\lambda_2}+\frac{y_1^2}{b^2+\lambda_2}=1$$

or
$$x_1^2(b^2+\lambda_1)+y_1^2(a^2+\lambda_1)-(a^2+\lambda_1)(b^2+\lambda_1)=0 \quad \text{...(1)}$$

and
$$x_1^2(b^2+\lambda_2)+y_1^2(a^2+\lambda_2)-(a^2+\lambda_2)(b^2+\lambda_2)=0 \quad \text{...(2)}$$

Solving the equations by cross multiplication, we have

$$\frac{x_1^2}{(a^2+\lambda_1)(a^2+\lambda_2)[b^2+\lambda_1-(b^2+\lambda_2)]}=\frac{y_1^2}{(b^2+\lambda_1)(b^2+\lambda_2)[a^2+\lambda_2-(a^2+\lambda_1)]}$$

$$=\frac{1}{(b^2+\lambda_1)(a^2+\lambda_2)-(a^2+\lambda_1)(b^2+\lambda_2)}$$

or
$$\frac{x_1^2}{(a^2+\lambda_1)(a^2+\lambda_2)(\lambda_1-\lambda_2)}=\frac{y_1^2}{(b^2+\lambda_1)(b^2+\lambda_2)(\lambda_2-\lambda_1)}=\frac{1}{(\lambda_1-\lambda_2)(a^2-b^2)}$$

$\therefore$
$$x_1^2=\frac{(a^2+\lambda_1)(a^2+\lambda_2)}{a^2-b^2}$$

and
$$y_1^2=-\frac{(b^2+\lambda_1)(b^2+\lambda_2)}{a^2-b^2}.$$

4. Confocal Parabolas

Two parabolas are said to be confocal if they have a common focus and common axis. If we take common focus as origin and x-axis as the common axis, then if 4λ is the latus rectum of one of the parabolas, its equation will be of the form

$$y^2 = 4\lambda(x+\lambda)$$

where λ is a parameter.

Example 1. *If the confocals through (x_1, y_1) to the ellipse $\frac{x^2}{a^2}+\frac{y^2}{b^2}=1$ are $\frac{x^2}{a^2+\lambda_1}+\frac{y^2}{b^2+\lambda_1}$ $=1$ and $\frac{x^2}{a^2+\lambda_2}+\frac{y^2}{b^2+\lambda_2}=1$,* *show that* (i) $\frac{x_1^2}{a^2}+\frac{y_1^2}{b^2}-1=-\frac{\lambda_1\lambda_2}{a^2b^2}$ *and*

(ii) $x_1^2+y_1^2-a^2-b^2=\lambda_1+\lambda_2$.

Sol. The given equation of the ellipse is $\frac{x^2}{a^2}+\frac{y^2}{b^2}=1$ ...(1)

Let the equation of any confocal to (1) be $\frac{x^2}{a^2+\lambda}+\frac{y^2}{b^2+\lambda}=1$ ...(2)

The two values of λ are given to be λ_1 and λ_2.

If the confocal (2) passes through (x_1, y_1), we have

$$\therefore \quad \frac{x_1^2}{a^2+\lambda}+\frac{y_1^2}{b^2+\lambda}=1$$

or $$(b^2+\lambda)x_1^2+(a^2+\lambda)y_1^2=(a^2+\lambda)(b^2+\lambda)$$

or $$\lambda^2+\lambda(a^2+b^2-x_1^2-y_1^2)+(a^2b^2-b^2x_1^2-a^2y_1^2)=0 \quad ...(3)$$

It is a quadratic equation in λ and hence has two values of λ which are given to be λ_1 and λ_2.

From equation (3), we have

$$\lambda_1+\lambda_2=-(a^2+b^2-x_1^2-y_1^2) \quad \text{or} \quad x_1^2+y_1^2-a^2-b^2=\lambda_1+\lambda_2$$

which proves the result (*ii*)

and $$\lambda_1\lambda_2=a^2b^2-b^2x_1^2-a^2y_1^2=a^2b^2-(b^2x_1^2+a^2y_1^2)$$

or $$\frac{x_1^2}{a^2}+\frac{y_1^2}{b^2}-1=-\frac{\lambda_1\lambda_2}{a^2b^2}$$

which is the result (*i*).

Example 2. *Show that the confocal hyperbola through the point on the ellipse* $\frac{x^2}{a^2}+\frac{y^2}{b^2}$ $= 1$ *whose eccentric angle is* α *has for equation* (M.D.U. 2007)

$$\frac{x^2}{\cos^2\alpha}-\frac{y^2}{\sin^2\alpha}=a^2-b^2.$$

Sol. The given equation of the ellipse is

$$\frac{x^2}{a^2}+\frac{y^2}{b^2}=1 \quad ...(1)$$

Let the equation of confocal to the ellipse (1) be

$$\frac{x^2}{a^2+\lambda}+\frac{y^2}{b^2+\lambda}=1 \quad ...(2)$$

The co-ordinates of the point whose eccentric angle is α on the ellipse (1) are $(a\cos\alpha, b\sin\alpha)$

The confocal (2) passes through the point $(a\cos\alpha, b\sin\alpha)$

$$\therefore \quad \frac{a^2\cos^2\lambda}{a^2+\lambda}+\frac{b^2\sin^2\alpha}{b^2+\lambda}=1$$

or $$(b^2 + \lambda)\, a^2 \cos^2 \alpha + (a^2 + \lambda) b^2 \sin^2 \alpha = (a^2 + \lambda)(b^2 + \lambda)$$

or $$\lambda^2 + \lambda[a^2(1 - \cos^2 \alpha) + b^2(1 - \sin^2 \alpha)] + a^2b^2 - a^2b^2(\cos^2 \alpha + \sin^2 \alpha) = 0$$

or $$\lambda^2 + \lambda(a^2 \sin^2 \alpha + b^2 \cos^2 \alpha) + a^2b^2 - a^2b^2 = 0$$

or $$\lambda[\lambda + a^2 \sin^2 \alpha + b^2 \cos^2 \alpha] = 0$$

or $$\lambda = -(a^2 \sin^2 \alpha + b^2 \cos^2 \alpha) \quad \text{or} \quad \lambda = 0$$

When $\lambda = 0$, equation (2) becomes equation (1).

Hence $\lambda = 0$ is rejected.

When $\lambda = -(a^2 \sin^2 \alpha + b^2 \cos^2 \alpha)$.

The equation of confocal (2) is given by

$$\frac{x^2}{a^2 - (a^2 \sin^2 \alpha + b^2 \cos^2 \alpha)} + \frac{y^2}{b^2 - (a^2 \sin^2 \alpha + b^2 \cos^2 \alpha)} = 1$$

or $$\frac{x^2}{a^2 (1 - \sin^2 \alpha) - b^2 \cos^2 \alpha} + \frac{y^2}{b^2 (1 - \cos^2 \alpha) - a^2 \sin^2 \alpha} = 1$$

or $$\frac{x^2}{(a^2 - b^2) \cos^2 \alpha} + \frac{y^2}{(b^2 - a^2) \sin^2 \alpha} = 1$$

or $$\frac{x^2}{\cos^2 \alpha} - \frac{y^2}{\sin^2 \alpha} = a^2 - b^2$$

which is the required equation of confocal hyperbola.

Example 3. *If ϕ be the angle between the tangents from a given point P to the ellipse $\frac{x^2}{a^2} + \frac{y^2}{b^2} = 1$, then show that $\tan \phi = \frac{2\sqrt{-\lambda_1\lambda_2}}{\lambda_1 + \lambda_2}$, where λ_1 and λ_2 are the parameters of the confocals through P.*

Sol. Let the equation of a conic confocal to

$$\frac{x^2}{a^2} + \frac{y^2}{b^2} = 1 \text{ be } \frac{x^2}{a^2 + \lambda} + \frac{y^2}{b^2 + \lambda} = 1 \qquad \text{...(1)}$$

Let P (x_1, y_1) be a given point which lies on confocal conic (1)

$\therefore$ $$\frac{x_1^2}{a^2 + \lambda} + \frac{y_1^2}{b^2 + \lambda} = 1$$

or $$(b^2 + \lambda)x_1^2 + (a^2 + \lambda)\, y_1^2 = (a^2 + \lambda)(b^2 + \lambda)$$

or $$\lambda^2 + (a^2 + b^2)\lambda + a^2b^2 - b^2x_1^2 - a^2y_1^2 - \lambda(x_1^2 + y_1^2) = 0$$

or $$\lambda^2 + \lambda(a^2 + b^2 - x_1^2 - y_1^2) + a^2b^2 \left[1 - \frac{x_1^2}{a^2} - \frac{y_1^2}{b^2}\right] = 0 \qquad \text{...(2)}$$

which is a quadratic equation in λ. If λ_1 and λ_2 be roots of (2)

then $$\lambda_1 + \lambda_2 = x_1^2 + y_1^2 - a^2 - b^2 \qquad \text{...(3)}$$

and $$\lambda_1\lambda_2 = -a^2b^2 \left(\frac{x_1^2}{a^2} + \frac{y_1^2}{b^2} - 1\right) \qquad \text{...(4)}$$

Again, the equation of pair of tangents from P (x_1, y_1) to the ellipse $\frac{x^2}{a^2}+\frac{y^2}{b^2}=1$ is given by $SS_1 = T^2$

i.e.,
$$\left(\frac{x^2}{a^2}+\frac{y^2}{b^2}-1\right)\left(\frac{x_1^{\,2}}{a^2}+\frac{y_1^{\,2}}{b^2}-1\right)=\left(\frac{xx_1}{a^2}+\frac{yy_1}{b^2}-1\right)^2$$

After simplification, we have

$$\frac{x^2}{a^2}\left(\frac{y_1^{\,2}}{b^2}-1\right)-2xy.\frac{x_1y_1}{a^2b^2}+\frac{y^2}{b^2}\left(\frac{x_1^{\,2}}{a^2}-1\right)+2x.\frac{x_1}{a^2}+2y.\frac{y_1}{b^2}-\left(\frac{x_1^{\,2}}{a^2}+\frac{y_1^{\,2}}{b^2}\right)=0$$

which is of the form $Ax^2 + 2Hxy + By^2 + 2Gx + 2Fy + c = 0$

where $H=-\frac{x_1y_1}{a^2b^2}$, $A=\frac{1}{a^2}\left(\frac{y_1^{\,2}}{b^2}-1\right)$ and $B=\frac{1}{b^2}\left(\frac{x_1^{\,2}}{a^2}-1\right)$

$$\therefore \qquad \tan\phi=\frac{2\sqrt{H^2-AB}}{A+B}=\frac{2ab\sqrt{\frac{x_1^{\,2}}{a^2}+\frac{y_1^{\,2}}{b^2}-1}}{x_1^{\,2}+y_1^{\,2}-a^2-b^2} \qquad ...(5)$$

From (4), $\frac{x_1^{\,2}}{a^2}+\frac{y_1^{\,2}}{b^2}-1=-\frac{\lambda_1\lambda_2}{a^2b^2}$

$$\therefore \qquad \tan\phi=\frac{2ab\sqrt{\frac{-\lambda_1\lambda_2}{a^2b^2}}}{\lambda_1+\lambda_2}=\frac{2\sqrt{-\lambda_1\lambda_2}}{\lambda_1+\lambda_2}.$$

Example 4. *Prove that the locus of the feet of the normals, drawn from the fixed point (h, k) to each of the series of confocals given by the equation*

$$\frac{x^2}{a^2+\lambda}+\frac{y^2}{b^2+\lambda}=1$$

is the cubic curve
$$\frac{x}{y-k}+\frac{y}{x-h}=\frac{a^2-b^2}{hy-kx}. \qquad \text{(M.D.U. 2003)}$$

Sol. Consider a series of confocals given by

$$\frac{x^2}{a^2+\lambda}+\frac{y^2}{b^2+\lambda}=1 \qquad ...(1)$$

Let (x_1, y_1) be any point on (1).

∴ The equation of the normal at (x_1, y_1) on (1) is

$$\frac{(a^2+\lambda)x}{x_1}-\frac{(b^2+\lambda)y}{y_1}=a^2-b^2$$

or
$$\frac{(a^2+\lambda)x}{x_1}-(a^2+\lambda)=\frac{(b^2+\lambda)y}{y_1}-(b^2+\lambda)$$

or
$$(a^2+\lambda)\left(\frac{x-x_1}{x_1}\right)=(b^2+\lambda)\left(\frac{y-y_1}{y_1}\right) \qquad ...(2)$$

It passes through the fixed point (h, k) (given)

$\therefore$ From (2), $$(a^2+\lambda)\left(\frac{h-x_1}{x_1}\right)=(b^2+\lambda)\left(\frac{k-y_1}{y_1}\right) \quad ...(3)$$

Since (x_1, y_1) lies on (1),

$$\therefore \quad \frac{x_1^2}{a^2+\lambda}+\frac{y_1^2}{b^2+\lambda}=1 \quad ...(4)$$

Now from (3), $$\frac{a^2+\lambda}{x_1(k-y_1)}=\frac{b^2+\lambda}{y_1(h-x_1)}=\frac{(a^2+\lambda)-(b^2+\lambda)}{x_1(k-y_1)-y_1(h-x_1)}=\frac{a^2-b^2}{kx_1-hy_1}$$

$$\therefore \quad a^2+\lambda=\frac{x_1(k-y_1)(a^2-b^2)}{kx_1-hy_1}$$

and $$b^2+\lambda=\frac{y_1(h-x_1)(a^2-b^2)}{kx_1-hy_1}$$

Putting these value of $a^2+\lambda$ and $b^2+\lambda$ in (4), we get

$$\frac{x_1^2(kx_1-hy_1)}{x_1(k-y_1)(a^2-b^2)}+\frac{y_1^2(kx_1-hy_1)}{y_1(h-x_1)(a^2-b^2)}=1$$

or $$\frac{x_1}{k-y_1}+\frac{y_1}{h-x_1}=\frac{a^2-b^2}{kx_1-hy_1}$$

$\therefore$ Locus of (x_1, y_1) is $\dfrac{x}{k-y}+\dfrac{y}{h-x}=\dfrac{a^2-b^2}{kx-hy}$

i.e., $$\frac{x}{y-k}+\frac{y}{x-h}=\frac{a^2-b^2}{hy-kx}$$

which is a cubic curve.

Example 5. *Prove that the locus of the points of contact of the tangents drawn from a given point to a system of confocals, is a cubic curve which passes through the given point and the foci. If the given points lies on major axis, the cubic curve reduces to a circle.*

Sol. Let the equation of the system of confocal be

$$\frac{x^2}{a^2+\lambda}+\frac{y^2}{b^2+\lambda}=1 \quad ...(1)$$

Let $P(x_1, y_1)$ be the given point.

$\therefore$ The equation of the chord of contact of the point $P(x_1, y_1)$ w.r.t. conic (1) is

$$\frac{xx_1}{a^2+\lambda}+\frac{yy_1}{b^2+\lambda}=1 \quad ...(2)$$

The points of contact of intersection of

$$\frac{x^2}{a^2+\lambda}+\frac{y^2}{b^2+\lambda}-1=0 \quad \text{and} \quad \frac{xx_1}{a^2+\lambda}+\frac{yy_1}{b^2+\lambda}-1=0$$

are the points of contact.

Solving the above equations by cross multiplication for

$$\frac{1}{a^2+\lambda} \text{ and } \frac{1}{b^2+\lambda}, \text{ we get}$$

$$\frac{\dfrac{1}{a^2+\lambda}}{yy_1 - y^2} = \frac{\dfrac{1}{b^2+\lambda}}{x^2 - xx_1} = \frac{1}{x^2yy_1 - y^2xx_1}$$

i.e.,
$$a^2 + \lambda = \frac{x(xy_1 - yx_1)}{y_1 - y} \qquad \text{...(3)}$$

and
$$b^2 + \lambda = \frac{y(xy_1 - yx_1)}{x - x_1} \qquad \text{...(4)}$$

To find the locus of points of contact, we eliminate λ, by subtracting (4) from (3).

$$\therefore \quad a^2 - b^2 = (xy_1 - yx_1)\left[\frac{x}{y_1 - y} - \frac{y}{x - x_1}\right]$$

$$= (xy_1 - yx_1)\frac{[xx_1 - x^2 + yy_1 - y^2]}{(y_1 - y)(x_1 - x)}$$

or
$$(a^2 - b^2)(y_1 - y)(x_1 - x) = -(xy_1 - yx_1)(x^2 + y^2 - xx_1 - yy_1)$$

or
$$(a^2 - b^2)(y_1 - y)(x_1 - x) + (xy_1 - yx_1)(x^2 + y^2 - xx_1 - yy_1) = 0 \qquad \text{...(5)}$$

which being an equation of third degree in x and y represents a cubic curve. It passes through the point (x_1, y_1) and the foci $(\pm\sqrt{a^2 - b^2}, 0)$ also lies on it.

Again, if (x_1, y_1) lies on the major axis, then $y_1 = 0$

Putting $y_1 = 0$ in equation (5), it reduces to

$$-y(a^2 - b^2)(x_1 - x) - yx_1(x^2 + y^2 - xx_1) = 0$$

or
$$x_1(x^2 + y^2 - xx_1) + (a^2 - b^2)(x_1 - x) = 0 \qquad | \because \; y \neq 0$$

which is the equation of a circle.

Example 6. *Prove that the two conics*

$$ax^2 + 2hxy + by^2 = 1 \text{ and } a'x^2 + 2h'xy + b'y^2 = 1$$

can be placed so as to be confocal if

$$\frac{(a-b)^2 + 4h^2}{(ab - h^2)^2} = \frac{(a' - b')^2 + 4h'^2}{(a'b' - h'^2)^2}.$$

Sol. The equations of the two conic are

$$ax^2 + 2hxy + by^2 = 1 \qquad \text{...(1)}$$

and
$$a'x^2 + 2h'xy + b'y^2 = 1 \qquad \text{...(2)}$$

The centres of both the conics lie at the origin.

We know that the lengths of semi-axes of a conic whose centre is the origin are given by

$$(ab - h^2)r^4 - (a + b)r^2 + 1 = 0 \qquad \text{...(3)}$$

Let r_1^2 and r_2^2 be the squares of the semi-axes of the conic (1), then r_1^2 and r_2^2 are the roots of equation (3).

$$\therefore \qquad r_1^{\,2} + r_2^{\,2} = \frac{a+b}{ab-h^2} \quad \text{and} \quad r_1^{\,2}\, r_2^{\,2} = \frac{1}{ab-h^2}$$

Now, we know that

$$(r_1^{\,2} - r_2^{\,2})^2 = (r_1^{\,2} + r_2^{\,2})^2 - 4r_1^{\,2}r_2^{\,2}$$

$$= \frac{(a+b)^2}{(ab-h^2)^2} - \frac{4}{(ab-h^2)} = \frac{(a+b)^2 - 4ab + 4h^2}{(ab-h^2)^2} = \frac{(a-b)^2 + 4h^2}{(ab-h^2)^2}$$

$$\therefore \qquad r_1^{\,2} - r_2^{\,2} = \frac{\sqrt{(a-b)^2 + 4h^2}}{ab-h^2} \qquad ...(4)$$

In a similar way, if $R_1^{\,2}$ and $R_2^{\,2}$ are the squares of the semi-axes of the conic (2), we have

$$R_1^{\,2} - R_2^{\,2} = \frac{\sqrt{(a'-b')^2 + 4h'^2}}{a'b' - h'^2} \qquad ...(5)$$

Since the conics (1) and (2) are confocal

$$\therefore \qquad r_1^{\,2} - r_2^{\,2} = R_1^{\,2} - R_2^{\,2} \quad i.e., \quad (r_1^{\,2} - r_2^{\,2})^2 = (R_1^{\,2} - R_2^{\,2})^2$$

From the results (4) and (5) the required condition is

$$\frac{(a-b)^2 - 4h^2}{(ab-h^2)^2} = \frac{(a'-b')^2 - 4h'^2}{(a'b'-h'^2)^2}.$$

EXERCISE 1

1. Prove that the conics $x^2 - y^2 - 4x + 2y + 2 = 0$ and $x^2 + 3y^2 - 4x - 6y + 4 = 0$ are confocal.

(K.U. 2007)

2. If two confocal ellipses

$$\frac{x^2}{a^2} + \frac{y^2}{b^2} = 1 \quad \text{and} \quad \frac{x^2}{a^2+\lambda} + \frac{y^2}{b^2+\lambda} = 1$$

are cut by a straight line $x \cos \alpha + y \sin \alpha = p$ and if T_1 T′ be the poles w.r.t. confocal ellipse of straight line then $TT' = \dfrac{\lambda}{p}$.

3. TQ, TP are tangents one to each of two confocal conics whose centre is C. Show that if the tangents are at right angles to one another CT will bisect PQ.

[**Hint.** $\dfrac{xx'}{a^2} + \dfrac{yy'}{b^2} = 1$ and $\dfrac{xx''}{a'^2} + \dfrac{yy''}{b'^2} = 1$ be two tangents then equation of CT is

$$x\left(\frac{x'}{a^2} - \frac{x''}{a'^2}\right) + y\left(\frac{y'}{b^2} - \frac{y''}{b'^2}\right) = 0$$

It passes through mid-point of PQ, if

$$(x' + x'')\left(\frac{x'}{a^2} - \frac{x''}{a'^2}\right) + (y' + y'')\left(\frac{y'}{b^2} - \frac{y''}{b'^2}\right) = 0$$

i.e., if
$$\frac{x'x''}{a^2a'^2} + \frac{y'y''}{b^2b'^2} = 1 \qquad \text{(since conics are confocal)}$$

i.e., if tangents are at rt. angles.]

4. If ϕ be the angle between the tangents from (x_1, y_1) to $\frac{x^2}{a^2}+\frac{y^2}{b^2}=1$, prove that

$$\tan\phi = \frac{2ab\sqrt{\frac{x_1^2}{a^2}+\frac{y_1^2}{b^2}-1}}{x_1^2+y_1^2-a^2-b^2}$$

Hence show that $\tan\frac{\phi}{2} = -\sqrt{\frac{\lambda_2}{\lambda_1}}$

where λ_1 and λ_2 are the parameters of the confocals through P.

5. Find the conics confocal with $x^2 + 2y^2 = 2$ which pass through (1, 1).

6. Prove that the confocal hyperbola through the point on the ellipse $\frac{x^2}{a^2}+\frac{y^2}{b^2}=1$ whose eccentric angle is 45°, is $2(x^2 - y^2) = a^2 - b^2$.

7. Two parabolas have a common focus and their axes are in the same straight lines. If TP and TQ be tangents one to each of the parabolas and if TP and TQ are at right angles, show that the locus of T is a straight line.

8. Find the locus of mid-point of the chords of the confocal $\frac{x^2}{a^2+\lambda}+\frac{y^2}{b^2+\lambda}=1$ which touch the ellipse $\frac{x^2}{a^2}+\frac{y^2}{b^2}=1$.

9. Prove that the normals to the confocals with an ellipse through any given point bisect the angles between the tangents drawn from that point to the ellipse. *(M.D.U. 2007)*

10. Prove that the locus of the point of intersection of two perpendicular tangents one to each of two given confocal conics is a circle.

11. If 2ϕ be the angle between the tangents from $P(x_1, y_1)$ to an ellipse $\frac{x^2}{a^2}+\frac{y^2}{b^2}=1$, prove that $\tan\phi = \sqrt{-\frac{\lambda_2}{\lambda_1}}$ and deduce that $\lambda_2\cos^2\phi + \lambda_1\sin^2\phi = 0$, where λ_1, λ_2 are the parameters of the two confocals through P.

Answers

5. $3x^2 - y^2 \pm \sqrt{5}(x^2 - y^2) = 2$.

8. $\frac{a^2x^2}{(a^2+\lambda)^2}+\frac{b^2y^2}{(b^2+\lambda)^2}=\left(\frac{x^2}{a^2+\lambda}+\frac{y^2}{b^2+\lambda}\right)^2$.

8

Polar Equation

1. Polar Co-ordinates

Definition. In Analytical Geometry, there is another system of expressing the co-ordinates of a point which are called polar-coordinates.

Let O be a fixed point and OX a fixed straight line through O. The positive direction of OX is indicated by the arrowhead. The fixed point is called the **pole** or the origin and the fixed straight line OX is called the **initial line** or the **polar axis**. Let P be any point in the plane through the line OX. Join OP then :

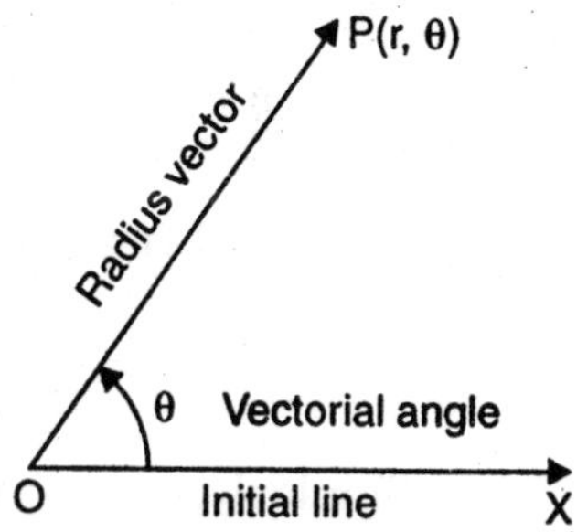

(*i*) The length OP is called the **radius vector** of the point P and is denoted by 'r'.

(*ii*) The angle XOP is called the **vectorial angle** of the point P and is generally denoted by 'θ'.

(*iii*) The numbers r and θ taken together in this very order are called the **polar co-ordinates** of the point P and it is written as P(r, θ). The vectorial angle is considered positive if it is measured in the anti-clockwise direction from OX. If PO be produced to P_1' so such that OP′ = OP in magnitude, and if the co-ordinates of P be (r, θ), then those of P′ will be either $(-r, \theta)$ or $(r, \theta + \pi)$.

(*iv*) If (x, y), are the co-ordinates of P referred to cartesian system, then it can be easily found that

$$x = r \cos \theta,\ y = r \sin \theta.$$

(*v*) Any equation in x, y can be transformed into an equation in polar co-ordinates by putting $x = r \cos \theta$ and $y = r \sin \theta$.

(*vi*) By putting $r = \sqrt{x^2 + y^2}$, $\theta = \tan^{-1} \dfrac{y}{x}$, we can transform equations in polar co-ordinates into equations in cartesian co-ordinates.

2. *To find the distance between two points* (r_1, θ_1) *and* (r_2, θ_2)

Let P and Q be two points whose co-ordinates are (r_1, θ_1) and (r_2, θ_2) respectively.

Then, by Trigonometry,

$$PQ^2 = OP^2 + OQ^2 - 2OP \cdot OQ \cos POQ$$

Here $\quad OP = r_1,\ OQ = r_2$

and $\quad \angle POQ = \angle XOQ - \angle XOP = \theta_2 - \theta_1$

$$\therefore \quad PO^2 = r_1^2 + r_2^2 - 2r_1 r_2 \cos(\theta_2 - \theta_1).$$

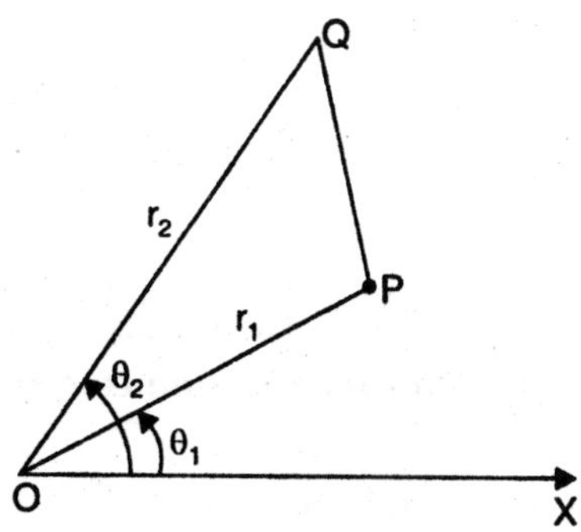

Example 1. *Find the distance between the points whose polar co-ordinates are (2, 10°) and (2, 40°).*

Sol. Let P(2, 10°) and Q(2, 40°) be the given point.

$\therefore$ Using the result,

$$PQ^2 = r_1^2 + r_2^2 - 2r_1r_2 \cos(\theta_2 - \theta_1), \text{ we have}$$

$$PQ^2 = 4 + 4 - 2 \times 2 \times 2 \cos(40 - 10)°$$

$$= 8 - 4 \times 2 \cos 30° = 8 - 4\sqrt{3} = 6 + 2 - 2\sqrt{12} = (\sqrt{6} - \sqrt{2})^2$$

$$\therefore \quad PQ = \sqrt{6} - \sqrt{2}.$$

3. *To find the area of a triangle having given the polar co-ordinates of its angular points*

Let angular points of the triangle P, Q, R be (r_1, θ_1), (r_2, θ_2), (r_3, θ_3) respectively.

Then area of ΔPQR

$$= \Delta OPQ + \Delta OQR - \Delta OPR.$$

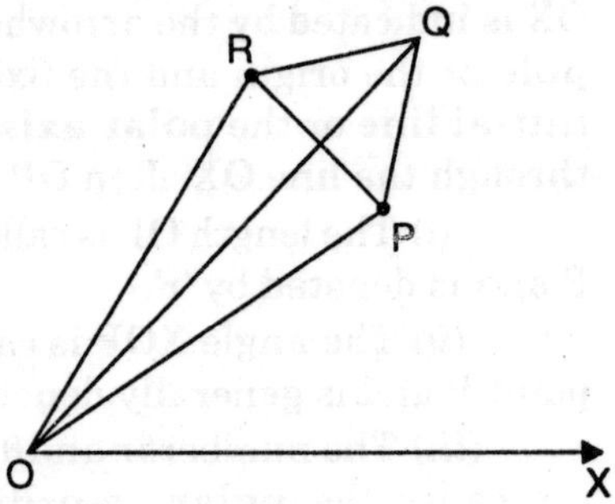

Then by Trigonometry

$$\Delta OPQ = \frac{1}{2} OP \,.\, OQ \sin POQ$$

$$= \frac{1}{2} r_1r_2 \sin(\theta_2 - \theta_1)$$

Similarly, $$\Delta OQR = \frac{1}{2} r_2r_3 \sin(\theta_3 - \theta_2)$$

and $$\Delta OPR = \frac{1}{2} r_3r_1 \sin(\theta_3 - \theta_1) = -\frac{1}{2} r_3r_1 \sin(\theta_1 - \theta_3)$$

$$\therefore \quad \Delta PQR = \frac{1}{2} \{r_1r_2 \sin(\theta_2 - \theta_1) + r_2r_3 \sin(\theta_3 - \theta_2) + r_3r_1 \sin(\theta_1 - \theta_3)\}$$

4. **(a)** *To find the equation of a straight line in polar equation*

Let AB be the given straight line and ON the perpendicular on it from O, let p be the length of the perpendicular ON and α the angle XON.

Let P(x, y) be any point on AB. Then we know that the equation of the straight line is

$$x \cos\alpha + y \sin\alpha = p \qquad ...(1)$$

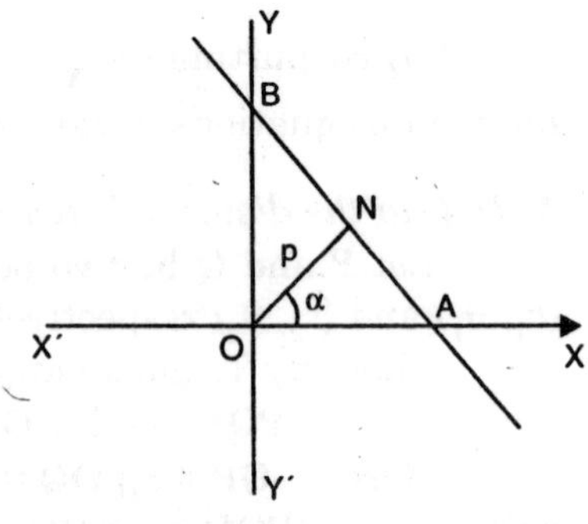

Now, take the origin as pole and OX as the initial line. To find the equation of the line in polar equation, we put

$$x = r\cos\theta,\ y = r\sin\theta \text{ in (1)},$$

$\therefore$ The equation of the line becomes

$$r\cos\theta\cos\alpha + r\sin\theta\sin\alpha = p$$

i.e., $$r\cos(\theta - \alpha) = p$$

This is the required polar equation.

(b) *In cartesian co-ordinates, the general equation of the straight line is of the form*

$$ax + by + c = 0 \qquad ...(1)$$

To find the equation of the line in polar equation, put $x = r\cos\theta$, $y = r\sin\theta$ in (1).

The polar equation of the line is

$$ar\cos\theta + br\sin\theta + c = 0$$

i.e.,
$$a\cos\theta + b\sin\theta = -\frac{c}{r}.$$

If the line does not pass through the origin *i.e.,* $c \neq 0$ then

$$-a\cos\theta - b\sin\theta = \frac{c}{r}.$$

Multiplying it by $\frac{k}{c}$, we have

$$-\frac{ak}{c}\cos\theta - \frac{bk}{c}\sin\theta = \frac{k}{r} \quad \text{or} \quad A\cos\theta + B\sin\theta = \frac{k}{r}$$

Thus, the general equation of the line not passing through the pole is

$$A\cos\theta + B\sin\theta = \frac{k}{r}.$$

where k is any constant.

5. (*a*) **Parallel straight lines**

The equation of a straight line parallel to $p = r\cos(\theta - \alpha)$ is

$$p' = r\cos(\theta - \alpha)$$

Also, a line parallel to

$$A\cos\theta + B\sin\theta = \frac{k}{r} \text{ is}$$

$$A\cos\theta + B\sin\theta = \frac{k'}{r}, \text{ where } k' \text{ is any constant.}$$

(*b*) **Perpendicular straight lines**

We know that the equation of a straight line perpendicular to $ax + by + c = 0$ is

$$bx - ay + k = 0.$$

Changing to polar co-ordinates, the equation of a line perpendicular to

$A\cos\theta + B\sin\theta = \frac{k}{r}$ is

$$B\cos\theta - A\sin\theta = \frac{k'}{r} \quad \text{or} \quad A\cos\left(\theta + \frac{\pi}{2}\right) + B\sin\left(\theta + \frac{\pi}{2}\right) = \frac{k'}{r}.$$

i.e., By changing θ into $\frac{\pi}{2} + \theta$ and k to k', we get the equation of a line perpendicular to

$$\mathbf{A\cos\theta + B\sin\theta = \frac{k}{r}.}$$

6. *To find the polar equation of the line joining the points* (r_1, θ_1) *and* (r_2, θ_2).

Let O be the pole and OX the initial line.

Let $A(r_1, \theta_1)$ and $B(r_2, \theta_2)$ be the given points. Take $P(r, \theta)$ any point on the line AB. Join OA, OB and OP.

Then $\quad \Delta OAB = \Delta OAP + \Delta OPB$

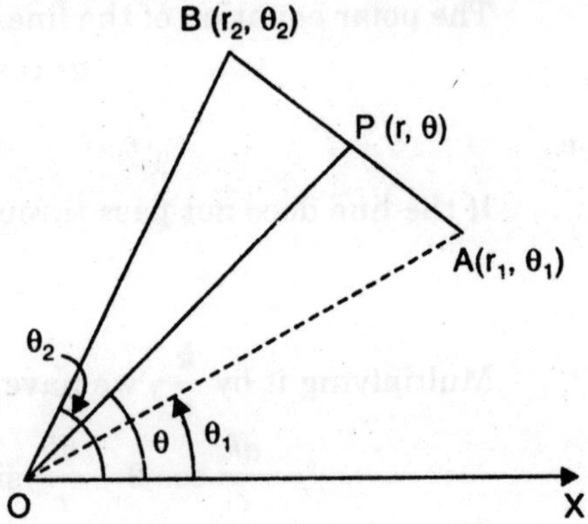

or $\quad \frac{1}{2}$ OA . OB sin A$\hat{O}$B

$$= \frac{1}{2} \text{OA . OP . sin A}\hat{\text{O}}\text{P} + \frac{1}{2} \text{OP . OB sin P}\hat{\text{O}}\text{B}$$

or $\quad r_1 r_2 \sin(\theta_2 - \theta_1) = r_1 r \sin(\theta - \theta_1) + r r_2 \sin(\theta_2 - \theta)$

or $\quad r_1 r_2 \sin(\theta_2 - \theta_1) + r r_1 \sin(\theta_1 - \theta)$

$$+ r r_2 \sin(\theta - \theta_2) = 0$$

Dividing by $r r_1 r_2$, we have

$$\frac{\sin(\theta_2 - \theta_1)}{r} + \frac{\sin(\theta_1 - \theta)}{r_2} + \frac{\sin(\theta - \theta_2)}{r_1} = 0$$

which is the required equation.

7. *To find length of the perpendicular drawn from (r_1, θ_1) on the straight line $r \cos(\theta - \alpha) = p$.*

Let AB be the given line $\quad r \cos(\theta - \alpha) = p$

and OL the perpendicular from O on it, so that

$$\text{OL} = p \text{ and } \angle\text{XOL} = \alpha.$$

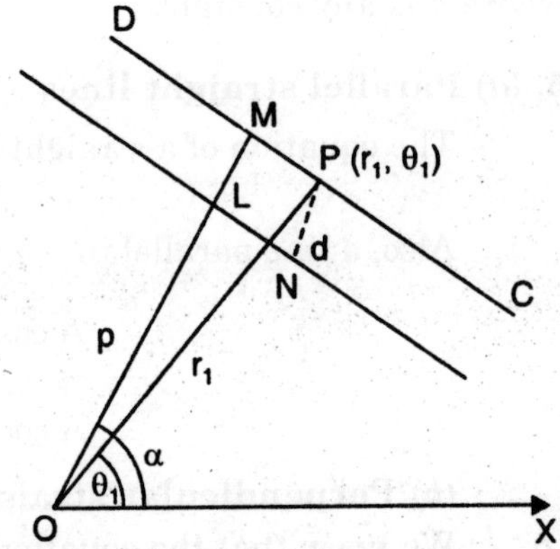

Let P(r_1, θ_1) be the given point. Draw PN ⊥ AB. Let PN = d. Through P draw a straight line CD parallel to AB, meeting OL produced in M.

Then $\quad$ OM = OL + LM = $p + d$

∴ Equation of the line CD is

$$r \cos(\theta - \alpha) = p + d$$

P(r_1, θ_1) lies on it.

∴ $\quad r_1 \cos(\theta_1 - \alpha) = p + d$

Hence $\quad \mathbf{d = r_1 \cos(\theta_1 - \alpha) - p}$

This is the required distance.

Example. *Find the polar co-ordinates of foot of the perpendicular from the pole on the line joining the points (r_1, θ_1) and (r_2, θ_2).*

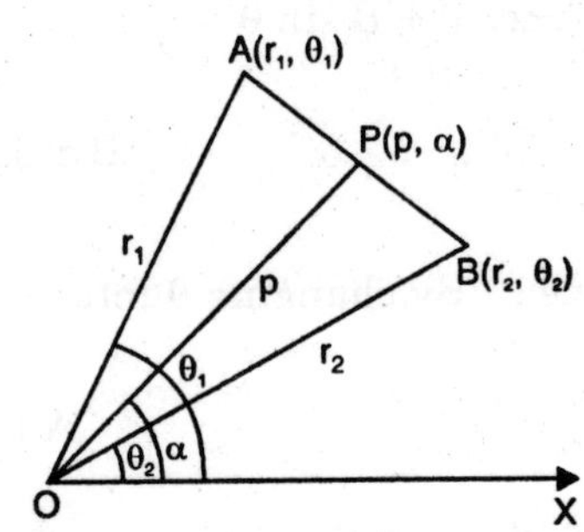

Sol. Let A(r_1, θ_1) and B(r_2, θ_2) be the given points and P(p, α) be the polar co-ordinates of the foot of the perpendicular from the pole O on AB. Join OA, OB.

∴ Equation of the line AB is

$$r \cos(\theta - \alpha) = p \qquad ...(i)$$

∵ A(r_1, θ_1) and B(r_2, θ_2) lie on (*i*)

$$\therefore \quad r_1 \cos(\theta_1 - \alpha) = p \qquad ...(ii)$$

and $$r_2 \cos(\theta_2 - \alpha) = p \qquad ...(iii)$$

∴ from (*ii*) and (*iii*), we have

$$r_1 \cos(\theta_1 - \alpha) = r_2 \cos(\theta_2 - \alpha)$$

or $\quad r_1 \cos\theta_1 \cos\alpha + r_1 \sin\theta_1 \sin\alpha = r_2 \cos\theta_2 \cos\alpha + r_2 \sin\theta_2 \sin\alpha$

or $\quad (r_1 \sin\theta_1 - r_2 \sin\theta_2) \sin\alpha = (r_2 \cos\theta_2 - r_1 \cos\theta_1) \cos\alpha$

$$\therefore \qquad \tan\alpha = \frac{r_2\cos\theta_2 - r_1\cos\theta_1}{r_1\sin\theta_1 - r_2\sin\theta_2}$$

Again $\qquad \Delta OAB = \frac{1}{2} OA\,.\,OB\sin A\hat{O}B = \frac{1}{2} r_1 r_2 \sin(\theta_1 - \theta_2)$

Also $\qquad \Delta OAB = \frac{1}{2} AB\,.\,p = \frac{1}{2}\,.\,p\sqrt{r_1^2 + r_2^2 - 2r_1r_2\cos(\theta_1 - \theta_2)}$

$$\therefore \quad \frac{1}{2} p\sqrt{r_1^2 + r_2^2 - 2r_1r_2\cos(\theta_1 - \theta_2)} = \frac{1}{2} r_1r_2\sin(\theta_1 - \theta_2)$$

$$\therefore \qquad p = \frac{r_1r_2\sin(\theta_1 - \theta_2)}{\sqrt{r_1^2 + r_2^2 - 2r_1r_2\cos(\theta_1 - \theta_2)}}$$

Hence the polar co-ordinates of the foot of perpendicular are

$$\left(\frac{r_1r_2\sin(\theta_1 - \theta_2)}{\sqrt{r_1^2 + r_2^2 - 2r_1r_2\cos(\theta_1 - \theta_2)}},\ \tan^{-1}\frac{r_2\cos\theta_2 - r_1\cos\theta_1}{r_1\sin\theta_1 - r_2\sin\theta_2}\right).$$

POLAR EQUATION OF A CIRCLE

8. *To find the polar equation of the circle, whose centre is* (C, α) *and radius* a.

Let OX be the initial line and C be the centre (c, α) of the circle.

Let $P(r, \theta)$ be any point on the circle. Join OC, OP and CP.

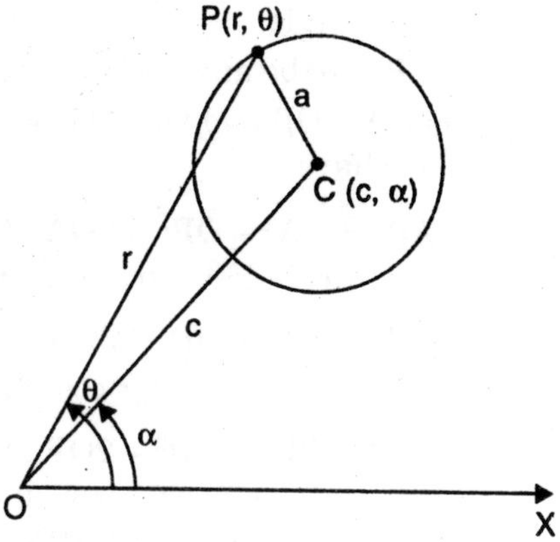

$$\therefore \qquad OC = c,\ OP = r,\ CP = a$$

and $\qquad \angle POC = \angle POX - \angle COX = \theta - \alpha$

From the ΔPOC, by cosine formula, we have

$$CP^2 = OP^2 + OC^2 - 2OP\,.\,OC\cos POC$$

i.e., $\qquad a^2 = r^2 + c^2 - 2rc\cos(\theta - \alpha)$

or $\qquad r^2 - 2cr\cos(\theta - \alpha) + c^2 - a^2 = 0$

This is the required equation of the circle.

9. *To find the polar equation of the circle with radius a and centre lying on the initial line at a distance c from the pole*

Let OX be the initial line passing through the centre $C(c, 0)$ of the circle of radius a. Let $P(r, \theta)$ by any point on the circle. Then, from ΔOCP, by cosine formula.

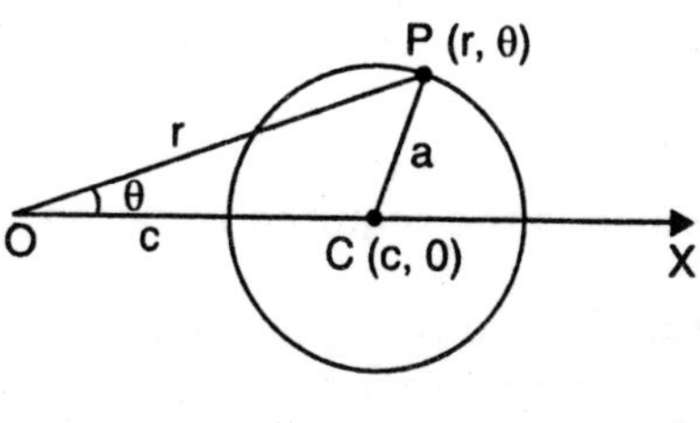

$$CP^2 = OP^2 + OC^2 - 2OP\,.\,OC\cos POC$$

$\therefore \qquad a^2 = r^2 + c^2 - 2rc\cos\theta$

or $\qquad r^2 - 2cr\cos\theta + c^2 - a^2 = 0$

This is the required equation.

[**Note.** We can also obtain the required equation from the general equation by putting $\alpha = 0$].

10. *To obtain the equation of the circle when the pole lies on the circumference, and the initial line does not pass through the centre.*

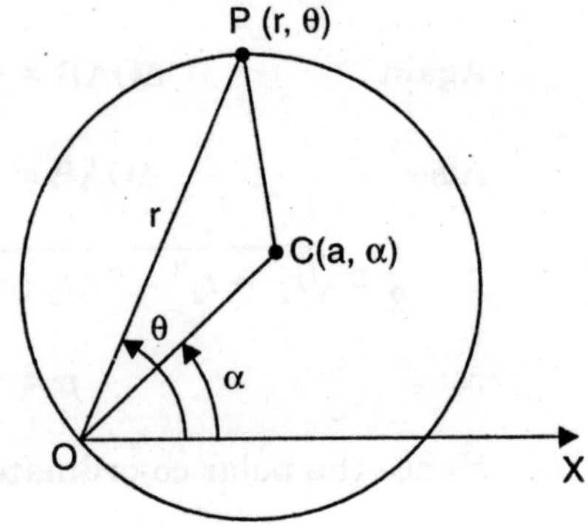

Here $\angle POC = \theta - \alpha$ and centre is (a, α).

$\therefore$ $$CP^2 = OP^2 + OC^2 - 2OP \cdot OC \cos (POC)$$

i.e., $$a^2 = r^2 + a^2 - 2ra \cos (\theta - \alpha)$$

or $$r^2 - 2ar \cos (\theta - \alpha) = 0$$

i.e., $$r - 2a \cos (\theta - \alpha) = 0$$

which is the required equation.

Cor 1. *When the pole is on the circumference and the initial line passes through the centre.*

Here co-ordinates of centre are $(a, 0)$ and radius $= a$.

$\therefore$ putting $\alpha = 0$ and $c = a$ in the general equation, we have

$$\mathbf{r = 2a \cos \theta}$$

Cor 2. *When the centre lies at the pole and radius is a.*

Here, we have $r = a$, the required equation of the circle.

11. *To find the polar equation of the circle on the join of* (r_1, θ_1) *and* (r_2, θ_2) *as diameter.*

Let A and B be the given points (r_1, θ_1) and (r_2, θ_2) respectively as the extremities of a diameter of a circle.

Let $P(r, \theta)$ be any point on the circle.

Join OA, OB and OP. Also join AP, BP and AB. From Δs OAB, OAP and OBP, by cosine formula, we have

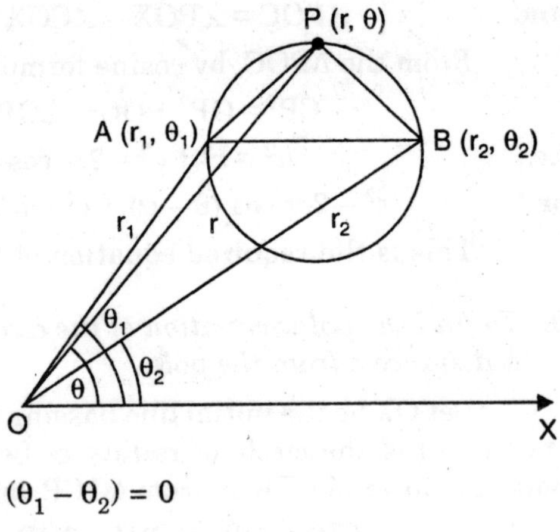

$$AB^2 = OA^2 + OB^2 - 2OA \cdot OB \cos A\hat{O}B$$
$$= r_1^2 + r_2^2 - 2r_1r_2 \cos (\theta_1 - \theta_2) \quad \ldots(1)$$
$$AP^2 = OA^2 + OP^2 - 2OA \cdot OP \cos A\hat{O}P$$
$$= r_1^2 + r^2 - 2r_1r \cos (\theta_1 - \theta) \quad \ldots(2)$$

and $$BP^2 = OP^2 + OB^2 - 2OP \cdot OB \cos P\hat{O}B$$
$$= r^2 + r_2^2 - 2rr_2 \cos (\theta - \theta_2) \quad \ldots(3)$$

From the right angled ΔPAB, we have

$$AB^2 = AP^2 + BP^2$$

i.e., $$r_1^2 + r_2^2 - 2r_1r_2 \cos (\theta_1 - \theta_2) = [r_1^2 + r^2 - 2r_1r \cos (\theta_1 - \theta)] + [r^2 + r_2^2 - 2rr_2 \cos (\theta - \theta_2)]$$

or $$r^2 - [r_1 \cos (\theta - \theta_1) + r_2 \cos (\theta - \theta_2)]r + r_1r_2 \cos (\theta_1 - \theta_2) = 0$$

which is the required equation of the circle.

12. *To find the equation to the chord joining the points whose vectorial angles are 'θ_1' and 'θ_2' on the circle $r = 2a \cos \theta$.*

Let P and Q be the points whose vectorial angles are 'θ_1' and 'θ_2' on the circle $r = 2a \cos \theta$. Hence these points are

$$P(2a \cos \theta_1, \theta_1) \text{ and } Q(2a \cos \theta_1, \theta_2)$$

$\therefore$ The equation of line joining these points is

$$4a^2 \cos\theta_1 \cos\theta_2 \sin(\theta_2 - \theta_1) + r \,.\, 2a\cos\theta_2 \sin(\theta - \theta_2) + r \,.\, 2a \cos\theta_1 \sin(\theta_1 - \theta) = 0$$

[Two point form $r_1r_2 \sin(\theta_2 - \theta_1) + r_2 r \sin(\theta - \theta_2) + rr_1 \sin(\theta_1 - \theta) = 0$]

or $$4a\cos\theta_1 \cos\theta_2 \sin(\theta_2 - \theta_1) + r[\sin\theta + \sin(\theta - 2\theta_2) + \sin(2\theta_1 - \theta) + \sin(-\theta)] = 0$$

or $$4a\cos\theta_1 \cos\theta_2 \sin(\theta_2 - \theta_1) + r[\sin(\theta - 2\theta_2) + \sin(2\theta_1 - \theta] = 0$$

or $$4a\cos\theta_1 \cos\theta_2 \sin(\theta_2 - \theta_1) + r[2\sin(\theta_1 - \theta_2)\cos(\theta - \theta_1 - \theta_2)] = 0$$

or $$2\sin(\theta_1 - \theta_2)\,[-2a\cos\theta_1\cos\theta_2 + r\cos(\theta - \theta_1 - \theta_2)] = 0$$

or $$r\cos(\theta - \theta_1 - \theta_2) = 2a\cos\theta_1 \cos\theta_2 \qquad | \because \theta_1 \neq \theta_2$$

which is the required equation.

Cor. *To deduce the equation of the tangent at the point 'θ_1'*

The equation of the chord joining the points 'θ_1' and 'θ_2' is

$$r\cos(\theta - \theta_1 - \theta_2) = 2a\cos\theta_1\cos\theta_2 \qquad ...(1)$$

The chord (1) becomes tangent at θ_1 when $\theta_2 \to \theta_1$. Then putting $\theta_2 = \theta_1$ in (1), the equation of tangent at θ_1 is

$$r\cos(\theta - 2\theta_1) = 2a\cos^2\theta_1.$$

13. Equation of Normal

To find the equation of the normal at the point 'θ_1' to the circle

$$r = 2a\cos\theta.$$

The equation of the circle is $r = 2a\cos\theta$ ...(1)

The equation of the tangent at the point 'θ_1' *i.e.*, at $(2a\cos\theta_1, \theta_1)$ is

$$r\cos(\theta - 2\theta_1) = 2a\cos^2\theta_1 \qquad ...(2)$$

The equation of any line $\perp$ to (2) is

$$r\sin(\theta - 2\theta_1) = p \qquad ...(3)$$

It will be a normal at θ_1 if $(2a\cos\theta_1, \theta_1)$ lies on (3)

$\therefore$ $$2a\cos\theta_1 \sin(-\theta_1) = p \text{ or } p = -a\sin 2\theta_1$$

$\therefore$ From (3), the equation of the normal is

$$r\sin(\theta - 2\theta_1) + a\sin 2\theta_1 = 0.$$

Example 1. *Find the equation of the circle whose centre is (4, π/3) and radius is 3.*

Sol. The equation of the circle whose centre is (c, α) is

$$r^2 - 2cr\cos(\theta - \alpha) + c^2 - a^2 = 0$$

Here $c = 4, \alpha = \dfrac{\pi}{3}$ and $a = 3$

$\therefore$ Required equation of the circle is

$$r^2 - 8r\cos(\theta - \pi/3) + 7 = 0.$$

Example 2. *Prove that the polar equation to the circle described on straight line joining the points (1, 60°), (2, 30°) as diameter is*

$$r^2 - r[\cos(\theta - 60°) + 2\cos(\theta - 30°)] + \sqrt{3} = 0.$$

Sol. The equation of the circle on the join of (r_1, θ_1) and (r_2, θ_2) as diameter is

$$r^2 - r[r_1\cos(\theta - \theta_1) + r_2\cos(\theta - \theta_2)] + r_1r_2\cos(\theta_1 - \theta_2) = 0 \qquad ...(1)$$

Here $r_1 = 1, \theta_1 = 60°$ and $r_2 = 2, \theta_2 = 30°$

$\therefore$ **from (1), the required equation of the circle is**

$$r^2 - r[1 . \cos(\theta - 60°) + 2\cos(\theta - 30°)] + 1.2\cos(60° - 30°) = 0$$

or $$r^2 - r[\cos(\theta - 60°) + 2\cos(\theta - 30°)] + \sqrt{3} = 0.$$

Example 3. *A circle passes through the point (r_1, θ_1) and touches the initial line at a distance c from the pole. Find the polar equation of the circle.*

Sol. Let $C(c_1, \alpha)$ be the centre and a the radius of the circle.

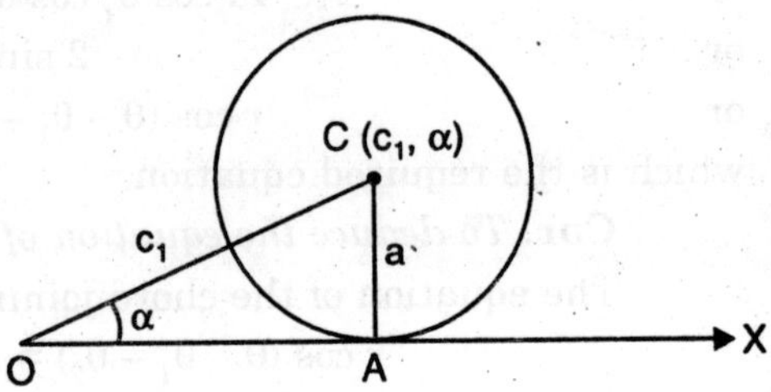

Let the circle touch the initial line at A.

Then $CA = a$, $OA = c$.

$OC = c_1$, and $\angle COA = \alpha$

$\therefore$ The equation of the circle is

$$r^2 - 2rc_1\cos(\theta - \alpha) + c_1^2 - a^2 = 0$$

or $$r^2 - 2r[c_1\cos\theta\cos\alpha + c_1\sin\theta\sin\alpha] + c_1^2 - a^2 = 0 \quad ...(1)$$

$\because$ $CA \perp OX$.

$\therefore$ From the right angled ΔOAC,

$$\cos\alpha = \frac{OA}{OC} = \frac{c}{c_1} \qquad \therefore \quad c_1\cos\alpha = c$$

and $$\sin\alpha = \frac{AC}{OC} = \frac{a}{c_1} \qquad \therefore \quad c_1\sin\alpha = a$$

Also $c_1^2 - a^2 = c^2$

Substituting these values in (1), we have

$$r^2 - 2r[c\cos\theta + a\sin\theta] + c^2 = 0$$

or $$r^2 - 2rc\cos\theta + c^2 = 2ar\sin\theta$$

or $$a = \frac{r^2 - 2rc\cos\theta + c^2}{2r\sin\theta} \quad ...(2)$$

Since, the circle passes through the point (r_1, θ_1)

$\therefore$ From (2), $$a = \frac{r_1^2 - 2r_1 c\cos\theta_1 + c^2}{2r_1\sin\theta_1} \quad ...(3)$$

Equating (2) and (3)

The required equation is $$\frac{r^2 - 2rc\cos\theta + c^2}{r\sin\theta} = \frac{r_1^2 - 2r_1 c\cos\theta_1 + c^2}{r_1\sin\theta_1}.$$

EXERCISE 1

1. Find the polar co-ordinates of the point whose rectangular co-ordinates are

   (*i*) $(-1, -1)$ (*ii*) $\left(-\frac{3}{\sqrt{2}}, \frac{3}{\sqrt{2}}\right)$.

2. Convert P $\left(2, \frac{\pi}{4}\right)$; Q $\left(3, \frac{\pi}{2}\right)$ into cartesian co-ordinates and hence write the length of PQ.

3. Prove that the line joining the points (0, 0), $\left(2, \frac{\pi}{3}\right)$, $\left(2, \frac{2\pi}{3}\right)$ form an equilateral triangle.

4. The vectorial angle of a point P on the join of the points $P_1(r_1, \theta_1)$ and $P_2(r_2, \theta_2)$ is $\frac{\theta_1 + \theta_2}{2}$. Show that the length of the radius vector of P is

$$\frac{2r_1r_2}{r_1 + r_2} \cos \frac{\theta_1 - \theta_2}{2}.$$

[**Hint.** Let (r_3, θ_3) be the co-ordinates of P $\therefore \theta_3 = \frac{\theta_1 + \theta_2}{2}$

$\therefore r_1r_2 \sin(\theta_2 - \theta_1) + r_2r_3 \sin\left(\frac{\theta_1 + \theta_2}{2} - \theta_2\right) + r_3r_1 \sin\left(\theta_1 - \frac{\theta_1 + \theta_2}{2}\right) = 0$. Find r_3 etc.]

5. Find the angle between the lines

$$r \cos \theta = 2 \quad \text{and} \quad r \cos\left(\theta - \frac{\pi}{6}\right) = 3.$$

6. Find the equation of any line passing through the intersection of

$$\frac{l}{r} = a \cos \theta + b \sin \theta \quad \text{and} \quad \frac{l'}{r} = a' \cos \theta + b' \sin \theta.$$

7. Prove that the equation of a circle which passes through the pole but the initial line does not pass through the centre can be put in the form

$$r = A \cos \theta + B \sin \theta.$$

8. Find the radius and the co-ordinates of the centre of the circle

$$r = A \cos \theta + B \sin \theta.$$

9. Find the co-ordinates of the centre and radius of the circle

$$r^2 - 6r \cos(\theta - \pi/6) - 16 = 0.$$

10. Any straight line is drawn from a fixed point O meeting a fixed circle in P and a point Q is taken on the line such that OP . OQ is constant. Show that locus of Q is a circle.

Answers

1. (i) $\left(\sqrt{2}, \frac{5\pi}{4}\right)$. (ii) $\left(3, \frac{3\pi}{4}\right)$

2. $P \leftrightarrow (\sqrt{2}, \sqrt{2})$ and $Q \leftrightarrow (0, 3)$ $\quad PQ = \sqrt{13 - 6\sqrt{2}}$

5. $\frac{\pi}{6}$

6. $\frac{l + kl'}{r} = (a + ka') \cos \theta + (b + kb') \sin \theta$

8. $\frac{1}{2}\sqrt{A^2 + B^2}; \frac{1}{2}\sqrt{A^2 + B^2} \tan^{-1} \frac{B}{A}$

9. $\left(3, \frac{\pi}{6}\right)$; 5.

9

Polar Equation of a Conic

1. *To find the polar equation of a conic, the focus being the pole.* (M.D.U. 2005)

Let S be the focus and ZM the directrix of the conic and let the eccentricity be e, SL ($= l$) the semi latus rectum of the conic.

Draw SZ perpendicular to the directrix and take SZ as the initial line and S as the pole.

Let the co-ordinates of any point P on the curve be (r, θ). So that $SP = r$ and $\angle ZSP = \theta$. Draw PM, PN perpendicular respectively to the directrix and to SZ. Draw LK perpendicular to ZM. Thus we have $SL = eLK = e \,.\, SZ$

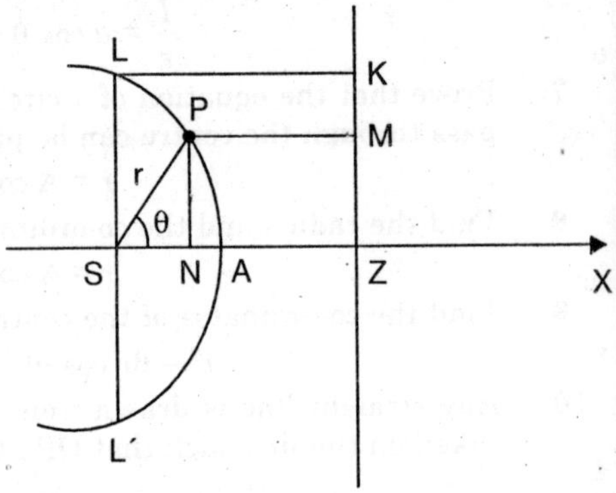

$$\therefore \quad SZ = \frac{SL}{e} = \frac{l}{e}$$

By the definition of a conic, $SP = ePM$

$$\therefore \quad r = eNZ = e(SZ - SN)$$

$$= e\left(\frac{l}{e} - r\cos\theta\right) = l - er\cos\theta$$

i.e., $\quad r[1 + e\cos\theta] = l$

$$\therefore \quad \frac{l}{r} = 1 + e\cos\theta.$$

which is the required polar equation of the conic.

Cor. If the axis of the conic SZ make an angle α with the initial line, SP makes an angle $(\theta - \alpha)$ with the initial line and then the equation of the conic will be

$$\frac{l}{r} = 1 + e\cos(\theta - \alpha).$$

2. *To find the equation of the directrix corresponding to the focus at the pole.*

Let the co-ordinates of any point P on the directrix ZM corresponding to focus S as pole be (r, θ).

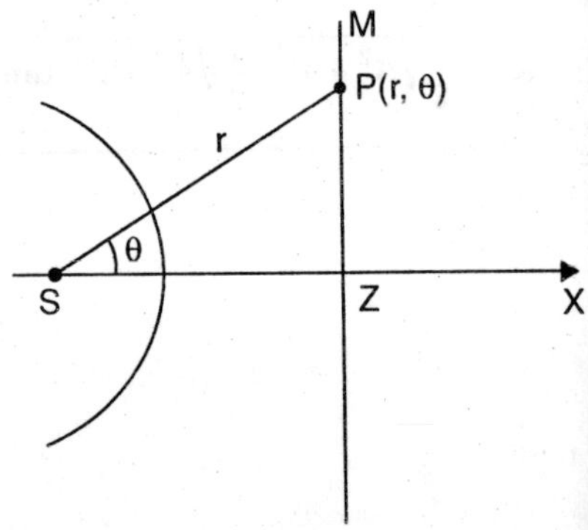

Then $\quad \dfrac{SZ}{SP} = \cos\theta \quad$ or $\quad SP\cos\theta = SZ$

i.e., $\quad r\cos\theta = \dfrac{l}{e} \quad$ or $\quad \dfrac{l}{r} = e\cos\theta$

which is the required equation of the directrix.

Cor. The equation of the directrix of the conic

$$\frac{l}{r} = 1 + e\cos(\theta - \alpha)$$

Corresponding to the focus S is $\dfrac{l}{r} = e\cos(\theta - \alpha)$.

3. *To find the equations of asymptotes of polar equation of conic*

Let the polar equation of conic be $\frac{l}{r} = 1 + e \cos \theta$ ...(1)

Let the co-ordinates of any point P on the conic (1) be (r_1, α)

Then, $$\frac{l}{r_1} = 1 + e \cos \alpha \quad \text{...(2)}$$

Now, the equation of a tangent to the conic (1) at $P(r_1, \alpha)$ is

$$\frac{l}{r} = e \cos \theta + \cos (\theta - \alpha) = e \cos \theta + \cos \theta \cos \alpha + \sin \theta \sin \alpha$$

i.e., $$\frac{l}{r} = (e + \cos \alpha) \cos \theta + \sin \theta \sin \alpha \quad \text{...(3)}$$

We know that the asymptote to a curve is the limiting position of the tangent whose point of contact tends to infinity *i.e.,* when P lies at infinity, $r_1 \to \infty$

∴ From (2), $0 = 1 + e \cos \alpha$

or $$\cos \alpha = -\frac{1}{e}. \text{ Also } \sin \alpha = \pm \sqrt{1 - \frac{1}{e^2}}$$

From (3), $$\frac{l}{r} = \left(e - \frac{1}{e}\right) \cos \theta \pm \sin \theta \sqrt{1 - \frac{1}{e^2}}$$

or $$\frac{l}{r} = \frac{\sqrt{e^2 - 1}}{e}\left[\sqrt{e^2 - 1} \cos \theta \pm \sin \theta\right]$$

which is the required equation of the asymptote.

4. To trace the conic $\frac{l}{r} = 1 + e \cos \theta$

The given conic is $\frac{l}{r} = 1 + e \cos \theta$ ...(1)

Case I. Let $e = 0$.

In this case, from (1), the equation becomes $r = l$ (a constant)

∴ The curve is a circle of radius l with its centre at the pole.

Case 2. Let $0 < e < 1$, then the curve is an ellipse when $\theta = 0$, then from (1), $r = \frac{l}{1+e}$.

Since the maximum value of $\cos \theta$ is 1

∴ The maximum value of $\frac{l}{r}$ is $1 + e$

i.e., the minimum value of r is $\frac{l}{1+e}$ when $\theta = 0$.

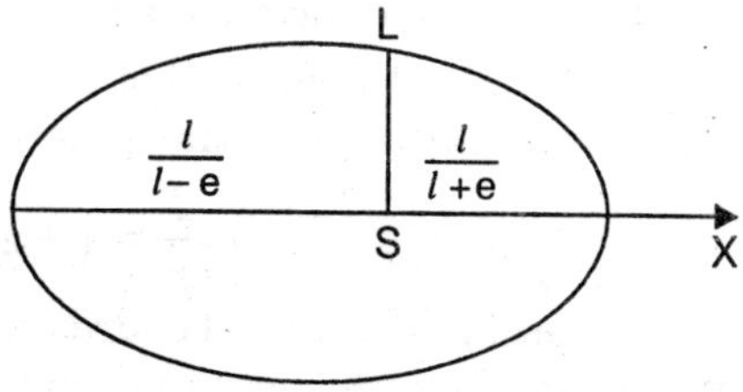

As θ increases from 0 to π, r increases and attains its maximum value $\frac{l}{1-e}$ when $\theta = \pi$. When $\theta = \frac{\pi}{2}$, $r = l$.

If θ be changed to $-\theta$, the equation of the curve does not change.

Hence the curve is symmetrical above the initial line.

Its shape is shown in the given figure.

Case 3. Let $e = 1$, then the conic is a parabola and the equation becomes $\frac{l}{r} = 1 + \cos\theta$

At the point A, where the curve cuts the axis, $\theta = 0$ and $r = \frac{l}{2}$ which is the minimum value of r. When θ increases, $\frac{l}{r}$ decreases and therefore r increases.

When $\theta \to \pi$, $r \to \infty$.

The curve is symmetrical above the initial line as proved in case 2.

Case 4. Let $e > 1$, then the curve is an hyperbola.

At the point A, $\theta = 0$ and $r = \frac{l}{1+e}$.

As θ increases, $\cos\theta$ decreases and therefore r increases. When $\theta = \frac{\pi}{2}$, then $r = l$. When θ increases beyond $\frac{\pi}{2}$, the expression $1 + e\cos\theta$ becomes zero for a value of θ lying between $\frac{\pi}{2}$ and π and the value of r will be infinitely great. Let α be the angle between $\frac{\pi}{2}$ and π, then

$$1 + e\cos\alpha = 0 \quad \therefore \quad \cos\alpha = -\frac{1}{e}$$

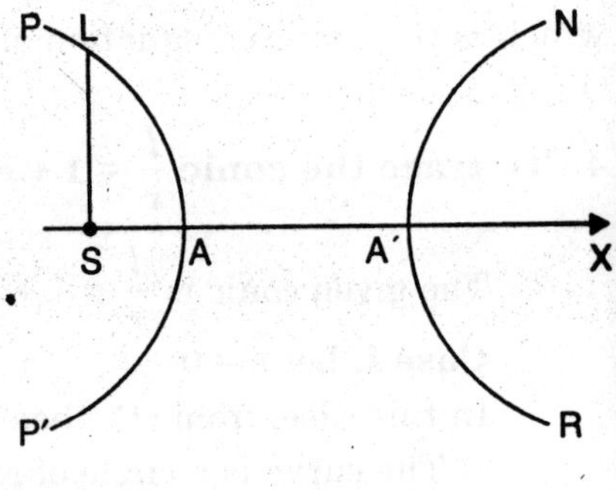

The part of the curve given by values of θ from 0 to α is the portion AP.

As θ increases beyond the value of α, $(1 + e\cos\theta)$ becomes negative and when $\theta = \pi$, $r = -\frac{l}{e-1} = SA'$ when θ goes on increasing from α to π, then r goes on diminishing numerically. These values of θ correspond to the portion RA′ of the curve. The portion A′N correspond to values of θ between π and $2\pi - \alpha$. The value of r will remain positive while θ changes from $2\pi - \alpha$ to 2π.

The curve consists of two separate branches as shown in the figure.

Example 1. *Show that in a conic the semi-latus rectum is the harmonic mean between the segments of a focal chord.*

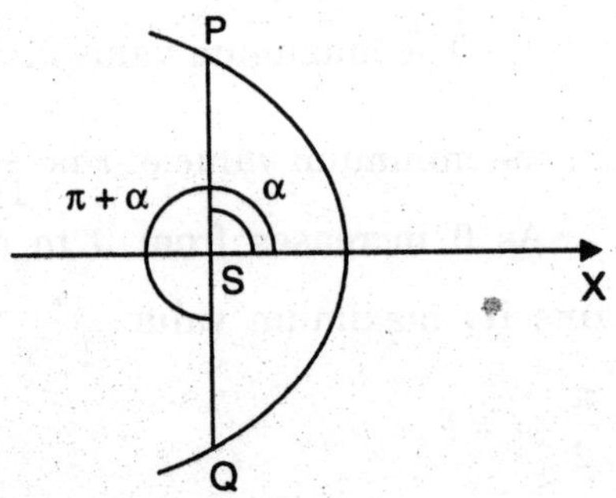

Sol. Let the focus S of the conic be pole and axis as the initial line, the equation of the conic is

$$\frac{l}{r} = 1 + e\cos\theta \qquad \text{...(1)}$$

Let PSQ be a focal chord and vectorial angle of P be α. Then the vectorial angle of Q is $\pi + \alpha$.

The co-ordinates of P and Q are (r_1, α) and $\left(\frac{r}{2}, \alpha + \pi\right)$ where $SP = r_1$ and $SQ = r_2$.

Since these points lie on the conic (1), we have

$$\frac{l}{r_1} = 1 + e\cos\alpha \quad \text{and} \quad \frac{l}{r_2} = 1 + e\cos(\pi + \alpha) = 1 - e\cos\alpha$$

$$\therefore \qquad \frac{l}{r_1} + \frac{l}{r_2} = 2$$

or
$$\frac{1}{r_1} + \frac{1}{r_2} = \frac{2}{l}, \qquad \text{where } l \text{ is the semi-latus rectum.}$$

Hence in any conic the semi latus rectum is a harmonic mean between the segments of any focal chord.

Example 2. *In any conic, prove that, the sum of the reciprocals of two perpendicular focal chords is constant.* (M.D.U. 2007)

Sol. Taking the focus S as pole and the axis as the initial line, we have the equation of the conic as

$$\frac{l}{r} = 1 + e\cos\theta \qquad ...(1)$$

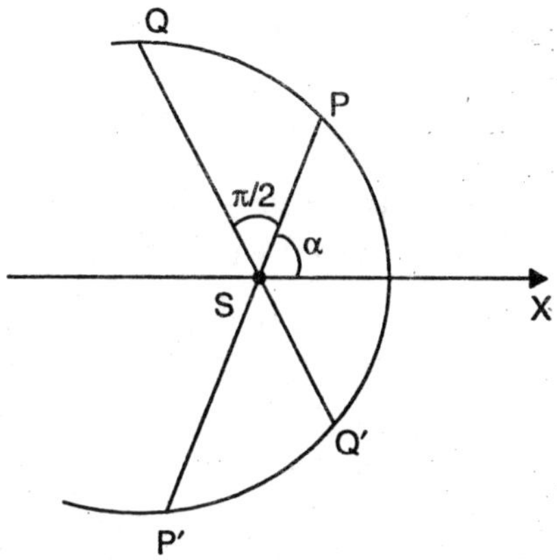

Let PSP′ and QSQ′ be two focal chords perpendicular to each other.

Let α be the vectorial angle of P.

Then the vectorial angles of Q, P′, Q′ are $\alpha + \frac{\pi}{2}$, $\alpha + \pi$, $\alpha + \frac{3\pi}{2}$ respectively.

Let SP = r_1, SP′ = r_2, SQ = r_3, SQ′ = r_4.

Since these points lie on the conic (1), we have

$$\frac{l}{r_1} = 1 + e\cos\alpha, \quad \frac{l}{r_2} = 1 + e\cos(\pi + \alpha) = 1 - e\cos\alpha$$

$$\frac{l}{r_3} = 1 + e\cos\left(\frac{\pi}{2} + \alpha\right) = 1 - e\sin\alpha$$

and
$$\frac{l}{r_4} = 1 + e\cos\left(\frac{3\pi}{2} + \alpha\right) = 1 + e\sin\alpha$$

Now, $$PP' = r_1 + r_2 = \frac{l}{1 + e\cos\alpha} + \frac{l}{1 - e\cos\alpha} = \frac{2l}{1 - e^2\cos^2\alpha}$$

and
$$QQ' = r_3 + r_4 = \frac{l}{1 - e\sin\alpha} + \frac{l}{1 + e\sin\alpha} = \frac{2l}{1 - e^2\sin^2\alpha}$$

Hence, $$\frac{1}{PP'} + \frac{1}{QQ'} = \frac{1 - e^2\cos^2\alpha}{2l} + \frac{1 - e^2\sin^2\alpha}{2l}$$

$$= \frac{2 - e^2(\cos^2\alpha + \sin^2\alpha)}{2l} = \frac{2 - e^2}{2l}$$

which is constant.

Example 3. *A circle passing through the focus of a conic whose latus rectum is 2l, meets the conic in four points whose distances from the focus are r_1, r_2, r_3, r_4, prove that*

$$\frac{1}{r_1} + \frac{1}{r_2} + \frac{1}{r_3} + \frac{1}{r_4} = \frac{2}{l}. \qquad \text{(M.D.U. 2003)}$$

Sol. Taking one of the foci S as the pole and axis as the initial line, then the equation of the conic is

$$\frac{l}{r} = 1 + e \cos \theta \quad \text{...(1)}$$

The equation of a circle of given radius passing through the focus S may be taken as

$$r = a \cos \theta + b \sin \theta \quad \text{...(2)}$$

which is of the form $r = \sqrt{a^2 + b^2} \cos (\theta - \alpha)$ whose radius $= \frac{1}{2}\sqrt{a^2 + b^2}$.

The focal distances of the common points of (1) and (2) *i.e.*, r_1, r_2, r_3, r_4 are the roots of the biquadratic which would be obtained by eliminating θ from (1) and (2).

From (1), $\cos \theta = \dfrac{l - r}{re}$ ...(3)

From (2), $(r - a \cos \theta)^2 = b^2 \sin^2 \theta = b^2(1 - \cos^2 \theta)$

$\therefore$ From (3), we have $\left[r - \dfrac{a(l-r)}{er}\right]^2 = b^2\left[1 - \left(\dfrac{l-r}{er}\right)^2\right]$

or $(er^2 + ar - al)^2 = b^2[e^2r^2 - (l^2 + r^2 - 2rl)] = b^2[(e^2 - 1)r^2 + 2lr - l^2]$

or $e^2r^4 + 2aer^3 + (a^2 - 2ael - b^2e^2 + b^2)r^2 - 2(a^2l + b^2l)r^2 + a^2l^2 + b^2l^2 = 0$ ...(4)

But $\dfrac{1}{r_1} + \dfrac{1}{r_2} + \dfrac{1}{r_3} + \dfrac{1}{r_4} = \dfrac{\Sigma r_2 r_3 r_4}{r_1 r_2 r_3 r_4}$

$$= \frac{-\text{ coeff. of } r/\text{coeff. of } r^4}{\text{constant term/coeff. of } r^4} \quad | \text{ From (4)}$$

$$= -\frac{\text{coeff. of } r}{\text{constant term}} = -\frac{-2l(a^2 + b^2)}{l^2(a^2 + b^2)} = \frac{2}{l}.$$

Example 4. *Trace the curve* $\dfrac{16}{r} = 4 + \sqrt{3} \cos \theta + 3 \sin \theta$.

Sol. The given curve is $\dfrac{16}{r} = 4 + \sqrt{3} \cos \theta + 3 \sin \theta$ which can be written as

$$\frac{4}{r} = 1 + \frac{\sqrt{3}}{4}\cos\theta + \frac{3}{4}\sin\theta = 1 + \frac{\sqrt{3}}{2}\left[\frac{1}{2}\cos\theta + \frac{\sqrt{3}}{2}\sin\theta\right]$$

$$= 1 + \frac{\sqrt{3}}{2}\left[\cos\frac{\pi}{3}\cos\theta + \sin\frac{\pi}{3}\sin\theta\right]$$

$$= 1 + \frac{\sqrt{3}}{2}\cos\left(\theta - \frac{\pi}{3}\right) \quad \text{...(1)}$$

which is of the form $\dfrac{l}{r} = 1 + e \cos (\theta - \alpha)$

$\therefore$ One of the foci is at the pole.

From (1), eccentricity of the conic is $\dfrac{\sqrt{3}}{2}$, semi latus rectum is 4 and the axis makes angle $\alpha = 60°$ with the initial line.

As the eccentricity $e = \dfrac{\sqrt{3}}{2}$ is less than one, the curve represents an ellipse.

$\therefore \quad l = \dfrac{b^2}{a} = a(1-e^2)$

$\therefore \quad 4 = a\left(1-\dfrac{3}{4}\right) \qquad \therefore a = 16$

and $\quad b^2 = al = 16 \times 4 = 64 \qquad \therefore b = 8$

Let S′ be the second focus

$\therefore$ Distance between the foci $SS' = 2ae = 2 \times 16 \times \dfrac{\sqrt{3}}{2} = 16\sqrt{3}$.

If the vertex near to S is A, then $SA = 16 - 8\sqrt{3} = 8[2-\sqrt{3}]$

The rough figure is as shown.

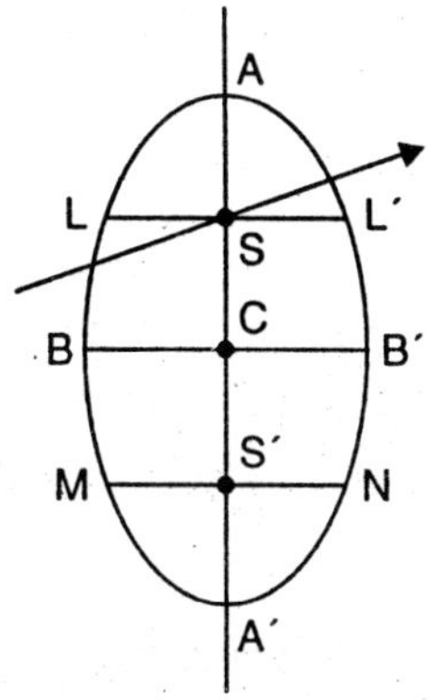

EXERCISE 1

1. Trace the conics

(a) $\dfrac{1}{r} = 1 + \cos\theta + \sqrt{3}\sin\theta$ (b) $\dfrac{10}{r} = 5 + 3\cos\theta + 4\sin\theta$.

2. In any conic prove that :
The sum of the reciprocals of the segments of any focal chord is constant.

3. If PSP′ and QSQ′ be two perpendicular focal chords of a conic, prove that : $\dfrac{1}{PS\,.\,SP'} + \dfrac{1}{QS\,.\,SQ'}$ is constant.

4. A circle of given radius passing through the focus of a given conic intersect it in A, B, C, D. Show that : SA . SB . SC . SD = constant.

5. Prove that the perpendicular focal chords of a rectangular hyperbola are equal.

6. A chord PQ of a conic whose eccentricity is e and semi latus rectum l, subtends a right angle at the focus S ; prove that :

$$\left(\frac{1}{SP}-\frac{1}{l}\right)^2 + \left(\frac{1}{SQ}-\frac{1}{l}\right)^2 = \frac{e^2}{l^2}.$$

7. Prove that perpendicular focal chords of a rectangular hyperbola are equal.

8. If PSP′ is a focal chord of a conic, prove that the locus of its middle point is a conic of the same kind as the original conic.

TANGENT AND NORMAL

5. *To find the polar equation of the straight line through two given points on a conic, and to find the equation of the tangent at any point.*

Let the vectorial angles of the two points P and Q be $(\alpha-\beta)$ and $(\alpha+\beta)$ respectively.

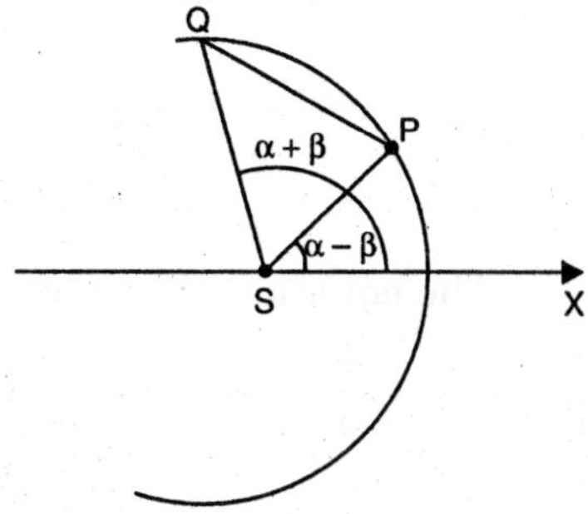

Let the equation of the conic be

$$\frac{l}{r} = 1 + e\cos\theta \qquad ...(1)$$

The general equation of a line not through the pole can be written as

$$\frac{l}{r} = A\cos\theta + B\sin\theta \qquad ...(2)$$

Since the points P and Q are on the conic, we have

$$\frac{l}{SP} = 1 + e\cos(\alpha - \beta) \qquad ...(3)$$

and
$$\frac{l}{SQ} = 1 + e\cos(\alpha + \beta) \qquad ...(4)$$

If the line (2) passes through the points P and Q, whose vectorial angles are $\alpha - \beta$ and $\alpha + \beta$, then we have

$$\frac{l}{SP} = A\cos(\alpha - \beta) + B\sin(\alpha - \beta) \qquad ...(5)$$

and
$$\frac{l}{SQ} = A\cos(\alpha + \beta) + B\sin(\alpha + \beta) \qquad ...(6)$$

From (3), (4), (5), (6), we get

$$1 + e\cos(\alpha - \beta) = A\cos(\alpha - \beta) + B\sin(\alpha - \beta)$$

and
$$1 + e\cos(\alpha + \beta) = A\cos(\alpha + \beta) + B\sin(\alpha + \beta)$$

These equation determine A and B.

We write these equation as $(A - e)\cos(\alpha - \beta) + B\sin(\alpha - \beta) = 1$

$$(A - e)\cos(\alpha + \beta) + B\sin(\alpha + \beta) = 1$$

Solving these, we have

$$\frac{A - e}{\sin(\alpha + \beta) - \sin(\alpha - \beta)} = \frac{B}{\cos(\alpha + \beta) - \cos(\alpha - \beta)} = \frac{1}{\sin 2\beta}$$

or
$$\frac{A - e}{\cos\alpha\sin\beta} = \frac{B}{\sin\alpha\sin\beta} = \frac{1}{\sin\beta\cos\beta}$$

$\therefore$
$$A - e = \cos\alpha\sec\beta,\ B = \sin\alpha\sec\beta$$

$\therefore$ The equation of the chord PQ is $\frac{l}{r} = (e + \cos\alpha\sec\beta)\cos\theta + \sin\alpha\sec\beta\sin\theta$

$$= e\cos\theta + (\cos\alpha\cos\theta + \sin\alpha\sin\theta)\sec\beta$$

i.e.,
$$\frac{l}{r} = e\cos\theta + \sec\beta\cos(\theta - \alpha) \qquad ...(7)$$

Equation of Tangent

To find the equation of the tangent, we put $\beta = 0$ in (7), the points P and Q coincide and the chord PQ becomes tangent at P whose vectorial angle becomes α and the equation of the tangent at P is $\frac{l}{r} = e\cos\theta + \cos(\theta - \alpha)$.

6. *To find the equation of the Normal at the point P whose vectorial angle is α.*

The equation of the tangent to the conic $\frac{l}{r} = 1 + e\cos\theta$ at the point α is

$$\frac{l}{r} = e\cos\theta + \cos(\theta - \alpha) \qquad ...(1)$$

The equation of any line perpendicular to the tangent is of the form

$$\frac{K}{r} = e\cos\left(\theta + \frac{\pi}{2}\right) + \cos\left(\theta + \frac{\pi}{2} - \alpha\right) \quad \text{or} \quad \frac{K}{r} = -e\sin\theta - \sin(\theta - \alpha)$$

This will be the equation of the normal if K is so chosen that it passes through the point whose polar co-ordinates are $\left(\frac{l}{1+e\cos\alpha}, \alpha\right)$.

That is $\quad K\left(\frac{1+e\cos\alpha}{l}\right) = -e\sin\theta \quad$ or $\quad K = \frac{-le\sin\alpha}{1+e\cos\alpha}$

Hence the equation of the normal is $\frac{le\sin\alpha}{1+e\cos\alpha}\cdot\frac{1}{r} = e\sin\theta + \sin(\theta-\alpha)$.

Example 1. *Find the condition that the line* $\frac{l}{r} = A\cos\theta + B\sin\theta$ *may be a tangent to the conic* $\frac{l}{r} = 1 + e\cos\theta$.

Sol. Let the line $\quad \frac{l}{r} = A\cos\theta + B\sin\theta \qquad$ ...(1)

be a tangent to the conic $\frac{l}{r} = 1 + e\cos\theta$ at the point P whose vectorial angle is α.

Then the tangent to the conic at α is

$$\frac{l}{r} = e\cos\theta + \cos(\theta-\alpha)$$
$$= e\cos\theta + \cos\theta\cos\alpha + \sin\theta\sin\alpha$$

or
$$\frac{l}{r} = \cos\theta\,(e+\cos\alpha) + \sin\theta\sin\alpha \qquad ...(2)$$

Identifying (1) and (2), we have $A = e + \cos\alpha$ and $B = \sin\alpha$

Eliminating α by squaring and adding we have $(A-e)^2 + B^2 = 1$, which is the required condition.

Example 2. *Find the point of intersection of the tangents at 'α' and 'β' to the conic*

$$\frac{l}{r} = 1 + e\cos\theta.$$

Sol. The equation of tangents at 'α' and 'β' are

$$\frac{l}{r} = e\cos\theta + \cos(\theta-\alpha) \qquad ...(1)$$

and
$$\frac{l}{r} = e\cos\theta + \cos(\theta-\beta) \qquad ...(2)$$

To find the value of r and θ, we have by subtraction,

$$\cos(\theta-\alpha) = \cos(\theta-\beta) \quad \therefore \quad \theta-\alpha = \pm(\theta-\beta)$$

Taking – ve sign, $\quad \therefore \quad \alpha \neq \beta$

$$\theta = \frac{\alpha+\beta}{2} \qquad ...(3)$$

Hence if T be the point of intersection of the tangents at P and Q of the conic, ST bisects $\angle PSQ$.

Substituting the value of θ in (2), we get $\frac{l}{r} = e\cos\frac{\alpha+\beta}{2} + \cos\frac{\alpha-\beta}{2}$,

or
$$r = \frac{l}{\cos\frac{\alpha-\beta}{2} + e\cos\frac{\alpha+\beta}{2}} \qquad ...(4)$$

The required co-ordinates are given by (3) and (4).

Example 3. *Prove that the slope of the tangent to the conic $\frac{l}{r} = 1 + e \cos \theta$ at the point α is $\frac{-(\cos \alpha + e)}{\sin \alpha}$.*

Sol. The equation of the tangent at α is $\frac{l}{r} = e \cos \theta + \cos (\theta - \alpha)$

$$= e \cos \theta + \cos \theta \cos \alpha + \sin \theta \sin \alpha$$

or $$l = (\cos \alpha + e) . r \cos \theta + (\sin \alpha) r \sin \theta$$

Changing into cartesian equation $(\cos \alpha + e)x + \sin \alpha . y = l$

$$\therefore \quad \text{Slope} = -\frac{(\cos \alpha + e)}{\sin \alpha}.$$

Example 4. *PSP′ is a focal chord of an ellipse. Prove that the angle between the tangents at P and P′ is $\tan^{-1}\left(\frac{2e \sin \alpha}{1-e^2}\right)$, where α is the angle between the chord and the major axis.*

Sol. Let the equation of the ellipse be $\frac{l}{r} = 1 + e \cos \theta$

Then the vectorial angle of P is α and that of P′ is $\pi + \alpha$

Now the slope of the tangent at P

$$m_1 = -\frac{\cos \alpha + e}{\sin \alpha} \quad \text{...(1)}$$

and slope of the tangent at P′

$$m_2 = -\frac{\cos (\pi + \alpha) + e}{\sin (\pi + \alpha)} = -\frac{\cos \alpha - e}{\sin \alpha} \quad \text{...(2)}$$

| Changing α into $\pi + \alpha$ in (1)

If θ be the angle between the tangents, then

$$\tan \theta = \frac{m_2 - m_1}{1 + m_1 m_2} = \frac{-\left(\frac{\cos \alpha - e}{\sin \alpha}\right) + \frac{\cos \alpha + e}{\sin \alpha}}{1 + \frac{\cos^2 \alpha - e^2}{\sin^2 \alpha}} = \frac{2e \sin \alpha}{1 - e^2}$$

$$\therefore \quad \theta = \tan^{-1}\left(\frac{2e \sin \alpha}{1 - e^2}\right).$$

Example 5. *Show that the two conics $\frac{l_1}{r} = 1 + e_1 \cos \theta$ and $\frac{l_2}{r} = 1 + e_2 \cos (\theta - \alpha)$ will touch each other if $l_1^2(1 - e_2^2) + l_2^2(1 - e_1^2) = 2l_1 l_2(1 - e_1 e_2 \cos \alpha)$.*

Sol. Let the two given conics touch each other at the point whose vectorial angle is β. Then, the equations of the tangents at the common point β to the given conics are

$$\frac{l_1}{r} = e_1 \cos \theta + \cos (\theta - \beta) \quad \text{...(1)}$$

and $$\frac{l_2}{r} = e_2 \cos (\theta - \alpha) + \cos (\theta - \beta) \quad \text{...(2)}$$

or from equations (1) and (2), we have

$$\frac{l_1}{r} = (e_1 + \cos\beta)\cos\theta + \sin\theta\sin\beta \qquad ...(3)$$

and

$$\frac{l_2}{r} = (e_2\cos\alpha + \cos\beta)\cos\theta + (\sin\beta + e_2\sin\alpha)\sin\theta \qquad ...(4)$$

Since equations (3) and (4) are identical, therefore comparing the coefficients of like terms, we get

$$\frac{l_1}{l_2} = \frac{e_1 + \cos\beta}{e_2\cos\alpha + \cos\beta} = \frac{\sin\beta}{e_2\sin\alpha + \sin\beta}$$

or $$(l_1 - l_2)\cos\beta = e_1 l_2 - l_1 e_2\cos\alpha \qquad ...(5)$$

and $$(l_1 - l_2)\sin\beta = -\,l_1 e_2\sin\alpha \qquad ...(6)$$

Squaring (5) and (6) and adding, we get

$$(l_1 - l_2)^2 = e_1^2 l_2^2 + e_2^2 l_1^2 - 2e_1 e_2 l_1 l_2\cos\alpha$$

or $l_1^2(1 - e_1^2) + l_2^2(1 - e_1^2) = 2l_1 l_2(1 - e_1 e_2\cos\alpha)$, which is the required result.

Example 6. *A chord of a conic subtends a constant angle at a focus of the conic. Show that the chord touches another conic.*

Sol. Let the equation of the conic be

$$\frac{l}{r} = 1 + e\cos\theta \qquad ...(1)$$

Let PQ be a chord of the conic (1). Let it subtend a constant angle 2β at the focus. Let the vectorial angles of the points P and Q be $\alpha - \beta$ and $\alpha + \beta$ respectively.

∴ The equation of the chord PQ is $\frac{l}{r} = e\cos\theta + \sec\beta\cos(\theta - \alpha)$

$$= e\cos\theta + \frac{\cos(\theta - \alpha)}{\cos\beta}$$

or $$\frac{(l\cos\beta)}{r} = e\cos\beta\cos\theta + \cos(\theta - \alpha) \qquad ...(2)$$

Clearly the straight line (2) is the equation of the tangent to the conic

$$\frac{l\cos\beta}{r} = 1 + e\cos\beta\cos\theta \qquad ...(3)$$

at the point whose vectorial angle is α.

∴ The chord PQ touches the conic (3).

Hence the result.

Example 7. *Prove that two points on the conic* $\frac{l}{r} = 1 + e\cos\theta$ *whose vectorial angles are* α *and* β *respectively will be the extremities of a diameter if* $\frac{e+1}{e-1} = \tan\frac{\alpha}{2}\tan\frac{\beta}{2}$.

Sol. The equation of the given conic is

$$\frac{l}{r} = 1 + e\cos\theta \qquad ...(1)$$

Let P and P′ be the extremities of the diameter of the conics whose vectorial angles are α and β respectively.

$\therefore$ The equation of the tangent at α is

$$\frac{l}{r} = e\cos\theta + \cos(\theta - \alpha) = e\cos\theta + \cos\theta\cos\alpha + \sin\theta\sin\alpha$$

or $$l = (\cos\alpha + e).r\cos\theta + (\sin\alpha)\, r\sin\theta$$

Changing into cartesian equation, we have $(\cos\alpha + e)x + \sin\alpha y = l$

$\therefore$ Slope of the tangent at P

$$m_1 = -\frac{\cos\alpha + e}{\sin\alpha} \qquad \text{...(2)}$$

Similarly, the slope of the tangent at P′

$$m_2 = -\frac{\cos\beta + e}{\sin\beta} \qquad \text{...(3)}$$

If the line PP′ is a diameter, then the tangents at P and P′ are parallel and hence their slopes are equal.

$\therefore$ From (2) and (3), $m_1 = m_2$

i.e., $$-\frac{\cos\alpha + e}{\sin\alpha} = -\frac{\cos\beta + e}{\sin\beta}$$

or $$\cos\alpha\sin\beta + e\sin\beta = \sin\alpha\cos\beta + e\sin\alpha$$

or $$e(\sin\beta - \sin\alpha) = \sin(\alpha - \beta)$$

or $$2e\cos\frac{\beta+\alpha}{2}\sin\frac{\beta-\alpha}{2} = 2\sin\frac{\alpha-\beta}{2}\cos\frac{\alpha-\beta}{2}$$

or $$e\cos\frac{\beta+\alpha}{2} = -\cos\frac{\alpha-\beta}{2}$$

or $$\frac{e}{1} = \frac{-\cos\frac{\alpha-\beta}{2}}{\cos\frac{\beta+\alpha}{2}}$$

Using componendo and dividendo, we get

$$\frac{e+1}{e-1} = \frac{-\cos\frac{\alpha-\beta}{2} + \cos\frac{\beta+\alpha}{2}}{-\cos\frac{\alpha-\beta}{2} - \cos\frac{\beta+\alpha}{2}} = \frac{\cos\frac{\alpha-\beta}{2} - \cos\frac{\beta+\alpha}{2}}{\cos\frac{\alpha-\beta}{2} + \cos\frac{\beta+\alpha}{2}} = \frac{2\sin\frac{\alpha}{2}\sin\frac{\beta}{2}}{2\cos\frac{\alpha}{2}\cos\frac{\beta}{2}}$$

$$\therefore \quad \frac{e+1}{e-1} = \tan\frac{\alpha}{2}\tan\frac{\beta}{2}$$

which is the required result.

7. A few properties of the general conic

Property 1. *The exterior angle between any two tangents to a parabola is equal to half the difference of the vertical angles of their points of contact.*

Sol. Let the equation of the parabola be $\frac{l}{r} = 1 + \cos\theta$

Let the vectorial angles of the points of contact of the two tangents be α and β ($\beta > \alpha$). Now, the slopes of the two tangents are

$$m_1 = -\frac{\cos\alpha + 1}{\sin\alpha} = -\frac{2\cos^2\frac{\alpha}{2}}{2\sin\frac{\alpha}{2}\cos\frac{\alpha}{2}} = -\cot\frac{\alpha}{2}.$$

and
$$m_2 = -\frac{\cos\beta + 1}{\sin\beta} = -\cot\frac{\beta}{2}$$

Let, ϕ be the exterior angle between the tangents measured positively from tangent at α to tangent at β, we have

$$\tan\phi = \frac{m_2 - m_1}{1 + m_1 m_2} = \frac{\cot\frac{\alpha}{2} - \cot\frac{\beta}{2}}{1 + \cot\frac{\alpha}{2}\cot\frac{\beta}{2}}$$

$$= \frac{\tan\frac{\beta}{2} - \tan\frac{\alpha}{2}}{1 + \tan\frac{\alpha}{2}\tan\frac{\beta}{2}} = \tan\left(\frac{\beta - \alpha}{2}\right)$$

$$\therefore \quad \phi = \frac{\beta - \alpha}{2}$$

Hence the required result.

Property 2. *The tangent at any point P of a conic meets a directrix in K. If S be the corresponding focus, show that angle PSK is a right angle.*

Or

The portion of a tangent to a conic intercepted between the directrix and the curve subtends a right angle at the focus.

Sol. Let the equation of the conic whose focus S as pole be

$$\frac{l}{r} = 1 + e\cos\theta \qquad ...(1)$$

Let α be the vectorial angle of any point P on the conic.

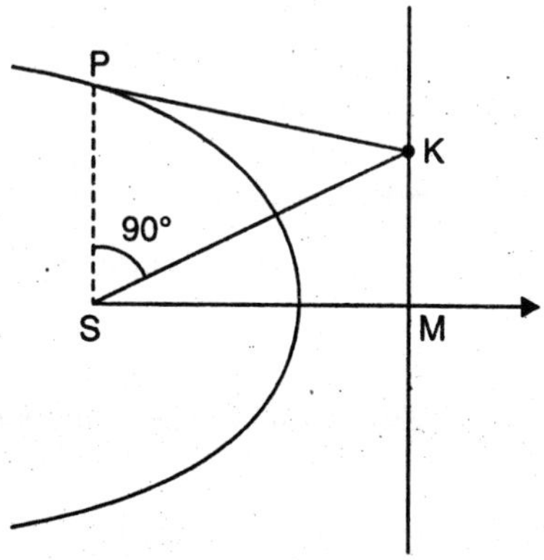

$\therefore \quad \angle MSP = \alpha$

The equation of the tangent at P is

$$\frac{l}{r} = e\cos\theta + \cos(\theta - \alpha) \qquad ...(2)$$

Also, the equation of the directrix is

$$\frac{l}{r} = e\cos\theta \qquad ...(3)$$

The lines (2) and (3) intersect at K.

Equating (2) and (3), the vectorial angle θ is given by

$$e\cos\theta + \cos(\theta - \alpha) = e\cos\theta$$

or
$$\cos(\theta - \alpha) = 0$$

$$\therefore \quad \theta - \alpha = \pm\frac{\pi}{2}$$

Now $\angle KSP = \angle MSP - \angle MSK$

$$= \alpha - \left(\alpha \pm \frac{\pi}{2}\right) = \mp \frac{\pi}{2}.$$

Property 3. *The tangents at the extremities of any focal chord of a conic intersect on the corresponding directrix.*

Sol. Let α be the vectorial angle of the extremity of P of the focal chord PSP′ of the conic $\frac{l}{r} = 1 + e \cos \theta$.

Then the vectorial angle of P′ is $\pi + \alpha$.

Now, the equation of tangents at P and P′ are

$$\frac{l}{r} = e \cos \theta + \cos (\theta - \alpha) \quad ...(1)$$

and
$$\frac{l}{r} = e \cos \theta + \cos (\theta - \alpha - \pi) \quad ...(2)$$

Let $K(r_1, \theta_1)$ be the point of intersection of (1) and (2).

$$\therefore \quad \frac{l}{r_1} = e \cos \theta_1 + \cos (\theta_1 - \alpha) \quad ...(3)$$

and
$$\frac{l}{r_1} = e \cos \theta_1 + \cos (\theta_1 - \alpha - \pi)$$

i.e.,
$$\frac{l}{r_1} = e \cos \theta_1 - \cos (\theta_1 - \alpha) \quad ...(4)$$

Adding (3) and (4), we have

$$\frac{2l}{r_1} = 2e \cos \theta_1 \quad \text{or} \quad \frac{l}{r_1} = e \cos \theta_1$$

$\therefore$ Locus of (r_1, θ_1) is $\frac{l}{r} = e \cos \theta$, which is the equation of the directrix.

Property 4. *The feet of the perpendiculars drawn from a focus to the tangent of a conic lie on a circle.*

Sol. Let α be the vectorial angle of any point P on the conic $\frac{l}{r} = 1 + e \cos \theta$.

then, tangent to the conic at P is

$$\frac{l}{r} = e \cos \theta + \cos (\theta - \alpha) \quad ...(1)$$

Equation of a line perpendicular to it passing through the pole is

$$0 = e \cos \left(\theta + \frac{\pi}{2}\right) + \cos \left(\theta + \frac{\pi}{2} - \alpha\right)$$

$\left(\text{By replacing } l \text{ to zero and changing } \theta \text{ into } \theta + \frac{\pi}{2}\right)$

or
$$0 = e \sin \theta + \sin (\theta - \alpha)$$

i.e.,
$$\sin (\theta - \alpha) = - e \sin \theta \quad ...(2)$$

From (1), $\cos(\theta - \alpha) = \dfrac{l}{r} - e\cos\theta$ ...(3)

To find the locus of the point of intersection of (1) and (2), we eliminate α from (2) and (3).

Squaring (2) and (3), and adding we have $\left(\dfrac{l}{r} - e\cos\theta\right)^2 + e^2\sin^2\theta = 1$

or $$\frac{l^2}{r^2} - \frac{2el}{r}\cos\theta + e^2\cos^2\theta + e^2\sin^2\theta = 1$$

or $$(e^2 - 1)r^2 - 2elr\cos\theta + l^2 = 0 \text{ which is a circle.}$$

Cor. When $e = 1$, the conic becomes the parabola.

$$\frac{l}{r} = 1 + \cos\theta \text{ and the locus becomes } 2r\cos\theta = l$$

i.e., $$\frac{l}{r} = 2\cos\theta \text{ which is the tangent to the parabola at } \theta = 0. \text{ i.e., at the vertex.}$$

Property 5. *The locus of the intersection of perpendicular tangents to a conic is a circle (called the director circle).*

Sol. Let the equation of the conic be

$$\frac{l}{r} = 1 + e\cos\theta \quad \text{...(1)}$$

Let (r_1, θ_1) be the co-ordinates of the point of intersection of a pair of perpendicular tangents and α and β be the vectorial angles of the points of contact of the tangents

$\therefore$ Tangents at α and β are

$$\frac{l}{r} = \cos(\theta - \alpha) + e\cos\theta \quad \text{...(2)}$$

and $$\frac{l}{r} = \cos(\theta - \beta) + e\cos\theta \quad \text{...(3)}$$

The point (r_1, θ_1) is a point on these tangents

i.e., $$\frac{l}{r_1} = \cos(\theta_1 - \alpha) + e\cos\theta_1 \quad \text{...(4)}$$

$$\frac{l}{r_1} = \cos(\theta_1 - \beta) + e\cos\theta_1 \quad \text{...(5)}$$

Solving (4) and (5), we have

$$\theta_1 - \alpha = \pm(\theta_1 - \beta)$$

$\therefore$ $$\theta_1 = \frac{1}{2}(\alpha + \beta) \quad \text{...(6)} \mid \because \alpha \neq \beta$$

Substituting this value of θ in (4), we get

$$\frac{l}{r_1} = e\cos\frac{\alpha + \beta}{2} + \cos\frac{\beta - \alpha}{2} \quad \text{...(7)}$$

Slopes of the tangents (1) and (2) are

$$-\left(\frac{\cos\alpha + e}{\sin\alpha}\right) \text{ and } -\left(\frac{\cos\beta + e}{\sin\beta}\right) \text{ respectively.}$$

Since the tangents are at right angles

$$\therefore \quad -\left(\frac{\cos\alpha + e}{\sin\alpha}\right) \times -\left(\frac{\cos\beta + e}{\sin\beta}\right) = -1$$

or $\quad e^2 + e(\cos\alpha + \cos\beta) + \cos(\alpha - \beta) = 0$

or $\quad e^2 + 2e\cos\frac{\alpha+\beta}{2}\cos\frac{\alpha-\beta}{2} + 2\cos^2\frac{\alpha-\beta}{2} - 1 = 0 \quad \text{...(8)}$

Substituting for $\frac{\alpha+\beta}{2}$ and $\cos\frac{\alpha-\beta}{2}$ from (6) and (7) in (8), we get

$$e^2 + 2e\cos\theta_1\left(\frac{l}{r_1} - e\cos\theta_1\right) + 2\left(\frac{l}{r_1} - e\cos\theta_1\right)^2 - 1 = 0$$

or $\quad e^2 - 1 + \frac{2el\cos\theta_1}{r_1} - 2e^2\cos^2\theta_1 + \frac{2l^2}{r_1^{\,2}} - \frac{4le\cos\theta_1}{r_1} + 2e^2\cos^2\theta_1 = 0$

or $\quad (e^2 - 1) - \frac{2el\cos\theta_1}{r_1} + \frac{2l^2}{r_1^{\,2}} = 0$

or $\quad (1 - e^2)\,r_1^{\,2} + 2elr_1\cos\theta_1 - 2l^2 = 0$

Hence locus of (r_1, θ_1) is $(1 - e^2)r^2 + 2elr\cos\theta - 2l^2 = 0$ which is a circle with (ae, π) as its centre.

8. *Equation of the chord of contact of the tangents drawn from a point (r_1, θ_1) to the conic*

$$\frac{l}{r} = 1 + e\cos\theta.$$

The equation of the chord joining the two points of contact of the tangents whose vectorial angles are α and β is

$$\frac{l}{r} = \sec\frac{\alpha-\beta}{2}\cos\left(\theta - \frac{\alpha+\beta}{2}\right) + e\cos\theta \quad \text{...(1)}$$

The point of intersection of two tangents is given by

$$\theta = \frac{1}{2}(\alpha+\beta),\ \frac{l}{r} - e\cos\theta = \cos\frac{\alpha-\beta}{2}$$

Since tangents intersect in (r_1, θ_1)

$$\therefore \quad \frac{1}{2}(\alpha+\beta) = \theta_1,\ \cos\frac{\alpha-\beta}{2} = \frac{l}{r_1} - e\cos\theta_1$$

To eliminate α and β, substituting the above values in (1), we get

$$\frac{l}{r} = \frac{\cos(\theta-\theta_1)}{\frac{l}{r_1} - e\cos\theta_1} + e\cos\theta$$

or $\quad \frac{l}{r} - e\cos\theta = \frac{\cos(\theta-\theta_1)}{\frac{l}{r_1} - e\cos\theta_1} \quad$ or $\quad \left(\frac{l}{r} - e\cos\theta\right)\left(\frac{l}{r_1} - e\cos\theta_1\right) = \cos(\theta - \theta_1)$

which is the equation of the chord of contact.

9. *To find the equation of the polar of any point* (r_1, θ_1) *with respect to the conic*

$$\frac{l}{r} = 1 + e \cos\theta.$$

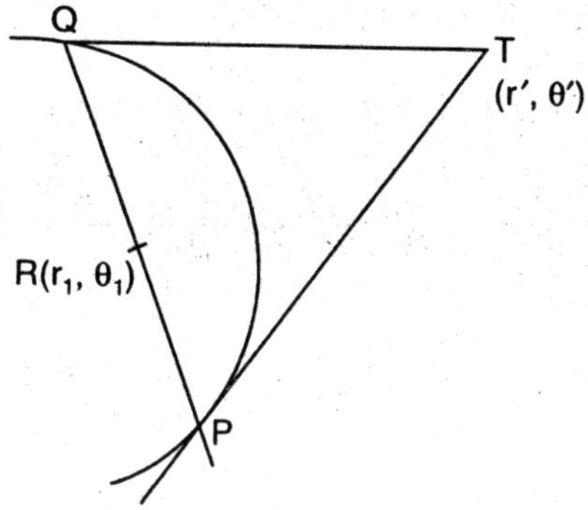

Let R be the point (r_1, θ_1).

Let any chord through $R(r_1, \theta_1)$ meet the conic in P and Q. Let the tangent at P and Q intersects in $T(r', \theta')$.

Therefore PQ is the chord of contact of tangents drawn from $T(r', \theta')$.

Hence the equation of chord of contact of tangents drawn from the point (r', θ') to the conic is

$$\left(\frac{l}{r} - e\cos\theta\right)\left(\frac{l}{r'} - e\cos\theta'\right) = \cos(\theta - \theta')$$

As this line passes through the point $R(r_1, \theta_1)$

$$\therefore \quad \left(\frac{l}{r_1} - e\cos\theta\right)\left(\frac{l}{r'} - e\cos\theta'\right) = \cos(\theta_1 - \theta')$$

Hence the locus of (r', θ') is $\left(\frac{l}{r_1} - e\cos\theta\right)\left(\frac{l}{r} - e\cos\theta\right) = \cos(\theta_1 - \theta) = \cos(\theta - \theta_1)$ which is the required equation of the polar.

EXERCISE 2

1. Show that the conic $l\sqrt{3} = r(\sqrt{3} + \cos\theta)$ and $l\sqrt{3} = 2r\left\{\sqrt{3} + \cos\left(\theta + \frac{\pi}{3}\right)\right\}$ touch where $\theta = \frac{\pi}{2}$.
2. A point moves so that, the sum of its distances from two fixed points S, S′ is constant and equal to 2α.

   Show that P lies on the conic $\frac{a(1 - e^2)}{r} = 1 - e\cos\theta$ referred to S as pole and SS′ as initial line, SS′ being equal to $2ae$.
3. If PSQ and PS′R be two chords of an ellipse through the foci S and S′, prove that $\frac{SP}{SQ} + \frac{S'P}{S'R}$ is independent of the position of P.
4. Prove that the locus of the middle points of a system of focal chords of a conic is a conic which is a parabola, ellipse or hyperbola according as the original conic is a parabola, ellipse or hyperbola.
5. If two conic have a common focus, show that two of their common chords will pass through the point of intersection of their directrices.
6. Show that the equations $\frac{l}{r} = 1 + e\cos\theta$ and $\frac{l}{r} = -1 + e\cos\theta$ represents the same conic.
7. If the tangents at P and Q to a conic intersect in T and the chord PQ meets the directrix in K, prove that angle KST is a right angle.

8. Two equal ellipses of eccentricity e, are placed with their axes at right angles and they have one focus S in common. If PQ be a common tangent, show that the angle PSQ is equal to $2\sin^{-1}\left(\frac{e}{\sqrt{2}}\right)$.

9. A conic touches the conic $\frac{l}{r} = 1 + e\cos\theta$ at α and has the same focus and eccentricity. Prove that the length of its latus rectum is $\frac{2l(1-e^2)}{(e^2+2e\cos\alpha+1)}$.

10. A focal chord P′SP of an ellipse is inclined at an angle α to the major axis. Show that the perpendicular from the focus on the tangent at P makes an angle $\tan^{-1}\left[\frac{\sin\alpha}{e+\cos\alpha}\right]$.

MISCELLANEOUS EXERCISE

1. Trace the conic $\frac{2}{r} = 1 + \cos\theta + \sin\theta$.

2. Find the condition that the line $\frac{l}{r} = A\cos\theta + B\sin\theta$ may touch the conic $\frac{l}{r} = 1 + e\cos(\theta - \alpha)$.

3. If the tangents at P and Q to a conic intersects in T and the chord PQ meets the directrix in K, prove that angle KST is a right angle.

4. Find the polar equation of the auxiliary circle of the conic $\frac{l}{r} = 1 + e\cos\theta$.

5. Prove that the chords of a rectangular hyperbola which subtend a right angle at a focus touch a fixed parabola.

6. If PQ is the chord of contact of the tangents drawn from a point T to a parabola whose focus is S, prove that $ST^2 = SP.SQ$.

7. If the tangents at P and Q of a conic meet at a point T, and S be the focus, prove that

$$\frac{1}{SP.SQ} - \frac{1}{ST^2} = \frac{1}{b^2}\sin^2\left(\frac{1}{2}PSQ\right).$$

8. A variable chord AB of a conic subtends a constant angle 2γ at the focus S. Show that the locus of the intersection of the tangents at A and B is a conic section whose focus is S.

9. A variable chord of a conic subtends a constant angle at a focus of the conic. Show that the chord touches another conic having the same focus.

10. If A, B, C be three points on a parabola and the tangents at these points meet at A′, B′, C′, prove that SA.SB.SC = SA′.SB′.SC′, S being the focus of the parabola.

11. If the normal at L, one of the extremities of the latus rectum of the conic $\frac{l}{r} = 1 + e\cos\theta$, meets the curve again at Q, show that $SQ = \frac{l(1+3e^2+e^4)}{1+e^2-e^4}$.

12. Given the focus and directrix of a conic, show that the polar of a given point with respect to it passes through a fixed point.

13. If the normals at the points whose vectorial angles are α, β, γ on the parabola $\frac{l}{r} = 1 + \cos\theta$ meet in the point (P, ϕ), then prove that $2\phi = \alpha + \beta + \gamma$. *(M.D.U. 2007)*

14. Prove that the locus of the pole of a chord of the conic $\frac{l}{r} = 1 + e\cos\theta$ which subtends a constant angle 2α at the focus is $\frac{l}{r} = \cos\alpha + e\cos\theta$.

Answers

2. $A^2 + B^2 - 2e(A\cos\alpha + B\sin\alpha) + e^2 - 1 = 0$
4. $(e^2 - 1)r^2 - 2elr\cos\theta + l^2 = 0$; $2r\cos\theta = 1$.

Review III: Important Results and Formulae

(from Solid Geometry learnt in Class XII)

1. Co-ordinates. If P(x, y, z) is any point in space, then

x = PL = ⊥ distance of P from YZ plane.
y = PM = ⊥ distance of P from ZX plane.
z = PN = ⊥ distance of P from XY plane.

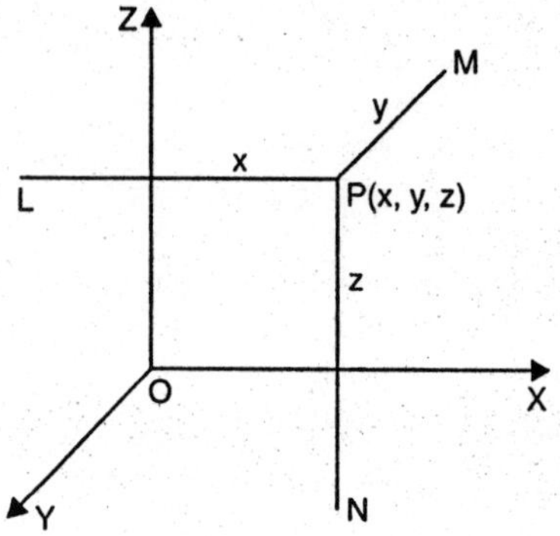

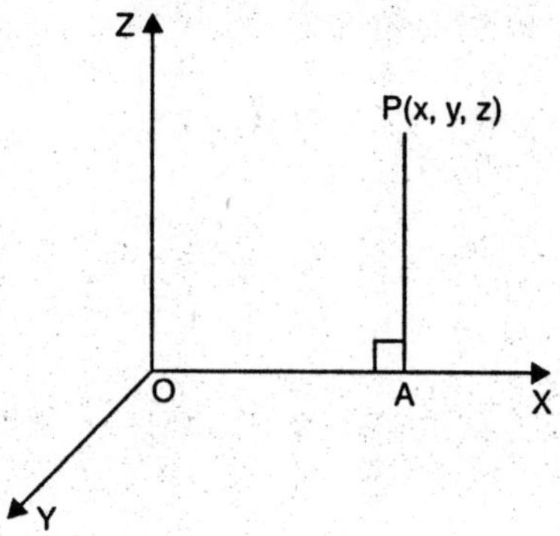

Also, if PA ⊥ OX, then OA = x,
Similarly, if PB ⊥ OY, then OB = y,
and PC ⊥ OZ, then OC = z.

2. Distance Formula. Distance between two points (x_1, y_1, z_1) and (x_2, y_2, z_2), then

$$d = \sqrt{(x_1 - x_2)^2 + (y_1 - y_2)^2 + (z_1 - z_2)^2}.$$

3. Section Formula. If (x, y, z) be the point which divides the join of (x_1, y_1, z_1) and (x_2, y_2, z_2) in the ratio $m_1 : m_2$, then

$$x = \frac{m_1x_2 + m_2x_1}{m_1 + m_2}$$

$$y = \frac{m_1y_2 + m_2y_1}{m_1 + m_2}$$

$$z = \frac{m_1z_2 + m_2z_1}{m_1 + m_2}$$

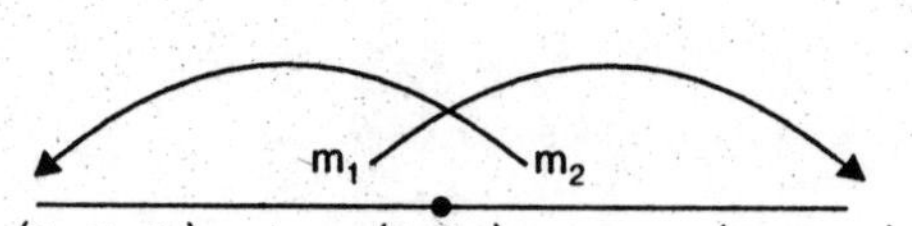

Rule. *Multiply m_1 by the x co-ordinate of the point remote from m_1 and m_2 by the x co-ordinate of the point remote from m_2 as shown by arrows, add these products, and divide the sum by $m_1 + m_2$.*

This gives x co-ordinate of the point of section.

Similarly for the y and z co-ordinates.

Mid-point Formula. The mid-point of the join of (x_1, y_1, z_1) and (x_2, y_2, z_2) is

$$\left(\frac{x_1 + x_2}{2}, \frac{y_1 + y_2}{2}, \frac{z_1 + z_2}{2}\right)$$

i.e., $$\left(\frac{\text{sum of } x \text{ co-ordinates}}{2}, \frac{\text{sum of } y \text{ co-cordinates}}{2}, \frac{\text{sum of } z \text{ co-ordinates}}{2}\right)$$

4. Centroid. The centroid of the Δ whose vertices are (x_1, y_1, z_1), (x_2, y_2, z_2), (x_3, y_3, z_3) is

$$\left(\frac{x_1 + x_2 + x_3}{3}, \frac{y_1 + y_2 + y_3}{3}, \frac{z_1 + z_2 + z_3}{3}\right)$$

i.e., $$\left(\frac{\text{sum of } x \text{ co-ordinates}}{3}, \frac{\text{sum of } y \text{ co-ordinates}}{3}, \frac{\text{sum of } z \text{ co-ordinates}}{3}\right)$$

Centroid of the tetrahedron whose vertices are (x_1, y_1, z_1), (x_2, y_2, z_2), (x_3, y_3, z_3) and (x_4, y_4, z_4) by

$$\left(\frac{x_1 + x_2 + x_3 + x_4}{4}, \frac{y_1 + y_2 + y_3 + y_4}{4}, \frac{z_1 + z_2 + z_3 + z_4}{4}\right)$$

i.e., $$\left(\frac{\text{sum of } x \text{ co-ordinates}}{4}, \frac{\text{sum of } y \text{ co-ordinates}}{4}, \frac{\text{sum of } z \text{ co-ordinates}}{4}\right)$$

Remember. The centroid of a Δ divides each median of the ratio 2 : 1 and the centroid of the tetrahedron divides the join of each vertex to the centroid of the opposite face in the ratio 3 : 1.

5. If l, m, n are the direction-cosines (d.c's) of a line OP and OP $= r$, then the co-ordinates of P are **(lr, mr, nr)**.

6. If l, m, n are the d.c's of a line, then $\mathbf{l^2 + m^2 + n^2 = 1}$.

7. If a, b, c are the direction ratios (or proportional d.c's) of a line, then the actual d.c's

are $$\frac{a}{\sqrt{\Sigma a^2}}, \frac{b}{\sqrt{\Sigma a^2}}, \frac{c}{\sqrt{\Sigma a^2}}, \quad \text{where } \Sigma a^2 = a^2 + b^2 + c^2.$$

Rule. To find the actual d.c's of a line whose proportional d.c's are a, b, c we divide each of these by $\sqrt{a^2 + b^2 + c^2}$, then the resulting ratios $\frac{a}{\sqrt{\Sigma a^2}}, \frac{b}{\sqrt{\Sigma a^2}}, \frac{c}{\sqrt{\Sigma a^2}}$ are the actual d.c's.

8. Direction cosines of the join of two points (x_1, y_1, z_1) and (x_2, y_2, z_2) are proportional to $\mathbf{x_2 - x_1, y_2 - y_1, z_2 - z_1}$,

i.e., difference of x co-ordinates, difference of y co-ordinates, difference of z co-ordinates.

9. Angle formulae :

(*i*) If l_1, m_1, n_1 and l_2, m_2, n_2 are the actual d.c's of two lines, then the angle θ between the two lines is given by

$$\cos\theta = l_1 l_2 + m_1 m_2 + n_1 n_2 = \Sigma l_1 l_2$$

$$\sin\theta = \sqrt{(m_1 n_2 - n_1 m_2)^2 + (n_1 l_2 - l_1 n_2)^2 + (l_1 m_2 - l_2 m_1)^2}$$

$$= \sqrt{\Sigma(m_1 n_2 - m_2 n_1)^2}$$

$$\tan\theta = \frac{\sin\theta}{\cos\theta}$$

$$= \frac{\sqrt{\Sigma(m_1 n_2 - n_1 m_2)^2}}{\Sigma l_1 l_2}$$

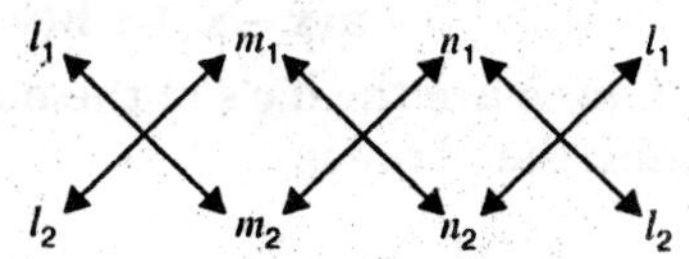

(*ii*) If a_1, b_1, c_1 and a_2, b_2, c_2 are the proportional d.c's of two lines inclined at an angle **θ**, then

$$\cos\theta = \frac{a_1a_2 + b_1b_2 + c_1c_2}{\sqrt{a_1^2 + b_1^2 + c_1^2} + \sqrt{a_2^2 + b_2^2 + c_2^2}} = \frac{\Sigma a_1a_2}{\sqrt{\Sigma a_1^2}\,\sqrt{\Sigma a_2^2}}$$

$$\sin\theta = \frac{\sqrt{\Sigma(a_1b_2 - b_1a_2)^2}}{\sqrt{\Sigma a_1^2}\,\sqrt{\Sigma a_2^2}}$$

and
$$\tan\theta = \frac{\sqrt{\Sigma(a_1b_2 - b_1a_2)^2}}{\Sigma a_1a_2}.$$

For **complete formula,** use ± sign with the radical sign in the formula for sine and tangent.

10. Condition of perpendicularity of two lines whose d.c's are l_1, m_1, n_1 and l_2, m_2, n_2 (or proportional d.c's a_1, b_1, c_1 and a_2, b_2, c_2) is

$$l_1l_2 + m_1m_2 + n_1n_2 = 0 \ (\textit{or}\ a_1a_2 + b_1b_2 + c_1c_2 = 0).$$

11. The condition for two lines to be parallel, whose d.c's are l_1, m_1, n_1 and l_2, m_2, n_2 (*or* proportional d.c's are a_1, b_1, c_1 and a_2, b_2, c_2) is

$$l_1 = l_2,\ m_1 = m_2,\ n_1 = n_2 \left(\textit{or}\ \frac{a_1}{a_2} = \frac{b_1}{b_2} = \frac{c_1}{c_2}\right).$$

12. Length of projection of the join of two points (x_1, y_1, z_1), (x_2, y_2, z_2) on a line whose d.c's are l, m, n is given by

$$l(x_2 - x_1) + m(y_2 - y_1) + n(z_2 - z_1).$$

13. Eqn. of the pl. ‖ to YZ pl. and at a distance a from it is **x = a** *i.e., missing co-ordinate* (in YZ pl.) = a.

Eqn. of the pl. ‖ to ZX pl. and at a distance b from it is **y = b** *or missing co-ordinate* = **b.**

Eqn. of the pl. ‖ to XY pl. and at a distance c from it is **z = c** *or missing co-ordinates* = **c.**

14. Equations of co-ordinate planes

The equations of the YZ, ZX, XY planes are $x = 0$, $y = 0$, $z = 0$ respectively. (**Remember,** *missing co-ordinate* = 0).

15. Equations of the co-ordinates axes :

Eqns. of X-axis are **y = 0, z = 0**

Eqns. of Y-axis are **z = 0, x = 0**

Eqns. of Z-axis are **x = 0, y = 0.**

[**Remember.** Missing co-ordinates = 0]

16. Equations of a plane in various forms :

(*i*) *Every equation of first degree in x, y, z represents a plane.*

$$ax + by + cz + d = 0 \qquad |\textit{General form}$$

(*ii*) Any plane through (x_1, y_1, z_1) is

$$a(x - x_1) + b(y - y_1) + c(z - z_1) = 0 \qquad |\textit{One-point form}$$

(*iii*) If l, m, n are the d.c's of the normal to the plane and p, the length of ⊥ from orgin to the plane, then the plane is

$$lx + my + nz = p \qquad |\textit{Normal form}$$

(*iv*) If the plane makes intercepts *a, b, c* on the axes, then the equation of the plane is

$$\frac{x}{a}+\frac{y}{b}+\frac{z}{c}=1. \qquad | \textit{Intercept form}$$

17. The d.c's of the normal to the plane are proportional to **co-eff. of x, co-eff. of y, co-eff. of z,** in the eqn. of plane.

18. Angle between two planes whose equations are

$$a_1x+b_1y+c_1z+d_1=0 \quad \text{and} \quad a_2x+b_2y+c_2z+d_2=0$$

is given by

$$\cos\theta=\frac{\Sigma a_1a_2}{\sqrt{\Sigma a_1^2}\sqrt{\Sigma a_2^2}},$$

$$\sin\theta=\frac{\sqrt{\Sigma(a_1b_2-b_1a_2)^2}}{\sqrt{\Sigma a_1^2}\sqrt{\Sigma a_2^2}} \quad \text{and} \quad \tan\theta=\frac{\sqrt{\Sigma(a_1b_2-b_1a_2)^2}}{\Sigma a_1a_2}.$$

19. Condition for the two planes $a_rx+b_ry+c_rz+d_r=0$ $[r=1, 2]$

(*i*) to be **perpendicular** is $a_1a_2+b_1b_2+c_1c_2=0$

and (*ii*) to be **parallel** is $\frac{a_1}{a_2}=\frac{b_1}{b_2}=\frac{c_1}{c_2}$.

20. Any plane parallel to the pl. $ax+by+cz+d=0$ is

$$ax+by+cz+K=0.$$

Rule to write down the equation of any plane || to a given plane : "*In the equation of the given plane, change* **only** *the constant term to a new constant.*"

21. Perpendicular Distance Formula

The perpendicular distance of the point (x_1, y_1, z_1) from the plane $ax+by+cz+d=0$ (equation in the *general form*) is

$$\frac{ax_1+by_1+cz_1+d}{\sqrt{a^2+b^2+c^2}}.$$

Rule. *Substitute the co-ordinates of the point of the L.H.S. of the equation of the plane (with R.H.S = 0), and divide the result by*

$$\sqrt{(\text{co-eff. of } x)^2+(\text{co-eff. of } y)^2+(\text{co-eff. of } z)^2}.$$

22. Equations of a st. line in various forms :

(*i*) $a_1x+b_1y+c_1z+d_1=0, \quad a_2x+b_2y+c_2z+d_2=0$ | *General form*

(*ii*) $\frac{x-x_1}{l}=\frac{y-y_1}{m}=\frac{z-z_1}{n}\ (=r)$ | *Distance form or Symmetrical form*

where (x_1, y_1, z_1) is a point on the line and *l, m, n* the d.c's of the line.

(*iii*) The line through (x_1, y_1, z_1) and with d.c's proportional to *a, b, c* are

$$\frac{x-x_1}{a}=\frac{y-y_1}{b}=\frac{z-z_1}{c}. \qquad | \textit{Symmetrical form}$$

23. A line and a plane

Let the line be $\frac{x-x_1}{l}=\frac{y-y_1}{m}=\frac{z-z_1}{n}$ ...(1)

and the plane be $ax+by+cz+d=0$ ...(2)

(*i*) The conditions for the line (1) to be ∥ to the plane (2) are [*that it should be* ⊥ *to the normal to the plane and its point* (x_1, y_1, z_1) *should not lie on the plane; i.e.,*]

$$\mathbf{al + bm + cn = 0 \text{ and } ax_1 + by_1 + cz_1 + d \neq 0.}$$

(*ii*) The line (1) will be ⊥ to the plane (2) *if it is* ∥ *to the normal to the plane, i.e., if* $\dfrac{\mathbf{l}}{\mathbf{a}} = \dfrac{\mathbf{m}}{\mathbf{b}} = \dfrac{\mathbf{n}}{\mathbf{c}}$.

(*iii*) The line (1) will lie in the plane (2) if [*it is* ⊥ *to the normal to the plane and a point* (x_1, y_1, z_1) *on the line lies on the plane, i.e., if*]

$$\mathbf{al + bm + cn = 0 \quad \text{and} \quad ax_1 + by_1 + cz_1 + d = 0.}$$

24. Any plane through the line intersection of two planes

(*a*) Any plane through the line of intersection of two planes

$$a_1x + b_1y + c_1z + d_1 = 0, \quad a_2x + b_2y + c_2z + d_2 = 0 \qquad | \textit{Equations in general form}$$

is $$\mathbf{a_1x + b_1y + c_1z + d_1 + K(a_2x + b_2y + c_2z + d_2) = 0}$$

i.e., **one plane + K (other plane) = 0.**

(*b*) Any plane through a given line

$$\frac{x - x_1}{l} = \frac{y - y_1}{m} = \frac{z - z_1}{n} \qquad | \textit{Equations in the symmetrical form}$$

is $$\left.\begin{aligned} \mathbf{A(x - x_1) + B(y - y_1) + C(z - z_1) = 0} \\ \mathbf{Al + Bm + Cn = 0} \end{aligned}\right].$$

where

Review IV: *Change of Axes

1. Introduction

In this chapter, we wish to learn to express the co-ordinates of a point on space referred to one set of axes in terms of the co-ordinates of the same point referred to another set of axes.

The reader will see for himself or herself that this process of change of axes simplifies some problems.

We shall discuss here

(*i*) **Shift of origin**

(*ii*) **Rotation and then those expressions which remain invariant under the change of axes.**

2. To change the origin of co-ordinates without changing the directions of the axes

Let OX, OY, OZ and O′X′, O′Y′, O′Z′ be two systems of parallel rectangular axes having O and O′ as their origins.

Let (f, g, h) be the co-ordinates of O′ referred to OX, OY, OZ.

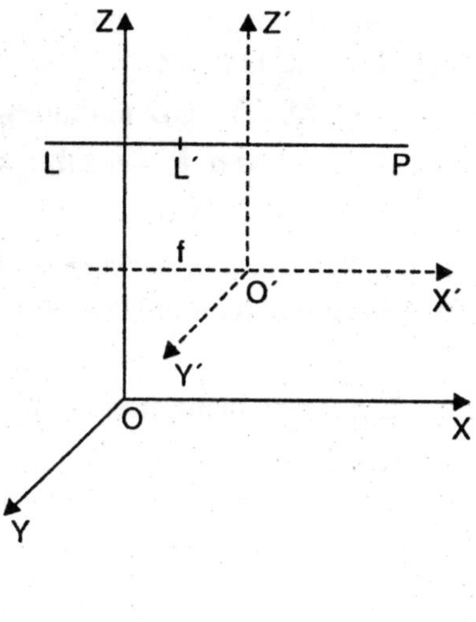

Let (x, y, z) be the co-ordinates of P referred OX, OY, OZ and (x', y', z') be the new co-ordinates of the point P referred to O′X′, O′Y′, O′Z′.

Draw the straight line PL ⊥ to the planes YOZ and Y′O′Z′ meeting them in L and L′ respectively.

Then $x = LP = LL' + L'P$ ...(1)

But LL′ = Distance between parallel planes YOZ and Y′O′Z′

$= \perp$ distance of O′ from plane YOZ $= f$

and $L'P = x'$ (By def. of co-ordinates)

$\therefore$ (1) becomes $x = f + x'$

or $x = x' + f$

Similarly $y = y' + g$

and $z = z' + h.$

Rule of Transformation when origin is shifted to (f, g, h)

Put $\mathbf{x = x' + f, y = y' + g, z = z' + h}$ ...(2)

---

*This chapter is not a part of syllabus but will be prove to be useful for learning this subject (Solid Geometry).

Cor. 1. From equations (2),

$$x' = x - f,\ y' = y - g,\ z' = z - h.$$

Cor. 2. *By shift of origin, degree of an equation remains unchanged.*

∵ The equations (2) of transformation are linear equations, therefore degree of an equation remains unchanged.

Example 1. *Find the co-ordinates of (1, 2, 3), (– 1, – 4, 0) referred to parallel axes through (– 2, – 3, – 4).*

Sol. Here $O'\ (f, g, h) = (-2, -3, -4)$.

$\therefore\quad f = -2,\ g = -3,\ h = -4$

(*i*) Let P be (1, 2, 3). $\therefore\quad x = 1,\ y = 2,\ z = 3$

Let P be (x', y', z') referred to parallel axes through $(-2, -3, -4)$

$\therefore\quad x' = x - f,\ y' = y - g,\ z' = z - h$

or $\quad x' = 1 + 2,\ y' = 2 + 3,\ z' = 3 + 4$

i.e., P is (3, 5, 7)

Similarly, the point (– 1, – 4, 0) transforms to the point (1, – 1, 4).

3. Formulae for change of directions of axes

To change the directions of the axes without changing the origin, both systems being rectangular.

Let OX, OY, OZ be the old axes and OX′, OY′, OZ′ the new.

Let $< l_1, m_1, n_1 >$; $< l_2, m_2, n_2 >$; $< l_3, m_3, n_3 >$ be the direction cosines of OX′, OY′, OZ′ referred to OX, OY, OZ.

Let OX′, OY′, OZ′ make angles α, β, γ respectively with OX.

$\therefore\quad l_1 = \cos\alpha,\ l_2 = \cos\beta,\ l_3 = \cos\gamma.$

∴ d.c's of OX referred to OX′, OY′, OZ′ are

$$< \cos\alpha, \cos\beta, \cos\gamma > \text{ i.e., } < l_1, l_2, l_3 >$$

Similarly d.c's of OY and OZ referred to OX′, OY′, OZ′ are

$$< m_1, m_2, m_3 > \text{ and } < n_1, n_2, n_3 >.$$

Let the co-ordinates of any point P in space be (x, y, z) and (x', y', z') referred to the two systems of axes respectively.

(*a*) *To express x, y, z in terms of x′, y′, z′*

Let A be the foot of ⊥ from P on OX. Join OP.

Then $\quad x = OA =$ Projection of OP on OX

$= (x' - 0)l_1 + (y' - 0)l_2 + (z' - 0)l_3$

$\mid (x_2 - x_1)\, l + (y_2 - y_1)\, m + (z_2 - z_1)\, n$

[∵ P is (x', y', z') and d.c's of OX referred to OX′, OY′, OZ′ are $< l_1, l_2, l_3 >$]

or $\quad \mathbf{x = l_1x' + l_2y' + l_3z'}$

Similarly, $\quad \mathbf{y = m_1x' + m_2y' + m_3z'}$

$\mathbf{z = n_1x' + n_2y' + n_3z'}$

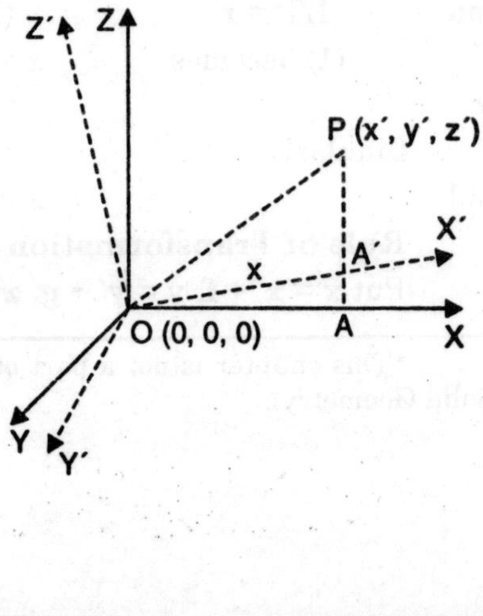

(*b*) *To express x', y', z' in terms of x, y, z*

Let A′ be the foot of ⊥ from P on OX′. Join OP.

Then $x' = OA' =$ Projection of OP on OX′

$$= (x - 0)l_1 + (y - 0)m_1 + (z - 0)n_1$$

$$| (x_2 - x_1)l + (y_2 - y_1)m + (z_2 - z_1)n$$

[∵ P is (x, y, z) and d.c's of OX referred to OX, OY, OZ are $< l_1, m_1, n_1 >$]

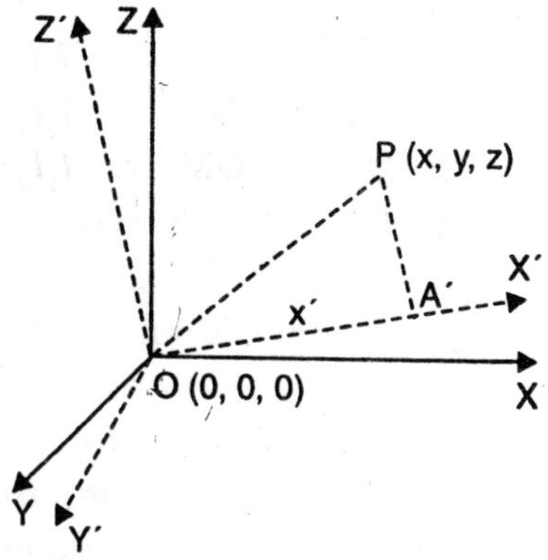

or $$\mathbf{x' = l_1x + m_1y + n_1z}$$

Similarly $$\mathbf{y' = l_2x + m_2y + n_2z,}$$

$$\mathbf{z' = l_3x + m_3y + n_3z}$$

Rule to write

The results of Art. 3 can easily be written with the help of the following table :

	x	y	z
x'	l_1	m_1	n_1
y'	l_2	m_2	n_2
z'	l_3	m_3	n_3

To find x. *Multiply the numbers in the x-column by x', y', z' respectively and add.*

Thus $$x = l_1x' + l_2y' + l_3z'.$$

Similarly, y and z.

To find x′. *Multiply the numbers in the x-row by x, y, z respectively and add.*

Thus $$x' = l_1x + m_1y + n_1z.$$

Similarly, y' and z'.

Cor. *If the axes are changed without changing the origin ; the degree of the equation remains unchanged.*

∵ The equations of transformations in Art. 3 are linear equations, therefore degree of the equation remains unchanged.

4. Relations between the direction-cosines of three mutually perpendicular lines

To find the relations between the direction-cosines of three mutually perpendicular lines.

Let OX, OY, OZ be the axes.

Let $< l_1, m_1, n_1 >$; $< l_2, m_2, n_2 >$; $<l_3, m_3, n_3>$ be the direction cosines of three mutually perpendicular lines OX′, OY′, OZ′ (referred to OX, OY, OZ).

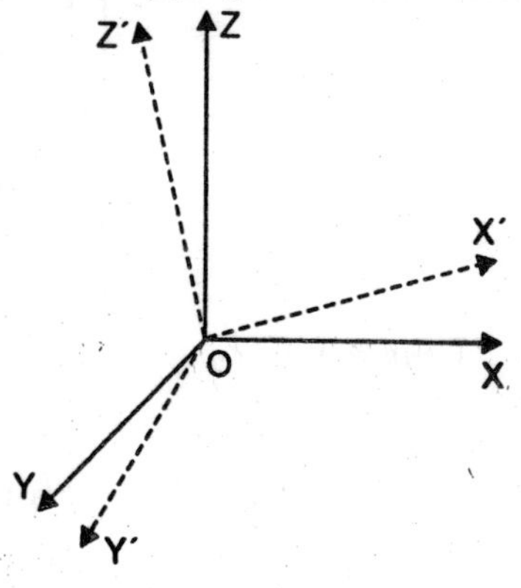

∵ $$< l_1, m_1, n_1 >,$$
$$< l_2, m_2, n_2 >,$$
$$< l_3, m_3, n_3 >$$

are d.c's of the lines

∵ $$\left.\begin{aligned} l_1^2 + m_1^2 + n_1^2 = 1 \\ l_2^2 + m_2^2 + n_2^2 = 1 \\ l_3^2 + m_3^2 + n_3^2 = 1 \end{aligned}\right] \quad \ldots(1)$$

$\because$ OX′ is ⊥ OY′ $\therefore$ $l_1 l_2 + m_1 m_2 + n_1 n_2 = 0$

$\because$ OY′ is ⊥ OZ′ $\therefore$ $l_2 l_3 + m_2 m_3 + n_2 n_3 = 0$ ...(2)

$\because$ OZ′ is ⊥ OX′ $\therefore$ $l_1 l_3 + m_1 m_3 + n_1 n_3 = 0$

Let OX′, OY′, OZ′ make angles α, β, γ with OX respectively.

$$\therefore \quad l_1 = \cos\alpha,\ l_2 = \cos\beta,\ l_3 = \cos\gamma.$$

$\therefore$ d.c's of OX referred to OX′, OY′, OZ′ are $< \cos\alpha, \cos\beta, \cos\gamma >$

i.e., $< l_1, l_2, l_3 >$

Similarly d.c's of OY and OZ referred to OX′, OY′, OZ′ are

$$< m_1, m_2, m_3 > \text{ and } < n_1, n_2, n_3 >$$

$\therefore$ $<l_1, l_2, l_3>$, $<m_1, m_2, m_3>$, $<n_1, n_2, n_3>$ are d.c's of the lines.

$$\therefore \quad \left.\begin{aligned} l_1^2 + l_2^2 + l_3^2 &= 1 \\ m_1^2 + m_2^2 + m_3^2 &= 1 \\ n_1^2 + n_2^2 + n_3^2 &= 1 \end{aligned}\right] \quad ...(3)$$

$\because$ OX ⊥ OY $\therefore$ $l_1 m_1 + l_2 m_2 + l_3 m_3 = 0$

$\therefore$ OY ⊥ OZ $\therefore$ $m_1 n_1 + m_2 n_2 + m_3 n_3 = 0$ ...(4)

$\because$ OZ ⊥ OX $\therefore$ $n_1 l_1 + n_2 l_2 + n_3 l_3 = 0$

(1), (2), (3) and (4) give the twelve relations among the direction cosines cf three mutually perpendicular lines.

Note. Rule to write down the above twelve relations.

Write down in three rows the direction-cosines of the three lines as :

$$\begin{matrix} l_1 & m_1 & n_1 \\ l_2 & m_2 & n_2 \\ l_3 & m_3 & n_3 \end{matrix}$$

(*i*) sum of the squares of the elements in each row is 1.

(*ii*) sum of the squares of the elements in each column is 1.

(*iii*) sum of the products of elements in two rows is zero.

(*iv*) sum of the products of elements in two columns is zero.

5. *If $< l_1, m_1, n_1 >$; $< l_2, m_2, n_2 >$; $< l_3, m_3, n_3 >$ are the direction-cosines of three mutually perpendicular lines, prove that*

$$\begin{vmatrix} l_1 & m_1 & n_1 \\ l_2 & m_2 & n_2 \\ l_3 & m_3 & n_3 \end{vmatrix} = \pm 1$$

$$\begin{vmatrix} l_1 & m_1 & n_1 \\ l_2 & m_2 & n_2 \\ l_3 & m_3 & n_3 \end{vmatrix}^2 = \begin{vmatrix} l_1 & m_1 & n_1 \\ l_2 & m_2 & n_2 \\ l_3 & m_3 & n_3 \end{vmatrix} \times \begin{vmatrix} l_1 & m_1 & n_1 \\ l_2 & m_2 & n_2 \\ l_3 & m_3 & n_3 \end{vmatrix}$$

$$= \begin{vmatrix} l_1^2 + m_1^2 + n_1^2 & l_1 l_2 + m_1 m_2 + n_1 n_2 & l_1 l_3 + m_1 m_3 + n_1 n_3 \\ l_2 l_1 + m_2 m_1 + n_2 n_1 & l_2^2 + m_2^2 + n_2^2 & l_2 l_3 + m_2 m_3 + n_2 n_3 \\ l_3 l_1 + m_3 m_1 + n_3 n_1 & l_3 l_2 + m_3 m_2 + n_3 n_2 & l_3^2 + m_3^2 + n_3^2 \end{vmatrix}$$

(Row by Row Multiplication)

Using 12 relations of Art. 4

$$= \begin{vmatrix} 1 & 0 & 0 \\ 0 & 1 & 0 \\ 0 & 0 & 1 \end{vmatrix}$$

Expanding by first row, it is 1.

Taking square roots.

$$\begin{vmatrix} l_1 & m_1 & n_1 \\ l_2 & m_2 & n_2 \\ l_3 & m_3 & n_3 \end{vmatrix} = \pm 1 \quad ...(1)$$

Cor. We know that an interchange of rows and columns of a determinant does not change the value of a determinant.

Therefore interchanging rows and columns in the L.H.S. of Eqn. (1) above, we have

$$\begin{vmatrix} l_1 & l_2 & l_3 \\ m_1 & m_2 & m_3 \\ n_1 & n_2 & n_3 \end{vmatrix} = \pm 1.$$

6. If l_1, m_1, n_1 ; l_2, m_2, n_2 ; and l_3, m_3, n_3 are d.c's of three mutually perpendicular lines, then $l_1 = (m_2n_3 - m_3n_2)$ etc.

First and third equations of the set of equations (4) of Art. 4 are

$$l_1m_1 + l_2m_2 + l_3m_3 = 0$$
$$l_1n_1 + l_2n_2 + l_3n_3 = 0$$

Solving these two equations for l_1, l_2, l_3, we have

$$\frac{l_1}{m_2n_3 - m_3n_2} = \frac{l_2}{m_3n_1 - m_1n_3} = \frac{l_3}{m_1n_2 - m_2n_1}$$

$\therefore$ d.r's of one of these three perpendicular lines (*i.e.*, d.r's of one of the transformed axes) are

$$m_2n_3 - m_3n_2,\ m_3n_1 - m_1n_3,\ m_1n_2 - m_2n_1.$$

Dividing each by $\sqrt{(m_2n_3 - m_3n_2)^2 + (m_3n_1 - m_1n_3)^2 + (m_1n_2 - m_2n_1)^2}$

$$= \sqrt{\sin\theta}$$

where θ = angle between the lines having d.c's m_1, m_2, m_3 and $n_1, n_2, n_3 = 90°$ (given)

$$= \sqrt{\sin 90°} = 1$$

d.c's l_1, m_1, n_1 of this line are

$$m_2n_3 - m_3n_2,\ m_3n_1 - m_1n_3,\ m_1n_2 - m_2n_1$$

i.e., $$l_1 = m_2n_3 - m_3n_2 \text{ etc.}$$

7. If by any change of rectangular axes, $ax^2 + by^2 + cz^2 + 2fyz + 2gzx + 2hxy$ be transformed, then the expressions $a + b + c$, $A + B + C$ and D remain unaltered in value, where A, B, C are co-factors of a, b, c in the determinant

$$D = \begin{vmatrix} a & h & g \\ h & b & f \\ g & f & c \end{vmatrix}$$

i.e., $\mathbf{A = bc - f^2, B = ca - g^2, C = ab - h^2,}$

$\mathbf{D = abc + 2fgh - af^2 - bg^2 - ch^2}$

(*i*) **If only the origin is changed** to the point (α, β, γ) ; the axes remaining parallel to original directions, then x, y, z change to $x + \alpha, y + \beta, z + \gamma$ respectively, so that the given expression transforms to

$$a(x+\alpha)^2 + b(y+\beta)^2 + c(z+\gamma)^2 + 2f(y+\beta)(z+\gamma) + 2g(z+\gamma)(x+\alpha) + 2h(x+\alpha)(y+\beta)$$

or $$ax^2 + by^2 + cz^2 + 2fyz + 2gzx + 2hxy + 2x(a\alpha + h\beta + g\gamma) + ...$$

i.e., the second degree terms remain unaltered by the transformation and since the expressions $a + b + c$,

$$A + B + C = ab + bc + ca - f^2 - g^2 - h^2$$
$$D = abc + 2fgh - af^2 - bg^2 - ch^2$$

contain only the co-efficients of second degree terms, therefore they remain unaltered.

(*ii*) **If the axes are changed without changing the origin**

Let $ax^2 + by^2 + cz^2 + 2fyz + 2gzx + 2hxy$ transform to

$$a'x^2 + b'y^2 + c'z^2 + 2f'yz + 2g'zx + 2h'xy$$

(The reader should keep in mind that (x, y, z) in the above two expressions are w.r.t. different axes).

Since O is fixed, therefore OP remains unaltered

Hence $\quad OP^2 = x^2 + y^2 + z^2$

$\therefore$ $$ax^2 + by^2 + cz^2 + 2fyz + 2gzx + 2hxy + \lambda(x^2 + y^2 + z^2) \quad ...(1)$$

transforms to

$$a'x^2 + b'y^2 + c'z^2 + 2f'yz + 2g'zx + 2h'xy + \lambda(x^2 + y^2 + z^2) \quad ...(2)$$

(1) and (2) (if equated to zero) represent the same surface. If (1) is the product of two linear factors (*i.e.,* (1) equated to 0 represents a pair of planes) for some value of λ ; then the same value of λ will also make (2) as the product of two linear factors.

(1) is the product of two linear factors if

$$(a+\lambda)(b+\lambda)(c+\lambda) + 2fgh - (a+\lambda)f^2 - (b+\lambda)g^2 - (c+\lambda)h^2 = 0$$

[Using $abc + 2fgh - af^2 - bg^2 - ch^2 = 0$]

i.e., if $$\lambda^3 + \lambda^2(a+b+c) + \lambda(ab+bc+ca-f^2-g^2-h^2) + (abc+2fgh-af^2-bg^2-ch^2) = 0$$

i.e., if $$\lambda^3 + \lambda^2(a+b+c) + \lambda(A+B+C) + D = 0 \quad ...(3)$$

(See the statement of the Article)

Similarly, (2) is the product of two linear factors if

$$\lambda^3 + \lambda^2(a'+b'+c') + \lambda(A'+B'+C') + D' = 0 \quad ...(4)$$

Since (3) and (4) hold for same values of λ,

$\therefore$ Equating co-effs. of λ^2, λ and constants (as co-efficient of λ^3 is 1 in both (3) and (4)) ;

$$a + b + c = a' + b' + c'$$
$$A + B + C = A' + B' + C'$$
$$D = D'$$

Hence the result.

10

Sphere

1. Definition of Sphere

A sphere is the locus of a point **in space** *which is at a constant distance from a fixed point.*

The constant distance is called *radius* of the sphere and the *fixed point* is called the *centre* of the sphere.

Remark. In Analytical Solid Geometry, 'Sphere' means 'hollow sphere' of negligible thickness.

2. Equation of a Sphere (Central form)

To find the equation of a sphere whose centre is (a, b, c) and radius is r.

Let C(a, b, c) be the centre of the sphere and r be the radius of the sphere.

Let P(x, y, z) be any point on the sphere.

∴ By definition of sphere,

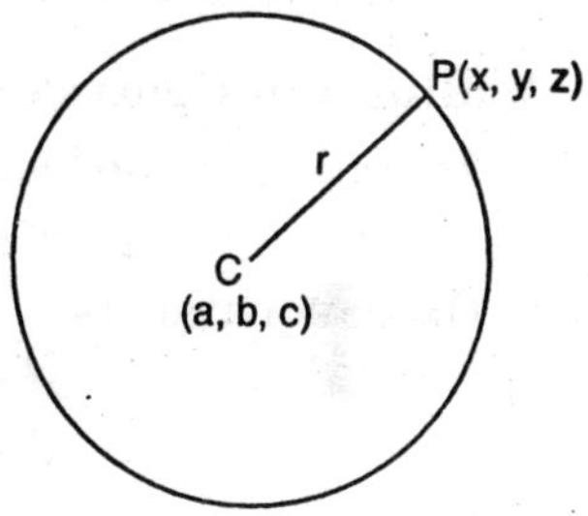

$$CP = r$$

or $$\sqrt{(x-a)^2 + (y-b)^2 + (z-c)^2} = r$$

or squaring,

$$\mathbf{(x - a)^2 + (y - b)^2 + (z - c)^2 = r^2} \qquad ...(1)$$

which is the required equation of the sphere.

Cor. 1. Simplest Form. *To find the equation of a sphere whose centre is origin (0, 0, 0) and radius r.*

Putting $a = 0, b = 0, c = 0$ in (1), equation of sphere becomes

$$\mathbf{x^2 + y^2 + z^2 = r^2}$$

Note. $\mathbf{x^2 + y^2 + z^2 = r^2}$ is the equation of a sphere with centre at origin and radius r.

Remark. Equation (1) can be written as

$$x^2 - 2ax + a^2 + y^2 - 2by + b^2 + z^2 - 2cz + c^2 = r^2$$

or $$x^2 + y^2 + z^2 - 2ax - 2by - 2cz + (a^2 + b^2 + c^2 - r^2) = 0 \qquad ...(2)$$

We note the following characteristics about the equation :

(i) *It is of second degree in x, y, z.*

(ii) *Co-efficients of x^2, y^2* and *z^2* are equal.

(iii) *The product terms involving xy, yz, zx are absent.*

Cor. 2. Equation (2) is of the form

$$\mathbf{x^2 + y^2 + z^2 + 2ux + 2vy + 2wz + d = 0}$$

where $u = -a, v = -b, w = -c,$

$d = a^2 + b^2 + c^2 - r^2$

Since this equation possesses all the three characteristics listed in the above remark, therefore it is called **general equation** of the sphere or equation of the sphere **in general form.**

Example. *Find the equation to the sphere whose centre is (2, 3, – 4) and radius 5.*

Sol. Equation of sphere with centre (2, 3, – 4) and radius 5 is

$$(x-2)^2 + (y-3)^2 + [z-(-4)]^2 = 5^2 \qquad | \; (x-a)^2 + (y-b)^2 + (z-c)^2 = r^2$$

Simplifying, $x^2 + y^2 + z^2 - 4x - 6y + 8z + 4 = 0.$

3. To find the equation of a sphere, the ends of whose diameter are (x_1, y_1, z_1) and (x_2, y_2, z_2)

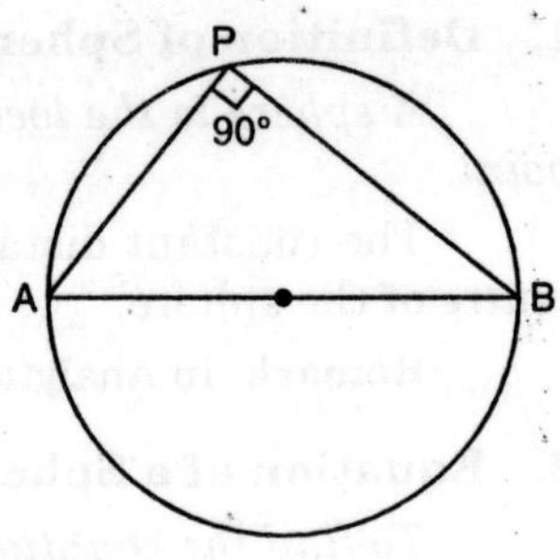

Let $A(x_1, y_1, z_1)$ and $B(x_2, y_2, z_2)$ be the ends of a diameter of a sphere.

Let P(x, y, z) be any point of the sphere.

$\therefore \quad \angle APB = 90°$ *i.e.*, $AP \perp BP$ ...(1)

Now direction ratios of AP are $< x - x_1, y - y_1, z - z_1 >$ and direction ratios of BP are $< x - x_2, y - y_2, z - z_2 >$

$$\therefore \quad (x - x_1)(x - x_2) + (y - y_1)(y - y_2) + (z - z_1)(z - z_2) = 0$$

$(\because$ From (1) $AP \perp BP \quad \therefore \quad aa' + bb' + cc' = 0)$

which is the required equation of sphere.

Note. It is called **diameter form** of the equation of a sphere.

4. Radius and Centre of the Sphere

To find the co-ordinates of centre and radius of sphere

$$x^2 + y^2 + z^2 + 2ux + 2vy + 2wz + d = 0$$

Given equation is

$$x^2 + y^2 + z^2 + 2ux + 2vy + 2wz + d = 0 \qquad ...(1)$$

or $$(x^2 + 2ux) + (y^2 + 2vy) + (z^2 + 2wz) = -d$$

On completing squares in x, y and z, we have

$$(x^2 + 2ux + u^2) + (y^2 + 2vy + v^2) + (z^2 + 2wz + w^2) = u^2 + v^2 + w^2 - d$$

or $$(x + u)^2 + (y + v)^2 + (z + w)^2 = u^2 + v^2 + w^2 - d$$

or $$[x - (-u)]^2 + [y - (-v)]^2 + [z - (-w)]^2$$

$$= \sqrt{(u^2 + v^2 + w^2 - d)^2}$$

Comparing this equation with the equation of sphere (centre at (a, b, c) and radius r) namely

$$(x - a)^2 + (y - b)^2 + (z - c)^2 = r^2, \text{ we have}$$

$$a = -u, \; b = -v, \; c = -w,$$

$$r = \sqrt{u^2 + v^2 + w^2 - d}.$$

$\therefore$ **Centre of sphere (1) is (– u, – v, – w) and radius is $\sqrt{u^2 + v^2 + w^2 - d}$.**

Note. *u, v, w* and *d* are $\frac{1}{2}$ co-eff. of *x*, $\frac{1}{2}$ co-eff. of *y*, $\frac{1}{2}$ co-eff. of *z* and constant term respectively.

$\therefore$ Centre is $\left(-\frac{1}{2} \text{ co-eff. of x}, \; -\frac{1}{2} \text{ co-eff. of y}, \; -\frac{1}{2} \text{ co-eff. of z}\right)$

Remark. If $u^2 + v^2 + w^2 - d$ is negative, then radius

$$\sqrt{u^2 + v^2 + w^2 - d}$$

is imaginary, but centre $(-u, -v, -w)$ is real.

In this situation, the sphere is called **virtual** sphere.

Caution. While finding centre and radius of sphere, **make the co-efficients of x^2, y^2, z^2 as unity** in the equation of the sphere.

Example 1. *Find the radius and centre of the sphere*

$$x^2 + y^2 + z^2 - 2x + 4y - 6z = 2.$$

Sol. Equation of sphere is

$$x^2 + y^2 + z^2 - 2x + 4y - 6z = 2$$

or
$$x^2 + y^2 + z^2 - 2x + 4y - 6z - 2 = 0$$

Comparing with

$$x^2 + y^2 + z^2 + 2ux + 2vy + 2wz + d = 0\ ;$$

we have

$$2u = -2, \quad 2v = 4, \quad 2w = -6, \quad d = -2$$

i.e.,
$$u = -1, \quad v = 2, \quad w = -3, \quad d = -2$$

Hence the centre $(-u, -v, -w)$ is $(1, -2, 3)$

and
$$\text{radius} = \sqrt{u^2 + v^2 + w^2 - d} = \sqrt{1+4+9+2} = 4.$$

Example 2. *Show that the equation*

$$ax^2 + ay^2 + az^2 + 2ux + 2vy + 2wz + d = 0$$

represents a sphere. Find its centre and radius.

Sol. The given equation is

$$ax^2 + ay^2 + az^2 + 2ux + 2vy + 2wz + d = 0 \quad \text{...(1)}$$

Equation (1) represents a sphere

[$\because$ (1) satisfies all the three conditions listed in Remark Art. 2]

Dividing both sides of (1) by a, we get

$$x^2 + y^2 + z^2 + \frac{2u}{a}x + \frac{2v}{a}y + \frac{2w}{a}z + \frac{d}{a} = 0$$

Grouping the terms of x, y and z, we get

$$\left(x^2 + \frac{2u}{a}x\right) + \left(y^2 + \frac{2v}{a}y\right) + \left(z^2 + \frac{2w}{a}z\right) = \frac{-d}{a}$$

Completing squares in x, y, z by adding ($\frac{1}{2}$ co-eff. of x)^2 etc., we have

$$\left[x^2 + \frac{2u}{a}x + \left(\frac{u}{a}\right)^2\right] + \left[y^2 + \frac{2v}{a}y + \left(\frac{v}{a}\right)^2\right] + \left[z^2 + \frac{2w}{a}z + \left(\frac{w}{a}\right)^2\right] = \frac{u^2}{a^2} + \frac{v^2}{a^2} + \frac{w^2}{a^2} - \frac{d}{a}$$

or
$$\left(x + \frac{u}{a}\right)^2 + \left(y + \frac{v}{a}\right)^2 + \left(z + \frac{w}{a}\right)^2 = \frac{u^2}{a^2} + \frac{v^2}{a^2} + \frac{w^2}{a^2} - \frac{d}{a}$$

or
$$\left[x - \left(\frac{-u}{a}\right)\right]^2 + \left[y - \left(\frac{-v}{a}\right)\right]^2 + \left[z - \left(\frac{-w}{a}\right)\right]^2 = \sqrt{\left(\frac{u^2}{a^2} + \frac{v^2}{a^2} + \frac{w^2}{a^2} - \frac{d}{a}\right)}^2$$

Comparing with

$$(x-a)^2+(y-b)^2+(z-c)^2=r^2, \text{ we have}$$

$$a=\frac{-u}{a},\ b=\frac{-v}{a},\ c=\frac{-w}{a}$$

$$r=\sqrt{\frac{u^2}{a^2}+\frac{v^2}{a^2}+\frac{w^2}{a^2}-\frac{d}{a}}$$

$\therefore$ Centre is $(a, b, c)=\left(\frac{-u}{a},\frac{-v}{a},\frac{-w}{a}\right)$

and radius is $\sqrt{\frac{u^2}{a^2}+\frac{v^2}{a^2}+\frac{w^2}{a^2}-\frac{d}{a}}$.

Example 3. ***A point moves so that the sum of the squares of its distances from the six faces of a cube is constant, show that its locus is a sphere.***

Sol. Let us take one corner of the cube as origin and the three edges through it as axes. Let the edge of the cube be a.

$\therefore$ Equations of the six faces of the cube are

$$x=0,\quad x=a,\quad y=0,\quad y=a,\quad z=0 \text{ and } z=a$$

i.e., $$x=0,\quad x-a=0,\quad y=0,\quad y-a=0,\quad z=0,\quad z-a=0$$

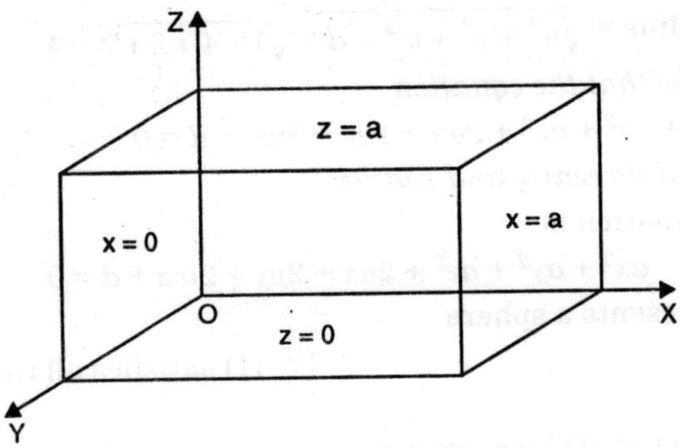

Let the moving point be (x, y, z).

The distances of (x, y, z) from the six faces are

$$x,\ x-a,\ y,\ y-a,\ z,\ z-a \qquad \left|\pm\frac{(ax_1+by_1+cz_1+d)}{\sqrt{a^2+b^2+c^2}}\right|$$

$$\therefore \quad x^2+(x-a)^2+y^2+(y-a)^2+z^2+(z-a)^2 = a \text{ constant} = k \text{ (say)}$$

or $2x^2+2y^2+2z^2-2ax-2ay-2az=k$ which is the locus of the moving points (x, y, z) and is a sphere. [See Remarks Art. 2]

EXERCISE 1

1. (*i*) Find the equation of the sphere if centre is (– 1, 2, – 3) and radius 5.

   (*ii*) Find the equation of the sphere passing through (2, 6, – 1) and having centre at the point (– 1, 2, – 3).

   (*iii*) Find the equation of the sphere on the join of (– 1, 3, 2) and (5, 7, – 6) as diameter.

(*iv*) Find the co-ordinates of the centre and radius of the sphere

$$2x^2 + 2y^2 + 2z^2 - 2x + 4y + 2z - 15 = 0.$$

(*v*) One end of the diameter of the sphere

$$x^2 + y^2 + z^2 + 4x - 6y - 2z - 22 = 0$$

is at the point (– 6, 7, 3). Find the co-ordinates of the other end.

2. Prove that a point, the sum of squares of whose distances from the planes $x + y + z = 0$, $x - z = 0$, $x - 2y + z = 0$ is 9, must lie on a sphere.

3. A is a point (1, 3, 4) and B is the point (1, – 2, – 1). A point P moves so that 3PA = 2PB. Prove that locus of P is the sphere $x^2 + y^2 + z^2 - 2x - 14y - 16z + 12 = 0$. Verify that the sphere divides AB internally and externally in the same ratio.

4. A point moves so that the ratio of its distances from two fixed points is constant, show that its locus is a sphere.

[**Hint.** *Let us take the two fixed points as* $(-a, 0, 0)$ and $(a, 0, 0)$.]

5. A plane passes through a fixed point (a, b, c). Show that the locus of the foot of perpendicular to it from the origin is the sphere $x^2 + y^2 + z^2 - ax - by - cz = 0$.

[**Hint.** *Let* $P(x_1, y_1, z_1)$ *be the foot of perpendicular. Therefore d.r's of OP are* $<x_1 - 0, y_1 - 0, z_1 - 0>$ *and hence equation of the plane is*

$$x_1(x - x_1) + y_1(y - y_1) + z_1(z - z_1) = 0.$$

It passes through (a, b, c).]

6. Show that spheres $x^2 + y^2 + z^2 = 25$, $x^2 + y^2 + z^2 - 18x - 24y - 40z + 225 = 0$ touch and find the co-ordinates of their common point.

[**Def.** *The spheres with centres* C_1, C_2 *and radii* r_1, r_2 *touch each other externally if* $C_1C_2 = r_1 + r_2$ *and internally if* $C_1C_2 = |r_1 - r_2|$. *The point of contact divides* C_1C_2 *in the ratio* $r_1 : r_2$ *internally or externally according as the spheres touch externally or internally.*]

7. Show that the spheres $x^2 + y^2 + z^2 = 64$ and

$$x^2 + y^2 + z^2 - 12x - 4y - 6z + 48 = 0$$

touch internally and find the point of contact.

Answers

1. (*i*) $x^2 + y^2 + z^2 + 2x - 4y + 6z - 11 = 0$ (*ii*) $x^2 + y^2 + z^2 + 2x - 4y + 6z - 15 = 0$

(*iii*) $x^2 + y^2 + z^2 - 4x - 10y + 4z + 4 = 0$ (*iv*) $\left(\frac{1}{2}, -1, -\frac{1}{2}\right)$; 3 (*v*) (2, – 1, – 1)

6. $\left(\frac{9}{5}, \frac{12}{5}, 4\right)$ **7.** $\left(\frac{48}{7}, \frac{-16}{7}, \frac{24}{7}\right)$.

5. Number of Constants in the Equation of a Sphere

The general equation of a sphere is

$$x^2 + y^2 + z^2 + 2ux + 2vy + 2wz + d = 0$$

This equation contains four arbitrary constants u, v, w and d.

Hence a sphere can be uniquely determined when four conditions giving four independent relations between u, v, w and d are known.

For example, a sphere can be determined to pass through four non-coplanar points.

6. Four Point Form

To find the equation of a sphere passing through four given points.

Let the four given points be (x_1, y_1, z_1), (x_2, y_2, z_2), (x_3, y_3, z_3) and (x_4, y_4, z_4).

Let the equation of the sphere through these four points be

$$(x^2 + y^2 + z^2) + 2ux + 2vy + 2wz + d = 0 \qquad ...(1)$$

∵ The four points lie on sphere (1)

∴
$$(x_1^2 + y_1^2 + z_1^2) + 2ux_1 + 2vy_1 + 2wz_1 + d = 0 \qquad ...(2)$$
$$(x_2^2 + y_2^2 + z_2^2) + 2ux_2 + 2vy_2 + 2wz_2 + d = 0 \qquad ...(3)$$
$$(x_3^2 + y_3^2 + z_3^2) + 2ux_3 + 2vy_3 + 2wz_3 + d = 0 \qquad ...(4)$$
$$(x_4^2 + y_4^2 + z_4^2) + 2ux_4 + 2vy_4 + 2wz_4 + d = 0 \qquad ...(5)$$

Eliminating u, v, w, d from (1), (2), (3), (4) and (5) with the help of determinants, we have

$$\begin{vmatrix} x^2+y^2+z^2 & x & y & z & 1 \\ x_1^2+y_1^2+z_1^2 & x_1 & y_1 & z_1 & 1 \\ x_2^2+y_2^2+z_2^2 & x_2 & y_2 & z_2 & 1 \\ x_3^2+y_3^2+z_3^2 & x_3 & y_3 & z_3 & 1 \\ x_4^2+y_4^2+z_4^2 & x_4 & y_4 & z_4 & 1 \end{vmatrix} = 0$$

Note. In numerical problems, find the values of u, v, w, d from (2), (3), (4), (5) and substitute them in (1).

Example 1. *Find the equation of the sphere through the origin and making intercepts a, b, c on the co-ordinates axes.*

Sol. Let the sphere make intercepts

OA = a, OB = b and OC = c on the axes.

∴ It passes through the points

A(a, 0, 0), B(0, b, 0) and C(0, 0, c).

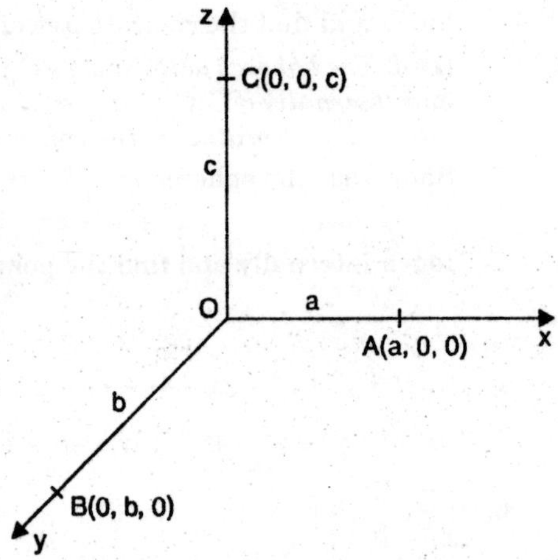

Let the equation of the sphere be

$$x^2 + y^2 + z^2 + 2ux + 2vy + 2wz + d = 0 \qquad ...(1)$$

It passes through origin (0, 0, 0) ∴ $d = 0$

Again because (a, 0, 0) lies on (1)

∴
$$a^2 + 2ua + d = 0$$

i.e.,
$$a^2 + 2ua = 0 \qquad (\because d = 0)$$

∴
$$2ua = -a^2 \quad \text{or} \quad u = \frac{-a}{2}$$

Similarly, $v = \frac{-b}{2}$ and $w = \frac{-c}{2}$

Putting these values of u, v, w, d in (1), equation of the sphere

$$x^2 + y^2 + z^2 - ax - by - cz = 0.$$

Example 2. *Find the equation of the sphere through the points*

(4, – 1, 2), (0, – 2, 3), (1, 5, – 1), (2, 0, 1).

Sol. Let the required equation of the sphere be

$$x^2 + y^2 + z^2 + 2ux + 2vy + 2wz + d = 0 \qquad ...(1)$$

∵ It passes through (4, – 1, 2)

∴
$$16 + 1 + 4 + 8u - 2v + 4w + d = 0$$

or
$$21 + 8u - 2v + 4w + d = 0 \qquad ...(2)$$

Similarly, as the sphere passes through

$$(0, -2, 3), (1, 5, -1), (2, 0, 1)$$

$\therefore$ $$0 + 4 + 9 + 0 - 4v + 6w + d = 0$$

or $$13 - 4v + 6w + d = 0 \quad ...(3)$$

and $$1 + 25 + 1 + 2u + 10v - 2w + d = 0$$

or $$27 + 2u + 10v - 2w + d = 0 \quad ...(4)$$

and $$4 + 0 + 1 + 4u + 0 + 2w + d = 0$$

or $$5 + 4u + 2w + d = 0 \quad ...(5)$$

(2) – (3) gives | To eliminate d

$$8 + 8u + 2v - 2w = 0 \quad \text{or} \quad 4 + 4u + v - w = 0 \quad ...(6)$$

(3) – (4) gives | To eliminate d

$$-14 - 2u - 14v + 8w = 0 \quad \text{or} \quad 7 + u + 7v - 4w = 0 \quad ...(7)$$

(4) – (5) gives | To eliminate d

$$22 - 2u + 10v - 4w = 0 \quad \text{or} \quad 11 - u + 5v - 2w = 0 \quad ...(8)$$

(7) – 4(6) gives

$$-9 - 15u + 3v = 0 \quad \text{or} \quad -5u + v = 3 \quad ...(9)$$

(8) – 2(6) gives

$$3 - 9u + 3v = 0 \quad \text{or} \quad -3u + v = -1 \quad ...(10)$$

(9) – (10) gives

$$-2u = 4 \quad \text{or} \quad u = -2$$

Putting $u = -2$ in (9), $v = -7$

Putting $u = -2$, $v = -7$ in (6), $w = -11$

Putting these values of u, v, w in (2), $d = 25$

Putting these values of u, v, w, d in (1), the required equation of the sphere is

$$x^2 + y^2 + z^2 - 4x - 14y - 22z + 25 = 0.$$

Example 3. *A sphere of constant radius 2k passes through the origin and meets the axes in A, B, C. Prove that the locus of the centroid of the tetrahedron OABC is the sphere*

$$x^2 + y^2 + z^2 = k^2.$$

Sol. Let OA = a, OB = b, OC = c

$\therefore$ Co-ordinates of the points O, A, B, C are

(0, 0, 0), (a, 0, 0), (0, b, 0), (0, 0, c) respectively.

$\therefore$ Equation of the sphere OABC is

$$x^2 + y^2 + z^2 - ax - by - cz = 0 \quad \text{|See Example 1}$$

Radius of this sphere $= \sqrt{u^2 + v^2 + w^2 - d}$

$$\sqrt{\left(\frac{a}{2}\right)^2 + \left(\frac{b}{2}\right)^2 + \left(\frac{c}{2}\right)^2} = 2k \quad \text{(given)}$$

Squaring, $$\frac{a^2}{4} + \frac{b^2}{4} + \frac{c^2}{4} = 4k^2$$

or $$a^2 + b^2 + c^2 = 16k^2 \quad ...(1)$$

Let (x, y, z) be the centroid of the tetrahedron OABC.

$\therefore \quad x = \dfrac{0+a+0+0}{4} = \dfrac{a}{4}$. Similarly, $y = \dfrac{b}{4}$, $z = \dfrac{c}{4}$

$\therefore \quad a = 4x, \quad b = 4y, \quad c = 4z$

Eliminating the unknown quantities a, b, c

[*i.e.*, Putting $a = 4x$, $b = 4y$, $c = 4z$ in (1)] ;

$$16x^2 + 16y^2 + 16z^2 = 16k^2$$

Cancelling 16, $x^2 + y^2 + z^2 = k^2$ which is the required locus.

Example 4. *Obtain the equation of the sphere* **inscribed** *in the tetrahedron whose faces are $x = 0$, $y = 0$, $z = 0$ and $x + 2y + 2z = 1$.*

Sol. Equations of the four plane faces of the tetrahedron are

$$x = 0 \quad ...(1)$$

$$y = 0 \quad ...(2)$$

$$z = 0 \quad ...(3)$$

$$x + 2y + 2z = 1 \quad ...(4)$$

Let us solve these four equations taken three at a time to get the vertices of the tetrahedron.

Solving (2), (3) and (4), we get the vertex A (1, 0, 0)

Solving (1), (3) and (4), we get the vertex B$(0, \frac{1}{2}, 0)$

Solving (1), (2) and (4), we get the vertex C $(0, 0, \frac{1}{2})$

Solving (1), (2) and (3), we get the vertex O(0, 0, 0)

Let (x, y, z) be the centre of the sphere inscribed in the tetrahedron.

$\therefore \quad \perp$ distance of the centre (x, y, z) from each of the four plane faces is the same.

$\perp$ distance of the centre (x, y, z) from plane (1).

$$= \pm \frac{x}{1}$$

$$\left(\pm \frac{(ax_1 + by_1 + cz_1 + d)}{\sqrt{a^2 + b^2 + c^2}} \right)$$

But the centre (x, y, z) and intersection of planes (2), (3), (4) *i.e.*, vertex A(1, 0, 0) lie on the same side of face (1) namely $x = 0$.

$\therefore \quad x$ and 1 are of same sign [substituting the co-ordinates of the two points in the L.H.S. of plane (1)]

$\therefore \quad x$ is positive. [$\because$ 1 is positive]

$\therefore \quad \perp$ distance of the centre (x, y, z) from plane (1) $= \dfrac{+x}{1}$.

Similarly, perpendicular distances of the centre (x, y, z) from planes (2) and (3) are $+\dfrac{y}{1}$ and $+\dfrac{z}{1}$.

Again perpendicular distance of the centre (x, y, z) from plane (4)

$$= \pm \frac{(x + 2y + 2z - 1)}{\sqrt{1+4+4}} = \pm \frac{(x + 2y + 2z - 1)}{3}$$

But centre (x, y, z) and vertex O (0, 0, 0) intersection of planes (1), (2), (3) lie on the same side of plane (4) namely.

$$x + 2y + 2z - 1 = 0$$

$\therefore$ $x + 2y + 2z - 1$ and -1 are of the same sign [substituting the co-ordinates of the two points (x, y, z) and origin in L.H.S. of (4)]

$\therefore$ $x + 2y + 2z - 1$ is negative.

$\therefore$ $\perp$ distance of centre (x, y, z) from plane (4)

$$= -\frac{(x + 2y + 2z - 1)}{3}$$

Equating the four perpendicular distances,

$$x = y = z = -\frac{(x + 2y + 2z - 1)}{3}$$

From first and fourth members

$$3x = -x - 2y - 2z + 1$$

or $$3x = -x - 2x - 2x + 1 \qquad \mid \because \; y = x, z = x$$

or $$8x = 1$$

$\therefore$ $$x = \frac{1}{8}, \; y = \frac{1}{8}, \; z = \frac{1}{8}$$

$\therefore$ Centre of the sphere is

$$(x, y, z) = \left(\frac{1}{8}, \frac{1}{8}, \frac{1}{8}\right)$$

and radius of sphere is

$$x = \frac{1}{8}$$

$\therefore$ Equation of the sphere is

$$\left(x - \frac{1}{8}\right)^2 + \left(y - \frac{1}{8}\right)^2 + \left(z - \frac{1}{8}\right)^2 = \left(\frac{1}{8}\right)^2$$

or $$32(x^2 + y^2 + z^2) - 8(x + y + z) + 1 = 0$$

which is the required equation.

Remark. To understand the above problem the reader is suggested to draw an Analogy by drawing *a triangle inscribed in a circle.*

EXERCISE 2

1. (*a*) Find the equation of the sphere through the four points (0, 0, 0), (1, 0, 0), (0, 2, 0) and (0, 0, 3).
   (*b*) Find the equation of the sphere through the four points
   (0, 0, 0), (0, 1, – 1), (– 1, 2, 0), (1, 2, 3).
2. (*a*) Find the equation of the sphere which passes through the points (1, – 4, 3), (1, – 5, 2), (1, – 3, 0) and whose centre lies on the plane $x + y + z = 0$.
   (*b*) Find the equation of the sphere which passes through the points $(a, 0, 0)$, $(0, a, 0)$, $(0, 0, a)$ and whose centre lies on the plane $x + y + z = a$.
   (*c*) Find the equation of the sphere which passes through the points (3, 0, 2), (– 1, 1, 1), (2, – 5, 4) and has its centre on the plane $2x + 3y + 4z = 6$. *(M.D.U. 1997)*
3. Obtain the sphere having its centre on the line $5y + 2z = 0$, $2x - 3y = 0$ and passing through the points (0, – 2, – 4) and (2, – 1, – 1).

4. Find the equation of the sphere circumscribing the tetrahedron whose faces are $x = 0$, $y = 0$, $z = 0$,

$$\frac{x}{a} + \frac{y}{b} + \frac{z}{c} = 1.$$

[**Hint.** *A sphere circumscribing the tetrahedron means a sphere passing through the four vertices of the tetrahedron.*]

5. (*a*) Find the equation of the sphere through the four points (0, 0, 0), (– *a*, *b*, *c*), (*a*, – *b*, *c*), (*a*, *b*, – *c*) and determine its radius.

(*b*) Prove that the equation to the sphere circumscribing the tetrahedron whose sides are

$$\frac{y}{b} + \frac{z}{c} = 0, \qquad \frac{z}{c} + \frac{x}{a} = 0,$$

$$\frac{x}{a} + \frac{y}{b} = 0, \quad \frac{x}{a} + \frac{y}{b} + \frac{z}{c} = 1$$

is
$$\frac{x^2 + y^2 + z^2}{a^2 + b^2 + c^2} - \frac{x}{a} - \frac{y}{b} - \frac{z}{c} = 0.$$

6. Find the equation of the sphere which passes through the points (1, 0, 0), (0, 1, 0), (0, 0, 1) and has its radius as small as possible.

[**Hint.** $R^2 = \frac{1}{4}(3d^2 + 2d + 3) = \frac{3}{4}\left(d^2 + \frac{2}{3}d + 1\right) = \frac{3}{4}\left(d^2 + \frac{2}{3}d + \frac{1}{9} + \frac{8}{9}\right) = \frac{3}{4}\left[\left(d + \frac{1}{3}\right)^2 + \frac{8}{9}\right]$

$\therefore$ *R is least when* $d + \frac{1}{3} = 0$ *i.e.,* $d = \frac{-1}{3}$.]

7. OA, OB, OC are three mutually perpendicular lines through the origin whose direction cosines are l_1, m_1, n_1 ; l_2, m_2, n_2 ; l_3, m_3, n_3. If OA = *a*, OB = *b*, OC = *c* ; show that the equation of the sphere OABC is

$$x^2 + y^2 + z^2 - x(al_1 + bl_2 + cl_3) - y(am_1 + bm_2 + cm_3) - z(an_1 + bn_2 + cn_3) = 0.$$

[**Hint.** Use the 12 relations of Art. 4. Review IV]

8. (*a*) A plane passes through a fixed point (*a*, *b*, *c*) and cuts the axes in A, B, C. Show that the locus of centre of the sphere OABC is

$$\frac{a}{x} + \frac{b}{y} + \frac{c}{z} = 2.$$ *(M.D.U. 2003)*

(*b*) A plane through a fixed point (1, 1, 1) cuts the axes in A, B, C. Find the locus of the centre of the sphere OABC, where O is the origin.

9. A sphere of constant radius *k* passes through the origin and meets the axes in A, B, C. Prove that the centroid of the triangle ABC lies on the sphere $9(x^2 + y^2 + z^2) = 4k^2$.

10. A variable sphere passes through the origin O and meets the axes in A, B, C so that the volume of the tetrahedron OABC is constant. Find the locus of the centre of the sphere.

11. A sphere of constant radius *r* passes through the origin O and cuts the axes in A, B, C. Prove that the locus of the foot of perpendicular from O to the plane ABC is given by

$$(x^2 + y^2 + z^2)^2 (x^{-2} + y^{-2} + z^{-2}) = 4r^2.$$

12. A sphere of radius R passes through the origin. Show that the ends of the diameter which is parallel to *x*-axis lie on each of the spheres $x^2 + y^2 + z^2 \pm 2Rx = 0$.

Answers

1. (*a*) $x^2 + y^2 + z^2 - x - 2y - 3z = 0$ (*b*) $7(x^2 + y^2 + z^2) - 15x - 25y - 11z = 0$
2. (*a*) $x^2 + y^2 + z^2 - 4x + 7y - 3z + 15 = 0$ (*b*) $3(x^2 + y^2 + z^2) - 2a(x + y + z) - a^2 = 0$
   (*c*) $x^2 + y^2 + z^2 + 4y - 6z = 1$
3. $x^2 + y^2 + z^2 - 6x - 4y + 10z + 12 = 0$ 4. $x^2 + y^2 + z^2 - ax - by - cz = 0$

5. (a) $\dfrac{x^2+y^2+z^2}{a^2+b^2+c^2} - \dfrac{x}{a} - \dfrac{y}{b} - \dfrac{z}{c} = 0,\ \dfrac{(a^2+b^2+c^2)}{2}\sqrt{\dfrac{1}{a^2}+\dfrac{1}{b^2}+\dfrac{1}{c^2}}$

6. $3(x^2+y^2+z^2) - 2(x+y+z) - 1 = 0$

8. (b) $\dfrac{1}{x}+\dfrac{1}{y}+\dfrac{1}{z} = 2$ 10. $xyz = \pm k$.

7. Intersections of a Straight Line and a Sphere

To find the points where the line

$$\frac{x-x_1}{l} = \frac{y-y_1}{m} = \frac{z-z_1}{n}$$

meets the sphere

$$x^2+y^2+z^2+2ux+2vy+2wz+d=0$$

Equation of the sphere is

$$x^2+y^2+z^2+2ux+2vy+2wz+d=0 \quad \text{...(1)}$$

Equations of the line are

$$\frac{x-x_1}{l} = \frac{y-y_1}{m} = \frac{z-z_1}{n} \ (= r) \quad \text{...(2)}$$

Any point on line (2) is

$$(lr+x_1,\ mr+y_1,\ nr+z_1) \quad \text{...(3)}$$

If it lies on sphere (1), then

$$(x_1+lr)^2+(y_1+mr)^2+(z_1+nr)^2+2u(x_1+lr)+2v(y_1+mr)+2w(z_1+nr)+d=0$$

or

$$r^2(l^2+m^2+n^2)+2r(lx_1+my_1+nz_1+lu+mv+nw) + (x_1^2+y_1^2+z_1^2+2ux_1+2vy_1+2wz_1+d)=0 \quad \text{...(4)}$$

which is a quadratic in r giving two values of r, real coincident or imaginary.

Putting these values of r one by one in (3), we get the two points of intersection which may be real, coincident or imaginary.

Cor. Power of a point w.r.t. a sphere

If in equation (2) l, m, n are the actual directions of the line (2), then equation (4) reduces to

$$r^2+2r(lx_1+my_1+nz_1+lu+mv+nw) + (x_1^2+y_1^2+z_1^2+2ux_1+2vy_1+2wz_1+d)=0 \quad \text{...(5)}$$

Let the roots of this quadratic be r_1 and r_2.

Now r_1, r_2 are the **distances** ($\because$ $<l, m, n>$ are the d.c's) of the two points of intersection P and Q (say) from a given point $A(x_1, y_1, z_1)$ on the line (2)

i.e., r_1, r_2 are the distances AP and AQ.

$\therefore$ From (5) $\quad AP\,.\,AQ = r_1r_2 = \dfrac{x_1^2+y_1^2+z_1^2+2ux_1+2vy_1+2wz_1+d}{1}$

or

$$\mathbf{AP\,.\,AQ = x_1^2+y_1^2+z_1^2+2ux_1+2vy_1+2wz_1+d}$$

which is constant because it is independent of l, m, n (the direction cosines of the line (2)].

This constant is called the power of the point A w.r.t. the given sphere.

Note. Rule to write the power of a point w.r.t. a sphere.

Substitute the co-ordinates of the point in the L.H.S. of the equation of the sphere (after making the R.H.S. zero). This expression obtained is called power of the point w.r.t. the sphere.

8. (*a*) **Tangent Line**

Def. A straight line which intersects a sphere in two **coincident** points is called a *tangent line* of the sphere.

(*b*) **Length of the tangent**

To find the length of the tangent from the point (x_1, y_1, z_1) *to the sphere* $x^2 + y^2 + z^2 + 2ux + 2vy + 2wz + d = 0$.

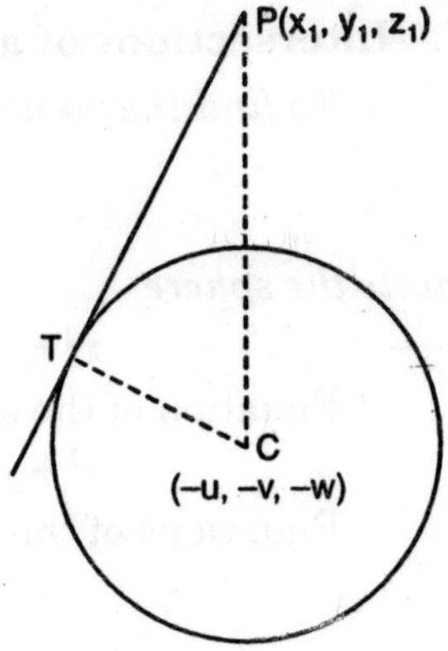

Let P be the point (x_1, y_1, z_1) and C the centre of the sphere and PT a tangent from P to the sphere.

The co-ordinates of C are

$$(-u, -v, -w).$$

CT is the radius of the sphere and

$$= \sqrt{u^2 + v^2 + w^2 - d}$$

CT is perpendicular to PT. [See Cor. 1, Art. 9 (*b*)]

$$\therefore \quad PT^2 = CP^2 - CT^2$$

$$= (x_1 + u)^2 + (y_1 + v)^2 + (z_1 + w)^2 - [u^2 + v^2 + w^2 - d]$$

$$= x_1^2 + u^2 + 2ux_1 + y_1^2 + v^2 + 2vy_1 + z_1^2 + w^2 + 2wz_1 - u^2 - v^2 - w^2 + d$$

or $$\mathbf{PT^2 = x_1^2 + y_1^2 + z_1^2 + 2ux_1 + 2vy_1 + 2wz_1 + d.}$$

Cor. The point (x_1, y_1, z_1) lies outside, on or inside the sphere according as

$$x_1^2 + y_1^2 + z_1^2 + 2ux_1 + 2vy_1 + 2wz_1 + d >=< 0.$$

Note. Rule to find the square of length of tangent from a point to the sphere is the same as given in Note Art. 7.

9. (*a*) **Definitions**

(*i*) **Conicoid.** A conicoid is a surface whose equation is of the second degree in *x*, *y*, *z*.

(*ii*) **Tangent Plane.** The locus of the tangent lines at a point P of a sphere is a plane called **the tangent plane** to the sphere at P.

P is called the **point of contact.**

(*b*) **Equation of the tangent plane to a sphere**

To find the equation of the tangent plane at the point (x_1, y_1, z_1) *of the sphere*

$$x^2 + y^2 + z^2 + 2ux + 2vy + 2wz + d = 0.$$

Equation of the sphere is

$$x^2 + y^2 + z^2 + 2ux + 2vy + 2wz + d = 0 \quad \text{...(1)}$$

$\because$ The point $P(x_1, y_1, z_1)$ lies on the sphere.

$$\therefore \quad x_1^2 + y_1^2 + z_1^2 + 2ux_1 + 2vy_1 + 2wz_1 + d = 0 \quad \text{...(2)}$$

Equations of *any* line through (x_1, y_1, z_1) are

$$\frac{x - x_1}{l} = \frac{y - y_1}{m} = \frac{z - z_1}{n} (= r) \quad \text{...(3)}$$

Any point on line (3) is $(lr + x_1, mr + y_1, nr + z_1)$.

If it lies on sphere (1), then

$$(lr + x_1)^2 + (mr + y_1)^2 + (nr + z_1)^2 + 2u(lr + x_1) + 2v(mr + y_1) + 2w(nr + z_1) + d = 0$$

or $$r^2(l^2 + m^2 + n^2) + 2r(lx_1 + my_1 + nz_1 + lu + mv + nw) + (x_1^2 + y_1^2 + z_1^2 + 2ux_1 + 2vy_1 + 2wz_1 + d) = 0$$

or $$r^2(l^2 + m^2 + n^2) + 2r(lx_1 + my_1 + nz_1 + lu + mv + nw) = 0 \quad ...(4) \quad | \text{ Using (2)}$$

which is a quadratic in r.

If the line (3) is a tangent to sphere (1), then the two values of r from (4) are equal *i.e.*, disc of (4) is 0.

i.e., $$4(lx_1 + my_1 + nz_1 + lu + mv + nw)^2 - 4(l^2 + m^2 + n^2) \cdot 0 = 0$$

or $$lx_1 + my_1 + nz_1 + lu + mv + nw = 0 \quad ...(5)$$

Equation (5) is the condition for the line (3) to be tangent to the sphere (1).

To find the locus of tangent lines (3), at the point (x_1, y_1, z_1) ; we are to eliminate l, m, n from (3) and (5).

Eliminating l, m, n from (3) and (5), we have

$$(x - x_1)x_1 + (y - y_1)y_1 + (z - z_1)z_1 + (x - x_1)u + (y - y_1)v + (z - z_1)w = 0.$$

or $$xx_1 + yy_1 + zz_1 + ux + vy + wz = x_1^2 + y_1^2 + z_1^2 + ux_1 + vy_1 + wz_1$$

Adding $ux_1 + vy_1 + wz_1 + d$ to both sides,

$$xx_1 + yy_1 + zz_1 + u(x + x_1) + v(y + y_1) + w(z + z_1) + d$$
$$= x_1^2 + y_1^2 + z_1^2 + 2ux_1 + 2vy_1 + 2wz_1 + d = 0 \quad | \text{ Using (2)}$$

or $$\mathbf{xx_1 + yy_1 + zz_1 + u(x + x_1) + v(y + y_1) + w(z + z_1) + d = 0} \quad ...(6)$$

which is the required equation of the tangent plane.

Cor. 1. *The radius of the sphere through the point of contact of the tangent line is perpendicular to the tangent line.*

Proof. Centre C of the sphere (1) is $(-u, -v, -w)$

d.r's of radius CP are $< x_1 - (-u), y_1 - (-v), z_1 - (-w) >$

i.e., $$< x_1 + u, y_1 + v, z_1 + w >$$

d.r's of the tangent line (3) are $< l, m, n >$

Radius CP is $\perp$ to tangent line (3) if

$$l(x_1 + u) + m(y_1 + v) + n(z_1 + w) = 0 \quad | \; aa' + bb' + cc' = 0$$

i.e., if $$lx_1 + my_1 + nz_1 + lu + mv + nw = 0$$

which is true by (5).

Cor. 2. *The radius of the sphere through the point of contact of the tangent plane is perpendicular to the tangent plane.*

Proof. Centre C of sphere (1) is $(-u, -v, -w)$

d.r's of radius CP are $< x_1 - (-u), y_1 - (-v), z_1 - (-w) >$

i.e., $$< x_1 + u, y_1 + v, z_1 + w >$$

From (6) equation of tangent plane is

$$xx_1 + yy_1 + zz_1 + u(x + x_1) + v(y + y_1) + w(z + z_1) + d = 0$$

or $$x(x_1 + u) + y(y_1 + v) + z(z_1 + w) + (ux_1 + vy_1 + wz_1 + d) = 0$$

Co-efficients of x, y, z in the equation of tangent plane are $x_1 + u, y_1 + v, z_1 + w$ which are the same as d.r's of radius CP.

$\therefore$ Radius CP is $\perp$ to the tangent plane at P.

Remark. $\therefore$ Radius CP through the point of contact is **Normal** to the sphere at P.

Note 1. Method to write down the equation of tangent plane at a point (x_1, y_1, z_1) of the sphere or a conicoid.

In the given equation of the surface, change x^2 *to* xx_1, y^2 *to* yy_1, z^2 *to* zz_1 *and* x *to* $\frac{1}{2}(x + x_1)$, y *to* $\frac{1}{2}(y + y_1)$, z *to* $\frac{1}{2}(z + z_1)$ *and* yz *to* $\frac{1}{2}(yz_1 + zy_1)$ *etc.*

Note. Tangent plane property.

If a plane touches a sphere, then the ⊥ distance of the centre from the plane = radius.

Similarly if a line touches a sphere, then the ⊥ distance of the centre from the line = radius of the sphere.

Example 1. *Find the condition that the plane* $lx + my + nz = p$ *may touch the sphere* $x^2 + y^2 + z^2 + 2ux + 2vy + 2wz + d = 0$.

Sol. The given sphere is

$$x^2 + y^2 + z^2 + 2ux + 2vy + 2wz + d = 0 \quad ...(1)$$

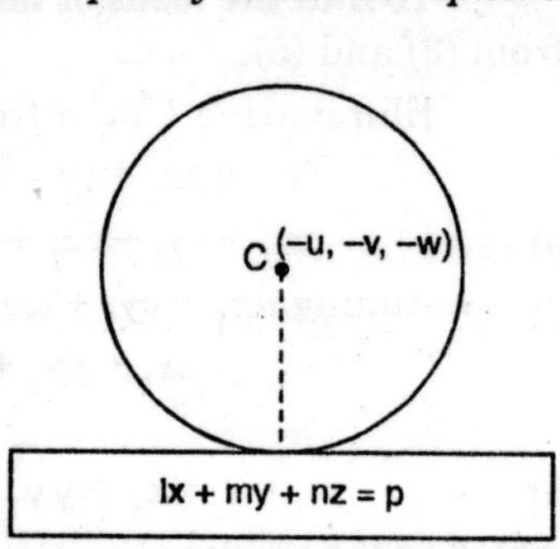

Its centre is $(-u, -v, -w)$ and radius is $\sqrt{u^2 + v^2 + w^2 - d}$.

The given plane is

$$lx + my + nz - p = 0 \quad ...(2)$$

Since the plane (2) touches the sphere (1).

∴ ⊥ distance of the centre $(-u, -v, -w)$ from the plane (2)

$$= \text{radius of sphere} = \sqrt{u^2 + v^2 + w^2 - d}$$

or
$$\pm \frac{l(-u) + m(-v) + n(-w) - p}{\sqrt{l^2 + m^2 + n^2}} = \sqrt{u^2 + v^2 + w^2 - d}$$

or
$$\mp (lu + mv + nw + p) = \sqrt{l^2 + m^2 + n^2}\ \sqrt{u^2 + v^2 + w^2 - d}$$

Squaring, $\quad (lu + mv + nw + p)^2 = (l^2 + m^2 + n^2)(u^2 + v^2 + w^2 - d)$.

Example 2. *Find the condition that the sphere* $x^2 + y^2 + z^2 + 2ux + 2vy + 2wz + d = 0$ *may touch the line* $\frac{x - \alpha}{l} = \frac{y - \beta}{m} = \frac{z - \gamma}{n}$.

Sol. The given sphere is

$$x^2 + y^2 + z^2 + 2ux + 2vy + 2wz + d = 0 \quad ...(1)$$

Its centre is $(-u, -v, -w)$ and radius is $\sqrt{u^2 + v^2 + w^2 - d}$.

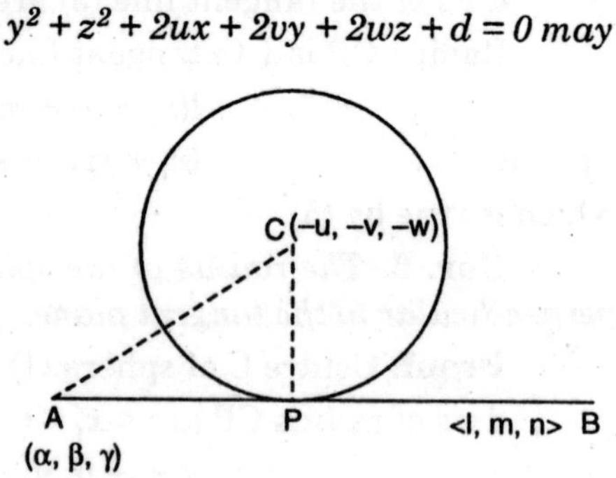

The given line is

$$\frac{x - \alpha}{l} = \frac{y - \beta}{m} = \frac{z - \gamma}{n} \quad ...(2)$$

(l, m , n are d.r's of the line)

Since the line (2) touches the sphere (1).

∴ ⊥ distance CP of centre $(-u, -v, -w)$ from line (2)

$$= \text{radius of sphere} = \sqrt{u^2 + v^2 + w^2 - d} \quad ...(3)$$

By Pythagorus Theorem in right angled ΔAPC,

$$AC^2 = AP^2 + CP^2 \quad ...(4)$$

$$AC^2 = (\alpha + u)^2 + (\beta + v)^2 + (\gamma + w)^2$$ | Distance Formula

AP = Projection of AC on the line AB whose d.c's are

$$\left\langle \frac{l}{\sqrt{\Sigma l^2}}, \frac{m}{\sqrt{\Sigma l^2}}, \frac{n}{\sqrt{\Sigma l^2}} \right\rangle = \frac{l(\alpha + u)}{\sqrt{\Sigma l^2}} + \frac{m(\beta + v)}{\sqrt{\Sigma l^2}} + \frac{n(\gamma + w)}{\sqrt{\Sigma l^2}}$$

| Using $l(x_2 - x_1) + m(y_2 - y_1) + n(z_2 - z_1)$

$$= \frac{l(\alpha + u) + m(\beta + v) + n(\gamma + w)}{\sqrt{l^2 + m^2 + n^2}}$$

Putting the values of CP, AC² and AP in (4),

$$(\alpha + u)^2 + (\beta + v)^2 + (\gamma + w)^2 = \frac{[l(\alpha + u) + m(\beta + v) + n(\gamma + w)]^2}{l^2 + m^2 + n^2} + (u^2 + v^2 + w^2 - d)$$

which is the required condition.

Example 3. *Show that the plane $2x - 2y + z + 12 = 0$ touches the sphere $x^2 + y^2 + z^2 - 2x - 4y + 2z - 3 = 0$ and find the point of contact.*

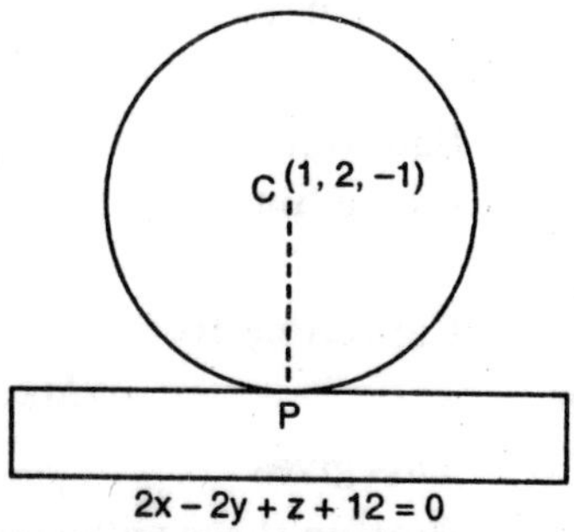

Sol. Equation of the sphere is

$$x^2 + y^2 + z^2 - 2x - 4y + 2z - 3 = 0 \quad ...(1)$$

Its centre is (1, 2, – 1) and radius is

$$\sqrt{1 + 4 + 1 - (-3)} = 3$$

Equation of the plane is

$$2x - 2y + z + 12 = 0 \quad ...(2)$$

⊥ distance of centre C(1, 2, – 1) from plane (2)

$$= \pm \frac{(2 - 4 - 1 + 12)}{\sqrt{4 + 4 + 1}} \quad \Bigg| \ \pm \frac{(ax_1 + by_1 + cz_1 + d)}{\sqrt{a^2 + b^2 + c^2}}$$

$$= \pm \frac{9}{3} = 3 = \text{Radius of the sphere}$$

∴ plane (2) touches the sphere (1).

Now let us find the point of contact P.

Point of contact is the foot of perpendicular from centre of the sphere on the tangent plane (2)

d.r's of CP (⊥ to tangent plane (2) are < 2, – 2, 1 > | Co-effs. of x, y, z in plane (2)

∴ Equations of CP are $\frac{x - 1}{2} = \frac{y - 2}{-2} = \frac{z + 1}{1} = r$

Any point on CP is $(2r + 1, - 2r + 2, r - 1)$ ...(3)

If it lies on plane (2), then

$$2(2r + 1) - 2(-2r + 2) + r - 1 + 12 = 0$$

or $9r + 9 = 0 \quad \therefore \quad r = -1$

Putting $r = -1$ in (3), the co-ordinates of the point of contact P are (– 1, 4, – 2).

Example 4. *Find the equations of the tangent planes to the sphere $x^2 + y^2 + z^2 + 2x - 4y + 6z - 7 = 0$ which intersect in the line*

$$6x - 3y - 23 = 0 = 3z + 2.$$ (K.U. 1999 S)

Sol. Equations of the line are

$$6x - 3y - 23 = 0$$
$$3z + 2 = 0$$

Equation of any plane through this line is

$$6x - 3y - 23 + k(3z + 2) = 0 \qquad [\text{L.K.S.} + k(\text{L.H.S.}) = 0]$$

or
$$6x - 3y + 3kz + (2k - 23) = 0 \qquad ...(1)$$

But this plane (1) is a tangent plane to the sphere

$$x^2 + y^2 + z^2 + 2x - 4y + 6z - 7 = 0$$

$\therefore$ $\perp$ distance of the centre $(-1, 2, -3)$ from plane (1)

$$= \text{Radius } \sqrt{1+4+9-(-7)}$$

i.e.,
$$\pm\frac{(-6-6-9k+2k-23)}{\sqrt{36+9+9k^2}} = \sqrt{21}$$

i.e.,
$$\pm(-7k - 35) = \sqrt{21}\sqrt{9k^2 + 45}$$

i.e.,
$$\mp 7(k+5) = \sqrt{21}\sqrt{9k^2 + 45}$$

Squaring $\qquad 49(k+5)^2 = 21(9k^2 + 45)$

Cancelling 7, $\qquad 7(k^2 + 10k + 25) = 3(9k^2 + 45)$

or $\qquad 20k^2 - 70k - 40 = 0$

Cancelling 10, $\qquad 2k^2 - 7k - 4 = 0$ or $2k^2 - 8k + k - 4 = 0$

or $\qquad 2k(k-4) + 1(k-4) = 0$ or $(k-4)(2k+1) = 0$

This gives $\qquad k = 4, -\frac{1}{2}$

Putting these values of k in (1), equations of the tangent planes are

$$2x - y + 4z - 5 = 0, \quad 4x - 2y - z - 16 = 0.$$

Example 5. *Find the centres of the two spheres which touch the plane $x + 2y + 2z - 5 = 0$ at the point (1, 1, 1) and the sphere*

$$x^2 + y^2 + z^2 + 2x + 4y + 6z - 11 = 0. \qquad \text{(K.U. 2007)}$$

Sol. Let C be the centre of the required sphere which touches the plane

$$x + 2y + 2z - 5 = 0 \qquad ...(1)$$

at the point P(1, 1, 1).

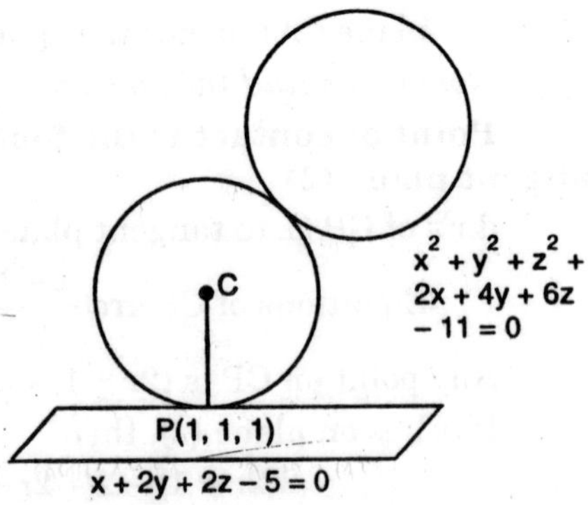

$\therefore$ CP is $\perp$ to tangent plane (1).

$\therefore$ D.R.'s of CP are <1, 2, 2>

| Co-effs. of x, y, z in (1)

$\therefore$ Equation of CP are

$$\frac{x-1}{1} = \frac{y-1}{2} = \frac{z-1}{2} = r$$

$\therefore$ Any point on CP is $C(r+1, 2r+1, 2r+1)$ $\qquad ...(2)$

R = Radius of required sphere = CP

$$= \sqrt{(r+1-1)^2 + (2r+1-1)^2 + (2r+1-1)^2}$$

$$= \sqrt{r^2 + 4r^2 + 4r^2} = \sqrt{9r^2} = 3r$$

Equation of the given sphere is

$$x^2 + y^2 + z^2 + 2x + 4y + 6z - 11 = 0 \quad ...(3)$$

Centre of this sphere is C′(– 1, – 2, – 3) [$-\frac{1}{2}$ co-eff. of x, ...]

Radius of this sphere is

$$R' = \sqrt{1+4+9+11} \quad | \sqrt{u^2 + v^2 + w^2 - d}$$

$$= 5.$$

Because the required sphere touches the sphere (3) (given) [externally or internally] [In the figure, we have shown only externally touching spheres].

$\therefore$ $$CC' = R \pm R'.$$

$$\Rightarrow \quad \sqrt{(r+1+1)^2 + (2r+1+2)^2 + (2r+1+3)^2} = 3r \pm 5$$

$$\Rightarrow \quad \sqrt{(r+2)^2 + (2r+3)^2 + (2r+4)^2} = 3r \pm 5.$$

S.B.S. $$(r + 2)^2 + (2r + 3)^2 + (2r + 4)^2 = (3r \pm 5)^2$$

or $$r^2 + 4 + 4r + 4r^2 + 9 + 12r + 4r^2 + 16 + 16r = 9r^2 + 25 \pm 30r$$

or $$32r + 4 = \pm 30r.$$

Taking positive sign, $$32r + 4 = 30r$$

or $$2r = -4 \quad \therefore \quad r = -2$$

Putting $r = -2$ in (2), centre of one required sphere is

$$C(-2 + 1, -4 + 1, -4 + 1) = C(-1, -3, -3).$$

Taking negative sign, $32r + 4 = -30r$

or $$62r = -4 \quad \therefore \quad r = \frac{-2}{31}.$$

Putting $r = \frac{-2}{31}$ in (2), centre of the second required sphere is

$$C\left(\frac{-2}{31}+1, \frac{-4}{31}+1, \frac{-4}{31}+1\right) \quad \text{or} \quad C\left(\frac{29}{31}, \frac{27}{31}, \frac{27}{31}\right)$$

$\therefore$ Centres of the two required spheres are

$$C(-1, -3, -3) \text{ and } C\left(\frac{29}{31}, \frac{27}{31}, \frac{27}{31}\right).$$

Example 6. *Prove that the centres of the sphere which touch the lines $y = mx$; $z = c$; $y = -mx$, $z = -c$ lie upon the conicoid*

$$mxy + cz(l + m^2) = 0.$$ (M.D.U. 2007 ; K.U. 1998 A)

Sol. Let the sphere be

$$x^2 + y^2 + z^2 + 2ux + 2vy + 2wz + d = 0 \quad ...(1)$$

Let $C(x_1, y_1, z_1)$ be the centre of the sphere.

$\therefore$ $$x_1 = -u, y_1 = -v, z_1 = -w \quad ...(2)$$

$\because$ Sphere (1) touches the line $y = mx$, $z = c$ ...(3)

$\therefore$ Line (3) meets the sphere (1) in two coincident points.

Solving (3) and (1) *i.e.*, putting $y = mx$ and $z = c$ in (1),

$$x^2 + m^2x^2 + c^2 + 2ux + 2vmx + 2wc + d = 0$$

or $$x^2(1 + m^2) + 2x(u + vm) + (c^2 + 2wc + d) = 0 \quad ...(4)$$

Two values of x from (4) are equal

[$\because$ Line (3) meets the sphere (1) in two coincident points]

$\therefore$ Disc. of (4) = 0.

$\therefore$ $4(u + vm)^2 - 4(1 + m^2)(c^2 + 2wc + d) = 0.$ | "$b^2 - 4ac$" = 0

Cancelling 4, $(u + vm)^2 = (1 + m^2)(c^2 + 2wc + d)$ ...(5)

Also sphere (1) touches the line $y = -mx, z = -c$

$\therefore$ Changing m to $-m$ and c to $-c$ in (5), we have

$$(u - vm)^2 = (1 + m^2)(c^2 - 2wc + d) \quad ...(6)$$

(5) – (6) gives (To eliminate d)

$$(u + vm)^2 - (u - vm)^2 = (1 + m^2)[c^2 + 2wc + d - c^2 + 2wc - d]$$

or $$4uvm = (1 + m^2)4wc$$

Cancelling 4, $uvm = cw(1 + m^2)$ ...(7)

Eliminating u, v, w [*i.e.*, putting $u = -x_1$, $v = -y_1$, $w = -z_1$ from (2) in (7)],

we have $$x_1y_1m = -cz_1(1 + m^2)$$

or $$mx_1y_1 + cz_1(1 + m^2) = 0$$

$\therefore$ Locus of centres (x_1, y_1, z_1) is

$$mxy + cz(1 + m^2) = 0$$

which being a second degree ($\because$ of xy) represents a conicoid.

EXERCISE 3

1. Find the equation of the tangent plane at a point (x_1, y_1, z_1) of the sphere $x^2 + y^2 + z^2 = a^2$.

2. Find the condition that the plane $lx + my + nz = p$ is a tangent plane of the sphere
$$x^2 + y^2 + z^2 = a^2.$$

3. Find the value of a for which $x + y + z = a\sqrt{3}$ touches the sphere
$$x^2 + y^2 + z^2 - 2x - 2y - 2z - 6 = 0.$$

4. (*a*) Show that the plane $3x - 4y - 25 = 0$ touches the sphere $x^2 + y^2 + z^2 - 25 = 0$. Find the point of contact.

   (*b*) Show that $2x - y - 2z = 4$ is a tangent plane to the sphere
$$x^2 + y^2 + z^2 + 2x - 6y + 1 = 0.$$
   Also find the point of contact.

5. Find the equations of the tangents planes to the sphere $x^2 + y^2 + z^2 - 2x - 8 = 0$ perpendicular to the line with direction ratios <1, 2, 2>.

   [**Hint.** Equation of any plane perpendicular, to the line with direction ratios <1, 2, 2> is $x + 2y + 2z + k = 0$.]

6. (*a*) Find the two tangent planes to the sphere
$$x^2 + y^2 + z^2 - 4x + 2y - 6z + 5 = 0$$
   which are parallel to the plane $2x + 2y - z = 0$.

   [**Hint.** Equation of any plane $\parallel$ to $2x + 2y - z = 0$ is $2x + 2y - z + k = 0$.]

   (*b*) Find the equations of the two tangent planes to the sphere $x^2 + y^2 + z^2 - 4x + 2y - 6z + 5 = 0$ which are parallel to the plane $2x + 2y = 5$.

7. **(a) Obtain the equations of the tangent planes to the sphere $x^2 + y^2 + z^2 = 9$, which can be drawn through the line**

$$\frac{x-5}{2} = -\frac{(y-1)}{2} = \frac{z-1}{1}$$

or $$x + y = 6,\ x - 2z = 3.$$

[Hint. Change the equations of the line to general form.]

(b) Find the equations of the tangent planes to the sphere

$$x^2 + y^2 + z^2 - 2x - 2y - 2z = 0$$

which contain the line given by $x - z = 0,\ y + 2 = 0$. *(M.D.U. 1996)*

8. **Find the centres of the two spheres which touch the plane $3x + 2y + 6z - 7 = 0$ at the point $(1, -1, 1)$ and also touch the sphere $x^2 + y^2 + z^2 - 2x - 4y - 6z + 10 = 0$.**

9. **Find the equation of the sphere with centre at the origin and which touches the line**

$$2(x + 1) = 2 - y = z + 3.$$

10. **Find the equation of a sphere touching the three co-ordinate planes. How many such spheres can be drawn ?**

[Hint. x co-ordinate of centre of sphere

$= \perp$ distance of centre from plane YOZ

$= \pm$ Radius λ ($\because$ The sphere touches the plane YOZ)

$\therefore$ Centre of sphere is $(\pm\lambda, \pm\lambda, \pm\lambda)$ and radius is λ.]

11. **Prove that the equation to a sphere, which lies in the octant OXYZ and touches the co-ordinate planes is of the form $x^2 + y^2 + z^2 - 2\lambda(x + y + z) + 2\lambda^2 = 0$.**

Prove that in general two spheres can be drawn through a given point to touch the co-ordinate planes.

12. **Find the equations of the two spheres whose centres lie in the positive octant and which touch the planes $x = 0, y = 0, z = 0, 2x + 2y + z = 8a$.**

13. **Find the equation of the sphere which touches the three co-ordinate axes. How many spheres can be so drawn.**

[Hint. Let the sphere be $x^2 + y^2 + z^2 + 2ux + 2vy + 2wz + d = 0$

Solve it with x-axis ($y = 0, z = 0$) to get $x^2 + 2ux + d = 0$.

Now put discriminant $= 0$ *i.e.*, $4u^2 - 4d = 0$ $\therefore$ $u^2 = d$

$\therefore$ $u = \pm\sqrt{d}$ etc.]

14. **Show that the sum of squares of the intercepts made by a given sphere on three mutually perpendicular straight lines through a fixed point is constant.**

[Hint. Let us take the three mutually perpendicular straight lines as axes. Let the equation of the sphere be $x^2 + y^2 + z^2 + 2ux + 2vy + 2wz + d = 0$. For intercept on x-axis, put $y = 0, z = 0$.

$\therefore$ $x^2 + 2ux + d = 0$.

Let the roots be x_1 and x_2.

$$\therefore \quad x_1 + x_2 = -2u\left(-\frac{b}{a}\right) \quad \text{and} \quad x_1x_2 = d\left(\frac{c}{a}\right)$$

Intercept on x-axis $= x_2 - x_1 = \sqrt{(x_1 + x_2)^2 - 4x_1x_2}$.]

15. **Tangent plane at any point of the sphere $x^2 + y^2 + z^2 = r^2$ meets the co-ordinate axes in A, B, C. Show that the locus of the point of intersection of planes drawn parallel to the co-ordinate planes through A, B, C is the surface $x^{-2} + y^{-2} + z^{-2} = r^{-2}$.**

$\left[\textbf{Hint.} \text{ Let the tangent plane be } \frac{x}{a} + \frac{y}{b} + \frac{z}{c} = 1.\right]$

16. Prove that the locus of the centres of spheres which pass through a given point and touch a given plane is a conicoid.

[**Hint.** Let the given plane be XY-plane (*i.e.*, $z = 0$) and the given point be $(0, 0, c)$ on the Z-axis.]

17. Find the locus of the centres of spheres which pass through a given point and touch a given line.

[**Hint.** Let the given line be x-axis ($y = 0$, $z = 0$) and the given point be $(0, 0, c)$ on the z-axis.]

Answers

1. $xx_1 + yy_1 + zz_1 = a^2$ 2. $a^2(l^2 + m^2 + n^2) = p^2$ 3. $\sqrt{3} \pm 3$
4. (a) $(3, -4, 0)$ (b) $(1, 2, -2)$ 5. $x + 2y + 2z + 8 = 0$, $x + 2y + 2z - 10 = 0$
6. (a) $2x + 2y - z + 10 = 0$, $2x + 2y - z - 8 = 0$ (b) $x + y \pm 3\sqrt{2} - 1 = 0$
7. (a) $x + 2y + 2z - 9 = 0$, $2x + y - 2z - 9 = 0$ (b) $x + y - z + 2 = 0$, $x - y - z - 2 = 0$
8. $\left(\frac{91}{64}, \frac{-46}{64}, \frac{118}{64}\right), \left(\frac{35}{8}, \frac{10}{8}, \frac{62}{8}\right)$ 9. $9(x^2 + y^2 + z^2) = 5$
10. $x^2 + y^2 + z^2 \pm 2\lambda(x + y + z) + 2\lambda^2 = 0$, where λ is the radius of the sphere. Infinite.
12. $x^2 + y^2 + z^2 - 2a(x + y + z) + 2a^2 = 0$, $x^2 + y^2 + z^2 - 8a(x + y + z) + 32a^2 = 0$.
13. $x^2 + y^2 + z^2 \pm 2ax \pm 2ay \pm 2az + a^2 = 0$. Infinite. 17. $x^2 - 2cz + c^2 = 0$.

---

10. (a) Plane Section of a Sphere

To prove that the section of a sphere by a plane is a circle i.e., the points common to a sphere and a plane lie on a circle.

Let P be *any* point common to a sphere and the plane.

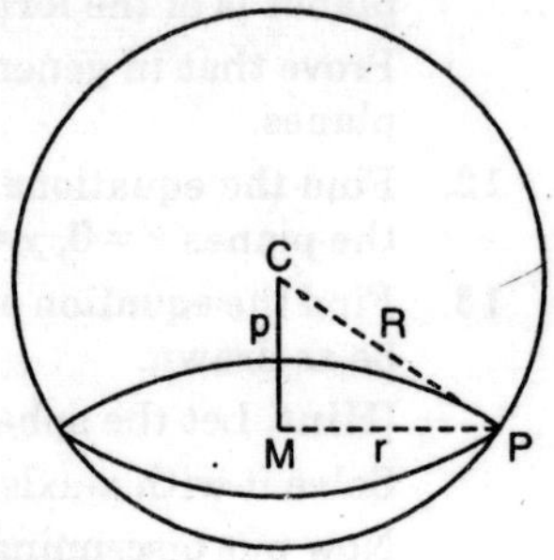

From C, the centre of the sphere, draw CM perpendicular to the plane. ($\therefore$ M is fixed)

$\therefore$ CM is $\perp$ MP also. Let CM $= p$

$\therefore$ $MP^2 = CP^2 - CM^2 = R^2 - p^2$

$\therefore$ $MP = \sqrt{R^2 - p^2}$

which is independent of P.

Also M is fixed.

$\therefore$ Locus of P is a circle whose centre is M and radius is $\sqrt{R^2 - p^2}$.

We thus arrive at the following results :

(i) The equations of a circle are the equations of a sphere and a plane taken together.

(ii) The centre of a circle is the foot of the perpendicular from the centre of the sphere on the plane of circle.

(iii) r, the radius of the circle $= \sqrt{R^2 - p^2}$, *where R is the radius of the sphere and p is the length of the perpendicular from the centre of the sphere on the plane.*

Note. The tangent plane at a point to the sphere cuts the sphere in a point circle.

$\therefore$ Equations of the point of contact are the equations of the sphere and the tangent plane taken together.

Def. Great Circle

The section of a sphere by a plane passing through its centre is called a great circle.

Note. $\because$ $p = 0$

$\therefore$ M coincides with C and $r = \sqrt{R^2 - p^2} = \sqrt{R^2 - 0^2} = R$.

Remark. Plane of the great circle passes through the centre of the sphere. Centre and radius of the great circle are the same as centre and radius of the sphere.

(b) Intersection of Two Spheres

Let the two spheres be

$$x^2 + y^2 + z^2 + 2u_1x + 2v_1y + 2w_1z + d_1 = 0 \qquad ...(1)$$

and

$$x^2 + y^2 + z^2 + 2u_2x + 2v_2y + 2w_2z + d_2 = 0 \qquad ...(2)$$

The points common to (1) and (2) will also lie on the surface [(1) – (2) = 0]

$$x^2 + y^2 + z^2 + 2u_1x + 2v_1y + 2w_1z + d_1 - (x^2 + y^2 + z^2 + 2u_2x + 2v_2y + 2w_2z + d_2) = 0$$

i.e., On $\quad 2(u_1 - u_2)x + 2(v_1 - v_2)y + 2(w_1 - w_2)z + (d_1 - d_2) = 0 \qquad ...(3)$

which is a plane.

Thus the common points of (1) and (2) are also the common points of (1) and (3) [or (2) and (3)] and ; therefore, they lie on a circle.

The equations of this circle are either (1) and (2) or (1) and (3) or (2) and (3).

(c) Plane Section of a Right Circular Cylinder

The section of a right circular cylinder by any plane perpendicular to its axis is a circle.

Now the equation of the right circular cylinder whose axis is the z-axis and radius a is

$$x^2 + y^2 = a^2 \qquad ...(1)$$ [See Art. 5 (*a*) Chapter 12]

Any plane ⊥ to its axis (*i.e.*, z-axis) is $z = k$ ...(2)

These equations (1) and (2) together represent a circle.

Adding square of equation (2) to equation (1), we get

$$x^2 + y^2 + z^2 = a^2 + k^2 \qquad ...(3)$$

which is the equation of a sphere.

Then the circle given by cylinder (1) and plane (2) is also given by sphere (3) and plane (2).

Example 1. *Find the centre and the radius of the circle in which the sphere $x^2 + y^2 + z^2 + 2x - 2y - 4z - 19 = 0$ is cut by the plane*

$$x + 2y + 2z + 7 = 0.$$

Sol. Equations of the circle are

$$x^2 + y^2 + z^2 + 2x - 2y - 4z - 19 = 0 \qquad ...(1)$$

and

$$x + 2y + 2z + 7 = 0 \qquad ...(2)$$

Centre of sphere (1) is C(– 1, 1, 2) | $(-u, -v, -w)$

Radius of sphere (1) is $\quad R = \sqrt{1+1+4+19} = 5$ | $\sqrt{u^2 + v^2 + w^2 - d}$

p = ⊥ distance from the centre C(– 1, 1, 2) on plane (2)

$$= \pm\frac{(-1+2+4+7)}{\sqrt{1+4+4}} = \frac{12}{3} = 4$$ | $\pm\dfrac{(ax_1 + by_1 + cz_1 + d)}{\sqrt{a^2 + b^2 + c^2}}$

∴ If r be the radius of the circle, then

$$r = \sqrt{R^2 - p^2} = \sqrt{5^2 - 4^2} = 3$$

Let us find the centre of the circle

It is the foot of perpendicular from centre of sphere on plane (2)

d.r's of the normal to plane (2) are <1, 2, 2>.

$\therefore$ Equations of the line through the centre $(-1, 1, 2)$ of the sphere and normal to plane (2) are

$$\frac{x+1}{1}=\frac{y-1}{2}=\frac{z-2}{2}(=r)$$

Any point on this line is $(r-1, 2r+1, 2r+2)$

If it lies on plane (2), then

$$r-1+2(2r+1)+2(2r+2)+7=0$$

or $$9r+12=0 \quad \therefore \quad r=\frac{-4}{3}$$

$\therefore$ Foot of $\perp$ or centre of the circle is $(r-1, 2r+1, 2r+2)$

$$=\left(\frac{-4}{3}-1, \frac{-8}{3}+1, \frac{-8}{3}+2\right)=\left(\frac{-7}{3}, \frac{-5}{3}, \frac{-2}{3}\right).$$

Example 2. *Find the equations of the circle circumscribing the triangle formed by the three points (a, 0, 0), (0, b, 0), (0, 0, c). Obtain the co-ordinates of the centre of this circle.*

Sol. We know that a circle in Solid Geometry is represented by the equations of a sphere and a plane taken together.

Now (*i*) The equation of the plane passing through the three points $(a, 0, 0)$, $(0, b, 0)$, $(0, 0, c)$ is

$$\frac{x}{a}+\frac{y}{b}+\frac{z}{c}=1 \qquad \textbf{(Intercept Form)}$$

(*ii*) Let us find the equation of **any** sphere through the three points ($\because$ An infinite number of spheres can pass through three points).

To find the equation of this sphere, a fourth point is necessary.

For the sake of convenience, let us take the origin as fourth point.

Equation of the sphere through the points $O(0, 0, 0)$; $(a, 0, 0)$, $(0, b, 0)$ and $(0, 0, c)$ is $x^2+y^2+z^2-ax-by-cz=0$.

Hence $x^2+y^2+z^2-ax-by-cz=0, \ \frac{x}{a}+\frac{y}{b}+\frac{z}{c}=1$

and the equations of the circle.

Now let us find the centre of this circle

It is the foot of perpendicular from centre of the sphere on the plane.

d.r's of the normal to the plane are $<\frac{1}{a}, \frac{1}{b}, \frac{1}{c}>$

$\therefore$ Equations of the line through the centre $\left(\frac{a}{2}, \frac{b}{2}, \frac{c}{2}\right)$ of the sphere and normal to the plane are

$$\frac{x-\frac{a}{2}}{\frac{1}{a}}=\frac{y-\frac{b}{2}}{\frac{1}{b}}=\frac{z-\frac{c}{2}}{\frac{1}{c}}(=r)$$

Any point on this Normal is $\left(\frac{a}{2}+\frac{r}{a}, \frac{b}{2}+\frac{r}{b}, \frac{c}{2}+\frac{r}{c}\right)$

i.e., $$\left(\frac{a}{2}+ra^{-1}, \frac{b}{2}+rb^{-1}, \frac{c}{2}+rc^{-1}\right) \qquad ...(1)$$

If it lies on the plane $\frac{x}{a}+\frac{y}{b}+\frac{z}{c}=1$.

Then $\frac{1}{a}\left(\frac{a}{2}+ra^{-1}\right)+\frac{1}{b}\left(\frac{b}{2}+rb^{-1}\right)+\frac{1}{c}\left(\frac{c}{2}+rc^{-1}\right)=1$

or $$\tfrac{1}{2}+ra^{-2}+\tfrac{1}{2}+rb^{-2}+\tfrac{1}{2}+rc^{-2}=1$$

or $$r(a^{-2}+b^{-2}+c^{-2})=\frac{-1}{2} \quad \therefore \quad r=\frac{-1}{2(a^{-2}+b^{-2}+c^{-2})}$$

Putting this value of r in (1), the foot of the $\perp$ *i.e.*, the centre of the circle is

$$\left[\frac{a}{2}-\frac{a^{-1}}{2(a^{-2}+b^{-2}+c^{-2})}, \frac{b}{2}-\frac{b^{-1}}{2(a^{-2}+b^{-2}+c^{-2})}, \frac{c}{2}-\frac{c^{-1}}{2(a^{-2}+b^{-2}+c^{-2})}\right]$$

i.e., $$\left[\frac{a(b^{-2}+c^{-2})}{2(a^{-2}+b^{-2}+c^{-2})}, \frac{b(c^{-2}+a^{-2})}{2(a^{-2}+b^{-2}+c^{-2})}, \frac{c(a^{-2}+b^{-2})}{2(a^{-2}+b^{-2}+c^{-2})}\right].$$

EXERCISE 4

1. (*a*) Find the centre and radius of the circle given by
$$x^2+y^2+z^2=49,\ 2x+3y+6z=14.$$ (*M.D.U. 2000*)

   (*b*) Find the centre and radius of the circle
$$x^2+y^2+z^2-4x-y+3z+2=0$$
$$2x+3y-6z-10=0.$$

2. If r is the radius of the circle
$$x^2+y^2+z^2+2ux+2vy+2wz+d=0,\ lx+my+nz=0\ ;$$
prove that $$(r^2+d)(l^2+m^2+n^2)=(mw-nv)^2+(nu-lw)^2+(lv-mu)^2.$$

3. Find the equations of the tangent line in the symmetrical form to the circle
$$x^2+y^2+z^3+5x-7y+2z-8=0,\ 3x-2y+4z+3=0$$
at the point $(-3, 5, 4)$.
[Hint. The tangent line to a circle is the line of intersection of the tangent plane to the sphere at the given point and the plane of the circle.]

4. Find the equation of that plane which cuts the sphere
$x^2+y^2+z^2=a^2$ in a circle whose centre is (α, β, γ).

5. Find the equation of the circle lying on the sphere $x^2+y^2+z^2-2y-4z=11$ and having its centre at $(1, 3, 4)$.

Answers

1. (*a*) $\left(\frac{4}{7}, \frac{6}{7}, \frac{12}{7}\right)$; $3\sqrt{5}$ (*b*) $\left(\frac{178}{98}, \frac{22}{98}, \frac{-93}{98}\right)$; $\frac{\sqrt{801}}{14}$ 3. $\frac{x+3}{32}=\frac{y-5}{34}=\frac{z-4}{-7}$

4. $\alpha(x-\alpha)+\beta(y-\beta)+\gamma(z-\gamma)=0$ 5. $x^2+y^2+z^2-2y-4z=11,\ x+2y+2z=15$.

---

11. Any Sphere through a given Circle

To prove that the equation of any sphere through the circle
$$x^2+y^2+z^2+2ux+2vy+2wz+d=0,\ Ax+By+Cz+D=0$$
is $$\mathbf{x^2+y^2+z^2+2ux+2vy+2wz+d+k(Ax+By+Cz+D)=0}$$
where k is any constant.

The equations of the given circle are

$$x^2 + y^2 + z^2 + 2ux + 2vy + 2wz + d = 0 \quad ...(1)$$

$$Ax + By + Cz + D = 0 \quad ...(2)$$

Consider the equation,

$$x^2 + y^2 + z^2 + 2ux + 2vy + 2wz + d + k(Ax + By + Cz + D) = 0 \quad ...(3)$$

(i) This is a second degree in x, y, z.

(ii) Co-eff. of x^2 = Co-eff. of y^2 = Co-eff. of z^2.

(iii) There is no mixed term (*i.e.*, terms of yz, zx or xy).

∴ (3) represents a sphere.

Also any point which satisfies (1) and (2) both, also satisfies (3),

[∵ $0 + k(0) = 0$ for all k]

Hence (3) represents a family of sphere passing through the circle determined by (1) and (2).

Note. Equation of any sphere through a given circle is **Sphere + k (plane) = 0** where 'sphere' stands for the L.H.S. of the equation of the sphere (R.H.S. being zero) and so for the 'plane'.

Example 1. *Find the equation of the sphere through the origin and the circle with equations*

$$x^2 + y^2 + z^2 = 1,$$

and

$$x^2 + y^2 + z^2 + x + 2y + 3z - 5 = 0.$$

Sol. Equations of the circle are (see Art. 10 (b)]

$$x^2 + y^2 + z^2 - 1 = 0 \quad ...(1)$$

$$x^2 + y^2 + z^2 + x + 2y + 3z - 5 = 0 \quad ...(2)$$

The plane of common points of spheres (1) and (2) is [(2) – (1) gives]

$$x + 2y + 3z - 4 = 0 \quad ...(3)$$

Equation of any sphere through the circle given by (1) and (2) or given by (1) and (3) [see Art. 10 (b)] is

$$x^2 + y^2 + z^2 - 1 + k(x + 2y + 3z - 4) = 0 \quad ...(4)$$

(Sphere + k(plane) = 0)

It *i.e.*, (4) passes through the origin (0, 0, 0).

$$\therefore \quad -1 - 4k = 0 \quad \text{or} \quad k = \frac{-1}{4}$$

Putting this value of k in (4), we have

$$x^2 + y^2 + z^2 - 1 - \tfrac{1}{4}(x + 2y + 3z - 4) = 0$$

or

$$4x^2 + 4y^2 + 4z^2 - x - 2y - 3z = 0$$

as the required equation of the sphere.

Example 2. *Find the equation of the sphere which touches the sphere $x^2 + y^2 + z^2 - x + 3y + 2z - 3 = 0$ at the point (1, 1 – 1) and passes through the origin.*

Sol. Equation of the given sphere is

$$x^2 + y^2 + z^2 - x + 3y + 2z - 3 = 0 \quad ...(1)$$

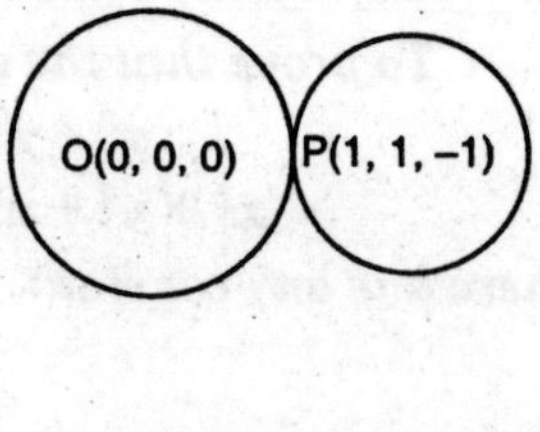

Equation of the tangent plane at (1, 1, – 1) to sphere (1) is (changing x^2 to xx_1 etc. changing x to $\frac{x + x_1}{2}$ etc.)

$$x \, . \, 1 + y \, . \, 1 + z(-1) - \frac{1}{2}(x+1) + \frac{3}{2}(y+1) + 2\left(\frac{z-1}{2}\right) - 3 = 0$$

i.e., $$x + 5y - 6 = 0 \qquad ...(2)$$

Tangent plane (2) cuts the sphere (1) in a point circle P (the point of contact) [see Note, Art. 10 (*a*)].

$\therefore$ Equation of any sphere through the point circle (P) given by (1) and (2) is

$$x^2 + y^2 + z^2 - x + 3y + 2z - 3 + k(x + 5y - 6) = 0 \qquad ...(3)$$

But this sphere passes through the origin (0, 0, 0)

$$\therefore \qquad -3 - 6k = 0 \quad \text{or} \quad k = \frac{-1}{2}$$

Putting $k = \frac{-1}{2}$ in (3), the required equation of the sphere is

$$2(x^2 + y^2 + z^2) - 3x + y + 4z = 0.$$

Example 3. *Obtain the equation of the sphere having the circle*

$$x^2 + y^2 + z^2 + 10y - 4z - 8 = 0, \; x + y + z = 3 \textit{ as great circle.}$$

Sol. Equations of the given circle are

$$x^2 + y^2 + z^2 + 10y - 4z - 8 = 0 \qquad ...(1)$$

$$x + y + z - 3 = 0 \qquad ...(2)$$

Equation of any sphere through the circle given by (1) and (2) is

$$(x^2 + y^2 + z^2 + 10y - 4z - 8) + k(x + y + z - 3) = 0 \qquad ...(3) \mid \text{Sphere} + k\text{ (plane)} = 0$$

$\because$ The given circle is a great circle lying on sphere (3).

$\therefore$ **The plane of the circle passes through the centre of (3)**

i.e., centre $\left(-\frac{k}{2}, -5 - \frac{k}{2}, 2 - \frac{k}{2}\right)$ of sphere (3) lies on plane (2).

$$\therefore \qquad \frac{-k}{2} - 5 - \frac{k}{2} + 2 - \frac{k}{2} - 3 = 0$$

i.e., $$\frac{-3k}{2} = 6 \quad \text{or} \quad k = \frac{-12}{3} = -4.$$

Putting this value of k in (3), the equation of the required sphere is

$$x^2 + y^2 + z^2 + 10y - 4z - 8 - 4(x + y + z - 3) = 0$$

or $$x^2 + y^2 + z^2 - 4x + 6y - 8z + 4 = 0.$$

Example 4. *Find the equations of the spheres passing through the circle*

$$x^2 + y^2 + z^2 - 5 = 0, \quad 2x + 3y + z - 3 = 0$$

and touching the plane $$3x + 4z - 15 = 0.$$

Sol. Equations of the given circle are

$$x^2 + y^2 + z^2 - 5 = 0, \quad 2x + 3y + z - 3 = 0 \qquad ...(1)$$

Equation of any sphere through circle (1) is

$$x^2 + y^2 + z^2 - 5 + k(2x + 3y + z - 3) = 0 \qquad ...(2)$$

But sphere (2) touches the plane

$$3x + 4z - 15 = 0 \qquad ...(3)$$

$\therefore$ $\perp$ distance of centre $\left(-k, \frac{-3k}{2}, \frac{-k}{2}\right)$ of sphere (2) from plane (3) is equal to radius

$$\sqrt{k^2+\frac{9k^2}{4}+\frac{k^2}{4}+5+3k} \quad \text{or} \quad \frac{\pm\left[-3k+4\left(\frac{-k}{2}\right)-15\right]}{\sqrt{9+16}}=\sqrt{\frac{14k^2+20+12k}{4}}$$

i.e.,
$$\mp\frac{5(k+3)}{5}=\sqrt{\frac{14k^2+20+12k}{4}}$$

Squaring both sides,

$$(k+3)^2=\frac{14k^2+20+12k}{4} \quad \text{or} \quad 4(k^2+6k+9)=14k^2+20+12k$$

i.e.,
$$10k^2-12k-16=0 \quad \text{or} \quad 5k^2-6k-8=0$$

$\therefore$
$$k=\frac{6\pm\sqrt{36+160}}{10}=\frac{6\pm 14}{10}$$

i.e.,
$$k=2,\ \frac{-4}{5}.$$

Putting these values of k in (2), the required equations of the spheres are

$$x^2+y^2+z^2+4x+6y+2z-11=0$$

and
$$5(x^2+y^2+z^2)-8x-12y-4z-13=0.$$

Example 5. ***Show that the two circles***

$$x^2+y^2+z^2-y+2z=0,\ x-y+z-2=0\ ;$$

and
$$x^2+y^2+z^2+x-3y+z-5=0,\ 2x-y+4z-1=0$$

lie on the same sphere and find its equation.

Sol. Equations of the two given circles are

$$x^2+y^2+z^2-y+2z=0,\ x-y+z-2=0 \quad \text{...(1)}$$

$$x^2+y^2+z^2+x-3y+z-5=0,\ 2x-y+4z-1=0 \quad \text{...(2)}$$

Equation of any sphere through circle (1) is

$$x^2+y^2+z^2-y+2z+k_1(x-y+z-2)=0 \quad \text{...(3)}$$

Equation of any sphere through circle (2) is

$$x^2+y^2+z^2+x-3y+z-5+k_2(2x-y+4z-1)=0 \quad \text{...(4)}$$

Since circles (1) and (2) lie on the same sphere.

$\therefore$ **Equations (3) and (4) must be identical for some values of k_1 and k_2.**

$\therefore$ **Co-efficients in (3) and (4) should be proportional.**

But co-efficients of x^2, y^2, z^2 are equal, so other co-efficients are also equal.

$\therefore$ **Equating co-efficients of x, y, z and constant terms in (3) and (4), we have**

$$k_1=1+2k_2 \quad \text{...(5)}$$

$$-1-k_1=-3-k_2 \quad \text{...(6)}$$

$$2+k_1=1+4k_2 \quad \text{...(7)}$$

$$-2k_1=-5-k_2 \quad \text{...(8)}$$

Let us solve (5) and (6) for k_1 and k_2

(5) + (6) gives,

$$-1=-2+k_2 \quad \text{or} \quad k_2=1$$

Putting $k_2=1$ **in (5),** $k_1=3$

These values of k_1, k_2 will satisfy, (7) and (8) if

$$2 + 3 = 1 + 4$$

and $$-6 = -5 - 1 \text{ which are both true.}$$

$\therefore$ Equations (3) and (4) are identical for $k_1 = 3$ and $k_2 = 1$.

$\therefore$ Circles (1) and (2) lie on the same sphere.

Putting $k_1 = 3$ in (3) or $k_2 = 1$ in (4), required equation of the sphere upon which circles (1) and (2) lie is $x^2 + y^2 + z^2 + 3x - 4y + 5z - 6 = 0$.

Example 6. *Obtain the equation of the sphere which passes through the circle $x^2 + y^2 = 4$, $z = 0$ and is cut by the plane $x + 2y + 2z = 0$ in a circle of radius 3.* (K.U. 2003)

Sol. Equations of the circle are $x^2 + y^2 = 4$, $z = 0$.

This circle is the section of the cylinder by a plane. [See Art. 10 (c)]

These equations of the circle can be rewritten as

$$\left.\begin{aligned} x^2 + y^2 + z^2 = 4 \\ z = 0 \end{aligned}\right] \quad (\because \; z = 0) \text{ (Note the inclusion of } z^2\text{)}$$

Equation of any sphere through this circle is

$$x^2 + y^2 + z^2 - 4 + kz = 0 \qquad \text{...(1)}$$

The plane $$x + 2y + 2z = 0 \qquad \text{...(2)}$$

cuts the sphere (1) in a circle of radius 3.

Centre of sphere (1) is

$$\left(0, 0, \frac{-k}{2}\right) \text{ and radius R} = \sqrt{0 + 0 + \frac{k^2}{4} + 4}$$

$p = \perp$ distance of the centre $\left(0, 0, \frac{-k}{2}\right)$ from plane (2)

$$= \frac{0 + 0 + 2\left(\frac{-k}{2}\right)}{\sqrt{1 + 4 + 4}} = \frac{-k}{3}$$

$\therefore$ 3 = Radius of the circle $= \sqrt{R^2 - p^2}$

or $$3 = \sqrt{\frac{k^2}{4} + 4 - \frac{k^2}{9}}$$

Squaring, $$9 = \frac{k^2}{4} + 4 - \frac{k^2}{9}$$

or $$324 = 9k^2 + 144 - 4k^2 \quad \text{or} \quad 5k^2 = 180$$

$\therefore$ $$k^2 = 36 \quad \therefore \quad k = \pm 6$$

Putting these values of k in (1), the equations of the spheres are

$$x^2 + y^2 + z^2 \pm 6z - 4 = 0.$$

EXERCISE 5

1. (a) Find the equation of the sphere through the circle $x^2 + y^2 + z^2 = 9$, $2x + 3y + 4z = 5$ and the point (1, 2, 3). *(M.D.U. 2007)*

   (b) Find the equation of the sphere through the circle $x^2 + y^2 + z^2 = 1$, $x + 2y + 3z = 4$ and the point (1, − 1, 2). *(M.D.U. 1998)*

2. Find the equation of the sphere through the circle $x^2 + y^2 + z^2 + 2x + 3y + 6 = 0, x - 2y + 4z - 9 = 0$ and the centre of the sphere $x^2 + y^2 + z^2 - 2x + 4y - 6z + 5 = 0$.

3. (a) Find the equation of the sphere having the circle

$$x^2 + y^2 + z^2 = 9, x - 2y + 2z = 5$$

for a great circle ; determine its centre and radius. *(K.U. 1997 S)*

(b) Find the equation of the sphere for which the circle

$$x^2 + y^2 + z^2 + 7y - 2z + 2 = 0, 2x + 3y + 4z - 8 = 0$$

is a great circle.

4. Prove that the plane $x + 2y - z = 4$ cuts the sphere $x^2 + y^2 + z^2 - x + z - 2 = 0$ in a circle of a radius unity ; and find the equation of the sphere which has this circle for one of the great circles.

5. (a) Find the equations of the spheres which pass through the circle

$$x^2 + y^2 + z^2 = 1, 2x + 4y + 5z = 6$$

and touch the plane $z = 0$. *(K.U. 1995 A)*

(b) Find the equations of the spheres which pass through the circle $x^2 + y^2 + z^2 = 5$, $x + 2y + 3z = 3$ and touch the plane $4x + 3y - 15 = 0$.

6. Prove that the circles

$$x^2 + y^2 + z^2 - 2x + 3y + 4z - 5 = 0, 5y + 6z + 1 = 0,$$

and $$x^2 + y^2 + z^2 - 3x - 4y + 5z - 6 = 0, x + 2y - 7z = 0$$

lie on the same sphere and find its equation. *(K.U. 1997 A)*

7. Find the equation to the sphere which passes through the point (α, β, γ) and the circle $z = 0, x^2 + y^2 = a^2$.

8. Find the equations of the spheres which pass through the circle $x^2 + z^2 - 2x + 2z = 2, y = 0$ and touch the plane $y - z = 7$. *(M.D.U. 1995)*

Answers

1. (a) $3(x^2 + y^2 + z^2) - 2x - 3y - 4z - 22 = 0$ (b) $x^2 + y^2 + z^2 - 5x - 10y - 15z + 19 = 0$
2. $x^2 + y^2 + z^2 + 7y - 8z + 24 = 0$.
3. (a) $9(x^2 + y^2 + z^2) - 10x + 20y - 20z - 31 = 0$; $\left(\frac{5}{9}, \frac{-10}{9}, \frac{10}{9}\right)$; $\frac{2}{3}\sqrt{14}$

   (b) $x^2 + y^2 + z^2 - 2x + 4y - 6z + 10 = 0$
4. $x^2 + y^2 + z^2 - 2x - 2y + 2z + 2 = 0$
5. (a) $x^2 + y^2 + z^2 - 2x - 4y - 5z + 5 = 0$; $5(x^2 + y^2 + z^2) - 2x - 4y - 5z - 1 = 0$

   (b) $x^2 + y^2 + z^2 + 2x + 4y + 6z - 11 = 0$; $5(x^2 + y^2 + z^2) - 4x - 8y - 12z - 13 = 0$
6. $x^2 + y^2 + z^2 - 2x - 2y - 2z - 6 = 0$
7. $\gamma(x^2 + y^2 + z^2 - a^2) = z(\alpha^2 + \beta^2 + \gamma^2 - a^2)$
8. $x^2 + y^2 + z^2 - 2x + 28y + 2z - 2 = 0$

   $x^2 + y^2 + z^2 - 2x - 4y + 2z - 2 = 0$.

ANGLES OF INTERSECTION OF TWO SPHERES

12. (a) Definition

The angle of intersection of two spheres is the angle between their tangent planes at a point of intersection. Now we know that the tangent plane at a point to the sphere is ⊥ to the radius through the point and

The angle between two planes = Angle between their normals

$\therefore$ Angle between two spheres at a point of intersection P

= Angle between the tangent planes at P
(*not shown in the figure*)
= Angle between their radii at P.

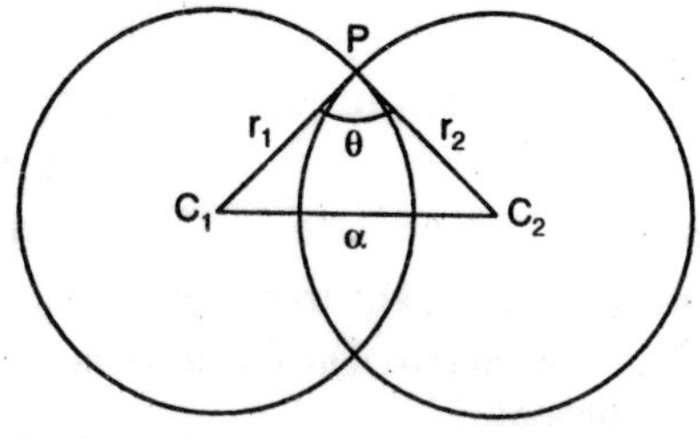

i.e., $\angle C_1PC_2$ is the angle between the spheres at P where C_1, C_2 are the centres of the two spheres.

Note. If the angle between the spheres *i.e.*, $\angle C_1PC_2 = 90°$, then the spheres are said to be **orthogonal**.

(*b*) Angle of Intersection of Two Spheres

If r_1, r_2 are the radii of two spheres, d the distance between their centres, and θ their angle of intersection, then

$$\cos\theta = \frac{r_1^2 + r_2^2 - d^2}{2r_1r_2}$$

Reproduce Art. 12 (*a*), $\angle C_1PC_2 = \theta$

$\therefore$ From the ΔPC_1C_2, by the cosine formula,

$$\cos\theta = \frac{C_1P^2 + C_2P^2 - C_1C_2^2}{2C_1P \cdot C_2P} \qquad \because \cos A = \frac{b^2 + c^2 - a^2}{2bc}$$

or
$$\cos\theta = \frac{r_1^2 + r_2^2 - d^2}{2r_1r_2}$$

13. Condition of Orthogonality of Two Spheres

To find the condition that the spheres

$$x^2 + y^2 + z^2 + 2u_1x + 2v_1y + 2w_1z + d_1 = 0$$

and
$$x^2 + y^2 + z^2 + 2u_2x + 2v_2y + 2w_2z + d_2 = 0$$

may cut orthogonally.

Equations of the given spheres are

$x^2 + y^2 + z^2 + 2u_1x + 2v_1y + 2w_1z + d_1 = 0$ ...(1)

$x^2 + y^2 + z^2 + 2u_2x + 2v_2y + 2w_2z + d_2 = 0$ ...(2)

Centre C_1 of (1) is $(-u_1, -v_1, -w_1)$

Centre C_2 of (2) is $(-u_2, -v_2, -w_2)$

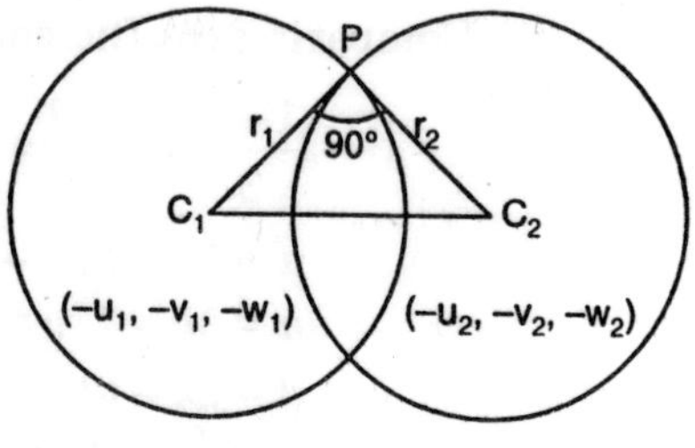

Radius r_1 (= C_1P) of (1) is = $\sqrt{u_1^2 + v_1^2 + w_1^2 - d_1}$

Radius r_2 (= C_2P) of (2) is = $\sqrt{u_2^2 + v_2^2 + w_2^2 - d_2}$

$\because$ The spheres cut orthogonally

$\therefore$ $\angle C_1PC_2$ (the angle between the spheres) = 90° [See Note Art. 12 (*a*)]

$\therefore$
$$C_1C_2^2 = C_1P^2 + C_2P^2$$

or
$$C_1C_2^2 = r_1^2 + r_2^2$$

or
$$(-u_1 + u_2)^2 + (-v_1 + v_2)^2 + (-w_1 + w_2)^2 = u_1^2 + v_1^2 + w_1^2 - d_1 + u_2^2 + v_2^2 + w_2^2 - d_2$$

or $$u_1^2 + u_2^2 - 2u_1u_2 + v_1^2 + v_2^2 - 2v_1v_2 + w_1^2 + w_2^2 - 2w_1w_2$$
$$= u_1^2 + v_1^2 + w_1^2 - d_1 + u_2^2 + v_2^2 + w_2^2 - d_2$$

Cancelling common terms on both sides,
$$-2u_1u_2 - 2v_1v_2 - 2w_1w_2 = -d_1 - d_2$$
or $$2u_1u_2 + 2v_1v_2 + 2w_1w_2 = d_1 + d_2$$
which is the required condition.

Remark. The converse is also true *i.e.*, if the above condition is satisfied, then the spheres are orthogonal.

Note. Rule to Remember the condition of orthogonality

(*i*) Write equations of spheres so as to make co-efficients of x^2, y^2, z^2 unity.

(*ii*) Then the condition of orthogonality is :

$\frac{1}{2}$ (Product of the co-efficients of x) + $\frac{1}{2}$ (Product of the co-efficients of y) + $\frac{1}{2}$ (Product of the co-efficients of z) = Sum of constant terms.

Example 1. *Find the equation of the sphere which touches the plane $3x + 2y - z + 2 = 0$ at the point $(1, -2, 1)$ and cuts orthogonally the sphere $x^2 + y^2 + z^2 - 4x + 6y + 4 = 0$.*

(K.U. 1999 A, 98 S)

Sol. Given sphere is
$$x^2 + y^2 + z^2 - 4x + 6y + 4 = 0 \quad ...(1)$$
Given equation of the plane is
$$3x + 2y - z + 2 = 0 \quad ...(2)$$

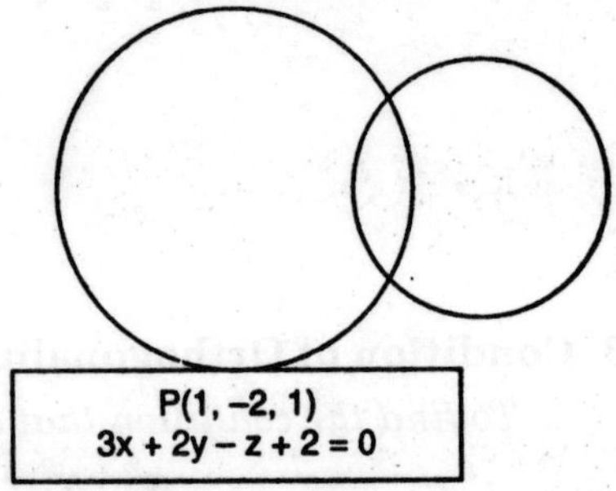

Let the equation of the required sphere be
$$x^2 + y^2 + z^2 + 2ux + 2vy + 2wz + d = 0 \quad ...(3)$$
Equation of the tangent plane at $(1, -2, 1)$ to (3) is
$$x(1) + y(-2) + z(1) + u(x + 1) + v(y - 2) + w(z + 1) + d = 0$$
i.e., $$x(1 + u) + y(-2 + v) + z(1 + w) + (u - 2v + w + d) = 0 \quad ...(4)$$
But (2) is also given to be tangent plane at $(1, -2, 1)$.

$\therefore$ (2) and (4) represent the same plane.

$\therefore$ Comparing co-efficients in (2) and (4), we have
$$\frac{1+u}{3} = \frac{v-2}{2} = \frac{1+w}{-1} = \frac{u-2v+w+d}{2} = k \text{ (say)}$$
$\therefore$ $$u + 1 = 3k,\ v - 2 = 2k,\ 1 + w = -k,\ u - 2v + w + d = 2k$$
$\therefore$ $$u = 3k - 1,\ v = 2k + 2,\ w = -k - 1,\ d = 2k - u + 2v - w$$
or $$d = 2k - (3k - 1) + 2(2k + 2) - (-k - 1) = 4k + 6$$
Again spheres (1) and (3) cut orthogonally.

$\therefore$ $$\tfrac{1}{2}(2u)(-4) + \tfrac{1}{2}(2v)(6) + \tfrac{1}{2}(2w)(0) = d + 4$$ (See Note Art. 13)

or $$-4u + 6v = d + 4$$
Putting $u = 3k - 1$, $v = 2k + 2$, $d = 4k + 6$ in the above relation,
$$-4(3k - 1) + 6(2k + 2) = 4k + 6 + 4$$
or $$-12k + 4 + 12k + 12 = 4k + 6 + 4$$
or $$4k = 6 \quad \therefore \quad k = \tfrac{3}{2}.$$

Putting this value of k in $u = 3k - 1$, $v = 2k + 2$, $w = -k - 1$,

and $d = 4k + 6$, we have

$$u = \frac{7}{2}, \quad v = 5, \quad w = -\frac{5}{2}, \quad d = 12.$$

Putting these values in (3), the required equation of the sphere is

$$x^2 + y^2 + z^2 + 7x + 10y - 5z + 12 = 0.$$

Example 2. *Two spheres of radii r_1 and r_2 cut orthogonally. Prove that the radius of the common circle is*

$$\frac{r_1 r_2}{\sqrt{r_1^2 + r_2^2}}. \qquad \text{(K.U. 1996 A)}$$

Sol. Let the equations of the common circle be $x^2 + y^2 + a^2$, $z = 0$

It is a circle of radius a and is in the XY plane.

The equations of this circle can be put as :

$$x^2 + y^2 + z^2 = a^2, z = 0 \qquad \text{(Note the inclusion of } z^2\text{)}$$

Equations of any two spheres through this circle are

$$x^2 + y^2 + z^2 - a^2 + k_1 z = 0 \qquad ...(1)$$

and

$$x^2 + y^2 + z^2 - a^2 + k_2 z = 0 \qquad ...(2)$$

$\because$ Spheres (1) and (2) cut orthogonally

$\therefore$

$$\tfrac{1}{2}(0) + \tfrac{1}{2}(0) + \tfrac{1}{2} k_1 k_2 = -a^2 - a^2 \qquad \text{(See Note Art. 13)}$$

or

$$k_1 k_2 = -4a^2 \qquad ...(3)$$

$$r_1 = \text{Radius of spheres (1)}$$

$$= \sqrt{0^2 + 0^2 + \frac{k_1^2}{4} + a^2} = \sqrt{\frac{k_1^2}{4} + a^2} \qquad ...(4)$$

Similarly,

$$r_2 = \sqrt{\frac{k_2^2}{4} + a^2} \qquad ...(5)$$

Let us eliminate k_1, k_2 from (3), (4) and (5) to find a.

Squaring (3), (4) and (5), we have

$$k_1^2 k_2^2 = 16a^4 \qquad ...(6)$$

$$r_1^2 = \frac{k_1^2}{4} + a^2 \quad \therefore \quad k_1^2 = 4(r_1^2 - a^2) \qquad ...(7)$$

$$r_2^2 = \frac{k_2^2}{4} + a^2 \quad \therefore \quad k_2^2 = 4(r_2^2 - a^2) \qquad ...(8)$$

Putting the values of k_1^2 and k_2^2 from (7) and (8) in (6), we have

$$16(r_1^2 - a^2)(r_2^2 - a^2) = 16a^4$$

Cancelling 16,

$$r_1^2 r_2^2 - a^2(r_1^2 + r_2^2) + a^4 = a^4$$

or

$$-a^2(r_1^2 + r_2^2) = -r_1^2 r_2^2.$$

$\therefore$

$$a^2 = \frac{r_1^2 r_2^2}{r_1^2 + r_2^2}$$

$\therefore$ a (The radius of the common circle) $= \dfrac{r_1 r_2}{\sqrt{r_1^2 + r_2^2}}$.

EXERCISE 6

1. Find the angle between the spheres

$$x^2 + y^2 + z^2 + 2u_1x + 2v_1y + 2w_1z + d_1 = 0$$

and $$x^2 + y^2 + z^2 + 2u_2x + 2v_2y + 2w_2z + d_2 = 0$$

and hence find the condition that they may cut orthogonally. ***(M.D.U. 1995 S)***

[Hint. See Art. 12 (*b*) and 13.]

2. Find the condition that the spheres $x^2 + y^2 + z^2 = a^2$ and $x^2 + y^2 + z^2 + 2ux + 2vy + 2wz + d = 0$ may cut orthogonally.

3. Show that every sphere through the circle

$$x^2 + y^2 - 2ax + r^2 = 0 \ ; \ z = 0$$

cuts orthogonally every sphere through the circle

$$x^2 + z^2 = r^2, \ y = 0.$$

4. Two spheres of radius 3 cm and 4 cm cut each other orthogonally. Find the radius of the common circle.

[Hint. Proceed as in Example 2 ; $r_1 = 3$, $r_2 = 4$.]

5. Find the equation of a sphere which passes through the circle $x^2 + y^2 + z^2 - 2x + 3y - 4z + 6 = 0$, $3x - 4y + 5z - 15 = 0$ and cuts orthogonally the sphere $x^2 + y^2 + z^2 + 2x + 4y - 6z + 11 = 0$. ***(K.U. 2007, 2000)***

6. Find the equation of the sphere that passes through the two points (0, 3, 0), (– 2, – 1, – 4) and cuts orthogonally the two spheres $x^2 + y^2 + z^2 + x - 3z - 2 = 0$ and $2(x^2 + y^2 + z^2) + x + 3y + 4 = 0$.

Answers

2. $a^2 - d = 0$ 4. $\dfrac{12}{5}$

5. $5(x^2 + y^2 + z^2) - 13x + 19y - 25z + 45 = 0$ 6. $x^2 + y^2 + z^2 + 2x - 2y + 4z - 3 = 0$.

---

14. (*a*) Radical Plane of Two Spheres

Definition. *The locus of all points whose powers with respect to two spheres are equal is a plane called the radical plane of two spheres.*

(*b*) *To find the equation of the radical plane of two given spheres.*

Let the equations of two given spheres be

$$x^2 + y^2 + z^2 + 2ux + 2vy + 2wz + d = 0 \quad ...(1)$$

$$x^2 + y^2 + z^2 + 2u'x + 2v'y + 2w'z + d' = 0 \quad ...(2)$$

Let P(*x*, *y*, *z*) be **any** point whose powers w.r.t. two spheres are equal (*i.e.*, P(*x*, *y*, *z*) is any point on the radical plane).

i.e., Power of P w.r.t. (1) = Power of P w.r.t. (2)

$$\therefore \quad x^2 + y^2 + z^2 + 2ux + 2vy + 2wz + d$$

$$= x^2 + y^2 + z^2 + 2u'x + 2v'y + 2w'z + d'$$

[By definition of the power of a point, **Cor.** Art. 7]

or $$2ux + 2vy + 2wz + d = 2u'x + 2v'y + 2w'z + d'$$

or $$2(u - u')x + 2(v - v')y + 2(w - w')z + (d - d') = 0 \quad ...(3)$$

which being a first degree in *x*, *y*, *z* represents a plane and is the required equation of the radical plane.

Cor. Radical plane of two spheres is perpendicular to the line joining the centres of the spheres.

d.r's of the normal to the radical plane (3) are

$$< 2(u - u'), (v - v'), 2(w - w') >$$ | co-effs. x, y, z in (3)

or $$< u - u', v - v', w - w' >$$ ...(4)

Centres of the spheres (1) and (2) are

$$C(-u, -v, -w) \text{ and } C'(-u', -v', -w')$$

$\therefore$ d.r's of CC′ are $< -u' + u, -v' + v, -w' + w >$ ...(5)

As (4) and (5) are the same.

$\therefore$ the line CC′ is $\perp$ to the radical plane.

Hence the radical plane is $\perp$ to the line of centres.

Working Rule to find the equation of radical plane

(*i*) In the equations of the spheres, make co-efficients of x^2, y^2, z^2 unity.

(*ii*) **Then subtract the equations of the two spheres to get the equation of the radical plane.**

Remark. *If the two spheres intersect, then the radical plane passes through their points of intersection.*

Any common point of spheres (1) and (2) will satisfy both (1) and (2).

Hence it will satisfy (1) – (2) *i.e.*, (3) also.

$\therefore$ Common points of (1) and (2) also lie on Radical Plane (3).

i.e., Radical plane (3) passes through the points of intersection of (1) and (2).

15. Equations of Two Spheres in the Simplest Form

To prove that the equations of two spheres can be put in the form

$$x^2 + y^2 + z^2 + 2u_1x + d = 0, \; x^2 + y^2 + z^2 + 2u_2x + d = 0.$$ (M.D.U. 2000 A)

Let us take the line joining the centres of the spheres as X-axis. Now let us take YZ plane as the radical plane. ($\because$ Radical Plane is $\perp$ to the line joining the centres)

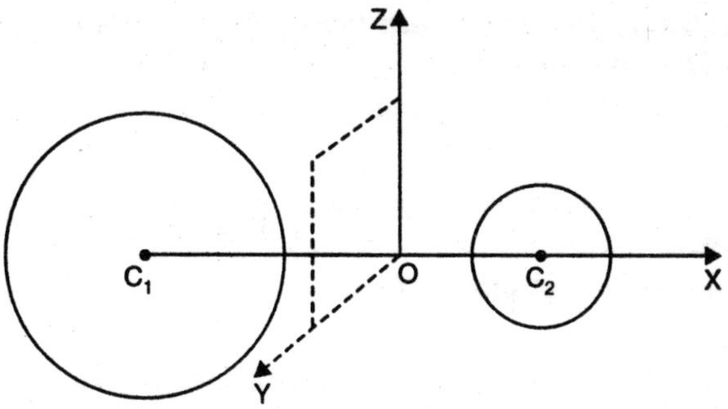

Let the equations of the two spheres (w.r.t. the above choice of axes)

be $$x^2 + y^2 + z^2 + 2u_1x + 2v_1y + 2w_1z + d_1 = 0$$ ...(1)

and $$x^2 + y^2 + z^2 + 2u_2x + 2v_2y + 2w_2z + d_2 = 0$$ ...(2)

$\therefore$ Centres $C_1(-u_1, -v_1, -w_1)$ and $C_2(-u_2, -v_2, -w_2)$ of spheres (1) and (2) lie on X-axis

i.e., $$y = 0, z = 0.$$

(Because of our circle of X-axis)

$\therefore$ $v_1 = 0, w_1 = 0, v_2 = 0$ and $w_2 = 0$

$\therefore$ Equations (1) and (2) reduce to

$$x^2 + y^2 + z^2 + 2u_1x + d_1 = 0$$
$$x^2 + y^2 + z^2 + 2u_2x + d_2 = 0$$

Subtracting these equations, the equation of the radical plane of two spheres is [By Art. 14 (*b*)]

$$2(u_1 - u_2)x + (d_1 - d_2) = 0 \quad ...(3)$$

But equation of the radical plane (because of our choice) is

$$\text{Y} - \text{Z plane } i.e., \text{ plane } x = 0 \quad ...(4)$$

Comparing (3) and (4), we have

$$\frac{2(u_1 - u_2)}{1} = \frac{d_1 - d_2}{0}$$

or $$d_1 - d_2 = 0 \quad \therefore \quad d_1 = d_2 = d \text{ (say)}$$

Putting these values in (1) and (2), the equations of the two spheres become

$$x^2 + y^2 + z^2 + 2u_1x + d = 0$$
$$x^2 + y^2 + z^2 + 2u_2x + d = 0.$$

16. (*a*) **Definition : System of Spheres**

We know that the equation of the sphere in general is

$$x^2 + y^2 + z^2 + 2ux + 2vy + 2wz + d = 0.$$

It contains four arbitrary constants u, v, w and d which can uniquely be determined if the sphere satisfies four independent conditions.

However, if the sphere satisfies less than four conditions, then some of the four constants u, v, w and d remain undetermined. These undetermined constants are called **parameters.** For different values of these parameters, we get different spheres. All these spheres are said to form a **system of spheres.**

(*b*) **Co-axal System of Spheres**

Definition. *A system of spheres is said to be a co-axal system if every two spheres of the system have the same radical plane.*

(*a*) *The centres of a Co-axal system of spheres are collinear.*

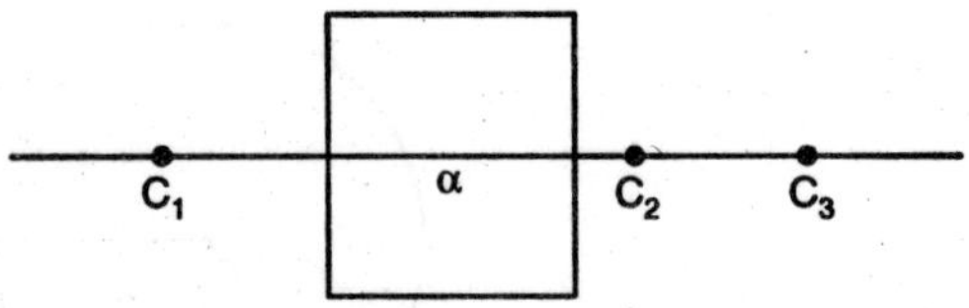

Let C_1, C_2, C_3, ... etc. be the centres of a co-axal system of spheres.

Let α be the common radical plane of this co-axal system.

$\because$ α is the radical plane of the spheres having centres C_1, C_2.

$\therefore$ C_1C_2 is $\perp$ to the plane α.

($\because$ By Cor. Art. 14, Line of Centres is $\perp$ to the radical plane). Similarly, C_1C_3 is $\perp$ to plane $\perp$.

$\therefore$ St. lines C_1C_2, C_1C_3 etc. pass through the same point C_1 and are $\perp$ to plane α.

But this is possible only if C_1, C_2, C_3, ... all lie in one straight line.

($\because$ Through a point one and only one line can be drawn $\perp$ to a plane.)

17. Equation of a Co-axal System of Spheres

To prove that the equations of a co-axal system of spheres can be put in the form $x^2 + y^2 + z^2 + 2ux + d = 0$, where u is a parameter.

Let us take the line joining the centres of the spheres of the co-axal system as x-axis.

($\because$ By Art. 16 (c), centres of a co-axal system are collinear)

Now let us taken YZ plane as the radical plane of co-axal system.

($\because$ Radical plane is $\perp$ to the line joining the centres)

Let the equation to the co-axal system of spheres be

$$x^2 + y^2 + z^2 + 2ux + 2vy + 2wz + d = 0$$

Centres of the spheres are $(-u, -v, -w)$

$\because$ They lie on x-axis *i.e.*, on $y = 0$ and $z = 0$

$\therefore$ $-v = 0$ and $-w = 0$

i.e., $v = 0$ and $w = 0$

Putting $v = 0$ and $w = 0$ in (1), equation of the co-axal system is

$$x^2 + y^2 + z^2 + 2ux + d = 0 \quad \text{...(1)}$$

$\therefore$ Equations of any two spheres of this co-axal system are

$$x^2 + y^2 + z^2 + 2u_1x + d_1 = 0$$
$$x^2 + y^2 + z^2 + 2u_2x + d_2 = 0$$

Subtracting these equations, the equation of the radical plane of two spheres is

$$2(u_1 - u_2)x + (d_1 - d_2) = 0 \quad \text{...(2)}$$

But equation of the radical plane (because of our choice) is Y – Z plane

i.e., plane $x = 0$ ...(3)

Comparing (2) and (3), we have

$$\frac{2(u_1 - u_2)}{1} = \frac{d_1 - d_2}{0} \quad \text{or} \quad d_1 - d_2 = 0$$

$\therefore$ $d_1 = d_2 = d$ (say)

i.e., d is same for all spheres.

Hence the equation of a co-axal system can be put in the form

$$x^2 + y^2 + z^2 + 2ux + d = 0$$

where u is a parameter.

Exercise. Show that the equation of a co-axal system of spheres can be put in the form $x^2 + y^2 + z^2 + 2wz + d = 0$, where w is a parameter.

18. (a) Definition. Limiting Points of a Co-axal System

The centres of two spheres of the co-axal system which have zero radius are called the limiting points of the co-oral system.

Note. Each limiting point is a sphere of *zero* radius and centre is limiting point.

(b) To find the limiting points of the Co-axal system of spheres given by $x^2 + y^2 + z^2 + 2ux + d = 0$, where u is a parameter.

The given co-axal system is $x^2 + y^2 + z^2 + 2ux + d = 0$, where u is a parameter.

Its centre is $(-u, 0, 0)$ and radius is $\sqrt{u^2 - d}$.

Since for limiting points, radius = 0.

$\therefore$ $$\sqrt{u^2-d}=0 \quad \text{or} \quad u^2-d=0 \quad \text{or} \quad u=\pm\sqrt{d}$$

Thus the centre $(-u, 0, 0)$ becomes

$$(-\sqrt{d}, 0, 0) \quad \text{and} \quad (\sqrt{d}, 0, 0)$$

which are the required limiting points.

Working Rule to find the Limiting points of a co-axal system.

(i) Write down the co-ordinates of the centre of the co-axal system and radius of co-axal system.

(ii) Put radius equal to zero and find the values of the parameter.

(iii) Substitute the values of the parameter in the co-ordinates of the centre to get the limiting points.

19. *If $S_1 = 0$, $S_2 = 0$ be two spheres, then equation*

$$S_1 + \lambda S_2 = 0,$$

λ being a parameter, represent a co-axal system of spheres.

Proof. $\because$ $$S_1 = 0 \quad \text{and} \quad S_2 = 0$$

are the equations of the two spheres, therefore,

$$S_1 + \lambda S_2 = 0$$

also represents a system of spheres where λ is a parameter.

Any two members of this system are (putting $\lambda = \lambda_1, \lambda_2$)

$$S_1 + \lambda_1 S_2 = 0 \quad \text{and} \quad S_1 + \lambda_2 S_2 = 0$$

Dividing two equations by $1 + \lambda_1$ and $1 + \lambda_2$ (so as to make the co-efficients of x^2, y^2, z^2 unity), the equations of the two spheres are

$$\frac{S_1 + \lambda_1 S_2}{1+\lambda_1} = 0 \quad \text{and} \quad \frac{S_1 + \lambda_2 S_2}{1+\lambda_2} = 0$$

where $\lambda_1 \neq \lambda_2$.

The equation of the radical plane of the two spheres is

$$\frac{S_1 + \lambda_1 S_2}{1+\lambda_1} - \frac{S_1 + \lambda_2 S_2}{1+\lambda_2} = 0 \qquad \text{(Subtracting to the equations)}$$

or $$(1+\lambda_2)(S_1 + \lambda_1 S_2) - (1+\lambda_1)(S_1 + \lambda_2 S_2) = 0$$

or $$S_1 + \lambda_1 S_2 + \lambda_2 S_1 + \lambda_1\lambda_2 S_2 - S_1 - \lambda_2 S_2 - \lambda_1 S_1 - \lambda_1\lambda_2 S_2 = 0$$

or $$\lambda_1 S_2 + \lambda_2 S_1 - \lambda_2 S_2 - \lambda_1 S_1 = 0$$

or $$\lambda_1(S_2 - S_1) - \lambda_2(S_2 - S_1) = 0$$

or $$(\lambda_1 - \lambda_2)(S_2 - S_1) = 0$$

But $$\lambda_1 - \lambda_2 \neq 0 \qquad |\because \lambda_1 \neq \lambda_2$$

$\therefore$ $$S_2 - S_1 = 0$$

is the equation of radical plane of the two spheres.

This being independent of λ_1 and λ_2 will be the same for any two members of the system

$$S_1 + \lambda S_2 = 0.$$

Hence the equation

$$S_1 + \lambda S_2 = 0$$

represents a co-axal system of spheres.

Note 1. Co-axal system determined by two spheres S_1 and S_2 is the same as co-axal system determined by one sphere say S_1 and their common radical plane *i.e.*, co-axal system is

$$S_1 + \lambda(S_2 - S_1) = 0.$$

Note 2. Equation of a co-axal system, determined by one sphere S and the radical plane P is

$$S + \lambda P = 0.$$

Example 1. *Find the equation of the sphere through the origin co-axal with the spheres*

$$x^2 + y^2 + z^2 = 1$$

and $$x^2 + y^2 + z^2 + x + 2y + 3z - 5 = 0.$$

Sol. The equations of the two spheres are

$$S_1 \equiv x^2 + y^2 + z^2 - 1 = 0 \quad \text{...(1)}$$

$$S_2 \equiv x^2 + y^2 + z^2 + x + 2y + 3z - 5 = 0 \quad \text{...(2)}$$

Equation of the radical plane of the spheres (1) and (2) is

$$P \equiv S_1 - S_2 \equiv -x - 2y - 3z + 4 = 0$$

or $$x + 2y + 3z - 4 = 0 \quad \text{...(3)}$$

Equation of any sphere co-axal with the spheres (1) and (2) is

$$S_1 + \lambda(S_2 - S_1) = 0 \quad \text{| See Note 1, Art. 19}$$

or $$x^2 + y^2 + z^2 - 1 + \lambda(x + 2y + 3z - 4) = 0 \quad \text{...(4)}$$

It *i.e.*, (4) passes through O(0, 0, 0).

$$\therefore \quad -1 - 4\lambda = 0 \quad \text{or} \quad \lambda = -\tfrac{1}{4}$$

Putting this value of λ in (4), the equaton of the required sphere is

$$x^2 + y^2 + z^2 - 1 - \tfrac{1}{4}(x + 2y + 3z - 4) = 0$$

or $$4x^2 + 4y^2 + 4z^2 - x - 2y - 3z = 0.$$

Example 2. *Find the limiting points of the co-axal system defined by the spheres*

$$x^2 + y^2 + z^2 + 3x - 3y + 6 = 0$$

$$x^2 + y^2 + z^2 - 6y - 6z + 6 = 0.$$

Sol. The equations of the two spheres are

$$x^2 + y^2 + z^2 + 3x - 3y + 6 = 0$$

and $$x^2 + y^2 + z^2 - 6y - 6z + 6 = 0$$

Equation of the radical plane of these two spheres is

$$S_1 - S_2 = 0 \quad \text{or} \quad 3x + 3y + 6z = 0 \quad \text{or} \quad x + y + 2z = 0$$

$\therefore$ Equation of the co-axal system defined by the two given spheres is

$$S_1 + \lambda(S_2 - S_1) = 0 \quad \text{or} \quad x^2 + y^2 + z^2 + 3x - 3y + 6 + \lambda(x + y + 2z) = 0$$

or $$x^2 + y^2 + z^2 + (3 + \lambda)x + (-3 + \lambda)y + 2\lambda z + 6 = 0 \quad \text{...(1)}$$

λ being a parameter.

The centre of (1) is

$$\left(-\frac{3+\lambda}{2}, \frac{-(\lambda-3)}{2}, -\lambda\right) \quad \text{...(2)}$$

and its radius is

$$\sqrt{\left(\frac{3+\lambda}{2}\right)^2 + \left(\frac{\lambda-3}{2}\right)^2 + \lambda^2 - 6}$$

Equating this radius to zero, we get

$$\left(\frac{\lambda+3}{2}\right)^2+\left(\frac{\lambda-3}{2}\right)^2+\lambda^2-6=0$$

or $$\lambda^2+6\lambda+9+\lambda^2-6\lambda+9+4\lambda^2-24=0 \quad \text{or} \quad 6\lambda^2-6=0$$

or $$\lambda^2=1 \quad \therefore \quad \lambda=\pm 1.$$

Putting these values of λ in (2), the limiting points are $(-2, 1, -1)$ and $(-1, 2, 1)$.

EXERCISE 7

1. Prove that the equations of the two spheres can be put in the simplified form :

$$x^2+y^2+z^2+2u_1z+d=0$$

and $$x^2+y^2+z^2+2u_2z+d=0.$$

[**Hint.** Take the line joining the centres of the two spheres a **z-axis** and let us take **XY-plane** as the radical plane. Now proceed as in Art. 15.]

2. Prove that the spheres

$$x^2+y^2+z^2+2x+4y+6z+10=0,$$
$$x^2+y^2+z^2+x+2y+3z+6=0,$$

and $x^2+y^2+z^2+2=0$ are co-axal.

3. Find the limiting points of the co-axal system of spheres

$$x^2+y^2+z^2-20x+30y-40z+29+\lambda(2x-3y+4z)=0.$$

4. (*a*) Find the limiting points of the co-axal system determined by the two spheres whose equations are

$$x^2+y^2+z^2-8x+2y-2z+32=0$$

and $$x^2+y^2+z^2-7x+z+23=0.$$ (*M.D.U. 2003*)

(*b*) Find the limiting points of the co-axial system defined by the spheres

$$x^2+y^2+z^2+3x-3y+6=0 \text{ and } x^2+y^2+z^2-6y+6z+6=0.$$

5. Find the equations to the two spheres of the co-axal system $x^2+y^2+z^2-5+\lambda(2x+y+3z-3)=0$ which touch the plane $3x+4y=15$.

6. The point $(-1, 2, 1)$ is a limiting point of a co-axal system of spheres of which

$$x^2+y^2+z^2+3x-3y+6=0 \text{ is a member.}$$

Find the equation of the radical plane of this system and the co-ordinates of the other limiting point.

[**Hint.** *$(-1, 2, 1)$ is a limiting point*

$\Rightarrow$ *Equation of one sphere of the co-axal system is*

$$(x+1)^2+(y-2)^2+(z-1)^2=0$$ [*See note. Def. Art. 18 (a)*]

i.e., $$x^2+y^2+z^2+2x-4y-2z+6=0.]$$

7. A, B are two fixed point and P moves so that PA = n PB ; show that the locus of P is a sphere ($n \neq 1$). Show also that all such spheres for different values of n, have a common radical plane.
[**Hint.** Take the fixed points as $(-a, 0, 0)$ and $(a, 0, 0)$].

8. Prove that the locus of the points whose powers w.r.t. two given spheres are in a constant ratio ($\neq 1$) is a sphere co-axal with two given spheres.
[**Hint.** Take the equations of the given spheres as obtained in Art. 15.]

9. Show that the equation $x^2+y^2+z^2+2\lambda x+2\mu y-d=0$, where λ and μ are parameters, represents a system of spheres passing through the limiting points of the system $x^2+y^2+z^2+2ux+d=0$, where u is a parameter, and cutting every member of that system at right angles.

Answers

3. (2, – 3, 4) and (– 2, 3, – 4)
4. (*a*) (3, 1, – 2) and (5, – 3, 4) (*b*) (– 1, 2, 1) and (– 2, 1, – 1)
5. $x^2 + y^2 + z^2 + 4x + 2y + 6z - 11 = 0$;
 $5x^2 + 5y^2 + 5z^2 - 8x - 4y - 12z - 13 = 0$
6. $x + y + 2z = 0$; (– 2, 1, – 1).

---

20. Diametral Plane Property

To find the locus of the middle points of a system of parallel chords of the sphere

$$x^2 + y^2 + z^2 = a^2.$$

Let $< l, m, n >$ be the **d.c's** of the system of parallel chords of the sphere

$$x^2 + y^2 + z^2 = a^2 \quad \text{...(1)}$$

Let (x_1, y_1, z_1) be the **mid**-point of one of these parallel chords. Then the equations of one of these parallel chords are $\frac{x - x_1}{l} = \frac{y - y_1}{m} = \frac{z - z_1}{n}$ ...(2)

Any point on this line is $(lr + x_1, mr + y_1, nr + z_1)$

If it lies on sphere (1), then

$$(lr + x_1)^2 + (mr + y_1)^2 + (nr + z_1)^2 = a^2$$

or
$$r^2(l^2 + m^2 + n^2) + 2r(lx_1 + my_1 + nz_1) + (x_1^2 + y_1^2 + z_1^2 - a^2) = 0 \quad \text{...(3)}$$

which is a quadratic in r.

$\because$ (x_1, y_1, z_1) is the mid-point of chord (1) and $< l, m, n >$ are d.c's of line (2), therefore, the quadratic (3) has equal and opposite roots.

i.e., the sum of roots is zero or the coefficient of $r = 0$

$$\therefore \quad lx_1 + my_1 + nz_1 = 0$$

$\therefore$ Locus of the mid-point (x_1, y_1, z_1) is [Changing (x_1, y_1, z_1) to (x, y, z)]

$lx + my + nz = 0$ which is a plane through the centre (0, 0, 0) of sphere (1) and $\perp$ to **line** (2).

Exercise. *Prove that the locus of the mid-points of a system of parallel chords of the sphere $x^2 + y^2 + z^2 + 2ux + 2vy + 2wz + d = 0$ is a plane through its centre perpendicular to the given chords.*

21. Plane of Contact

To find the locus of the plane of contact of the tangent planes to the sphere $x^2 + y^2 + z^2 = a^2$ which pass through a given point (x_1, y_1, z_1).

The given sphere is
$$x^2 + y^2 + z^2 = a^2 \quad \text{...(1)}$$

Let (x', y', z') be the point of contact of *any* tangent plane to the sphere (1).

Then the equation of the tangent plane at (x', y', z') to sphere (1) is $xx' + yy' + zz' = a^2$.

Since it passes through the given point (x_1, y_1, z_1)

$$\therefore \quad x_1x' + y_1y' + z_1z' = a^2$$

$\therefore$ Locus of the points of contact (x', y', z') is [changing (x', y', z') to (x, y, z)]

$$xx_1 + yy_1 + zz_1 = a^2$$

which being a first degree equation in x, y, z represents a plane and is known as the **plane of contact** of the point (x_1, y_1, z_1) w.r.t. sphere (1).

Exercise. *Find the locus of the points of contact of the tangent planes to the sphere $x^2 + y^2 + z^2 + 2ux + 2vy + 2wz + d = 0$ which pass through a given point (x_1, y_1, z_1).*

[**Ans.** $xx_1 + yy_1 + zz_1 + u(x + x_1) + v(y + y_1) + w(z + z_1) + d = 0$]

22. (a) Definition : Polar Plane of a Point w.r.t. a Sphere

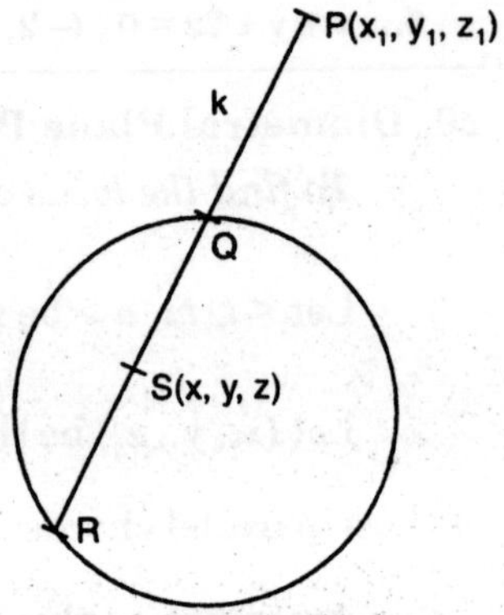

Let P be a given point.

Through P draw any straight line to meet the sphere in Q and R.

Mark a point S on this line such that PS is divided internally at Q and externally at R in the same ratio.

Then locus of S is called the **polar plane** of P w.r.t. the sphere.

P is called **pole** of the **polar plane.**

(b) Equation of the Polar Plane

To find the equation of the polar plane of the point (x_1, y_1, z_1) w.r.t. the sphere $x^2 + y^2 + z^2 = a^2$

Firstly, write the definition of the polar plane given above.

Let P be the point (x_1, y_1, z_1).

Equation of the sphere is $x^2 + y^2 + z^2 = a^2$ ...(1)

$\therefore$ Polar plane of $P(x_1, y_1, z_1)$ is the locus of $S(x, y, z)$.

Let S be the point (x, y, z) whose locus is required. Any point which divides PS in the ratio $k : 1$ is

$$\left(\frac{kx + x_1}{k+1}, \frac{ky + y_1}{k+1}, \frac{kz + z_1}{k+1}\right).$$

If it lies on the sphere (1), (to become Q or R), then

$$\left(\frac{kx + x_1}{k+1}\right)^2 + \left(\frac{ky + y_1}{k+1}\right)^2 + \left(\frac{kz + z_1}{k+1}\right)^2 = a^2$$

or $$(kx + x_1)^2 + (ky + y_1)^2 + (kz + z_1)^2 = a^2 (k + 1)^2$$

or $$k^2(x^2 + y^2 + z^2 - a^2) + 2k(xx_1 + yy_1 + zz_1 - a^2) + (x_1^2 + y_1^2 + z_1^2 - a^2) = 0 \quad ...(2)$$

which is a quadratic in k.

Since PS is divided internally and externally in the same ratio at Q and R, the quadratic (2) must have equal and opposite roots.

$\therefore$ Sum of roots = 0 or the co-efficient of k is 0.

$\therefore$ $\mathbf{xx_1 + yy_1 + zz_1 - a^2 = 0}$ which is the required locus of $S(x, y, z)$ and hence is the equation of the polar plane of $P(x_1, y_1, z_1)$ w.r.t. the sphere.

Exercise. *Find the equation of the polar plane of the point (x_1, y_1, z_1) w.r.t. the sphere $x^2 + y^2 + z^2 + 2ux + 2vy + 2wz + d = 0$.*

[**Ans.** $xx_1 + yy_1 + zz_1 + u(x + x_1) + v(y + y_1) + w(z + z_1) + d = 0$]

23. Pole of the given Plane

To find the pole of the plane $lx + my + nz = p$ with respect to the sphere $x^2 + y^2 + z^2 = a^2$.

The given sphere is $x^2 + y^2 + z^2 = a^2$ ...(1)

and the given plane is $lx + my + nz = p$ ...(2)

Let $P(x_1, y_1, z_1)$ be the pole of plane (2) w.r.t. sphere (1)

∴ By def. Art. 22 (*a*) ; plane (2) is the polar plane of (x_1, y_1, z_1) w.r.t. sphere (1).

But equation of the polar plane of $P(x_1, y_1, z_1)$ w.r.t. sphere (1) is

$$xx_1 + yy_1 + zz_1 = a^2 \qquad ...(3)$$

Now equations (2) and (3) represent the same plane (polar plane of the same point w.r.t. the sphere).

∴ Comparing co-efficients in (3) and (2), we have

$$\frac{x_1}{l} = \frac{y_1}{m} = \frac{z_1}{n} = \frac{a^2}{p}$$

$$\therefore \qquad x_1 = \frac{la^2}{p}, \quad y_1 = \frac{ma^2}{p}, \quad z_1 = \frac{na^2}{p}$$

Thus the required pole is $\left(\frac{la^2}{p}, \frac{ma^2}{p}, \frac{na^2}{p}\right)$.

24. Reciprocal Properties

(*a*) **If the polar plane of a point P passes through another point Q, then polar plane of Q passes through P.**

Let the sphere be

$$x^2 + y^2 + z^2 = a^2 \qquad ...(1)$$

and let P, Q be the points (x_1, y_1, z_1), (x_2, y_2, z_2).

The polar plane of $P(x_1, y_1, z_1)$ w.r.t. sphere (1) is

$$xx_1 + yy_1 + zz_1 = a^2$$

If it passes through $Q(x_2, y_2, z_2)$, then

$$x_2x_1 + y_2y_1 + z_2z_1 = a^2$$

or

$$x_1x_2 + y_1y_2 + z_1z_2 = a^2 \qquad ...(2)$$

The polar plane of $Q(x_2, y_2, z_2)$ w.r.t. sphere (1) is

$$xx_2 + yy_2 + zz_2 = a^2$$

It passes through $P(x_1, y_1, z_1)$ if

$$x_1x_2 + y_1y_2 + z_1z_2 = a^2$$

which is true by (2)

Hence the polar plane of Q passes through P.

Conjugate points (Def.). Two points such that the polar plane of either passes through the other are called **conjugate points.**

(*b*) **If the pole of a plane α lies on another plane β, then the pole of plane β lies on α**

Let the plane α and β be $\qquad lx + my + nz = p \qquad ...(1)$

and $\qquad l'x + m'y + n'z = p' \qquad ...(2)$

Let the sphere be $\qquad x^2 + y^2 + z^2 = a^2 \qquad ....(3)$

Pole of plane α, *i.e.*, (1) w.r.t. sphere (3) is

$$\left(\frac{la^2}{p}, \frac{ma^2}{p}, \frac{na^2}{p}\right) \qquad \text{| Reproduce Art. 23}$$

Since it lies on plane β, *i.e.*, (2)

$$\therefore \qquad l'\left(\frac{la^2}{p}\right)+m'\left(\frac{ma^2}{p}\right)+n'\left(\frac{na^2}{p}\right)=p'$$

or $$a^2(ll'+mm'+nn')=pp' \qquad ...(4)$$

Symmetry of the result (4) shows that the pole of plane (2) w.r.t. sphere (3) lies on plane (1).

Conjugate Planes (Def.). *Two planes such that the pole of either w.r.t. a sphere lies on the other are called* **conjugate planes** *w.r.t. a sphere.*

Example. Salmon's Theorem. *Prove that the distances of two points from the centre of a sphere are proportional to the distances of each from the polar of the other.*

Sol. Let the sphere be $x^2+y^2+z^2=a^2$ ...(1)

Centre of the sphere is O(0, 0, 0).

Let the two given points be $P(x_1, y_1, z_1)$, $Q(x_2, y_2, z_2)$.

Polar planes of P and Q w.r.t. sphere (1) are

$$xx_1+yy_1+zz_1=a^2 \qquad ...(2)$$

and $$xx_2+yy_2+zz_2=a^2 \qquad ...(3)$$

$$\therefore \quad \frac{\text{Distance of P from polar plane of Q } i.e., \text{ plane (3)}}{\text{Distance of Q from polar plane of P } i.e., \text{ plane (2)}}$$

$$=\frac{(x_1x_2+y_1y_2+z_1z_2-a^2)\Big/\sqrt{x_2^2+y_2^2+z_2^2}}{(x_1x_2+y_1y_2+z_1z_2-a^2)\Big/\sqrt{x_1^2+y_1^2+z_1^2}}$$

$$=\frac{\sqrt{x_1^2+y_1^2+z_1^2}}{\sqrt{x_2^2+y_2^2+z_2^2}}=\frac{OP}{OQ}.$$

25. To prove that the polar planes of all the points on a line w.r.t. a sphere pass through another line

Let the given line be $\dfrac{x-x_1}{l}=\dfrac{y-y_1}{m}=\dfrac{z-z_1}{n}=r$ (say) ...(1)

and the given sphere be $x^2+y^2+z^2=a^2$ ...(2)

Any point on line (1) is $(lr+x_1, mr+y_1, nr+z_1)$

Polar plane of this point w.r.t. sphere (2) is

$$x(x_1+lr)+y(y_1+mr)+z(z_1+nr)=a^2 \qquad | \ \mathbf{xx_1+yy_1+zz_1=a^2}$$

or $$(xx_1+yy_1+zz_1-a^2)+r(lx+my+nz)=0$$

This equation is of the form (one plane) + k (other plane) = 0.

∴ This plane passes through the line

$$xx_1+yy_1+zz_1-a^2=0,\ lx+my+nz=0 \text{ for all values of } r.$$

This line is called **polar line** of the line (1) w.r.t. sphere (2).

Def. : Polar Lines. (Two lines are called **polar lines w.r.t.** a sphere if the polar plane of any point on one line w.r.t. a sphere passes through the other).

Example. *Find the equations of the polar line of the line*

$$\frac{x-1}{2}=\frac{y-2}{3}=\frac{z-3}{4} \text{ w.r.t. the sphere } x^2+y^2+z^2=16.$$

Sol. Any point on the given line is $(2r + 1, 3r + 2, 4r + 3)$

Polar plane of this point w.r.t. the sphere $x^2 + y^2 + z^2 = 16$ is

$$x(2r + 1) + y(3r + 2) + z(4r + 3) = 16 \qquad | xx_1 + yy_1 + zz_1 = a^2$$

or
$$(x + 2y + 3z - 16) + r(2x + 3y + 4z) = 0$$

This equation is of the form one plane + k(other plane) = 0.

$\therefore$ This plane passes through the line

$$x + 2y + 3z - 16 = 0, 2x + 3y + 4z = 0 \text{ for all values of } r.$$

These equations of the line obtained above are the equations of the polar line of the given line w.r.t. the given sphere.

EXERCISE 8

1. Prove that the polar plane of a point P w.r.t. a sphere (centre O) is perpendicular to the line OP.
2. If the polar of a point P w.r.t. a sphere (centre O) meets OP in P′ ; then OP.OP′ = (radius)^2.
3. Find the equatons of the polar line of the line

$$\frac{x-1}{2} = \frac{y-2}{2} = z + 3 \text{ w.r.t. the sphere } x^2 + y^2 + z^2 = 1.$$

Answers

3. $x - 2y + 3z + 1 = 0, 2x + 3y + z = 0.$

MISCELLANEOUS EXERCISE

1. If L, M, N are the feet of the perpendiculars from (1, 2, 3) on the co-ordinate planes, find the equation of the sphere through O, L, M, N where O is the origin.
2. Find the co-ordinates of the centre of the sphere through the points (3, 4, – 12), (0, – 5, 12), (– 4, 12, – 3), (13, 0, 0).
3. A point moves so that the sum of squares of its distances from any number of given points is constant. Prove that its locus is a sphere.
4. Find the locus of the centres of the spheres of constant radius which pass through a given point and touch a given plane.
   [**Hint.** *Let the given point be (0, 0, c) and the given plane be XY-plane (i.e., z = 0).*]
5. Show that the following points are concyclic : (5, 0, 2), (2, – 6, 0), (7, – 3, 8) and (4, – 9, 6).
   [**Hint.** *Find the equations of the circle through any three points and prove that the fourth point satisfies the equations of the circle.*]
6. What is the equation of a sphere whose radius is 4 and whose centre is (– 6, 1, 3) ? Find the area of the section in which the sphere is cut by the plane $x - y + 2z + 5 = 0$.
7. Prove that a sphere which cuts the two spheres S = 0 and S′ = 0 at right angles, will also cut lS + mS′ = 0 at right angles.
8. P is a variable point on a given line and A, B, C are its projections on the axes. Show that the sphere OABC passes through a fixed circle.
9. Show that the centres of all sections of the sphere $x^2 + y^2 + z^2 = r^2$ by planes through a point (x', y', z') lie on the sphere $x(x - x') + y(y - y') + z(z - z') = 0$.
10. Show that the spheres which cut two given spheres along great circles all pass through two fixed points.
11. Find the equation of the sphere which has its centre at (a, b, c) and which touches the line

$$\frac{x-f}{l} = \frac{y-g}{m} = \frac{z-h}{n}.$$

12. At what angle does the sphere
$$x^2 + y^2 + z^2 - 2x - 4y - 6z + 10 = 0$$
intersect the sphere which has the points (1, 2, – 3) and (5, 0, 1) as the extremities of a diameter.

13. If the sphere
$$S_1 = x^2 + y^2 + z^2 + 2u_1x + 2v_1y + 2w_1z + d_1 = 0$$
cuts the sphere $S_2 = x^2 + y^2 + z^2 + 2u_2x + 2v_2y + 2w_2z + d_2 = 0$
in a great circle, then prove that
$$2(u_2^2 + v_2^2 + w_2^2) - d_2 = 2(u_1u_2 + v_1v_2 + w_1w_2) - d_1.$$

14. Prove that the centres of spheres which cut both the spheres
$$x^2 + y^2 + z^2 + 2u_1x + d = 0,\ x^2 + y^2 + z^2 + 2u_2x + d = 0$$
in great circles lie on the plane $x + u_1 + u_2 = 0$.

15. Prove that the sum of the squares of the intercepts made by a given sphere on any three mutually ⊥ lines through a fixed point is constant.

16. Find the areas of projection of the circle $x^2 + y^2 + z^2 = 25$, $2x + y + 2z = 9$ on the co-ordinates planes and find the area of the curve.

17. POP′ is a variable diameter for the ellipse
$$z = 0, \frac{x^2}{a^2} + \frac{y^2}{b^2} = 1$$
and a circle is described in the plane PP′ ZZ′ on PP′ as diameter. Prove that as PP′ varies, the circle generates the surface
$$(x^2 + y^2 + z^2)\left(\frac{x^2}{a^2} + \frac{y^2}{b^2}\right) = x^2 + y^2.$$

18. If O be the centre of a sphere of radius unity and A, B be two points in a line with O such that OA . OB = 1, and if P be a variable point on the sphere, show that PA : PB = constant.

19. Find the centres of the two spheres which touch the plane $4x + 3y = 47$ at the point (8, 5, 4) and the sphere $x^2 + y^2 + z^2 = 1$.

20. Two points P, Q are conjugate w.r.t. a sphere S, show that the sphere on PQ as diameter cuts S orthogonally.

21. If two spheres S_1 and S_2 are orthogonal, the polar plane of any point on S_1 w.r.t. S_2 passes through the other end of the diameter of S_1 through P.

22. Show that the locus of a point from which equal tangents may be drawn to the three spheres
$$x^2 + y^2 + z^2 + 2x + 2y + 2z + 2 = 0,\ x^2 + y^2 + z^2 + 4x + 4z + 4 = 0,$$
$$x^2 + y^2 + z^2 + x + 6y - 4z - 2 = 0 \text{ is the straight line } \frac{x}{2} = \frac{y-1}{5} = \frac{z}{3}.$$

Answers

1. $x^2 + y^2 + z^2 - x - 2y - 3z = 0$
2. (0, 0, 0)
4. The curve of intersection of surfaces
$$x^2 - 2cz + c^2 = 0 \quad \text{and} \quad y^2 + z^2 = \lambda^2$$
where λ is the radius of the sphere
6. $x^2 + y^2 + z^2 + 12x - 2y - 6z + 30 = 0,\ \frac{40\pi}{3}$
11. $(l^2 + m^2 + n^2)[(x - a)^2 + (y - b)^2 + (z - c)^2] = \Sigma[(b - g)n - (c - h)m^2]$
12. $\cos^{-1}\frac{2}{3}$
16. $\frac{32\pi}{3}, \frac{16\pi}{3}, \frac{32\pi}{3}$; 16π
19. (4, 2, 4), $\left(\frac{64}{21}, \frac{27}{21}, 4\right)$.

11
Cone

1. Definition

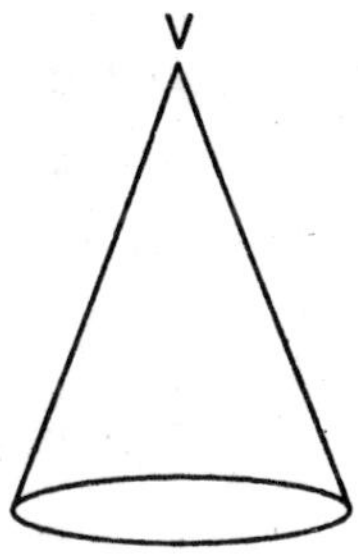

A Cone is a surface generated by a (moving) straight line which passes through a fixed point and is subject to some other condition ; for example it may intersect a given curve, or touch a given surface or make a given angle with a given line through the fixed point.

The moving line in any position is called the **generator** of the cone, the fixed point is called the **vertex** and if the generators cut a fixed curve, **then** the curve is called the **guiding curve.**

2. What is Homogeneous Equation ?

An equation $f(x, y, z) = 0$ is said to be homogeneous in x, y, z if $f(tx, ty, tz) = 0$ for all values of t.

For example,

(*i*) $f(x, y, z) \equiv ax^2 + by^2 + cz^2 + 2fyz + 2gzx + 2hxy = 0$ is a homogeneous equation, because

$$\begin{aligned} f(tx, ty, tz) &= a(tx)^2 + b(ty)^2 + c(tz)^2 + 2f(ty)(tz) + 2g(tz)(tx) + 2h(tx)(ty) \\ &= t^2(ax^2 + by^2 + cz^2 + 2fyz + 2gzx + 2hxy) \\ &= t^2(0) = 0 \text{ for all values of } t. \end{aligned}$$

or Equivalently : An equation in x, y, z is said to be **homogeneous** if the sum of powers of x, y, z in each term is the same, *i.e.,* each term is of the same degree.

3. Equation of a Cone whose vertex is the origin

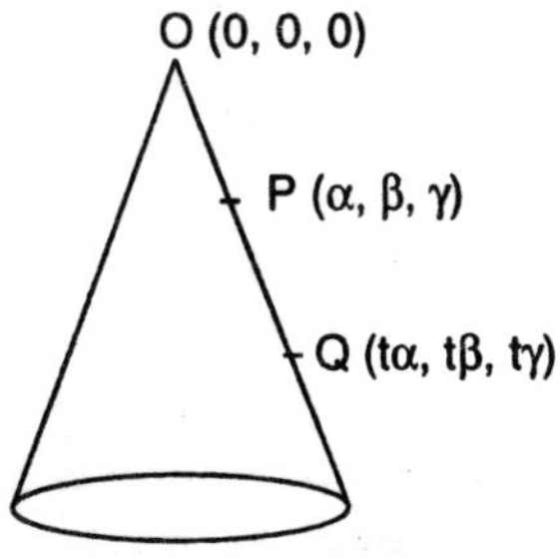

To show that the equation of a Cone, whose vertex is the origin, is homogeneous in x, y, z.

Let the equation of the cone with vertex as origin be

$$f(x, y, z) = 0 \qquad ...(1)$$

Let $P(\alpha, \beta, \gamma)$ be any point on the cone (1), then

$$f(\alpha, \beta, \gamma) = 0 \qquad ...(2)$$

D.R's of the generator OP are $\alpha - 0, \beta - 0, \gamma - 0$, *i.e.,* α, β, γ.

Equations of the generator OP are

$$\frac{x-0}{\alpha} = \frac{y-0}{\beta} = \frac{z-0}{\gamma}$$

i.e.,

$$\frac{x}{\alpha} = \frac{y}{\beta} = \frac{z}{\gamma} = (t) \qquad ...(3)$$

Any point Q on this generator is $(t\alpha, t\beta, t\gamma)$.

Since the generator completely lies on the Cone (By def. of Cone), therefore every point $(t\alpha, t\beta, t\gamma)$ of this generator must lie on Cone (1).

$$\therefore \qquad f(t\alpha, t\beta, t\gamma) = 0 \qquad \text{...(4)}$$

Form (2) and (4), we conclude that $f(x, y, z) = 0$ is a homogeneous equation in x, y, z.(See Art. 2).

Note. In this book, we shall discuss equations of second degree in x, y, z.

Remark. We have already proved in the chapter on "Plane" that $ax^2 + by^2 + cz^2 + 2fyz + 2gzx + 2hxy = 0$ (a homogeneous of second degree in x, y, z) represents a pair of planes if $abc + 2fgh - af^2 - bg^2 - ch^2 = 0$.

Here below, we will discuss a 'partial' converse of Art. 3 (in the light of above remark) :

4. Every homogeneous equation in x, y, z represents a cone whose vertex is the origin (provided it does not represent planes).

Let $$f(x, y, z) = 0 \qquad \text{...(1)}$$

be a homogeneous equation of degree $n \geq 2$ in x, y, z.

Let us suppose that $f(x, y, z)$ can't be decomposed into linear factors (*i.e.*, does not represent planes).

Let $P(\alpha, \beta, \gamma)$ be ANY point on the locus of (1).

$$\therefore \qquad f(\alpha, \beta, \gamma) = 0 \qquad \text{...(2)}$$

$$\therefore \qquad f(t\alpha, t\beta, t\gamma) = 0 \qquad \text{...(3)}$$

for all values of t. [$\because$ $f(x, y, z) = 0$ is a homogeneous equation in x, y, z]

In particular, taking $t = 0$; $f(0, 0, 0) = 0$ which shows that the origin O lies on (1).

Equations of OP are $$\frac{x}{\alpha} = \frac{y}{\beta} = \frac{z}{\gamma}.$$

Any point on OP is Q $(t\alpha, t\beta, t\gamma)$.

(3) Shows that every point Q of the line OP lies on surface (1) *i.e.*, the line OP lies completely on surface (1).

Hence the surface (1) is being generated by lines through O, and therefore represents a cone with vertex at the origin. Hence the result.

Note. The second degree homogeneous equation

$$ax^2 + by^2 + cz^2 + 2fyz + 2gzx + 2hxy = 0$$

represent a cone (called a **quadric** cone) with vertex at the origin.

5. Working rule to find the equation of a cone whose vertex is origin and guiding curve is given

Step I. *Make both the given equations of the guiding curve homogeneous in x, y, z and t by introducing proper powers of $t(t = 1)$.*

Step II. *Eliminate t from the above two equations to get the equation of the cone.*

Example 1. *Find the equation to the cone with vertex at the origin and which passes through the curve.* (M.D.U. 1998)

(*i*) $ax^2 + by^2 + cz^2 = 1,\ lx + my + nz = p$

(*ii*) $ax^2 + by^2 + cz^2 - 1 = 0 = \alpha x^2 + \beta y^2 - 2z$.

Sol. (*i*) Equations of the guiding curve are

$$ax^2 + by^2 + cz^2 = 1 \quad \text{...(1)}$$

$$lx + my + nz = p \quad \text{...(2)}$$

We are to find the equation of a cone with vertex at the origin and the guiding curve given by (1) and (2).

Homogeneousing (1) and (2) with the help of t, we have

$$ax^2 + by^2 + cz^2 = 1.\, t^2 \quad \text{...(3)}$$

$$lx + my + nz = p.\, t \text{ where } t = 1 \quad \text{...(4)}$$

Let us eliminate t from (3) and (4),

From (4), $$t = \frac{lx + my + nz}{p}$$

Putting this value of t in (3), we have

$$ax^2 + by^2 + cz^2 = \frac{(lx + my + nz)^2}{p^2}$$

or $$p^2(ax^2 + by^2 + cz^2) = (lx + my + nz)^2$$

which is the required equation of the cone.

(*ii*) Equations of the guiding curve are

$$ax^2 + by^2 + cz^2 - 1 = 0 \quad \text{and} \quad \alpha x^2 + \beta y^2 - 2z = 0$$

Homogeneousing the two equations with the help of t, we have

$$ax^2 + by^2 + cz^2 - t^2 = 0 \quad \text{...(1)}$$

and $$\alpha x^2 + \beta y^2 - 2zt = 0 \quad \text{...(2)}$$

where $$t = 1$$

Let us eliminate t from (1) and (2)

From (2), $$t = \frac{\alpha x^2 + \beta y^2}{2z}$$

Putting this value of t in (1), we have

$$ax^2 + by^2 + cz^2 - \left(\frac{\alpha x^2 + \beta y^2}{2z}\right)^2 = 0$$

or $$4z^2(ax^2 + by^2 + cz^2) - (\alpha x^2 + \beta y^2)^2 = 0$$

which is the required equation of the cone.

Example 2. *The plane* $\frac{x}{a} + \frac{y}{b} + \frac{z}{c} = 1$ *meets the co-ordinate axes in A, B, C. Prove that the equation of the cone generated by lines drawn from O to meet the circle ABC is*

$$yz\left(\frac{b}{c} + \frac{c}{b}\right) + zx\left(\frac{c}{a} + \frac{a}{c}\right) + xy\left(\frac{a}{b} + \frac{b}{a}\right) = 0. \quad \text{(K.U. 1997 A)}$$

Sol. Equation of the plane ABC is $$\frac{x}{a} + \frac{y}{b} + \frac{z}{c} = 1 \quad \text{...(1)}$$

where A, B, C are the points in which the plane (1) meets the co-ordinate axes at A,B,C.

$\therefore$ A is $(a, 0, 0)$; B$(0, b, 0)$ and C$(0, 0, c)$.

To find the equations of the circle ABC, let us find the equation of the sphere through A, B, C.

*For convenience sake, let us take the sphere as passing through the fourth point O(0, 0, 0).

Equation of the sphere OABC is

$$x^2 + y^2 + z^2 - ax - by - cz = 0 \quad \text{...(2)}$$

$\therefore$ Equations (1) and (2) are equations of circle ABC.

We are to find the equation of the Cone having origin as vertex and circle ABC as guiding curve.

Homogeneousing equations (1) and (2) of the guiding curve with the help of t, we have

$$\frac{x}{a} + \frac{y}{b} + \frac{z}{c} = 1 \cdot t \quad \text{...(3)}$$

and

$$x^2 + y^2 + z^2 - axt - byt - czt = 0 \quad \text{...(4)}$$

where

$$t = 1$$

Let us eliminate t from (3) and (4).

Putting the value fo t from (3) in (4), we have

$$x^2 + y^2 + z^2 - ax\left(\frac{x}{a} + \frac{y}{b} + \frac{z}{c}\right) - by\left(\frac{x}{a} + \frac{y}{b} + \frac{z}{c}\right) - cz\left(\frac{x}{a} + \frac{y}{b} + \frac{z}{c}\right) = 0$$

or

$$x^2 + y^2 + z^2 - x^2 - \frac{a}{b}xy - \frac{a}{c}zx - \frac{b}{a}xy - y^2 - \frac{b}{c}yz - \frac{c}{a}zx - \frac{c}{b}yz - z^2 = 0$$

or

$$-\left(\frac{a}{b} + \frac{b}{a}\right)xy - \left(\frac{a}{c} + \frac{c}{a}\right)zx - \left(\frac{b}{c} + \frac{c}{b}\right)yz = 0$$

or

$$\left(\frac{a}{b} + \frac{b}{a}\right)xy + \left(\frac{a}{c} + \frac{c}{a}\right)zx + \left(\frac{b}{c} + \frac{c}{b}\right)yz = 0$$

which is the required equation of the cone.

EXERCISE 1

1. Find the equation to the cones with vertex at the origin and which pass through the curves given by
   (*i*) $ax^2 + by^2 = 2z,\ lx + my + nz = p$
   (*ii*) $x^2 + y^2 + z^2 + 2ux + d = 0,\ lx + my + nz = p$.
2. Find the equation of the cone whose vertex is at the origin and guiding curve is
$$\frac{x^2}{4} + \frac{y^2}{9} + \frac{z^2}{1} = 1,\ x + y + z = 1.$$
3. Find the equation of the cone with vertex at the origin and which passes through the curve
$$z = 2,\ x^2 + y^2 = 4.$$
4. Find the equation of the cone whose vertex is the origin and which passes through the intersection of the surfaces $4x^2 + y^2 + z^2 + 2x + 4z = 8$ and $4x^2 + y^2 + z^2 + x + 2z = 4$.
   [**Hint.** Subtract the two equations.]
5. Find the equation to the cone whose vertex is the origin and base the circle $x = a,\ y^2 + z^2 = b^2$ and show that the section of the cone by a plane parallel to the plane XOY is a hyperbola.

*Through three points A, B, C an infinite number of spheres can pass, but circle ABC is **unique.**

6. A variable plane is parallel to the given plane $\frac{x}{a}+\frac{y}{b}+\frac{z}{c}=0$ and meets the axes in A, B, C. Prove that the circle ABC lies on the cone

$$yz\left(\frac{b}{c}+\frac{c}{b}\right)+zx\left(\frac{c}{a}+\frac{a}{c}\right)+xy\left(\frac{a}{b}+\frac{b}{a}\right)=0.$$ *(M.D.U. 2003 S)*

7. Planes through OX, OY include an angle α. Show that their line of intersection lies on the cone $z^2(x^2+y^2+z^2)=x^2y^2\tan^2\alpha$.

Answers

1. (i) $p(ax^2+by^2)=2z(lx+my+nz)$
   (ii) $p^2(x^2+y^2+z^2)+2pux(lx+my+nz)+d(lx+my+nz)^2=0$
2. $27x^2+32y^2+72yz+72zx+72xy=0$
3. $x^2+y^2-z^2=0$
4. $4x^2+y^2+z^2=0$.

6. Equation of a Cone with Vertex (α, β, γ)

To find the equation of the cone whose vertex is (α, β, γ) *and base conic is* $f(x, y) \equiv ax^2+2hxy+by^2+2gx+2fy+c=0,\ z=0.$ (K.U. 1999 A)

The equations of the given base conic are

$$ax^2+2hxy+by^2+2gx+2fy+c=0,\ z=0 \quad ...(1)$$

The required cone is the locus of lines drawn through $P(\alpha, \beta, \gamma)$ to intersect the curve given by (1).

Equations of any (generating) line through (α, β, γ) are

$$\frac{x-\alpha}{l}=\frac{y-\beta}{m}=\frac{z-\gamma}{n} \quad ...(2)$$

Line (2) meets the curve (1) where $z=0$

$$\therefore \quad \frac{x-\alpha}{l}=\frac{y-\beta}{m}=\frac{-\gamma}{n}$$

or
$$x-\alpha=-\frac{l\gamma}{n},\ y-\beta=-\frac{m\gamma}{n}$$

or
$$x=\alpha-\frac{l\gamma}{n},\ y=\beta-\frac{m\gamma}{n}$$

If the point lies on the given conic [$\because$ Generator (2) must intersect guiding curve (1)], then putting these values of x, y in (1) ;

$$a\left(\alpha-\frac{l\gamma}{n}\right)^2+2h\left(\alpha-\frac{l\gamma}{n}\right)\left(\beta-\frac{m\gamma}{n}\right)+b\left(\beta-\frac{m\gamma}{n}\right)^2+2g\left(\alpha-\frac{l\gamma}{n}\right)+2f\left(\beta-\frac{m\gamma}{n}\right)+c=0 \quad ...(3)$$

Equations (3) is the condition for the line (2) to intersect conic (1).

To find the locus of lines (2), we have to eliminate l, m, n from (2) and (3)

[$\because$ For different generators l, m, n are different]

Now putting the values of l, m, n from (2) in (3), we have

$$a\left(\alpha-\frac{x-\alpha}{z-\gamma}\cdot\gamma\right)^2+2h\left(\alpha-\frac{x-\alpha}{z-\gamma}\cdot\gamma\right)\left(\beta-\frac{y-\beta}{z-\gamma}\cdot\gamma\right)+b\left(\beta-\frac{y-\beta}{z-\gamma}\cdot\gamma\right)^2+2g\left(\alpha-\frac{x-\alpha}{z-\gamma}\cdot\gamma\right)$$
$$+2f\left(\beta-\frac{y-\beta}{z-\gamma}\cdot\gamma\right)+c=0$$

or $$a\frac{(\alpha z-\gamma x)^2}{(z-\gamma)^2}+2h\frac{(\alpha z-\gamma x)(\beta z-\gamma y)}{(z-\gamma)^2}+b\frac{(\beta z-\gamma y)^2}{(z-\gamma)^2}+2g\frac{(\alpha z-\gamma x)}{z-\gamma}+2f\frac{(\beta z-\gamma y)}{z-\gamma}+c=0$$

or $$a(\alpha z-x\gamma)^2+2h(\alpha z-\gamma x)(\beta z-\gamma y)+b(\beta z-\gamma y)^2+2g(\alpha z-\gamma x)(z-\gamma)$$
$$+2f(\beta z-\gamma y)(z-\gamma)+c(z-\gamma)^2=0 \quad ...(4)$$

which is the required equation of the cone.

Remark. Examples 1 and 2 given under Art. 5 and every question of Exercise 2 (*a*) namely equation of the cone with vertex at origin and a given guiding curve) can also be done by the general method given in Art. 6.

Example 1. *The section of a cone whose vertex is P and guiding curve the ellipse $\frac{x^2}{a^2}+\frac{y^2}{b^2}=1$, $z=0$ by the plane $x=0$ is a rectangular hyperbola. Show that the locus of P is*

$$\frac{x^2}{a^2}+\frac{y^2+z^2}{b^2}=1.$$

Sol. Let the point P whose locus is required be $P(x_1, y_1, z_1)$.

Let us find the equation of a cone whose vertex is $P(x_1, y_1, z_1)$ and guiding curve the ellipse

$$\frac{x^2}{a^2}+\frac{y^2}{b^2}=1,\ z=0. \quad ...(1)$$

Equations of any generating line through $P(x_1, y_1, z_1)$ are

$$\frac{x-x_1}{l}=\frac{y-y_1}{m}=\frac{z-z_1}{n} \quad ...(2)$$

Line (2) meets the curve (1) where $z=0$

$$\therefore \quad \frac{x-x_1}{l}=\frac{y-y_1}{m}=-\frac{z_1}{n}$$

or $$x-x_1=-\frac{l}{n}z_1,\ y-y_1=-\frac{m}{n}z_1 \text{ or } x=x_1-\frac{l}{n}z_1,\ y=y_1-\frac{m}{n}z_1$$

If this point lies on the given conic.

[$\because$ Generator (2) must intersect guiding curve (1)], then putting these values of x, y in (1) ;

$$\frac{1}{a^2}\left(x_1-\frac{l}{n}z_1\right)^2+\frac{1}{b^2}\left(y_1-\frac{m}{n}z_1\right)^2=1 \quad ...(3)$$

Putting the values of l, m, n from (2) in (3) [To eliminate l, m, n],

$$\frac{1}{a^2}\left(x_1-\frac{x-x_1}{z-z_1}z_1\right)^2+\frac{1}{b^2}\left(y_1-\frac{y-y_1}{z-z_1}z_1\right)^2=1$$

or $$\frac{1}{a^2}\frac{(zx_1-xz_1)^2}{(z-z_1)^2}+\frac{1}{b^2}\frac{(zy_1-yz_1)^2}{(z-z_1)^2}=1$$

or $$\frac{1}{a^2}(zx_1-xz_1)^2+\frac{1}{b^2}(zy_1-yz_1)^2=(z-z_1)^2 \quad ...(4)$$

which is the equation of the cone.

Cone (4) is met by the plane $x=0$ in a rectangular hyperbola (given).

Putting $x = 0$ in (4),

$$\frac{1}{a^2}(zx_1)^2 + \frac{1}{b^2}(zy_1 - yz_1)^2 = (z - z_1)^2$$

or $$\frac{z^2 x_1^2}{a^2} + \frac{z^2 y_1^2 + y^2 z_1^2 - 2yzy_1z_1}{b^2} - (z^2 + z_1^2 - 2zz_1) = 0 \quad ...(5)$$

$\because$ Equation (5) is to represent a rectangular hyperbola in the plane $x = 0$.

$\therefore$ *Co-eff. of* y^2 + *co-eff. of* $z^2 = 0$

or $$\frac{z_1^2}{b^2} + \frac{x_1^2}{a^2} + \frac{y_1^2}{b^2} - 1 = 0$$

or $$\frac{x_1^2}{a^2} + \frac{y_1^2 + z_1^2}{b^2} = 1$$

$\therefore$ Locus of $P(x_1, y_1, z_1)$ is $\dfrac{x^2}{a^2} + \dfrac{y^2 + z^2}{b^2} = 1.$

Example 2. *Find the equation of the cone with vertex at (1, 1, 1) and passing through the curve of intersection of* $x^2 + y^2 + z^2 = 1$ *and* $x + y + z = 1$. (K.U. 2006)

Sol. Equations of the given curve are

$$x^2 + y^2 + z^2 = 1 \quad ...(1)$$

$$x + y + z = 1 \quad ...(2)$$

Equations of any generating line through (1, 1, 1) are

$$\frac{x-1}{l} = \frac{y-1}{m} = \frac{z-1}{n} \quad ...(3)$$

Any point on generator (3) is $(1 + lr, 1 + mr, 1 + nr)$.

Since generator (3) must meet the curve given by (1) and (2), therefore, this point must satisfy (1) and (2) for some value of r.

$\therefore$ $$(1 + lr)^2 + (1 + mr)^2 + (1 + nr)^2 = 1$$

and $$1 + lr + 1 + mr + 1 + nr = 1$$

or $$r^2(l^2 + m^2 + n^2) + 2r(l + m + n) + 2 = 0 \quad ...(4)$$

and $$r(l + m + n) = -2 \quad ...(5)$$

Let us eliminate r from (4) and (5).

Putting the value of $r = \dfrac{-2}{l + m + n}$ from (5) in (4), we have

$$\frac{4}{(l + m + n)^2}(l^2 + m^2 + n^2) - 4 + 2 = 0$$

or $$4(l^2 + m^2 + n^2) - 2(l + m + n)^2 = 0$$

Simplifying $l^2 + m^2 + n^2 - 2lm - 2mn - 2nl = 0 \quad ...(6)$

Equation (6) is the condition for line (3) to be a generator of the required cone.

Eliminating l, m, n from (3) and (6), we have

$$(x - 1)^2 + (y - 1)^2 + (z - 1)^2 - 2(x - 1)(y - 1) - 2(y - 1)(z - 1) - 2(z - 1)(x - 1) = 0$$

Simplifying $x^2 + y^2 + z^2 - 2xy - 2yz - 2zx + 2x + 2y + 2z - 3 = 0$ which is the required equation of the cone.

EXERCISE 2

1. Find the equation of the cone whose vertex is (α, β, γ) and whose base is
   (i) $y^2 = 4ax, z = 0$
   (ii) $z^2 = 4ax, y = 0$
   (iii) $ax^2 + by^2 = 1, z = 0$.
2. Find the equation of the cone whose vertex is the point (α, β, γ) and whose generating lines pass through the conic $\frac{x^2}{a^2} + \frac{y^2}{b^2} = 1, z = 0$. *(M.D.U. 2003, 2007)*
   [**Hint.** It is Example 1 upto equation (4).]
3. (i) Find the equation of the cone whose vertex is the point $(-1, 1, 2)$ and whose guiding curve is $3x^2 - y^2 = 1, z = 0$. *(M.D.U 1996)*
   (ii) Find the equation of the cone whose vertex is the point $(3, 1, 2)$ and whose guiding lines pass through the ellipse $2x^2 + 3y^2 = 1, z = 0$.
   (iii) Find the equation of the cone whose vertex is at the point $(0, 0, 5)$ and whose guiding curve is $z = 3, x^2 + 2xy + 3y^2 = 1$.
4. (a) Find the equation of the cone with vertex at $(5, 4, 3)$ and the guiding curve
   $$3x^2 - 2y^2 = 6, \; y + z = 0.$$
   (b) Find the equation of the cone with vertex at $(2, 3, 1)$ and passing through the curve of intersection of $x^2 + y^2 + z^2 - 2x + 4y - 6z + 7 = 0$ and $x + 2y + 2z = 5$. *(M.D.U. 2000)*
5. Show that the equation to the cone whose vertex is origin and the base curve is $z = c, f(x, y) = 0$ is $f\left(\frac{xc}{z}, \frac{yc}{z}\right) = 0$.

Answers

1. (i) $(\beta z - \gamma y)^2 = 4a(\alpha z - \gamma x)(z - \gamma)$ (ii) $(\beta z - \gamma y)^2 = 4a(\beta x - \alpha y)(\beta - y)$
   (iii) $a(\alpha z - \gamma x)^2 + b(\beta z - \gamma y)^2 = (z - \gamma)^2$
2. $\frac{1}{a^2}(\alpha z - \gamma x)^2 + \frac{1}{b^2}(\beta z - \gamma y)^2 = (z - \gamma)^2$
3. (i) $12x^2 - 4y^2 + z^2 + 4yz + 12zx + 4z - 4 = 0$ (ii) $2x^2 + 3y^2 + 5z^2 - 3yz - 6zx + z - 1 = 0$
   (iii) $4x^2 + 12y^2 - z^2 + 8xy + 10z - 25 = 0$
4. (a) $3(-7x + 5y + 5z)^2 + 2(-3y + 4z)^2 = 6(y + z - 7)^2$
   (b) $38x^2 + 17y^2 + 157z^2 + 22xy + 124yz + 92zx - 310x - 270y - 870z + 1150 = 0$.

7. The Right Circular Cone

A right circular cone is the locus of a line (i.e., generated by a st. line) which passes through a fixed point and makes a constant angle with a fixed line through the fixed point.

The moving line in any position is called the **generator** of the cone, the fixed point is called **vertex** and the constant angle the **semi-vertical angle** of the cone.

Note. The semi-vertical angle is the angle between the axis and any generator of the cone.

Remark. The section of a right circular cone by a plane perpendicular to its axis is a circle.

8. Working Rule to find the Equation of a Right circular Cone

1. We must find (if not given) first
   (a) d.r's of axis of cone.
   (b) Semi-vertical angle α of the cone.

2. Take any point P(x, y, z) on the required right circular cone.
3. Find cos α where α is the angle between the axis say VN of the cone and generator VP where V is the vertex of the cone.
4. Square both sides to get the equation of the cone.

9. (a) To show that the equation of the right circular cone whose vertex is the origin, axis the z-axis and semi-vertical angle α is $x^2 + y^2 = z^2 \tan^2\alpha$. *(M.D.U. 1999; K.U. 2003, 2007)*

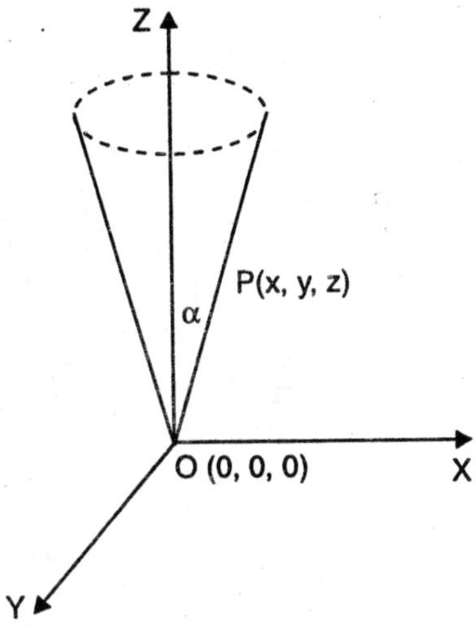

Proof. Let P(x, y, z) be *any* point on the right circular cone.

∴ By def. of right circular cone ∠ZOP = semi-vertical angle α. *i.e.*, angle between axis of the cone and any generator of the cone is α.

d.r's of the generator OP are $x - 0,\ y - 0,\ z - 0$ *i.e.*, $< x, y, z >$.

d.c's of the axis of the cone (here z-axis) are < 0, 0, 1 >.

$$\therefore \quad \cos\alpha = \frac{x(0) + y(0) + z(1)}{\sqrt{x^2 + y^2 + z^2}}$$

$$\left|\ \because \cos\theta = \frac{a_1a_2 + b_1b_2 + c_1c_2}{\sqrt{a_1^2 + b_1^2 + c_1^2}\sqrt{a_2^2 + b_2^2 + c_2^2}}\right.$$

or
$$\cos\alpha = \frac{z}{\sqrt{x^2 + y^2 + z^2}}$$

Squaring and cross-multiplying, $(x^2 + y^2 + z^2)\cos^2\alpha = z^2$.

$$\therefore \quad x^2 + y^2 + z^2 = \frac{z^2}{\cos^2\alpha} = z^2\sec^2\alpha$$

$$\therefore \quad x^2 + y^2 = z^2\sec^2\alpha - z^2 = z^2(\sec^2\alpha - 1)$$

or
$$x^2 + y^2 = z^2\tan^2\alpha$$

which is the required equation of the cone.

(b) Find the equation of a right circular cone whose vertex is (α, β, γ), semi-vertical angle θ and axis has the direction cosines < l, m, n >.

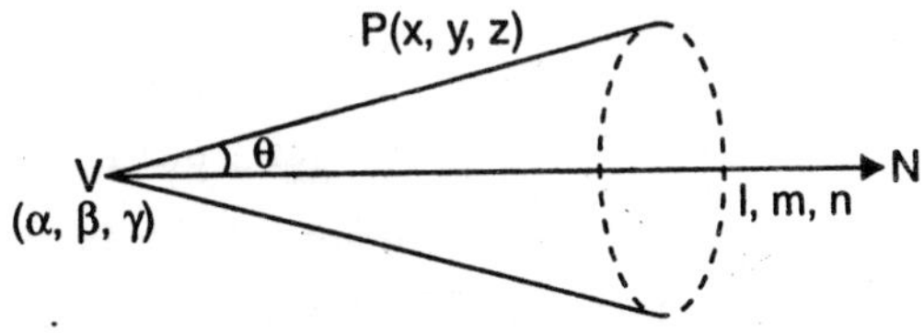

Let P(x, y, z) be *any* point on the right circular cone.

∴ By def. of right circular cone.

∠NVP = Semi-vertical angle θ

i.e., the angle between the axis of the cone and any generator the cone is θ.

d.r's of the generator VP are $< x - \alpha, y - \beta, z - \gamma >$.

d.c's of the axis of the cone are $< l, m, n >$.

$$\therefore \quad \cos\theta = \frac{(x-\alpha)l + (y-\beta)m + (z-\gamma)n}{\sqrt{(x-\alpha)^2 + (y-\beta)^2 + (z-\gamma)^2}\sqrt{l^2 + m^2 + n^2}}$$

Squaring and cross-multiplying , (Also using that $l^2 + m^2 + n^2 = 1$)

$$[(x-\alpha)^2 + (y-\beta)^2 + (z-\gamma)^2] \cos^2\theta = [l(x-\alpha) + m(y-\beta) + n(z-\gamma)]^2 \quad ...(1)$$

which is the required equation of the cone.

Cor. If in (1), we put $\alpha = \beta = \gamma = 0$, then the equation of the right circular cone whose vertex is origin and axis with d.c's $< l, m, n >$ and semi-vertical angle θ is

$$(lx + my + nz)^2 = (x^2 + y^2 + z^2) \cos^2\theta.$$

Example 1. *The axis of a right cone, vertex O, makes equal angles with the co-ordinate axes, and the cone passes through the line drawn from O with direction cosines proportional to (1, – 2, 2). Find the equation of the Cone.*

Sol. For figure, see Fig. of Art. 9 (*b*).

∵ The axis of the cone makes equal angles say θ with the co-ordinate axes.

∴ d.c's of the axis of the cone are $< \cos\theta, \cos\theta, \cos\theta >$

i.e., d.r's of the axis are < 1, 1, 1 >.

Let α be the semi-vertical angle of the cone.

⇒ α is the angle between the axis of the cone and any generator of the cone.

d.r's of one of the generators of the cone are < 1, – 2, 2 > (given)

$$\therefore \quad \cos\alpha = \frac{(1)(1) + (1)(-2) + (1)(2)}{\sqrt{1+1+1}\sqrt{1+4+4}} = \frac{1-2+2}{3\sqrt{3}} = \frac{1}{3\sqrt{3}} \quad ...(1)$$

Now let P(x, y, z) be any point on the right circular cone.

$$\therefore \quad \angle NOP = \alpha$$

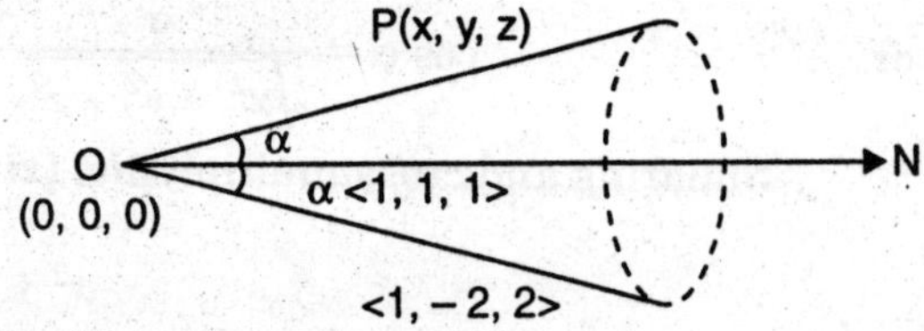

d.r's of the generator OP are $< x - 0, y - 0, z - 0 >$

i.e., $< x, y, z >$.

d.r's of axis of the cone are < 1, 1, 1 >.

$$\therefore \quad \cos\alpha = \frac{x.1 + y.1 + z.1}{\sqrt{x^2+y^2+z^2}\sqrt{1+1+1}}$$

or $$\frac{1}{3\sqrt{3}} = \frac{x+y+z}{\sqrt{3}\sqrt{x^2+y^2+z^2}} \quad | \text{ By (1)}$$

Squaring and cross-multiplying,

$$9(x+y+z)^2 = x^2 + y^2 + z^2$$

or $$9(x^2 + y^2 + z^2 + 2xy + 2yz + 2zx) = x^2 + y^2 + z^2$$

or $$8x^2 + 8y^2 + 8z^2 + 18xy + 18yz + 18zx = 0$$

Cancelling 2, we get $4x^2 + 4y^2 + 4z^2 + 9xy + 9yz + 9zx = 0$

which is the required equation of the cone.

Example 2. *Find the equation to the right circular cone whose vertex is P(2, – 3, 5), axis PQ which makes equal angles with the axes, and which passes through A(1, – 2, 3).*

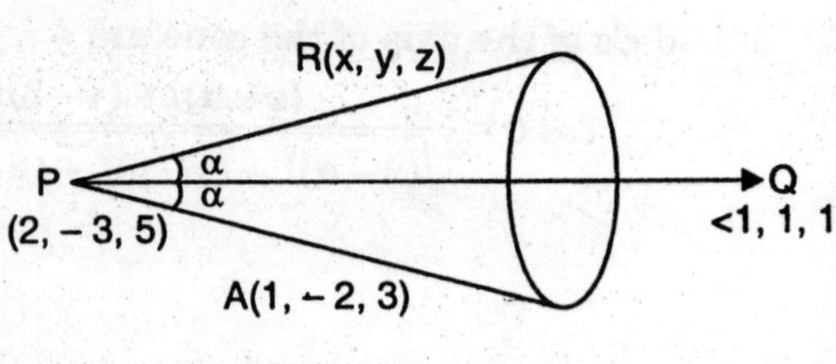

Sol. ∵ The axis PQ of the cone makes equal angles with the axes.

∴ d.r's of axis PQ are < 1, 1, 1 >

(See Example 1)

Let α be the semi-vertical angle of the cone.

d.r's of one of the generators PA of the cone are

$$< 1-2, -2+3, 3-5 > \quad i.e., \quad < -1, 1, -2 > \quad i.e., \quad < 1, -1, 2 >$$

$$\therefore \qquad \cos\alpha = \frac{(1)(1)+(1)(-1)+(1)(2)}{\sqrt{1+1+1}\sqrt{1+1+4}} = \frac{1-1+2}{\sqrt{3}\sqrt{6}} = \frac{2}{\sqrt{18}} = \frac{2}{3\sqrt{2}} = \frac{\sqrt{2}}{3}$$

Now let R(x, y, z) be any point on the right circular cone.

$$\therefore \qquad \angle QPR = \alpha$$

d.r's of the generator PR are $<x-2, y+3, z-5>$

d.r's of the axis of the cone are <1, 1, 1>

$$\therefore \qquad \cos\alpha = \frac{(x-2)\,1+(y+3)\,.\,1+(z-5)\,.\,1}{\sqrt{(x-2)^2+(y+3)^2+(z-5)^2}\ \sqrt{1+1+1}}$$

or
$$\frac{\sqrt{2}}{3} = \frac{x+y+z-4}{\sqrt{(x-2)^2+(y+3)^2+(z-5)^2}\ \sqrt{3}} \qquad | \text{ Using (1)}$$

Squaring and cross-multiplying $6[(x-2)^2+(y+3)^2+(z-5)^2] = 9[x+y+z-4]^2$.

Cancelling 3, $2(x^2-4x+4+y^2+6y+9+z^2+25-10z)$

$$= 3(x^2+y^2+z^2+16+2xy+2xz-8x+2yz-8y-8z)$$

$[\because \quad (\mathbf{a}+\mathbf{b}+\mathbf{c}+\mathbf{d})^2 = \mathbf{a}^2+\mathbf{b}^2+\mathbf{c}^2+\mathbf{d}^2+2\mathbf{ab}+2\mathbf{ac}+2\mathbf{ad}+2\mathbf{bc}+2\mathbf{bd}+2\mathbf{cd}]$

Simplifying, $x^2+y^2+z^2+6(yz+zx+xy)-16x-36y-4z-28=0$

which is the required equation of the cone.

EXERCISE 3

1. Obtain the equation of the right circular cone whose vertex is the point (x_1, y_1, z_1) and axis is the line $\frac{x-x_1}{l} = \frac{y-y_1}{m} = \frac{z-z_1}{n}$ and θ is the semi-vertical angle.
   [**Hint.** It is Art. 9 (*b*).]
2. (*a*) Find the equation of the right circular cone whose vertex is at the origin, whose axis is $\frac{x}{1} = \frac{y}{2} = \frac{z}{3}$ and whose vertical angle is 60°.
   (*b*) Find the equation to the right circular cone whose vertex is P(2, – 3, 5), axis PQ which makes equal angles with the axes, and semi-vertical angle is 30°.
3. Prove that $x^2-y^2+z^2-2x+4y+6z+6=0$ represents a right circular cone whose vertex is the point (1, 2, – 3), whose axis is parallel to OY and whose semi-vertical angle is 45°.
   [**Hint.** *The d.c's of the axis of the cone which is parallel to OY are* $< 0, 1, 0 >$.]
4. Find the equation of the right circular cone which passes through the point (1, 1, 2) and has vertex at the origin and axis $\frac{x}{2} = \frac{y}{-4} = \frac{z}{3}$.
5. Find the equation of the right circular cone whose vertex is at the origin, whose axis is the line $\frac{x}{1} = \frac{y}{2} = \frac{z}{3}$ and which has a vertical angle of 60°. Show that the sections by planes parallel to $x = 0$ are hyperbolas and by planes parallel to $y = 0$ and $z = 0$ are ellipse.
   [**Hint.** *A second degree equation (in two dimensions) namely* $ax^2+2hxy+by^2+2gx+2fy+c=0$ *represents a hyperbola, parabola, or an ellipse according as* $h^2-ab > 0 = 0$ *or* < 0.]

6. Lines are drawn from the origin, with direction cosines proportional to <1, 2, 2 >, < 2, 3, 6 >, < 3, 4, 12 >. Find the direction cosines of the axis of the right circular cone through them and prove that the semi-vertical angle of the cone is $\cos^{-1}\dfrac{1}{\sqrt{3}}$.

[**Hint.** *Axis of the right circular cone makes the same angle with every generator.*]

7. If α is the semi-vertical angle of the right circular cone which passes through the lines OX, OY, $x = y = z$; show that $\cos\alpha = (9 - 4\sqrt{3})^{-1/2}$.

Answers

1. $[l(x - x_1) + m(y - y_1) + n(z - z_1)]^2 = (l^2 + m^2 + n^2)[(x - \alpha)^2 + (y - \beta)^2 + (z - \gamma)^2]\cos^2\theta$
2. (*a*) $19x^2 + 13y^2 + 3z^2 - 24yz - 12zx - 8xy = 0$
   (*b*) $5x^2 + 5y^2 + 5z^2 - 8xy - 8yz - 8zx - 4x + 86y - 58z + 278 = 0$
4. $4x^2 + 40y^2 + 19z^2 - 72yz + 36zx - 48xy = 0$ **5.** $19x^2 + 13y^2 + 3z^2 - 8xy - 24yz - 12zx = 0$
6. $\dfrac{1}{\sqrt{3}}, \dfrac{-1}{\sqrt{3}}, \dfrac{-1}{\sqrt{3}}$.

---

10. (*a*) Enveloping Cone

Def. *The locus of the tangent lines from a given point to a sphere is a cone called the* **enveloping cone** *or the* **tangent cone** *from the point to the sphere.*

(*b*) **Equation of the Enveloping Cone.** *To find the equation of the enveloping cone from the point* (x_1, y_1, z_1) *to the sphere* $x^2 + y^2 + z^2 = a^2$.

The equation of the sphere is $x^2 + y^2 + z^2 = a^2$ ...(1)

Let $P(x_1, y_1, z_1)$ be the given point.

Let $Q(x, y, z)$ be *any* point on the tangent line from P to the sphere.

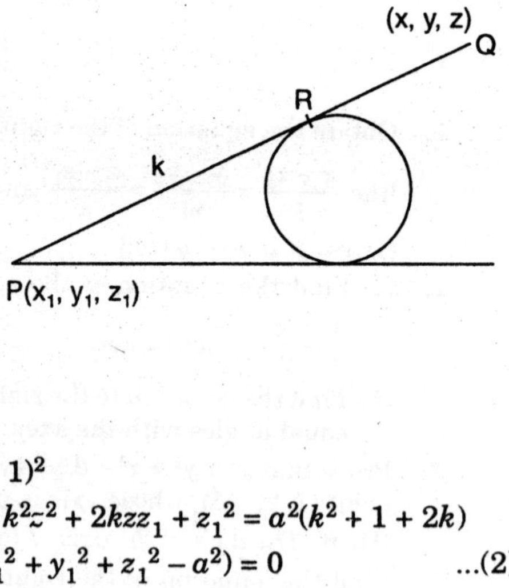

Let PQ be divided by the point of contact R in the ratio $k : 1$.

Then the co-ordinates of R are

$$\left(\frac{kx + x_1}{k+1}, \frac{ky + y_1}{k+1}, \frac{kz + z_1}{k+1}\right).$$

But R lies on the sphere (1).

$$\therefore \quad \left(\frac{kx + x_1}{k+1}\right)^2 + \left(\frac{ky + y_1}{k+1}\right)^2 + \left(\frac{kz + z_1}{k+1}\right)^2 = a^2$$

or $$(kx + x_1)^2 + (ky + y_1)^2 + (kz + z_1)^2 = a^2(k + 1)^2$$

or $$k^2x^2 + 2kxx_1 + x_1^2 + k^2y^2 + 2kyy_1 + y_1^2 + k^2z^2 + 2kzz_1 + z_1^2 = a^2(k^2 + 1 + 2k)$$

or $$k^2(x^2 + y^2 + z^2 - a^2) + 2k(xx_1 + yy_1 + zz_1 - a^2) + (x_1^2 + y_1^2 + z_1^2 - a^2) = 0 \quad ...(2)$$

which is a quadratic in k.

$\because$ The line PQ is a tangent to a sphere (1) ; therefore it meets the sphere in two coincident points R.

$\therefore$ Discriminant of (2) is 0 *i.e.*, "$b^2 - 4ac$" = 0

or $$4(xx_1 + yy_1 + zz_1 - a^2)^2 - 4(x^2 + y^2 + z^2 - a^2)(x_1^2 + y_1^2 + z_1^2 - a^2) = 0$$

Cancelling 4, $(x^2 + y^2 + z^2 - a^2)(x_1^2 + y_1^2 + z_1^2 - a^2) = (xx_1 + yy_1 + zz_1 - a^2)^2$

which is the required equation of the Enveloping Cone.

Rule to Remember. *The equation of the enveloping cone obtained above is* $SS_1 = T^2$

where $S \equiv (x^2 + y^2 + z^2 - a^2)$ *is the L.H.S. of the equation of the sphere (R.H.S. being zero).*

$S_1 \equiv (x_1^2 + y_1^2 + z_1^2 - a^2)$ *is the result of substituting the point* (x_1, y_1, z_1) *in S.*

$T = xx_1 + yy_1 + zz_1 - a^2$ *is the L.H.S of equation of the tangent plane at* (x_1, y_1, z_1) *to the sphere.* *(R.H.S. being zero)*

Example. *Show that the plane* $z = 0$ *cuts the enveloping cone of the sphere* $x^2 + y^2 + z^2 = 11$ *which has its vertex at (2, 4, 1) in a rectangular hyperbola.*

Sol. The equation of the sphere is $x^2 + y^2 + z^2 - 11 = 0$ | Make R.H.S. = 0

and the vertex is (2, 4, 1).

By the above Rule,

Here $S = x^2 + y^2 + z^2 - 11$ or $S_1 = 2^2 + 4^2 + 1^2 - 11 = 10$

$$T = x(2) + y(4) + z(1) - 11 = 2x + 4y + z - 11$$

$\therefore$ Equation of the enveloping cone is $SS_1 = T^2$

or $$(x^2 + y^2 + z^2 - 11)(10) = (2x + 4y + z - 11)^2.$$

This cone meets the plane $z = 0$ in the curve $(x^2 + y^2 - 11)(10) = (2x + 4y - 11)^2$ ...(1)

[Putting $z = 0$]

In this equation, co-eff. of x^2 + co-eff. of $y^2 = 10 - 4 + 10 - 16 = 0$

$\therefore$ (1) represents a rectangular Hyperbola.

EXERCISE 4

1. Find the enveloping cone of the sphere $x^2 + y^2 + z^2 + 2x - 2y = 2$ with its vertex at (1, 1, 1).
2. Find the equation of the cone with vertex at the origin and generators touching the sphere $x^2 + y^2 + z^2 - 2x + 4z = 1$.
3. Prove that the lines drawn from the origin so as to touch the sphere $x^2 + y^2 + z^2 + 2ux + 2vy + 2wz + d = 0$ lie on the cone $d(x^2 + y^2 + z^2) = (ux + vy + wz)^2$.

Answers

1. $3x^2 - y^2 + 4zx - 10x + 2y - 4z + 6 = 0$ 2. $2x^2 + y^2 + 5z^2 - 4zx = 0$.

CONE REPRESENTED BY THE GENERAL EQUATION OF THE SECOND DEGREE

11. *To find the condition that the equation*

$$F(x, y, z) \equiv ax^2 + by^2 + cz^2 + 2fyz + 2gzx + 2hxy + 2ux + 2vy + 2wz + d = 0$$

may represent a cone.

The given equation is

$$ax^2 + by^2 + cz^2 + 2fyz + 2gzx + 2hxy + 2ux + 2vy + 2wz + d = 0 \quad ...(1)$$

If (1) represents a cone with vertex (α, β, γ) say, then shifting the origin to the point (α, β, γ) *i.e.*, changing

x to x + α, y to y + β, z to z + γ : (1) becomes

$$a(x+\alpha)^2 + b(y+\beta)^2 + c(z+\gamma)^2 + 2f(y+\beta)(z+\gamma) + 2g(z+\gamma)(x+\alpha) + 2h(x+\alpha)(y+\beta) + 2u(x+\alpha) + 2v(y+\beta) + 2w(z+\gamma) + d = 0$$

or $$ax^2 + by^2 + cz^2 + 2fyz + 2gzx + 2hxy + 2x(a\alpha + h\beta + g\gamma + u) + 2y(h\alpha + b\beta + f\gamma + v) + 2z(g\alpha + f\beta + c\gamma + w) + (a\alpha^2 + b\beta^2 + c\gamma^2 + 2f\beta\gamma + 2g\gamma\alpha + 2h\alpha\beta + 2u\alpha + 2v\beta + 2w\gamma + d) = 0 \quad ...(2)$$

Since (2) represents a cone with vertex at the origin, therefore (2) must be homogeneous in x, y, z (By Art. 3).

$\therefore$ Co-eff. of $x = 0$, co-eff. of $y = 0$, co-eff. of $z = 0$, and constant term = 0

i.e., $$a\alpha + h\beta + g\gamma + u = 0 \quad ...(3)$$

$$h\alpha + b\beta + f\gamma + v = 0 \quad ...(4)$$

$$g\alpha + f\beta + c\gamma + w = 0 \quad ...(5)$$

$$a\alpha^2 + b\beta^2 + c\gamma^2 + 2f\beta\gamma + 2g\gamma\alpha + 2h\alpha\beta + 2u\alpha + 2v\beta + 2w\gamma + d = 0 \quad ...(6)$$

Now (6) can be written as

$$\alpha(a\alpha + h\beta + g\gamma + u) + \beta(h\alpha + b\beta + f\gamma + v) + \gamma(g\alpha + f\beta + c\gamma + w) + (u\alpha + v\beta + w\gamma + d) = 0$$

i.e., $$\alpha(3) + \beta(4) + \gamma(5) + u\alpha + v\beta + w\gamma + d = 0$$

or $$u\alpha + v\beta + w\gamma + d = 0 \quad ...(7) \mid \text{Using (3), (4), (5)}$$

Eliminating α, β, γ from (3), (4), (5) and (7) with the help of determinants, we have

$$\begin{vmatrix} a & h & g & u \\ h & b & f & v \\ g & f & c & w \\ u & v & w & d \end{vmatrix} = 0$$

which is the required condition.

Cor. The vertex of the cone is obtained by solving any three of the four equations (3), (4), (5) and (7) for α, β, γ.

Remark. The equations (3), (4), (5) and (7) of the above Article can be expressed in a very neat form by the use of Calculus. Let us introduce proper powers of t to make the equation (1) homogeneous.

Then $F(x, y, z, t) \equiv ax^2 + by^2 + cz^2 + 2fyz + 2gzx + 2hxy + 2uxt + 2vyt + 2wzt + dt^2 = 0$

$$\therefore \quad \frac{\partial F}{\partial x} = 2(ax + hy + gz + ut), \qquad \frac{\partial F}{\partial y} = 2(hx + by + fz + vt),$$

$$\frac{\partial F}{\partial z} = 2(gx + fy + cz + wt), \qquad \frac{\partial F}{\partial t} = 2(ux + vy + wz + dt).$$

Now equating $\frac{\partial F}{\partial x}, \frac{\partial F}{\partial y}, \frac{\partial F}{\partial z}, \frac{\partial F}{\partial t}$ each to zero and putting $t = 1$, we get equations (3), (4), (5), (7).

METHOD OF NUMERICAL PROBLEMS

1. Make the given equation homogeneous in x, y, z, t by introducing proper powers t where $t = 1$ and denote the L.H.S. of this equation by $F(x, y, z, t)$ (after making R.H.S. zero).

2. Then the four equations (3), (4), (5), (7) are obtained by equating $\frac{\partial F}{\partial x}, \frac{\partial F}{\partial y}, \frac{\partial F}{\partial z}, \frac{\partial F}{\partial t}$ to zero and putting $t = 1$. (See Remark above)

3. Solve any three of the above four equations in x, y, z.

4. Substitute these values of x, y, z in the fourth equation, and if it is satisfied, then the given equation represents a cone and values of x, y, z found in Step III are the co-ordinates of the **VERTEX.**

Example. *Prove that the equation $7x^2 + 2y^2 + 2z^2 - 10zx + 10xy + 26x - 2y + 2z - 17 = 0$ represents a cone whose vertex is (1, – 2, 2).*

Sol. The given equation is

$$7x^2 + 2y^2 + 2z^2 - 10zx + 10xy + 26x - 2y + 2z - 17 = 0 \quad \text{...(1)}$$

Making it homogeneous in x, y, z, t by introducing proper powers of t, we have

$$F(x, y, z, t) \equiv 7x^2 + 2y^2 + 2z^2 - 10zx + 10xy + 26xt - 2yt + 2zt - 17t^2.$$

$$\therefore \quad \frac{\partial F}{\partial x} = 14x - 10z + 10y + 26t = 14x - 10z + 10y + 26 \quad | \because t = 1$$

$$\frac{\partial F}{\partial y} = 4y + 10x - 2t = 4y + 10x - 2 \quad | \because t = 1$$

$$\frac{\partial F}{\partial z} = 4z - 10x + 2t = 4z - 10x + 2 \quad | \because t = 1$$

and

$$\frac{\partial F}{\partial t} = 26x - 2y + 2z - 34t = 26x - 2y + 2z - 34 \quad | \because t = 1$$

Putting $\frac{\partial F}{\partial x}, \frac{\partial F}{\partial y}, \frac{\partial F}{\partial z}$ and $\frac{\partial F}{\partial t}$ each equal to zero, we have

$$14x + 10y - 10z + 26 = 0 \quad \text{...(2)}$$

$$10x + 4y - 2 = 0 \quad \text{...(3)}$$

$$-10x + 4z + 2 = 0 \quad \text{...(4)}$$

$$26x - 2y + 2z - 34 = 0 \quad \text{...(5)}$$

Let us solve (2), (3), (4) for x, y, z

2(2) + 5(4) gives (to eliminate z), $-22x + 20y + 62 = 0$

Also $\quad 10x + 4y - 2 = 0 \quad$ | From (3)

$$\therefore \quad \frac{x}{-40-248} = \frac{y}{620-44} = \frac{1}{-88-200} \quad \text{or} \quad \frac{x}{-288} = \frac{y}{576} = \frac{1}{-288}$$

$$\therefore \quad x = 1, y = -2$$

From (4), $\quad -10 + 4z + 2 = 0 \quad$ or $\quad 4z = 8 \quad \therefore \quad z = 2$

Putting these values of x, y, z in (5), we have

$$26(1) - 2(-2) + 2(2) - 34 = 0$$

or $\quad 26 + 4 + 4 - 34 = 0 \quad$ or $\quad 0 = 0$ which is true.

Hence equation (1) represents a cone with vertex at (1, – 2, 2).

12. Condition that the general homogeneous equation of the second degree should represent a right circular cone

The conditions that the equation $ax^2 + by^2 + cz^2 + 2fyz + 2gzx + 2hxy = 0$ may represent a right circular cone are

$$\frac{F}{f} = \frac{G}{g} = \frac{H}{h}, \text{ where F, G, H are the co-factors of f, g, h in the determinant } \begin{vmatrix} a & h & g \\ h & b & f \\ g & f & c \end{vmatrix}$$

i.e., $\quad F = -(af - gh), G = hf - bg, H = -(ch - fg).$

Proof. The given equation is $ax^2 + by^2 + cz^2 + 2fyz + 2gzx + 2hxy = 0 \quad$ ...(1)

We know that (1) represent a cone with vertex at origin.

[$\because$ (1) is a homogeneous equation (of the second degree) in $x, y\ z$]

If (1) is to represent a right circular cone, then let us compare (1) with

$$(lx + my + nz)^2 = (x^2 + y^2 + z^2) \cos^2 \alpha$$

[Equation of right circular cone with vertex at origin (See cor. Art. 9 (*b*))]

i.e., $(l^2 - \cos^2 \alpha)\, x^2 + (m^2 - \cos^2 \alpha)\, y^2 + (n^2 - \cos^2 \alpha) z^2 + 2mnyz + 2nlzx + 2lmxy = 0$ ...(2)

Comparing (1) and (2),

$$\frac{a}{l^2 - \cos^2 \alpha} = \frac{b}{m^2 - \cos^2 \alpha} = \frac{c}{n^2 - \cos^2 \alpha} = \frac{f}{mn} = \frac{g}{nl} = \frac{h}{lm} = k \text{ (say)}$$

$$\therefore \quad a = k(l^2 - \cos^2 \alpha),\ b = k(m^2 - \cos^2 \alpha),\ c = k(n^2 - \cos^2 \alpha)$$

$$f = kmn, \qquad g = knl, \qquad h = klm$$

$$\therefore \quad \frac{F}{f} = -\frac{(af - gh)}{f} = \frac{gh - af}{f} = \frac{k^2 l^2\, mn - k(l^2 - \cos^2 \alpha) kmn}{kmn}$$

$$= \frac{kmn[kl^2 - kl^2 + k \cos^2 \alpha]}{kmn} = k \cos^2 \alpha$$

Similarly, $\dfrac{G}{g} = k \cos^2 \alpha$ and $\dfrac{H}{h} = k \cos^2 \alpha$

$$\therefore \quad \frac{F}{f} = \frac{G}{g} = \frac{H}{h}.$$

EXERCISE 5

1. (*a*) Prove that the equation $x^2 - 2y^2 + 3z^2 + 5yz - 6zx - 4xy + 8x - 19y - 2z - 20 = 0$ represents a cone whose vertex is $(1, -2, 3)$.

   (*b*) Prove that the equation $2x^2 + 2y^2 + 7z^2 - 10yz - 10zx + 2x + 2y + 26z - 17 = 0$ represents a cone whose vertex is $(2, 2, 1)$.

2. Prove that equation $2y^2 - 8yz - 4zx - 8xy + 6x - 4y - 2z + 5 = 0$ represents a cone whose vertex is $\left(-\dfrac{7}{6}, \dfrac{1}{3}, \dfrac{5}{6}\right)$.

3. Prove that $ax^2 + by^2 + cz^2 + 2ux + 2vy + 2wz + d = 0$ represents a cone if $\dfrac{u^2}{a} + \dfrac{v^2}{b} + \dfrac{w^2}{c} = d.$ (*K.U. 1997 S, 96 A*)

4. Show that $33x^2 + 13y^2 - 95z^2 - 144yz - 96zx - 48xy = 0$ represents a right circular cone whose axis is the line $3x = 2y = z$. Find the semi-vertical angle.

Answers

4. 60°.

---

13. To prove that the direction cosines of any generator of a cone, whose equation is homogeneous, satisfy the equation of the cone

Proof. Let the Homogeneous equation of the cone be $f(x, y, z) = 0$ ...(1)

Vertex of cone (1) is O(0, 0, 0) [$\because$ (1) is homogeneous in x, y, z]

Let $<l, m, n>$ be the d.c's of any generator of the cone.

$\therefore$ Equations of the generator are $\dfrac{x-0}{l} = \dfrac{y-0}{m} = \dfrac{z-0}{n}$

[$\because$ A generator of the cone must pass through the vertex]

Any point on this generator is (lr, mr, nr)

It must lie on cone (1) for all values of r.

$\therefore$ $f(lr, mr, nr) = 0$ for all values of r.

Taking $r = 1$; $f(l, m, n) = 0$ ...(2)

Comparing (1) and (2) we can say that d.c's $<l, m, n>$ of a generator of cone (1) satisfy the cone.

Remark. The converse of the above result is also true.

14. Equation of a quadric cone through the axes

To show that the general equation of the cone of the second degree, which passes through the axes, is **fyz + gzx + hxy = 0.**

Proof. The general equation of a cone of the second degree, whose vertex is the origin ($\because$ the three co-ordinate axes which are three generators here pass through the origin), is

$$ax^2 + by^2 + cz^2 + 2fyz + 2gzx + 2hxy = 0 \qquad ...(1)$$

$\because$ Cone (1) passes through x-axis

i.e., x -axis is a generator of cone (1).

$\therefore$ d.c's of x-axis *i.e.*, $<1, 0, 0>$ must satisfy (1) | By Art. 13

$\therefore$ $a = 0$

Similarly, $b = 0$ and $c = 0$

Substituting these values of a, b, c in (1), the equation of the cone is $2fyz + 2gzx + 2hxy = 0$

or **fyz + gzx + hxy = 0** which is the required equation.

Example. *Show that a cone of second degree can be found to pass through any two sets of rectangular axes through the same origin.*

Sol. Let OX, OY, OZ ; OX′, OY′, OZ′ be two sets of rectangular axes through the same origin.

Let us take OX, OY, OZ as co-ordinate axes and let the d.c's of the other set OX′, OY′, OZ′ referred to them be $<l_1, m_1, n_1>$; $<l_2, m_2, n_2>$ and $<l_3, m_3, n_3>$.

$\therefore$ There exist 12 relations among these d.c's [See Art. 4, Review IV]

The equation of any cone through the co-ordinate axes OX, OY, OZ is

$$fyz + gzx + hxy = 0 \qquad ...(1)$$

$\because$ (1) passes through OX′.

$\therefore$ Its d.c's $<l_1, m_1, n_1>$ satisfy (1)

$$\therefore \qquad fm_1n_1 + gn_1l_1 + hl_1m_1 = 0 \qquad ...(2)$$

Again because (1) passes through OY′.

$$\therefore \qquad fm_2n_2 + gn_2l_2 + hl_2m_2 = 0 \qquad ...(3)$$

Adding (2) and (3) vertically,

$$f(m_1n_1 + m_2n_2) + g(n_1l_1 + n_2l_2) + h(l_1m_1 + l_2m_2) = 0 \qquad ...(4)$$

But $<l_1, m_1, n_1>, <l_2, m_2, n_2>, <l_3, m_3, n_3>$
are the d.c's of three mutually $\perp$ lines.

$$\therefore \quad m_1n_1 + m_2n_2 + m_3n_3 = 0 \quad \text{or} \quad m_1n_1 + m_2n_2 = -m_3n_3$$
$$n_1l_1 + n_2l_2 + n_3l_3 = 0 \quad \text{or} \quad n_1l_1 + n_2l_2 = -n_3l_3$$
$$l_1m_1 + l_2m_2 + l_3m_3 = 0 \quad \text{or} \quad l_1m_1 + l_2m_2 = -l_3m_3$$

Putting these values in (4), we have

$$-fm_3n_3 - gn_3l_3 - hl_3m_3 = 0 \quad \text{or} \quad fm_3n_3 + gn_3l_3 + hl_3m_3 = 0$$

i.e., (1) is satisfied by the d.c's $<l_3, m_3, n_3>$ of OZ′.

$\therefore$ The cone (1) passes through OZ′.

$\therefore$ A cone of second degree can be found to pass through any two sets of rectangular axes.

Note. For numerical problems, we should solve (2) and (3) for f, g, h and substitute these values of f, g, h in (1) to get the equation of the cone.

EXERCISE 6

1. Show that the line $\frac{x}{l} = \frac{y}{m} = \frac{z}{n}$, where $l^2 + 2m^2 - 3n^2 = 0$, is a generator of the cone $x^2 + 2y^2 - 3z^2 = 0$.
2. Show that the lines through the point (α, β, γ) whose direction ratios satisfy $al^2 + bm^2 + cn^2 = 0$ generate the cone $a(x - \alpha)^2 + b(y - \beta)^2 + c(z - \gamma)^2 = 0$.
3. Find the equation to the quadric cone which passes through the three co-ordinate axes and the three mutually perpendicular lines $\frac{x}{1} = \frac{y}{-2} = \frac{z}{3}, \frac{x}{1} = \frac{y}{-1} = \frac{z}{-1}, \frac{x}{5} = \frac{y}{4} = \frac{z}{1}$.
4. (*a*) Find the equation of the cone which passes through three co-ordinate axes as well as the two lines $\frac{x}{1} = \frac{y}{-2} = \frac{z}{3}, \frac{x}{3} = \frac{y}{-1} = \frac{z}{1}$.

   (*b*) Find the equation of a cone which passes through the three co-ordinate axes as well as the two lines $6x = -3y = 2z$ and $x = -3y = 3z$.
5. Find the equation to the quadric cone which contains the three co-ordinate axes and the lines through the origin having direction cosines l_1, m_1, n_1 and l_2, m_2, n_2.

Answers

3. $5yz + 8zx - 3xy = 0$ 4. (*a*) $3yz + 16zx + 15xy = 0$ (*b*) $3yz + 16zx + 15xy = 0$

5. $l_1l_2(m_1n_2 - n_1m_2)yz + m_1m_2(n_1l_2 - n_2l_1)zx + n_1n_2(l_1m_2 - l_2m_1)xy = 0$.

---

15. Angle between two lines in which a plane through the vertex of a cone cuts the cone

Example 1. *Find the equations to the lines in which the plane $2x + y - z = 0$ cuts the cone $4x^2 - y^2 + 3z^2 = 0$. Also find the angle between the lines.* (K.U. 1999 S)

Sol. The given plane is $2x + y - z = 0$ ...(1)

and the cone is $4x^2 - y^2 + 3z^2 = 0$ ...(2)

Plane (1) is passing through the vertex (0, 0, 0) of the cone (2)

$\therefore$ It cuts the cone (2) in two lines through (0, 0, 0)

$\therefore$ Equations of the lines of section are $\frac{x}{l} = \frac{y}{m} = \frac{z}{n}$ ...(3)

$\because$ Line (3) lies on plane (1)

$\therefore$ It is $\perp$ to the normal to plane (1) with d.r's <2, 1, – 1>

$$\therefore \qquad 2l + m - n = 0 \qquad ...(4)$$

Also the line (3) lies on the cone (2)

$\therefore$ Its d.c's satisfy the equation of the cone.

Thus*
$$4l^2 - m^2 + 3n^2 = 0 \qquad ...(5)$$

Now let us solve (4) and (5) for l, m, n.

From (4), $n = 2l + m$. Putting this value in (5), we have

$$4l^2 - m^2 + 3(2l + m)^2 = 0 \quad \text{or} \quad 4l^2 - m^2 + 3(4l^2 + m^2 + 4lm) = 0$$

or
$$16l^2 + 2m^2 + 12lm = 0$$

or
$$8l^2 + 6lm + m^2 = 0 \quad \text{or} \quad (2l + m)(4l + m) = 0$$

$\therefore$ Either $2l + m = 0$

i.e., $2l + m + 0.n = 0$

Also from (4), $2l + m - n = 0$

Solving for l, m, n ;

$$\frac{l}{-1-0} = \frac{m}{0+2} = \frac{n}{2-2}$$

i.e.,
$$\frac{l}{-1} = \frac{m}{2} = \frac{n}{0}$$

or $4l + m = 0$

i.e., $4l + m + 0.n = 0$

Also $2l + m - n = 0$

Solving for l, m, n :

$$\frac{l}{-1-0} = \frac{m}{0+4} = \frac{n}{4-2}$$

i.e.,
$$\frac{l}{-1} = \frac{m}{4} = \frac{n}{2}$$

Putting these values of l, m, n in (3), the required lines of section are

$$\frac{x}{-1} = \frac{y}{2} = \frac{z}{0} \quad \text{and} \quad \frac{x}{-1} = \frac{y}{4} = \frac{z}{2}$$

If θ is the angle between these two lines

then
$$\cos\theta = \frac{(-1)(-1) + 2(4) + 0(2)}{\sqrt{1+4+0}\,\sqrt{1+16+4}} = \frac{9}{\sqrt{5}\,\sqrt{21}}$$

$$\therefore \qquad \theta = \cos^{-1}\frac{9}{\sqrt{105}}.$$

Example 2. *Prove that the angle between the lines given by*

$$x + y + z = 0,\ ayz + bzx + cxy = 0$$

is (i) $\frac{\pi}{2}$ *if* $a + b + c = 0$ (K.U. 1998 S)

(ii) $\frac{\pi}{3}$ *if* $\frac{1}{a} + \frac{1}{b} + \frac{1}{c} = 0.$ (K.U. 1998 S)

Sol. The given plane is
$$x + y + z = 0 \qquad ...(1)$$

and the cone is
$$ayz + bzx + cxy = 0 \qquad ...(2)$$

Plane (1) is passing through the vertex (0, 0, 0) of the cone (2)

*The reader is suggested to observe that (4) and (5) have been obtained by replacing (x, y, z) by (l, m, n) in (1) and (2).

$\therefore$ It cuts the cone (2) in two lines through (0, 0, 0)

$\therefore$ Equations of the lines of section are $\frac{x}{l} = \frac{y}{m} = \frac{z}{n}$ ...(3)

$\because$ Line (3) lies on plane (1)

$\therefore$ It is $\perp$ to the normal to plane (1) with d.r's <1, 1, 1>

$$\therefore \quad l + m + n = 0 \quad \text{...(4)}$$

Again the line (3) lies on the cone (2), $\therefore$ Its d.c's satisfy the equation of the cone.

Thus, $$amn + bnl + clm = 0 \quad \text{...(5)}$$

From (4), $n = -l - m$. Putting this value in (5), we have

$$am(-l - m) + bl(-l - m) + clm = 0$$

i.e., $$-alm - am^2 - bl^2 - blm + clm = 0 \quad \text{or} \quad -bl^2 - (a + b - c)lm - am^2 = 0$$

Cancelling -1, $bl^2 + (a + b - c)lm + am^2 = 0$

Dividing by m^2, $$b\left(\frac{l}{m}\right)^2 + (a + b - c)\frac{l}{m} + a = 0 \quad \text{...(6)}$$

It is a quadratic in $\frac{l}{m}$.

Its roots are $\frac{l_1}{m_1}$ and $\frac{l_2}{m_2}$, where $<l_1, m_1, n_1>$ and $<l_2, m_2, n_2>$ are d.c's of the two lines in which plane (1) cuts cone (2).

Product of roots of (6) $= \frac{l_1}{m_1} \cdot \frac{l_2}{m_2} = \frac{a}{b}$

or $$\frac{l_1 l_2}{a} = \frac{m_1 m_2}{b} = \frac{n_1 n_2}{c} \quad \text{(By Symmetry)}$$
$$= k \text{ (Say)}$$

$$\therefore \quad l_1 l_2 = ak, \ m_1 m_2 = bk, \ n_1 n_2 = ck \quad \text{...(7)}$$

Let θ be the angle between the two lines.

(*i*) $\theta = 90°$ *i.e.*, the two lines are perpendicular to each other

$$\therefore \quad l_1 l_2 + m_1 m_2 + n_1 n_2 = 0$$

Putting values from Eqn. (7),

$$ak + bk + ck = 0$$

Dividing by k, $$a + b + c = 0$$

which is the required condition.

(*ii*) We know that

$$\tan\theta = \frac{\sqrt{(m_1 n_2 - m_2 n_1)^2 + (n_1 l_2 - n_2 l_1)^2 + (l_1 m_2 - l_2 m_1)^2}}{l_1 l_2 + m_1 m_2 + n_1 n_2} \quad \text{...(8)}$$

(We have not found cos θ because the unknown k would have created problems).

Here $$\theta = \frac{\pi}{3} \quad \text{(given)} \quad \text{...(9)}$$

Again from (6),

Sum of roots of (6) $= \frac{l_1}{m_1} + \frac{l_2}{m_2} = -\frac{(a + b - c)}{b}$

or $$\frac{l_1m_2 + l_2m_1}{m_1m_2} = -\frac{(a+b-c)}{b}$$

or $$l_1m_2 + l_2m_1 = -\frac{(a+b-c)}{b} m_1m_2 = -\frac{(a+b-c)}{b} bk \qquad \text{| By (7)}$$

$$\left.\begin{aligned} &= -k(a+b-c) \\ \text{Similarly} \quad m_1n_2 + m_2n_1 &= -k(b+c-a) \\ \text{and} \quad n_1l_2 + n_2l_1 &= -k(c+a-b) \end{aligned}\right\} \qquad \text{...(10)}$$

$\therefore$
$$\begin{aligned} (m_1n_2 - m_2n_1)^2 &= (m_1n_2 + m_2n_1)^2 - 4m_1m_2n_1n_2 \\ &= k^2(b+c-a)^2 - 4k^2bc \qquad \text{| Using (10) and (7)} \\ &= k^2[b^2 + c^2 + a^2 + 2bc - 2ac - 2ab - 4bc] \\ &= k^2[a^2 + b^2 + c^2 - 2ab - 2bc - 2ac] \end{aligned}$$

Similarly $$(n_1l_2 - n_2l_1)^2 = k^2(a^2 + b^2 + c^2 - 2ab - 2bc - 2ac)$$

and $$(l_1m_2 - l_2m_1)^2 = k^2(a^2 + b^2 + c^2 - 2ab - 2bc - 2ac).$$

Putting these values and also values from Equation (7) and Eqn. (9) in Eqn. (8), we have

$$\tan\frac{\pi}{3} = \frac{\sqrt{3k^2(a^2 + b^2 + c^2 - 2ab - 2bc - 2ca)}}{k(a+b+c)}$$

$$\sqrt{3} = \frac{\sqrt{3}\, k\sqrt{a^2 + b^2 + c^2 - 2ab - 2bc - 2ca}}{k(a+b+c)}$$

Dividing both sides by $\sqrt{3}$, and cross-multiplying

$$a + b + c = \sqrt{a^2 + b^2 + c^2 - 2ab - 2bc - 2ac}$$

S.B.S. $a^2 + b^2 + c^2 + 2ab + 2bc + 2ac = a^2 + b^2 + c^2 - 2ab - 2bc - 2ac$

or $$4bc + 4ac + 4ab = 0$$

Dividing every term by $4abc$, we have

$$\frac{1}{a} + \frac{1}{b} + \frac{1}{c} = 0$$

which is the required condition.

EXERCISE 7

1. Find the angles between the lines of section of the following planes and cones :
   (a) $3x + y + 5z = 0, \quad 6yz - 2zx + 5xy = 0$
   (b) $6x - y - 2z = 0, \quad 108x^2 - 7y^2 - 20z^2 = 0$
   (c) $6x - 10y - 7z = 0, \; 108x^2 - 20y^2 - 7z^2 = 0$
   (d) $x + 2y - 6z = 0, \quad 7x^2 + 20y^2 - 108z^2 = 0.$
2. Find the equations to the lines in which the plane $x + y + z = 0$, cuts the cone $x^2 + 2y^2 - z^2 = 0$. Also find the angle between the lines.
3. Prove that the plane $ax + by + cz = 0$ cuts the cone $yz + zx + xy = 0$ in perpendicular lines if
$$\frac{1}{a} + \frac{1}{b} + \frac{1}{c} = 0.$$
[Hint. *Proceed as in Example 2, part* (*i*)] [*For an Alternative Solution See Example 2, Art. 16*]

4. Find the condition that the plane $ux + vy + wz = 0$ may cut the cone $ax^2 + by^2 + cz^2 = 0$ in perpendicular lines.
5. Find the condition that the plane $ux + vy + wz = 0$ may meet the cone

$$f(x, y, z) \equiv ax^2 + by^2 + cz^2 + 2fyz + 2gzx + 2hxy = 0$$

in two mutually $\perp$ lines.

Answers

1. (a) $\cos^{-1}\frac{1}{6}$ (b) $\cos^{-1}\frac{20}{21}$ (c) $\cos^{-1}\frac{16}{21}$ (d) $\cos^{-1}\frac{20}{21}$.
2. $\frac{x}{1} = \frac{y}{0} = \frac{z}{-1}, \frac{x}{1} = \frac{y}{2} = \frac{z}{-3}$; $\cos^{-1}\frac{2}{\sqrt{7}}$ 4. $(b + c)u^2 + (c + a)v^2 + (a + b)w^2 = 0$
5. $a(v^2 + w^2) + b(w^2 + u^2) + c(u^2 + v^2) = 2fvw + 2guw + 2huv$.

---

16. Condition for a Cone to have Three Mutually Perpendicular Generators

To show that the necessary and sufficient condition for a cone $ax^2 + by^2 + cz^2 + 2fyz + 2gzx + 2hxy = 0$ to have three mutually perpendicular generators is that $a + b + c = 0$.

Necessary Condition. Let the cone

$$ax^2 + by^2 + cz^2 + 2fyz + 2gzx + 2hxy = 0 \quad ...(1)$$

have three mutually perpendicular generators.

Taking them as axes, let equation (1) of cone transform to

$$a'x^2 + b'y^2 + c'z^2 + 2f'yz + 2g'zx + 2h'xy = 0 \quad ...(2) \text{ [By Art. 3]}$$

$\because$ x-axis (new) is a generator of cone (2)

$\therefore$ Its d.c's <1, 0, 0> must satisfy cone (2)

$\therefore$ $a' = 0$

Similarly, $b' = 0$ and $c' = 0$

$\because$ Equations (1) and (2) represent the same cone

$\therefore$ By Invariants, $a + b + c = a' + b' + c' = 0 + 0 + 0 = 0$.

Sufficient Condition

The equation of the cone (having origin as vertex) is

$$ax^2 + by^2 + cz^2 + 2fyz + 2gzx + 2hxy = 0 \quad ...(1)$$

where $a + b + c = 0$.

To prove that the cone (1) has three mutually $\perp$ generators.

Let us take any generator as the z-axis and two perpendicular lines through the vertex (origin) as the axis of x and y in a plane perpendicular to z-axis.

(These x-axis and y-axis need not be generators of the cone.)

Taking these as axes, let equation (1) of the cone transform to

$$a'x^2 + b'y^2 + c'z^2 + 2f'yz + 2g'zx + 2h'xy = 0 \quad ...(2)$$

$\because$ z-axis (new) is a generator of cone (2).

$\therefore$ Its d.c's <0, 0, 1> must satisfy (2) $\therefore$ $c = 0$.

By invariants, $a' + b' + c' = a + b + c = 0$ (given)

$\therefore$ $a' + b' = 0$ | $\because$ $c' = 0$ ...(3)

Now the plane through the vertex and ⊥ to generator z-axis namely the plane $z = 0$ meets the cone (2) in two lines (also generators of the cone) given by

$$a'x^2 + b'y^2 + 2h'xy = 0$$

Using Plane Co-ordinate Geometry these two lines of section (in the plane $z = 0$ ⊥ to z-axis) are ⊥. [∵ By (3) $a' + b' = 0$]

These two lines of section (also generators of the cone) are ⊥ to z-axis (a generator).

[∵ These lines of section lie in a plane ⊥ to z-axis]

∴ The cone has three mutually ⊥ generators.

Remark. Further since we have taken *any* generator as the z-axis.

∴ The cone has an Infinite number of sets of three mutually ⊥ generators.

Rule to remember the condition for three mutually perpendicular generators

co-eff. of x² + co-eff. of y² + co-eff. of z² = 0.

Example 1. *If a right circular cone has three mutually perpendicular generators, the semi-vertical angle is* $\tan^{-1}\sqrt{2}$.

Sol. Let the equation of the right circular cone be [By Art. 9 (*a*) Page 317]

$$x^2 + y^2 = z^2 \tan^2 \alpha$$

or $$x^2 + y^2 - z^2 \tan^2 \alpha = 0 \quad ...(1)$$

∵ Cone (1) has three mutually ⊥ generators.

∴ Co-eff. of x^2 + co-eff. of y^2 + co-eff. of $z^2 = 0$

or $$1 + 1 - \tan^2 \alpha = 0$$

or $$\tan^2 \alpha = 2 \quad \therefore \quad \tan \alpha = \sqrt{2}$$

∴ $$\alpha = \tan^{-1} \sqrt{2}.$$

Example 2. *Prove that the plane* $ax + by + cz = 0$ *cuts the cone* $yz + zx + xy = 0$ *is perpendicular lines if* $\frac{1}{a} + \frac{1}{b} + \frac{1}{c} = 0$.

Sol. Let the plane $$ax + by + cz = 0 \quad ...(1)$$

cut the cone $$yz + zx + xy = 0 \quad ...(2)$$

in two perpendicular lines say OA and OB.

[OA and OB are two lines in plane (1) and are generators of cone (2)].

Now in cone (2),

Co-efficient of x^2 + co-eff. of y^2 + co-eff. of $z^2 = 0 + 0 + 0 = 0$.

∴ Cone (2) has an infinite number of sets of three mutually ⊥ generators (See Remark Art. 16).

Let one such set be OA, OB, OC. [∵ OA, OB are already ⊥ generators]

∵ OC is ⊥ to both OA and OB which lie in plane (1).

∴ OC is ⊥ to plane (1) and hence d.r's of OC are $<a, b, c>$

∴ d.r's of OC must satisfy cone (2) [∵ OC is a generator of cone (2)]

∴ $$bc + ca + ab = 0$$

Dividing throughout by abc [∵ $abc \neq 0$]

$$\frac{1}{a} + \frac{1}{b} + \frac{1}{c} = 0.$$

EXERCISE 8

1. Prove that the plane $lx + my + nz = 0$ cuts the cone $(b - c)x^2 + (c - a)y^2 + (a - b)z^2 + 2fyz + 2gzx + 2hxy = 0$ in perpendicular lines if $(b - c)l^2 + (c - a)m^2 + (a - b)n^2 + 2fmn + 2gnl + 2hlm = 0$.
2. (*i*) If $\frac{x}{1} = \frac{y}{2} = \frac{z}{3}$ represents one of a set of three mutually perpendicular generators of the cone $5yz - 8zx - 3xy = 0$, find the equations to the other two.

   [**Hint.** The other two generators lie in a plane through vertex perpendicular to the generator $\frac{x}{1} = \frac{y}{2} = \frac{z}{3}$. $\therefore$ Equation of this plane is $x + 2y + 3z = 0$. Now proceed by the method of Art. 15.]

   (*ii*) If $x = \frac{1}{2}y = z$ represents one of the sets of three mutually perpendicular generators of the cone $11yz + 6zx - 14xy = 0$, find the equations of the other two. *(K.U. 1998)*
3. Show that the cone whose vertex is the origin and which passes through the curve of intersection of the surface $2x^2 - y^2 + 2z^2 = 3a^2$ and any plane at a distance a from the origin, has three mutually perpendicular generators.

   [**Hint.** Equation of any plane at a distance a from the origin is $lx + my + nz = a$. Use the Method of Art. 4 to write the equation of the cone.]
4. Show that three mutually perpendicular tangent lines can be drawn to the sphere $x^2 + y^2 + z^2 = a^2$ from any point on the sphere $2(x^2 + y^2 + z^2) = 3a^2$.

   [**Hint.** Write the equation of enveloping cone of a point (x_1, y_1, z_1) on the sphere $2(x^2 + y^2 + z^2) = 3a^2$ w.r.t. the sphere $x^2 + y^2 + z^2 = a^2$.]
5. Find the locus of the points from which three mutually perpendicular lines can be drawn to intersect the conic $z = 0$, $ax^2 + by^2 = 1$.

Answers

2. (*i*) $\frac{x}{1} = \frac{y}{1} = \frac{z}{-1}; \frac{x}{5} = \frac{y}{-4} = \frac{z}{1}$ (*ii*) $\frac{x}{11} = \frac{y}{-2} = \frac{z}{-7}; \frac{x}{2} = \frac{y}{-3} = \frac{z}{4}$.

5. $ax^2 + by^2 + (a + b)z^2 = 1$.

17. Equation of the Tangent Plane

To find the equation of the tangent plane at the point (x_1, y_1, z_1) of the cone $ax^2 + by^2 + cz^2 + 2fyz + 2gzx + 2hxy = 0$. (K.U. 2000)

We know that tangent plane at a point is the locus of the tangent lines at that point.

The given equation of the cone is

$$ax^2 + by^2 + cz^2 + 2fyz + 2gzx + 2hxy = 0 \quad ...(1)$$

$\because$ The point (x_1, y_1, z_1) lies on the cone (1)

$$\therefore \quad ax_1^2 + by_1^2 + cz_1^2 + 2fy_1z_1 + 2gz_1x_1 + 2hx_1y_1 = 0 \quad ...(2)$$

Equations of any line through (x_1, y_1, z_1) are

$$\frac{x - x_1}{l} = \frac{y - y_1}{m} = \frac{z - z_1}{n} \quad ...(3)$$

Any point on line (3) is $(lr + x_1, mr + y_1, nr + z_1)$

If it lies on cone (1), then

$$a(lr + x_1)^2 + b(mr + y_1)^2 + c(nr + z_1)^2 + 2f(mr + y_1)(nr + z_1) + 2g(nr + z_1)(lr + x_1) + 2h(lr + x_1)(mr + y_1) = 0$$

or $$r^2(al^2 + bm^2 + cn^2 + 2fmn + 2gnl + 2hlm) + 2r[l(ax_1 + hy_1 + gz_1) + m(hx_1 + by_1 + fz_1) + n(gx_1 + fy_1 + cz_1)] + (ax_1^2 + by_1^2 + cz_1^2 + 2fy_1z_1 + 2gz_1x_1 + 2hx_1y_1) = 0 \quad ...(4)$$

which being a quadratic in r gives two values of r.

$\therefore$ Line (3) meets the cone (1) in two points.

If the line (3) is to be a tangent to cone (1), these values of r from (4) must be equal *i.e.*, discriminant $B^2 - 4AC = 0$.

But $C = 0$ by (2) $\quad \therefore \quad B^2 = 0$ or $B = 0$ *i.e.*, Co-eff. of $r = 0$

$$\therefore \quad l(ax_1 + hy_1 + gz_1) + m(hx_1 + by_1 + fz_1) + n(gx_1 + fy_1 + cz_1) = 0 \quad ...(5)$$

Equation (5) is the condition for the line (3) to be a tangent to cone (1).

To find the locus of tangent lines given by (3), let us eliminate l, m, n from (3) and (5).

Thus putting the values of l, m, n from (3) in (5), we have

$$(x - x_1)(ax_1 + hy_1 + gz_1) + (y - y_1)(hx_1 + by_1 + fz_1) + (z - z_1)(gx_1 + fy_1 + cz_1) = 0$$

or $$x(ax_1 + hy_1 + gz_1) + y(hx_1 + by_1 + fz_1) + z(gx_1 + fy_1 + cz_1)$$
$$= x_1(ax_1 + hy_1 + gz_1) + y_1(hx_1 + by_1 + fz_1) + z_1(gx_1 + fy_1 + cz_1)$$
$$= ax_1^2 + by_1^2 + cz_1^2 + 2fy_1z_1 + 2gz_1x_1 + 2hx_1y_1 = 0 \quad | \text{ Using (2)}$$

or $$\mathbf{x(ax_1 + hy_1 + gz_1) + y(hx_1 + by_1 + fz_1) + z(gx_1 + fy_1 + cz_1) = 0} \quad ...(6)$$

which is the required equation of the tangent plane.

> **Method to write down the Tangent Plane at (x_1, y_1, z_1)**
> *In the equation of the given cone, change*
> x^2 to xx_1, y^2 to yy_1, z^2 to zz_1 ; yz to $\frac{1}{2}(yz_1 + zy_1)$,
> zx to $\frac{1}{2}(zx_1 + xz_1)$ etc., x to $\frac{1}{2}(x + x_1)$ etc.

Cor. 1. The tangent plane at any point of the cone passes through its vertex.

Proof. The vertex of the cone (1) is (0, 0, 0) and it clearly lies on the tangent plane (6).

Cor. 2. The tangent plane at any point P of a cone touches the cone along the generator through P.

Proof. Let P (x_1, y_1, z_1) be any point.

The equation of the tangent plane at P (x_1, y_1, z_1) to cone (1) (In above Art.)

is $$x(ax_1 + hy_1 + gz_1) + y(hx_1 + by_1 + fz_1) + z(gx_1 + fy_1 + cz_1) = 0 \quad ...(7)$$

(By Equation (6) of above Article)

d.r's of generator OP are $<x_1, y_1, z_1>$

$\therefore$ Equations of OP, the generator through P are $\frac{x}{x_1} = \frac{y}{y_1} = \frac{z}{z_1}$.

Any point on this generator OP is $Q(rx_1, ry_1, rz_1)$.

Equation of the tangent plane at $Q(rx_1, ry_1, rz_1)$ to cone (1) is

[Changing x_1 to rx_1, y_1 to ry_1, z_1 to rz_1 in equation (7)]

$$x(arx_1 + hry_1 + grz_1) + y(hrx_1 + bry_1 + frz_1) + z(grx_1 + fry_1 + crz_1) = 0$$

Cancelling r, $\quad x(ax_1 + hy_1 + gz_1) + y(hx_1 + by_1 + fz_1) + z(gx_1 + fy_1 + cz_1) = 0$

which is the same as the equation (7) of the tangent plane at P.

Thus the tangent plane at P also touches the cone at every Q of generator OP.

i.e., it touches the cone along the generator through P.

Remark. The tangent plane at a point P of the cone cuts the cone in two coincident lines OP where O is the vertex of the cone.

18. Condition of Tangency of a Plane and a Cone

To find the condition that the plane $lx + my + nz = 0$ may touch the cone $ax^2 + by^2 + cz^2 + 2fyz + 2gzx + 2hxy = 0$

Equation of the cone is $ax^2 + by^2 + cz^2 + 2fyz + 2gzx + 2hxy = 0$

Equation of the tangent plane to cone (1) is $lx + my + nz = 0$ ...(2)

Let the plane (2) touch the cone (1) at (x_1, y_1, z_1).

Equation of the tangent plane at (x_1, y_1, z_1) to cone (1) is

$$x(ax_1 + hy_1 + gz_1) + y(hx_1 + by_1 + fz_1) + z(gx_1 + fy_1 + cz_1) = 0 \quad ...(3)$$

[By Equation (6) of Art. 17]

Equations (2) and (3) represent the same plane.

Comparing co-efficients in (3) and (2), we have

$$\frac{ax_1 + hy_1 + gz_1}{l} = \frac{hx_1 + by_1 + fz_1}{m} = \frac{gx_1 + fy_1 + cz_1}{n} = k \text{ (say)}$$

$$\therefore \quad ax_1 + hy_1 + gz_1 - lk = 0 \quad ...(4)$$

$$hx_1 + by_1 + fz_1 - mk = 0 \quad ...(5)$$

$$gx_1 + fy_1 + cz_1 - nk = 0 \quad ...(6)$$

Also the point (x_1, y_1, z_1) lies on tangent plane (2)

$$\therefore \quad lx_1 + my_1 + nz_1 - 0.k = 0 \quad ...(7)$$

Eliminating x_1, y_1, z_1, k from (4), (5), (6) and (7) with the help of determinants, we have

$$\begin{vmatrix} a & h & g & l \\ h & b & f & m \\ g & f & c & n \\ l & m & n & 0 \end{vmatrix} = 0$$

or $\quad -(Al^2 + Bm^2 + Cn^2 + 2Fmn + 2Gnl + 2Hlm) = 0$ (From Higher Algebra)

or $\quad \mathbf{Al^2 + Bm^2 + Cn^2 + 2Fmn + 2Gnl + 2Hlm = 0}$ ...(8)

where A, B, C, F, G, H are cofactors of corresponding small letters in the determinant $\begin{vmatrix} a & h & g \\ h & b & f \\ g & f & c \end{vmatrix}$.

Thus (8) is the required condition for the plane (2) to touch the cone (1).

RECIPROCAL CONE

19. (*a*) Reciprocal Cone. Def.

The locus of the lines through the vertex of the cone perpendicular (normal) to the tangent planes to the cone is another cone called the **Reciprocal Cone.**

(*b*) To find the equation of the cone reciprocal to the cone

$$\mathbf{ax^2 + by^2 + cz^2 + 2fyz + 2gzx + 2hxy = 0}$$

The equation of the cone is $ax^2 + by^2 + cz^2 + 2fyz + 2gzx + 2hxy = 0$ ...(1)

Let $$lx + my + nz = 0 \qquad ...(2)$$

be tangent plane to cone (1).

$$\therefore \quad Al^2 + Bm^2 + Cn^2 + 2Fmn + 2Gnl + 2Hlm = 0 \qquad ...(3)$$

where A, B, C, F, G, H are co-factors of the corresponding small letters in $\begin{vmatrix} a & h & g \\ h & b & f \\ g & f & c \end{vmatrix}$.

d.r's of the normal to the tangent plane (2) are $<l, m, n>$.

$\therefore$ Equations of the line through the vertex (0, 0, 0) of the cone (1) and normal to the tangent plane (2) are

$$\frac{x-0}{l} = \frac{y-0}{m} = \frac{z-0}{n} \quad i.e., \quad \frac{x}{l} = \frac{y}{m} = \frac{z}{n} \qquad ...(4)$$

(3) is the condition for the plane (2) to be a tangent plane or line (4) to be a normal.

To find the locus of the lines (4), let us eliminate $<l, m, n>$ from (3) and (4).

Putting the values of l, m, n from (4) in (3), the locus of the normals to the tangent planes to cone (1) is

$$\mathbf{Ax^2 + By^2 + Cz^2 + 2Fyz + 2Gzx + 2Hxy = 0}$$

which is the required equation of the reciprocal cone.

Note. A, B, C, F, G, H are co-factors of a, b, c, f, g, h in

$$D = \begin{vmatrix} a & h & g \\ h & b & f \\ g & f & c \end{vmatrix} = abc + 2fgh - af^2 - bg^2 - ch^2$$

$$\therefore \quad \mathbf{A = bc - f^2, \quad B = ca - g^2, \quad C = ab - h^2,}$$
$$\mathbf{F = gh - af, \quad G = hf - bg, \quad H = fg - ch.}$$

Remark. We have learnt in Art. 19 that reciprocal cone of the cone

$$ax^2 + by^2 + cz^2 + 2fyz + 2gzx + 2hxy = 0 \qquad ...(1)$$

is $$Ax^2 + By^2 + Cz^2 + 2Fyz + 2Gzx + 2Hxy = 0 \qquad ...(2)$$

where $A = bc - f^2, \quad B = ca - g^2, \quad C = ab - h^2$

$F = gh - af, \quad G = hf - bg, \quad H = fg - ch$

Let us find the reciprocal cone of the cone (2)

Let the reciprocal cone of (2) be

$$A'x^2 + B'y^2 + C'z^2 + 2F'yz + 2G'zx + 2H'xy = 0 \qquad ...(3)$$

where $A' = BC - F^2 = (ca - g^2)(ab - h^2) - (gh - af)^2$

$= a^2bc - ach^2 - abg^2 + g^2h^2 - (g^2h^2 + a^2f^2 - 2afgh)$

$= a^2bc - ach^2 - abg^2 - a^2f^2 + 2afgh = a(abc + 2fgh - af^2 - bg^2 - ch^2)$

$= aD$ | See Note above

Similarly, $B' = bD, \quad C' = cD, \quad F' = fD, G' = gD, \quad H' = hD$

Putting these value of A′, B′, C′, F′, G′, H′ in (3).

$$aDx^2 + bDy^2 + cDz^2 + 2fDyz + 2gDzx + 2hDxy = 0$$

Dividing throughout by D [which $\neq 0$ because we have assumed that (1) does not represent a pair of planes].

$ax^2 + by^2 + cz^2 + 2fyz + 2gzx + 2hxy = 0$ which is same as the cone (1).

Thus the cones (1) and (2) are **Reciprocal** of each other.

Reciprocal Cones. Two cones which are such that each is the locus of the normals through the vertex to the tangent planes to the other are called **Reciprocal Cones.**

20. Condition that a cone may have three mutually perpendicular tangent planes

To prove that the condition, that the cone

$$ax^2 + by^2 + cz^2 + 2fyz + 2gzx + 2hxy = 0$$

may have three mutually perpendicular tangent planes is

$$\mathbf{A + B + C = 0 \text{ where } A = bc - f^2, B = ca - g^2, C = ab - h^2.}$$

Equation of the given cone is $ax^2 + by^2 + cz^2 + 2fyz + 2gzx + 2hxy = 0$ ...(1)

∴ By Art. 19, equation of the Reciprocal cone is

$$Ax^2 + By^2 + Cz^2 + 2Fyz + 2Gzx + 2Hxy = 0 \quad ...(2)$$

∵ The cone (1) has three mutually ⊥ tangent planes.

∴ Normals to these tangent planes through the vertex of cone (1) are generators of cone (2). [By def. of Reciprocal cone in Art. 19 (*a*)]

Again these Normals are mutually ⊥ (∵ The tangent planes are mutually ⊥)

∴ Cone (2) has three mutually ⊥ generators.

∴ Co-eff. of x^2 + Co-eff. of y^2 + Co-eff. of $z^2 = 0$ (By Art. 16)

i.e.,

$$\mathbf{A + B + C = 0.}$$

Example 1. *Find the equation of the reciprocal cone of the cone*

$$ax^2 + by^2 + cz^2 = 0.$$

Sol. The equation of the cone is $ax^2 + by^2 + cz^2 = 0$

Comparing it with $ax^2 + by^2 + cz^2 + 2fyz + 2gzx + 2hxy = 0$,

we have "a" $= a$, "b" $= b$, "c" $= c$;

$f = 0, \quad g = 0, \quad h = 0$

∴ $A = bc - f^2 = bc - 0 = bc$

$B = ca - g^2 = ca - 0 = ca$

$C = ab - h^2 = ab - 0 = ab$

$F = gh - af = 0$

$G = hf - bg = 0$

$H = fg - ch = 0$

Putting these values of A, B, C, F, G, H in

$$Ax^2 + By^2 + Cz^2 + 2Fyz + 2Gzx + 2Hxy = 0,$$

the equation of the reciprocal cone is

$$bcx^2 + cay^2 + abz^2 = 0$$

or Dividing throughout by abc, $\dfrac{x^2}{a} + \dfrac{y^2}{b} + \dfrac{z^2}{c} = 0.$

Example 2. *Show that the general equation of the cone which touches the three co-ordinate planes is* $\sqrt{fx} \pm \sqrt{gy} \pm \sqrt{hz} = 0$.

Sol. We know that normals to the tangent planes to a given cone through the vertex are generators of the reciprocal cone.

Here co-ordinates planes are tangent planes to the required cone.

$\therefore$ Their normals *i.e.*, the co-ordinate axes are generators of the reciprocal cone.

$\therefore$ Equation of the reciprocal cone having co-ordinates axes as generators is

$$fyz + gzx + hxy = 0$$

or $$2fyz + 2gzx + 2hxy = 0$$

$\therefore$ The required cone is also reciprocal cone of cone (1). [See remark, Art. 19]

Comparing (1) with $ax^2 + by^2 + cz^2 + 2fyz + 2gzx + 2hxy = 0$:

$$a = 0, b = 0, c = 0, f = f, g = g, h = h.$$

$\therefore$ $$A = bc - f^2 = -f^2, B = ca - g^2 = -g^2 ; C = ab - h^2 = -h^2$$

$$F = gh - af = gh, G = hf, H = fg$$

Putting these values of A, B, C, F, G, H in

$$Ax^2 + By^2 + Cz^2 + 2Fyz + 2Gzx + 2Hxy = 0,$$

the equation of required cone touching the co-ordinate planes is

$$-f^2x^2 - g^2y^2 - h^2z^2 + 2ghyz + 2hfzx + 2fgxy = 0$$

Cancelling -1, $$f^2x^2 + g^2y^2 + h^2z^2 - 2ghyz - 2hfzx - 2fgxy = 0 \quad ...(1)$$

Adding $4fgxy$ to both sides, to make L.H.S. a perfect square,

$$f^2x^2 + g^2y^2 + h^2z^2 + 2fgxy - 2ghyz - 2hfzx = 4fgxy$$

or $$(fx + gy - hz)^2 = 4fgxy$$

Taking square root, we have

$$fx + gy - hz = \pm 2\sqrt{fgxy} \quad \text{or} \quad fx + gy \mp 2\sqrt{fgxy} = hz$$

or $$(\sqrt{fx} \mp \sqrt{gy})^2 = hz$$

Taking square root, $\sqrt{fx} \mp \sqrt{gy} = \pm\sqrt{hz}$

or $$\sqrt{fx} \mp \sqrt{gy} \mp \sqrt{hz} = 0 \quad \text{or} \quad \sqrt{fx} \pm \sqrt{gy} \pm \sqrt{hz} = 0$$

which is the required equation.

Example 3. *If a right circular cone has three mutually perpendicular tangent planes, the semi-vertical angle is* $\tan^{-1}\frac{1}{\sqrt{2}}$. (K.U. 1996 S)

Sol. Let the equation of the right circular cone be

$$x^2 + y^2 = z^2 \tan^2 \alpha \quad \text{or} \quad x^2 + y^2 - z^2 \tan^2 \alpha = 0 \quad ...(1)$$

Comparing (1) with $ax^2 + by^2 + cz^2 + 2fyz + 2gzx + 2hxy = 0$, we have

$$a = 1, b = 1, c = -\tan^2 \alpha ;$$
$$f = 0, g = 0, h = 0.$$

$\because$ The cone (1) has three mutually $\perp$ tangent planes

$\therefore$ $$A + B + C = 0 \quad \text{(By Art. 20)}$$

or $$bc - f^2 + ca - g^2 + ab - h^2 = 0$$

$\therefore$ $$-\tan^2 \alpha - 0 - \tan^2 \alpha - 0 + 1 - 0 = 0$$

or $$-2\tan^2 \alpha = -1 \quad \therefore \quad \tan^2 \alpha = \tfrac{1}{2}$$

or $$\tan \alpha = \frac{1}{\sqrt{2}} \quad \therefore \quad \alpha = \tan^{-1}\frac{1}{\sqrt{2}}.$$

Hence the result.

EXERCISE 9

1. Find the locus of the lines through the vertex of the cone $ax^2 + by^2 + cz^2 + 2fyz + 2gzx + 2hxy = 0$ perpendicular to its tangent-planes.
[**Hint.** It is Art 19.]

2. Prove that the cones $ax^2 + by^2 + cz^2 = 0$ and $\frac{x^2}{a} + \frac{y^2}{b} + \frac{z^2}{c} = 0$ are reciprocal.
(M.D.U. 1997)
[**Hint.** It is example 1].

3. Prove that tangent planes to the cone $fyz + gzx + hxy = 0$ are perpendicular to generators of the cone $f^2x^2 + g^2y^2 + h^2z^2 - 2ghyz - 2hfzx - 2fgxy = 0$.
[**Hint.** Find the reciprocal cone of the cone $fyz + gzx + hxy = 0$. It is Ex. 2, upto equation (1).]

4. Show that the general equation to a cone which touches the co-ordinate planes is
$$a^2x^2 + b^2y^2 + c^2z^2 - 2bcyz - 2cazx - 2abxy = 0 \text{ or } \sqrt{ax} \pm \sqrt{by} \pm \sqrt{cz} = 0.$$
[**Hint.** It is Example 2]

5. (*a*) Find the condition that the plane $lx + my + nz = 0$ may touch the cone $2x^2 - 3y^2 + z^2 = 0$ and hence find the equation of the reciprocal cone.
(*b*) Find the condition that the plane $lx + my + nz = 0$ may touch the cone $4x^2 - y^2 + 3z^2 = 0$.

Answers

1. $Ax^2 + By^2 + Cz^2 + 2Fyz + 2Gzx + 2Hxy = 0$
5. (*a*) $3l^2 - 2m^2 + 6n^2 = 0$; $3x^2 - 2y^2 + 6z^2 = 0$.
(*b*) $3l^2 - 12m^2 + 4n^2 = 0$.

MISCELLANEOUS EXERCISE

1. A is a point on OX and B on OY so that the angle OAB is constant (= α). On AB as diameter a circle is described whose plane is parallel to OZ. Prove that as AB varies, the circle generates the cone $2xy - z^2 \sin 2\alpha = 0$.

2. Show that the equation to the right circular cone whose vertex is at the origin, has direction-cosines $\cos\alpha$, $\cos\beta$, $\cos\gamma$ and whose semi-vertical angle θ is
$$(y\cos\gamma - z\cos\beta)^2 + (z\cos\alpha - x\cos\gamma)^2 + (x\cos\beta - y\cos\alpha)^2 = (x^2 + y^2 + z^2)\sin^2\theta.$$

3. Show that the straight lines $x = 2y = 8z$; $x = y = 2z$; $4x = 7y = 7z$ lie on a circular cone of semi-vertical angle $\cos^{-1}\left(\frac{11}{\sqrt{126}}\right)$.

4. Find the equation of the right circular cone passing through the co-ordinate axes. Obtain the semi-vertical angle and the equations of the axis of the cone.

5. The straight line $x = y = z$ rotates round the straight line $x = -y = -z$ about the origin, to generate a right circular cone. Find the equation of the cone.

6. The vertex of a cone is (a, b, c) and the y-z plane meets it in the curve $f(y, z) = 0$, $x = 0$.
Show that the zx-plane meets it in the curve $y = 0$, $f\left(\frac{bx}{x-a}, \frac{cx-az}{x-a}\right) = 0$.

7. Find the angle between the lines of section of the plane $x + y + z = 0$ and the cone

$$\frac{yz}{q-r} + \frac{zx}{r-p} + \frac{xy}{p-q} = 0.$$

8. Show that the plane $ax + by + cz = 0$ cuts the cone $yz + zx + xy = 0$ in the two lines inclined at an angle $\tan^{-1}\left[\frac{\{(a^2+b^2+c^2)(a^2+b^2+c^2-2bc-2ca-2ab)\}^{1/2}}{bc+ca+ab}\right]$.

9. Find he equation of the quadric cone which touches the three co-ordinate planes and the planes $x + y + z = 0$, $2x - 3y + z = 0$.
[**Hint.** The required cone is reciprocal to the cone which passes through the normals through the origin to the given planes.]

10. Find the equation of the cone whose semi-vertical angle is 45°, and which has its vertex at the origin and its axis along the line $x = -2y = z$, and show that the plane $z = 0$ cuts the cone in two lines inclined at an angle $\cos^{-1}\frac{4}{5}$.

11. Find the equation of the cone generated by straight lines drawn from the origin to cut the circle through the three points (1, 0, 0), (0, 2, 0), (2, 1, 1) and prove that the acute angle between the two lines in which the plane $x = 2y$ cuts the cone is $\cos^{-1}\sqrt{\frac{5}{4}}$.

12. Prove that the perpendiculars drawn from the origin to the tangent planes to the cone
$2x^2 + 3y^2 + 4z^2 + 2yz + 4zx + 6xy = 0$ lie on the cone
$11x^2 + 4y^2 - 3z^2 + 8yz - 6zx - 20xy = 0.$

13. Show that the equation of the cone through the intersection of the cones
$x^2 - 2y^2 + 3z^2 - 4yz + 5zx - 6xy = 0$
and $2x^2 - 3y^2 + 4z^2 - 5yz + 6zx - 10xy = 0$
and the line with direction cosines proportional to 1, 1, 1 is $y^2 - 2z^2 + 3yz - 4zx + 2xy = 0$.

Answers

4. $yz + zx + xy = 0$; $\cos^{-1}\frac{1}{\sqrt{3}}$, $x = y = z$

5. $x^2 + y^2 + z^2 + 3yz - 3zx - 3xy = 0$

6. $64x^2 + 9y^2 + 25z^2 - 30yz + 80zx + 48xy = 0$

7. $\frac{\pi}{3}$

9. $x^2 + 7y^2 + z^2 + 8yz - 16zx + 8xy = 0$

11. $8z^2 + 4yz - zx - 5xy = 0$.

12

Cylinder

1. Cylinder

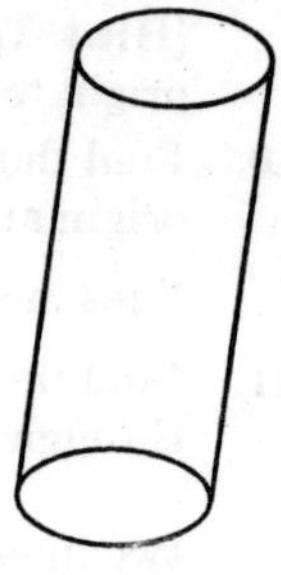

Definition. A cylinder is a surface generated by a (moving) straight line which is parallel to a fixed line and is subject to one more condition *e.g.*, it may intersect a given curve, or touch a given surface or the moving line is at a constant distance from the fixed line.

The moving line in any position is called the **generator** of the cylinder and if the generators intersect a fixed curve, then the curve is called the **guiding curve.**

2. To find the equation of the cylinder whose generators are parallel to the line $\frac{x}{l}=\frac{y}{m}=\frac{z}{n}$ and the base conic is

$$f(x, y) \equiv ax^2 + 2hxy + by^2 + 2gx + 2fy + c = 0,\ z = 0.$$

$\because$ All generators of the cylinder are parallel to the line

$$\frac{x}{l}=\frac{y}{m}=\frac{z}{n}$$

$\therefore$ d.r's of every generators are $<l, m, n>$.

Let (x_1, y_1, z_1) be ANY point on the cylinder.

$\therefore$ Equations of the generator of cylinder through (x_1, y_1, z_1) are

$$\frac{x-x_1}{l}=\frac{y-y_1}{m}=\frac{z-z_1}{n} \qquad ...(1)$$

Equations of the base conic (*i.e.*, guiding curve) are

$$ax^2 + 2hxy + by^2 + 2gx + 2fy + c = 0,\ z = 0 \qquad ...(2)$$

$\because$ Line (1) is a generator of the required cylinder.

$\therefore$ Line (1) must intersect the guiding curve (2).

Line (1) meets the plane $z = 0$ where

$$\frac{x-x_1}{l}=\frac{y-y_1}{m}=-\frac{z_1}{n}$$

or

$$x-x_1=-\frac{l}{n}z_1,\ y-y_1=-\frac{m}{n}z_1$$

or

$$x=x_1-\frac{l}{n}z_1,\ y=y_1-\frac{m}{n}z_1.$$

$\because$ This point lies on the given conic.

$\therefore$ Substituting these values of x, y in (2) ;

$$a\left(x_1-\frac{l}{n}z_1\right)^2+2h\left(x_1-\frac{l}{n}z_1\right)\left(y_1-\frac{m}{n}z_1\right)$$
$$+b\left(y_1-\frac{m}{n}z_1\right)^2+2g\left(x_1-\frac{l}{n}z_1\right)+2f\left(y_1-\frac{m}{n}z_1\right)+c=0$$

$\therefore$ The locus of (x_1, y_1, z_1) is [changing (x_1, y_1, z_1) to (x, y, z)]

$$a\left(x-\frac{l}{n}z\right)^2+2h\left(x-\frac{l}{n}z\right)\left(y-\frac{m}{n}z\right)+b\left(y-\frac{m}{n}z\right)^2+2g\left(x-\frac{l}{n}z\right)+2f\left(y-\frac{m}{n}z\right)+c=0$$

or multiplying throughout by n^2,

$$a(nx-lz)^2+2h(nx-lz)(ny-mz)+b(ny-mz)^2+2g(nx-lz)n+2f(ny-mz)n+cn^2=0$$

which is the required equation of the cylinder.

Example 1. *Find the equation of the cylinder whose generators are parallel to the line* $\frac{x}{1}=\frac{y}{2}=\frac{z}{3}$ *and which passes through the curve* $x^2+y^2=16$, $z=0$.

Sol. $\because$ All generators are parallel to the line

$$\frac{x}{1}=\frac{y}{2}=\frac{z}{3}$$

$\therefore$ d.r's of every generator are $<1, 2, 3>$.

Let (x_1, y_1, z_1) be ANY point on the cylinder.

$\therefore$ Equations of the generator of the cylinder through (x_1, y_1, z_1) are

$$\frac{x-x_1}{1}=\frac{y-y_1}{2}=\frac{z-z_1}{3} \quad \text{...(1)}$$

Equations of the base conic (*i.e.*, guiding curve) are

$$x^2+y^2=16,\ z=0 \quad \text{...(2)}$$

$\because$ Line (1) is a generator of the required cylinder.

$\therefore$ Line (1) must intersect the guiding curve (2).

Line (1) meets the plane $z=0$ where

$$\frac{x-x_1}{1}=\frac{y-y_1}{2}=-\frac{z_1}{3}$$

$\therefore$
$$x-x_1=\frac{-1}{3}z_1,\quad y-y_1=\frac{-2}{3}z_1$$

or
$$x=x_1-\frac{1}{3}z_1,\quad y=y_1-\frac{2}{3}z_1$$

$\because$ This point lies on the given conic $x^2+y^2=16$

$\therefore$ Putting these values of x and y in $x^2+y^2=16$, we have

$$\left(x_1-\frac{1}{3}z_1\right)^2+\left(y_1-\frac{2}{3}z_1\right)^2=16$$

$\therefore$ Locus of (x_1, y_1, z_1) is

$$\left(x-\frac{1}{3}z\right)^2+\left(y-\frac{2}{3}z\right)^2=16$$

or $(3x-z)^2+(3y-2z)^2=144$ or $9x^2+z^2-6zx+9y^2+4z^2-12yz=144$

or $9x^2+9y^2+5z^2-6xz-12yz-144=0$

which is the required equation of the cylinder.

3. To find the locus of the equation f(x, y) = 0, where f(x, y) contains only positive integral powers of x and y

The equation is $f(x, y) = 0$ ...(1)

where $f(x, y)$ contains only positive integral powers of x and y.

Let (1) represent the curve C in the x-y plane (referred to OX, OY as axes)

Let $P(x_1, y_1)$ be any point on this curve C in the x-y plane represented by $f(x, y) = 0$.

$$\therefore \quad f(x_1, y_1) = 0 \quad ...(2)$$

Through P draw a straight line PQ || to the z-axis.

$\therefore$ Any point Q on this line is (x_1, y_1, z_1)

This point $Q(x_1, y_1, z_1)$ will lie on the locus of (1) if $f(x_1, y_1) = 0$

which is true by (2).

$\therefore$ Every point Q on the line through P || to the z-axis lies on the locus of (1)

and $\therefore$ the line itself lies on the locus of (1).

$\therefore$ (1) represents a cylinder whose generators are || to z-axis and intersect the curve whose equation in the xy-plane (referred to OX, OY as axes) is $f(x, y) = 0$.

Note. Similarly $f(y, z) = 0$ represents a cylinder whose generators are parallel to x-axis and intersect the curve whose equation in the yz-plane (referred to OY, OZ as axes) is $f(y, z) = 0$.

Similarly we can talk about $f(x, y) = 0$.

Aid to memory. *The equation $f(x, y) = 0$ from which z is absent, represents a cylinder whose generators are parallel to z-axis.*

Example 2. *Find the equation of the quadric cylinder with generators parallel to x-axis and passing through the curve $ax^2 + by^2 + cz^2 = 1$, $lx + my + nz = p$.*

Sol. Equations of the guiding curve are $ax^2 + by^2 + cz^2 = 1$ ...(1)

$$lx + my + nz = p \quad ...(2)$$

Since the generators of the cylinder are parallel to x-axis therefore, the equation of the cylinder should be free from the x co-ordinate (By Art. 3).

$\therefore$ To find the equation of the cylinder, *let us eliminate x from (1) and (2).*

From equation (2), $lx = p - my - nz$ or $x = \dfrac{p - my - nz}{l}$

Putting this value of x in (1),

$$a\left(\frac{p - my - nz}{l}\right)^2 + by^2 + cz^2 = 1$$

or $$a(p - my - nz)^2 + bl^2y^2 + cl^2z^2 = l^2$$

or $$a(p^2 + m^2y^2 + n^2z^2 - 2mpy - 2npz + 2mnyz) + bl^2y^2 + cl^2z^2 - l^2 = 0$$

or $$(bl^2 + am^2)y^2 + (cl^2 + an^2)z^2 + 2amnyz - 2ampy - 2anpz + (ap^2 - l^2) = 0$$

which is the required equation of the cylinder.

EXERCISE 1

1. Find the equation of the cylinder whose generators are parallel to the line $\dfrac{x}{1} = \dfrac{-y}{2} = \dfrac{z}{3}$ and whose guiding curve is the ellipse $x^2 + 2y^2 = 1$, $z = 3$. *(K.U. 1999 S, 97 S)*

2. Find the equation of the cylinder whose generators are parallel to the line $3x = 3y = z$ and whose guiding curve is $2x^2 + y^2 = 1, z = 1$.
3. Find the equation of the cylinder whose generators are parallel to the line $\frac{x}{3} = \frac{y-4}{5} = \frac{z+1}{-4}$ and which has for its guiding curve the hyperbola $3x^2 - 4y^2 = 5, z = 2$.
4. Find the equation of the cylinder whose generating lines have the d.c's l, m, n and which passes through the circumference of the fixed circle $x^2 + z^2 = a^2$ in the ZOX-plane.
5. Find the equation of a cylinder, whose generators intersect the conic $ax^2 + 2hxy + by^2 = 1, z = 0$ and are parallel to a line with d.c's $<l, m, n>$.
6. Find the equation of the surface generated by a straight line which is parallel to the line $y = mx$, $z = nx$ and intersects the ellipse $\frac{x^2}{a^2} + \frac{y^2}{b^2} = 1, z = 0$.

   $\left[\textbf{Hint.} \text{ The line } y = mx, z = nx \text{ can be put as } \frac{x}{l} = \frac{y}{m} = \frac{z}{n}.\right]$
7. Find the equation of the cylinder whose generators are parallel to z-axis and which passes through the curve of intersection of $x^2 + y^2 + z^2 = 1$ and $x + y + z = 1$.

   [**Hint.** Eliminate z.]

Answers

1. $3x^2 + 6y^2 + 3z^2 + 8yz - 2zx + 6x - 24y - 18z + 24 = 0$
2. $6x^2 + 3y^2 + z^2 - 4xz - 2yz + 4x + 2y - 2z - 2 = 0$
3. $48x^2 - 64y^2 - 73z^2 - 160yz + 72zx - 144x + 320y + 292z - 372 = 0$
4. $(mx - ly)^2 + (mz - ny)^2 = m^2a^2$
5. $a(nx - lz)^2 + 2h(nx - lz)(ny - mz) + b(ny - mz)^2 = n^2$
6. $\frac{1}{a^2}(nx - z)^2 + \frac{1}{b^2}(ny - mz)^2 = n^2$
7. $x^2 + y^2 + xy - x - y = 0$.

---

4. Right Circular Cylinder

Def. *A right circular cylinder is a surface generated by a straight line which is parallel to a fixed line and is at a constant distance from it.*

The fixed line is called the **axis** of the cylinder. The constant distance is called the **radius** of the cylinder. (In fact, **perpendicular distance of any point on the cylinder from the axis of cylinder = Radius).**

Note. The section of a right circular cylinder by any plane perpendicular to its axis is a circle.

5. Equation of a Right Circular Cylinder

(*a*) **Standard Form.** *To show that the equation of the right circular cylinder whose axis is the z-axis and radius a, is $x^2 + y^2 = a^2$.*

Let P(x, y, z) be any point on the cylinder.

Draw PM ⊥ on z-axis (the axis of the cylinder).

∴ PM = Radius of cylinder = a.

O(0, 0, 0) is given point on the axis of cylinder (z-axis) whose d.c.'s are < 0, 0, 1 >.

In ΔOMP, $OP^2 = OM^2 + MP^2 = OM^2 + a^2$ ...(1)

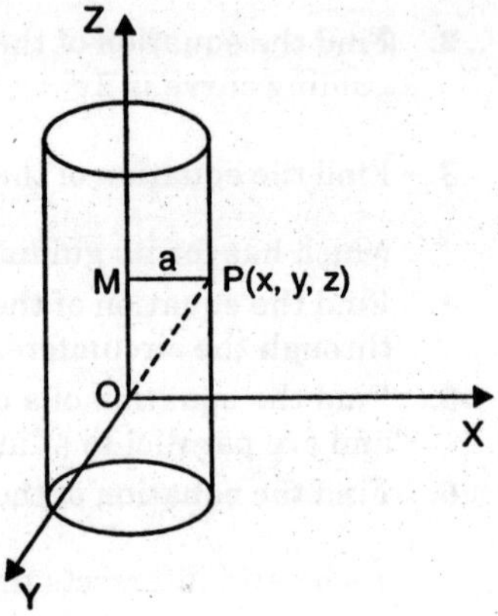

$$OP = \sqrt{(x-0)^2 + (y-0)^2 + (z-0)^2}$$
$$= \sqrt{x^2 + y^2 + z^2}$$

OM = projection of OP on z-axis
$$= (x - 0).\,0 + (y - 0).0 + (z - 0).\,1 = z$$

[∵ Length of projection
$$= l(x_2 - x_1) + m(y_2 - y_1) + n(z_2 - z_1)]$$

Putting these values of OP and OM in (1)
$$x^2 + y^2 + z^2 = z^2 + a^2$$
or
$$\mathbf{x^2 + y^2 = a^2}$$
which is the required equation of the cylinder.

(b) **General Form.** *To find the equation to the right circular cylinder whose radius is r and axis is the line* $\dfrac{x-\alpha}{l} = \dfrac{y-\beta}{m} = \dfrac{z-\gamma}{n}$.

Let P(x, y, z) be any point on the cylinder.

Draw PM ⊥ on the axis of the cylinder given by

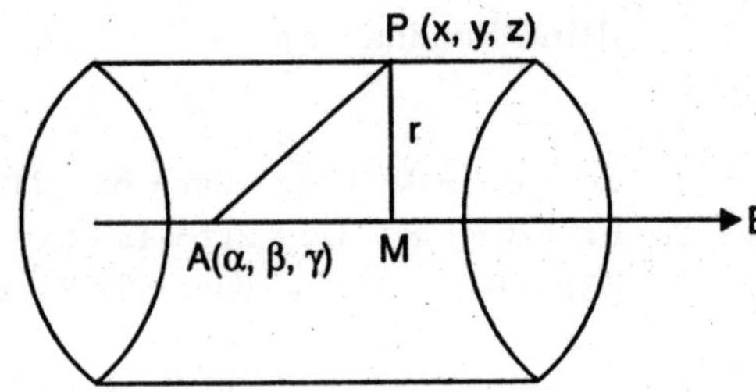

$$\frac{x-\alpha}{l} = \frac{y-\beta}{m} = \frac{z-\gamma}{n} \quad ...(1)$$

∴ PM = Radius of the cylinder = r.

A(α, β, γ) is a given point on the axis (1) of the cylinder and d.r's of the axis are < l, m, n >

In right angled ΔAMP, $AP^2 = AM^2 + MP^2 = AM^2 + r^2$ ...(2)

$$AP = \sqrt{(x-\alpha)^2 + (y-\beta)^2 + (z-\gamma)^2}$$

AM = Projection of AP on the axis of the cylinder having d.r's < l, m, n >

$$= (x-\alpha)\frac{l}{\sqrt{l^2+m^2+n^2}} + (y-\beta)\frac{m}{\sqrt{l^2+m^2+n^2}} + (z-\gamma)\frac{n}{\sqrt{l^2+m^2+n^2}}$$

[∵ Length of projection $= l(x_2 - x_1) + m(y_2 - y_1) + n(z_2 - z_1)$]

$$= \frac{l(x-\alpha) + m(y-\beta) + n(z-\gamma)}{\sqrt{l^2+m^2+n^2}}.$$

Putting these values of AP and AM in (2) ;

$$[(x-\alpha)^2 + (y-\beta)^2 + (z-\gamma)^2] = \frac{[(l(x-\alpha) + m(y-\beta) + n(z-\gamma)]^2}{l^2+m^2+n^2} + r^2$$

which is the required equation of the cylinder.

Example 1. *Find the equation of the right circular cylinder of radius 2 whose axis passes through (1, 2, 3) and has direction cosines proportional to 2, 1, 2.*

Sol. Let P(x, y, z) be any point on the cylinder.

Equations of the axis of the cylinder passing through (1, 2, 3) and having d.c's proportional to 2, 1, 2 are $\dfrac{x-1}{2} = \dfrac{y-2}{1} = \dfrac{z-3}{2}$ ...(1)

Draw PM ⊥ on the axis of the cylinder.

∴ PM = Radius of the cylinder = 2.

A(1, 2, 3) is a given point on the axis and d.r's of axis are 2, 1, 2.

∴ d.c's of axis are (Dividing each by $\sqrt{2^2+1^2+2^2}=3$) $\frac{2}{3}, \frac{1}{3}, \frac{2}{3}$.

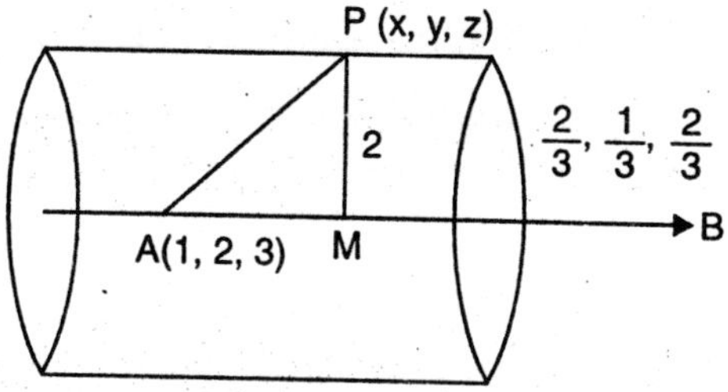

In ΔAMP, $AP^2 = AM^2 + MP^2$

$$= AM^2 + 2^2 = AM^2 + 4 \quad ...(2)$$

$$AP = \sqrt{(x-1)^2+(y-2)^2+(z-3)^2}$$

AM = Projection of AP on the axis of the cylinder having d.c's $< \frac{2}{3}, \frac{1}{3}, \frac{2}{3} >$

$$= \frac{2}{3}(x-1) + \frac{1}{3}(y-2) + \frac{2}{3}(z-3)$$

$$= \frac{2x+y+2z-10}{3}$$

Putting these values of AP and AM in (2),

$$(x-1)^2+(y-2)^2+(z-3)^2 = \left(\frac{2x+y+2z-10}{3}\right)^2 + 4$$

or
$$9[x^2-2x+1+y^2-4y+4+z^2-6z+9]$$
$$= 4x^2+y^2+4z^2+100+4xy+8xz-40x+4yz-20y-40z+36$$

or
$$5x^2+8y^2+5z^2-4yz-8zx-4xy+22x-16y-14z-10=0$$

which is the required equation of the cylinder.

Example 2. *Find the equation of the right circular cylinder whose guiding circle is $x^2+y^2+z^2=9$, $x-y+z=3$.* (K.U. 2007)

Sol. The axis of the cylinder is a line through the centre of the sphere (or the centre of the circle) **and ⊥ to the plane of circle**

Radius of cylinder = Radius of circle.

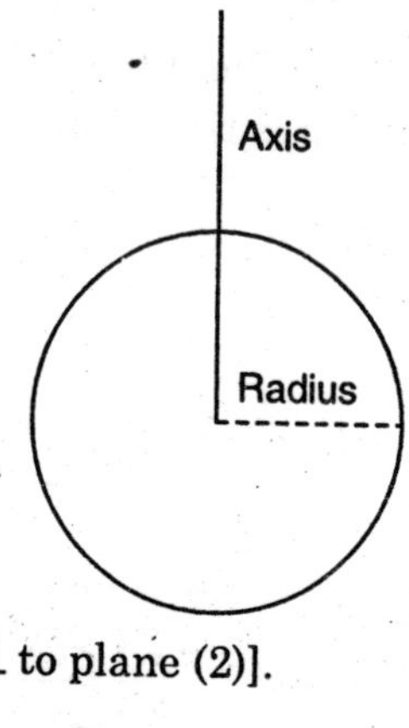

The sphere is $x^2+y^2+z^2=9$ ...(1)

and the plane is $x-y+z=3$ ...(2)

Centre of the sphere (1) is O(0, 0, 0) and d.r's of normal to plane (2) are < 1, – 1, 1 >. | Co-efficients of x, y, z in plane (2)

∴ Equations of the axis of the cylinder are

$$\frac{x-0}{1} = \frac{y-0}{-1} = \frac{z-0}{1}$$

i.e.,
$$\frac{x}{1} = \frac{y}{-1} = \frac{z}{1} \quad ...(3)$$

[Axis of the cylinder is a line through centre of the sphere and ⊥ to plane (2)].

To find radius of cylinder = (Radius of circle)

R = Radius of sphere (1) = 3

p = ⊥ distance of centre (0, 0, 0) of sphere (1) from plane (2)

$$= \frac{0-0+0-3}{\sqrt{1+1+1}} = -\frac{3}{\sqrt{3}} = -\sqrt{3}$$

$\therefore$ Radius of circle $= \sqrt{R^2 - p^2} = \sqrt{9-3} = \sqrt{6}$

$\therefore$ Radius of cylinder $= \sqrt{6}$.

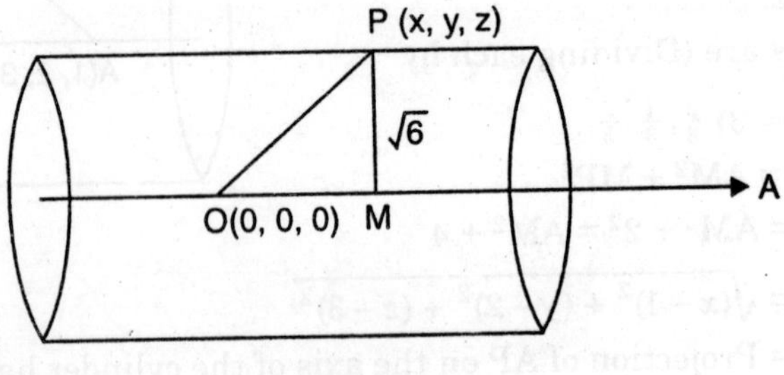

Let P(x, y, z) be ANY point on the cylinder.

Draw PM ⊥ on the axis of the cylinder.

$\therefore$ PM = Radius of the cylinder $= \sqrt{6}$

O(0, 0, 0) is a given point on the axis and d.r's of the axis are $< 1, -1, 1 >$.

$\therefore$ d.c.'s of the axis are $< \frac{1}{\sqrt{3}}, -\frac{1}{\sqrt{3}}, \frac{1}{\sqrt{3}} >$.

In ΔOMP, $OP^2 = OM^2 + MP^2 = OM^2 + 6$ ...(4)

Now $OP = \sqrt{(x-0)^2 + (y-0)^2 + (z-0)^2} = \sqrt{x^2 + y^2 + z^2}$

OM = Projection of OP on the axis of cylinder

$$= \frac{1}{\sqrt{3}}(x-0) - \frac{1}{\sqrt{3}}(y-0) + \frac{1}{\sqrt{3}}(z-0) = \frac{x-y+z}{\sqrt{3}}$$

$\therefore$ From (4), $x^2 + y^2 + z^2 = \left(\frac{x-y+z}{\sqrt{3}}\right)^2 + 6$

or $3(x^2 + y^2 + z^2) = (x - y + z)^2 + 18$.

Simplifying $x^2 + y^2 + z^2 + xy + yz - zx - 9 = 0$

which is the required equation of the cylinder.

EXERCISE 2

1. (*a*) Find the equation of the right circular cylinder of radius 2 whose axis passes through (1, 2, 3) and has direction cosines proportional to $< 2, -3, 6 >$.

(*M.D.U. 1995*)

or Find the equation of a right circular cylinder of radius 2 units and whose axis is the line $\frac{x-1}{2} = \frac{2-y}{3} = \frac{z-3}{6}$ (*K.U. 2003*)

[**Hint.** Equation of axis is $\frac{x-1}{2} = \frac{y-2}{-3} = \frac{z-3}{6}$.]

(*b*) Find the equation of the right circular cylinder of radius 2 whose axis passes through (1, 0, 3) and has direction cosines proportional to 3, 3, 1.

2. (a) Find the equation of the right circular cylinder of radius 2 whose axis is the line $\frac{x-1}{2} = y - 2 = \frac{z-3}{2}$ or $x - 1 = 2(y - 2) = z - 3$. (K.U. 2000 A, 1995 A)

(b) Find the equation of the right circular cylinder whose radius is 2 and axis is $\frac{x-1}{2} = \frac{y}{3} = \frac{z-3}{1}$. (M.D.U. 1998)

3. Find the equation of the right circular cylinder whose axis is $x = 2y = -z$ and radius 4.

4. Find the equation of the right circular cylinder whose generating circle is $x^2 + y^2 + z^2 = 4$, $x + y + z = 3$.

5. Find the equation of the right circular cylinder whose guiding circle is

$$x^2 + y^2 + z^2 - 2x + 4y - 6z - 2 = 0$$
$$2x + 3y + 6z = 0.$$ (K.U. 2006 ; M.D.U. 2000)

6. Obtain the equation of the right circular cylinder whose guiding curve is the circle through the points (1, 0, 0), (0, 1, 0) and (0, 0, 1).

Answers

1. (a) $45x^2 + 40y^2 + 13z^2 + 36yz - 24zx + 12xy - 42x - 280y - 126z + 294 = 0$

(b) $5x^2 + 5y^2 + 9z^2 - 9xy - 3xz - 3yz - x + 18y - 51z + 39 = 0$

2. (a) $5x^2 + 8y^2 + 5z^2 - 4yz - 8zx - 4xy + 22x - 16y - 14z - 10 = 0$

(b) $10x^2 + 5y^2 + 13z^2 - 12xy - 6yz - 4zx - 8x + 30y - 74z + 59 = 0$

3. $5x^2 + 8y^2 + 5z^2 + 4yz + 8zx - 4xy - 144 = 0$

4. $x^2 + y^2 + z^2 - yz - zx - xy - \frac{3}{2} = 0$

5. $45x^2 + 40y^2 + 13z^2 - 12xy - 24xy - 36yz - 42x + 280y - 126z - 98 = 0$

6. $x^2 + y^2 + z^2 - yz - zx - xy - 1 = 0.$

---

6. (a) Enveloping Cylinder

Def. The locus of the tangent to a sphere which are parallel to a given line is a cylinder known as the **enveloping cylinder.**

(b) Equation of the Enveloping Cylinder

To find the equation of the enveloping cylinder of the sphere $x^2 + y^2 + z^2 = a^2$, whose generators are parallel to the line

$$\frac{x}{l} = \frac{y}{m} = \frac{z}{n}.$$

The given sphere is $x^2 + y^2 + z^2 = a^2$ ...(1)

Let (x_1, y_1, z_1) be any point on the enveloping cylinder.

$\therefore$ Equations of the generator of the enveloping cylinder through (x_1, y_1, z_1) are

$$\frac{x - x_1}{l} = \frac{y - y_1}{m} = \frac{z - z_1}{n} \qquad ...(2)$$

$\left(\because \text{ All generators are} \parallel \text{the line } \frac{x}{l} = \frac{y}{m} = \frac{z}{n}\right)$

By def. of Enveloping cylinder, a generator (2) must be a tangent to sphere (1).

Any point on line (2) is $(lr + x_1, mr + y_1, nr + z_1)$.

If it lies on sphere (1), then

$$(lr + x_1)^2 + (mr + y_1)^2 + (nr + z_1)^2 = a^2$$

or $$r^2(l^2 + m^2 + n^2) + 2r(lx_1 + my_1 + nz_1) + (x_1^2 + y_1^2 + z_1^2 - a^2) = 0 \quad ...(3)$$

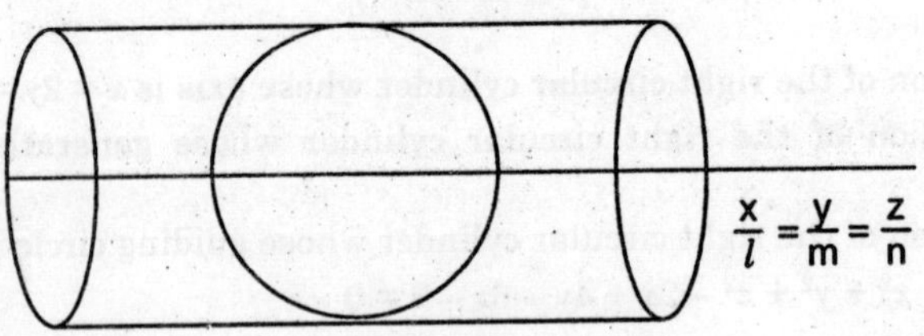

Since the line (2) touches the sphere (1), the two values of r must be equal *i.e.*, quadratic (3) must have two equal roots for which discriminant = 0.

i.e., $$b^2 - 4ac = 0$$

$\therefore$ $$4(lx_1 + my_1 + nz_1)^2 - 4(l^2 + m^2 + n^2)(x_1^2 + y_1^2 + z_1^2 - a^2) = 0$$

or $$(lx_1 + my_1 + nz_1)^2 = (l^2 + m^2 + n^2)(x_1^2 + y_1^2 + z_1^2 - a^2)$$

$\therefore$ Locus of (x_1, y_1, z_1) is [changing (x_1, y_1, z_1) to (x, y, z)]

$$(lx + my + nz)^2 = (l^2 + m^2 + n^2)(x^2 + y^2 + z^2 - a^2)$$

which is the required equation of the enveloping cylinder.

Example. *Find the equation of the enveloping cylinder of the sphere $x^2 + y^2 + z^2 - 2x + 4y = 1$ having its generators parallel to the line $x = y = z$.*

(M.D.U. 2007 ; K.U. 1996, 1999)

Sol. For figure, see Art. 6 (*b*).

The equation of the given sphere is $x^2 + y^2 + z^2 - 2x + 4y = 1$. ...(1)

Let (x_1, y_1, z_1) be any point on the enveloping cylinder

$\therefore$ All generators are parallel to the line $x = y = z$

i.e., $$\frac{x}{1} = \frac{y}{1} = \frac{z}{1}$$

$\therefore$ d.r's of every generator are 1, 1, 1.

$\therefore$ Equations of the generator of the enveloping cylinder through (x_1, y_1, z_1) are

$$\frac{x - x_1}{1} = \frac{y - y_1}{1} = \frac{z - z_1}{1} \quad ...(2)$$

By Def. of enveloping cylinder, generator (2) must be a tangent to sphere (1).

Any point on line (2) is $(r + x_1, r + y_1, r + z_1)$.

If it lies on the sphere (1), then

$$(r + x_1)^2 + (r + y_1)^2 + (r + z_1)^2 - 2(r + x_1) + 4(r + y_1) - 1 = 0$$

or $$3r^2 + 2r(x_1 + y_1 + z_1 + 1) + (x_1^2 + y_1^2 + z_1^2 - 2x_1 + 4y_1 - 1) = 0 \quad ...(3)$$

Since the line (2) touches the sphere (1), the two values of r must be equal *i.e.*, quadratic (3) must have two equal roots for which discriminant = 0.

i.e., $$b^2 - 4ac = 0$$

$\therefore$ $$4(x_1 + y_1 + z_1 + 1)^2 - 12(x_1^2 + y_1^2 + z_1^2 - 2x_1 + 4y_1 - 1) = 0$$

Dividing by 4,

$$(x_1 + y_1 + z_1 + 1)^2 - 3(x_1^2 + y_1^2 + z_1^2 - 2x_1 + 4y_1 - 1) = 0$$

$\therefore$ Locus of (x_1, y_1, z_1) [changing (x_1, y_1, z_1) to (x, y, z)]

$$(x + y + z + 1)^2 - 3(x^2 + y^2 + z^2 - 2x + 4y - 1) = 0$$

i.e., $x^2 + y^2 + z^2 + 1 + 2xy + 2xz + 2x + 2yz + 2y + 2z - 3x^2 - 3y^2 - 3z^2 + 6x - 12y + 3 = 0$

i.e., $$-2x^2 - 2y^2 - 2z^2 + 2xy + 2xz + 2yz + 8x - 10y + 2z + 4 = 0$$

Dividing by -2,

$$x^2 + y^2 + z^2 - xy - xz - yz - 4x + 5y - z - 2 = 0$$

which is the required equation of the enveloping cylinder.

EXERCISE 3

1. Find the equation of the enveloping cylinder of the sphere $x^2 + y^2 + z^2 = 25$ whose generators are parallel to the line $\frac{x}{1} = \frac{y}{2} = \frac{z}{3}$. (*M.D.U. 1998, 96*)
2. (*a*) Find the equation of the enveloping cylinder of the sphere $x^2 + y^2 + z^2 + 2x + 2y + 2z + 2 = 0$ and whose generators are parallel to the line $\frac{x}{1} = \frac{y}{-1} = \frac{z}{1}$.

   (*b*) Find the equation of enveloping cylinder of the sphere $x^2 + y^2 + z^2 - 4x + 6y - 8z - 4 = 0$ having its generators parallel to the line $\frac{x}{-3} = \frac{y}{5} = \frac{z}{1}$.
3. Find the equation of the right circular cylinder which envelopes a sphere of centre (a, b, c) and radius r and has its generators parallel to the direction $< l, m, n >$. (*K.U. 1998 A*)

Answers

1. $13x^2 + 10y^2 + 5z^2 - 4xy - 6xz - 12yz - 350 = 0$
2. (*a*) $x^2 + y^2 + z^2 + 2yz - 2zx + 2xy + 4x + 8y + 4z + 5 = 0$

   (*b*) $26x^2 + 10y^2 + 34z^2 + 30xy + 6xz - 10yz - 38x + 40y - 314z - 429 = 0$
3. $[l(x - a) + m(y - b) + n(z - c)]^2 = [(l^2 + m^2 + n^2)][(x - a)^2 + (y - b)^2 + (z - c)^2 - r^2]$.

MISCELLANEOUS EXERCISE

1. Find the equation of the quadratic cylinder whose generators intersect the curve $ax^2 + by^2 = 2z$, $lx + my + nz = p$ and are parallel to z-axis.
2. A cylinder cuts the plane $z = 0$ in the curve $x^2 + \frac{y^2}{4} = \frac{1}{4}$ and has its axis parallel to $3x = -6y = 2z$. Find its equation.
3. A straight line is always parallel to the YZ-plane and intersects the curves $x^2 + y^2 = a^2, z = 0$ and $x^2 = az, y = 0$; prove that it generates the surface $x^4y^2 = (x^2 - az)^2 (a^2 - x^2)$.
4. Show that the equation of the tangent plane at the point (x_1, y_1, z_1) to the cylinder $ax^2 + 2hxy + by^2 + 2gx + 2fy + c = 0$ is $x(ax_1 + hy_1 + g) + y(hx_1 + by_1 + f) + gx_1 + fy_1 + c = 0$.

Answers

1. $n(ax^2 + by^2) + 2lx + 2my - 2p = 0$
2. $36x^2 + 9y^2 + 17z^2 + 6yz - 48zx - 9 = 0$.

SECTION IV

13

Central Conicoids, Paraboloids

1. Introduction

The locus of the general equation of the second degree

$ax^2 + by^2 + cz^2 + 2fyz + 2gzx + 2hxy + 2ux + 2vy + 2wz + d = 0$

is called a *conicoid* or a *quadric.*

Why the name conicoid ? $\because$ Every plane section of it is a conic and every straight line meets it in two points.

Sphere, cone and cylinder are three conicoids which we have already discussed in chapters 2, 3 and 4.

By suitable transformations of the co-ordinates axes, this equation can be reduced to very simple forms as given below in Art. 2

Remark. The above equation can also represent a pair of planes. But in that case the locus of this equation is called a *degenerate conicoid.*

2. Conicoids in their Simplest Forms

These conicoids can be divided into two groups :

I. (*i*) $\dfrac{x^2}{a^2} + \dfrac{y^2}{b^2} + \dfrac{z^2}{c^2} = 1$ **Ellipsoid**

(*ii*) $\dfrac{x^2}{a^2} + \dfrac{y^2}{b^2} - \dfrac{z^2}{c^2} = 1$ **Hyperboloid of one sheet**

(*iii*) $\dfrac{x^2}{a^2} - \dfrac{y^2}{b^2} - \dfrac{z^2}{c^2} = 1$ **Hyperboloid of two sheets**

(*iv*) $\dfrac{x^2}{a^2} + \dfrac{y^2}{b^2} - \dfrac{z^2}{c^2} = 0$ **Cone**

II. (*i*) $\dfrac{x^2}{a^2} + \dfrac{y^2}{b^2} = \dfrac{2z}{c}$ **Elliptic paraboloid**

(*ii*) $\dfrac{x^2}{a^2} - \dfrac{y^2}{b^2} = \dfrac{2z}{c}$ **Hyperbolic paraboloid**

Note. For tracing of these conicoids see Appendix given after this chapter.

3. Definitions

(1) **Centre** (*Centre of symmetry*). *Centre of a figure is a point (C say) such that every line through C cuts the figure in a pair of points equidistant from C.*

If the origin is **the centre,** then to every point (x, y, z) on the surface, there corresponds a point $(-x, -y, -z)$ on the surface.

i.e., If the equation of a surface remains unchanged on changing **x, y, z to – x –y, – z** respectively ; then origin is the centre of this surface.

(2) **Axis of symmetry.** An axis of symmetry is a line such that every line which cuts this axis at right angles cuts the figure in a pair of points equidistant from this axis *e.g.* if the axis of x is an axis of symmetry, then to every point (x, y, z) on the surface, there corresponds a point $(x, -y, -z)$ on the surface.

i.e., if the equation of a surface remains unchanged on changing **y** to **– y and z** to **– z,** then x-axis is said to be an *axis of symmetry* for the surface.

Similarly if on changing x to $-x$ and z to $-z$, the equation remain unchanged, then y-axis is an axis of symmetry. Similarly, for z-axis.

(3) **Plane of Symmetry.** A plane of symmetry is a plane such that every line perpendicular to this plane cuts the figure in a pair of points equidistant from the plane.

e.g., if the YZ-plane is a plane of symmetry, then to every point (x, y, z) on the surface, there corresponds a point $(-x, y, z)$ on the surface.

i.e., if on changing x to $-x$, the equation remains unchanged ; then the surface is symmetrical about the YZ-plane.

Similarly, if on changing y to $-y$, the equation remains unchanged, then the surface is symmetrical about the ZX-plane.

Similarly for the XY-plane.

SECTION I : CENTRAL CONICOIDS

4. Central Conicoid

Def. *A conicoid whose all chords through the origin are bisected at the origin is called a central conicoid.*

i.e., A conicoid whose centre is the origin (*i.e.*, A conicoid whose equation remains unchanged on changing x to $-x$, y to $-y$ and z to $-z$) is called a central conicoid.

The equation
$$\mathbf{ax^2 + by^2 + cz^2 = 1} \quad \ldots(1)$$
in general, represents a central conicoid [$\because$ (1) remains unchanged when x, y, z are changed to $-x, -y, -z$ respectively].

[Equations (*i*), (*ii*), (*iii*) of group I listed in Art. 2 namely
$$\frac{x^2}{a^2}+\frac{y^2}{b^2}+\frac{z^2}{c^2}=1,\ \frac{x^2}{a^2}+\frac{y^2}{b^2}-\frac{z^2}{c^2}=1,\ \frac{x^2}{a^2}-\frac{y^2}{b^2}-\frac{z^2}{c^2}=1$$
are covered by equation (1) above.]

(*i*) When a, b, c are + ve, (1) represents an *ellipsoid.*

(*ii*) When two of a, b, c are + ve and one –ve, then (1) represents a *hyperboloid of one sheet.*

(*iii*) When one of a, b, c is + ve and two are –ve, then (1) represents a *hyperboloid of two sheets.*

The equation $ax^2 + by^2 + cz^2 = 1$ is called the **standard form** of the central conicoid.

Remark. If a, b, c are all negative, then equation (1) represents a **virtual** (or Imaginary) **ellipsoid.**

5. Intersections of a Line and a Conicoid

To find the points of intersection of the line $\dfrac{x-x_1}{l}=\dfrac{y-y_1}{m}=\dfrac{z-z_1}{n}$ *with the conicoid* $ax^2 + by^2 + cz^2 = 1$.

The given line is

$$\frac{x - x_1}{l} = \frac{y - y_1}{m} = \frac{z - z_1}{n} \qquad ...(1)$$

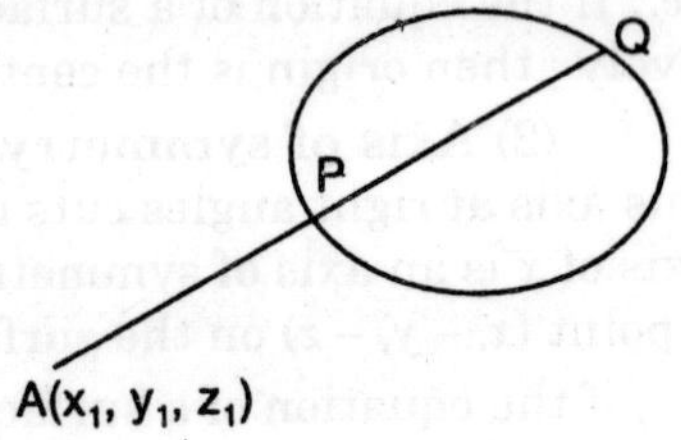

and the conicoid is

$$ax^2 + by^2 + cz^2 = 1 \qquad ...(2)$$

Any point on the line (1) is

$$(lr + x_1,\ mr + y_1,\ nr + z_1) \qquad ...(3)$$

If it lies on the conicoid (2) ; then

$$a(lr + x_1)^2 + b(mr + y_1)^2 + c(nr + z_1)^2 = 1$$

or $\quad r^2(al^2 + bm^2 + cn^2) + 2r(alx_1 + bmy_1 + cnz_1) + (ax_1^{\ 2} + by_1^{\ 2} + cz_1^{\ 2} - 1) = 0 \qquad ...(4)$

which is a quadratic in a giving two values of r.

Putting these values of r in (3), we get the two points of inter-section P and Q.

Cor. Condition of tangency of a line and a conicoid. If the line (1) touches the conicoid (2), then the two values of r from (3) are equal.

$\therefore$ Disc. of (3) is 0 *i.e.*, $\quad b^2 = 4ac$

or $\quad 4(alx_1 + bmy_1 + cnz_1)^2 = 4(al^2 + bm^2 + cn^2)(ax_1^{\ 2} + by_1^{\ 2} + cz_1^{\ 2} - 1)$

Cancelling 4, $\quad (alx_1 + bmy_1 + cnz_1)^2 = (al^2 + bm^2 + cn^2)(ax_1^{\ 2} + by_1^{\ 2} + cz_1^{\ 2} - 1)$

which is the required condition.

6. Equation of the Tangent Plane

To find the equation of the tangent plane at the point (x_1, y_1, z_1) of the central conicoid $ax^2 + by^2 + cz^2 = 1$. (K.U. 1996 S)

We know that tangent plane at a point is the locus of the tangent lines to the surface at that point.

Equation of the conicoid is $ax^2 + by^2 + cz^2 = 1 \qquad ...(1)$

$\because$ The point (x_1, y_1, z_1) lies on (1)

$$\therefore \qquad ax_1^{\ 2} + by_1^{\ 2} + cz_1^{\ 2} = 1 \qquad ...(2)$$

Equations of *any* line through (x_1, y_1, z_1) are

$$\frac{x - x_1}{l} = \frac{y - y_1}{m} = \frac{z - z_1}{n} = r \text{ (say)} \qquad ...(3)$$

Any point on the line (3) is $(lr + x_1, mr + y_1, nr + z_1)$.

If it lies on conicoid (1), then

$$a(lr + x_1)^2 + b(mr + y_1)^2 + c(nr + z_1)^2 = 1$$

i.e., $\quad r^2(al^2 + bm^2 + cn^2) + 2r(alx_1 + bmy_1 + cnz_1) + (ax_1^{\ 2} + by_1^{\ 2} + cz_1^{\ 2} - 1) = 0 \qquad ...(4)$

If line (3) is a tangent to conicoid (1), then this line cuts the conicoid in two coincident points and therefore the two values of r from (4) must be equal.

i.e., Disc. of (4) is 0 $\qquad$ *i.e.*, $\quad B^2 - 4AC = 0$

But C of (4) is $\qquad ax_1^{\ 2} + by_1^{\ 2} + cz_1^{\ 2} - 1 = 0 \qquad$ [By (2)]

$$\therefore \qquad B^2 = 0 \text{ or } B = 0$$

or $$alx_1 + bmy_1 + cnz_1 = 0 \qquad ...(5)$$

Equation (5) is the condition for the line (3) to be a tangent to conicoid (1).

To find the equation of the tangent plane (*i.e.*, the locus of the tangent lines), let us **eliminate l, m, n from (3) and (5).**

Putting the values of l, m, n from (3) in (5), we have

$$ax_1(x - x_1) + by_1(y - y_1) + cz_1(z - z_1) = 0$$

or $$axx_1 + byy_1 + czz_1 = ax_1^2 + by_1^2 + cz_1^2 = 1 \qquad | \text{ Using (2)}$$

or $\mathbf{axx_1 + byy_1 + czz_1 = 1}$ which is the required equation of the tangent plane.

Method to write the equation of the tangent plane at a point (x_1, y_1, z_1) of the central conicoid.

Change x^2 to xx_1, y^2 to yy_1 and z^2 to zz_1.

Exercise. *Show that the equation of the tangent plane at (x_1, y_1, z_1) to the ellipsoid*

$$\frac{x^2}{a^2} + \frac{y^2}{b^2} + \frac{z^2}{c^2} = 1 \text{ is } \frac{xx_1}{a^2} + \frac{yy_1}{b^2} + \frac{zz_1}{c^2} = 1.$$

7. Condition of Tangency

To find the condition that the plane $lx + my + nz = p$, should touch the conicoid $ax^2 + by^2 + cz^2 = 1$.

The given plane is $$lx + my + nz = p \qquad ...(1)$$

and the conicoid is $$ax^2 + by^2 + cz^2 = 1 \qquad ...(2)$$

Let the plane (1) touch the conicoid (2) at the point (x_1, y_1, z_1).

Equation of the tangent plane at the point (x_1, y_1, z_1) to

Conicoid (2) is $$axx_1 + byy_1 + czz_1 = 1 \qquad ...(3)$$

Comparing (1) and (3). [$\because$ They represent the same tangent plane],

we have $$\frac{ax_1}{l} = \frac{by_1}{m} = \frac{cz_1}{n} = \frac{1}{p}$$

$\therefore$ $$x_1 = \frac{l}{ap},\ y_1 = \frac{m}{bp},\ z_1 = \frac{n}{cp} \qquad ...(4)$$

But (x_1, y_1, z_1) being the point of contact lies on conicoid (2),

$\therefore$ $$ax_1^2 + by_1^2 + cz_1^2 = 1$$

Putting the value of x_1, y_1, z_1 from (4),

$\therefore$ $$a\left(\frac{l}{ap}\right)^2 + b\left(\frac{m}{bp}\right)^2 + c\left(\frac{n}{cp}\right)^2 = 1$$

or $$\frac{l^2}{ap^2} + \frac{m^2}{bp^2} + \frac{n^2}{cp^2} = 1 \quad \text{or} \quad \mathbf{\frac{l^2}{a} + \frac{m^2}{b} + \frac{n^2}{c} = p^2}$$

which is the required condition of **tangency.**

Cor. 1. From (4) the co-ordinates of the point of contact are

$$\left(\frac{l}{ap}, \frac{m}{bp}, \frac{n}{cp}\right).$$

Cor. 2. Equation of any tangent plane to the conicoid.

To find the equations of the tangents planes to the conicoid $ax^2 + by^2 + cz^2 = 1$ which are parallel to the plane $lx + my + nz = 0$.

The conicoid is $$ax^2 + by^2 + cz^2 = 1 \qquad ...(1)$$

Equation of any plane parallel to the plane $lx + my + nz = 0$ is

$$lx + my + nz = p \qquad ...(2)$$

If plane (2) is to be tangent plane to conicoid (1), then

$$\frac{l^2}{a}+\frac{m^2}{b}+\frac{n^2}{c}=p^2 \qquad \text{(By Art. 7 above)}$$

or
$$p=\pm\sqrt{\frac{l^2}{a}+\frac{m^2}{b}+\frac{n^2}{c}}$$

Putting these values of p in (2), equations of the tangent planes are

$$lx+my+nz=\pm\sqrt{\frac{l^2}{a}+\frac{m^2}{b}+\frac{n^2}{c}} \qquad ...(3)$$

Note. The equation (3) represents tangent planes to the conicoid (1) for all values of l, m, n. Thus any tangent plane to the conicoid (1) is

$$\mathbf{lx+my+nz=\sqrt{\frac{l^2}{a}+\frac{m^2}{b}+\frac{n^2}{c}}.}$$

Exercise. *Show that the condition for the plane $lx+my+nz=p$ to be a tangent plane to the ellipsoid $\frac{x^2}{a^2}+\frac{y^2}{b^2}+\frac{z^2}{c^2}=1$ is $a^2l^2+b^2m^2+c^2n^2=p^2$.*

Also prove that the equation of any tangent plane to the ellipsoid $\frac{x^2}{a^2}+\frac{y^2}{b^2}+\frac{z^2}{c^2}=1$ is

$$\mathbf{lx+my+nz=\sqrt{a^2l^2+b^2m^2+c^2n^2}.}$$

[**Hint.** Proceed as in Art. 7]

8. Director Sphere

To find locus of the point of intersection of three mutually perpendicular tangent planes to the ellipsoid

$$\frac{x^2}{a^2}+\frac{y^2}{b^2}+\frac{z^2}{c^2}=1.$$

(By the above exercise) equation of any tangent plane to the ellipsoid $\frac{x^2}{a^2}+\frac{y^2}{b^2}+\frac{z^2}{c^2}=1$ is $lx+my+nz=\sqrt{a^2l^2+b^2m^2+c^2n^2}$, where l, m, n are d.c's of normal to the plane.

So let the equations of any three mutually perpendicular tangent planes to the ellipsoid be

$$l_1x+m_1y+n_1z=\sqrt{a^2l_1^2+b^2m_1^2+c^2n_1^2} \qquad ...(1)$$

$$l_2x+m_2y+n_2z=\sqrt{a^2l_2^2+b^2m_2^2+c^2n_2^2} \qquad ...(2)$$

$$l_3x+m_3y+n_3z=\sqrt{a^2l_3^2+b^2m_3^2+c^2n_3^2} \qquad ...(3)$$

$\because$ The three tangents planes are mutually $\perp$, therefore their normals are also mutually $\perp$.

$\therefore$ $l_1, m_1, n_1, l_2, m_2, n_2$; l_3, m_3, n_3 are the d.c's of three mutually $\perp$ lines.

$\therefore$ (By Art. 4, Review IV) there exist 12 relations among these d.c's

namely $l_1l_2+m_1m_2+n_1n_2=0$ etc. ; $l_1m_1+l_2m_2+l_3m_3=0$ etc.

and $l_1^2+m_1^2+n_1^2=1$ etc. ; $l_1^2+l_2^2+l_3^2=1$ etc. $\qquad$ ...(4)

Let (x, y, z) be the point of intersection of three mutually ⊥ tangent planes (1), (2), (3).

To find the locus of the point (x, y, z) we are to eliminate $l_1, m_1, n_1 ; l_2, m_2, n_2 ; l_3, m_3, n_3$ from (1), (2), and (3).

Squaring and adding (1), (2) and (3), (To eliminate the above quantities)

we have
$$(l_1x + m_1y + n_1z)^2 + (l_2x + m_2y + n_2z)^2 + (l_3x + m_3y + n_3z)^2$$
$$= (a^2l_1^2 + b^2m_1^2 + c^2n_1^2) + (a^2l_2^2 + b^2m_2^2 + c^2n_2^2) + (a^2l_3^2 + b^2m_3^2 + c^2n_3^2)$$

or
$$x^2(l_1^2 + l_2^2 + l_3^2) + y^2(m_1^2 + m_2^2 + m_3^2) + z^2(n_1^2 + n_2^2 + n_3^2)$$
$$+ 2xy(l_1m_1 + l_2m_2 + l_3m_3) + 2yz(m_1n_1 + m_2n_2 + m_3n_3) + 2zx(n_1l_1 + n_2l_2 + n_3l_3)$$
$$= a^2(l_1^2 + l_2^2 + l_3^2) + b^2(m_1^2 + m_2^2 + m_3^2) + c^2(n_1^2 + n_2^2 + n_3^2)$$

or $x^2(1) + y^2(1) + z^2(1) + 2xy(0) + 2yz(0) + 2zx(0) = a^2(1) + b^2(1) + c^2(1)$ | Using (4)

or $x^2 + y^2 + z^2 = a^2 + b^2 + c^2$ which is the required locus and is a sphere concentric with the ellipsoid and is known as the **Director Sphere.**

Exercise. *Prove that the locus of the point of intersection of three mutually ⊥ tangents planes to the central conicoid*

$$ax^2 + by^2 + cz^2 = 1 \text{ is}$$
$$x^2 + y^2 + z^2 = \frac{1}{a} + \frac{1}{b} + \frac{1}{c}.$$ (K.U. 2000 A)

[**Hint.** Proceed as in Art. 8.]

Example 1. *Prove that the sum of the square of the reciprocals of any three mutually perpendicular diameters of an ellipsoid is constant.*

(M.D.U. 1997 ; K.U. 1999 A, 99 S)

Sol. Let the ellipsoid be $\frac{x^2}{a^2} + \frac{y^2}{b^2} + \frac{z^2}{c^2} = 1$ ...(1)

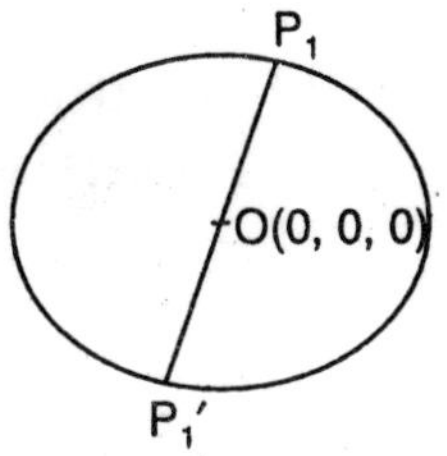

having O(0, 0, 0) as centre.

Let $<l_1, m_1, n_1>$, $<l_2, m_2, n_2>$, $<l_3, m_3, n_3>$ be the d.c's of three mutually ⊥ diameters P_1OP_1', P_2OP_2', P_3OP_3'.

One of the diameters P_1OP_1' has been shown in the adjoining figure.

Let $OP_1 = r_1$

∴ P_1 is (l_1r_1, m_1r_1, n_1r_1) | (lr, mr, nr)

But P_1 lies on the ellipsoid (1)

$$\therefore \quad \frac{l_1^2r_1^2}{a^2} + \frac{m_1^2r_1^2}{b^2} + \frac{n_1^2r_1^2}{c^2} = 1$$

or
$$\frac{1}{r_1^2} = \frac{l_1^2}{a^2} + \frac{m_1^2}{b^2} + \frac{n_1^2}{c^2}$$

$$\therefore \quad \frac{1}{(P_1P_1')^2} = \frac{1}{(2r_1)^2} = \frac{1}{4r_1^2} = \frac{1}{4}\left(\frac{l_1^2}{a^2} + \frac{m_1^2}{b^2} + \frac{n_1^2}{c^2}\right)$$

Similarly,
$$\frac{1}{(P_2P_2')^2} = \frac{1}{4}\left(\frac{l_2^2}{a^2} + \frac{m_2^2}{b^2} + \frac{n_2^2}{c^2}\right)$$

and
$$\frac{1}{(P_3P_3')^2} = \frac{1}{4}\left(\frac{l_3^2}{a^2} + \frac{m_3^2}{b^2} + \frac{n_3^2}{c^2}\right)$$

Adding vertically, we get $\frac{1}{(P_1P_1')^2}+\frac{1}{(P_2P_2')^2}+\frac{1}{(P_3P_3')^2}$

$$=\frac{1}{4}\left[\frac{1}{a^2}(l_1^2+l_2^2+l_3^2)+\frac{1}{b^2}(m_1^2+m_2^2+m_3^2)+\frac{1}{c^2}(n_1^2+n_2^2+n_3^2)\right]$$

$$=\frac{1}{4}\left[\frac{1}{a^2}+\frac{1}{b^2}+\frac{1}{c^2}\right]$$

($\because$ l_1, m_1, n_1 ; l_2, m_2, n_2 ; l_3, m_3, n_3 are the d.c's of three mutually $\perp$ diameters)

$\therefore$ Sum of the square of the reciprocals of any three mutually $\perp$ diameters

$$=\frac{1}{4}\left[\frac{1}{a^2}+\frac{1}{b^2}+\frac{1}{c^2}\right]=\text{Constant.}$$

Example 2. *Show that the plane $x + 2y + 3z = 2$ touches the conicoid $x^2 - 2y^2 + 3z^2 = 2$. Find also the point of contact.* (K.U. 2003)

Sol. Equation of the plane is $x + 2y + 3z = 2$ ...(1)

Comparing with $lx + my + nz = p$, we have $l = 1, m = 2, n = 3, p = 2$

Equation of the conicoid is $x^2 - 2y^2 + 3z^2 = 2$ ...(2)

Dividing every term by 2 (To make R.H.S. = 1)

$$\frac{1}{2}x^2 - y^2 + \frac{3}{2}z^2 = 1$$

Comparing with $ax^2 + by^2 + cz^2 = 1$, we have

$$a=\frac{1}{2},\ b=-1,\ c=\frac{3}{2}$$

Condition of tangency

Plane (1) will touch conicoid (2), if

$$\frac{\mathbf{l^2}}{\mathbf{a}}+\frac{\mathbf{m^2}}{\mathbf{b}}+\frac{\mathbf{n^2}}{\mathbf{c}}=\mathbf{p^2} \qquad \text{(By Art. 7)}$$

i.e., if
$$\frac{1}{\left(\frac{1}{2}\right)}+\frac{4}{-1}+\frac{9}{\frac{3}{2}}=(2)^2$$

i.e., if $2 - 4 + 6 = 4$ *i.e.,* if $4 = 4$

which is true.

$\therefore$ Plane (1) touches conicoid (2).

To find point of contact

Let plane (1) touch conicoid (2) at (x_1, y_1, z_1).

Equation of the tangent plane at (x_1, y_1, z_1) to conicoid (2) is $xx_1 - 2yy_1 + 3zz_1 = 2$ ...(3)

[Change x^2 to xx_1, y^2 to yy_1 and z^2 to zz_1]

Comparing (3) and (1), we have

$$\frac{x_1}{1}=-\frac{2y_1}{2}=\frac{3z_1}{3}=\frac{2}{2} \quad \text{or} \quad x_1=-y_1=z_1=1$$

$\therefore$ $x_1 = 1, y_1 = -1, z_1 = 1$

$\therefore$ Point of contact is $(x_1, y_1, z_1) = (1, -1, 1)$.

Example 3. *Find the equations to the two tangent planes which contain the line given by $7x + 10y - 30 = 0$, $5y - 3z = 0$ and touch the conicoid $7x^2 + 5y^2 + 3z^2 = 60$.* (M.D.U. 2001)

Sol. Equations of the given line are $7x + 10y - 30 = 0$, $5y - 3z = 0$.

Any plane through this line is $7x + 10y - 30 + k(5y - 3z) = 0$ [L.H.S. + k(L.H.S.) = 0]

or
$$7x + 5(k + 2)y - 3kz = 30 \qquad ...(1)$$

[(1) is the form $lx + my + nz = p$]

Comparing, $l = 7$, $m = 5(k + 2)$, $n = -3k$, $p = 30$.

The given conicoid is $7x^2 + 5y^2 + 3z^2 = 60$

Dividing both sides by 60 (to get the equation in the standard form $ax^2 + by^2 + cz^2 = 1$).

$$\frac{7}{60}x^2 + \frac{5}{60}y^2 + \frac{3}{60}z^2 = 1 \qquad ...(2)$$

Comparing with $ax^2 + by^2 + cz^2 = 1$, we have

$$a = \frac{7}{60},\ b = \frac{5}{60},\ c = \frac{3}{60}$$

But plane (1) touches the conicoid (2).

$$\therefore \quad \frac{\mathbf{l^2}}{\mathbf{a}} + \frac{\mathbf{m^2}}{\mathbf{b}} + \frac{\mathbf{n^2}}{\mathbf{c}} = \mathbf{p^2} \qquad \text{(By Art. 7)}$$

$$\therefore \quad \frac{(7)^2}{\left(\frac{7}{60}\right)} + \frac{25(k+2)^2}{\frac{5}{60}} + \frac{(-3k)^2}{\frac{3}{60}} = (30)^2$$

or
$$\frac{49 \times 60}{7} + \frac{25(k+2)^2 \times 60}{5} + \frac{9k^2 \times 60}{3} = 900$$

or
$$420 + 300(k^2 + 4k + 4) + 180k^2 = 900$$

or
$$480k^2 + 1200k + 720 = 0$$

Cancelling 240,
$$2k^2 + 5k + 3 = 0$$

or
$$(2k + 3)(k + 1) = 0$$

$$\therefore \quad k = -\frac{3}{2} \quad \text{or} \quad k = -1$$

Putting these values of k in (1), the required tangent planes are

$$7x + 5\left(-\frac{3}{2} + 2\right)y + \frac{9}{2}z = 30 \text{ and } 7x + 5(-1 + 2)y + 3z = 30$$

or $\quad 7x + \frac{5}{2}y + \frac{9}{2}z = 30 \quad$ and $\quad 7x + 5y + 3z = 30$

or $\quad 14x + 5y + 9z = 60 \quad$ and $\quad 7x + 5y + 3z = 30$.

Example 4. *Tangent planes are drawn to the ellipsoid*

$$\frac{x^2}{a^2} + \frac{y^2}{b^2} + \frac{z^2}{c^2} = 1$$

through the point (α, β, γ). *Prove that the perpendiculars to them through the origin generate the cone* $(\alpha x + \beta y + \gamma z)^2 = a^2x^2 + b^2y^2 + c^2z^2$.

Sol. Equation of any plane through (α, β, γ) is $l(x - \alpha) + m(y - \beta) + n(z - \gamma) = 0$

(one point form)

or $$lx + my + nz = l\alpha + m\beta + n\gamma \quad ...(1)$$

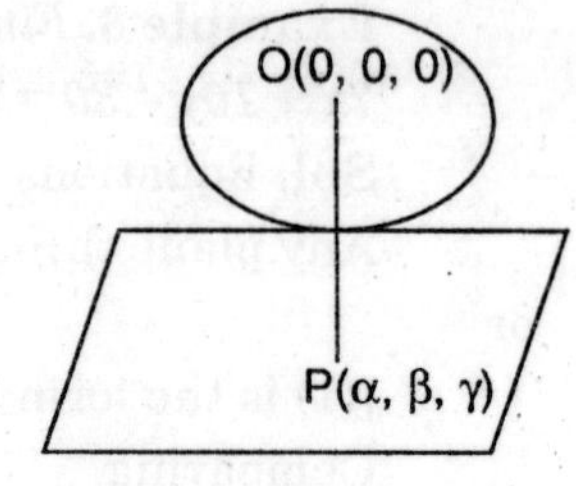

(Equation (1) is of the form $lx + my + nz = p$)

If plane (1) is to be a tangent plane to the ellipsoid

$$\frac{x^2}{a^2} + \frac{y^2}{b^2} + \frac{z^2}{c^2} = 1, \text{ then}$$

$$a^2l^2 + b^2m^2 + c^2n^2 = (l\alpha + m\beta + n\gamma)^2 \quad ...(2)$$

(By Exercise, Art. 7 ; $a^2l^2 + b^2m^2 + c^2n^2 = p^2$)

Now equations of the perpendicular drawn from the origin to the tangent plane (1) are

$$\frac{x}{l} = \frac{y}{m} = \frac{z}{n} \quad ...(3)$$

We are to find the surface generated by perpendiculars given by (3) *i.e.*, we are to find the locus of the lines (3).

Eliminating l, m, n from (2) and (3) *i.e.*, putting the values of l, m, n from (3) in (2), the required locus is $a^2x^2 + b^2y^2 + c^2z^2 = (\alpha x + \beta y + \gamma z)^2$ which being a second degree homogeneous in x, y, z represents a cone.

Example 5. *Prove that the locus of the foot of the perpendicular drawn from the centre of the ellipsoid* $\frac{x^2}{a^2} + \frac{y^2}{b^2} + \frac{z^2}{c^2} = 1$ *to any of its tangent planes is* $a^2x^2 + b^2y^2 + c^2z^2 = (x^2 + y^2 + z^2)^2$.

(K.U. 1997 A)

Sol. The ellipsoid is $\frac{x^2}{a^2} + \frac{y^2}{b^2} + \frac{z^2}{c^2} = 1 \quad ...(1)$

Let $P(x_1, y_1, z_1)$ be the foot of the perpendicular from centre $O(0, 0, 0)$ of (1) of any tangent plane to (1).

∴ d.r's of OP are x_1, y_1, z_1.

Since OP is ⊥ to the tangent plane, d.r's of OP are the coefficients of x, y, z in the equation of the tangent plane.

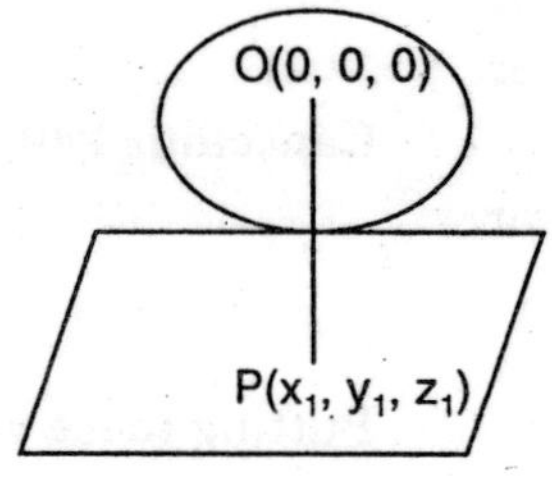

∴ Equation of the tangent plane is

$$x_1(x - x_1) + y_1(y - y_1) + z_1(z - z_1) = 0$$

[One-point Form]

or $$xx_1 - x_1^2 + yy_1 - y_1^2 + zz_1 - z_1^2 = 0$$

or $$xx_1 + yy_1 + zz_1 = x_1^2 + y_1^2 + z_1^2 \quad ...(2)$$

$(lx + my + nz = p)$

Because plane (2) touches ellipsoid (1).

∴ $a^2x_1^2 + b^2y_1^2 + c^2z_1^2 = (x_1^2 + y_1^2 + z_1^2)^2$ [**Condition of tangency :** $a^2l^2 + b^2m^2 + c^2n^2 = p^2$]

∴ Locus of (x_1, y_1, z_1) *i.e.*, foot of perpendicular is $a^2x^2 + b^2y^2 + c^2z^2 = (x^2 + y^2 + z^2)^2$.

EXERCISE 1

1. OP, OQ, OR is any set of three mutually perpendicular semi-diameters of the central conicoid $ax^2 + by^2 + cz^2 = 1$. Prove that $\frac{1}{OP^2} + \frac{1}{OQ^2} + \frac{1}{OR^2}$ is a constant quantity.

2. Show that the length of the perpendicular p from the origin on the tangent plane at the point (x', y', z') of the ellipsoid $\frac{x^2}{a^2}+\frac{y^2}{b^2}+\frac{z^2}{c^2}=1$ is given by $\frac{1}{p^2}=\frac{x'^2}{a^4}+\frac{y'^2}{b^4}+\frac{z'^2}{c^4}$.

3. (a) Show that the plane $3x + 12y - 6z - 17 = 0$ touches the conicoid $3x^2 - 6y^2 + 9z^2 + 17 = 0$. Find also the point of contact.
   (b) Show that the plane $2x - 4y - z = -3$ touches the paraboloid $x^2 - 2y^2 = 3z$. Also, find the point of contact. (K.U. 2006)

4. Find the equations to the tangent planes to the surface
   (a) $4x^2 - 5y^2 + 7z^2 + 13 = 0$ parallel to the plane $4x + 20y - 21z = 0$.
   (b) $x^2 - 2y^2 + 3z^2 = 2$ parallel to the plane $x - 2y + 3z = 0$. (M.D.U. 2000)

5. (a) Find the equations of the tangent planes to $2x^2 - 6y^2 + 3z^2 = 5$ which pass through the line $x + 9y - 3z = 0 = 3x - 3y + 6z - 5$. (K.U. 2007)
   (b) Find the equations to the tangent planes to $7x^2 - 3y^2 - z^2 + 21 = 0$ which pass through the line $7x + 6y + 9 = 0$, $z = 3$.
   (c) Find the equations of the tangent planes to $3x^2 + 2y^2 - 6z^2 - 5 = 0$ which pass through the line $3x - y - 9z = 0 = 6x + 3y - 3z - 5$.

6. Prove that the equation to the two tangent planes to the conicoid $ax^2 + by^2 + cz^2 = 1$ which pass through the line

$$u \equiv lx + my + nz - p = 0.$$

$$u' \equiv l'x + m'y + n'z - p' = 0 \text{ is}$$

$$u^2\left(\frac{l'^2}{a}+\frac{m'^2}{b}+\frac{n'^2}{c}-p'^2\right)-2uu'\left(\frac{ll'}{a}+\frac{mm'}{b}+\frac{nn'}{c}-pp'\right)+u'^2\left(\frac{l^2}{a}+\frac{m^2}{b}+\frac{n^2}{c}-p^2\right)=0$$

7. A tangent plane to the ellipsoid $\frac{x^2}{a^2}+\frac{y^2}{b^2}+\frac{z^2}{c^2}=1$ meets the co-ordinates axes in A, B and C. Find the locus of the centroid of the (i) triangle ABC (ii) tetrahedron OABC.

8. Tangent planes are drawn to the conicoid $ax^2 + by^2 + cz^2 = 1$ through the point (α, β, γ). Prove that the perpendiculars to them from the origin generate the cone

$$(\alpha x + \beta y + \gamma z)^2 = \frac{x^2}{a}+\frac{y^2}{b}+\frac{z^2}{c}.$$

9. Obtain the tangent planes for the ellipsoid $\frac{x^2}{a^2}+\frac{y^2}{b^2}+\frac{z^2}{c^2}=1$.
   which are parallel to $lx + my + nz = 0$.
   If $2r$ is the distance between the planes, show that a line through the origin and perpendicular to the planes lies on the cone $x^2(a^2 - r^2) + y^2(b^2 - r^2) + z^2(c^2 - r^2) = 0$.

Answers

3. $(-1, 2, \frac{2}{3})$

4. (a) $4x + 20y - 21z = \pm 13$ (b) $x - 2y + 3z = \pm 2$

5. (a) $4x + 6y + 3z - 5 = 0$, $2x - 12y + 9z - 5 = 0$
   (b) $7x + 6y - 4z + 21 = 0$, $14x + 12y - z + 21 = 0$
   (c) $9x + 2y - 12z = 5$, $3x + 4y + 6z = 5$

7. (i) $\frac{a^2}{x^2}+\frac{b^2}{y^2}+\frac{c^2}{z^2}=9$ (ii) $\frac{a^2}{x^2}+\frac{b^2}{y^2}+\frac{c^2}{z^2}=16$.

NORMALS TO A CONICOID

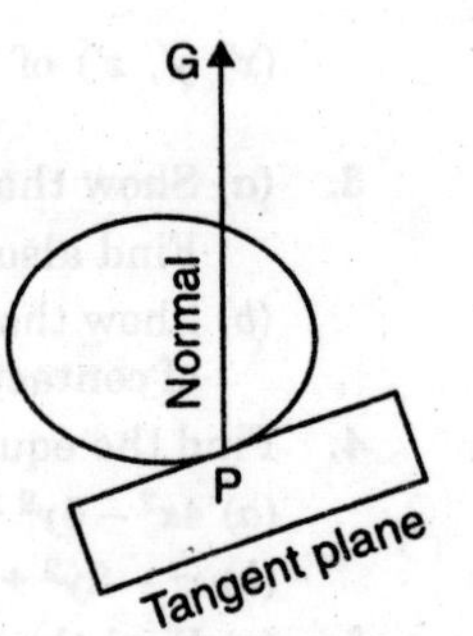

9. (a) Def. Normal

The **normal** *at any point P of a surface is the straight line passing through P and perpendicular to the tangent plane at P.*

(b) Equations of the normal

To find the equations of the normal at the point (x_1, y_1, z_1) of the conicoid $ax^2 + by^2 + cz^2 = 1$.

The equation of the conicoid is $ax^2 + by^2 + cz^2 = 1$

The equation of the tangent plane at (x_1, y_1, z_1) to the conicoid is

$$axx_1 + byy_1 + czz_1 = 1$$

∴ d.r's of the normal to this tangent plane are

$$ax_1,\ by_1,\ cz_1 \qquad \text{| Co-effs. of } x, y, z \text{ in (1)}$$

∴ Equations of the normal at (x_1, y_1, z_1) are

$$\frac{x - x_1}{ax_1} = \frac{y - y_1}{by_1} = \frac{z - z_1}{cz_1}.$$

Cor. *To find the equations of the normal in the actual direction cosines form.*

If p is the length of the ⊥ from the centre (0, 0, 0) of the conicoid on the tangent plane (1) namely.

$$ax_1x + by_1y + cz_1z - 1 = 0 \text{ ; then}$$

$$p = \pm \frac{(0 + 0 + 0 - 1)}{\sqrt{a^2x_1^2 + b^2y_1^2 + c^2z_1^2}} = \frac{1}{\sqrt{a^2x_1^2 + b^2y_1^2 + c^2z_1^2}}.$$

Now d.r's of the normal to the tangent plane (1) are ax_1, by_1, cz_1

Dividing each by $\sqrt{a^2x_1^2 + b^2y_1^2 + c^2z_1^2} = \frac{1}{p}$, the actual direction cosines of the normal are

$$\frac{ax_1}{(1/p)}, \frac{by_1}{(1/p)}, \frac{cz_1}{(1/p)}$$

i.e.,
$$ax_1p,\ by_1p,\ cz_1p$$

∴ The equations of the normal at (x_1, y_1, z_1) to the conicoid $ax^2 + by^2 + cz^2 = 1$ in the actual direction cosines form, are

$$\frac{x - x_1}{ax_1p} = \frac{y - y_1}{by_1p} = \frac{z - z_1}{cz_1p}.$$

Note. When and where to use the above forms.

(*i*) For problems relating to distances measured from (x_1, y_1, z_1) along the normal at this point, use the equations of the normal in the **actual direction cosines form**

$$\frac{x - x_1}{ax_1p} = \frac{y - y_1}{by_1p} = \frac{z - z_1}{cz_1p}$$

(*ii*) And for other problems, use equations of the normal in the simpler form namely

$$\frac{x - x_1}{ax_1} = \frac{y - y_1}{by_1} = \frac{z - z_1}{cz_1}.$$

Exercise. *Prove that the equations of the normal at a point (x_1, y_1, z_1) to the ellipsoid*

$$\frac{x^2}{a^2}+\frac{y^2}{b^2}+\frac{z^2}{c^2}=1$$

are
$$\frac{x-x_1}{\frac{x_1}{a^2}}=\frac{y-y_1}{\frac{y_1}{b^2}}=\frac{z-z_1}{\frac{z_1}{c^2}}$$

Also show that these equations can be put in the actual direction-cosines form as

$$\frac{x-x_1}{\frac{px_1}{a^2}}=\frac{y-y_1}{\frac{py_1}{b^2}}=\frac{z-z_1}{\frac{pz_1}{c^2}}$$

where p is the length of ⊥ from centre (0, 0, 0) of the ellipsoid on the tangent plane at (x_1, y_1, z_1) to the ellipsoid.

Example. *The normal at a point P of the ellipsoid* $\frac{x^2}{a^2}+\frac{y^2}{b^2}+\frac{z^2}{c^2}=1$ *meets the principal planes in* G_1, G_2, G_3;

(*i*) *Show that* $PG_1 : PG_2 : PG_3 = a^2 : b^2 : c^2$ (M.D.U. 2003 S)

(*ii*) *If* $PG_1^2 + PG_2^2 + PG_3^2 = k^2$, *find the locus of P.*

Sol. The principal planes are the planes YOZ, ZOX and XOY.

The given ellipsoid is $\frac{x^2}{a^2}+\frac{y^2}{b^2}+\frac{z^2}{c^2}=1$ ...(1)

Let P be the point (x_1, y_1, z_1).

The equations of the normal at $P(x_1, y_1, z_1)$ to (1) in the actual direction cosines form are [By Exercise Art. 9 (*b*)]

$$\frac{x-x_1}{\frac{px_1}{a^2}}=\frac{y-y_1}{\frac{py_1}{b^2}}=\frac{z-z_1}{\frac{pz_1}{c^2}}=r \quad ...(2)$$

where r is the distance of the point $P(x_1, y_1, z_1)$ from any point (x, y, z).

This normal (2) meets the principal plane YOZ *i.e.*, $x = 0$ in the point G_1.

$$\therefore \quad \frac{0-x_1}{\frac{px_1}{a^2}}=r=PG_1 \text{ or } -\frac{a^2}{p}=PG_1 \text{ i.e., } PG_1=-\frac{a^2}{p}.$$

Similarly, $PG_2=-\frac{b^2}{p}$, $PG_3=-\frac{c^2}{p}$

(*i*) $\therefore$ $PG_1 : PG_2 : PG_3 = -\frac{a^2}{p} : -\frac{b^2}{p} : -\frac{c^2}{p} = a^2 : b^2 : c^2$.

(*ii*) We are given that $PG_1^2 + PG_2^2 + PG_3^2 = k^2$

or
$$\frac{a^4}{p^2}+\frac{b^4}{p^2}+\frac{c^4}{p^2}=k^2$$

or
$$\frac{(a^4+b^4+c^4)}{p^2}=k^2 \quad ...(3)$$

But p = length of ⊥ from centre (0, 0, 0) on the tangent plane $\frac{xx_1}{a^2}+\frac{yy_1}{b^2}+\frac{zz_1}{c^2}=1$ at (x_1, y_1, z_1) to ellipsoid (1)

$$= \left| \frac{-1}{\sqrt{\frac{x_1^2}{a^4}+\frac{y_1^2}{b^4}+\frac{z_1^2}{c^4}}} \right| = \frac{-1}{\sqrt{\frac{x_1^2}{a^4}+\frac{y_1^2}{b^4}+\frac{z_1^2}{c^4}}}$$

Putting this value of p in (3), we have

$$\left(\frac{x_1^2}{a^4}+\frac{y_1^2}{b^4}+\frac{z_1^2}{c^4}\right)(a^4+b^4+c^4)=k^2$$

∴ Locus of $P(x_1, y_1, z_1)$ is [changing (x_1, y_1, z_1) to (x, y, z)]

$$\left(\frac{x^2}{a^4}+\frac{y^2}{b^4}+\frac{z^2}{c^4}\right)(a^4+b^4+c^4)=k^2 \quad ...(4)$$

Also P lies on (1). Thus P lies on the curve of intersection of ellipsoids (1) and (4).

10. Number of Normals From a Given Point

To prove that there are six points on an ellipsoid the normals at which pass through a given point (α, β, γ).

Or

To prove that six normals can be drawn from a point (α, β, γ) *to the ellipsoid.*

Let the ellipsoid be

$$\frac{x^2}{a^2}+\frac{y^2}{b^2}+\frac{z^2}{c^2}=1 \quad ...(1)$$

Let (x_1, y_1, z_1) be one of the points on the ellipsoid (1), the normal at which passes through (α, β, γ).

Equations of the nomal at (x_1, y_1, z_1) are

$$\frac{x-x_1}{\frac{x_1}{a^2}}=\frac{y-y_1}{\frac{y_1}{b^2}}=\frac{z-z_1}{\frac{z_1}{c^2}}$$

∵ It passes through (α, β, γ)

∴ $$\frac{\alpha-x_1}{\frac{x_1}{a^2}}=\frac{\beta-y_1}{\frac{y_1}{b^2}}=\frac{\gamma-z_1}{\frac{z_1}{c^2}}=k \text{ (say)}$$

From first and fourth members

$$\alpha-x_1=\frac{kx_1}{a^2} \quad \text{or} \quad \alpha=x_1\left(1+\frac{k}{a^2}\right)=x_1\left(\frac{a^2+k}{a^2}\right)$$

∴ $$x_1=\frac{a^2\alpha}{a^2+k}$$

Similarly, $$y_1=\frac{b^2\beta}{b^2+k} \text{ and } z_1=\frac{c^2\gamma}{c^2+k} \quad ...(2)$$

But (x_1, y_1, z_1) lies on ellipsoid (1),

therefore, $$\frac{x_1^2}{a^2}+\frac{y_1^2}{b^2}+\frac{z_1^2}{c^2}=1$$

or $$\frac{1}{a^2}\cdot\frac{a^4\alpha^2}{(a^2+k)^2}+\frac{1}{b^2}\cdot\frac{b^4\beta^2}{(b^2+k)^2}+\frac{1}{c^2}\cdot\frac{c^4\gamma^2}{(c^2+k)^2}=1$$ | Using (2) or $$\frac{a^2\alpha^2}{(a^2+k)^2}+\frac{b^2\beta^2}{(b^2+k)^2}+\frac{c^2\gamma^2}{(c^2+k)^2}=1$$

or $$a^2\alpha^2(b^2+k)^2(c^2+k)^2+b^2\beta^2(a^2+k)^2(c^2+k)^2+c^2\gamma^2(a^2+k)^2(b^2+k)^2$$
$$=(a^2+k)^2\,(b^2+k)^2\,(c^2+k)^2$$

It is a sixth degree equation in k and gives six value of k corresponding to which six points (x_1, y_1, z_1) are obtained from (2), the normals at which pass through the given point (α, β, γ).

Cor. Foot of normal. (*i.e., the point of the ellipsoid where normal is drawn* or the point of contact) is

$$(x_1, y_1, z_1)=\left(\frac{a^2\alpha}{a^2+k}, \frac{b^2\beta}{b^2+k}, \frac{c^2\gamma}{c^2+k}\right) \qquad | \text{ From (2)}$$

11. Cubic Curve Through the Feet of Six Normals From a Point

To show that the feet of the normals from (α, β, γ) to the ellipsoid are the six points of intersection of the ellipsoid and a certain cubic curve.

Let the ellipsoid be $$\frac{x^2}{a^2}+\frac{y^2}{b^2}+\frac{z^2}{c^2}=1 \qquad ...(1)$$

If the normal at (x_1, y_1, z_1) to the ellipsoid passes through the point (α, β, γ) ; then

$$x_1=\frac{a^2\alpha}{a^2+k},\ y_1=\frac{b^2\beta}{b^2+k},\ z_1=\frac{c^2\gamma}{c^2+k}$$

[Reproduce Art. 10 upto equation (2)]

∴ Locus of the feet (x_1, y_1, z_1) of the normals is [changing (x_1, y_1, z_1) to (x, y, z)]

$$x=\frac{a^2\alpha}{a^2+k},\ y=\frac{b^2\beta}{b^2+k},\ z=\frac{c^2\gamma}{c^2+k} \qquad ...(2)$$

where k is a parameter.

Let us prove that the curve (2) is a cubic curve.

Let us find the points of intersection of curve (2) and the plane

$$Ax+By+Cz+D=0 \qquad ...(3)$$

Putting the values of x, y, z from (2) in (3), we have

$$A\frac{a^2\alpha}{a^2+k}+B.\frac{b^2\beta}{b^2+k}+C.\frac{c^2\gamma}{c^2+k}+D=0$$

or $$a^2\alpha A(b^2+k)(c^2+k)+b^2\beta B(a^2+k)(c^2+k)+c^2\gamma C(a^2+k)(b^2+k)$$
$$+D(a^2+k)(b^2+k)(c^2+k)=0$$

which is a cubic in k giving three values of k. Thus the curve (2) meets the plane (3) in three points.

∴ Curve (2) is a cubic curve.

Exercise. *Show that in general six normals can be drawn from a given point (f, g, h) to the conicoid $ax^2+by^2+cz^2=1$. Prove also that the six feet of the normals from (f, g, h) to the conicoid are the intersections of the conicoid with a cubic curve.* (M.D.U. 2003)

[Hint. Proceed as in Art. 10 and Art. 11.]

12. Quadric Cone Through the Six Feet of the Normals

To prove that the feet of the six normals from (α, β, γ) *to the ellipsoid* $\frac{x^2}{a^2}+\frac{y^2}{b^2}+\frac{z^2}{c^2}=1$ *lie on the curve of intersection of the ellipsoid and the cone*

$$\frac{a^2(b^2-c^2)\alpha}{x}+\frac{b^2(c^2-a^2)\beta}{y}+\frac{c^2(a^2-b^2)\gamma}{z}=0$$

Reproduce Art 10 upto equation (2), we get

$$x_1=\frac{a^2\alpha}{a^2+k},\ y_1=\frac{b^2\beta}{b^2+k},\ z_1=\frac{c^2\gamma}{c^2+k}$$

From these equations, we have

$$a^2+k=\frac{a^2\alpha}{x_1},\ b^2+k=\frac{b^2\beta}{y_1},\ c^2+k=\frac{c^2\gamma}{z_1}$$

Multiplying these equations by (b^2-c^2), (c^2-a^2), (a^2-b^2) and adding, we have

$$(a^2+k)(b^2-c^2)+(b^2+k)(c^2-a^2)+(c^2+k)(a^2-b^2)$$
$$=\frac{a^2\alpha}{x_1}(b^2-c^2)+\frac{b^2\beta}{y_1}(c^2-a^2)+\frac{c^2\gamma}{z_1}(a^2-b^2)$$

or
$$0=\frac{a^2\alpha(b^2-c^2)}{x_1}+\frac{b^2\beta(c^2-a^2)}{y_1}+\frac{c^2\gamma(a^2-b^2)}{z_1}$$

($\because$ L.H.S. become zero on simplification)

$\therefore$ The feet of normals lie on *i.e.*, locus of the feet of normals is

$$\frac{a^2\alpha(b^2-c^2)}{x}+\frac{b^2\beta(c^2-a^2)}{y}+\frac{c^2\gamma(a^2-b^2)}{z}=0$$

or
$$a^2\alpha(b^2-c^2)yz+b^2\beta(c^2-a^2)zx+c^2\gamma(a^2-b^2)xy=0$$

which being a homogeneous equation of the second degree in x, y, z represents a cone. Thus the feet of the normals lie on the curve of intersection of the ellipsoid and the above cone.

Note. In the equation of the above cone through the feet of six normals from a point to an ellipsoid.

Co-eff. of $x^2 = 0$, Co-eff. of $y^2 = 0$, Co-eff. of $z^2 = 0$, constant term = 0.

Example. *If P, Q, R ; P′, Q′, R′ are the feet of six normals from a point to the ellipsoid* $\frac{x^2}{a^2}+\frac{y^2}{b^2}+\frac{z^2}{c^2}=1$, *and the plane PQR is given by* $lx+my+nz=p$, *then the plane P′ Q′ R′ is given by*

$$\frac{x}{a^2l}+\frac{y}{b^2m}+\frac{z}{c^2n}+\frac{1}{p}=0.$$

Sol. The equation of the ellipsoid is $\frac{x^2}{a^2}+\frac{y^2}{b^2}+\frac{z^2}{c^2}-1=0$ ...(1)

and that of the plane PQR is $lx+my+nz-p=0$ ...(2)

Let the required equation of the plane P′ Q′ R′ is $l'x+m'y+n'z-p'=0$ ...(3)

$\therefore$ The joint equation of the planes PQR and P′ Q′ R′ is

$$(lx+my+nz-p)(l'x+m'y+n'z-p')=0 \qquad ...(4)$$

The equation of any second degree surface through the six feet of normals namely P, Q, R ; P′, Q′, R′ which are the intersections of the ellipsoid (1) and the pair of planes (4) is

$$\left(\frac{x^2}{a^2}+\frac{y^2}{b^2}+\frac{z^2}{c^2}-1\right)+k(lx+my+nz-p)(l'x+m'y+n'z-p')=0 \quad ...(5)$$

For some value of k equation (5) represent a cone.

($\because$ By Art. 12, six feet of the normals also lie on a cone)

$\therefore$ By Note, Art. 12,

Co-eff of x^2 in (5) is 0, i.e., $\frac{1}{a^2}+kll'=0 \quad \therefore \quad l'=-\frac{1}{kla^2}$

Similarly, $m'=\frac{-1}{kmb^2}$ and $n'=\frac{-1}{knc^2}$

Also *constant term in (5) is 0 i.e.,* $-1+kpp'=0$ or $p'=\frac{1}{kp}$

Putting these values of l', m', n', p' in (3), the required plane P′ Q′ R′ is

$$\frac{-x}{kla^2}-\frac{y}{kmb^2}-\frac{z}{knc^2}-\frac{1}{kp}=0$$

Cancelling $\frac{-1}{k}$, we get $\frac{x}{la^2}+\frac{y}{mb^2}+\frac{z}{nc^2}+\frac{1}{p}=0.$

13. Quadric Cone Through Six Concurrent Normals

To show that the six normals from a point (α, β, γ) to the ellipsoid lie on a cone of second degree.

Let (x_1, y_1, z_1) be one of the six feet of the normals drawn from (α, β, γ) to the ellipsoid.

We know that the actual direction cosines of the normal at (x_1, y_1, z_1) to the ellipsoid are

$$\frac{px_1}{a^2}, \frac{py_1}{b^2}, \frac{pz_1}{c^2} \quad | \text{By Exercise Art. 9 } (b)$$

Let these be denoted by l, m, n

$$\therefore \quad l=\frac{px_1}{a^2}, m=\frac{py_1}{b^2}, n=\frac{pz_1}{c^2} \quad ...(1)$$

Equations of the normal at (x_1, y_1, z_1) to ellipsoid are

$$\frac{x-x_1}{\frac{x_1}{a^2}}=\frac{y-y_1}{\frac{y_1}{b^2}}=\frac{z-z_1}{\frac{z_1}{c^2}}$$

It passes through (α, β, γ)

$$\therefore \quad \frac{\alpha-x_1}{\frac{x_1}{a^2}}=\frac{\beta-y_1}{\frac{y_1}{b^2}}=\frac{\gamma-z_1}{\frac{z_1}{c^2}}=k \text{ (say)}$$

From first and fourth members

$$\alpha-x_1=\frac{kx_1}{a^2} \quad \text{or} \quad \alpha=x_1\left(\frac{a^2+k}{a^2}\right)$$

$$\therefore \quad x_1=\frac{a^2\alpha}{a^2+k}$$

Similarly, $y_1 = \dfrac{b^2\beta}{b^2+k}$ and $z_1 = \dfrac{c^2\gamma}{c^2+k}$

Putting these values of x_1, y_1, z_1 in (1)

$$l = \frac{p}{a^2}\cdot\frac{a^2\alpha}{a^2+k} = \frac{p\alpha}{a^2+k}$$

Similarly, $$m = \frac{p\beta}{b^2+k},\quad n = \frac{p\gamma}{c^2+k}$$

$$\therefore \quad a^2+k = \frac{p\alpha}{l},\ b^2+k = \frac{p\beta}{m},\ c^2+k = \frac{p\gamma}{n}$$

Multiplying these equations by $b^2 - c^2$, $c^2 - a^2$ and $(a^2 - b^2)$ and adding, we get

$$(a^2+k)(b^2-c^2) + (b^2+k)(c^2-a^2) + (c^2+k)(a^2-b^2)$$
$$= \frac{p\alpha}{l}(b^2-c^2) + \frac{p\beta}{m}(c^2-a^2) + \frac{p\gamma}{n}(a^2-b^2)$$

or $$0 = \frac{p\alpha(b^2-c^2)}{l} + \frac{p\beta(c^2-a^2)}{m} + \frac{p\gamma(a^2-b^2)}{n}$$

($\because$ L.H.S. become 0 on simplification)

Cancelling *p*, $$\frac{\alpha(b^2-c^2)}{l} + \frac{\beta(c^2-a^2)}{m} + \frac{\gamma(a^2-b^2)}{n} = 0 \qquad ...(2)$$

Now equations of any normal through (α, β, γ) having d.c's l, m, n are

$$\frac{x-\alpha}{l} = \frac{y-\beta}{m} = \frac{z-\gamma}{n} \qquad ...(3)$$

To find the locus of normals given by (3), let us eliminate l, m, n from (2) and (3).

Putting the values of l, m, n from (3) in (2), the required locus of concurrent normals is

$$\frac{\alpha(b^2-c^2)}{x-\alpha} + \frac{\beta(c^2-a^2)}{y-\beta} + \frac{\gamma(a^2-b^2)}{z-\gamma} = 0.$$

14. Polar Plane of a Point

To find the equation of the polar plane of the point (x_1, y_1, z_1) w.r.t. the conicoid

$$ax^2 + by^2 + cz^2 = 1.$$

The given conicoid is $ax^2 + by^2 + cz^2 = 1$ ...(1)

Let $P(x_1, y_1, z_1)$ be the given point.

Through P draw any straight line to meet the conicoid in Q and R.

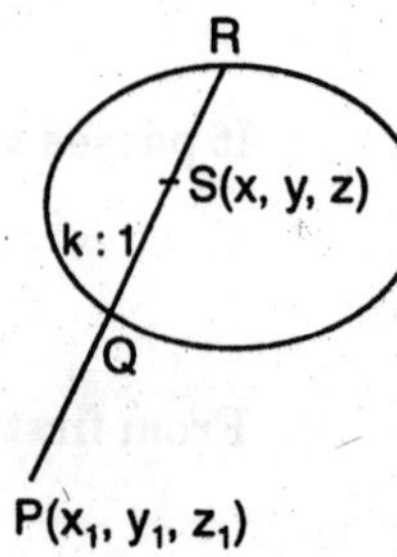

Let S be a point on this line such that PS is divided internally at Q and externally at R in the same ratio.

Then locus of S is called *polar plane* of P and P is called **pole** of polar plane.

Let S be (x, y, z).

Any point on PS which divides PS in the ratio $k : 1$ is

$$\left(\frac{kx+x_1}{k+1}, \frac{ky+y_1}{k+1}, \frac{kz+z_1}{k+1}\right)$$

If it lies on the conicoid (1),

$$\therefore \quad a\left(\frac{kx+x_1}{k+1}\right)^2 + b\left(\frac{ky+y_1}{k+1}\right)^2 + c\left(\frac{kz+z_1}{k+1}\right)^2 = 1$$

or $$a(kx + x_1)^2 + b(ky + y_1)^2 + c(kz + z_1)^2 = (k + 1)^2$$

or $$k^2(ax^2 + by^2 + cz^2 - 1) + 2k(axx_1 + byy_1 + czz_1 - 1) + (ax_1^2 + by_1^2 + cz_1^2 - 1) = 0 \quad ...(2)$$

These values of k will give the points Q and R.

These values of k are equal in magnitude, but opposite in sign.

($\because$ PS is divided at Q and R internally and externally in the same ratio).

$\therefore$ Sum of values of k in (2) is 0

i.e., co-eff. of $k = 0$

$$\therefore \quad axx_1 + byy_1 + czz_1 - 1 = 0$$

or $$\mathbf{axx_1 + byy_1 + czz_1 = 1}$$

which is the locus of S *i.e.*, polar plane of $P(x_1, y_1, z_1)$ w.r.t. conicoid (1).

Cor. If P lies on the conicoid, then polar plane of P becomes the tangent plane at P.

Rule to remember the equation of the polar plane

Change x^2 to xx_1, y^2 to yy_1, z^2 to zz_1.

15. Pole of a given Plane

To find the pole of the plane $lx + my + nz = p$ w.r.t. the conicoid $ax^2 + by^2 + cz^2 = 1$.

Equation of the conicoid is $ax^2 + by^2 + cz^2 = 1$ ...(1)

Equation of the plane is $lx + my + nz = p$ ...(2)

Let $P(x_1, y_1, z_1)$ be the pole of the plane (2) w.r.t. conicoid (1).

$\therefore$ Plane (2) is the polar plane of the point P w.r.t. conicoid (1).

Also equation of the polar plane of $P(x_1, y_1, z_1)$ w.r.t. conicoid (1) is

$$axx_1 + byy_1 + czz_1 = 1 \quad ...(3)$$

Equations (2) and (3) are identical

($\because$ They represent the polar plane of the same point)

Comparing (3) and (2), we have

$$\frac{ax_1}{l} = \frac{by_1}{m} = \frac{cz_1}{n} = \frac{1}{p}$$

$$\therefore \quad x_1 = \frac{l}{ap}, \; y_1 = \frac{m}{bp}, \; z_1 = \frac{n}{cp}$$

Thus the pole is $\left(\frac{\mathbf{l}}{\mathbf{ap}}, \frac{\mathbf{m}}{\mathbf{bp}}, \frac{\mathbf{n}}{\mathbf{cp}}\right)$.

Example. *Prove that the locus of the poles of the tangent planes to $a^2x^2 + b^2y^2 - c^2z^2 = 1$ with respect to $\alpha^2x^2 + \beta^2y^2 + \gamma^2z^2 = 1$ is the hyperboloid of one sheet. Find its equation.*

Sol. Let $P(x_1, y_1, z_1)$ be the pole of a tangent plane to the conicoid

$$a^2x^2 + b^2y^2 - c^2z^2 = 1 \quad ...(1)$$

w.r.t. the conicoid $$\alpha^2x^2 + \beta^2y^2 + \gamma^2z^2 = 1 \quad ...(2)$$

$\therefore$ Equation of the polar plane of $P(x_1, y_1, z_1)$ w.r.t. conicoid (2) is

[This polar plane is also a tangent plane to conicoid (1)]

$$\alpha^2 xx_1 + \beta^2 yy_1 + \gamma^2 zz_1 = 1$$

or
$$(\alpha^2 x_1)x + (\beta^2 y_1)y + (\gamma^2 z_1)z = 1 \qquad ...(3)$$

But plane (3) is a tangent plane to conicoid (1)

$$\therefore \quad \frac{l^2}{a} + \frac{m^2}{b} + \frac{n^2}{c} = p^2$$

$$\therefore \quad \frac{(\alpha^2 x_1)^2}{a^2} + \frac{(\beta^2 y_1)^2}{b^2} + \frac{(\gamma^2 z_1)^2}{-c^2} = 1$$

or
$$\frac{\alpha^4 x_1^2}{a^2} + \frac{\beta^4 y_1^2}{b^2} - \frac{\gamma^4 z_1^2}{c^2} = 1$$

$\therefore$ **Locus of the pole** (x_1, y_1, z_1) **is** $\dfrac{\alpha^4 x^2}{a^2} + \dfrac{\beta^4 y^2}{b^2} - \dfrac{\gamma^4 z^2}{c^2} = 1$

which is a hyperboloid of one sheet.

[$\because$ Co-efficients of x^2 and y^2 are positive but co-efficient of z^2 is negative]

16. Polar of a Line w.r.t. a Conicoid

To find the equations of the polar line of the line

$$\frac{x - x_1}{l} = \frac{y - y_1}{m} = \frac{z - z_1}{n}$$

w.r.t. the conicoid $ax^2 + by^2 + cz^2 = 1$. **(K.U. 2007)**

Def. **A line l_2 is said to be polar line of a line l_1 w.r.t. the conicoid if polar plane of every point on the line l_1 passes through the line l_2.**

Now the given conicoid is $ax^2 + by^2 + cz^2 = 1$ **...(1)**

and the given line is $\dfrac{x - x_1}{l} = \dfrac{y - y_1}{m} = \dfrac{z - z_1}{n}$ **...(2)**

Any point on the line (2) is $(lr + x_1, mr + y_1, nr + z_1)$

Polar plane of this point w.r.t. conicoid (1) is

$$ax(lr + x_1) + by(mr + y_1) + cz(nr + z_1) = 1$$

or
$$(axx_1 + byy_1 + czz_1 - 1) + r(alx + bmy + cnz) = 0$$

This plane passes through the line

$$\left.\begin{aligned} axx_1 + byy_1 + czz_1 - 1 = 0 \\ alx + bmy + cnz = 0 \end{aligned}\right] \qquad ...(3)$$

for all values of *r*.

Hence line (3) is the *polar line* of line (2) w.r.t. conicoid (1).

Example 1. *Show that the equations of the polar of the line* $\dfrac{x-1}{2} = \dfrac{y-2}{3} = \dfrac{z-3}{4}$ *w.r.t. the conicoid* $x^2 - 2y^2 + 3z^2 = 4$ *are* $\dfrac{x+6}{3} = \dfrac{y-2}{3} = z - 2$. **(M.D.U. 2007, 2003 S)**

Sol. **The given conicoid is** $x^2 - 2y^2 + 3z^2 = 4$ **...(1)**

and the given line is $\dfrac{x-1}{2} = \dfrac{y-2}{3} = \dfrac{z-3}{4}$ **...(2)**

Any point on line (2) is $(2r + 1, 3r + 2, 4r + 3)$

Polar plane of this point w.r.t. conicoid (1) is $x(2r + 1) - 2y(3r + 2) + 3z(4r + 3) = 4$

or $$(x - 4y + 9z - 4) + r(2x - 6y + 12z) = 0$$

This plane passes through the line

$$\left.\begin{aligned} x - 4y + 9z - 4 = 0 \\ 2x - 6y + 12z = 0 \end{aligned}\right] \quad ...(3)$$

for all values of r.

Hence line (3) is the polar line of line (2) w.r.t. conicoid (1).

Let us reduce the line (3) (which is in the general form) to Symmetrical Form.

To find d.r's of this line (3) [omitting constant terms in (3)],

we have $$x - 4y + 9z = 0$$
$$2x - 6y + 12z = 0$$

$$\therefore \quad \frac{x}{-48+54} = \frac{y}{18-12} = \frac{z}{-6+8}$$

or $$\frac{x}{6} = \frac{y}{6} = \frac{z}{2} \quad i.e., \quad \frac{x}{3} = \frac{y}{3} = \frac{z}{1}$$

$\therefore$ The d.r's of the polar line (3) are 3, 3, 1.

To get a point on the line (3), put $z = 2$ [As suggested by the answer]

$$\therefore \quad x - 4y + 14 = 0$$
$$2x - 6y + 24 = 0.$$

Solving for x and y, we have $x = -6, y = 2$

Also $z = 2$.

Hence one point on the polar line (3) is $(-6, 2, 2)$.

$\therefore$ The equations of the polar line (3) in symmetrical form are

$$\frac{x+6}{3} = \frac{y-2}{3} = \frac{z-2}{1}.$$

Example 2. *Find the locus of the straight lines drawn through a fixed point* (α, β, γ) *at right angles to its polars with respect to* $ax^2 + by^2 + cz^2 = 1$.

Sol. Equation of the conicoid is $ax^2 + by^2 + cz^2 = 1$ ...(1)

Equations of any line through (α, β, γ) are

$$\frac{x-\alpha}{l} = \frac{y-\beta}{m} = \frac{z-\gamma}{n} \quad ...(2)$$

Equation of the polar of line (2) w.r.t. conicoid (1) are (proceeding as in Art. 16).

$$\left.\begin{aligned} ax\alpha + by\beta + cz\gamma = 1 \\ alx + bmy + cnz = 0 \end{aligned}\right] \quad ...(3)$$

It is given that line (2) is $\perp$ to its polar line (3).

Now to find d.r's of line (3), let us omit constant terms in (3).

$$\therefore \quad a\alpha x + b\beta y + c\gamma z = 0$$
$$alx + bmy + cnz = 0$$

$$\therefore \quad \frac{x}{bc(n\beta - m\gamma)} = \frac{y}{ca(l\gamma - n\alpha)} = \frac{z}{ab(m\alpha - l\beta)}$$

$\therefore$ d.r's of line (3) are $bc(n\beta - m\gamma), ca(l\gamma - n\alpha), ab(m\alpha - l\beta)$

But lines (1) and (3) are ⊥

$$\therefore \quad l.bc(n\beta - m\gamma) + m\,.\,ca(l\gamma - n\alpha) + n.ab(m\alpha - l\beta) = 0$$

or
$$\alpha mna(b-c) + \beta nlb(c-a) + \gamma lmc(a-b) = 0$$

Dividing by lmn,
$$\frac{\alpha a(b-c)}{l} + \frac{\beta b(c-a)}{m} + \frac{\gamma c(a-b)}{n} = 0 \quad ...(4)$$

To find the locus of the lines (2), let us eliminate l, m, n from (2) and (4).

$$\therefore \quad \frac{\alpha a(b-c)}{x-\alpha} + \frac{\beta b(c-a)}{y-\beta} + \frac{\gamma c(a-b)}{z-\gamma} = 0$$

which is the required locus.

EXERCISE 2

1. Find the equation of the polar plane of P(x_1, y_1, z_1) w.r.t. the ellipsoid
$$\frac{x^2}{a^2} + \frac{y^2}{b^2} + \frac{z^2}{c^2} = 1.$$
2. Find the condition for the two points (x_1, y_1, z_1) and (x_2, y_2, z_2) to be conjugate w.r.t. the conicoid
$$ax^2 + by^2 + cz^2 = 1.$$
[**Hint.** Two points are said to be **conjugate** if the polar plane of each passes through the other.]
3. Prove that the condition for the planes $l_1x + m_1y + n_1z = p_1$, $l_2x + m_2y + n_2z = p_2$ to be conjugate with respect to the conicoid $ax^2 + by^2 + cz^2 = 1$, is
$$\frac{l_1l_2}{a} + \frac{m_1m_2}{b} + \frac{n_1n_2}{c} = p_1p_2.$$
[**Hint.** Two planes are said to be **conjugate** planes w.r.t. a conicoid if the pole of each lies on the other.
Find the **pole** of the first plane w.r.t. the conicoid (Reproduce Art. 15) and put the co-ordinates of this pole in the equation of second plane.]
4. If A, B are the points (x_1, y_1, z_1) and (x_2, y_2, z_2) ; prove that the polar line of AB w.r.t. $ax^2 + by^2 + cz^2 = 1$ is given by $axx_1 + byy_1 + czz_1 = 1$, $axx_2 + byy_2 + czz_2 = 1$.
5. Find the equations of the polar of the line $\frac{x-1}{5} = \frac{y-3}{7} = \frac{z+5}{2}$ w.r.t. the conicoid $x^2 + 3y^2 - 7z^2 - 21 = 0$ in symmetrical form. *(M.D.U. 1996)*

Answers

1. $\frac{xx_1}{a^2} + \frac{yy_1}{b^2} + \frac{zz_1}{c^2} = 1$ 2. $ax_1x_2 + by_1y_2 + cz_1z_2 = 1$ 5. $\frac{x + \frac{441}{24}}{861} = \frac{y - \frac{105}{24}}{-189} = \frac{z}{24}$.

17. Enveloping Cone

To find the equation of the enveloping cone from the point (x_1, y_1, z_1) to the ellipsoid
$$\frac{x^2}{a^2} + \frac{y^2}{b^2} + \frac{z^2}{c^2} = 1.$$

Def. Enveloping Cone is the locus of the tangent lines drawn from a given point to a surface (here ellipsoid).

The equation of the ellipsoid is $\frac{x^2}{a^2}+\frac{y^2}{b^2}+\frac{z^2}{c^2}=1$ ...(1)

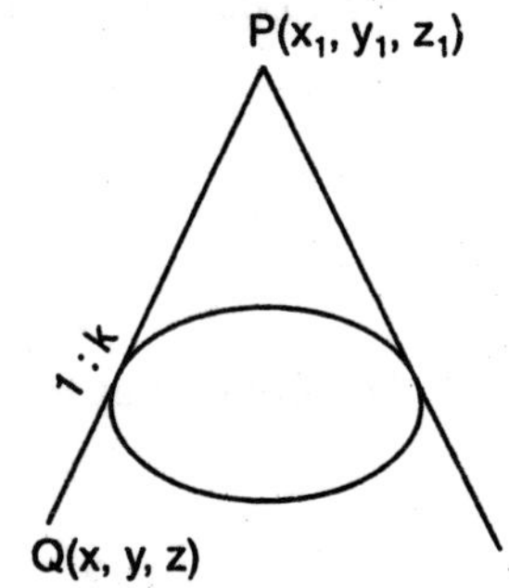

Let P be the point (x_1, y_1, z_1).

Let $Q(x, y, z)$ be *any* point on a tangent line from P to the ellipsoid.

Any point on tangent PQ is (dividing PQ in the ratio $k : 1$)

$$\left(\frac{kx+x_1}{k+1}, \frac{ky+y_1}{k+1}, \frac{kz+z_1}{k+1}\right).$$

If it lies on ellipsoid (1), then

$$\frac{1}{a^2}\left(\frac{kx+x_1}{k+1}\right)^2+\frac{1}{b^2}\left(\frac{ky+y_1}{k+1}\right)^2+\frac{1}{c^2}\left(\frac{kz+z_1}{k+1}\right)^2=1$$

or $$\frac{1}{a^2}(kx+x_1)^2+\frac{1}{b^2}(ky+y_1)^2+\frac{1}{c^2}(kz+z_1)^2=(k+1)^2$$

or $$k^2\left(\frac{x^2}{a^2}+\frac{y^2}{b^2}+\frac{z^2}{c^2}-1\right)+2k\left(\frac{xx_1}{a^2}+\frac{yy_1}{b^2}+\frac{zz_1}{c^2}-1\right)+\left(\frac{x_1^2}{a^2}+\frac{y_1^2}{b^2}+\frac{z_1^2}{c^2}-1\right)=0 \quad ...(2)$$

Since the line PQ touches the conicoid (1)

$\therefore$ (2) must have equal roots

$\therefore$ Disc = 0 *i.e.,* $b^2 = 4ac$

$$\therefore \quad 4\left(\frac{xx_1}{a^2}+\frac{yy_1}{b^2}+\frac{zz_1}{c^2}-1\right)^2=4\left(\frac{x^2}{a^2}+\frac{y^2}{b^2}+\frac{z^2}{c^2}-1\right)\left(\frac{x_1^2}{a^2}+\frac{y_1^2}{b^2}+\frac{z_1^2}{c^2}-1\right)$$

or cancelling 4, $$\left(\frac{x^2}{a^2}+\frac{y^2}{b^2}+\frac{z^2}{c^2}-1\right)\left(\frac{x_1^2}{a^2}+\frac{y_1^2}{b^2}+\frac{z_1^2}{c^2}-1\right)=\left(\frac{xx_1}{a^2}+\frac{yy_1}{b^2}+\frac{zz_1}{c^2}-1\right)^2$$

which is the required equation of the enveloping cone.

Rule to Remember the equation of Enveloping Cone $SS_1 = T^2$

where $S \equiv \frac{x^2}{a^2}+\frac{y^2}{b^2}+\frac{z^2}{c^2}-1$, $S_1 \equiv \frac{x_1^2}{a^2}+\frac{y_1^2}{b^2}+\frac{z_1^2}{c^2}-1$, $T \equiv \frac{xx_1}{a^2}+\frac{yy_1}{b^2}+\frac{zz_1}{c^2}-1$.

Exercise. Show that the equation of the enveloping cone from the point (x_1, y_1, z_1) to the conicoid $ax^2 + by^2 + cz^2 = 1$ is

$$(ax^2 + by^2 + cz^2 - 1)\,(ax_1^2 + by_1^2 + cz_1^2 - 1) = (axx_1 + byy_1 + czz_1 - 1)^2.$$

(*K.U. 1997*)

Or

Find the locus of the tangents drawn from a given point (α, β, γ) to the conicoid $ax^2 + by^2 + cz^2 = 1$. (*M.D.U. 2003*)

Example 1. *The enveloping cone of the ellipsoid* $\frac{x^2}{a^2}+\frac{y^2}{b^2}+\frac{z^2}{c^2}=1$ *is cut by the plane* $z = 0$ *in a rectangular hyperbola. Prove that the locus of the vertex of the cone is*

$$\frac{x^2+y^2}{a^2+b^2}+\frac{z^2}{c^2}=1.$$

Sol. Let $P(x_1, y_1, z_1)$ be the vertex of an enveloping cone of the ellipsoid

$$\frac{x^2}{a^2}+\frac{y^2}{b^2}+\frac{z^2}{c^2}=1$$

or
$$S=\frac{x^2}{a^2}+\frac{y^2}{b^2}+\frac{z^2}{c^2}-1=0$$

We know that equation of the enveloping cone is $SS_1 = T^2$

or
$$\left(\frac{x^2}{a^2}+\frac{y^2}{b^2}+\frac{z^2}{c^2}-1\right)\left(\frac{x_1^2}{a^2}+\frac{y_1^2}{b^2}+\frac{z_1^2}{c^2}-1\right)=\left(\frac{xx_1}{a^2}+\frac{yy_1}{b^2}+\frac{zz_1}{c^2}-1\right)^2 \quad ...(1)$$

To take the section of enveloping cone (1) by plane $z = 0$ (given), let us put $z = 0$ in (1).

$$\therefore \quad \left(\frac{x^2}{a^2}+\frac{y^2}{b^2}-1\right)\left(\frac{x_1^2}{a^2}+\frac{y_1^2}{b^2}+\frac{z_1^2}{c^2}-1\right)=\left(\frac{xx_1}{a^2}+\frac{yy_1}{b^2}-1\right)^2 \quad ...(2)$$

Because Eqn. (2) represents a rectangular hyperbola (given), therefore

coeff. of x^2 + coeff. of $y^2 = 0$

$$\therefore \quad \frac{1}{a^2}\left(\frac{x_1^2}{a^2}+\frac{y_1^2}{b^2}+\frac{z_1^2}{c^2}-1\right)-\frac{x_1^2}{a^4}+\frac{1}{b^2}\left(\frac{x_1^2}{a^2}+\frac{y_1^2}{b^2}+\frac{z_1^2}{c^2}-1\right)-\frac{y_1^2}{b^4}=0$$

or
$$\frac{x_1^2}{a^4}+\frac{y_1^2}{a^2b^2}+\frac{z_1^2}{a^2c^2}-\frac{1}{a^2}-\frac{x_1^2}{a^4}+\frac{x_1^2}{a^2b^2}+\frac{y_1^2}{b^4}+\frac{z_1^2}{b^2c^2}-\frac{1}{b^2}-\frac{y_1^2}{b^4}=0$$

or
$$\frac{x_1^2}{a^2b^2}+\frac{y_1^2}{a^2b^2}+\frac{z_1^2}{a^2c^2}+\frac{z_1^2}{b^2c^2}=\frac{1}{a^2}+\frac{1}{b^2}$$

or
$$\frac{x_1^2+y_1^2}{a^2b^2}+z_1^2\left(\frac{b^2+a^2}{a^2b^2c^2}\right)=\frac{a^2+b^2}{a^2b^2}$$

$\therefore$ Locus of (x_1, y_1, z_1) is

$$\frac{x^2+y^2}{a^2b^2}+z^2\frac{(a^2+b^2)}{a^2b^2c^2}=\frac{a^2+b^2}{a^2b^2}$$

Dividing by $(a^2 + b^2)$ and multiplying by a^2b^2,

$$\frac{x^2+y^2}{a^2+b^2}+\frac{z^2}{c^2}=1$$

which is the required equation of the locus of the vertex of the cone.

Example 2. *Find the locus of points from which three mutually perpendicular tangents can be drawn to the surface $ax^2 + by^2 + cz^2 = 1$.*

Sol. Let $P(x_1, y_1, z_1)$ be the point whose locus is required. Equation of the enveloping cone of w.r.t. conicoid $ax^2 + by^2 + cz^2 = 1$.

(Each generator of this enveloping cone is a tangent from P to the conicoid) is $(ax^2 + by^2 + cz^2 - 1)(ax_1^2 + by_1^2 + cz_1^2 - 1) = (axx_1 + byy_1 + czz_1 - 1)^2$ ...(1)

$(SS_1 = T^2)$

But three of the tangents drawn from P are mutually ⊥ (given).

∴ Cone (1) has three mutually ⊥ generators.

∴ **Co-eff. of x² + co-eff. of y² + co-eff. of z² = 0.**

i.e., $a(by_1^2 + cz_1^2 - 1) + b(ax_1^2 + cz_1^2 - 1) + c(ax_1^2 + by_1^2 - 1) = 0$

i.e., $a(b + c)\, x_1^2 + b(c + a)y_1^2 + c(a + b)\, z_1^2 = a + b + c$

∴ Locus of $P(x_1, y_1, z_1)$ is [changing (x_1, y_1, z_1) to (x, y, z)]

$$a(b + c)x^2 + b(c + a)y^2 + c(a + b)z^2 = a + b + c.$$

18. Enveloping Cylinder

To find the equation of the enveloping cylinder of the conicoid $ax^2 + by^2 + cz^2 = 1$ whose generators are parallel to the line $\frac{x}{l} = \frac{y}{m} = \frac{z}{n}$.

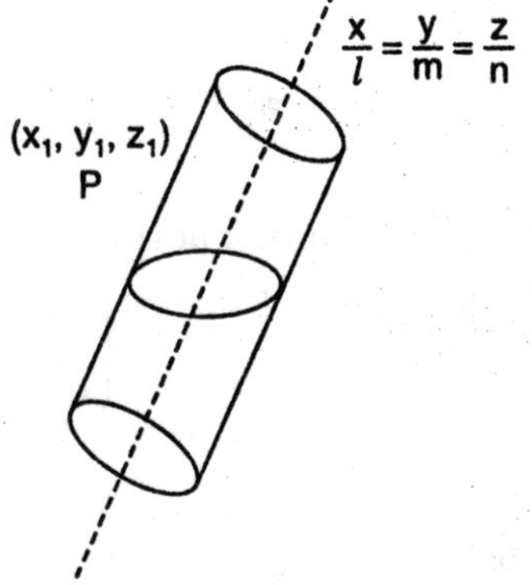

The Enveloping cylinder of the conicoid is the locus of the tangent lines to the conicoid drawn parallel to a given line which here is $\frac{x}{l} = \frac{y}{m} = \frac{z}{n}$.

Let P (x_1, y_1, z_1) be ANY point on the cylinder.

∴ Equations of the generator of the enveloping cylinder through $P(x_1, y_1, z_1)$ parallel to the line

$$\frac{x}{l} = \frac{y}{m} = \frac{z}{n} \quad \text{are} \quad \frac{x - x_1}{l} = \frac{y - y_1}{m} = \frac{z - z_1}{n} \qquad ...(1)$$

∴ Line (1) is a tangent line (being a generator of the enveloping cylinder) to the conicoid

$$ax^2 + by^2 + cz^2 = 1 \qquad ...(2)$$

Any point on line (1) is $(lr + x_1, mr + y_1, nr + z_1)$

If it lies on conicoid (2), then $a(lr + x_1)^2 + b(mr + y_1)^2 + c(nr + z_1)^2 = 1$

or $r^2(al^2 + bm^2 + cn^2) + 2r(alx_1 + bmy_1 + cnz_1) + (ax_1^2 + by_1^2 + cz_1^2 - 1) = 0$ ...(3)

Since the line (1) touches the conicoid (2).

∴ (3) has equal roots

∴ $4(alx_1 + bmy_1 + cnz_1)^2 - 4(al^2 + bm^2 + cn^2)(ax_1^2 + by_1^2 + cz_1^2 - 1) = 0$

| Using $b^2 - 4ac = 0$

or cancelling 4, $(alx_1 + bmy_1 + cnz_1)^2 = (al^2 + bm^2 + cn^2)(ax_1^2 + by_1^2 + cz_1^2 - 1)$

∴ Locus of (x_1, y_1, z_1) is

$$(ax^2 + by^2 + cz^2 - 1)\,(al^2 + bm^2 + cn^2) = (alx + bmy + cnz)^2 \qquad ...(4)$$

which is the required equation of the enveloping cylinder.

Method to write down the equation of the enveloping cylinder. Let S denote the L.H.S. of the equation of the conicoid (after making the R.H.S. zero)

i.e., $S \equiv ax^2 + by^2 + cz^2 - 1$;

then $s_1 = al^2 + bm^2 + cn^2$

i.e., s_1 is obtained by putting l, m, n in S and omitting the constant term.

$$t = alx + bmy + cnz,$$

where t is the expression for the tangent plane at (l, m, n) after omitting the constant term.

$\therefore$ Equation (4) of the enveloping cylinder can be put as $Ss_1 = t^2$.

Example. *Prove that the enveloping cylinder of the ellipsoid* $\frac{x^2}{a^2}+\frac{y^2}{b^2}+\frac{z^2}{c^2} = 1$ *whose generators are parallel to the lines* $\frac{x}{0} = \frac{y}{\pm\sqrt{a^2-b^2}} = \frac{z}{c}$ *meets the plane* $z = 0$ *in a circle.*

Sol. $\because$ The generators of the enveloping cylinder of the ellipsoid $\frac{x^2}{a^2}+\frac{y^2}{b^2}+\frac{z^2}{c^2} = 1$ are parallel to the lines

$$\frac{x}{0} = \frac{y}{\pm\sqrt{a^2-b^2}} = \frac{z}{c}.$$

$$\therefore \qquad l = 0, \quad m = \pm\sqrt{a^2-b^2}, \quad n = c.$$

Equation of the enveloping cylinder of ellipsoid is $Ss_1 = t^2$

or $$\left(\frac{x^2}{a^2}+\frac{y^2}{b^2}+\frac{z^2}{c^2}-1\right)\left(\frac{0^2}{a^2}+\frac{a^2-b^2}{b^2}+\frac{c^2}{c^2}\right) = \left(\frac{x\,.\,0}{a^2} \pm y\frac{\sqrt{a^2-b^2}}{b^2}+\frac{zc}{c^2}\right)^2$$

or $$\left(\frac{x^2}{a^2}+\frac{y^2}{b^2}+\frac{z^2}{c^2}-1\right)\left(\frac{a^2-b^2}{b^2}+1\right) = \left(\pm\frac{\sqrt{a^2-b^2}}{b^2}\,y+\frac{z}{c}\right)^2$$

This meets the plane $z = 0$

where $$\left(\frac{x^2}{a^2}+\frac{y^2}{b^2}-1\right)\left(\frac{a^2}{b^2}\right) = \left(\pm\frac{\sqrt{a^2-b^2}}{b^2}\,y+0\right)^2$$

or $$\frac{x^2}{b^2}+\frac{a^2y^2}{b^4}-\frac{a^2}{b^2} = \frac{(a^2-b^2)}{b^4}\,y^2$$

or $$\frac{x^2}{b^2}+\frac{a^2y^2}{b^4}-\frac{a^2}{b^2} = \frac{a^2y^2}{b^4}-\frac{y^2}{b^2}$$

or $$\frac{x^2}{b^2}+\frac{y^2}{b^2} = \frac{a^2}{b^2}$$

or $$x^2+y^2 = a^2 \quad \text{which is a circle}$$

19. Section with a Given Centre

To find the locus of the chords of the conicoid $ax^2 + by^2 + cz^2 = 1$ *which are bisected at a given point* (x_1, y_1, z_1). (K.U. 1995 A)

The given conicoid is $\quad ax^2 + by^2 + cz^2 = 1$ ...(1)

Let $< l, m, n >$ be the **direction cosines** of one of the chords of the conicoid which are bisected at $P(x_1, y_1, z_1)$.

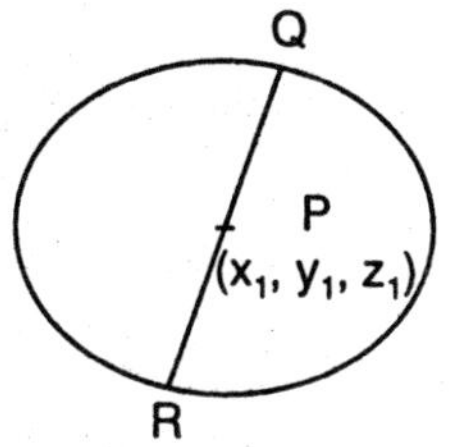

∴ Equations of any line through P are

$$\frac{x-x_1}{l}=\frac{y-y_1}{m}=\frac{z-z_1}{n}=r \text{ (say)} \quad ...(2)$$

where r is the distance of the point P(x_1, y_1, z_1) from any point (x, y, z).

Any point (x, y, z) on this line is $(lr + x_1, mr + y_1, nr + z_1)$

If it lies on (1), then $a(lr + x_1)^2 + b(mr + y_1)^2 + c(nr + z_1)^2 = 1$

or $r^2(al^2 + bm^2 + cn^2) + 2r(alx_1 + bmy_1 + cnz_1) + (ax_1^2 + by_1^2 + cz_1^2 - 1) = 0$ ...(3)

Equation (3) gives two values of r corresponding to which there are two points of intersection (say Q and R) of the conicoid and the line.

These two values of r namely PQ and PR are equal and opposite.

[∵ P is the mid-point of QR]

∴ Sum of values of r given by equation (3) is 0

∴ Co-eff. of $r = 0$ giving $alx_1 + bmy_1 + cnz_1 = 0$ ...(4)

To find the locus of the chords given by (2), let us eliminate l, m, n from (2) and (4).

Putting the values of l, m, n from (2) in (4), the locus of the chords is

$$ax_1(x - x_1) + by_1(y - y_1) + cz_1(z - z_1) = 0$$

or
$$\mathbf{axx_1 + byy_1 + czz_1 = ax_1^2 + by_1^2 + cz_1^2}$$

or
$$axx_1 + byy_1 + czz_1 - 1 = ax_1^2 + by_1^2 + cz_1^2 - 1$$

which is a plane and is of the form $\mathbf{T = S_1}$.

Note. ∴ The locus of chords of the conicoid $ax^2 + by^2 + cz^2 = 1$ which are bisected at the point (x_1, y_1, z_1) is the plane

$$\mathbf{axx_1 + byy_1 + czz_1 = ax_1^2 + by_1^2 + cz_1^2}$$

This plane is called the **plane section** of the conicoid with centre at (x_1, y_1, z_1).

The point (x_1, y_1, z_1) is called the **centre** of this plane section. This plane meets the conicoid in a conic whose centre is (x_1, y_1, z_1).

Example 1. *Find the locus of centres of all plane sections of a conicoid which pass through a given line.*

Sol. Let (x_1, y_1, z_1) be the centre of one of plane sections of the conicoid

$$ax^2 + by^2 + cz^2 = 1 \quad ...(1)$$

which pass through a given line

$$\frac{x-\alpha}{l}=\frac{y-\beta}{m}=\frac{z-\gamma}{n} \quad ...(2)$$

The equation of the plane with (x_1, y_1, z_1) as centre is

$$axx_1 + byy_1 + czz_1 - 1 = ax_1^2 + by_1^2 + cz_1^2 - 1 \quad | \text{ Using } T = S_1$$

or
$$axx_1 + byy_1 + czz_1 = ax_1^2 + by_1^2 + cz_1^2 \quad ...(3)$$

Now plane section (3) passes through the line (2)

∴ $(ax_1)l + (by_1)m + (cz_1)n = 0$ ...(4)

| $al + bm + cn = 0$

and also one point (α, β, γ) of line (2) (In fact, every point) lies on the plane (3).

$$\therefore \qquad a\alpha x_1 + b\beta y_1 + c\gamma z_1 = ax_1^{\,2} + by_1^{\,2} + cz_1^{\,2} \qquad \text{...(5)}$$

From (4) and (5), the locus of (x_1, y_1, z_1) is

$$alx + bmy + cnz = 0$$

$$ax^2 + by^2 + cz^2 = a\alpha x + b\beta y + c\gamma z$$

which being the intersection of a plane and a conicoid represents a conic.

Example 2. ***Prove that the section of the ellipsoid $\frac{x^2}{a^2}+\frac{y^2}{b^2}+\frac{z^2}{c^2}=1$ whose centre is the point $\left(\frac{a}{3},\frac{b}{3},\frac{c}{3}\right)$ passes through the three extremities of its principal axes.*** **(M.D.U. 2003)**

Sol. **Equation of the ellipsoid is** $\quad \frac{x^2}{a^2}+\frac{y^2}{b^2}+\frac{z^2}{c^2}=1 \qquad$ **...(1)**

We know that principal axes of ellipsoid (1) are x-axis ; y-axis and z-axis respectively.

(See Art. "Appendix for Chapter 4" at the end of this chapter)

The ellipsoid (1) meets the x-axis where putting

$$y = 0 \text{ and } z = 0 \text{ in (1)}, \quad \frac{x^2}{a^2}=1 \quad \text{or} \quad x^2 = a^2 \quad \text{or} \quad x = \pm a$$

$\therefore$ **The two extremities of one principal axis (x-axis) of ellipsoid (1) are A(a, 0, 0) and A′ ($-a$, 0, 0).**

Similarly, the other four extremities of the other two principal axis (y-axis and z-axis) of ellipsoid (1) are B(0, b, 0), B′(0, $-b$, 0), C(0, 0, c), C′(0, 0, $-c$).

Again Equation of plane section of ellipsoid (1) with centre at $\left(\frac{a}{3},\frac{b}{3},\frac{c}{3}\right)$ is

$$T = S_1 \qquad \text{(By Art. 19)}$$

or
$$\frac{x\cdot\frac{a}{3}}{a^2}+\frac{y\cdot\frac{b}{3}}{b^2}+\frac{z\cdot\frac{c}{3}}{c^2}-1=\frac{\left(\frac{a}{3}\right)^2}{a^2}+\frac{\left(\frac{b}{3}\right)^2}{b^2}+\frac{\left(\frac{c}{3}\right)^2}{c^2}-1$$

or
$$\frac{x}{3a}+\frac{y}{3b}+\frac{z}{3c}=\frac{1}{9}+\frac{1}{9}+\frac{1}{9}=\frac{3}{9}=\frac{1}{3}$$

Multiplying by 3, $\quad \frac{x}{a}+\frac{y}{b}+\frac{z}{c}=1 \qquad$ **...(2)**

Plane section (2) is clearly passing through three extremities A(a, 0, 0), B(0, b, 0) and C(0, 0, c) out of the six extremities.

20. Diameter Plane Property

To find the locus of the mid-points of a system of parallel chords of the conicoid

$$ax^2 + by^2 + cz^2 = 1.$$

The given conicoid is

$$ax^2 + by^2 + cz^2 = 1 \qquad \text{...(1)}$$

Let <l, m, n> be the direction cosines of the system of parallel chords.

Let $P(x_1, y_1, z_1)$ be the mid-point of one of these parallel chords.

Equations of any line through P are

$$\frac{x-x_1}{l} = \frac{y-y_1}{m} = \frac{z-z_1}{n} = r \text{ (say)} \qquad ...(2)$$

where r is the distance of the point $P(x_1, y_1, z_1)$ from any point (x, y, z).

Any point (x, y, z) on this line is $(lr + x_1, mr + y_1, nr + z_1)$

If it lies on (1), then

$$a(lr + x_1)^2 + b(mr + y_1)^2 + c(nr + z_1)^2 = 1$$

or
$$r^2(al^2 + bm^2 + cn^2) + 2r(alx_1 + bmy_1 + cnz_1) + (ax_1^2 + by_1^2 + cz_1^2 - 1) = 0 \qquad ...(3)$$

Equation (3) gives two values of r corresponding to which there are two points of intersection say Q and R of the conicoid and the line.

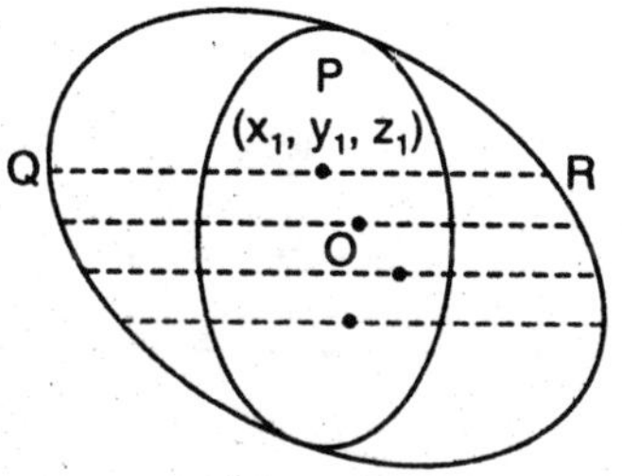

These two values of r namely PQ and PR are equal and opposite. [∵ P is the mid-point of QR]

∴ Sum of values of r gives by equation (3) is 0.

∴ Co-eff. of $r = 0$ giving $alx_1 + bmy_1 + cnz_1 = 0$

∴ Locus of $P(x_1, y_1, z_1)$ is $alx + bmy + cnz = 0$ ...(4)

Note. It is a plane through the centre (0, 0, 0) of the conicoid (1).

EXERCISE 3

1. The section of the enveloping cone of the ellipsoid $\frac{x^2}{a^2} + \frac{y^2}{b^2} + \frac{z^2}{c^2} = 1$ whose vertex is P, by the plane $z = 0$ is a parabola. Find the locus of P.

   [**Hint.** For a **parabola** $AB = H^2$ where A is co-efficient of x^2, B is the co-eff. of y^2 and $H = \frac{1}{2}$ co-eff. of xy.]

2. Find the locus of a luminous point if the ellipsoid $\frac{x^2}{a^2} + \frac{y^2}{b^2} + \frac{z^2}{c^2} = 1$ casts a circular shadow on the plane $z = 0$.

3. Find the equation of the plane section with centre (x_1, y_1, z_1) of the conicoid $ax^2 + by^2 + cz^2 = 1$.
   [**Hint.** It is Art. 19.]

4. Find the equation of the plane section with centre (x_1, y_1, z_1) of the ellipsoid $\frac{x^2}{a^2} + \frac{y^2}{b^2} + \frac{z^2}{c^2} = 1$.

   [**Hint.** Do as Art. 19.]

5. (*a*) Find the equation to the plane which cuts the surface $2x^2 + 3y^2 + 5z^2 = 4$ in a conic whose centre is at the point (1, 2, 3).

   (*b*) Show that the equation of the plane which cuts the conicoid $x^2 + 4y^2 - 5z^2 = 1$ in a conic with centre at (2, 3, 4) is $x + 6y - 10z + 20 = 0$.

6. Find the centre of the conic given by the equations $2x - 2y - 5z + 5 = 0$, $3x^2 + 2y^2 - 15z^2 = 4$.

7. Show that the centre of the conic $lx + my + nz = p$, $ax^2 + by^2 + cz^2 = 1$ is the point $\left(\frac{lp}{ap_0^2}, \frac{mp}{bp_0^2}, \frac{np}{cp_0^2}\right)$

where $p_0^2 = \frac{l^2}{a} + \frac{m^2}{b} + \frac{n^2}{c}$.

8. Prove that the locus of the centres of all plane sections of a conicoid which pass through a fixed point is a conicoid.

9. Find the locus of the centres of section of the conicoid $ax^2 + by^2 + cz^2 = 1$ which touch the conicoid $\alpha x^2 + \beta y^2 + \gamma z^2 = 1$.

Answers

1. $z = \pm c$ 2. $x = 0, \frac{y^2}{b^2 - a^2} + \frac{z^2}{c^2} = 1$ or $y = 0, \frac{x^2}{a^2 - b^2} + \frac{z^2}{c^2} = 1$

5. (a) $2x + 6y + 15z = 59$ 6. $(-2, 3, -1)$

9. $(ax^2 + by^2 + cz^2)^2 = \frac{a^2x^2}{\alpha} + \frac{b^2y^2}{\beta} + \frac{c^2z^2}{\gamma}$.

SECTION II. PARABOLOID

21. Introduction

We have already discussed in Art. 2, that the equation **$ax^2 + by^2 = 2z$** represents a **paraboloid.**

(For tracing of "Paraboloid" see Appendix at the end of this Chapter).

If a and b are both +ve, then $ax^2 + by^2 = 2z$ represents an **elliptic paraboloid** (standard equation of **elliptic paraboloid is** $\left(\frac{x^2}{a^2} + \frac{y^2}{b^2} = 2z\right)$.

If one of the two a and b is positive and the other is negative, then the equation $ax^2 + by^2 = 2z$ represents a **hyperbolic paraboloid** (Standard equation of **hyperboloid** is $\frac{x^2}{a^2} - \frac{y^2}{b^2} = 2z$).

22. Intersections of a Line and a Paraboloid

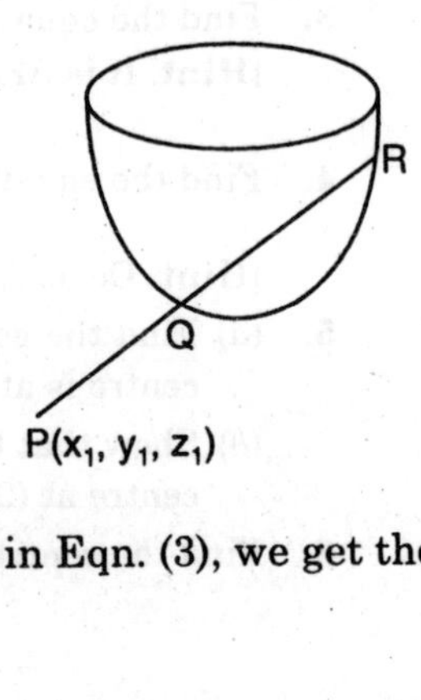

Let the line be $\frac{x - x_1}{l} = \frac{y - y_1}{m} = \frac{z - z_1}{n}$ ...(1)

and the paraboloid be $ax^2 + by^2 = 2z$ ...(2)

Any point on the line (1) is $(lr + x_1, mr + y_1, nr + z_1)$ ...(3)

If it lies on paraboloid (2), then

$$a(lr + x_1)^2 + b(mr + y_1)^2 = 2(nr + z_1)$$

i.e., $r^2(al^2 + bm^2) + 2r(alx_1 + bmy_1 - n) + (ax_1^2 + by_1^2 - 2z_1) = 0$ ...(4)

Eqn. (4) gives two values of r and putting these values of r in Eqn. (3), we get the co-ordinates of the two points of intersection.

Remark. $\therefore$ Every plane section of the paraboloid is a conic.

Note. If the line (1) is parallel to z-axis, then

$$l = 0,\ m = 0,\ n = 1.$$

$\therefore$ The quadratic (4) in r reduces to $2r(-1) + (ax_1^2 + by_1^2 - 2z_1) = 0$ ...(5)
(which is a linear in r ; showing that one value of r is infinite.)

$\therefore$ Any line parallel to z-axis meets the paraboloid in one point at an infinite distance from P and in the other point Q whose distance from P is given by

$$r = \frac{ax_1^2 + by_1^2 - 2z_1^2}{2} \qquad \text{| From (5)}$$

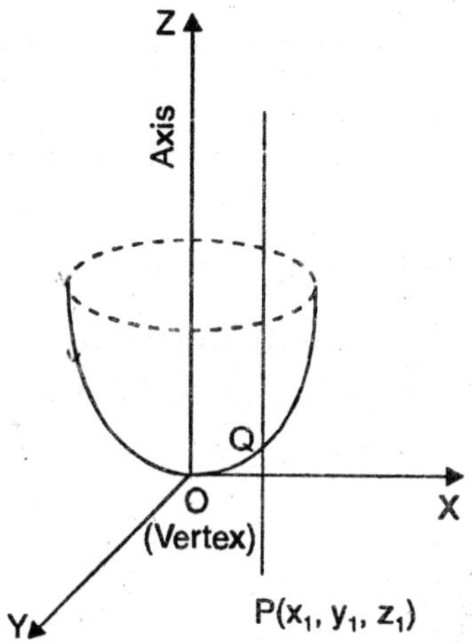

Def. *If one point of intersection of a line and a paraboloid extends to infinity, then the line is called a* **diameter.**

Thus a line || to OZ is a diameter of the paraboloid

$$ax^2 + by^2 = 2z.$$

23. Some Standard Results about the Paraboloid

Following are some of the results about a paraboloid which can be easily proved by seeing the corresponding articles in section I.

These results are important and the **reader is suggested to prove these results for himself or herself.**

Let the paraboloid be $ax^2 + by^2 = 2z$ ...(1)

(*i*) The equation of the **tangent plane** to (1) at (x_1, y_1, z_1) is

$$\mathbf{axx_1 + byy_1 = z + z_1}$$

$$\left(\text{Changing } x^2 \text{ to } xx_1,\ y^2 \text{ to } yy_1,\ z \text{ to } \frac{z+z_1}{2}\right)$$

(*ii*) The condition for the plane $lx + my + nz = p$ to be a **tangent plane** to paraboloid (1) is

$$\frac{\mathbf{l^2}}{\mathbf{a}} + \frac{\mathbf{m^2}}{\mathbf{b}} = -\,\mathbf{2np}$$

and the point of contact is $\left(-\dfrac{l}{an}, -\dfrac{m}{bn}, -\dfrac{p}{n}\right)$

(*iii*) The equation of the **polar plane** of the point (x_1, y_1, z_1) w.r.t. (1) is

$$\mathbf{axx_1 + byy_1 = z + z_1}.$$

(*iv*) **The enveloping cone** of the point (x_1, y_1, z_1) w.r.t. paraboloid (1) is

$$\mathbf{SS_1 = T^2}$$

or $$(ax^2 + by^2 - 2z)(ax_1^2 + by_1^2 - 2z_1) = [axx_1 + byy_1 - (z + z_1)]^2.$$

(*v*) The **plane section** of paraboloid (1) with **centre** (x_1, y_1, z_1) is

$$\mathbf{T = S_1}$$

i.e., $$axx_1 + byy_1 - (z + z_1) = ax_1^2 + by_1^2 - 2z_1.$$

Example 1. *Show that the plane $2x - 4y - z + 3 = 0$ touches the paraboloid $x^2 - 2y^2 = 3z$ and find the co-ordinates of the point of contact.* **(K.U. 2007)**

Sol. Equation of the plane is $2x - 4y - z + 3 = 0$

i.e., $$2x - 4y - z = -3 \quad \text{...(1)}$$

Comparing with $lx + my + nz = p$, we have

$$l = 2,\ m = -4,\ n = -1,\ p = -3$$

Equation of the paraboloid is $x^2 - 2y^2 = 3z$ ...(2)

Dividing both sides by 3, $\frac{1}{3}x^2 - \frac{2}{3}y^2 = z$

Multiplying both sides by 2, $\frac{2}{3}x^2 - \frac{4}{3}y^2 = 2z$ **(Note)**

Comparing with $ax^2 + by^2 = 2z$, we have $a = \frac{2}{3}, b = -\frac{4}{3}$

Condition of tangency

Plane (1) will touch paraboloid (2) if

$$\frac{l^2}{a} + \frac{m^2}{b} = -2np \quad | \text{ By Art. 23}$$

i.e., if $$\frac{4}{\left(\frac{2}{3}\right)} + \frac{16}{\left(-\frac{4}{3}\right)} = -2(-1)(-3)$$

i.e., if $$6 - 12 = -6$$

i.e., if $$-6 = -6 \text{ which is true.}$$

$\therefore$ Plane (1) touches paraboloid (2)

To find point of contact

Let plane (1) touch paraboloid (2) at (x_1, y_1, z_1).

Equation of the tangent plane at (x_1, y_1, z_1) to paraboloid (2) is

$$xx_1 - 2yy_1 = 3\left(\frac{z + z_1}{2}\right)$$

$$\left[\text{Change } x^2 \text{ to } xx_1,\ y^2 \text{ to } yy_1,\ x \text{ to } \frac{x + x_1}{2}\right]$$

Cross-multiplying, $$2xx_1 - 4yy_1 = 3z + 3z_1$$

or $$2xx_1 - 4yy_1 - 3z = 3z_1 \quad \text{...(3)}$$

Comparing (3) and (1), we have

$$\frac{2x_1}{2} = -\frac{4y_1}{-4} = \frac{-3}{-1} = \frac{3z_1}{-3}$$

or $$x_1 = y_1 = 3 = -z_1$$

$\therefore$ $$x_1 = 3,\ y_1 = 3,\ -z_1 = 3 \text{ i.e., } z_1 = -3$$

$\therefore$ Point of contact is $(3, 3, -3)$.

Example 2. *Show that the equations to the two tangent planes to the surface $ax^2 + by^2 = 2z$ which pass through the line*

$$u \equiv lx + my + nz - p = 0, \quad u' \equiv l'x + m'y + n'z - p' = 0$$

is
$$u^2\left(\frac{l'^2}{a}+\frac{m'^2}{b}+2n'p'\right)-2uu'\left(\frac{ll'}{a}+\frac{mm'}{b}+np'+n'p\right)+u'^2\left(\frac{l^2}{a}+\frac{m^2}{b}+2np\right)=0.$$

(M.D.U. 2003 S)

Sol. Equation of any plane through the line

$$u \equiv lx + my + nz - p = 0, \quad u' \equiv l'x + m'y + n'z - p' = 0$$

is
$$u + ku' = 0 \qquad \text{...(1)}$$

i.e.,
$$lx + my + nz - p + k(l'x + m'y + n'z - p') = 0$$

or
$$(l + kl')x + (m + km')y + (n + kn')z = p + kp$$

If this plane touches the paraboloid $ax^2 + by^2 = 2z$,

then
$$\frac{\mathbf{l^2}}{\mathbf{a}}+\frac{\mathbf{m^2}}{\mathbf{b}}=-\mathbf{2np}$$

or
$$\frac{(l+kl')^2}{a}+\frac{(m+km')^2}{b}=-2(n+kn')(p+kp')$$

or
$$k^2\left(\frac{l'^2}{a}+\frac{m'^2}{b}+2n'p'\right)+2k\left(\frac{ll'}{a}+\frac{mm'}{b}+np'+n'p\right)+\left(\frac{l^2}{a}+\frac{m^2}{b}+2np\right)=0 \qquad \text{...(2)}$$

Putting $k = -\frac{u}{u'}$ from (1) in (2) ; equations of the required planes are

$$\frac{u^2}{u'^2}\left(\frac{l'^2}{a}+\frac{m'^2}{b}+2n'p'\right)-\frac{2u}{u'}\left(\frac{ll'}{a}+\frac{mm'}{b}+np'+n'p\right)+\left(\frac{l^2}{a}+\frac{m^2}{b}+2np\right)=0.$$

Multiplying every term by u'^2.

$$u^2\left(\frac{l'^2}{a}+\frac{m'^2}{b}+2n'p'\right)-2uu'\left(\frac{ll'}{a}+\frac{mm'}{b}+np'+n'p\right)+\left(\frac{l^2}{a}+\frac{m^2}{b}+2np\right)u'^2=0.$$

EXERCISE 4

1. (*a*) Show that the plane $2x - 4y - z + 3 = 0$ touches the paraboloid $x^2 - 2y^2 = 3z$. (*K.U. 1999 S*)

   (*b*) Show that the plane $8x - 6y - z = 5$ touches the paraboloid $\frac{x^2}{2} - \frac{y^2}{3} = z$ and find the co-ordinates of the point of contact. (*M.D.U. 1999*)

2. Find the condition that the plane $lx + my + nz = 1$ may be a tangent plane to the paraboloid $x^2 + y^2 = 2z$.

3. Prove that the paraboloids $\frac{x^2}{a_1^2}+\frac{y^2}{b_1^2}=\frac{2z}{c_1}, \frac{x^2}{a_2^2}+\frac{y^2}{b_2^2}=\frac{2z}{c_2}, \frac{x^2}{a_3^2}+\frac{y^2}{b_3^2}=\frac{2z}{c_3}$ have a common tangent plane if $\begin{vmatrix} a_1^2 & a_2^2 & a_3^2 \\ b_1^2 & b_2^2 & b_3^2 \\ c_1 & c_2 & c_3 \end{vmatrix}=0.$

[**Hint.** The first paraboloid is $\frac{x^2}{a_1^2}+\frac{y^2}{b_1^2}=\frac{2z}{c_1}$

Multiplying by c_1, $\frac{c_1}{a_1^2}x^2+\frac{c_1}{b_1^2}y^2=2z.$

The plane $lx+my+nz=p$ will touch this paraboloid if

$$\frac{l^2}{\left(\frac{c_1}{a_1^2}\right)}+\frac{m^2}{\left(\frac{c_1}{b_1^2}\right)}=-2np \text{ or } a_1^2l^2+b_1^2m^2+(2np)c_1=0.$$]

4. Find the locus of the point from which three mutually perpendicular tangent lines can be drawn to the paraboloid
   (*i*) $ax^2+by^2+2z=0$ (*M.D.U. 2003*)
   (*ii*) $ax^2+by^2=2z.$
5. Find the equation of the diametral plane of the paraboloid $ax^2+by^2=2z$ which bisects chords parallel to the line $\frac{x}{l}=\frac{y}{m}=\frac{z}{n}.$
6. Find the equation of the plane which cuts the paraboloid $x^2-2y^2=z$ in a conic with its centre at $\left(2, \frac{3}{2}, 4\right).$
7. Find the centre of the conic $ax^2+by^2=2z,\ lx+my+nz=p.$

Answers

1. (*b*) (8, 9, 5)
2. $l^2+m^2=-2n$
4. (*i*) $ab(x^2+y^2)+2(a+b)z-1=0$ (*ii*) $ab(x^2+y^2)-2(a+b)z-1=0$
5. $alx+bmy-n=0$
6. $4x-6y-z+5=0$
7. $\left(-\frac{l}{an}, -\frac{m}{bn}, \frac{l^2}{an^2}+\frac{m^2}{bn^2}+\frac{p}{n}\right).$

---

24. To find the locus of the point of intersection of three mutually perpendicular tangent planes to the paraboloid

$$\frac{x^2}{a^2}+\frac{y^2}{b^2}=2z.$$

Let $lx+my+nz=p$ (where $l, m, n,$ are actual direction cosines of the normal to the plane) be a tangent plane to the paraboloid

$$\frac{x^2}{a^2} + \frac{y^2}{b^2} = 2z.$$

$$\therefore \qquad \frac{l^2}{\left(\frac{1}{a^2}\right)} + \frac{m^2}{\left(\frac{1}{b^2}\right)} = -2np \quad \text{or} \quad a^2l^2 + b^2m^2 = -2np$$

$$\therefore \qquad p = \frac{-1}{2n}(a^2l^2 + b^2m^2)$$

Putting this value of p in $lx + my + nz = p$, equation of the tangent plane is

$$lx + my + nz = \frac{-1}{2n}(a^2l^2 + b^2m^2)$$

or $$nlx + nmy + n^2z = -\tfrac{1}{2}(a^2l^2 + b^2m^2)$$

where l, m, n are d.c's of the normal to the plane.

Let $$n_1l_1x + n_1m_1y + n_1^2z = -\tfrac{1}{2}[a^2l_1^2 + b^2m_1^2] \qquad ...(1)$$

$$n_2l_2x + n_2m_2y + n_2^2z = -\tfrac{1}{2}[a^2l_2^2 + b^2m_2^2] \qquad ...(2)$$

and $$n_3l_3x + n_3m_3y + n_3^2z = -\tfrac{1}{2}[a^2l_3^2 + b^2m_3^2] \qquad ...(3)$$

be three mutually ⊥ tangents planes to the paraboloid.

∴ Normals to these tangents planes with d.c's l_1, m_1, n_1 ; l_2, m_2, n_2 ; l_3, m_3, n_3 are also mutually ⊥.

∴ By Art. 4, Review IV, there exist 12 relations among these d.c's.

Let (x, y, z) be the point of intersection of planes (1), (2), (3).

To find the locus of the point (x, y, z), let us eliminate l_1, m_1, n_1 ; l_2, m_2, n_2 ; l_3, m_3, n_3.

Adding (1), (2), (3) ;

$$x(n_1l_1 + n_2l_2 + n_3l_3) + y(n_1m_1 + n_2m_2 + n_3m_3) + z(n_1^2 + n_2^2 + n_3^2)$$
$$= -\tfrac{1}{2}[a^2(l_1^2 + l_2^2 + l_3^2) + b^2(m_1^2 + m_2^2 + m_3^2)]$$

or $$x(0) + y(0) + z(1) = -\tfrac{1}{2}[a^2 + b^2]$$

or $$z = -\tfrac{1}{2}(a^2 + b^2).$$

Exercise. Prove that the locus of the point of intersection of three mutually perpendicular tangent planes to the paraboloid

$$ax^2 + by^2 = 2z \text{ is } z = -\tfrac{1}{2}\left(\frac{1}{a} + \frac{1}{b}\right).$$

25. Equation of Normal to the Paraboloid

To find the equation of the normal at the point (x_1, y_1, z_1) to the paraboloid $ax^2 + by^2 = 2z$.

Equation of the paraboloid is $$ax^2 + by^2 = 2z \qquad ...(1)$$

Equation of the tangent plane at (x_1, y_1, z_1) to (1) is

$$axx_1 + byy_1 = z + z_1$$

or $$axx_1 + byy_1 - z = z_1.$$

d.r's of the normal to this tangent plane are $ax_1, by_1, -1$.

$\therefore$ Equations of the normal at (x_1, y_1, z_1)

to the given paraboloid are $$\boxed{\frac{x - x_1}{ax_1} = \frac{y - y_1}{by_1} = \frac{z - z_1}{-1}}$$

Exercise. Prove that the equations of the normal at the point (x_1, y_1, z_1) to the paraboloid $\frac{x^2}{a^2} + \frac{y^2}{b^2} = 2z$ are

$$\frac{x - x_1}{\frac{x_1}{a^2}} = \frac{y - y_1}{\frac{y_1}{b^2}} = \frac{z - z_1}{-1}.$$

26. Number of Normals

To prove that there are five points on an elliptic paraboloid $\frac{x^2}{a^2} + \frac{y^2}{b^2} = 2z$, *the normals at which pass through a given point* (α, β, γ). (M.D.U. 2007)

The equation of the paraboloid is $\frac{x^2}{a^2} + \frac{y^2}{b^2} = 2z$ ...(1)

Let (x_1, y_1, z_1) be one of the points on the paraboloid (1), the normal at which passes through (α, β, γ).

Equations of the normal at (x_1, y_1, z_1), to the paraboloid (1) are

$$\frac{x - x_1}{x_1 / a^2} = \frac{y - y_1}{y_1 / b^2} = \frac{z - z_1}{-1}$$

But this normal passes through (α, β, γ)

$$\therefore \quad \frac{\alpha - x_1}{x_1 / a^2} = \frac{\beta - y_1}{y_1 / b^2} = \frac{\gamma - z_1}{-1} = k \text{ (say)}$$

From first and fourth members

$$\alpha - x_1 = \frac{kx_1}{a^2} \quad \text{or} \quad \alpha = x_1 + \frac{kx_1}{a^2}$$

or $$\alpha = x_1\left(\frac{a^2 + k}{a^2}\right) \quad \therefore \quad x_1 = \frac{a^2\alpha}{a^2 + k}$$

From second and fourth members $$y_1 = \frac{b^2\beta}{b^2 + k}$$ ...(2)

From third and fourth members $$\gamma - z_1 = -k$$

or $$z_1 = \gamma + k$$

But the point (x_1, y_1, z_1) lies on paraboloid (1),

$$\therefore \quad \frac{x_1^2}{a^2} + \frac{y_1^2}{b^2} = 2z_1$$ ...(3)

Putting these values of x_1, y_1, z_1 from (2) in (3), we have

$$\frac{a^4\alpha^2}{a^2(a^2 + k)^2} + \frac{b^4\beta^2}{b^2(b^2 + k)^2} = 2(\gamma + k)$$

or
$$\frac{a^2\alpha^2}{(a^2+k)^2}+\frac{b^2\beta^2}{(b^2+k)^2}=2(\gamma+k)$$

or
$$a^2\alpha^2(b^2+k)^2+b^2\beta^2(a^2+k)^2=2(\gamma+k)\,(a^2+k)^2\,(b^2+k)^2 \qquad \ldots(4)$$

Equation (4) is a fifth degree in k and hence it gives five values of k.

Putting these values of k in (2), we get five points on the paraboloid, the normals at which pass through (α, β, γ).

Cor. From (2), the foot of the normal is

$$\left(\frac{a^2\alpha}{a^2+k}, \frac{b^2\beta}{b^2+k}, \gamma+k\right).$$

Example. *Prove that the normals from (α, β, γ) to the paraboloid $\frac{x^2}{a^2}+\frac{y^2}{b^2}=2z$ lies on the cone*

$$\frac{\alpha}{x-\alpha}-\frac{\beta}{y-\beta}+\frac{a^2-b^2}{z-\gamma}=0.$$

Or

Find the locus of the normals drawn from (α, β, γ) to the paraboloid

$$\frac{x^2}{a^2}+\frac{y^2}{b^2}=2z.$$ (K.U. 2003, 1998 S)

Sol. Reproduce Art. 26 upto equation (2)

Let l, m, n be d.c's of any of the normals drawn from (α, β, γ) to the paraboloid.

$\therefore$ l, m, n are proportional to $\frac{x_1}{a^2}, \frac{y_1}{b^2}, -1$

$$\therefore \qquad l=\frac{px_1}{a^2}, \quad m=\frac{py_1}{b^2}, \quad n=-p$$

where p is the constant of proportionality,

or
$$l=\frac{p}{a^2}\,\frac{a^2\alpha}{a^2+k}, \; m=\frac{p}{b^2}\,\frac{b^2\beta}{b^2+k}, \; n=-p \qquad \mid \text{Using equation (2)}$$

or
$$l=\frac{p\alpha}{a^2+k}, \; m=\frac{p\beta}{b^2+k}, \; n=-p$$

$$\therefore \qquad a^2+k=\frac{p\alpha}{l} \qquad \ldots(3)$$

$$b^2+k=\frac{p\beta}{m} \qquad \ldots(4)$$

$$p=-n \qquad \ldots(5)$$

(3) – (4) gives [To eliminate k]

$$a^2-b^2=p\left(\frac{\alpha}{l}-\frac{\beta}{m}\right) \quad \text{or} \quad \frac{a^2-b^2}{p}=\frac{\alpha}{l}-\frac{\beta}{m}$$

Putting $p = -n$ from (5), [To eliminate p], we have

$$-\left(\frac{a^2-b^2}{n}\right)=\frac{\alpha}{l}-\frac{\beta}{m}$$

$$\frac{\alpha}{l}-\frac{\beta}{m}+\left(\frac{a^2-b^2}{n}\right)=0 \qquad ...(6)$$

Now equations of any normal from (α, β, γ) are

$$\frac{x-\alpha}{l}=\frac{y-\beta}{m}=\frac{z-\gamma}{n} \qquad ...(7)$$

Putting the values of l, m, n from (7) in (6), the locus of the normals is the cone

$$\frac{\alpha}{x-\alpha}-\frac{\beta}{y-\beta}+\frac{(a^2-b^2)}{z-\gamma}=0.$$

EXERCISE 5

1. Prove that the feet of the normals from a given point (α, β, γ) to an elliptic paraboloid are the five points of intersection of the elliptic paraboloid and a certain cubic curve.
2. Prove that in general three normals can be drawn from a given point to the paraboloid of revolution $x^2 + y^2 = 2az$. Prove also that if that point lies on the surface $27a(x^2 + y^2) + 8(a - z)^3 = 0$, then two of the three normals coincide.

MISCELLANEOUS EXERCISE

1. Find the locus of the feet of the perpendicular from the origin to the tangent planes to the surface $\frac{x^2}{a^2}+\frac{y^2}{b^2}+\frac{z^2}{c^2}=1$ which cut off from the axes intercepts the sum of whose reciprocals is equal to constant $\frac{1}{K}$.
2. Prove that the locus of the centres of sections of a central conicoid which are parallel to a given line is a plane.
3. Find the locus of the centres of sections of a conicoid that are at a constant distance from the centre.
4. Prove that the locus of the centres of parallel plane sections of a conicoid is a diameter. Prove also that the tangent plane at an extremity of a diameter is parallel to the plane sections.
5. Show that the locus of the centres of a system of parallel plane sections of a paraboloid is a diameter. Prove also that the tangent plane at the extremity of the diameter is parallel to the plane sections.
6. Show that the feet of the normals from the point (α, β, γ) to the paraboloid $x^2 + y^2 = 2az$ lie on the sphere $x^2 + y^2 + z^2 - z(a + \gamma) - \frac{y}{2\beta}(\alpha^2 + \beta^2) = 0$.
7. If the line of intersection of two perpendicular tangent planes to the ellipsoid $\frac{x^2}{a^2}+\frac{y^2}{b^2}+\frac{z^2}{c^2}=1$ passes through the fixed point $(0, 0, k)$, show that it lies on the cone
$$x^2(b^2 + c^2 - k^2) + y^2(c^2 + a^2 - k^2) + (z - k)^2(a^2 + b^2) = 0.$$

8. Find the length of the normal chord through $P(x_1, y_1, z_1)$ of the ellipsoid $\frac{x^2}{a^2}+\frac{y^2}{b^2}+\frac{z^2}{c^2}=1$ and prove that if it is equal to $4PG_3$, where G_3 is the point in which the normal chord meets the plane XOY, then P lies on the cone $\frac{x^2}{a^6}(2c^2-a^2)+\frac{y^2}{b^6}(2c^2-b^2)+\frac{z^2}{c^4}=0$.

9. The normals to an ellipsoid at the points P and P′ meet a principal plane in G and G′. Show that the plane which bisects PP′ at right angles also bisects GG′.

10. Show that the locus of the point of intersection of three mutually perpendicular tangent planes to the conicoid $a^2x^2+b^2y^2+c^2z^2=1$ is the sphere $x^2+y^2+z^2=\frac{1}{a^2}+\frac{1}{b^2}+\frac{1}{c^2}$.

11. If P be the point of contact of a tangent plane to the ellipsoid $\frac{x^2}{a^2}+\frac{y^2}{b^2}+\frac{z^2}{c^2}=1$, which meets the axes in A, B, C and PD, PE, PF are perpendiculars drawn from P to the axes, prove that

$$OD.OA = a^2,\ OE.OB = b^2,\ OF.\ OC = c^2.$$

12. If P is the point on the ellipsoid $x^2+2y^2+\frac{1}{3}z^2=1$, such that the perpendicular from the origin on the tangent plane at P is of unit length, show that P lies on one or other of the planes $3y=\pm z$.

13. The normal at a variable point P of the ellipsoid $\frac{x^2}{a^2}+\frac{y^2}{b^2}+\frac{z^2}{c^2}=1$ meets the plane XOY in A and AQ is drawn parallel to OZ and equal to AP. Prove that the locus of Q is $\frac{x^2}{a^2-c^2}+\frac{y^2}{b^2-c^2}+\frac{z^2}{c^2}=1$.

14. Prove that two normals to the ellipsoid $\frac{x^2}{a^2}+\frac{y^2}{b^2}+\frac{z^2}{c^2}=1$ lie in the plane $lx+my+nz=0$ and the line joining their feet has direction cosines proportional to $a^2(b^2-c^2)mn$, $b^2(c^2-a^2)nl$, $c^2(a^2-b^2)lm$. Also find the co-ordinates of there points. *(K.U. 1998)*

Answers

1. $x^2+y^2+z^2=k(x+y+z)$ and $a^2x^2+b^2y^2+c^2z^2=k^2(x+y+z)^2$
3. $(ax^2+by^2+cz^2)^2=k^2(a^2x^2+b^2y^2+c^2z^2)$ which k is the constant distance
8. $$\frac{2\left[\frac{x_1^2}{a^4}+\frac{y_1^2}{b^4}+\frac{z_1^2}{c^4}\right]^{3/2}}{\frac{x_1^2}{a^6}+\frac{y_1^2}{b^6}+\frac{z_1^2}{c^6}}.$$

Appendix

1. To trace the ellipsoid $\frac{x^2}{a^2}+\frac{y^2}{b^2}+\frac{z^2}{c^2}=1.$

Sol. **The given ellipsoid is** $\frac{x^2}{a^2}+\frac{y^2}{b^2}+\frac{z^2}{c^2}=1$ ...(1)

(*i*) Centre. **If (α, β, γ) is a point on the ellipsoid (1), then $(-\alpha, -\beta, -\gamma)$ is also a point on (1).**

The middle point of the joint of these points is (0, 0, 0), the origin. Thus (α, β, γ), $(-\alpha, -\beta, -\gamma)$ are the points on a straight line through the origin and are equidistant from the origin. Thus origin bisects all chords which pass through it and is therefore the *centre* of the surface.

(*ii*) Symmetry. **Since the equation (1) remains unchanged when x is changed to $-x$, $\therefore$ the ellipsoid is symmetrical about the YZ-plane.**

Similarly, the ellipsoid is also symmetrical about the ZX and XY-planes.

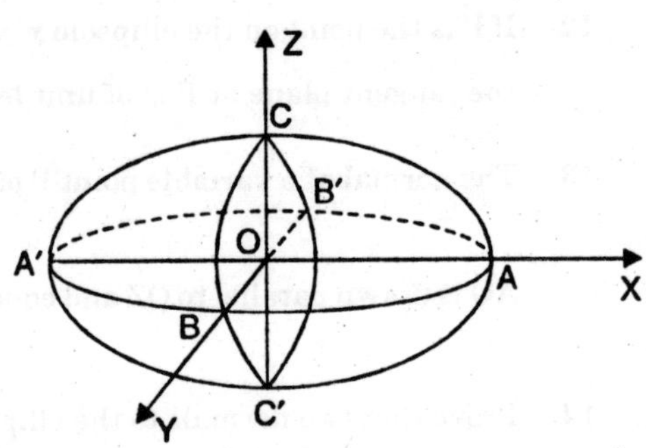

If the point (α, β, γ) satisfies (1), then $(\alpha, \beta, -\gamma)$ also satisfies it, that means the XY-plane bisects all chords perpendicular to it. Similarly, other co-ordinate planes also bisect all chords $\perp$ to them.

The three co-ordinate planes by virtue of their *property that they bisect chords $\perp$ to them are* called principal planes of the surface. And the three co-ordinate axes which are the lines of intersection of the three principal planes taken in pairs are called principal axes.

(*iii*) Axes intersection. **The surface (1) meets the x-axis where putting $y = 0$, $z = 0$ in (1), $\frac{x^2}{a^2}=1$ or $x^2 = a^2$ as $x = \pm a$, *i.e.*, the surface meets the x-axis in the points A(a, 0, 0) and A′($-a$, 0, 0).**

Similarly, it meets Y-axis ($x = 0, z = 0$) at B(0, b, 0), B′(0, $-b$, 0) and Z-axis ($x = 0, y = 0$) at the points C(0, 0, c) and C′(0, 0, $-c$). These six points A, A′ ; B, B′ and C, C′ are called the extremities of the principal axes.

(*iv*) Sections by co-ordinate planes. **The surface meets the YZ-plane, where putting $x = 0$ in (1), $\frac{y^2}{b^2}+\frac{z^2}{c^2}=1$ which is an ellipse *in that plane*.**

Similarly, it meets the ZX-plane ($y = 0$) in the ellipse $\frac{x^2}{a^2}+\frac{z^2}{c^2}=1$ *in that plane* and it meets XY-plane ($z = 0$) in the ellipse $\frac{x^2}{a^2}+\frac{y^2}{b^2}=1$, in that plane.

(*v*) **Generated by a variable curve.** The surface (1) meets the plane $z = k$ [where putting $z = k$ in (1)]

$$\frac{x^2}{a^2}+\frac{y^2}{b^2}+\frac{k^2}{c^2}=1 \quad \text{or} \quad \frac{x^2}{a^2}+\frac{y^2}{b^2}=1-\frac{k^2}{c^2}$$

$\therefore$ The surface is generated by a variable ellipse

$$\frac{x^2}{a^2}+\frac{y^2}{b^2}=1-\frac{k^2}{c^2},\ z=k \qquad ...(2)$$

as k varies and whose plane is || to the XY-plane and centre (0, 0, k) moves on the Z-axis.

The ellipse (2) is real only if $1-\frac{k^2}{c^2}>0$ or if $\frac{k^2}{c^2}<1$

or if $k^2 < c^2$ or k is numerically less than c

i.e., k lies between c and $-c$.

But $z = k$ from (2). Thus the surface lies between the planes $z = c$ and $z = -c$. Similarly, it lies between the planes $x = a$ and $x = -a$ and between the planes $y = b$ and $y = -b$. *Hence the surface is closed.*

From (2), $$\frac{x^2}{a^2\left(1-\frac{k^2}{c^2}\right)}+\frac{y^2}{b^2\left(1-\frac{k^2}{c^2}\right)}=1,\ z=k.$$

The semi-axes of the ellipse are $a\sqrt{1-\frac{k^2}{c^2}}$ and $b\sqrt{1-\frac{k^2}{c^2}}$ which decrease as k (numerically $< c$) increases numerically and become zero when $k = \pm c$, in which case the ellipse reduces to a point. Hence the shape of the surface is as shown in the figure. Its shape is like an egg.

Note. If in the equation of the ellipsoid $\frac{x^2}{a^2}+\frac{y^2}{b^2}+\frac{z^2}{c^2}=1$; $a^2 = b^2$ or $b^2 = c^2$ or $c^2 = a^2$, it is called an **ellipsoid of revolution.**

2. Hyperboloid of One Sheet

To trace the surface $\frac{x^2}{a^2}+\frac{y^2}{b^2}-\frac{z^2}{c^2}=1$.

The equation of the surface is $\frac{x^2}{a^2}+\frac{y^2}{b^2}-\frac{z^2}{c^2}=1$ ...(1)

(*i*) **Origin is the centre.** The origin bisects all chords which pass through it and is therefore the centre of the surface.

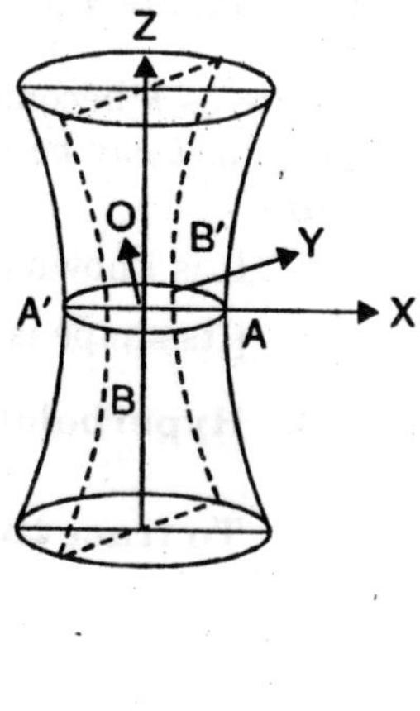

(*ii*) **Symmetry.** Since the equation (1) contains even powers of x, y, z *i.e.*, the equation is unaltered on changing the sign of x or y or z, the surface (2) is symmetrical about all the co-ordinate planes.

Also the co-ordinate planes bisect all chords $\perp$ to them and are therefore the **principal planes** of the surface. The co-ordinate axes which are the lines of intersection of principal planes taken in pairs are called **principal axes.**

(iii) **Axes Intersection.** The surface (1) meets X-axis where [putting $y = 0, z = 0$ in (1)].

$$\frac{x^2}{a^2} = 1 \text{ or } x = \pm a.$$

Thus the surface meets X-axis in the points A(a, 0, 0) and A′($-a$, 0, 0).

Also it meets Y-axis ($z = x = 0$) in the points B(0, b, 0) and B′(0, $-b$, 0).

But it meets Z-axis, where [putting $x = 0, y = 0$ in (1)]

where $-\frac{z^2}{c^2} = 1$ or $z^2 = -c^2$, which gives imaginary values of z. Thus the surface does not meet the Z-axis.

(iv) **Sections by co-ordinate planes.** The surface meets the YZ-plane, where [putting $x = 0$ in (1)],

$$\frac{y^2}{b^2} - \frac{z^2}{c^2} = 1 \text{ which is a hyperbola } \textit{in that plane.}$$

Similarly, the surface meets the ZX-plane ($y = 0$) in the hyperbola $\frac{x^2}{a^2} - \frac{z^2}{c^2} = 1$ *in that plane*, and it meets the XY-plane ($z = 0$) in the ellipse $\frac{x^2}{a^2} + \frac{y^2}{b^2} = 1$ *in that plane.*

(v) **Generated by a variable curve.** The surface meets the plane $z = k$, where [putting $z = k$ in (1)]

$$\frac{x^2}{a^2} + \frac{y^2}{b^2} - \frac{k^2}{c^2} = 1 \quad \text{or} \quad \frac{x^2}{a^2} + \frac{y^2}{b^2} = 1 + \frac{k^2}{c^2}.$$

Thus the surface is generated by the variable ellipse (as k varies)

$$\frac{x^2}{a^2} + \frac{y^2}{b^2} = 1 + \frac{k^2}{c^2},\ z = k$$

or

$$\frac{x^2}{a^2\left(1 + \frac{k^2}{c^2}\right)} + \frac{y^2}{b^2\left(1 + \frac{k^2}{c^2}\right)} = 1,\ z = k \qquad \text{...(2)}$$

whose plane is ∥ to XY-plane, and centre (0, 0, k) moves on the Z-axis.

The ellipse (1) is real for all values of k, whether, +ve or – ve. The semi-axes of the ellipse (2) are $a\sqrt{1 + \frac{k^2}{c^2}}$ and $b\sqrt{1 + \frac{k^2}{c^2}}$ which increase as k increases numerically and become infinite as k become infinite. Thus the surface extends to infinity ($\because\ z = k$) both above and below (as k can be +ve or – ve) the XY-plane. Hence the shape of the surface is as shown in the figure.

It is known as **hyperboloid of one sheet.**

[Its shape is like that of **Murha** (मूढ़ा) with elliptic ends.]

3. Hyperboloid of Two Sheets

To trace the surface $\frac{x^2}{a^2} - \frac{y^2}{b^2} - \frac{z^2}{c^2} = 1.$

The equation of surface is $\frac{x^2}{a^2}-\frac{y^2}{b^2}-\frac{z^2}{c^2}=1$ ...(1)

(*i*) **Origin is the centre.** Here also the origin bisects all chords through it [$\because$ If (α, β, γ) lies on the surface, then $(-\alpha, -\beta, -\gamma)$ also lies on it] and is therefore called the centre of the surface.

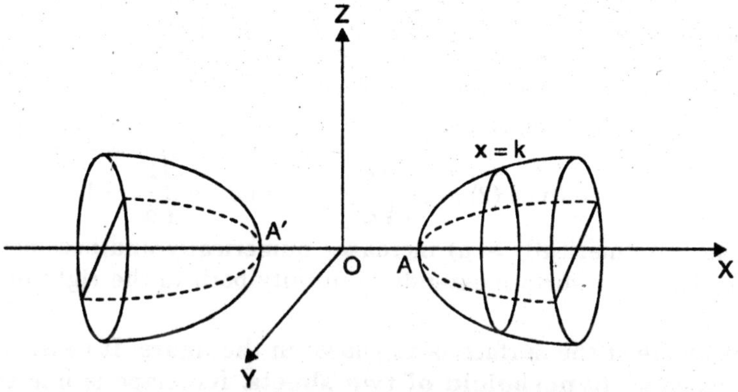

(*ii*) **Symmetry.** Since the equation (1) contains only even powers of x, y and z, so the surface is symmetrical about the YZ, ZX and XY planes.

Moreover, co-ordinate planes bisect chord $\perp$ to them and are therefore called the **principal planes** and the co-ordinate axes are **principal axes.**

(*iii*) **Intersection with the axes.** The surface meets the X-axis where [putting $y = 0$, $z = 0$ in (1)], $\frac{x^2}{a^2}=1$ or $x = \pm a$.

i.e., at the points A$(a, 0, 0)$ and A′$(-a, 0, 0)$. It meets y-axis, $(x = 0, z = 0)$ at points where $y^2 = -b^2$ *i.e.*, at imaginary points. Similarly it meets the Z-axis at imaginary points.

(*iv*) **Sections by the co-ordinate planes.** The surface meets the YZ-plane $(x = 0)$ where $-\frac{y^2}{b^2}-\frac{z^2}{c^2}=1$ or $\frac{y^2}{b^2}+\frac{z^2}{c^2}=-1$ which is an imaginary ellipse *in that plane*. Thus, it does not meet the YZ-plane.

(1) meets the ZX-plane $(y = 0)$, where $\frac{x^2}{a^2}-\frac{z^2}{c^2}=1$ which is a hyperbola in that plane.

Similarly, it meets the XY-plane $(z = 0)$ in the hyperbola $\frac{x^2}{a^2}-\frac{y^2}{b^2}=1$ *in that plane.*

(*v*) **Generated by a variable curve.** The surface meets the plane $x = k$ **(Note this step** $x = k$ and not $z = k$ as before) where

$$\frac{k^2}{a^2}-\frac{y^2}{b^2}-\frac{z^2}{c^2}=1 \quad \text{or} \quad \frac{y^2}{b^2}+\frac{z^2}{c^2}=\frac{k^2}{a^2}-1. \qquad \text{[putting } x = k \text{ in (1)]}$$

$\therefore$ The surface is generated by a variable ellipse (as k varies)

$$\frac{y^2}{b^2}+\frac{z^2}{c^2}=\frac{k^2}{a^2}-1,\ x=k$$

or $$\frac{y^2}{b^2\left(\dfrac{k^2}{a^2}-1\right)}+\frac{z^2}{c^2\left(\dfrac{k^2}{a^2}-1\right)}=1,\ x=k \qquad ...(2)$$

whose plane is ∥ to the YZ-plane and centre $(k, 0, 0)$ moves on the X-axis.

The ellipse (2) is real if $\frac{k^2}{a^2}-1>0$ or $\frac{k^2}{a^2}>1$ or $k^2>a^2$

i.e., k is numerically $> a$ *i.e.*, k does not lie between a and $-a$.

But $x = k$ from (2). Thus no portion of the surface lies between the planes $x = a$ and $x = -a$.

The semi-axes of the ellipse (2) are $b\sqrt{\frac{k^2}{a^2}-1}$ and $c\sqrt{\frac{k^2}{a^2}-1}$

which increase as k (numerically $> a$) increases numerically and becomes infinite when k becomes infinite. Thus the surface extends to infinity both to the right of A and to the left of A′.

Hence the shape of the surface is as shown in the figure. It consists of two detached portions. It is known as **hyperboloid of two sheets.** Its shape is like that of two *tablas* (तबले) placed as shown in the figure.

4. To Trace the Cone. $\frac{x^2}{a^2}+\frac{y^2}{b^2}-\frac{z^2}{c^2}=0.$

The given surface is $\frac{x^2}{a^2}+\frac{y^2}{b^2}-\frac{z^2}{c^2}=0 \qquad ...(1)$

(*i*) **Symmetry.** Since the equation (1) contains only even powers of x, y, z, so the surface is symmetrical about the YZ, ZX and XY planes.

(*ii*) **Axes intersection.** The cone meets X-axis ($y = 0, z = 0$) where $\frac{x^2}{a^2}=0$ or $x = 0, 0$ *i.e.*, in two coincident points.

Thus cone meets X-axis at the origin. Similarly it meets Y and Z-axis also at the origin.

(*iii*) **Sections by co-ordinate planes.** The cone (1) meets the YZ plane ($x = 0$), where $\frac{y^2}{b^2}-\frac{z^2}{c^2}=0$ or $y=\pm\frac{b}{c}z$ which are two straight lines in that plane [on opposite sides of Z-axis and making equal angles with it].

Similarly, the cone (1) meet ZX plane ($y = 0$) in two lines $x=\pm\frac{a}{c}z$ which are equally inclined to Z-axis and on opposite sides of it.

Again it meets the XY-plane ($z = 0$), where $\frac{x^2}{a^2}+\frac{y^2}{b^2}=0$ which is a point ellipse in *that plane*.

(*iv*) **Generated by a variable curve.** The surface (1) meets the plane $z = k$, [where putting $z = k$ in (1)].

$$\frac{x^2}{a^2}+\frac{y^2}{b^2}-\frac{k^2}{c^2}=0 \quad \text{or} \quad \frac{x^2}{a^2}+\frac{y^2}{b^2}=\frac{k^2}{c^2}$$

Thus the cone (1) is generated by the variables ellipse

$$z = k,\ \frac{x^2}{a^2} + \frac{y^2}{b^2} = \frac{k^2}{c^2} \qquad \text{...(2)}$$

(k varies) whose plane is || to XY plane and centre (0, 0, k) moves on Z-axis. The ellipse (2) is real for all values of k ; + ve or – ve and the semi-axes $\frac{ak}{c}, \frac{bk}{c}$ increase as k increases numerically and $\to \infty$ as $k \to \infty$.

∴ The cone extends to infinity both above and below the XY-plane.

Hence the shape is as shown in the figure.

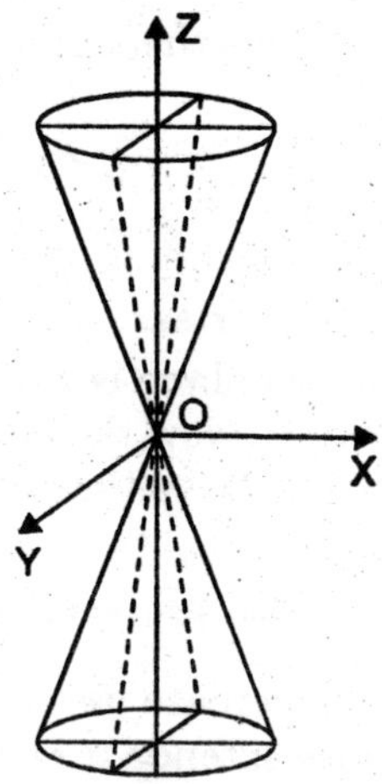

PARABOLOID

5. (a) To Trace Elliptic Paraboloid $\frac{x^2}{a^2} + \frac{y^2}{b^2} = \frac{2z}{c}$.

The given paraboloid is $\frac{x^2}{a^2} + \frac{y^2}{b^2} = \frac{2z}{c}$. ...(1)

(*i*) **Symmetry.** Since (1) contains only even powers of x and y, only so it is symmetrical about the YZ and ZX-planes.

(*ii*) **Intersections with axes.** The surface (1) meets X-axis ($y = 0$, $z = 0$) where $\frac{x^2}{a^2} = 0$ or $x^2 = 0$ ∴ $x = 0, 0$.

∴ The surface (1) touches the X-axis at O(0, 0, 0).

Similarly, it touches Y-axis at (0, 0, 0). Thus the surface (1) touches the XY-plane. Also (1) meets Z-axis at (0, 0, 0).

(*iii*) **Sections by co-ordinate planes.** (1) meets the YZ-plane ($x = 0$) where $\frac{y^2}{b^2} = \frac{2z}{c}$ or $y^2 = \frac{2b^2}{c} z$ which is an upward parabola *in that plane*.

Similarly, (1) meets the ZX-plane ($y = 0$) in the upward parabola

$$x^2 = \frac{2a^2}{c} z \textit{ in that plane.}$$

Again, (1) **meets the** XY-plane ($z = 0$) in $\frac{x^2}{a^2} + \frac{y^2}{b^2} = 0$ which is a point ellipse in that plane.

(*iv*) **Generated by a variable curve.** The surface (1) meets the plane $z = k$, where

$$\frac{x^2}{a^2} + \frac{y^2}{b^2} = \frac{2k}{c} \quad \text{or} \quad \frac{x^2}{\left(\frac{2a^2k}{c}\right)} + \frac{y^2}{\left(\frac{2b^2k}{c}\right)} = 1$$

Thus the surface is generated by a variable ellipse

$$z = k, \quad \frac{x^2}{\left(\frac{2a^2k}{c}\right)} + \frac{y^2}{\left(\frac{2b^2k}{c}\right)} = 1 \quad \text{...(1)}$$

where k varies.

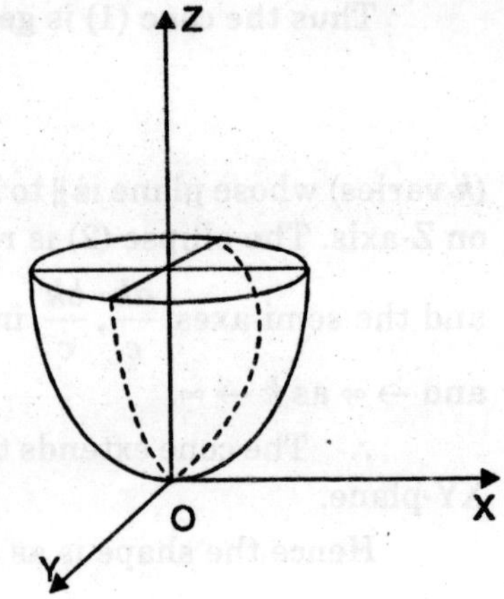

Its plane is ∥ to XY-plane and the centre (0, 0, k) moves on Z-axis. Now the ellipse (2) is real if k + ve. Thus the surface lies only above the XY-plane.

Also the semi-axis of the ellipse (2) are $a\sqrt{\frac{2k}{c}}$, $b\sqrt{\frac{2k}{c}}$ which increase as $k > 0$ increases and $\to \infty$ as $k \to \infty$. Thus the surface extends to ∞ above the XY-plane.

The shape is as shown in the figure.

(*b*) **To trace the hyperbolic paraboloid** $\frac{x^2}{a^2} - \frac{y^2}{b^2} = \frac{2z}{c}$

The equation of the surface is $\frac{x^2}{a^2} - \frac{y^2}{b^2} = \frac{2z}{c}$ ...(1)

(*i*) **Symmetry.** Since the equation (1) contains even powers of x and y, so the surface is symmetrical about the YZ and ZX-planes.

(*ii*) **Axes intersection.** The surface (1) meets X-axis ($y = 0$, $z = 0$), where $\frac{x^2}{a^2} = 0$ or $x^2 = 0$ or $x = 0, 0$. Thus the surface (1), touches X-axis at the origin. Similarly, it touches Y-axis at the origin.

Thus (1) touches XY-plane at O(0, 0, 0).

It meets Z-axis ($x = 0$, $y = 0$), where $\frac{2z}{c} = 0$ or $z = 0$ *i.e.*, at the origin.

(*iii*) **Sections by co-ordinate planes.** The surface (1) meets the YZ-plane ($x = 0$), where

$$\frac{-y^2}{b^2} = \frac{2z}{c} \quad \text{or} \quad y^2 = -2\frac{b^2}{c}z$$

which is a downward parabola *in that plane (assuming* c to be +ve).

Similarly (1) meets the ZX-plane in the upward parabola

$$x^2 = \frac{2a^2}{c}z \textit{ in that plane}$$

It meets the XY-plane ($z = 0$), where $\frac{x^2}{a^2} - \frac{y^2}{b^2} = 0$ or $y = \pm\frac{b}{a}x$

which are two straight lines in that plane equally inclined to X-axis.

(*iv*) **Generated by a variable curve.** The surface (1) meets the plane $z = k$, where [putting $z = k$ in (1)],

$$\frac{x^2}{a^2} - \frac{y^2}{b^2} = \frac{2k}{c} \quad \text{or} \quad \frac{x^2}{\left(\frac{2a^2k}{c}\right)} - \frac{y^2}{\left(\frac{2b^2k}{c}\right)} = 1.$$

Thus the surface is generated by a variable hyperbola

$$\frac{x^2}{\left(\frac{2a^2k}{c}\right)} - \frac{y^2}{\left(\frac{2b^2k}{c}\right)} = 1,\ z = k \qquad \text{...(2) [as } k \text{ varies]}$$

whose plane is || to the XY-plane, and centre (0, 0, k) moves on the Z-axis.

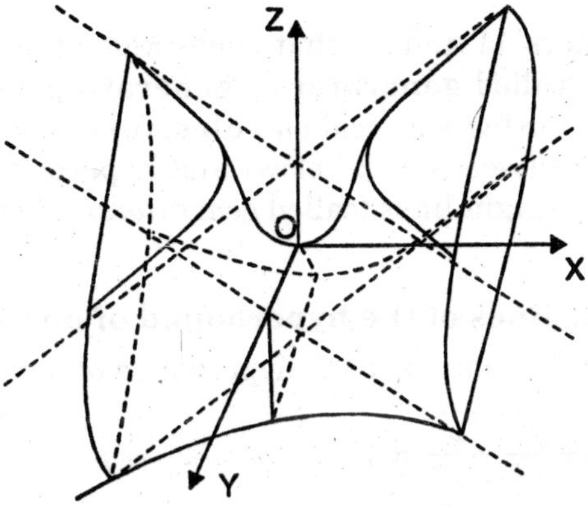

The hyperbola (2) has transverse axis || to X-axis if k is +ve and || to Y-axis if k is –ve.
[$\because$ *c is assumed to* be +ve]

Also the transverse semi-axis is $a\sqrt{\frac{2k}{c}}$ which increases as k (+ve) increases and $\to \infty$ as $k \to \infty$.

Thus surface extends to infinity above the XY-plane.

Similarly, the surface extends to infinity below the XY-plane.

The surface extends to infinity both above and below the XY-plane. Hence the shape of the surface is shown in the figure.

14

Generating Lines

1. Introduction

We know from Chapters 11 and 12 that cones and cylinders are surfaces which are generated by a moving line (called **generator** or **generating line**). (see Art. 1, chapter 11, and Art. 1, chapter 12) Such surfaces are called **ruled surfaces.** Hyperboloid of one sheet is also a ruled surface (For proof see Art. 2 below) and hyperbolic paraboloid is also a ruled surface *i.e.*, is generated by straight lines (called *generators*). We shall learn this aspect of the surfaces in this Chapter.

2. To find the generating lines of the hyperboloid of one sheet

We know that standard equation of the hyperboloid of one sheet is

$$\frac{x^2}{a^2}+\frac{y^2}{b^2}-\frac{z^2}{c^2}=1 \qquad \text{...(1)}$$

Rewriting Eqn. (1) as

$$\frac{x^2}{a^2}-\frac{z^2}{c^2}=1-\frac{y^2}{b^2} \quad \text{or} \quad \left(\frac{x}{a}-\frac{z}{c}\right)\left(\frac{x}{a}+\frac{z}{c}\right)=\left(1-\frac{y}{b}\right)\left(1+\frac{y}{b}\right)$$

$$\therefore \qquad \frac{\frac{x}{a}-\frac{z}{c}}{1-\frac{y}{b}}=\frac{1+\frac{y}{b}}{\frac{x}{a}+\frac{z}{c}} \qquad \text{...(2)}$$

and

$$\frac{\frac{x}{a}-\frac{z}{c}}{1+\frac{y}{b}}=\frac{1-\frac{y}{b}}{\frac{x}{a}+\frac{z}{c}} \qquad \text{...(3)}$$

Putting these fractions in Eqns. (2) and (3) equal to arbitrary constants λ and μ respectively, we have

From (2),
$$\frac{x}{a}-\frac{z}{c}=\lambda\left(1-\frac{y}{b}\right),\ 1+\frac{y}{b}=\lambda\left(\frac{x}{a}+\frac{z}{c}\right) \qquad \text{...(4)}$$

and from (3)
$$\frac{x}{a}-\frac{z}{c}=\mu\left(1+\frac{y}{b}\right),\ 1-\frac{y}{b}=\mu\left(\frac{x}{a}+\frac{z}{c}\right) \qquad \text{...(5)}$$

We know that each of the two equations (4) and (5) represents a family of lines (general form) for different values of λ and μ.

To each value of λ corresponds a member of family of lines (4) and to each value of μ corresponds a member of family of lines (5).

Now let us prove that **every point of each line of family (4) and every point of each line of family (5) lies on hyperboloid (1).**

Let (x_0, y_0, z_0) be any point of a member of family (4) obtained for some value λ_0 of λ.

$\therefore$ From (4), $\quad \frac{x_0}{a} - \frac{z_0}{c} = \lambda_0\left(1 - \frac{y_0}{b}\right)$ and $1 + \frac{y_0}{b} = \lambda_0\left(\frac{x_0}{a} + \frac{z_0}{c}\right)$

Dividing these two equations (to eliminate λ_0), we have

$$\frac{\frac{x_0}{a} - \frac{z_0}{c}}{1 + \frac{y_0}{b}} = \frac{1 - \frac{y_0}{b}}{\frac{x_0}{a} + \frac{z_0}{c}}$$

Cross-multiplying $\quad \frac{x_0^2}{a^2} - \frac{z_0^2}{c^2} = 1 - \frac{y_0^2}{b^2}$ or $\frac{x_0^2}{a^2} + \frac{y_0^2}{b^2} - \frac{z_0^2}{c^2} = 1.$

It shows that any point (x_0, y_0, z_0) of any member of family of lines (4) satisfies hyperboloid (1) *i.e.*, each member of family of lines (4) lies **Completely** (wholly) on hyperboloid (1).

$\therefore$ Lines (4) are generators of hyperboloid (1).

Similarly lines (5) are also generators of hyperboloid (1).

These two families of lines (4) and (5) are called **two systems of generating lines** (or generators) of the hyperboloid (1).

3. Through every point of the hyperboloid there passes one generator of each system

Let (x_0, y_0, z_0) be any point on hyperboloid (1) (Art. 2)

$$\therefore \quad \frac{x_0^2}{a^2} + \frac{y_0^2}{b^2} - \frac{z_0^2}{c^2} = 1 \quad ...(6)$$

Now the generators given by Eqn. (4) for some $\lambda = \lambda_0$ (say) of hyperboloid (1) will pass through the point (x_0, y_0, z_0) if (x_0, y_0, z_0) satisfies (4) for $\lambda = \lambda_0$

i.e., if $\quad \frac{x_0}{a} - \frac{z_0}{c} = \lambda_0\left(1 - \frac{y_0}{b}\right)$ and $1 + \frac{y_0}{b} = \lambda_0\left(\frac{x_0}{a} + \frac{z_0}{c}\right)$

i.e., if (on dividing the two equations) $\quad \frac{\frac{x_0}{a} - \frac{z_0}{c}}{1 + \frac{y_0}{b}} = \frac{1 - \frac{y_0}{b}}{\frac{x_0}{a} + \frac{z_0}{c}}$

i.e., if cross-multiplying $\quad \frac{x_0^2}{a^2} - \frac{z_0^2}{c^2} = 1 - \frac{y_0^2}{b^2}$

i.e., if $\quad \frac{x_0^2}{a^2} + \frac{y_0^2}{b^2} - \frac{z_0^2}{c^2} = 1$ which is true by (6).

Therefore a member of family (4) of generators is given by

$$\lambda (= \lambda_0) = \frac{\frac{x_0}{a} - \frac{z_0}{c}}{1 - \frac{y_0}{b}} \quad \text{or} \quad \frac{1 + \frac{y_0}{b}}{\frac{x_0}{a} + \frac{z_0}{c}}$$

passes through the given point (x_0, y_0, z_0) of hyperboloid (1).

Similarly it can be shown that a member of family (5) of generators will pass through the given point (x_0, y_0, z_0) of hyperboloid (1) for a value of

$$\mu = \mu_0 = \frac{\frac{x_0}{a} - \frac{z_0}{c}}{1 + \frac{y_0}{b}} \quad \text{or} \quad \frac{1 - \frac{y_0}{b}}{\frac{x_0}{a} + \frac{z_0}{c}}.$$

$\therefore$ **Through every point (x_0, y_0, z_0) of hyperboloid (1) there passes one generator of each system of generating lines (4) and (5).**

4. To prove that no two generators of the same system of the hyperboloid of one sheet intersect.

If possible, let two generators of the system (4) given by

$$\lambda = \lambda_1 \text{ and } \lambda = \lambda_2 \ (\lambda_1 \neq \lambda_2) \text{ intersect.}$$

Putting $\lambda = \lambda_1$ **and** $\lambda = \lambda_2$ **in(4), we have**

I. $\dfrac{x}{a} - \dfrac{z}{c} = \lambda_1\left(1 - \dfrac{y}{b}\right)$ ...(*i*) $\quad 1 + \dfrac{y}{b} = \lambda_1\left(\dfrac{x}{a} + \dfrac{z}{c}\right)$ ...(*ii*)

II. $\dfrac{x}{a} - \dfrac{z}{c} = \lambda_2\left(1 - \dfrac{y}{b}\right)$ ...(*iii*) $\quad 1 + \dfrac{y}{b} = \lambda_2\left(\dfrac{x}{a} + \dfrac{z}{c}\right)$ ...(*iv*)

as the equations of any two different generators of the λ-system.

Eqn. (*i*) – Eqn. (*iii*) gives

$$0 = (\lambda_1 - \lambda_2)\left(1 - \frac{y}{b}\right)$$

But $\lambda_1 - \lambda_2 \neq 0$ (as $\lambda_1 \neq \lambda_2$)

$$\therefore \quad 1 - \frac{y}{b} = 0 \quad \text{or} \quad 1 = \frac{y}{b} \quad \text{or} \quad y = b \qquad \text{...(v)}$$

Multiplying Eqn. (*ii*) by λ_2, Eqn (*iv*) by λ_1 and then subtracting, we have

$$\lambda_2\left(1 + \frac{y}{b}\right) - \lambda_1\left(1 + \frac{y}{b}\right) = \lambda_1\lambda_2\left(\frac{x}{a} + \frac{z}{c}\right) - \lambda_1\lambda_2\left(\frac{x}{a} + \frac{z}{c}\right)$$

or

$$(\lambda_2 - \lambda_1)\left(1 + \frac{y}{b}\right) = 0$$

But $\lambda_2 - \lambda_1 \neq 0$ as $\lambda_1 \neq \lambda_2$

$$\therefore \quad 1 + \frac{y}{b} = 0 \quad \text{or} \quad \frac{y}{b} = -1 \quad \text{or} \quad y = -b \qquad \text{...(vi)}$$

Eqn. (*v*) – Eqn. (*vi*) gives

$$0 = 2b \quad \text{or} \quad b = 0.$$

But this is impossible. ($\because$ Eqn. (1) of hyperboloid of one sheet does not exist for $b = 0$)

$\therefore$ **Equations (*i*), (*ii*), (*iii*), (*iv*) in x, y, z are inconsistent.**

Hence the two different generators of the λ-system given by Eqn. (4) do not intersect.

Similarly the two different generators of the μ-system given by Eqn. (5) do not intersect.

Example 1. *Find the condition that the line* $\dfrac{x-\alpha}{l} = \dfrac{y-\beta}{m} = \dfrac{z-\gamma}{n}$ *is a generator of the conicoid* $ax^2 + by^2 + cz^2 = 1$. **(K.U. 2007, 2003)**

Sol. **Equation of the conicoid is** $ax^2 + by^2 + cz^2 = 1$ ...(1)

Equation of the line is $\dfrac{x-\alpha}{l} = \dfrac{y-\beta}{m} = \dfrac{z-\gamma}{n}$ (= r say) ...(2)

Any point on line (2) is $x = lr + \alpha,\ y = mr + \beta,\ z = nr + \gamma$ ...(3)

Because line (2) is a generator of conicoid (1), therefore every point of line (2) satisfies conicoid (1).

Putting values of x, y, z from (3) in (1),

$$a(lr + \alpha)^2 + b(mr + \beta)^2 + c(nr + \gamma)^2 = 1$$

or $$a(l^2r^2 + 2\alpha lr + \alpha^2) + b(m^2r^2 + 2\beta mr + \beta^2) + c(n^2r^2 + 2\gamma nr + \gamma^2) - 1 = 0$$

or $$r^2(al^2 + bm^2 + cn^2) + 2r(al\alpha + bm\beta + cn\gamma) + (a\alpha^2 + b\beta^2 + c\gamma^2 - 1) = 0 \quad ...(4)$$

Equation (4) is true for all values of r.

[$\because$ Line (2) is a generator of conicoid (1) and hence every point of line (2) satisfies (1)]

$\therefore$ Equation (4) is an identity in r and hence all its coefficients are zero.

$\therefore$ $$al^2 + bm^2 + cn^2 = 0 \quad ...(5)$$

$$al\alpha + bm\beta + cn\gamma = 0 \quad ...(6)$$

$$a\alpha^2 + b\beta^2 + c\gamma^2 - 1 = 0 \quad ...(7)$$

Relations (5), (6) and (7) all conditions for line (2) to be generator of conicoid (1).

Example 2. *Find the equations of the generating lines of the hyperboloid* $\frac{x^2}{4} + \frac{y^2}{9} - \frac{z^2}{16}$ $= 1$ *which pass through the point* $(2, 3, -4)$.

Sol. We know that equations of any line through $(2, 3, -4)$ is

$$\frac{x-2}{l} = \frac{y-3}{m} = \frac{z+4}{n} \ (= (r \text{ say})) \quad ...(1)$$

Any point on line (1) is $(lr + 2, mr + 3, nr - 4)$.

This point for all values of r will satisfy the hyperboloid

$$\frac{x^2}{4} + \frac{y^2}{9} - \frac{z^2}{16} = 1 \quad ...(2)$$

[$\because$ Line (1) is a generator of hyperboloid (2)]

$\therefore$ $$\frac{(lr+2)^2}{4} + \frac{(mr+3)^2}{9} - \frac{(nr-4)^2}{16} = 1$$

or $$\frac{l^2r^2 + 4lr + 4}{4} + \frac{(m^2r^2 + 6mr + 9)}{9} - \frac{(n^2r^2 - 8nr + 16)}{16} = 1$$

or $$\left(\frac{l^2}{4} + \frac{m^2}{9} - \frac{n^2}{16}\right)r^2 + \left(l + \frac{2m}{3} + \frac{n}{2}\right)r = 0 \quad ...(3)$$

Equation (3) is true for all values of r and hence is an identity.

$\therefore$ Coeff. of r^2 is 0 and also coeff. of r is 0.

$\therefore$ $$\frac{l^2}{4} + \frac{m^2}{9} - \frac{n^2}{16} = 0$$

Multiplying by L.C.M. = 144, $36l^2 + 16m^2 - 9n^2 = 0 \quad ...(4)$

and $$l + \frac{2m}{3} + \frac{n}{2} = 0 \quad \text{or} \quad 6l + 4m + 3n = 0 \quad ...(5)$$

Let us solve Eqns. (4) and (5) for l, m, n.

From Eqn. (5), $6l = -(4m + 3n)$.

Putting this value of $6l$ in Eqn. (4),

$$(4m + 3n)^2 + 16m^2 - 9n^2 = 0 \quad \text{or} \quad 16m^2 + 9n^2 + 24mn + 16m^2 - 9n^2 = 0$$

or $$32m^2 + 24mn = 0 \quad \text{or} \quad 8m\,(4m + 3n) = 0$$

$\therefore$ Either $$8m = 0 \quad \text{or} \quad 4m + 3n = 0$$

i.e., $$m = 0 \quad \text{or} \quad 4m + 3n = 0$$

i.e., $$0\,.\,l + 1\,.\,m + 0\,.\,n = 0 \qquad \text{...(6)}$$

or $$0\,.\,l + 4m + 3n = 0 \qquad \text{...(7)}$$

Solving Eqns. (5) and (6) for l, m, n, we have

$$\frac{l}{4\,(0) - 3\,(1)} = \frac{m}{3\,(0) - 6\,(0)} = \frac{n}{6\,(1) - 4\,(0)} \quad \text{or} \quad \frac{l}{-3} = \frac{m}{0} = \frac{n}{6}$$

Dividing denominators by -3, $\dfrac{l}{1} = \dfrac{m}{0} = \dfrac{n}{-2}$

Putting these values of l, m, n in Eqn. (1), equation of one generator is

$$\frac{x-2}{1} = \frac{y-3}{0} = \frac{z+4}{-2}$$

Solving Eqns. (5) and (7) for l, m, n, we have

$$\frac{l}{12-12} = \frac{m}{0-18} = \frac{n}{24-0}$$

or $$\frac{l}{0} = \frac{m}{-18} = \frac{n}{24}$$

Dividing denominators by -6, $\dfrac{l}{0} = \dfrac{m}{3} = \dfrac{n}{-4}$

Putting these values of l, m, n in Eqn. (1), equations of the second generator are

$$\frac{x-2}{0} = \frac{y-3}{3} = \frac{z+4}{-4}.$$

5. *Prove that the equation of a generating line through any point of the hyperboloid of one sheet* $\dfrac{x^2}{a^2} + \dfrac{y^2}{b^2} - \dfrac{z^2}{c^2} = 1$ *can be written in the form* $\dfrac{x - a\cos\alpha}{a\sin\alpha} = \dfrac{y - b\sin\alpha}{-b\cos\alpha} = \dfrac{z}{\pm c}$.

Sol. Equation of the hyperboloid is $$\frac{x^2}{a^2} + \frac{y^2}{b^2} - \frac{z^2}{c^2} = 1 \qquad \text{...(1)}$$

Comparing with $\cos^2\alpha + \sin^2\alpha = 1$, *i.e.,* $(\cos\alpha)^2 + (\sin\alpha)^2 + 0^2 = 1$,

we have

$$\frac{x}{a} = \cos\alpha, \quad \frac{y}{b} = \sin\alpha \quad \text{and} \quad \frac{z}{c} = 0$$

$\therefore$ $x = a\cos\alpha$, $y = b\sin\alpha$ and $z = 0$

$\therefore$ Any point on hyperboloid (1) is $(a\cos\alpha, b\sin\alpha, 0)$.

$\therefore$ Equation of any generator of hyperboloid (1) through the point $(a\cos\alpha, b\sin\alpha, 0)$ is

$$\frac{x-a\cos\alpha}{l}=\frac{y-b\sin\alpha}{m}=\frac{z-0}{n}\ (=r \text{ say}) \quad ...(2)$$

Any point on generator (2) is $x = lr + a\cos\alpha$, $y = mr + b\sin\alpha$, $z = nr$.

Putting these values of x, y, z in Eqn. (2), we have

$$\frac{(lr+a\cos\alpha)^2}{a^2}+\frac{(mr+b\sin\alpha)^2}{b^2}-\frac{n^2r^2}{c^2}=1$$

or
$$\frac{l^2r^2+2lra\cos\alpha+a^2\cos^2\alpha}{a^2}+\frac{m^2r^2+b^2\sin^2\alpha+2mr\,b\sin\alpha}{b^2}-\frac{n^2r^2}{c^2}=1$$

or
$$r^2\left(\frac{l^2}{a^2}+\frac{m^2}{b^2}-\frac{n^2}{c^2}\right)+2r\left(\frac{l\cos\alpha}{a}+\frac{m\sin\alpha}{b}\right)+(\cos^2\alpha+\sin^2\alpha-1)=0$$

or
$$r^2\left(\frac{l^2}{a^2}+\frac{m^2}{b^2}-\frac{n^2}{c^2}\right)+2r\left(\frac{l\cos\alpha}{a}+\frac{m\sin\alpha}{b}\right)=0 \quad ...(3)$$

$\because$ Line (2) is a generator of hyperboloid (1) therefore equation (3) is true for all values of r.

$\therefore$ Equation (3) is an identity.

$\therefore$ Coeff. of r^2 is zero and coeff. of r is zero.

i.e.,
$$\frac{l^2}{a^2}+\frac{m^2}{b^2}-\frac{n^2}{c^2}=0 \quad ...(4)$$

and
$$\frac{l\cos\alpha}{a}+\frac{m\sin\alpha}{b}=0 \quad ...(5)$$

Let us solve Eqns. (4) and (5) for l, m, n.

From Eqn. (5),
$$l=-\frac{ma\sin\alpha}{b\cos\alpha} \quad ...(6)$$

Putting this value of l in equation (4),

$$\frac{m^2a^2\sin^2\alpha}{a^2b^2\cos^2\alpha}+\frac{m^2}{b^2}-\frac{n^2}{c^2}=0 \quad \text{or} \quad \frac{m^2}{b^2}(\tan^2\alpha+1)=\frac{n^2}{c^2}$$

or
$$\frac{m^2}{b^2}\sec^2\alpha=\frac{n^2}{c^2} \quad \text{or} \quad \frac{m^2}{b^2\cos^2\alpha}=\frac{n^2}{c^2} \quad \therefore \quad \frac{m}{b\cos\alpha}=\pm\frac{n}{c} \quad ...(7)$$

From (6),
$$\frac{-l}{a\sin\alpha}=\frac{-m}{b\cos\alpha}=\pm\frac{n}{c} \quad \text{[By (7)]}$$

or
$$\frac{l}{-a\sin\alpha}=\frac{m}{b\cos\alpha}=\frac{n}{\pm c} \quad \text{or} \quad \frac{l}{a\sin\alpha}=\frac{m}{-b\cos\alpha}=\frac{n}{\mp c}$$

Putting these values of l, m, n in Eqn. (2), equations of required generating lines are

$$\frac{x-a\cos\alpha}{a\sin\alpha}=\frac{y-b\sin\alpha}{-b\cos\alpha}=\frac{z}{\mp c}.$$

Example 1. *To show that the projections of the generators of a hyperboloid on any principal plane are tangents to the section of the hyperboloid by the principal plane.*

Sol. By Art. 5. equations of the generating lines through any point on the hyperboloid of one sheet are

$$\frac{x - a\cos\alpha}{a\sin\alpha} = \frac{y - b\sin\alpha}{-b\cos\alpha} = \frac{z}{\pm c}.$$

Taking positive sign, equations of one generator of hyperboloid of one sheet is

$$\frac{x - a\cos\alpha}{a\sin\alpha} = \frac{y - b\sin\alpha}{-b\cos\alpha} = \frac{z}{c} \quad \text{...(1)}$$

We know that equation of hyperboloid of one sheet is

$$\frac{x^2}{a^2} + \frac{y^2}{b^2} - \frac{z^2}{c^2} = 1 \quad \text{...(2)}$$

Now the projection of generator (1) on a principal plane say XOY *i.e.*, $z = 0$ (Putting $z = 0$ in (1)),

$$\frac{x - a\cos\alpha}{a\sin\alpha} = \frac{y - b\sin\alpha}{-b\cos\alpha}, z = 0$$

Cross-multiplying $-bx\cos\alpha + ab\cos^2\alpha = ay\sin\alpha - ab\sin^2\alpha$

or $\quad bx\cos\alpha + ay\sin\alpha = ab(\cos^2\alpha + \sin^2\alpha)$

or $\quad bx\cos\alpha + ay\sin\alpha = ab$

Dividing every term by ab,

$$\frac{x\cos\alpha}{a} + \frac{y\sin\alpha}{b} = 1, z = 0 \quad \text{...(3)}$$

Again the section of hyperboloid (2) by the same principal plane $z = 0$ is (putting $z = 0$ in (2))

$$\frac{x^2}{a^2} + \frac{y^2}{b^2} = 1, z = 0 \quad \text{...(4)}$$

Clearly line (3) is a tangent to ellipse (4) at the point $(a\cos\alpha, b\sin\alpha)$.

Again the projection of generating line (1) on the principal plane ZOX *i.e.*, $y = 0$ is

$$\frac{x - a\cos\alpha}{a\sin\alpha} = \frac{-b\sin\alpha}{-b\cos\alpha} = \frac{z}{c}$$

From first and third members

$$cx - ac\cos\alpha = az\sin\alpha$$

or $$cx - az\sin\alpha = ac\cos\alpha$$

Dividing by $ac\cos\alpha$,

$$\frac{x\sec\alpha}{a} - \frac{z}{c}\tan\alpha = 1 \quad \text{...(5)}$$

Again section of hyperboloid (2) by the principal plane $y = 0$ is

$$\frac{x^2}{a^2} - \frac{z^2}{c^2} = 1, y = 0 \quad \text{...(6)}$$

Clearly line (5) is a tangent to hyperbola (6) at the point $(a\sec\alpha, c\tan\alpha)$.

Similarly, we can show for the section $x = 0$.

6. To prove that any two generating lines of the hyperboloid of one sheet belonging to different systems intersect.

We known by equations (4) and (5) of Art. 2. that equations of two generators, one of each system are

I. $$\frac{x}{a}-\frac{z}{c}=\lambda\left(1-\frac{y}{b}\right) \quad ...(i) \qquad 1+\frac{y}{b}=\lambda\left(\frac{x}{a}+\frac{z}{c}\right) \quad ...(ii)$$

II. $$\frac{x}{a}-\frac{z}{c}=\mu\left(1+\frac{y}{b}\right) \quad ...(iii) \qquad 1-\frac{y}{b}=\mu\left(\frac{x}{a}+\frac{z}{c}\right) \quad ...(iv)$$

Let us firstly solve Eqns. (*i*), (*ii*) and (*iii*) simultaneously for x, y, z.

From (*i*) and (*iii*), equating the two values of $\frac{x}{a}-\frac{z}{c}$, we have

$$\lambda\left(1-\frac{y}{b}\right)=\mu\left(1+\frac{y}{b}\right) \quad \text{or} \quad \lambda-\lambda\frac{y}{b}=\mu+\mu\frac{y}{b}$$

or $$\lambda-\mu=\lambda\frac{y}{b}+\mu\frac{y}{b} \quad \text{or} \quad \lambda-\mu=(\lambda+\mu)\frac{y}{b} \quad \therefore \quad y=b\left(\frac{\lambda-\mu}{\lambda+\mu}\right)$$

Putting this value of y in Eqns. (*i*) and (*ii*),

$$\frac{x}{a}-\frac{z}{c}=\lambda\left(1-\frac{\lambda-\mu}{\lambda+\mu}\right)=\lambda\frac{(\lambda+\mu-\lambda+\mu)}{\lambda+\mu}=\frac{2\lambda\mu}{\lambda+\mu} \quad ...(v)$$

and from (*ii*), $$\frac{x}{a}+\frac{z}{c}=\frac{1}{\lambda}\left(1+\frac{y}{b}\right)=\frac{1}{\lambda}\left(1+\frac{\lambda-\mu}{\lambda+\mu}\right)=\frac{1}{\lambda}\left(\frac{\lambda+\mu+\lambda-\mu}{\lambda+\mu}\right)$$

$$=\frac{1}{\lambda}\cdot\frac{2\lambda}{\lambda+\mu}=\frac{2}{\lambda+\mu} \quad ...(vi)$$

Adding Eqns. (*v*) and (*vi*),

$$\frac{2x}{a}=\frac{2\lambda\mu+2}{\lambda+\mu}=\frac{2(\lambda\mu+1)}{\lambda+\mu} \quad \therefore \quad x=a\frac{(\lambda\mu+1)}{\lambda+\mu}=\frac{a(1+\lambda\mu)}{\lambda+\mu}$$

Similarly Eqn. (*vi*) – Eqn. (*v*) gives

$$z=\frac{c(1-\lambda\mu)}{\lambda+\mu}$$

Putting these values of x, y, z in Eqn. (*iv*), we have

$$1-\left(\frac{\lambda-\mu}{\lambda+\mu}\right)=\mu\left[\frac{\lambda\mu+1}{\lambda+\mu}+\frac{1-\lambda\mu}{\lambda+\mu}\right]$$

Multiplying by $\lambda+\mu$,

$$\lambda+\mu-\lambda+\mu=\mu(\lambda\mu+1+1-\lambda\mu) \quad \text{or} \quad 2\mu=2\mu \text{ which is true.}$$

Thus the lines I and II (two generating lines of hyperboloid of one sheet belonging to two different systems) intersect and their point of intersection is

$$\left(\frac{a(1+\lambda\mu)}{\lambda+\mu}, \frac{b(\lambda-\mu)}{\lambda+\mu}, \frac{c(1-\lambda\mu)}{\lambda+\mu}\right) \quad ...(vii)$$

Cor. Parametric Equations of the Hyperboloid

The co-ordinates given by (*vii*) (The point of intersection of any two generating lines of two different systems of hyperboloid of one sheet and hence is a point of hyperboloid) are

$$x = \frac{a(1+\lambda\mu)}{\lambda+\mu}, y = \frac{b(\lambda-\mu)}{\lambda+\mu}, z = \frac{c(1-\lambda\mu)}{\lambda+\mu}.$$

These are called **Parametric Equations** of the hyperboloid ; λ, μ being the two parameters.

Remark. Art. 6 is just converse of Art. 3.

7. To show that any plane through a generating line of a hyperboloid of one sheet is a tangent plane.

We know by equations (4) and (5) of Art. 2 that equations of two generators, one of each system are

I. $$\frac{x}{a} - \frac{z}{c} = \lambda\left(1 - \frac{y}{b}\right) \quad ...(i) \qquad 1 + \frac{y}{b} = \lambda\left(\frac{x}{a} + \frac{z}{c}\right) \quad ...(ii)$$

II. $$\frac{x}{a} - \frac{z}{c} = \mu\left(1 + \frac{y}{b}\right) \quad ...(iii) \qquad 1 - \frac{y}{b} = \mu\left(\frac{x}{a} + \frac{z}{c}\right) \quad ...(iv)$$

Firstly, let us prove that any plane through generator I passes through generator II and conversely also.

We know that equation of any plane through generator I is L.H.S. of (*i*) after making R.H.S. zero + *k* (L.H.S. of (*ii*) after making R.H.S. zero) = 0

i.e., $$\frac{x}{a} - \frac{z}{c} - \lambda\left(1 - \frac{y}{b}\right) + k\left[\left(1 + \frac{y}{b}\right) - \lambda\left(\frac{x}{a} + \frac{z}{c}\right)\right] = 0 \quad ...(v)$$

Similarly equation of any plane through generator II is

$$\frac{x}{a} - \frac{z}{c} - \mu\left(1 + \frac{y}{b}\right) + k'\left[1 - \frac{y}{b} - \mu\left(\frac{x}{a} + \frac{z}{c}\right)\right] = 0 \quad ...(vi)$$

Comparing coeffs. of *x*, *y*, *z* and constants in equations (*v*) and (*vi*),

we have $$\frac{1}{a} - \frac{k\lambda}{a} = \frac{1}{a} - \frac{k'\mu}{a} \quad \text{or} \quad \frac{-k\lambda}{a} = -\frac{k'\mu}{a} \quad \text{or} \quad k\lambda = k'\mu \quad ...(vii)$$

and $$\frac{\lambda}{b} + \frac{k}{b} = -\frac{\mu}{b} - \frac{k'}{b} \quad \text{or} \quad \lambda + k = -\mu - k' \quad \text{or} \quad k + k' = -\lambda - \mu \quad ...(viii)$$

and $$\frac{-1}{c} - \frac{k\lambda}{c} = \frac{-1}{c} - \frac{k'\mu}{c}$$ or $k\lambda + k'\mu$ which is same as Eqn. (*vii*).

and $$-\lambda + k = -\mu + k' \quad \text{or} \quad k - k' = \lambda - \mu \quad ...(ix)$$

Adding Eqns. (*viii*) and (*ix*), we have

$$2k = -2\mu \quad \text{or} \quad k = -\mu$$

Eqn. (*viii*) – Eqn. (*ix*) gives

$$2k' = -2\lambda \quad \text{or} \quad k' = -\lambda$$

Putting $k = -\mu$ and $k' = -\lambda$ in (*vii*), we have

$$-\lambda\mu = -\lambda\mu \quad \text{which is true.}$$

∴ Eqns. (*v*) and (*vi*) are identical for $k = -\mu$ and $k' = -\lambda$.

∴ Every plane through generator I contains generator II and conversely also.

Putting $k = -\mu$ in Eqn. (*v*) (or putting $k' = -\lambda$ in Eqn. (*vi*)), equation of any plane through generators I and II is

$$\frac{x}{a} - \frac{z}{c} - \lambda\left(1 - \frac{y}{b}\right) - \mu\left[\left(1 + \frac{y}{b}\right) - \lambda\left(\frac{x}{a} + \frac{z}{c}\right)\right] = 0$$

or
$$\frac{x}{a} - \frac{z}{c} - \lambda + \frac{\lambda y}{b} - \mu - \frac{\mu y}{b} + \frac{\lambda\mu x}{a} + \frac{\lambda\mu z}{c} = 0$$

or
$$\frac{x}{a}(1 + \lambda\mu) + \frac{y}{b}(\lambda - \mu) + \frac{z}{c}(-1 + \lambda\mu) = \lambda + \mu$$

Dividing every term by $\lambda + \mu$,

$$\frac{x}{a}\frac{(1+\lambda\mu)}{\lambda+\mu} + \frac{y}{b}\frac{(\lambda-\mu)}{\lambda+\mu} - \frac{z}{c}\frac{(1-\lambda\mu)}{\lambda+\mu} = 1 \qquad \ldots(x)$$

We know By Cor. Art. 6 that any point on hyperboloid of one sheet is

$$\left[\frac{a(1+\lambda\mu)}{\lambda+\mu}, \frac{b(\lambda-\mu)}{\lambda+\mu}, \frac{c(1-\lambda\mu)}{\lambda+\mu}\right] = (x_1, y_1, z_1)$$

We know that equation of the tangent plane at any point (x_1, y_1, z_1) to the hyperboloid of

one sheet
$$\frac{x^2}{a^2} + \frac{y^2}{b^2} - \frac{z^2}{c^2} = 1 \text{ is } \frac{xx_1}{a^2} + \frac{yy_1}{b^2} - \frac{zz_1}{c^2} = 1$$

i.e.,
$$\frac{x}{a^2}\frac{a(1+\lambda\mu)}{\lambda+\mu} + \frac{y}{b^2}\frac{b(\lambda-\mu)}{\lambda+\mu} - \frac{z}{c^2}\frac{c(1-\lambda\mu)}{\lambda+\mu} = 1$$

or
$$\frac{x}{a}\frac{(1+\lambda\mu)}{\lambda+\mu} + \frac{y}{b}\frac{(\lambda-\mu)}{\lambda+\mu} - \frac{z}{c}\frac{(1-\lambda\mu)}{\lambda+\mu} = 1$$

which is same as Eqn. (*x*).

$\therefore$ Any plane through a generating line of a hyperboloid of one sheet is a tangent plane.

Example 1. *Find the condition that the line $lx + my + nz + p = 0$, $l'x + m'y + n'z + p' = 0$ should be a generator of the hyperboloid* $\frac{x^2}{a^2} + \frac{y^2}{b^2} + \frac{z^2}{c^2} = 1$.

Sol. Equations of the given line are

and
$$\left.\begin{aligned} lx + my + nz + p &= 0 \\ l'x + m'y + n'z + p' &= 0 \end{aligned}\right\} \qquad \ldots(1)$$

Because line (1) is a generator of the hyperboloid

$$\frac{x^2}{a} + \frac{y^2}{b} + \frac{z^2}{c} = 1 \qquad \ldots(2)$$

therefore any plane through line (1) must be a tangent plane to hyperboloid (2) (by Art. 7) ...(3)

We know that equation of any plane through line (1) is

L.H.S. of I + K (L.H.S. of II) = 0

i.e.,
$$lx + my + nz + p + k(l'x + m'y + n'z + p') = 0$$

or
$$(l + kl')x + (m + km')y + (n + kn')z = -(p + kp')$$

By (3) this plane is a tangent plane to hyperboloid (2) for all values of k.

$$\therefore \quad \frac{(l+kl')^2}{\frac{1}{a}} + \frac{(m+km')^2}{\frac{1}{b}} + \frac{(n+kn')^2}{\frac{1}{c}} = (p+kp')^2$$

[Condition for the plane $lx + my + nz = p$ to be a tangent plane to conicoid

$$ax^2 + by^2 + cz^2 = 1 \text{ is } \frac{l^2}{a} + \frac{m^2}{b} + \frac{n^2}{c} = p^2 \Bigg]$$

or
$$a(l^2 + k^2l'^2 + 2ll'k) + b(m^2 + k^2m'^2 + 2mm'k) + c(n^2 + k^2n'^2 + 2nn'k) - (p^2 + k^2 p'^2 + 2kpp') = 0$$

or
$$k^2 (al'^2 + bm'^2 + cn'^2 - p'^2) + 2k(all' + bmm' + cnn' - pp') + (al^2 + bm^2 + cn^2 - p^2) = 0.$$

It is true for all values of k. Therefore it an identity. Therefore coeff. of k^2 is zero, coeff of k is zero and constant term is zero

i.e.,
$$al'^2 + bm'^2 + cn'^2 - p'^2 = 0$$
and
$$all' + bmm' + cnn' - pp' = 0$$
and
$$al^2 + bm^2 + cn^2 - p^2 = 0$$
which are the required conditions for line (1) to be a generator of hyperboloid (2).

8. Def. Point (θ, ϕ) on the hyperboloid $\frac{x^2}{a^2} + \frac{y^2}{b^2} - \frac{z^2}{c^2} = 1$ is the point whose co-ordinates are $(a \cos \theta \sec \phi, b \sin \theta \sec \phi, c \tan \phi)$

Putting $x = a \cos \theta \sec \phi$, $y = b \sin \theta \sec \phi$, $z = c \tan \phi$ in $\frac{x^2}{a^2} + \frac{y^2}{b^2} - \frac{z^2}{c^2} = 1$, we have

$$\frac{a^2 \cos^2 \theta \sec^2 \phi}{a^2} + \frac{b^2 \sin^2 \theta \sec^2 \phi}{b^2} - \frac{c^2 \tan^2 \phi}{c^2} = 1$$

or
$$(\cos^2 \theta + \sin^2 \theta) \sec^2 \phi - \tan^2 \phi = 1$$

or
$$\sec^2 \phi - \tan^2 \phi = 1$$

which is true.

$\therefore$ Any point (θ, ϕ) on the hyperboloid $\frac{x^2}{a^2} + \frac{y^2}{b^2} - \frac{z^2}{c^2} = 1$ is

$$(a \cos \theta \sec \phi, b \sin \theta \sec \phi, c \tan \phi).$$

9. Find the locus of the point of intersection of perpendicular generators of a hyperboloid of one sheet.

We know by Art. 2 that equations of λ and μ generators of hyperboloid of one sheet are

I. $$\frac{x}{a} - \frac{z}{c} = \lambda\left(1 - \frac{y}{b}\right) \quad ...(i) \qquad 1 + \frac{y}{b} = \lambda\left(\frac{x}{a} + \frac{z}{c}\right) \quad ...(ii)$$

i.e.,
$$\frac{x}{a} + \frac{\lambda y}{b} - \frac{z}{c} = \lambda \text{ and } \frac{\lambda x}{a} - \frac{y}{b} + \frac{\lambda z}{c} = 1$$

II. $$\frac{x}{a} - \frac{z}{c} = \mu\left(1 + \frac{y}{b}\right), \quad ...(iii) \qquad 1 - \frac{y}{b} = \mu\left(\frac{x}{a} + \frac{z}{c}\right) \quad ...(iv)$$

i.e., $$\frac{x}{a}-\frac{\mu y}{b}-\frac{z}{c}=\mu \quad \text{and} \quad \frac{\mu x}{a}+\frac{y}{b}+\frac{\mu z}{c}=1$$

Let l_1, m_1, n_1 be d.c.'s of λ generator I.

$\therefore$ This λ generator lies in both the planes (*i*) and (*ii*).

Using the condition $al + bm + cn = 0$ for a line to lie in a plane, we have

$$\frac{l_1}{a}+\lambda\frac{m_1}{b}-\frac{n_1}{c}=0 \quad \text{and} \quad \lambda\frac{l_1}{a}-\frac{m_1}{b}+\lambda\frac{n_1}{c}=0$$

(Dropping constants in (*i*) and (*ii*) and changing x, y, z to l_1, m_1, n_1 in (*i*) and (*ii*))

Solving for $\frac{l_1}{a}, \frac{m_1}{b}$ and $\frac{n_1}{c}$, we have

$$\frac{\frac{l_1}{a}}{\lambda^2-1}=\frac{\frac{m_1}{b}}{-\lambda-\lambda}=\frac{\frac{n_1}{c}}{-1-\lambda^2} \quad \text{or} \quad \frac{l_1}{a(\lambda^2-1)}=\frac{m_1}{-2b\lambda}=\frac{n_1}{-c(1+\lambda^2)}$$

$\therefore$ d.c.'s of λ-generator I are

$$a(\lambda^2-1), -2b\lambda, -c(1+\lambda^2) \quad (= l_1, m_1, n_1)$$

Let l_2, m_2, n_2 be d.c.'s of μ-generator II

Proceeding as above.

$$l_2, m_2, n_2 \text{ are } a(1-\mu^2), -2b\mu, c(1+\mu^2)$$

Because the two generators are perpendicular to each other, therefore

$$l_1l_2 + m_1m_2 + n_1n_2 = 0$$

$$\therefore \quad a^2(\lambda^2-1)(1-\mu^2) + 4b^2\lambda\mu - c^2(1+\lambda^2)(1+\mu^2) = 0$$

or $$a^2(\lambda^2-\lambda^2\mu^2-1+\mu^2) + 4b^2\lambda\mu - c^2(1+\mu^2+\lambda^2+\lambda^2\mu^2) = 0$$

or $$a^2[\lambda^2+\mu^2+2\lambda\mu-1-\lambda^2\mu^2-2\lambda\mu] + b^2[(\lambda+\mu)^2-(\lambda-\mu)^2]$$
$$- c^2[\lambda^2+\mu^2+2\lambda\mu+1+\lambda^2\mu^2-2\lambda\mu] = 0$$

or $$a^2[(\lambda+\mu)^2-(1+\lambda\mu)^2] + b^2[(\lambda+\mu)^2-(\lambda-\mu)^2] - c^2[(\lambda+\mu)^2+(1-\lambda\mu)^2] = 0$$

or $$a^2(\lambda+\mu)^2 - a^2(1+\lambda\mu)^2 + b^2(\lambda+\mu)^2 - b^2(\lambda-\mu)^2 - c^2(\lambda+\mu)^2 - c^2(1-\lambda\mu)^2 = 0$$

or $$(a^2+b^2-c^2)(\lambda+\mu)^2 = a^2(1+\lambda\mu)^2 + b^2(\lambda-\mu)^2 + c^2(1-\lambda\mu)^2$$

Dividing both sides by $(\lambda+\mu)^2$,

$$\left[a\left(\frac{1+\lambda\mu}{\lambda+\mu}\right)\right]^2+\left[b\left(\frac{\lambda-\mu}{\lambda+\mu}\right)\right]^2+\left[c\left(\frac{1-\lambda\mu}{\lambda+\mu}\right)\right]^2=a^2+b^2-c^2 \quad \text{...(v)}$$

Let (x, y, z) be the point of intersection of λ-generator I and μ-generator II.

$\therefore$ By Art. 6, $$x=\frac{a(1+\lambda\mu)}{\lambda+\mu},\ y=\frac{b(\lambda-\mu)}{\lambda+\mu},\ z=\frac{c(1-\lambda\mu)}{\lambda+\mu} \quad \text{...(vi)}$$

Putting values from Eqns. (*vi*) in (*v*) (To eliminate λ, μ),

required locus of point of intersection of λ and μ generators I and II is

$$x^2+y^2+z^2=a^2+b^2-c^2 \quad \text{...(vii)}$$

which is a sphere whose centre is origin and radius is $\sqrt{a^2+b^2-c^2}$.

Remark. Sphere (*vii*) is called **director sphere** of hyperboloid of one sheet.

10. To obtain the equations of the two systems of generators of the paraboloid

$$\frac{x^2}{a^2}-\frac{y^2}{b^2}=2z$$

We know that equation of paraboloid is $\frac{x^2}{a^2}-\frac{y^2}{b^2}=2z$ ...(1)

Rewriting Eqn. (1) as

$$\left(\frac{x}{a}-\frac{y}{b}\right)\left(\frac{x}{a}+\frac{y}{b}\right)=2z \quad \therefore \quad \frac{\frac{x}{a}-\frac{y}{b}}{z}=\frac{2}{\frac{x}{a}+\frac{y}{b}} \qquad \text{...(2)}$$

and

$$\frac{\frac{x}{a}-\frac{y}{b}}{2}=\frac{z}{\frac{x}{a}+\frac{y}{b}} \qquad \text{...(3)}$$

Putting these fractions in Eqns. (2) and (3) equal to arbitrary constants λ and μ respectively, we have

From (2), $\frac{x}{a}-\frac{y}{b}=\lambda z,\ 2=\lambda\left(\frac{x}{a}+\frac{y}{b}\right)$ ...(4)

and from (3), $\frac{x}{a}-\frac{y}{b}=2\mu,\ z=\mu\left(\frac{x}{a}+\frac{y}{b}\right)$ ...(5)

We know that each of the two equations (4) and (5) represents a family of lines (general form) for different values of λ and μ.

To each value of λ corresponds a member of family of lines (4) and to each value of μ corresponds a member of family of lines (5).

Now let us prove that every point of each line of family (4) and every point of each line of family (5) lies on paraboloid (1)

Let (x_0, y_0, z_0) be any point of a member of family (4) obtained for some value λ_0 of λ.

∴ From (4), $\frac{x_0}{a}-\frac{y_0}{b}=\lambda_0 . z_0$ and $2=\lambda_0\left(\frac{x_0}{a}+\frac{y_0}{b}\right)$

Dividing these two equations (To eliminate λ_0),

$$\frac{\frac{x_0}{a}-\frac{y_0}{b}}{2}=\frac{z_0}{\frac{x_0}{a}+\frac{y_0}{b}}$$

Cross-multiplying $\frac{x_0^2}{a^2}-\frac{y_0^2}{b^2}=2z_0$.

It shows that any point (x_0, y_0, z_0) of any member of family of lines (4) satisfies paraboloid (1) *i.e.*, each member of family of lines (4) lies completely (wholly) on paraboloid (1).

∴ Lines (4) are generators of paraboloid (1).

Similarly lines (5) are generators of paraboloid (1).

These two families of lines (4) and (5) are called two systems of generating lines (or generators) of the hyperboloid (1).

EXERCISE 1

1. Show that the equation to the generating lines through the point P(θ, ϕ) on the hyperboloid $\frac{x^2}{a^2}+\frac{y^2}{b^2}-\frac{z^2}{c^2}=1$ may be put in the form $\frac{x-a\cos\theta\sec\phi}{a\sin(\theta\pm\phi)}=\frac{y-b\sin\theta\sec\phi}{-b\cos(\theta\pm\phi)}=\frac{z-c\tan\phi}{\pm c}$.

[**Hint.** By Art. 5, any generator to the given hyperboloid is

$$\frac{x-a\cos\alpha}{a\sin\alpha}=\frac{y-b\sin\alpha}{-b\cos\alpha}=\frac{z}{c} \qquad ...(1)$$

It passes through the point (θ, ϕ) *i.e.*, the point

$$(a\cos\theta\sec\phi,\ b\sin\theta\sec\phi,\ c\tan\phi) \qquad \text{(By Art. 8)}$$

$\therefore$ It satisfies Eqn. (1).

$$\therefore \quad \frac{a\cos\theta\sec\phi-a\cos\alpha}{a\sin\alpha}=\frac{b\sin\theta\sec\phi-b\sin\alpha}{-b\cos\alpha}=\frac{c\tan\phi}{c}$$

Solve these equations for $\cos\alpha$ and $\sin\alpha$ and put the values in Eqn. (1).]

2. Find the equations of the generating lines of the hyperboloid $\frac{x^2}{4}+\frac{y^2}{9}-\frac{z^2}{16}=1$ which pass through the point $\left(2,-1,\frac{4}{3}\right)$.

3. Prove that the angle α between any two generating lines of two different systems of a hyperboloid of one sheet is given by

$$\tan\alpha=\frac{2abc}{p(a^2+b^2+c^2-\text{OP}^2)}$$

where p is the length of the perpendicular from the origin to the tangent plane at the point P to the hyperboloid of one sheet.

[**Hint.** Use the equations of generating lines obtained in Question 1.]

4. Prove that the perpendiculars from the origin on the generators of the hyperboloid $\frac{x^2}{a^2}+\frac{y^2}{b^2}-\frac{z^2}{c^2}=1$ lie on the curve $\frac{a^2(b^2+c^2)^2}{x^2}+\frac{b^2(a^2+c^2)^2}{y^2}=\frac{c^2(a^2-b^2)^2}{z^2}$. *(M.D.U. 2007)*

5. Show that the locus of the point of intersection of perpendicular generators to the paraboloid $\frac{x^2}{a^2}-\frac{y^2}{b^2}=2z$ is the plane $2z+a^2-b^2=0$.

6. Tangent planes to $\frac{x^2}{a^2}+\frac{y^2}{b^2}-\frac{z^2}{c^2}=1$ which are parallel to tangent planes to

$$\frac{b^2c^2x^2}{c^2-b^2}+\frac{c^2a^2y^2}{c^2-a^2}+\frac{a^2b^2z^2}{a^2+b^2}=0$$ cuts the surface in perpendicular generators.

7. Find the equation of a hyperboloid when x-axis is one of the generators.

[**Hint.** Let the equation of the hyperboloid (Quadric) be

$$ax^2+by^2+cz^2+2fyz+2gzx+2hxy+2ux+2vy+2wz+d=0 \qquad ...(1)$$

We know that equations of x-axis are $\frac{x-0}{1}=\frac{y-0}{0}=\frac{z-0}{0}=r$ (say) ...(2)

$\therefore \quad x = r, y = 0, z = 0$

Putting these values in (1), $ar^2 + 2ur + d = 0$ ...(3)

Equation (3) is an identity in r because line (2) *i.e.*, x-axis is a generator of conicoid (1).

$\therefore \quad a = 0, u = 0, d = 0.$

8. Find the equations of the generating lines of the hyperboloid $yz + 2zx + 3xy + 6 = 0$ which pass through the point $(-1, 0, 3)$. *(K.U. 2006)*

Answers

2. $\dfrac{x-2}{3} = \dfrac{y+1}{0} = \dfrac{z-\frac{4}{3}}{10}, \dfrac{x-2}{0} = \dfrac{y+1}{3} = \dfrac{z-\frac{4}{3}}{-4}.$

7. $by^2 + cz^2 + 2fyz + 2gzx + 2hxy + 2vy + 2wz = 0$

8. $x = -1, z = 3;\ \dfrac{x+1}{1} = \dfrac{y}{-1} = \dfrac{z-3}{3}.$

15

Confocal Conicoids

1. Introduction

We remind the reader that in co-ordinate geometry, we have defined **Confocal Conics.** All those conics which have the same two points as their foci are called **Confocal Conics.**

The equation $$\frac{x^2}{a^2-\lambda}+\frac{y^2}{b^2-\lambda}=1 \qquad ...(1)$$

represents a family of confocal conics because their foci are $(\pm\sqrt{A^2-B^2},0)$

i.e., $(\pm\sqrt{a^2-\lambda-(b^2-\lambda)},0) = (\pm\sqrt{a^2-b^2},0)$ which are independent of λ and hence are same for all the conics given by (1).

Def. Confocal Conicoids

Conicoids whose principal sections *i.e.,* sections by principal planes are confocal conics are called **Confocal Conicoids.**

For example consider the equation of central conicoids

$$\frac{x^2}{a^2-\lambda}+\frac{y^2}{b^2-\lambda}+\frac{z^2}{c^2-\lambda}=1 \qquad ...(2)$$

Let us consider the principal section of conicoid (2) by the principal plane XOY *i.e.,* $z=0$.

Putting $z=0$ in equation (2), we have

$\frac{x^2}{a^2-\lambda}+\frac{y^2}{b^2-\lambda}=1$ which is same as Eqn. (1) and represents a family of confocal conics as proved above.

$\therefore$ By def. given above Eqn. (2) represents a family of confocal conicoids for different values of λ.

Putting $\lambda=0$ in Eqn. (2), we have one member of family (2) as

$$\frac{x^2}{a^2}+\frac{y^2}{b^2}+\frac{z^2}{c^2}=1 \qquad ...(3)$$

We can also say that equation (2) represents confocal conicoids with conicoid of equation (3) for all values of λ. Similarly, we could prove it by taking sections by other principal planes.

Another example of confocal conicoids is

$$\frac{x^2}{a^2-\lambda}+\frac{y^2}{b^2-\lambda}=2z-\lambda$$

represents a family of paraboloids confocal with the paraboloid $\frac{x^2}{a^2}+\frac{y^2}{b^2}=2z$.

(The principal sections' are parabolic sections by the principal planes $x = 0$ and $y = 0$ only. See Art. 5 (*a*) (*iii*) Appendix Chapter 4)

Remark. If in the equation $\frac{x^2}{\alpha^2}+\frac{y^2}{\alpha^2-b^2}+\frac{z^2}{\alpha^2-c^2}=1$ ...(1)

where b and c are constants and α is **the parameter**, then equation (1) also represents a system of confocal conicoids (prove it as proved above). In this case α is called the **primary semi-axis.**

2. To prove that three conicoids confocal with a given central conicoid will pass through a given point, and one of the three is an ellipsoid, one is a Hyperboloid of one sheet and one is a Hyperboloid of two sheets.

Let the equation of given central conicoid be

$$\frac{x^2}{a^2}+\frac{y^2}{b^2}+\frac{z^2}{c^2}=1 \quad ...(1)$$

Let $a^2 > b^2 > c^2$ let the given point be (α, β, γ).

We know (By Art. 1) that equation of any conicoid confocal to (1) is

$$\frac{x^2}{a^2-\lambda}+\frac{y^2}{b^2-\lambda}+\frac{z^2}{c^2-\lambda}=1 \quad ...(2)$$

Because (2) is to pass through the given point (α, β, γ)

$$\therefore \quad \frac{\alpha^2}{a^2-\lambda}+\frac{\beta^2}{b^2-\lambda}+\frac{\gamma^2}{c^2-\lambda}=1$$

Multiplying by L.C.M. $= (a^2 - \lambda)(b^2 - \lambda)(c^2 - \lambda)$, we have

$$\alpha^2 (b^2 - \lambda)(c^2 - \lambda) + \beta^2 (a^2 - \lambda)(c^2 - \lambda) + \gamma^2 (a^2 - \lambda)(b^2 - \lambda)$$
$$= (a^2 - \lambda)(b^2 - \lambda)(c^2 - \lambda)$$

or
$$f(\lambda) = \alpha^2 (b^2 - \lambda)(c^2 - \lambda) + \beta^2 (a^2 - \lambda)(c^2 - \lambda) + \gamma^2 (a^2 - \lambda)(b^2 - \lambda)$$
$$- (a^2 - \lambda)(b^2 - \lambda)(c^2 - \lambda) = 0 \quad ...(3)$$

$f(\lambda)$ is a polynomial of degree 3 in λ. Highest degree term is λ^3.

Putting $\lambda = -\infty, c^2, b^2, a^2, \infty$ in Eqn. (3), we have

$$f(-\infty) = (-\infty)^3 = -\infty = \ominus$$

($\because$ A polynomial $f(x)$ behaves as its highest degree term as $x \to \pm\infty$)

$$f(c^2) = \gamma^2(a^2 - c^2)(b^2 - c^2) = \oplus \qquad | \because a^2 > b^2 > c^2$$
$$f(b^2) = \beta^2(a^2 - b^2)(c^2 - b^2) = \ominus \qquad | \because a^2 > b^2 > c^2$$
$$f(a^2) = \alpha^2(b^2 - a^2)(c^2 - a^2) = \oplus \qquad | \because a^2 > b^2 > c^2$$
$$f(\infty) = (\infty)^3 = \infty = \oplus$$

$f(-\infty)$ and $f(c^2)$, are of opposite signs.

$\therefore$ one real root say λ_1 of equation (3) lies between $-\infty$ and c^2.

($\because$ If $f(a)$ and $f(b)$ are of opposite signs for a polynomial $f(x)$, then one real root of $f(x) = 0$ lies between a and b.)

Similarly one real value of λ say λ_2 lies between c^2 and b^2 and one real value of λ say λ_3 lies between b^2 and a^2.

$\therefore \quad -\infty < \lambda_1 < c^2 < \lambda_2 < b^2 < \lambda_3 < a^2$

Because all the three roots $\lambda_1, \lambda_2, \lambda_3$ of equation (3) are real ; therefore three conicoids confocal with a given centroid conicoid (1) will pass through the given point (α, β, γ).

Putting $\lambda = \lambda_1, \lambda_2, \lambda_3$ in Eqn. (2) the equations of three confocal conicoids passing through a given point (α, β, γ) are

$$\frac{x^2}{a^2 - \lambda_1} + \frac{y^2}{b^2 - \lambda_1} + \frac{z^2}{c^2 - \lambda_1} = 1 \quad \text{...(4)}$$

and
$$\frac{x^2}{a^2 - \lambda_2} + \frac{y^2}{b^2 - \lambda_2} + \frac{z^2}{c^2 - \lambda_2} = 1 \quad \text{...(5)}$$

and
$$\frac{x^2}{a^2 - \lambda_3} + \frac{y^2}{b^2 - \lambda_3} + \frac{z^2}{c^2 - \lambda_3} = 1 \quad \text{...(6)}$$

Because $\lambda_1 < c^2 < b^2 < a^2 \quad \therefore \quad a^2 - \lambda_1, b^2 - \lambda_1$ and $c^2 - \lambda_1$ are all positive and hence equation (4) represents an ellipsoid.

Because $c^2 < \lambda_2$ and $\lambda_2 < b^2 < a^2 \quad \therefore \quad a^2 - \lambda_2, b^2 - \lambda_2$ are positive and $c^2 - \lambda_2$ is negative and hence equation (5) represents a hyperboloid of one sheet.

Because $c^2 < \lambda_3$ and $b^2 < \lambda_3$ and $\lambda_3 < a^2$, therefore $a^2 - \lambda_3$ is positive, $b^2 - \lambda_3$ is negative and $c^2 - \lambda_3$ is negative ; therefore Eqn. (6) represents a hyperboloid of two sheets.

$\therefore$ Three confocal conicoids ; one ellipsoid, one hyperboloid of one sheet and hyperboloid of two sheets pass through a given point.

3. To prove that three paraboloids confocal with a given paraboloid pass through a given point-two elliptic and one hyperbolic.

Let the equation of the given paraboloid be

$$\frac{x^2}{a^2} + \frac{y^2}{b^2} = 2z \quad \text{...(1)}$$

without loss of generality, let $a^2 > b^2$

Let the given point be (α, β, γ).

We know that (By Art. 1) equation of any conicoid confocal to (1) is

$$\frac{x^2}{a^2 - \lambda} + \frac{y^2}{b^2 - \lambda} = 2z - \lambda \quad \text{...(2)}$$

Because (2) is to pass through the given point (α, β, γ),

$$\therefore \quad \frac{\alpha^2}{a^2 - \lambda} + \frac{\beta^2}{b^2 - \lambda} = 2\gamma - \lambda$$

Multiplying by L.C.M. $= (a^2 - \lambda)(b^2 - \lambda)$, we have

$$\alpha^2 (b^2 - \lambda) + \beta^2 (a^2 - \lambda) = (a^2 - \lambda)(b^2 - \lambda)(2\gamma - \lambda)$$

or
$$f(\lambda) = \alpha^2 (b^2 - \lambda) + \beta^2 (a^2 - \lambda) - (a^2 - \lambda)(b^2 - \lambda)(2\gamma - \lambda) = 0 \quad \text{...(3)}$$

$f(\lambda)$ is a polynomial of degree 3 in λ. Highest degree term is λ^3

$$f(-\infty) = (-\infty)^3 = -\infty = \ominus$$
$$f(b^2) = \beta^2(a^2 - b^2) = \oplus \qquad | \because \; a^2 > b^2$$
$$f(a^2) = \alpha^2(b^2 - a^2) = \ominus \qquad | \because \; a^2 > b^2$$
$$f(\infty) = (\infty)^3 = \infty = \oplus$$

$f(-\infty)$ and $f(b^2)$ are of opposite signs,

therefore one real root say λ_1 of Eqn. (3) lies between $-\infty$ and b^2.

i.e., $$-\infty < \lambda_1 < b^2.$$

Similarly one real value of λ say λ_2 lies between b^2 and a^2 and one real value of λ say λ_3 lies between a^2 and ∞.

$\therefore$ $$-\infty < \lambda_1 < b^2 < \lambda_2 < a^2 < \lambda_3 < \infty.$$

Because all the three roots $\lambda_1, \lambda_2, \lambda_3$ of equation (3) are real, therefore three paraboloids confocal with given paraboloid (1) will pass through the given point (α, β, γ).

Putting $\lambda = \lambda_1, \lambda_2, \lambda_3$ in Eqn. (3), the equations of three confocal paraboloids are

$$\frac{x^2}{a^2-\lambda_1} + \frac{y^2}{b^2-\lambda_1} = 2z - \lambda_1 \qquad \text{...(4)}$$

and $$\frac{x^2}{a^2-\lambda_2} + \frac{y^2}{b^2-\lambda_2} = 2z - \lambda_2 \qquad \text{...(5)}$$

and $$\frac{x^2}{a^2-\lambda_3} + \frac{y^2}{b^2-\lambda_3} = 2z - \lambda_3 \qquad \text{...(6)}$$

Because $\lambda_1 < b^2 < a^2$, therefore $a^2 - \lambda_1$ and $b^2 - \lambda_1$ are both positive and hence equation (4) represents an elliptic paraboloid.

Because $b^2 < \lambda_2$ and $\lambda_2 < a^2$, therefore $a^2 - \lambda_2$ is positive and $(b^2 - \lambda_2)$ is negative and hence equation (5) represents a hyperbolic paraboloid.

Again because $b^2 < a^2 < \lambda_3$, therefore $a^2 - \lambda_3$ and $b^2 - \lambda_3$ are both negative and hence equation (6) represents an elliptic paraboloid. Therefore three paraboloids confocal with a given paraboloid pass through a given point, two elliptic and one hyperbolic.

4. To prove that one conicoid confocal with a given conicoid touches a given plane

Let the equation of the given conicoid be

$$\frac{x^2}{a^2} + \frac{y^2}{b^2} + \frac{z^2}{c^2} = 1 \qquad \text{...(1)}$$

We know that equation of conicoids confocal with (1) is

$$\frac{x^2}{a^2-\lambda} + \frac{y^2}{b^2-\lambda} + \frac{z^2}{c^2-\lambda} = 1 \qquad \text{...(2)}$$

Let the equation of the given plane be

$$lx + my + nz = p \qquad \text{...(3)}$$

If plane (3) touches confocal conicoid (2), then

$$\frac{l^2}{\left(\dfrac{1}{a^2-\lambda}\right)} + \frac{m^2}{\left(\dfrac{1}{b^2-\lambda}\right)} + \frac{n^2}{\left(\dfrac{1}{c^2-\lambda}\right)} = p^2$$

(Condition for the plane $lx + my + nz = p$ to be a tangent plane to conicoid $ax^2 + by^2 + cz^2 = 1$ is $\dfrac{l^2}{a} + \dfrac{m^2}{b} + \dfrac{n^2}{c} = p^2$)

or $$(a^2 - \lambda)\, l^2 + (b^2 - \lambda)\, m^2 + (c^2 - \lambda)\, n^2 = p^2$$

It is a first degree equation in λ and hence gives one and one value of λ. Hence only one confocal conicoid given by equation (2) will touch the given plane (3).

5. To prove that two conicoids confocal with a given conicoid touch a given line

(K.U. 2003, 2006)

Let the equation of the given conicoid be

$$\frac{x^2}{a^2}+\frac{y^2}{b^2}+\frac{z^2}{c^2}=1 \qquad ...(1)$$

We know that equation of conicoids confocal with (1) is

$$\frac{x^2}{a^2-\lambda}+\frac{y^2}{b^2-\lambda}+\frac{z^2}{c^2-\lambda}=1 \qquad ...(2)$$

Let the equation of the given line be

$$\frac{x-\alpha}{l}=\frac{y-\beta}{m}=\frac{z-\gamma}{n} \;(= r \text{ say}) \qquad ...(3)$$

Any point on line (3) is $x = lr + \alpha$, $y = mr + \beta$, $z = nr + \gamma$.

Putting these values of x, y, z in Eqn. (2), we have

$$\frac{(lr+\alpha)^2}{a^2-\lambda}+\frac{(mr+\beta)^2}{b^2-\lambda}+\frac{(nr+\gamma)^2}{c^2-\lambda}=1$$

or $$\frac{l^2r^2+\alpha^2+2lr\alpha}{a^2-\lambda}+\frac{m^2r^2+\beta^2+2mr\beta}{b^2-\lambda}+\frac{n^2r^2+\gamma^2+2nr\gamma}{c^2-\lambda}-1=0$$

or $$r^2\left(\frac{l^2}{a^2-\lambda}+\frac{m^2}{b^2-\lambda}+\frac{n^2}{c^2-\lambda}\right)+2r\left(\frac{l\alpha}{a^2-\lambda}+\frac{m\beta}{b^2-\lambda}+\frac{n\gamma}{c^2-\lambda}\right)$$
$$+\left(\frac{\alpha^2}{a^2-\lambda}+\frac{\beta^2}{b^2-\lambda}+\frac{\gamma^2}{c^2-\lambda}-1\right)=0 \qquad ...(4)$$

Equation (4) is a quadratic in r and hence two values of r and hence two points of intersection of line (3) and confocal conicoid (2).

Because line (3) is a tangent line to confocal conicoid (2), therefore the two values of r from equation (4) are equal.

$\therefore$ For Eqn. (4), $B^2 - 4AC = 0$ *i.e.,* $B^2 = 4AC$

i.e., $$4\left(\frac{l\alpha}{a^2-\lambda}+\frac{m\beta}{b^2-\lambda}+\frac{n\gamma}{c^2-\lambda}\right)^2=4\left(\frac{l^2}{a^2-\lambda}+\frac{m^2}{b^2-\lambda}+\frac{n^2}{c^2-\lambda}\right)$$
$$\left(\frac{\alpha^2}{a^2-\lambda}+\frac{\beta^2}{b^2-\lambda}+\frac{\gamma^2}{c^2-\lambda}-1\right)$$

Dividing both sides by 4,

$$\frac{l^2\alpha^2}{(a^2-\lambda)^2}+\frac{m^2\beta^2}{(b^2-\lambda)^2}+\frac{n^2\gamma^2}{(c^2-\lambda)^2}+\frac{2lm\alpha\beta}{(a^2-\lambda)(b^2-\lambda)}$$
$$+\frac{2mn\beta\gamma}{(b^2-\lambda)(c^2-\lambda)}+\frac{2ln\,\alpha\gamma}{(a^2-\lambda)(c^2-\lambda)}$$

$$= \frac{l^2\alpha^2}{(a^2-\lambda)^2} + \frac{l^2\beta^2}{(a^2-\lambda)(b^2-\lambda)} + \frac{l^2\gamma^2}{(a^2-\lambda)(c^2-\lambda)} - \frac{l^2}{a^2-\lambda} + \frac{m^2\alpha^2}{(b^2-\lambda)(a^2-\lambda)}$$

$$+ \frac{m^2\beta^2}{(b^2-\lambda)^2} + \frac{m^2\gamma^2}{(b^2-\lambda)(c^2-\lambda)} - \frac{m^2}{b^2-\lambda} + \frac{n^2\alpha^2}{(a^2-\lambda)(c^2-\lambda)}$$

$$+ \frac{n^2\beta^2}{(b^2-\lambda)(c^2-\lambda)} + \frac{n^2\gamma^2}{(c^2-\lambda)^2} - \frac{n^2}{c^2-\lambda}$$

or $$\frac{(-m^2\gamma^2 - n^2\beta^2 + 2mn\beta\gamma)}{(b^2-\lambda)(c^2-\lambda)} + \frac{(-l^2\gamma^2 - n^2\alpha^2 + 2ln\alpha\gamma)}{(a^2-\lambda)(c^2-\lambda)} + \frac{-l^2\beta^2 - m^2\alpha^2 + 2lm\alpha\beta)}{(a^2-\lambda)(b^2-\lambda)}$$

$$= -\frac{l^2}{a^2-\lambda} - \frac{m^2}{b^2-\lambda} - \frac{n^2}{c^2-\lambda}$$

Dividing every term by – 1 and multiplying by L.C.M. = $(a^2 - \lambda)(b^2 - \lambda)(c^2 - \lambda)$, we have

$$(m\gamma - n\beta)^2 (a^2 - \lambda) + (l\gamma - n\alpha)^2 (b^2 - \lambda) + (l\beta - m\alpha)^2 (c^2 - \lambda)$$
$$= l^2 (b^2 - \lambda)(c^2 - \lambda) + m^2 (a^2 - \lambda)(c^2 - \lambda) + n^2(a^2 - \lambda)(b^2 - \lambda) \quad ...(5)$$

This equation (5) is a quadratic in λ and hence gives the values of λ say λ_1 and λ_2. Hence two confocal conicoids given by Eqn. (2) namely

$$\frac{x^2}{a^2-\lambda_1} + \frac{y^2}{b^2-\lambda_1} + \frac{z^2}{c^2-\lambda_1} = 1 \text{ and } \frac{x^2}{a^2-\lambda_2} + \frac{y^2}{b^2-\lambda_2} + \frac{z^2}{c^2-\lambda_2} = 1$$

will touch the given line (3).

6. If a straight line touches two confocal conicoids, the tangent planes at the points of contact will be at right angles

By Art. 5, we know that a given straight line say l touches two **different** confocal conicoids namely

$$\frac{x^2}{a^2-\lambda_1} + \frac{y^2}{b^2-\lambda_1} + \frac{z^2}{c^2-\lambda_1} = 1 \quad ...(1)$$

and $$\frac{x^2}{a^2-\lambda_2} + \frac{y^2}{b^2-\lambda_2} + \frac{z^2}{c^2-\lambda_2} = 1 \quad ...(2)$$

where of course $\lambda_1 \neq \lambda_2$.

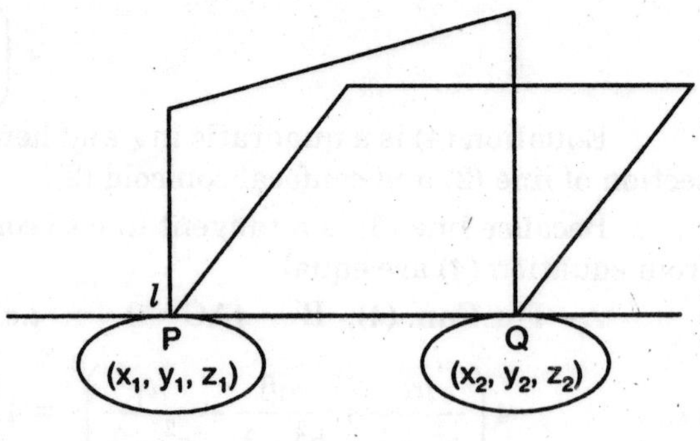

Let line l touch the confocal conicoids (1) and (2) at $P(x_1, y_1, z_1)$ and $Q(x_2, y_2, z_2)$ respectively.

We know that equation of tangent plane at $P(x_1, y_1, z_1)$ to confocal conicoid (1) is

$$\frac{xx_1}{a^2-\lambda_1} + \frac{yy_1}{b^2-\lambda_1} + \frac{zz_1}{c^2-\lambda_1} = 1 \quad ...(3)$$

(Changing x^2 to xx_1, y^2 to yy_1, z^2 to zz_1.)

Again equation of tangent plane at $Q(x_2, y_2, z_2)$ to confocal conicoid (2) is

$$\frac{xx_2}{a^2-\lambda_2} + \frac{yy_2}{b^2-\lambda_2} + \frac{zz_2}{c^2-\lambda_2} = 1 \quad ...(4)$$

We know by definition of tangent plane that tangent plane at a point is the locus of (*i.e.*, contains all) the tangent lines at that point to the conicoid.

Therefore point $Q(x_2, y_2, z_2)$ of common tangent line l also lies on tangent plane (3) and point $P(x_1, y_1, z_1)$ of common tangent line l also lies on tangent plane (4).

$$\therefore \quad \frac{x_1x_2}{a^2-\lambda_1}+\frac{y_1y_2}{b^2-\lambda_1}+\frac{z_1z_2}{c^2-\lambda_1}=1 \qquad \text{...(5)}$$

and

$$\frac{x_1x_2}{a^2-\lambda_2}+\frac{y_1y_2}{b^2-\lambda_2}+\frac{z_1z_2}{c^2-\lambda_2}=1 \qquad \text{...(6)}$$

Eqn. (5) – Eqn. (6) gives

$$\Sigma x_1x_2\left(\frac{1}{a^2-\lambda_1}-\frac{1}{a^2-\lambda_2}\right)=0 \quad \text{or} \quad \Sigma x_1x_2\frac{(a^2-\lambda_2-a^2+\lambda_1)}{(a^2-\lambda_1)(a^2-\lambda_2)}=0$$

$$\text{or} \quad \Sigma(\lambda_1-\lambda_2)\frac{x_1x_2}{(a^2-\lambda_1)(a^2-\lambda_2)}=0 \quad \text{or} \quad (\lambda_1-\lambda_2)\,\Sigma\frac{x_1x_2}{(a^2-\lambda_1)(a^2-\lambda_2)}=0$$

But $\lambda_1-\lambda_2\neq 0$ $\quad (\because \ \lambda_1\neq\lambda_2)$

$$\therefore \quad \Sigma\frac{x_1x_2}{(a^2-\lambda_1)(a^2-\lambda_2)}=0$$

i.e.,

$$\frac{x_1x_2}{(a^2-\lambda_1)(a^2-\lambda_2)}+\frac{y_1y_2}{(b^2-\lambda_1)(b^2-\lambda_2)}+\frac{z_1z_2}{(c^2-\lambda_1)(c^2-\lambda_2)}=0 \qquad \text{...(7)}$$

Now tangent planes (3) and (4) will be at right angles if

$$a_1a_2+b_1b_2+c_1c_2=0$$

i.e., if product of coeffs. of x + ... = 0

i.e., if

$$\frac{x_1}{a^2-\lambda_1}\,\frac{x_2}{a^2-\lambda_2}+\frac{y_1}{b^2-\lambda_1}\,\frac{y_2}{b^2-\lambda_2}+\frac{z_1}{c^2-\lambda_1}\,\frac{z_2}{c^2-\lambda_2}=0$$

i.e., if

$$\frac{x_1x_2}{(a^2-\lambda_1)(a^2-\lambda_2)}+\frac{y_1y_2}{(b^2-\lambda_1)(b^2-\lambda_2)}+\frac{z_1z_2}{(c^2-\lambda_1)(c^2-\lambda_2)}=0$$

which is true by Eqn. (7).

Hence the tangent planes at the two points of contact of a common tangent line are at right angles.

Def. Angle between two surfaces at their point of intersection

= Angle between their tangent planes at their point of intersection.

7. To prove that the confocal conicoids cut one another at right angles at all their common points i.e., the tangent planes at any common point are at right angles

Let the two different conicoids confocal to the conicoid $\frac{x^2}{a^2}+\frac{y^2}{b^2}+\frac{z^2}{c^2}=1$ be

$$\frac{x^2}{a^2-\lambda_1}+\frac{y^2}{b^2-\lambda_1}+\frac{z^2}{c^2-\lambda_1}=1 \qquad \text{...(1)}$$

and

$$\frac{x^2}{a^2-\lambda_2}+\frac{y^2}{b^2-\lambda_2}+\frac{z^2}{c^2-\lambda_2}=1 \qquad \text{...(2)}$$

where $\lambda_1\neq\lambda_2$.

Let (x_1, y_1, z_1) be any common point (*i.e.*, point of intersection) of conicoids (1) and (2).

$\therefore$ (x_1, y_1, z_1) satisfies both equations (1) and (2).

$$\therefore \quad \frac{x_1^2}{a^2-\lambda_1}+\frac{y_1^2}{b^2-\lambda_1}+\frac{z_1^2}{c^2-\lambda_1}=1 \quad ...(3)$$

and
$$\frac{x_1^2}{a^2-\lambda_2}+\frac{y_1^2}{b^2-\lambda_2}+\frac{z_1^2}{c^2-\lambda_2}=1 \quad ...(4)$$

Eqn. (3) – Eqn. (4) gives

$$\Sigma x_1^2\left(\frac{1}{a^2-\lambda_1}-\frac{1}{a^2-\lambda_2}\right)=0 \quad \text{or} \quad \Sigma x_1^2\left(\frac{a^2-\lambda_2-a^2+\lambda_1}{(a^2-\lambda_1)(a^2-\lambda_2)}\right)=0$$

or
$$\Sigma\frac{x_1^2(\lambda_1-\lambda_2)}{(a^2-\lambda_1)(a^2-\lambda_2)}=0 \quad \text{or} \quad (\lambda_1-\lambda_2)\,\Sigma\frac{x_1^2}{(a^2-\lambda_1)(a^2-\lambda_2)}=0$$

But $\lambda_1-\lambda_2\neq 0$ as $\lambda_1\neq\lambda_2$

$$\therefore \quad \Sigma\frac{x_1^2}{(a^2-\lambda_1)(a^2-\lambda_2)}=0$$

or
$$\frac{x_1^2}{(a^2-\lambda_1)(a^2-\lambda_2)}+\frac{y_1^2}{(b^2-\lambda_1)(b^2-\lambda_2)}+\frac{z_1^2}{(c^2-\lambda_1)(c^2-\lambda_2)}=0 \quad ...(5)$$

Equation of the tangent plane at (x_1, y_1, z_1) to confocal conicoid (1) is

$$\frac{xx_1}{a^2-\lambda_1}+\frac{yy_1}{b^2-\lambda_1}+\frac{zz_1}{c^2-\lambda_1}=1 \quad ...(6)$$

Equation of the tangent plane at (x_1, y_1, z_1) to confocal conicoid (2) is

$$\frac{xx_1}{a^2-\lambda_2}+\frac{yy_1}{b^2-\lambda_2}+\frac{zz_1}{c^2-\lambda_2}=1 \quad ...(7)$$

Tangent planes (6) and (7) will be at right angles

if
$$a_1a_2+b_1b_2+c_1c_2=0$$

i.e., if
$$\frac{x_1}{a^2-\lambda_1}\cdot\frac{x_1}{a^2-\lambda_2}+\frac{y_1}{b^2-\lambda_1}\cdot\frac{y_1}{b^2-\lambda_2}+\frac{z_1}{c^2-\lambda_1}\cdot\frac{z_1}{c^2-\lambda_2}=0$$

i.e., if
$$\frac{x_1^2}{(a^2-\lambda_1)(a^2-\lambda_2)}+\frac{y_1^2}{(b^2-\lambda_1)(b^2-\lambda_2)}+\frac{z_1^2}{(c^2-\lambda_1)(c^2-\lambda_2)}=0$$

which is true by Eqn. (5).

$\therefore$ Tangent planes (6) and (7) at the common point (x_1, y_1, z_1) to conicoids (1) and (2) are at right angles.

$\therefore$ By def. (given just above Art. 7), confocal conicoids (1) and (2) cut one another at right angles at all their common points.

8. Corresponding Points

Def. Two points (x, y, z) and (x', y', z') lying respectively on the conicoids

$$\frac{x^2}{a^2}+\frac{y^2}{b^2}+\frac{z^2}{c^2}=1 \quad \text{and} \quad \frac{x^2}{\alpha^2}+\frac{y^2}{\beta^2}+\frac{z^2}{\gamma^2}=1$$

are said to be *corresponding points* when

$$\frac{x}{a}=\frac{x'}{\alpha}, \qquad \frac{y}{b}=\frac{y'}{\beta}, \qquad \frac{z}{c}=\frac{z'}{\gamma}.$$

Example 1. *A given plane and parallel tangent plane to a conicoid are at distances p and p_0 from the centre. Prove that the parameter of the confocal conicoid which touches the plane is $p_0^2 - p^2$.*

Sol. Let the given conicoid be $\frac{x^2}{a^2}+\frac{y^2}{b^2}+\frac{z^2}{c^2}=1$ ...(1)

We know that centre of conicoid (1) is origin (0, 0, 0).

We also know that equation of the plane at a given distance p from the origin (centre) is

$$lx + my + nz = p \text{ (Normal Form)} \qquad ...(2)$$

where l, m, n are d.c.'s of Normal to plane (2) and hence

$$l^2 + m^2 + n^2 = 1 \qquad ...(3)$$

Again equation of any plane parallel to plane (2) at a given distance p_0 from the centre (origin) is $lx + my + nz = p_0$ ...(4)

Because plane (4) is a tangent plane to conicoid (1) (given)

$$\therefore \quad \frac{l^2}{\left(\frac{1}{a^2}\right)}+\frac{m^2}{\left(\frac{1}{b^2}\right)}+\frac{n^2}{\left(\frac{1}{c^2}\right)}=p_0^2 \qquad \left|\ \frac{l^2}{a}+\frac{m^2}{b}+\frac{n^2}{c}=p^2\right.$$

or
$$a^2l^2 + b^2m^2 + c^2n^2 = p_0^2 \qquad ...(5)$$

We know that equation of any conicoid confocal to conicoid (1) is

$$\frac{x^2}{a^2-\lambda}+\frac{y^2}{b^2-\lambda}+\frac{z^2}{c^2-\lambda}=1 \qquad ...(6)$$

where λ is the parameter.

Because plane (1) touches the confocal conicoid (6) (given),

$$\therefore \quad \frac{l^2}{\left(\frac{1}{a^2-\lambda}\right)}+\frac{m^2}{\left(\frac{1}{b^2-\lambda}\right)}+\frac{n^2}{\left(\frac{1}{c^2-\lambda}\right)}=p^2$$

or $$(a^2-\lambda)l^2 + (b^2-\lambda)m^2 + (c^2-\lambda)n^2 = p^2$$

or $$a^2l^2 - \lambda l^2 + b^2m^2 - \lambda m^2 + c^2n^2 - \lambda n^2 = p^2$$

or $$(a^2l^2 + b^2m^2 + c^2n^2) - \lambda(l^2 + m^2 + n^2) = p^2$$

or $$p_0^2 - \lambda(1) = p^2$$

[By (5)] [By (3)]

or $$p_0^2 - p^2 = \lambda \quad \text{or} \quad \lambda = p_0^2 - p^2.$$

Example 2. *Show that two confocal paraboloids cut every where at right angle.*

(K.U. 2007 ; M.D.U. 2003, 2007)

Sol. Let the two different paraboloids confocal to the paraboloid $\frac{x^2}{a^2}+\frac{y^2}{b^2}=2z$ be

$$\frac{x^2}{a^2-\lambda_1}+\frac{y^2}{b^2-\lambda_1}=2z-\lambda_1 \qquad ...(1)$$

and $$\frac{x^2}{a^2-\lambda_2}+\frac{y^2}{b^2-\lambda_2}=2z-\lambda_2 \qquad ...(2)$$

where $\lambda_1 \neq \lambda_2$.

Let (x_1, y_1, z_1) be any common point (*i.e.*, point of intersection) of paraboloids (1) and (2).

$\therefore$ (x_1, y_1, z_1) satisfies both equations (1) and (2).

$\therefore$ $$\frac{x_1^2}{a^2-\lambda_1}+\frac{y_1^2}{b^2-\lambda_1}=2z_1-\lambda_1 \qquad ...(3)$$

and $$\frac{x_1^2}{a^2-\lambda_2}+\frac{y_1^2}{b^2-\lambda_2}=2z_1-\lambda_2 \qquad ...(4)$$

Eqn. (3) – Eqn. (4) gives

$$x_1^2\left(\frac{1}{a^2-\lambda_1}-\frac{1}{a^2-\lambda_2}\right)+y_1^2\left(\frac{1}{b^2-\lambda_1}-\frac{1}{b^2-\lambda_2}\right)=-\lambda_1+\lambda_2$$

or $$x_1^2\left(\frac{a^2-\lambda_2-a^2+\lambda_1}{(a^2-\lambda_1)(a^2-\lambda_2)}\right)+y_1^2\left(\frac{b^2-\lambda_2-b^2+\lambda_1}{(b^2-\lambda_1)(b^2-\lambda_2)}\right)=-\lambda_1+\lambda_2$$

or $$\frac{x_1^2(\lambda_1-\lambda_2)}{(a^2-\lambda_1)(a^2-\lambda_2)}+\frac{y_1^2(\lambda_1-\lambda_2)}{(b^2-\lambda_1)(b^2-\lambda_2)}=-(\lambda_1-\lambda_2)$$

Dividing every term by $(\lambda_1-\lambda_2)\neq 0$ ($\because \lambda_1 \neq \lambda_2$),

$$\frac{x_1^2}{(a^2-\lambda_1)(a^2-\lambda_2)}+\frac{y_1^2}{(b^2-\lambda_1)(b^2-\lambda_2)}=-1 \qquad ...(5)$$

Equation of the tangent plane at (x_1, y_1, z_1) to paraboloid (1) is

$\left(\text{Change } x^2 \text{ to } xx_1 \text{ etc., } x \text{ to } \frac{x+x_1}{2} \text{ etc.}\right)$

$$\frac{xx_1}{a^2-\lambda_1}+\frac{yy_1}{b^2-\lambda_1}=2\frac{(z+z_1)}{2}-\lambda_1$$

or $$\frac{xx_1}{a^2-\lambda_1}+\frac{yy_1}{b^2-\lambda_1}=z+z_1-\lambda_1$$

or $$\frac{xx_1}{a^2-\lambda_1}+\frac{yy_1}{b^2-\lambda_1}-z=z_1-\lambda_1 \qquad ...(6)$$

Similarly equation of the tangent plane at (x_1, y_1, z_1) to paraboloid (2) is

$$\frac{xx_1}{a^2-\lambda_2}+\frac{yy_1}{b^2-\lambda_2}-z=z_1-\lambda_2 \qquad ...(7)$$

Tangent planes (6) and (7) will be at right angles

if $$a_1a_2+b_1b_2+c_1c_2=0$$

i.e., if $$\frac{x_1}{a^2-\lambda_1}\frac{x_1}{a^2-\lambda_2}+\frac{y_1}{b^2-\lambda_1}\frac{y_1}{b^2-\lambda_2}+(-1)(-1)=0$$

i.e., if $$\frac{x_1^2}{(a^2-\lambda_1)(a^2-\lambda_2)}+\frac{y_1^2}{(b^2-\lambda_1)(b^2-\lambda_2)}=-1$$

which is true by equation (5).

$\therefore$ tangent planes (6) and (7) at the common point (x_1, y_1, z_1) to paraboloids (1) and (2) are at right angles.

$\therefore$ By def. (given just above Art. 7), confocal paraboloids cut one another at right angles at all their common points.

Example 3. *Prove that the perpendiculars from the origin to the tangent planes to the ellipsoid which touch it along its curve of intersection with the confocal whose parameter is λ, lie on the cone* $\frac{a^2x^2}{a^2-\lambda}+\frac{b^2y^2}{b^2-\lambda}+\frac{c^2z^2}{c^2-\lambda}=0.$

Sol. We know that the equation of ellipsoid is $\frac{x^2}{a^2}+\frac{y^2}{b^2}+\frac{z^2}{c^2}=1$ ...(1)

$\therefore$ Equation of the confocal whose parameter is λ is

$$\frac{x^2}{a^2-\lambda}+\frac{y^2}{b^2-\lambda}+\frac{z^2}{c^2-\lambda}=1 \qquad ...(2)$$

Let (x_1, y_1, z_1) be any point on the curve of intersection of ellipsoid (1) and its confocal (2).

We know that equation of the tangent plane at (x_1, y_1, z_1) to ellipsoid (1) is

$$\frac{xx_1}{a^2}+\frac{yy_1}{b^2}+\frac{zz_1}{c^2}=1 \qquad ...(3)$$

$\therefore$ d.r.'s of the normal to the tangent plane (3) are $\frac{x_1}{a^2}, \frac{y_1}{b^2}, \frac{z_1}{c^2}$ | coeffs. of x, y, z in (3)

$\therefore$ Equations of the perpendicular from the origin (0, 0, 0) to the tangent plane (3) are

$$\frac{x-0}{\frac{x_1}{a^2}}=\frac{y-0}{\frac{y_1}{b^2}}=\frac{z-0}{\frac{z_1}{c^2}} \quad \text{or} \quad \frac{a^2x}{x_1}=\frac{b^2y}{y_1}=\frac{c^2z}{z_1} \qquad ...(4)$$

Because (x_1, y_1, z_1) is a common point of ellipsoid (1) and its confocal (2) ($\because$ (x_1, y_1, z_1) is a point on the curve of intersection of ellipsoid (1) and its confocal (2)), therefore (x_1, y_1, z_1) satisfies both equations (1) and (2).

$$\therefore \quad \frac{x_1^2}{a^2}+\frac{y_1^2}{b^2}+\frac{z_1^2}{c^2}=1 \qquad ...(5)$$

and

$$\frac{x_1^2}{a^2-\lambda}+\frac{y_1^2}{b^2-\lambda}+\frac{z_1^2}{c^2-\lambda}=1 \qquad ...(6)$$

Eqn. (5) – Eqn. (6) gives

$$x_1^2\left(\frac{1}{a^2}-\frac{1}{a^2-\lambda}\right)+y_1^2\left(\frac{1}{b^2}-\frac{1}{b^2-\lambda}\right)+z_1^2\left(\frac{1}{c^2}-\frac{1}{c^2-\lambda}\right)=0$$

or

$$\Sigma x_1^2\left[\frac{a^2-\lambda-a^2}{a^2(a^2-\lambda)}\right]=0 \quad \text{or} \quad \Sigma\frac{-\lambda x_1^2}{a^2(a^2-\lambda)}=0$$

or

$$-\frac{\lambda x_1^2}{a^2(a^2-\lambda)}-\frac{\lambda y_1^2}{b^2(b^2-\lambda)}-\frac{\lambda z_1^2}{c^2(c^2-\lambda)}=0$$

Dividing by $-\lambda$, $\dfrac{x_1^2}{a^2(a^2-\lambda)}+\dfrac{y_1^2}{b^2(b^2-\lambda)}+\dfrac{z_1^2}{c^2(c^2-\lambda)}=0$ ...(7)

We are to find the locus of lines (4)

(*i.e.*, lines drawn through the origin and perpendicular to tangent planes (3)).

Let us eliminate the unknowns x_1, y_1, z_1 from (4) and (7).

Putting the values of x_1, y_1, z_1 from (4) in (7), the required locus is

$$\frac{a^4x^2}{a^2(a^2-\lambda)}+\frac{b^4y^2}{b^2(b^2-\lambda)}+\frac{c^4z^2}{c^2(c^2-\lambda)}=0 \quad \text{or} \quad \frac{a^2x^2}{a^2-\lambda}+\frac{b^2y^2}{b^2-\lambda}+\frac{c^2z^2}{c^2-\lambda}=0$$

which being a homogenous of second degree in x, y, z represents a cone.

EXERCISE 1

1. **Find the locus of the point of intersection of three planes mutually at right angles, each of which touches one of the three given confocals, is a sphere.**
   [Hint. See Art. 8, Page 356, Chapter 13.]
2. **Prove that the locus of the poles of a given plane with respect to a system of confocal conicoids is a straight line.**
3. **A given plane and a parallel tangent plane to a conicoid are at distances d_1 and d_2 from the centre. Prove that the parameter of the confocal conicoid which touches the plane is $d_2^2-d_1^2$.**
4. **Show that the tangent planes to the two confocal paraboloids at their point of intersection are at right angles.**
   [Hint. It is Example 2, Art. 7, Page 421.]
5. **Prove that the perpendiculars from the origin to the tangent planes to the ellipsoid which touch it along its curve of intersection with the confocal whose parameter is λ, lie on the cone**
$$\frac{a^2x^2}{a^2-\lambda}+\frac{b^2y^2}{b^2-\lambda}+\frac{c^2z^2}{c^2-\lambda}=0.$$
   [Hint. It is Example 3.]
6. **If P is a point on an ellipsoid and P′ is a corresponding point on a confocal with parameter λ, prove that $OP^2-OP'^2=\lambda$, where O is the centre of the ellipsoid.**
   [Hint. See Def. Art. 8.]
7. **Prove that the equation of the confocal which has a system of circular sections parallel to the plane $x=y$ is**
$$\frac{x^2}{(c^2-a^2)(a^2-b^2)}+\frac{y^2}{(b^2-c^2)(a^2-b^2)}-\frac{z^2}{2(b^2-c^2)(c^2-a^2)}=\frac{1}{2c^2-(a^2+b^2)}.$$
   (M.D.U. 1993)

16

Reduction of Second Degree Equations

1. Introduction

We define a **quadric** as the locus of a point satisfying an equation of the second degree in x, y, z.

We know that a general equation of the second degree in x, y, z is

$$F(x, y, z) \equiv ax^2 + by^2 + cz^2 + 2fyz + 2gzx + 2hxy + 2ux + 2vy + 2wz + d = 0 \quad ...(1)$$

By the above definition, Eqn. (1) represents a **quadric**

Equation (1) can be rewritten as

$$\underbrace{\Sigma(ax^2 + 2fyz)}_{\text{(Homogeneous group of degree 2)}} + \underbrace{2\,\Sigma ux + d = 0}_{\text{(Homogeneous group of degree 1)}}$$

We know that sphere, cone, cylinder, central conicoids are all **quadrics** (see Chapters 10, 11, 12, 13) because their equations were of second degree in x, y, z.

Remark. Eqn. (1) above is

$$F(x, y, z) = ax^2 + by^2 + cz^2 + 2fyz + 2gzx + 2hxy + 2ux + 2vy + 2wz + d = 0 \quad ...(1)$$

Changing x, y, z to α, β, γ in (1),

$$F(\alpha, \beta, \gamma) = a\alpha^2 + b\beta^2 + c\gamma^2 + 2f\beta\gamma + 2g\gamma\alpha + 2h\alpha\beta + 2u\alpha + 2v\beta + 2w\gamma + d$$
$$= \Sigma\,(a\alpha^2 + 2f\beta\gamma) + 2\Sigma u\alpha + d$$

It may be noted that $F(\alpha, \beta, \gamma) \neq 0$ always.

$F(\alpha, \beta, \gamma) = 0$ only if the point (α, β, γ) satisfies (1).

Diff. both sides of Eqn. (1) partially w.r.t. x, y, z respectively.

$$\frac{\partial F}{\partial x} = 2ax + 2gz + 2hy + 2u = 2(ax + hy + gz + u)$$

$$\frac{\partial F}{\partial y} = 2by + 2fz + 2hx + 2v = 2(hx + by + fz + v)$$

$$\frac{\partial F}{\partial z} = 2cz + 2fy + 2gx + 2w = 2(gx + fy + cz + w)$$

At the point (α, β, γ)

$$\frac{\partial F}{\partial \alpha} = 2\,(a\alpha + h\beta + g\gamma + u)$$

$$\frac{\partial F}{\partial \beta} = 2(h\alpha + b\beta + f\gamma + v)$$

$$\frac{\partial F}{\partial \gamma} = 2(g\alpha + f\beta + c\gamma + w).$$

2. Intersections of any line with quadric (1)

Equation (1) of quadric is

$$F(x, y, z) = ax^2 + by^2 + cz^2 + 2fyz + 2gzx + 2hxy + 2ux + 2vy + 2wz + d = 0 \quad ...(1)$$

We know that equations of any line through a given point (α, β, γ) and having d.c's l, m, n are

$$\frac{x-\alpha}{l} = \frac{y-\beta}{m} = \frac{z-\gamma}{n} = (r \text{ say}) \quad ...(2)$$

We know that any point on line (2) is $(lr + \alpha, mr + \beta, nr + \gamma)$ ...(3)

If it lies on quadric (1), then

$$a(lr + \alpha)^2 + b(mr + \beta)^2 + c(nr + \gamma)^2 + 2f(mr + \beta)(nr + \gamma) + 2g(nr + \gamma)(lr + \alpha) + 2h(lr + \alpha)(mr + \beta) + 2u(lr + \alpha) + 2v(mr + \beta) + 2w(nr + \gamma) + d = 0$$

or
$$(al^2 + bm^2 + cn^2 + 2fmn + 2gnl + 2hlm)r^2 + 2r[al\alpha + bm\beta + cn\gamma + fm\gamma + fn\beta + gn\alpha + gl\gamma + hl\beta + hm\alpha + ul + vm + wn] + (a\alpha^2 + b\beta^2 + c\gamma^2 + 2f\beta\gamma + 2g\gamma\alpha + 2h\alpha\beta + 2u\alpha + 2v\beta + 2w\gamma + d) = 0$$

or
$$F(l, m, n)\, r^2 + 2r[l(a\alpha + h\beta + g\gamma + u) + m(h\alpha + b\beta + f\gamma + v) + n(g\alpha + f\beta + c\gamma + w)] + F(\alpha, \beta, \gamma) = 0$$

(Changing x, y, z to l, m, n in (1), and also x, y, z to α, β, γ in (1))

i.e.,
$$r^2 F(l, m, n) + 2r\left[l\frac{\partial F}{\partial \alpha} + m\frac{\partial F}{\partial \beta} + n\frac{\partial F}{\partial \gamma}\right] + F(\alpha, \beta, \gamma) = 0 \quad ...(4)$$

(By Remark Art. 1)

Equation (4) is a quadratic in r giving two values of r. Therefore line (2) meets quadric (1) in two points.

3. Equation of the tangent plane

To find the equation of the tangent plane at the point (α, β, γ) of the quadric

$$ax^2 + by^2 + cz^2 + 2fyz + 2gzx + 2hxy + 2ux + 2vy + 2wz + d = 0$$

We know that *tangent plane* at a point to a quadric is the locus of the tangent lines at that point. *Reproduce Art. 2.*

Because the point (α, β, γ) lies on the given quadric, (1), therefore

$$a\alpha^2 + b\beta^2 + c\gamma^2 + 2f\beta\gamma + 2g\gamma\alpha + 2h\alpha\beta + 2u\alpha + 2v\beta + 2w\gamma + d = 0$$

i.e.,
$$F(\alpha, \beta, \gamma) = 0 \quad ...(5)$$

Because line (2) is a tangent to conic (1), therefore the two values of r from equation (4) are equal.

$\therefore$ $B^2 - 4AC = 0$

But from Eqn. (4), $C = F(\alpha, \beta, \gamma) = 0$ [By (5)]

$\therefore$ $B^2 = 0$ or $B = 0$ *i.e.,* coeff. of $r = 0$

$$\therefore \quad l\frac{\partial F}{\partial \alpha} + m\frac{\partial F}{\partial \beta} + n\frac{\partial F}{\partial \gamma} = 0 \quad ...(6)$$

(Condition of tangency for line (2) to be a tangent to quadric (1)).

To find the locus of tangent lines given by (2) (i.e., To find Equation of tangent plane), let us eliminate l, m, n from (3) and (6).

Thus putting the values of l, m, n from (3) in (6), we have

$$(x-\alpha)\, 2(a\alpha + h\beta + g\gamma + u) + (y-\beta)\, 2(h\alpha + b\beta + f\gamma + v) + (z-\gamma)\, 2(g\alpha + f\beta + c\gamma + w) = 0$$

Dividing by 2,

$$x\,(a\alpha + h\beta + g\gamma + u) + y\,(h\alpha + b\beta + f\gamma + v) + z(g\alpha + f\beta + c\gamma + w)$$
$$= \alpha\,(a\alpha + h\beta + g\gamma + u) + \beta\,(h\alpha + b\beta + f\gamma + v) + \gamma(g\alpha + f\beta + c\gamma + w)$$

Adding $u\alpha + v\beta + w\gamma + d$ to both sides, we have

$$x\,(a\alpha + h\beta + g\gamma + u) + y(h\alpha + b\beta + f\gamma + v) + z(g\alpha + f\beta + c\gamma + w)$$
$$+ u\alpha + v\beta + w\gamma + d = a\alpha^2 + b\beta^2 + c\gamma^2 + 2f\beta\gamma$$
$$+ 2g\gamma\alpha + 2h\alpha\beta + 2u\alpha + 2v\beta + 2w\gamma + d$$
$$= F(\alpha, \beta, \gamma) = 0 \qquad \text{[By (5)]}$$

or $$x(a\alpha + h\beta + g\gamma + u) + y\,(h\alpha + b\beta + f\gamma + v) + z(g\alpha + f\beta + c\gamma + w) + u\alpha + v\beta + w\gamma + d = 0$$

or $$T = 0 \qquad ...(7)$$

which is the required equation of the tangent plane at the point (α, β, γ) to quadric (1).

Cor. d.r's of normal to tangent plane (7) are coefficients of x, y, z in (7) *i.e.*, $a\alpha + h\beta + g\gamma + u$, $h\alpha + b\beta + f\gamma + v$, $g\alpha + f\beta + c\gamma + w$.

$\therefore$ Equations of the normal at the point (α, β, γ) to tangent plane (7) are

$$\frac{x-\alpha}{a\alpha + h\beta + g\gamma + u} = \frac{y-\beta}{h\alpha + b\beta + f\gamma + v} = \frac{z-\gamma}{g\alpha + f\beta + c\gamma + w}.$$

4. Reduction of General Equation of Second Degree in x, y, z

The equation $F(x, y, z) = ax^2 + by^2 + cz^2 + 2fyz + 2gzx + 2hxy + 2ux + 2vy + 2wz + d = 0$ can be reduced to either of the two standard forms by transformation of axes :

(*i*) $\lambda_1 x^2 + \lambda_2 y^2 + \lambda_3 z^2 = \mu$ (*ii*) $\lambda_1 x^2 + \lambda_2 y^2 = 2\mu z$

We have already learnt the nature of the surfaces represented by (*i*) and (*ii*) in chapter 4 (central conicoids). For the convenience of the reader we write them here in the form of a table :

(*i*) $\lambda_1 \mathbf{x}^2 + \lambda_2 \mathbf{y}^2 + \lambda_3 \mathbf{z}^2 = \mu$

λ_1	λ_2	λ_3	μ	Surfaces
+	+	+	+	Ellipsoid
+	+	–	+	Hyperboloid of one sheet
+	–	–	+	Hyperboloid of two sheets
+	+	–	0	Cone
+	+	0	+	Elliptic cylinder
+	–	0	+	Hyperbolic cylinder
+	–	0	0	Pair of Intersecting planes
+	0	0	+	Pair of parallel planes

(*ii*) $\lambda_1 x^2 + \lambda_2 y^2 = 2\mu z$

λ_1	λ_2	μ	Surfaces
+	+	+	Elliptic paraboloid
+	−	+	Hyperbolic paraboloid
+	0	±	Parabolic cylinder

Remark. If λ_1, λ_2 are both negative and μ is positive, the equation still represents an elliptic paraboloid.

Note. Let $D = \begin{vmatrix} a & h & g \\ h & b & f \\ g & f & c \end{vmatrix}$

Expanding by first row $D = abc + 2fgh - af^2 - bg^2 - ch^2$ ...(1)

Let A, B, C, F, G, H denote the cofactors of a, b, c, f, g, h in D.

$\therefore$ $A = bc - f^2, B = ca - g^2, C = ab - h^2$

$F = gh - af, G = hf - bg, H = fg - ch.$

$$BC - F^2 = (ca - g^2)(ab - h^2) - (gh - af)^2$$
$$= a^2bc - ach^2 - abg^2 + g^2h^2 - g^2h^2 - a^2f^2 + 2afgh$$

or $\quad BC - F^2 = a(abc + 2fgh - af^2 - bg^2 - ch^2) = a\,D$

Similarly, $\quad CA - G^2 = bD, AB - H^2 = cD.$

Similarly, $\quad GH - AF = f\,D, HF - BG = gD$ etc.

Also we know that $\quad aA + hH + gG = D$ etc.

and $\quad aH + hB + gF = 0$

(We know that if the elements of a line of a determinant are multiplied by their corresponding factors, then their sum is D. Also if the elements of a line of a determinant are multiplied by co-factors of some other line, then the sum is zero).

5. To prove that by rotation of axes

$f(x, y, z) \equiv ax^2 + by^2 + cz^2 + 2fyz + 2gzx + 2hxy$ transforms $\lambda_1 x^2 + \lambda_2 y^2 + \lambda_3 z^2$, where $\lambda_1, \lambda_2, \lambda_3$ are roots of the equation

$$\begin{vmatrix} a-\lambda & h & g \\ h & b-\lambda & f \\ g & f & c-\lambda \end{vmatrix} = 0$$

(Expanding this determinant by first row and dividing by − 1)

i.e., $\quad \lambda^3 - \lambda^2(a + b + c) + \lambda(A + B + C) - D = 0$

where $\quad A = bc - f^2, B = ca - g^2, C = ab - h^2, D = abc + 2fgh - af^2 - bg^2 - ch^2$

By Art. 3 (*b*), Review IV, let us make the following transformations :

(Changing the axes without changing the origin)

$$X = l_1x + m_1y + n_1z$$
$$Y = l_2x + m_2y + n_2z$$
$$Z = l_3x + m_3y + n_3z$$

$\therefore$ The transformed expression

$$\lambda_1 x^2 + \lambda_2 y^2 + \lambda_3 z^2 = \lambda_1(l_1 x + m_1 y + n_1 z)^2 + \lambda_2(l_2 x + m_2 y + n_2 z)^2 + \lambda_3(l_3 x + m_3 y + n_3 z)^2$$
$$= \text{The given expression } ax^2 + by^2 + cz^2 + 2fyz + 2gzx + 2hxy$$

Comparing coefficients of x^2, y^2, z^2, yz, zx, xy on both sides, we have

$$\begin{array}{l|l} a = \lambda_1 l_1^2 + \lambda_2 l_2^2 + \lambda_3 l_3^2 & f = \lambda_1 m_1 n_1 + \lambda_2 m_2 n_2 + \lambda_3 m_3 n_3 \\ b = \lambda_1 m_1^2 + \lambda_2 m_2^2 + \lambda_3 m_3^2 & g = \lambda_1 n_1 l_1 + \lambda_2 n_2 l_2 + \lambda_3 n_3 l_3 \\ c = \lambda_1 n_1^2 + \lambda_2 n_2^2 + \lambda_3 n_3^2 & h = \lambda_1 l_1 m_1 + \lambda_2 l_2 m_2 + \lambda_3 l_3 m_3 \end{array} \quad ...(1)$$

$$\therefore \quad a + b + c = \lambda_1(l_1^2 + m_1^2 + n_1^2) + \lambda_2(l_2^2 + m_2^2 + n_2^2) + \lambda_3(l_3^2 + m_3^2 + n_3^2)$$
$$= \lambda_1 + \lambda_2 + \lambda_3 \quad ...(2) \text{ [By Art. 4, Review IV; Eqn. (1)]}$$

Again $\quad A = bc - f^2 = (\lambda_1 m_1^2 + \lambda_2 m_2^2 + \lambda_3 m_3^2)(\lambda_1 n_1^2 + \lambda_2 n_2^2 + \lambda_3 n_3^2) - (\lambda_1 m_1 n_1 + \lambda_2 m_2 n_2 + \lambda_3 m_3 n_3)^2$

$$= \lambda_1^2 m_1^2 n_1^2 + \lambda_1\lambda_2 m_1^2 n_2^2 + \lambda_1\lambda_3 m_1^2 n_3^2 + \lambda_1\lambda_2 m_2^2 n_1^2$$
$$+ \lambda_2^2 m_2^2 n_2^2 + \lambda_2\lambda_3 m_2^2 n_3^2 + \lambda_1\lambda_3 m_3^2 n_1^2 + \lambda_2\lambda_3 m_3^2 n_2^2 + \lambda_3^2 m_3^2 n_3^2$$
$$- \lambda_1^2 m_1^2 n_1^2 - \lambda_2^2 m_2^2 n_2^2 - \lambda_3^2 m_3^2 n_3^2 - 2\lambda_1\lambda_2\, m_1 m_2\, n_1 n_2 - 2\lambda_2\lambda_3\, m_2 m_3\, n_2 n_3$$
$$- 2\lambda_1\lambda_3\, m_1 m_3 n_1 n_3$$
$$= \Sigma\, \lambda_2\lambda_3(m_2^2 n_3^2 + m_3^2 n_2^2 - 2m_2 m_3 n_2 n_3) = \Sigma\lambda_2\lambda_3\,(m_2 n_3 - m_3 n_2)^2$$

or $\quad A = \Sigma\lambda_2\lambda_3 l_1^2 \quad$ (By Art. 6)

$$= \lambda_2\lambda_3 l_1^2 + \lambda_3\lambda_1 l_2^2 + \lambda_1\lambda_2 l_3^2$$

Similarly $\quad B = \lambda_2\lambda_3 m_1^2 + \lambda_3\lambda_1 m_2^2 + \lambda_1\lambda_2 m_3^2$

and $\quad C = \lambda_2\lambda_3 n_1^2 + \lambda_3\lambda_1 n_2^2 + \lambda_1\lambda_2 n_3^2$

Adding $\quad A + B + C = \lambda_2\lambda_3\,(l_1^2 + m_1^2 + n_1^2) + \lambda_3\lambda_1\,(l_2^2 + m_2^2 + n_2^2) + \lambda_1\lambda_2\,(l_3^2 + m_3^2 + n_3^2) = \lambda_2\lambda_3 + \lambda_3\lambda_1 + \lambda_1\lambda_2 \quad ...(3)$

Now to find $\lambda_1\lambda_2\lambda_3$, let us consider the product

$$\begin{vmatrix} l_1 & l_2 & l_3 \\ m_1 & m_2 & m_3 \\ n_1 & n_2 & n_3 \end{vmatrix} \begin{vmatrix} \lambda_1 l_1 & \lambda_2 l_2 & \lambda_3 l_3 \\ \lambda_1 m_1 & \lambda_2 m_2 & \lambda_3 m_3 \\ \lambda_1 n_1 & \lambda_2 n_2 & \lambda_3 n_3 \end{vmatrix}$$

Performing row by row multiplication,

$$\begin{vmatrix} \Sigma\lambda_1 l_1^2 & \Sigma\lambda_1 l_1 m_1 & \Sigma\lambda_1 n_1 l_1 \\ \Sigma\lambda_1 l_1 m_1 & \Sigma\lambda_1 m_1^2 & \Sigma\lambda_1 m_1 n_1 \\ \Sigma\lambda_1 n_1 l_1 & \Sigma\lambda_1 m_1 n_1 & \Sigma\lambda_1 n_1^2 \end{vmatrix}$$

Taking λ_1, λ_2, λ_3 common from C_1, C_2, C_3 respectively of the second determinant on L.H.S. and putting values from equation (1) in the determinant on R.H.S., we have

$$\lambda_1\lambda_2\lambda_3 \begin{vmatrix} l_1 & l_2 & l_3 \\ m_1 & m_2 & m_3 \\ n_1 & n_2 & n_3 \end{vmatrix}^2 = \begin{vmatrix} a & h & g \\ h & b & f \\ g & f & c \end{vmatrix}$$

Now Expanding by first row,

$$= abc + 2fgh - af^2 - bg^2 - ch^2$$

or $\quad \lambda_1\lambda_2\lambda_3\,(\pm 1)^2 = D \text{ (given)} \quad$ (By Cor. Art. 5, Review IV)

or $\quad \lambda_1\lambda_2\lambda_3 = D \quad ...(4)$

We know that $\lambda_1, \lambda_2, \lambda_3$ are roots of the equation

$$(\lambda - \lambda_1)(\lambda - \lambda_2)(\lambda - \lambda_3) = 0$$

or $$\lambda^3 - \lambda^2(\lambda_1 + \lambda_2 + \lambda_3) + \lambda(\lambda_1\lambda_2 + \lambda_2\lambda_3 + \lambda_3\lambda_1) - \lambda_1\lambda_2\lambda_3 = 0$$

Putting values from (2), (3) and (4),

$$\lambda^3 - (a + b + c)\lambda^2 + (A + B + C)\lambda - D = 0 \quad ...(5)$$

i.e., $\lambda_1, \lambda_2, \lambda_3$ are roots of the equation

$$\begin{vmatrix} a-\lambda & h & g \\ h & b-\lambda & f \\ g & f & c-\lambda \end{vmatrix} = 0 \quad ...(6)$$

(Expanding this determinant by first row and dividing by – 1, it becomes same as Eqn. (5))

Def. The cubic in λ derived in eqn. (5) or eqn. (6) of the above Article is called the **discriminating cubic** of the homogeneous expression

$$f(x, y, z) = ax^2 + by^2 + cz^2 + 2fyz + 2gzx + 2hxy$$

of the second degree in x, y, z.

6. To find the direction cosines l_1, m_1, n_1 ; l_2, m_2, n_2 and l_3, m_3, n_3 of the new-axes of transformations

For a proper understanding of this Art. 6, first read Art. 3 of Review IV.

Reproduce Art. 5 upto equation (1)

Putting values of a, b, c from equations (1),

$$al_1 + hm_1 + gn_1 = (\lambda_1 l_1^2 + \lambda_2 l_2^2 + \lambda_3 l_3^2)l_1 + (\lambda_1 l_1 m_1 + \lambda_2 l_2 m_2 + \lambda_3 l_3 m_3)m_1 + (\lambda_1 n_1 l_1 + \lambda_2 n_2 l_2 + \lambda_3 n_3 l_3)n_1$$

$$= \lambda_1 l_1^3 + \lambda_2 l_2^2 l_1 + \lambda_3 l_3^2 l_1 + \lambda_1 l_1 m_1^2 + \lambda_2 l_2 m_1 m_2 + \lambda_3 l_3 m_1 m_3 + \lambda_1 n_1^2 l_1 + \lambda_2 n_1 n_2 l_2 + \lambda_3 n_1 n_3 l_3$$

$$= \lambda_1 l_1 (l_1^2 + m_1^2 + n_1^2) + \lambda_2 l_2 (l_1 l_2 + m_1 m_2 + n_1 n_2) + \lambda_3 l_3 (l_1 l_3 + m_1 m_3 + n_1 n_3)$$

$$= \lambda_1 l_1 (1) + \lambda_2 l_2 (0) + \lambda_3 l_3 (0) \quad \text{[Using equations (1) and (2) of Art. 4 of Review IV]}$$

$$= \lambda_1 l_1$$

$$\therefore \quad \frac{al_1 + hm_1 + gn_1}{l_1} = \lambda_1$$

Similarly, $$\frac{hl_1 + bm_1 + fn_1}{m_1} = \lambda_1$$

and $$\frac{gl_1 + fm_1 + cn_1}{n_1} = \lambda_1$$

$\therefore$ d.c's l_1, m_1, n_1 of one of the new-axes satisfy the equations

$$\frac{\mathbf{al_1 + hm_1 + gn_1}}{\mathbf{l_1}} = \frac{\mathbf{hl_1 + bm_1 + fn_1}}{\mathbf{m_1}} = \frac{\mathbf{gl_1 + fm_1 + cn_1}}{\mathbf{n_1}} = \lambda_1$$

where λ_1 is the corresponding root of the discriminating cubic.

Similarly d.c's l_2, m_2, n_2 ; and l_3, m_3, n_3 of the other two new axes of transformations satisfy the equations

$$\therefore \quad \frac{al_2 + hm_2 + gn_2}{l_2} = \frac{hl_2 + bm_2 + fn_2}{m_2} = \frac{gl_2 + fm_2 + cn_2}{n_2} = \lambda_2$$

and $$\frac{al_3 + hm_3 + gn_3}{l_3} = \frac{hl_3 + bm_3 + fn_3}{m_3} = \frac{gl_3 + fm_3 + cn_3}{n_3} = \lambda_3$$

where λ_2 and λ_3 are the corresponding roots of the discriminating cubic.

7. General equation of a conicoid whose centre is origin

Let the equation of the conicoid (quadric) whose centre is origin be

$$ax^2 + by^2 + cz^2 + 2fyz + 2gzx + 2hxy + 2ux + 2vy + 2wz + d = 0 \quad ...(1)$$

Because centre of the conicoid is origin (given), and origin being centre will bisect every chord passes through it. Therefore if $P(x_1, y_1, z_1)$ is any point on quadric (1), then $Q(-x_1, -y_1, -z_1)$ is also a point on (1).

$$\because \frac{P+Q}{2} = 0 \Rightarrow P + Q = 0 \Rightarrow Q = -P = (-x_1, -y_1, -z_1).$$

Therefore co-ordinates (x_1, y_1, z_1) and $(-x_1, -y_1, -z_1)$ will both satisfy Equation (1).

$$\therefore \quad ax_1^2 + by_1^2 + cz_1^2 + 2fy_1z_1 + 2gz_1x_1 + 2hx_1y_1 + 2ux_1 + 2vy_1 + 2wz_1 + d = 0 \quad ...(2)$$

and $$ax_1^2 + by_1^2 + cz_1^2 + 2fy_1z_1 + 2gz_1x_1 + 2hx_1y_1 - 2ux_1 - 2vy_1 - 2wz_1 + d = 0 \quad ...(3)$$

Eqn. (2) – Eqn. (3) gives

$$4ux_1 + 4vy_1 + 4wz_1 = 0$$

Dividing by 4, $$ux_1 + vy_1 + wz_1 = 0 \quad ...(4)$$

Because equation (4) is true for all values of x_1, y_1, z_1 ; therefore it is an Identity.

$$\therefore \quad \mathbf{u = 0, v = 0, w = 0}$$

i.e., coeffs of x, y and z are all zero if the centre of the conicoid is origin.

8. To find the co-ordinates of the centre of the surface

$$\mathbf{f(x, y, z) = ax^2 + by^2 + cz^2 + 2fyz + 2gzx + 2hxy + 2ux + 2vy + 2wz + d = 0}$$

Let (x_1, y_1, z_1) be the centre of the quadric (conicoid)

$$f(x, y, z) = ax^2 + by^2 + cz^2 + 2fyz + 2gzx + 2hxy + 2ux + 2vy + 2wz + d = 0 \quad ...(1)$$

Let us shift the origin to centre (x_1, y_1, z_1) axes remaining parallel to original axes by putting $x = X + x_1$, $y = Y + y_1$, $z = Z + z_1$, where (X, Y, Z) are co-ordinates of the point (x, y, z) w.r.t. new axes. (By Art. 2, Review IV)

$\therefore$ Equation of the conicoid w.r.t. new axes is

$$a(X + x_1)^2 + b(Y + y_1)^2 + c(Z + z_1)^2 + 2f(Y + y_1)(Z + z_1) + 2g(Z + z_1)(X + x_1) + 2h(X + x_1)(Y + y_1) + 2u(X + x_1) + 2v(Y + y_1) + 2w(Z + z_1) + d = 0$$

i.e., $$aX^2 + bY^2 + cZ^2 + 2fYZ + 2gZX + 2hXY + 2X(ax_1 + hy_1 + gz_1 + u) + 2Y(hx_1 + by_1 + fz_1 + v) + 2Z(gx_1 + fy_1 + cz_1 + w) + (ax_1^2 + by_1^2 + cz_1^2 + 2fy_1z_1 + 2gz_1x_1 + 2hx_1y_1 + 2ux_1 + 2vy_1 + 2wz_1 + d) = 0 \quad ...(2)$$

Because centre of conicoid has been shifted to origin, therefore by Art. 7 ; coefficients of X, Y, Z in equation (2) are all zero.

$$\therefore \quad ax_1 + hy_1 + gz_1 + u = 0 \quad ...(3)$$

i.e., $$\frac{1}{2}\frac{\partial f}{\partial x} = 0 \text{ at } (x_1, y_1, z_1) \quad \text{(By Remark Art. 1)}$$

$$\therefore \quad \frac{\partial f}{\partial x} = 0 \text{ at centre } (x_1, y_1, z_1)$$

and $$hx_1 + by_1 + fz_1 + v = 0 \qquad ...(4)$$

$\therefore$ $$\frac{\partial f}{\partial y} = 0 \text{ at centre } (x_1, y_1, z_1) \qquad \text{(By Remark Art. 1)}$$

and $$gx_1 + fy_1 + cz_1 + w = 0 \qquad ...(5)$$

$\therefore$ $$\frac{\partial f}{\partial z} = 0 \text{ at centre } (x_1, y_1, z_1) \qquad \text{(By Remark Art. 1)}$$

$\therefore$ **Centre of the conicoid (1) is obtained by solving the equations**

$$\frac{\partial f}{\partial x} = 0, \frac{\partial f}{\partial y} = 0 \text{ and } \frac{\partial f}{\partial z} = 0 \text{ for x, y and z.}$$

And constant term of Eqn. (2)

$$= ax_1^2 + by_1^2 + cz_1^2 + 2fy_1z_1 + 2gz_1x_1 + 2hx_1y_1 + 2ux_1 + 2vy_1 + 2wz_1 + d$$
$$= (ax_1^2 + hx_1y_1 + 2gx_1z_1 + ux_1) + (hx_1y_1 + by_1^2 + fy_1z_1 + vy_1)$$
$$+ (gz_1x_1 + fy_1z_1 + cz_1^2 + wz_1) + ux_1 + vy_1 + wz_1 + d$$
$$= x_1(ax_1 + hy_1 + gz_1 + u) + y_1(hx_1 + by_1 + fz_1 + v)$$
$$+ z_1(gx_1 + fy_1 + cz_1 + w) + (ux_1 + vy_1 + wz_1 + d)$$
$$= x_1(0) + y_1(0) + z_1(0) + ux_1 + vy_1 + wz_1 + d$$

by (3) by (4) by (5)

$$= ux_1 + vy_1 + wz_1 + d = d' \text{ (say)}$$

$\therefore$ Equation (2) of the conicoid referred to centre as origin is

$$aX^2 + bY^2 + cZ^2 + 2fYZ + 2gZX + 2hXY + d' = 0. \qquad ...(6)$$

Remark 1. Comparing equations (1) and (6), we can say that we can remove first degree terms from the equation of the conicoid by shifting the origin to the centre and while doing so, the coefficients of the second degree terms remain unchanged.

Remark 2. Equation (6) can be further reduced to the form $\lambda_1X^2 + \lambda_2Y^2 + \lambda_3Z^2 = \mu$, where $\mu = -d'$ by changing the axes without changing the origin. (By Art. 3 (*b*), Review IV)

Remark 3. We can also say with the help of remark 1 that equations (1) and (3) have the same **discriminating cubic** given by Eqn. (5) or (6) of Art. 5 because this equation involves coefficients of only second degree terms.

9. To prove that all the roots of the discriminating cubic of the general equation of the second degree are real.

We know that discriminating cubic of general equation of second degree is

$$\begin{vmatrix} a-\lambda & h & g \\ h & b-\lambda & f \\ g & f & c-\lambda \end{vmatrix} = 0 \qquad \text{(By Art. 5 and Remark 3, Art. 8)}$$

Expanding by first row, we get

$$(a-\lambda)[(b-\lambda)(c-\lambda) - f^2] - h[h(c-\lambda) - fg] + g[hf - g(b-\lambda)] = 0$$

Dividing the equation by -1, we get

$$(\lambda - a)[(\lambda - b)(\lambda - c) - f^2] - h^2(\lambda - c) - fgh - fgh - g^2(\lambda - b) = 0$$
$$(\lambda - a)\, g(\lambda) - [h^2(\lambda - c) - g^2(\lambda - b) - 2fgh] = 0 \qquad ...(1)$$

where $g(\lambda) = (\lambda - b)(\lambda - c) - f^2$

Let us assume that $a > b > c$

Putting $\lambda = -\infty, c, b, \infty$, we get (See Art. 2, Chapter 15)

$$g(-\infty) = \infty \quad = \oplus \qquad \text{(See Art. 2, Chapter 15)}$$
$$g(c) = -f^2 \quad = \ominus$$
$$g(b) = -f^2 \quad = \ominus$$
$$g(\infty) = \infty \quad = \oplus$$

$\therefore$ $g(-\infty)$ and $g(c)$ are of opposite signs. Also $g(b)$ and $g(\infty)$ are of opposite signs.

Hence one real root say α of $g(\lambda) = 0$ lies between $-\infty$ and c (*i.e.*, $-\infty < \alpha < c$) and one real root say β of $g(\lambda) = 0$ lies between b and ∞

i.e., $b < \beta < \infty$. $\therefore$ $\alpha < c < b < \beta$.

Because α and β are roots of $g(\lambda) = 0$, we have

$$g(\alpha) = 0 \text{ and } g(\beta) = 0 \qquad ...(2)$$

$\therefore$ $(\alpha - b)(\alpha - c) - f^2 = 0$

or $(\alpha - b)(\alpha - c) = f^2$ ...(3)

Similarly, $(\beta - b)(\beta - c) = f^2$ ...(4)

Putting $\lambda = -\infty, \alpha, \beta, \infty$ in $\phi(\lambda)$ given by (1),

$$\phi(-\infty) = -\infty = \ominus$$

$$\phi(\alpha) = (\alpha - a)\, g(\alpha) - [h^2(\alpha - c) + g^2(\alpha - b) + 2fgh]$$

$$= 0 - [h^2(\alpha - c) + g^2(\alpha - b) \pm 2gh\sqrt{(\alpha - b)(\alpha - c)}\,] \qquad \text{(By (2) and (3))}$$

$$= [h^2(c - \alpha) + g^2(b - \alpha) \mp 2gh\sqrt{(b - \alpha)(c - \alpha)}\,]$$

$$= (h\sqrt{c - \alpha} \pm g\sqrt{b - \alpha}\,)^2 = \oplus \qquad (\because \ \alpha < c < b \Rightarrow c - \alpha > 0, b - \alpha > 0)$$

$$\phi(\beta) = (\beta - a)\, g(\beta) - [h^2(\beta - c) + g^2(\beta - b) + 2fgh]$$

$$= 0 - [h^2(\beta - c) + g^2(\beta - b) \pm 2gh\sqrt{(\beta - b)(\beta - c)}\,] \qquad \text{[By (2) and (4)]}$$

$$= -[h\sqrt{\beta - c} \pm g\sqrt{\beta - b}\,]^2 = \ominus \qquad (\because \ \alpha < c < b < \beta \Rightarrow \beta - c > 0, \beta - b > 0)$$

$$\phi(\infty) = \infty$$

$\therefore$ If $\lambda = -\infty, \alpha, \beta, \infty$, we get

$\phi(\lambda) = -$ ve, + ve, $-$ ve, + ve.

$\therefore$ The discriminating cubic has all three roots real one less than α $(-\infty < \lambda_1 < \alpha)$; one between α and β $(\alpha < \lambda_2 < \beta)$ and the third greater than β $(\beta < \lambda_3 < \infty)$.

10. Working rule for reducing a general equation of second degree in x, y, z to standard form if the terms of second degree donot form a perfect square

Step I. Solve the discriminating cubic

$$\lambda^3 - \lambda^2(a + b + c) + \lambda(A + B + C) - D = 0.$$

Sub-Case I

Step II. If all the three roots are different from zero say $\lambda_1, \lambda_2, \lambda_3$; then the centre of the quadric (conicoid) is obtained by solving the equations $\frac{\partial f}{\partial x} = 0, \frac{\partial f}{\partial y} = 0$ and $\frac{\partial f}{\partial z} = 0$ for x, y and z (Art. 8). Let the centre obtained be (x_1, y_1, z_1).

Step III. Shifting the origin to the centre and transforming the axes, the reduced equation will be

$$\lambda_1 x^2 + \lambda_2 y^2 + \lambda_3 z^2 = \mu$$

where $\mu = -d' = -(ux_1 + vy_1 + wz_1 + d)$ (Art. 8)

Step IV. The direction cosines l, m, n of one of the new axes are given by

$$\frac{al + hm + gn}{l} = \frac{hl + bm + fn}{m} = \frac{gl + fm + cn}{n} = \lambda \quad \text{...(1) (Art. 6)}$$

where λ is the corresponding root of the discriminating cubic.

Note. If two roots of the discriminating cubic are equal, then it is a surface of revolution and axis of rotation is obtained by putting that value of λ which is different from the two equal values.

Remark. The student is being suggested to use the following results without proof at this BA/B.Sc I level of learning.

Sub-Case II. If one root say $\lambda_3 = 0$; then find l_3, m_3, n_3 for third new axis from the equations

$$al_3 + hm_3 + gn_3 = 0, \; hl_3 + bm_3 + fn_3 = 0$$

and $gl_3 + fm_3 + cn_3 = 0$ (Putting $\lambda = 0$ in equation (1) and cross-multiplying)

Evaluate $k = ul_3 + vm_3 + wn_3$

If $k \neq 0$, reduced equation is the paraboloid $\lambda_1 x^2 + \lambda_2 y^2 + 2kz = 0$, and the **vertex** is obtained by solving following equations (2), (3), (4) and (5).

$$ax + hy + gz - l_3 k + u = 0 \quad \text{...(2)}$$

$$hx + by + fz - m_3 k + v = 0 \quad \text{...(3)}$$

$$gx + fy + cz - n_3 k + w = 0 \quad \text{...(4)}$$

and $k(l_3 x + m_3 y + n_3 z) + ux + vy + wz + d = 0$ ...(5)

In all cases equation (5) must be used to find the **Vertex**.

If k = 0, then there is a line of centres given by any two of equations

$$\frac{\partial f}{\partial x} = 0, \frac{\partial f}{\partial y} = 0, \frac{\partial f}{\partial z} = 0.$$

Choose any point (α, β, γ) on this line as the centre and reduced equation is the cylinder $\lambda_1 x^2 + \lambda_2 y^2 + u\alpha + v\beta + w\gamma + d = 0$ because one variable here z is absent. (See Art. 3, Chapter 12)

Type I. Illustrative examples when all the roots of the discriminating cubic are different from zero

Example 1. *Prove that the equation $x^2 + y^2 + z^2 + yz + zx + xy + 3x + y + 4z + 4 = 0$ represents an ellipsoid the squares of whose semi-axes are 2, 2, $\frac{1}{2}$. Show also that its principal axis is given by $x + 1 = y - 1 = z + 2$.*

Sol. The given equation of the quadric is

$$f(x, y, z) = x^2 + y^2 + z^2 + yz + zx + xy + 3x + y + 4z + 4 = 0 \quad \text{...(1)}$$

Comparing with $ax^2 + by^2 + cz^2 + 2fyz + 2gzx + 2hxy + 2ux + 2vy + 2wz + d = 0$, we have

$$a = 1, b = 1, c = 1, 2f = 1, 2g = 1, 2h = 1, 2u = 3, 2v = 1, 2w = 4, d = 4$$

or $$a = 1, b = 1, c = 1, f = \frac{1}{2}, g = \frac{1}{2}, h = \frac{1}{2}, u = \frac{3}{2}, v = \frac{1}{2}, w = 2, d = 4.$$

Step I. To solve discriminating cubic

$$a + b + c = 1 + 1 + 1 = 3$$

$$A + B + C = bc - f^2 + ca - g^2 + ab - h^2$$
$$= 1 - \frac{1}{4} + 1 - \frac{1}{4} + 1 - \frac{1}{4} = 3 - \frac{3}{4} = \frac{9}{4}$$
$$D = abc + 2fgh - af^2 - bg^2 - ch^2$$
$$= 1 + \frac{1}{4} - \frac{1}{4} - \frac{1}{4} - \frac{1}{4} = \frac{5}{4} - \frac{3}{4} = \frac{2}{4} = \frac{1}{2}$$

We know that discriminating cubic is

$$\lambda^3 - \lambda^2 (a + b + c) + \lambda (A + B + C) - D = 0$$

Putting values

$$\lambda^3 - 3\lambda^2 + \frac{9}{4}\lambda - \frac{1}{2} = 0 \quad \text{or} \quad 4\lambda^3 - 12\lambda^2 + 9\lambda - 2 = 0$$

$\lambda = 2$ is a root by trial method.

$(\because \ 4.2^3 - 12.2^2 + 9.2 - 2 = 0 \text{ or } 32 - 48 + 18 - 2 = 0 \text{ or } 0 = 0)$

Applying **Synthetic division**

$$\begin{array}{c|cccc} 2 & 4 & -12 & 9 & -2 \\ & & 8 & -8 & 2 \\ \hline & 4 & -4 & 1 & 0 \end{array}$$

The reduced equation is $4\lambda^2 - 4\lambda + 1 = 0$ or $(2\lambda - 1)^2 = 0$

$\therefore$ $$\lambda = \frac{1}{2}, \frac{1}{2}$$

$\therefore$ Roots of the discriminating cubic are

$$\lambda = \frac{1}{2}, \frac{1}{2}, 2. \qquad (\lambda_1 \le \lambda_2 \le \lambda_3)$$

Step II. To find centre of quadric I

We know by Art. 8 that centre is given by

$$\frac{\partial f}{\partial x} = 0, \frac{\partial f}{\partial y} = 0 \text{ and } \frac{\partial f}{\partial z} = 0$$

i.e., $ax + hy + gz + u = 0$ *i.e.,* $x + \frac{1}{2}y + \frac{1}{2}z + \frac{3}{2} = 0$

$hx + by + fz + v = 0$ *i.e.,* $\frac{1}{2}x + y + \frac{1}{2}z + \frac{1}{2} = 0$

$gx + fy + cz + w = 0$ *i.e.,* $\frac{1}{2}x + \frac{1}{2}y + z + 2 = 0$

or $2x + y + z + 3 = 0$ ...(2)

$x + 2y + z + 1 = 0$ ...(3)

$x + y + 2z + 4 = 0$ ...(4)

To find the centre, let us solve Eqns. (2), (3) and (4) for x, y, z.

Eqn. (2) – Eqn. (3) gives (to eliminate z) $x - y + 2 = 0$ ...(5)

2 × Eqn. (3) – Eqn. (4) gives (To eliminate z) $x + 3y - 2 = 0$ ...(6)

Eqn. (6) – Eqn. (5) gives $4y - 4 = 0$ or $4y = 4$ $\therefore$ $y = 1$.

Putting $y = 1$ in (5) $x - 1 + 2 = 0$ or $x = -1$

Putting $x = -1$ and $y = 1$ in (2), $-2 + 1 + z + 3 = 0$ or $z + 2 = 0$ $\therefore$ $z = -2$

$\therefore$ Centre of the conicoid (1) is $(-1, 1, -2) = (x_1, y_1, z_1)$ ...(7)

Step III. Shifting the origin to the centre and transforming the axes, the reduced equation will be

$$\lambda_1 X^2 + \lambda_2 Y^2 + \lambda_3 Z^2 = \mu \quad ..(8)$$

where $\mu = -d' = -(ux_1 + vy_1 + wz_1 + d)$ (By Art. 8 and Art. 5)

$$= -\left[\frac{3}{2}(-1) + \frac{1}{2}(1) + 2(-2) + 4\right]$$

$$= -\left(-\frac{3}{2} + \frac{1}{2} - 4 + 4\right) = \frac{3}{2} - \frac{1}{2} = 1.$$

Putting values of $\lambda_1, \lambda_2, \lambda_3$ and μ in (8), reduced equation of the conicoid is

$$\frac{1}{2}X^2 + \frac{1}{2}Y^2 + 2Z^2 = 1 \quad \text{or} \quad \frac{X^2}{2} + \frac{Y^2}{2} + \frac{Z^2}{\frac{1}{2}} = 1$$

This is an ellipsoid squares whose semi-axes are 2, 2, $\frac{1}{2}$.

Step IV. Because the two values of λ are equal, therefore this ellipsoid is a surface of revolution and d.c's of its principal axis (*i.e.*, axis of revolution)

(See **Note** sub-case I, Art. 10) are given by

$$\frac{al + hm + gn}{l} = \frac{hl + bm + fn}{m} = \frac{gl + fm + cn}{n} = \lambda\ (= 2) \quad \text{(By Art. 6)}$$

From first and fourth members $al + hm + gn = 2l$

or $(a - 2)l + hm + gn = 0$

or $-l + \frac{1}{2}m + \frac{1}{2}n = 0$ or $-2l + m + n = 0$ ...(9)

From second and fourth members $hl + bm + fn = 2m$

or $hl + (b - 2)m + fn = 0$ or $\frac{1}{2}l - m + \frac{1}{2}n = 0$

or $l - 2m + n = 0$ ...(10)

From third and fourth members

$$gl + fm + cn = 2n \quad \text{or} \quad gl + fm + (c - 2)n = 0$$

$\therefore$ $\frac{1}{2}l + \frac{1}{2}m - n = 0$ or $l + m - 2n = 0$ ...(11)

Solving Eqns. (9) and (10) for l, m, n [In fact any two of the three equations out of 9, 10, 11], we have

$$\frac{l}{1+2} = \frac{m}{1+2} = \frac{n}{4-1} \text{ or } \frac{l}{3} = \frac{m}{3} = \frac{n}{3} \quad \text{or} \quad \frac{l}{1} = \frac{m}{1} = \frac{n}{1}$$

$\therefore$ d.r's of principal axis are 1, 1, 1

$\therefore$ Equations of the principal axis (passing through of course $(-1, 1, -2)$ the centre of the conicoid) are

$$\frac{x+1}{1} = \frac{y-1}{1} = \frac{z+2}{1}.$$

Example 2. *Show that the equation $2y^2 + 4zx + 2x - 4y + 6z + 5 = 0$ represents a right circular cone. Show also that the semi-vertical angle of this cone is $\frac{\pi}{4}$ and that its axis is given by $x + z + 2 = 0$, $y = 1$.*

Sol. The given equation of the quadric is

$$f(x, y, z) = 2y^2 + 4zx + 2x - 4y + 6z + 5 = 0 \quad ...(1)$$

Comparing with $ax^2 + by^2 + cz^2 + 2fyz + 2gzx + 2hxy + 2ux + 2vy + 2wz + d = 0$, we have

$$a = 0, b = 2, c = 0, 2f = 0, 2g = 4, 2h = 0,$$
$$2u = 2, 2v = -4, 2w = 6 \text{ and } d = 5$$

i.e.,
$$a = 0, b = 2, c = 0, f = 0, g = 2, h = 0,$$
$$u = 1, v = -2, w = 3 \text{ and } d = 5.$$

Step I. To solve discriminating cubic

$\therefore$
$$a + b + c = 0 + 2 + 0 = 2$$
$$A + B + C = bc - f^2 + ca - g^2 + ab - h^2 = 0 - 0 + 0 - 4 + 0 - 0 = -4$$
$$D = abc + 2fgh - af^2 - bg^2 - ch^2 = 0 + 0 - 0 - 0 - 2(2)^2 - 0 = -8$$

We know that discriminating cubic is

$$\lambda^3 - \lambda^2 (a + b + c) + \lambda (A + B + C) - D = 0 \quad \text{or} \quad \lambda^3 - 2\lambda^2 - 4\lambda + 8 = 0$$

$\lambda = 2$ is a root of this equation by Trial method.

$$[\because \; 2^3 - 2 \cdot 2^2 - 4.2 + 8 = 0 \text{ or } 8 - 8 - 8 + 8 = 0 \text{ or } 0 = 0]$$

Applying synthetic division

2	1	−2	−4	8
		2	0	−8
	1	0	−4	0

The reduced equation is $\lambda^2 - 4 = 0$ or $\lambda^2 = 4$

$\therefore$
$$\lambda = \pm 2.$$

$\therefore$ Roots of the discriminating cubic are $-2, 2, 2$. $(\lambda_1 \le \lambda_2 \le \lambda_3)$

Step II. To find centre of quadric (1)

We know by Art. 8 that centre is given by

$$\frac{\partial f}{\partial x} = 0, \frac{\partial f}{\partial y} = 0 \text{ and } \frac{\partial f}{\partial z} = 0$$

i.e.,
$$ax + hy + gz + u = 0 \quad i.e., \quad 2z + 1 = 0 \quad ...(2)$$
$$hx + by + fz + v = 0 \quad i.e., \quad 2y - 2 = 0 \quad ...(3)$$

and
$$gx + fy + cz + w = 0 \quad i.e., \quad 2x + 3 = 0 \quad ...(4)$$

From Eqn. (2) $z = \frac{-1}{2}$. From Eqn. (3), $y = 1$

and from Eqn. (4) $x = \frac{-3}{2}$

$\therefore$ Centre of the conicoid is $\left(-\frac{3}{2}, 1, \frac{-1}{2}\right) = (x_1, y_1, z_1)$ (say) ...(5)

Step III. Shifting the origin to the centre and transforming the axes, the reduced equation will be

$$\lambda_1 X^2 + \lambda_2 Y^2 + \lambda_3 Z^2 = \mu \quad ...(6)$$

where $\quad \mu = -d' = -(ux_1 + vy_1 + wz_1 + d)$

or $\quad \mu = -\left[1\left(\frac{-3}{2}\right) - 2(1) + 3\left(\frac{-1}{2}\right) + 5\right]$

$$= -\left[\frac{-3}{2} - 2 - \frac{3}{2} + 5\right] = -[-5+5] = 0$$

Putting values of $\lambda_1, \lambda_2, \lambda_3$ and μ in Eqn. (6), reduced equation of the conicoid is

$$-2X^2 + 2Y^2 + 2Z^2 = 0 \quad \text{or} \quad X^2 - Y^2 - Z^2 = 0$$

or $$Y^2 + Z^2 = X^2 = X^2 \tan^2 45°.$$

which is of the form $Y^2 + Z^2 = X^2 \tan^2 \alpha$

and hence represents a right circular cone with semi-vertical $\alpha = 45°$ and vertex (new) origin *i.e.,* centre. [See Art. 9 (*a*), Chapter 11]

$\therefore$ Vertex of this right circular is the centre $\left(-\frac{3}{2}, 1, -\frac{1}{2}\right)$ given by (5).

Step IV. Because the two values of λ are equal, therefore this right circular cone is a surface of revolution and d.c's of its principal axis (*i.e.,* axis of revolution) (See Note Sub-case I, Art. 10) are given by

$$\frac{al + hm + gn}{l} = \frac{hl + bm + fn}{m} = \frac{gl + fm + cn}{n} = \lambda = -2 \qquad \text{(By Art. 6)}$$

Putting values of *a, b, c, f, g, h,* we have

$$\frac{2n}{l} = \frac{2m}{m} = \frac{2l}{n} = -2$$

Dividing every term by 2, $\quad \frac{n}{l} = \frac{m}{m} = \frac{l}{n} = -1$

$\therefore \quad \frac{n}{l} = -1 \quad \text{and} \quad \frac{m}{m} = -1$

$\therefore \quad n = -l \quad \text{and} \quad m = -m \quad \text{or} \quad 2m = 0 \quad \therefore \quad m = 0$

$\therefore$ d.r's of the axis of right circular cone $l, m, n\ (= -l)$ *i.e.,* $l, 0, -l$.

Dividing by l, 1, 0, – 1.

$\therefore$ Equations of axis of right circular cone (through the vertex $\left(-\frac{3}{2}, 1, -\frac{1}{2}\right)$ are

$$\frac{x + \frac{3}{2}}{1} = \frac{y-1}{0} = \frac{z + \frac{1}{2}}{-1}$$

From first and third members $x + \frac{3}{2} = -z - \frac{1}{2}$ or $x + z + 2 = 0$

From second and third members $y - 1 = 0$ or $y = 1$.

$\therefore$ Equations of the axis of right circular cone in general form are

$$x + z + 2 = 0 \quad \text{and} \quad y = 1.$$

Type II. When one Root of the discriminating cubic is zero and k ≠ 0.

Example 3. *Show that the surface whose equation is*

$$16x^2 + 4y^2 + 4z^2 + 4yz - 8zx + 8xy + 4x + 4y - 16z - 24 = 0$$

is an elliptic paraboloid. Find the co-ordinates of its vertex and equation to its axis.

Sol. The given equation of the quadratic is

$$f(x, y, z) = 16x^2 + 4y^2 + 4z^2 + 4yz - 8zx + 8xy + 4x + 4y - 16z - 24 = 0 \quad ...(1)$$

Comparing with $ax^2 + by^2 + cz^2 + 2fyz + 2gzx + 2hxy + 2ux + 2vy + 2wz + d = 0$, we have

$$a = 16, b = 4, c = 4, 2f = 4, 2g = -8, 2h = 8,$$
$$2u = 4, 2v = 4, 2w = -16 \quad \text{and} \quad d = -24$$

or

$$a = 16, b = 4, c = 4, f = 2, g = -4, h = 4,$$
$$u = 2, v = 2, w = -8 \quad \text{and} \quad d = -24.$$

Step I. To solve the discriminating cubic

$$a + b + c = 16 + 4 + 4 = 24$$
$$A + B + C = bc - f^2 + ca - g^2 + ab - h^2$$
$$= 16 - 4 + 64 - 16 + 64 - 16 = 12 + 48 + 48 = 108$$
$$D = abc + 2fgh - af^2 - bg^2 - ch^2$$
$$= 256 + 2(2)(-4)(4) - 16(2)^2 - 4(-4)^2 - 4(4)^2$$
$$= 256 - 64 - 64 - 64 - 64 = 0$$

We know that discriminating cubic is

$$\lambda^3 - \lambda^2 (a + b + c) + \lambda (A + B + C) - D = 0$$

Putting values $\lambda^3 - 24\lambda^2 + 108\lambda = 0$ or $\lambda(\lambda^2 - 24\lambda + 108) = 0$

$\therefore$ Either $\lambda = 0$ or $\lambda^2 - 24\lambda + 108 = 0$ or $\lambda^2 - 6\lambda - 18\lambda + 108 = 0$

$\therefore$ $\lambda = 6, 18$ or $\lambda = 0, 18, 6$

Here one value of λ say λ_3 is 0, therefore we proceed as in sub-case II, Art. 10.

$\therefore$ Eqn. (1) represents either a paraboloid or a cylinder or a pair of planes.

Step II. Let us find l, m, n the d.c's of the axis corresponding to $\lambda_3 = 0$ from the equations

$$al + hm + gn = 0$$
$$hl + bm + fn = 0$$
$$gl + fm + cn = 0$$

i.e., $16l + 4m - 4n = 0$ or $4l + m - n = 0$ ...(2)

and $4l + 4m + 2n = 0$ or $2l + 2m + n = 0$ ...(3)

and $-4l + 2m + 4n = 0$ or $2l - m - 2n = 0$ ...(4)

Solving Eqns. (2) and (3) for l, m, n

$$\frac{l}{1+2} = \frac{m}{-2-4} = \frac{n}{8-2} \text{ or } \frac{l}{3} = \frac{m}{-6} = \frac{n}{6}$$

Dividing denominators by 3, $\frac{l}{1} = \frac{m}{-2} = \frac{n}{2}$

These values of l, m, n clearly satisfy Eqn. (4).

$\therefore$ d.r's of the axis are $1, -2, 2$.

Dividing each by $\sqrt{(1)^2 + (-2)^2 + (2)^2} = \sqrt{1+4+4} = 3$,

d.c's of the axis are $\frac{1}{3}, \frac{-2}{3}, \frac{2}{3} = l, m, n$.

Step III. To find k

$$\therefore \quad k = ul + vm + wn = 2\left(\frac{1}{3}\right) + 2\left(\frac{-2}{3}\right) - 8\left(\frac{2}{3}\right) = \frac{2}{3} - \frac{4}{3} - \frac{16}{3} = -6 \neq 0$$

$\therefore$ Equation (1) represents a paraboloid and reduced equation is

$$\lambda_1 X^2 + \lambda_2 Y^2 + 2kZ = 0$$

i.e., $6X^2 + 18Y^2 + 2(-6)Z = 0$ or $6X^2 + 18Y^2 = 12Z$

Dividing by 6, $X^2 + 3Y^2 = 2Z$

which is an elliptic paraboloid.

Step IV. To find vertex of paraboloid

Let us solve the equations

$$ax + hy + gz - lk + u = 0$$
$$hx + by + fz - mk + v = 0$$
$$gx + fy + cz - nk + w = 0$$

and $k(lx + my + nz) + ux + vy + wz + d = 0$

(This last equation must be used for finding the vertex) for x, y, z to get the co-ordinates of the vertex.

Putting values $16x + 4y - 4z - \frac{1}{3}(-6) + 2 = 0$

or $16x + 4y - 4z + 4 = 0$ or $4x + y - z + 1 = 0$ ...(5)

and $4x + 4y + 2z - \left(-\frac{2}{3}\right)(-6) + 2 = 0$

or $4x + 4y + 2z - 2 = 0$ or $2x + 2y + z - 1 = 0$ ...(6)

and $-4x + 2y + 4z - \frac{2}{3}(-6) - 8 = 0$

or $-4x + 2y + 4z - 4 = 0$ or $2x - y - 2z + 2 = 0$ ...(7)

and $-6\left(\frac{1}{3}x - \frac{2}{3}y + \frac{2}{3}z\right) + 2x + 2y - 8z - 24 = 0$

or $-2x + 4y - 4z + 2x + 2y - 8z - 24 = 0$

or $6y - 12z - 24 = 0$

Dividing by 6, $y - 2z - 4 = 0$ ...(8)

Eqn. (6) – Eqn. (7) gives $3y + 3z - 3 = 0$ or $y + z - 1 = 0$ ...(9)

Eqn. (8) – Eqn. (9) gives $-3z - 3 = 0$ or $-3z = 3$ or $z = -1$

Putting $z = -1$ in Eqn. (9) $y - 1 - 1 = 0$ or $y = 2$

Putting values of y and z in Eqn. (6), $2x + 4 - 1 - 1 = 0$ or $2x = -2$ or $x = -1$

These values of $x = -1$, $y = 2$, $z = -1$ satisfy Eqn. (5).

$\therefore$ Co-ordinates of the **vertex** of paraboloid are $(-1, 2, -1)$.

$\therefore$ Equations of the axis of the paraboloid passing through the vertex $(-1, 2, -1)$ and having d.r's $1, -2, 2$ are $\frac{x+1}{1} = \frac{y-2}{-2} = \frac{z+1}{2}$.

Example 4. *Prove that $x^2 + y^2 + z^2 - yz - zx - xy - 3x - 6y - 9z + 21 = 0$ represents a paraboloid of revolution and find the equation of axis and co-ordinates of focus.*

Sol. The given equation of the quadric is

$$f(x, y, z) = x^2 + y^2 + z^2 - yz - zx - xy - 3x - 6y - 9z + 21 = 0 \quad ...(1)$$

Comparing with $ax^2 + by^2 + cz^2 + 2fyz + 2gzx + 2hxy + 2ux + 2vy + 2wz + d = 0$, we have

$$a = 1, b = 1, c = 1, 2f = -1, 2g = -1, 2h = -1, 2u = -3,$$
$$2v = -6, 2w = -9, d = 21$$

or
$$a = 1, b = 1, c = 1, f = \frac{-1}{2}, g = \frac{-1}{2}, h = \frac{-1}{2}, u = \frac{-3}{2},$$
$$v = -3, w = -\frac{9}{2}, d = 21.$$

Step I. To solve the discriminating cubic

$$a + b + c = 1 + 1 + 1 = 3$$
$$A + B + C = bc - f^2 + ca - g^2 + ab - h^2 = 1 - \frac{1}{4} + 1 - \frac{1}{4} + 1 - \frac{1}{4} = 3 - \frac{3}{4} = \frac{9}{4}$$
$$D = abc + 2fgh - af^2 - bg^2 - ch^2 = 1 - \frac{1}{4} - \frac{1}{4} - \frac{1}{4} - \frac{1}{4} = 0$$

We know that discriminating cubic is

$$\lambda^3 - \lambda^2 (a + b + c) + \lambda (A + B + C) - D = 0$$

or
$$\lambda^3 - 3\lambda^2 + \frac{9}{4}\lambda = 0 \quad \text{or} \quad 4\lambda^3 - 12\lambda^2 + 9\lambda = 0$$

or
$$\lambda(4\lambda^2 - 12\lambda + 9) = 0 \quad \text{or} \quad \lambda(2\lambda - 3)^2 = 0$$

∴ Either $\lambda = 0$ or $\lambda = \frac{3}{2}, \frac{3}{2}$.

Here one value of λ say λ_3 is 0, therefore we proceed as in sub-case II, Art. 10.

∴ Equation (1) represents either a paraboloid or a cylinder or a pair of planes.

Step II. Let us find l, m, n the d.c's of the axis corresponding to $\lambda_3 = 0$ from the equations

$$al + hm + gn = 0$$
$$hl + bm + fn = 0$$
$$gl + fm + cn = 0$$

i.e.,
$$l - \frac{1}{2}m - \frac{1}{2}n = 0 \quad \text{or} \quad 2l - m - n = 0 \qquad ...(2)$$

and
$$\frac{-1}{2}l + m - \frac{1}{2}n = 0 \quad \text{or} \quad -l + 2m - n = 0 \qquad ...(3)$$

and
$$\frac{-1}{2}l - \frac{1}{2}m + n = 0 \quad \text{or} \quad -l - m + 2n = 0 \qquad ...(4)$$

Solving Eqns. (2) and (3) for l, m, n

$$\frac{l}{1+2} = \frac{m}{1+2} = \frac{n}{4-1}$$

or
$$\frac{l}{3} = \frac{m}{3} = \frac{n}{3} \quad \text{or} \quad \frac{l}{1} = \frac{m}{1} = \frac{n}{1}$$

These values of l, m, n clearly satisfy Eqn. (4).

∴ d.r's of the axis are 1, 1, 1.

Dividing each by $\sqrt{(1)^2 + (1)^2 + (1)^2} = \sqrt{3}$.

d.c's of the axis are $\frac{1}{\sqrt{3}}, \frac{1}{\sqrt{3}}, \frac{1}{\sqrt{3}} = l, m, n$.

Step III. To find k

$$k = ul + vm + wn$$

$$= \left(\frac{-3}{2}\right)\frac{1}{\sqrt{3}} - 3\left(\frac{1}{\sqrt{3}}\right) - \frac{9}{2}\left(\frac{1}{\sqrt{3}}\right) = \frac{-3}{2\sqrt{3}} - \frac{3}{\sqrt{3}} - \frac{9}{2\sqrt{3}} = \frac{-3-6-9}{2\sqrt{3}} = \frac{-18}{2\sqrt{3}}$$

$$= \frac{-9}{\sqrt{3}} = \frac{-3\times 3}{\sqrt{3}} = -3\sqrt{3} \neq 0$$

$\therefore$ Equation (1) represents a paraboloid of revolution because two values of λ are equal and reduced equation is

$$\lambda_1 X^2 + \lambda_2 Y^2 + 2kZ = 0 \quad i.e., \quad \frac{3}{2}X^2 + \frac{3}{2}Y^2 - 6\sqrt{3}\,Z = 0$$

Dividing every term by 3 and multiplying by 2, $X^2 + Y^2 = 4\sqrt{3}\,Z$.

Step IV. To find vertex of paraboloid

Let us solve the equations

$$ax + hy + gz - lk + u = 0$$
$$hx + by + fz - mk + v = 0$$
$$gx + fy + cz - nk + w = 0$$

and $\quad k(lx + my + nz) + ux + vy + wz + d = 0$

(This last equation must be used for finding the vertex) for x, y, z to get the co-ordinates of the vertex.

Putting values $\quad x - \frac{1}{2}y - \frac{1}{2}z - \frac{1}{\sqrt{3}}(-3\sqrt{3}) - \frac{3}{2} = 0$

or $\quad 2x - y - z + 6 - 3 = 0 \quad \text{or} \quad 2x - y - z + 3 = 0$ ...(5)

and $\quad \frac{-1}{2}x + y - \frac{1}{2}z - \frac{1}{\sqrt{3}}(-3\sqrt{3}) - 3 = 0$

or $\quad x - 2y + z - 6 + 6 = 0 \quad \text{or} \quad x - 2y + z = 0$ ...(6)

and $\quad \frac{-1}{2}x - \frac{1}{2}y + z - \frac{1}{\sqrt{3}}(-3\sqrt{3}) - \frac{9}{2} = 0$

or $\quad x + y - 2z - 6 + 9 = 0 \quad \text{or} \quad x + y - 2z + 3 = 0$ ...(7)

and $\quad -3\sqrt{3}\left(\frac{1}{\sqrt{3}}x + \frac{1}{\sqrt{3}}y + \frac{1}{\sqrt{3}}z\right) - \frac{3}{2}x - 3y - \frac{9}{2}z + 21 = 0$

or $\quad -3x - 3y - 3z - \frac{3}{2}x - 3y - \frac{9}{2}z + 21 = 0$

or $\quad -6x - 6y - 6z - 3x - 6y - 9z + 42 = 0$

or $\quad -9x - 12y - 15z + 42 = 0 \quad \text{or} \quad 3x + 4y + 5z - 14 = 0$ ...(8)

Let us solve any two of equations (5), (6), (7); say (5), (6) here and equation (8) for x, y, z.

Eqn. (5) + Eqn. (6) gives $\quad 3x - 3y + 3 = 0 \quad \text{or} \quad x - y + 1 = 0$ ...(9)

Eqn (8) + 5 × Eqn. (5) gives $\quad 13x - y + 1 = 0$ ...(10)

Eqn. (10) – Eqn. (9) gives $12x = 0$ or $x = 0$

Putting $x = 0$ in (9) $y = 1$.

Putting $x = 0, y = 1$ in (6), $-2 + z = 0$ or $z = 2$

Putting $x = 0, y = 1, z = 2$ in Eqn. (7), it is satisfied.

∴ Vertex of paraboloid is (0, 1, 2).

∴ Equations of the axis of the parabola passing through the vertex (0, 1, 2) and having d.r's 1, 1, 1 are $\frac{x-0}{1} = \frac{y-1}{1} = \frac{z-2}{1}$.

To find focus. Let us use distance form of the equation of the axis (*i.e.*, in the denominator, let us use d.c's)

∴ Equation of axis are

$$\frac{x-0}{\frac{1}{\sqrt{3}}} = \frac{y-1}{\frac{1}{\sqrt{3}}} = \frac{z-2}{\frac{1}{\sqrt{3}}} = r \qquad ...(11)$$

We have derived above that the reduced equation is

$$X^2 + Y^2 = 4\sqrt{3}\,Z$$

Taking principal section by plane $X = 0$ (or $Y = 0$) ; $Y^2 = 4\sqrt{3}\,Z$

Comparing with $Y^2 = 4aZ$, we have

Length of L.R. $= 4a = 4\sqrt{3}$ or $a = \sqrt{3}$

Putting $r = a = \sqrt{3}$ in Eqn. (11)

(∵ Focus is always at a distance of one fourth of the latus rectum from the vertex)

Putting $r = \sqrt{3}$ in (1),

Co-ordinates of the focus are given by

$$\frac{x-0}{\frac{1}{\sqrt{3}}} = \frac{y-1}{\frac{1}{\sqrt{3}}} = \frac{z-2}{\frac{1}{\sqrt{3}}} = \sqrt{3}$$

∴ $x - 0 = 1, y - 1 = 1, z - 2 = 1$ or $x = 1, y = 2, z = 3$

∴ Co-ordinates of the focus of the paraboloid of revolution are (1, 2, 3).

Type III. When one root of the discriminating cubic is zero and k = 0

Example 5. *Prove that the equation $5x^2 + 5y^2 + 8z^2 + 8yz + 8zx - 2xy + 12x - 12y + 6 = 0$ represents a cylinder whose cross-section is an ellipse of eccentricity $\frac{1}{\sqrt{2}}$ and find the equation of its axis.* (K.U. 2007)

Sol. The given equation of the quadric is

$$f(x, y, z) = 5x^2 + 5y^2 + 8z^2 + 8yz + 8zx - 2xy + 12x - 12y + 6 = 0 \qquad ...(1)$$

Comparing with $ax^2 + by^2 + cz^2 + 2fyz + 2gzx + 2hxy + 2ux + 2vy + 2wz + d = 0$, we have

$$a = 5, b = 5, c = 8, 2f = 8, 2g = 8, 2h = -2, 2u = 12,$$
$$2v = -12, 2w = 0, d = 6$$

or $a = 5, b = 5, c = 8, f = 4, g = 4, h = -1, u = 6, v = -6, w = 0, d = 6.$

Step I. To solve the discriminating cubic

$$a + b + c = 5 + 5 + 8 = 18$$
$$A + B + C = bc - f^2 + ca - g^2 + ab - h^2 = 40 - 16 + 40 - 16 + 25 - 1 = 72$$

$$D = abc + 2fgh - af^2 - bg^2 - ch^2 = 200 - 32 - 80 - 80 - 8 = 200 - 200 = 0$$

We know that discriminating cubic is

$$\lambda^3 - \lambda^2 (a + b + c) + \lambda(A + B + C) - D = 0$$

or $\lambda^3 - 18\lambda^2 + 72\lambda = 0$ or $\lambda(\lambda^2 - 18\lambda + 72) = 0$

or $\lambda(\lambda^2 - 6\lambda - 12\lambda + 72) = 0$ or $\lambda(\lambda - 6)(\lambda - 12) = 0$

$\therefore$ $\lambda = 0, 6, 12$

Here one value of λ say λ_3 is zero, therefore we proceed as is sub-case II, Art. 10.

$\therefore$ Equation (1) represents either a paraboloid or a cylinder or a pair of planes.

Step II. Let us find l, m, n the d.c's of the axis corresponding to $\lambda (= \lambda_3) = 0$ from the equations

$$al + hm + gn = 0 \quad i.e., \quad 5l - m + 4n = 0 \qquad ...(2)$$

$$hl + bm + fn = 0 \quad i.e., \quad -l + 5m + 4n = 0 \qquad ...(3)$$

$$gl + fm + cn = 0 \quad i.e., \quad 4l + 4m + 8n = 0 \qquad ...(4)$$

Solving Eqns. (2) and (3) for l, m, n, we have

$$\frac{l}{-4-20} = \frac{m}{-4-20} = \frac{n}{25-1} \quad \text{or} \quad \frac{l}{-24} = \frac{m}{-24} = \frac{n}{24}$$

Dividing denominators by -24, $\frac{l}{1} = \frac{m}{1} = \frac{n}{-1}$

These values of l, m, n clearly satisfy Eqn. (4).

$\therefore$ d.r's of the axis are $1, 1, -1$.

Dividing each by $\sqrt{(1)^2 + (1)^2 + (-1)^2} = \sqrt{3}$,

d.c's of the axis are $\frac{1}{\sqrt{3}}, \frac{1}{\sqrt{3}}, \frac{-1}{\sqrt{3}}$.

Step III. To find k

$$k = ul + vm + wn = 6\left(\frac{1}{\sqrt{3}}\right) - 6\left(\frac{1}{\sqrt{3}}\right) + 0\left(\frac{-1}{\sqrt{3}}\right) = \frac{6}{\sqrt{3}} - \frac{6}{\sqrt{3}} = 0$$

$\therefore$ Equation (1) represents a cylinder or a pair of planes and there is a line of centres which is given by any two of the equations

$$\frac{\partial f}{\partial x} = 0, \frac{\partial f}{\partial y} = 0, \frac{\partial f}{\partial z} = 0. \text{ Let us take } \frac{\partial f}{\partial x} = 0, \frac{\partial f}{\partial y} = 0$$

i.e., $ax + hy + gz + u = 0$ and $hx + by + fz + v = 0$

or
$$5x - y + 4z + 6 = 0 \qquad ...(5)$$

$$-x + 5y + 4z - 6 = 0 \qquad ...(6)$$

Let us take a point (α, β, γ) on the line of centres given by equations (5) and (6).

Putting $z = 0$ (say) in equations (5) and (6), we have

$$5x - y + 6 = 0 \qquad ...(7)$$

and
$$-x + 5y - 6 = 0 \qquad ...(8)$$

Eqn. (7) + 5 × Eqn. (8) gives, $24y - 24 = 0$ or $24y = 24$

$\therefore$ $y = 1$.

Putting $y = 1$ in (7), $5x + 5 = 0$ or $x = -1$

$\therefore$ $(-1, 1, 0) = (\alpha, \beta, \gamma)$ is a point on the line of centres given by (5) and (6).

$\therefore$ Reduced equation of cylinder conicoid is

$$\lambda_1 X^2 + \lambda_2 Y^2 + u\alpha + v\beta + w\gamma + d = 0 \qquad \text{(By sub-case II, Art. 10)}$$

Putting values, $6X^2 + 12Y^2 + 6(-1) + (-6)(1) + 0(0) + 6 = 0$

or $6X^2 + 12Y^2 - 6 = 0$ or $6X^2 + 12Y^2 = 6$

Dividing by 6, $X^2 + 2Y^2 = 1$ or $\dfrac{X^2}{1} + \dfrac{Y^2}{\frac{1}{2}} = 1$

which is an elliptic cylinder (because $u\alpha + v\beta + w\gamma + d \neq 0$) (See Art. 3, Chapter 12)

Its cross-section by the plane $Z = 0$ is an ellipse

$$\frac{X^2}{1} + \frac{Y^2}{\frac{1}{2}} = 1, \ (Z = 0) \qquad \text{which is of the form } \frac{X^2}{a^2} + \frac{Y^2}{b^2} = 1.$$

Here $a^2 = 1, b^2 = \dfrac{1}{2}$.

$\therefore$ Eccentricity is given by $b^2 = a^2(1 - e^2)$

or $$\frac{1}{2} = 1 - e^2 \quad \therefore \quad e^2 = \frac{1}{2}$$

$\therefore$ $$e = \frac{1}{\sqrt{2}}.$$

Equation of the axis of cylinder, a line passing through $(\alpha, \beta, \gamma) = (-1, 1, 0)$ a point on the line of centres and having d.r's $1, 1, -1$ is

$$\frac{x+1}{1} = \frac{y-1}{1} = \frac{z-0}{-1}.$$

EXERCISE 1

1. Show that the equation $x^2 + y^2 + z^2 + yz + zx + xy = 1$ represents an ellipsoid the squares of whose semi-axes are $2, 2, \dfrac{1}{2}$. Show also that the equation of its axis of revolution is $x = y = z$. *(M.D.U. 2007)*
2. Prove that the equation $2x^2 + 5y^2 + z^2 - 4xy - 8x + 14y + 3 = 0$ is a surface of revolution. Determine the reduced equation and the equation of its principal axis.
3. Reduce the equation $11x^2 + 10y^2 + 6z^2 - 8yz + 4zx - 12xy + 72x - 72y + 36z + 150 = 0$ to the standard form. Show that it represents an ellipsoid and find the equations to the principal axes. *(M.D.U. 2003 S)*

   [**Hint.** Discriminating cubic is $\lambda^3 - 27\lambda^2 + 180\lambda - 324 = 0$; $\lambda = 3, 6, 18$. Centre is $(-2, 2, -1)$.]
4. Reduce the equation $2x^2 - 7y^2 + 2z^2 - 10yz - 8zx - 10xy + 6x + 12y - 6z + 5 = 0$ to the standard form. Write the standard form and show that it represents a cone. *(K.U. 2003)*

   [**Hint.** $\lambda = 3, 6, -12$.]
5. Show that $2x^2 + 2y^2 + z^2 + 2yz - 2zx - 4xy + x + y = 0$ represents a paraboloid. Find the reduced equation, the co-ordinates of the vertex and equation to the axis. *(M.D.U. 2003)*
6. Prove that the surface represented by the equation $3z^2 - 6yz - 6zx - 7x - 5y + 6z + 3 = 0$ is a hyperbolic paraboloid.

7. Reduce the following equation of a hyperbolic paraboloid to the standard form : $5x^2 - 16y^2 + 5z^2 + 8yz - 14zx + 8xy + 4x + 20y + 4z - 24 = 0$.
   Find also the co-ordinates of the vertex and the equation of its axis.
   $$\left[\textbf{Hint. } \lambda = 12, -18, 0; \frac{l}{2} = \frac{m}{1} = \frac{n}{2}.\right]$$
8. Show that the surface whose equation is $4x^2 - y^2 - z^2 + 2yz - 8x - 4y + 8z - 2 = 0$ is a hyperbolic paraboloid. Find the co-ordinates of its vertex and the equation to its axis.
9. Reduce to the standard form the equation
   $$2x^2 + 5y^2 + 2z^2 - 2yz + 4xz - 2xy + 14x - 16y + 14z + 26 = 0.$$
10. Show that $26x^2 + 20y^2 + 10z^2 - 4yz - 16zx - 36xy + 52x - 36y - 16z + 25 = 0$ represents an elliptic cylinder. Reduce it to its standard form and find the equation of its axis.
11. Prove that the equation $5x^2 - 4y^2 + 5z^2 + 4yz - 14zx + 4xy + 16x + 16y - 32z + 8 = 0$ represents a pair of planes which pass through the line $x + 3 = y = z + 1$. Find also the angle between the planes.
   [**Hint.** Proceed as in Example 5.
   You will find that $u\alpha + v\beta + w\gamma + d = 0$
   Reduced equation is $2X^2 - Y^2 = 0$ or $(\sqrt{2}X + Y)(\sqrt{2}X - Y) = 0$
   Planes are $\sqrt{2}X + Y = 0$ and $\sqrt{2}X - Y = 0$.
   Angle between the planes is given by $\cos\theta = \dfrac{a_1a_2 + b_1b_2 + c_1c_2}{\sqrt{a_1^2 + b_1^2 + c_1^2}\sqrt{a_2^2 + b_2^2 + c_2^2}}$.

Answers

2. $X^2 + Y^2 + 6Z^2 = 8$; $\dfrac{x-1}{1} = \dfrac{y+1}{-2} = \dfrac{z}{0}$
3. $\dfrac{X^2}{4} + \dfrac{Y^2}{2} + \dfrac{3Z^2}{2} = 1$, $\dfrac{x+2}{2} = \dfrac{y-2}{1} = \dfrac{z+1}{-2}$, $\dfrac{x+2}{1} = \dfrac{y-2}{2} = \dfrac{z+1}{2}$, $\dfrac{x+2}{-2} = \dfrac{y-2}{2} = \dfrac{z+1}{-1}$
4. $X^2 + 2Y^2 - 4Z^2 = 0$
5. $\left(\dfrac{5-\sqrt{17}}{2}\right)X^2 + \left(\dfrac{5+\sqrt{17}}{2}\right)Y^2 + \sqrt{2}\,Z = 0$
   Vertex is (0, 0, 0) and the equations to its axis is $\dfrac{x}{1} = \dfrac{y}{1} = \dfrac{z}{0}$.
7. $3X^2 - 2Y^2 - 2Z = 0$, vertex is (1, 1, 1), equation of its axis is $\dfrac{x-1}{2} = \dfrac{y-1}{1} = \dfrac{z-1}{2}$
8. $X^2 - 2Y^2 - \sqrt{2}Z = 0$, vertex $\left(1, -\dfrac{9}{4}, \dfrac{3}{4}\right)$, Equations of its axis are $\dfrac{x-1}{0} = \dfrac{y+\frac{9}{4}}{1} = \dfrac{z-\frac{3}{4}}{1}$
9. $X^2 + 2Y^2 = 1$; Elliptic cylinder
10. $14X^2 + 42Y^2 = 1$, Equations of its axis are $\dfrac{x+1}{1} = \dfrac{y}{1} = \dfrac{z}{1}$.
11. $\cos^{-1}\dfrac{1}{3}$.

11. Reduction of a second degree equation in x, y, z when the second degree terms form a perfect square

Because the second degree terms are given to be forming a perfect square so the equation of the conicoid is of the form

$$(px + qy + rz)^2 + 2ux + 2vy + 2wz + d = 0 \qquad ...(1)$$

Let us consider $(px + qy + rz + t)^2$, where t is any real number.

Opening the square using $(a + b)^2$, where $a = px + qy + rz$.

$$\therefore \quad (px + qy + rz + t)^2 = (px + qy + rz)^2 + t^2 + 2t\,(px + qy + rz)$$

$$= (-2ux - 2vy - 2wz - d) + t^2 + 2tpx + 2tqy + 2trz \quad \text{[By (1)]}$$

or
$$(px + qy + rz + t)^2 = 2x\,(pt - u) + 2y\,(tq - v) + 2z\,(tr - w) + t^2 - d \quad ...(2)$$

Let us choose t so that the two planes given by $px + qy + rz + t = 0$

(Equating L.H.S. to zero)

and $2x\,(pt - u) + 2y\,(tq - v) + 2z\,(tr - w) + t^2 - d = 0$ (obtained by equating R.H.S. to zero) are at right angles to each other.

$$\therefore \quad a_1a_2 + b_1b_2 + c_1c_2 = 0$$

i.e.,
$$p\,.\,2(pt - u) + q\,.\,2\,(tq - v) + r\,.\,2\,(tr - w) = 0.$$

Having thus choosen t, we rewrite (2) as

$$(p^2 + q^2 + r^2)\left(\frac{px + qy + rz + t}{\sqrt{p^2 + q^2 + r^2}}\right)^2 = k\left[\frac{2(pt - u)x + 2(tq - v)y + 2(tr - w)z + t^2 - d}{k}\right]$$

where $k = \sqrt{4(pt - u)^2 + 4(tq - v)^2 + 4(tr - w)^2}$

Putting
$$\frac{px + qy + rz + t}{\sqrt{p^2 + q^2 + r^2}} = Y$$

and
$$\frac{2(pt - u)x + 2(tq - v)y + 2(tr - w)z + t^2 - d}{k} = X$$

and dividing by $(p^2 + q^2 + r^2)$, we have

$$Y^2 = \frac{k}{p^2 + q^2 + r^2}\,X$$

and we know that this equation represents a parabolic cylinder.

Example 1. *Reduce the equation $36x^2 + 4y^2 + z^2 - 4yz - 12zx + 24xy + 4x + 16y - 26z - 3 = 0$ to the standard form and find the latus rectum of a normal section.*

Sol. The given equation is $36x^2 + 4y^2 + z^2 - 4yz - 12zx + 24xy + 4x + 16y - 26z - 3 = 0$

Here second degree terms are forming a perfect square

or
$$(6x)^2 + (2y)^2 + z^2 - 2\,.\,2y\,.\,z - 2\,.\,6x\,.\,z + 2\,.\,6x\,.\,2y + 4x + 16y - 26z - 3 = 0$$

or
$$(6x + 2y - z)^2 = -4x - 16y + 26z + 3 \quad ...(1)$$

Consider
$$(6x + 2y - z + t)^2 = (6x + 2y - z)^2 + t^2 + 2t\,(6x + 2y - z)$$

$$= -4x - 16y + 26z + 3 + t^2 + 12tx + 4ty - 2tz \quad \text{[By (1)]}$$

or
$$(6x + 2y - z + t)^2 = (12t - 4)x + (4t - 16)y + (26 - 2t)z + t^2 + 3 \quad ...(2)$$

Let us choose t such that the two planes given by

$$6x + 2y - z + t = 0 \quad \text{and} \quad (12t - 4)x + (4t - 16)y + (26 - 2t)z + t^2 + 3 = 0$$

are at right angles.

$$\therefore \quad a_1a_2 + b_1b_2 + c_1c_2 = 0$$

i.e., Product of coeffs. of x + Product of coeffs. of y + Product of coeff. of $z = 0$

i.e.,
$$6(12t - 4) + 2(4t - 16) - (26 - 2t) = 0$$

i.e.,
$$72t - 24 + 8t - 32 - 26 + 2t = 0$$

or
$$82t = 82 \text{ or } t = 1$$

Putting $t = 1$ in Eqn. (2), equation (2) transforms as

$$(6x + 2y - z + 1)^2 = 8x - 12y + 24z + 4 = 4(2x - 3y + 6z + 1)$$

Multiplying and dividing both L.H.S. and R.H.S. by $\sqrt{a^2 + b^2 + c^2}$, we have

$$41\left(\frac{6x+2y-z+1}{\sqrt{36+4+1}=\sqrt{41}}\right)^2 = 4\times 7\left(\frac{2x-3y+6z+1}{\sqrt{4+9+36}=7}\right)$$

Put $\dfrac{6x+2y-z+1}{\sqrt{41}} = Y$ and $\dfrac{2x-3y+6z+1}{7} = X$,

the above equation becomes $41Y^2 = 28X$ or $Y^2 = \dfrac{28}{41}X$

which is a parabolic cylinder.

A normal section of this cylinder by the plane $Z = 0$ is the parabola

$$Y^2 = \frac{28}{41}X$$

Comparing with $Y^2 = 4ax$, we have

Length of latus rectum $= 4a = \dfrac{28}{41}$.

EXERCISE 2

1. Show that the equation $x^2 + 4y^2 + 9z^2 + 12yz + 6zx + 4xy - 54x - 52y + 62z + 113 = 0$ represents a parabolic cylinder and the length of latus rectum of a normal section of the parabola is $4\sqrt{3}$.
2. Find the nature of the conicoid whose equation is
$$4x^2 + y^2 + 4z^2 - 4yz + 8zx - 4xy + 2x - 4y + 5z + 1 = 0.$$
3. Show that the equation $9x^2 + 4y^2 + 4z^2 + 8yz + 12zx + 12xy + 4x + y + 10z + 1 = 0$ represents a parabolic cylinder. *(M.D.U. 2007)*

Answer

2. Parabolic cylinder ; $Y^2 = \dfrac{1}{3}X$.

17

Plane Sections of Conicoids

1. Introduction

We know that the locus of the general equation of the second degree $ax^2 + by^2 + cz^2 + 2fyz + 2gzx + 2hxy + 2ux + 2vy + 2wz + d = 0$ is called a **conicoid.**

Also the locus of general equation of the first degree

$$ax + by + cz + d = 0 \text{ is a plane.}$$

Also we know that every plane section of a conicoid is a plane figure.

Def. If the conicoid is **central** (*i.e.*, has a centre), then a plane section through the centre is known as a **central section.** A plane section which does not pass through the centre is called a **non-central section.**

We can easily understand the above statements if we imagine

(*i*) Cutting spherical fruit by a plane knife is a plane circle and this circular plane will have greatest radius if the spherical fruit is cut by the plane knife passing through the centre of the spherical fruit.

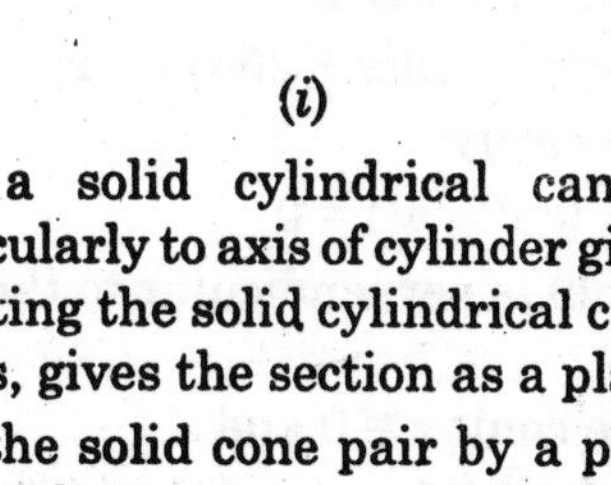

(*i*)

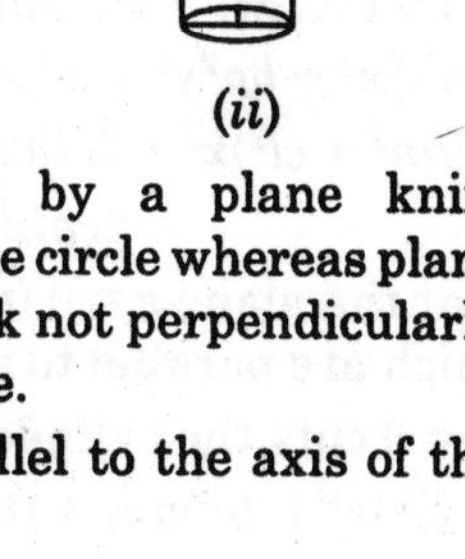

(*ii*)

(*ii*) Cutting a solid cylindrical candle stick by a plane knife perpendicularly to axis of cylinder gives a plane circle whereas plane knife cutting the solid cylindrical candle stick not perpendicularly to its axis, gives the section as a plane ellipse.

(*iii*) Cutting the solid cone pair by a plane parallel to the axis of the cone gives the section as a plane hyperbola.

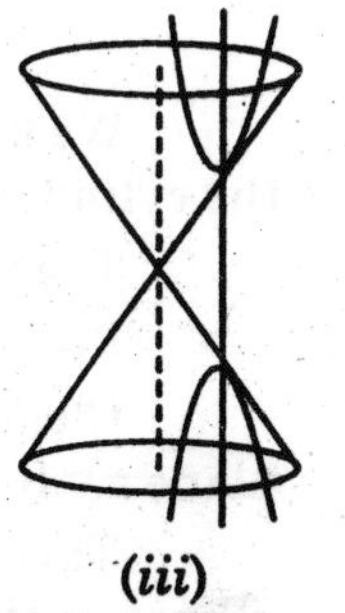

(*iii*)

Remark 1. When two persons meet each other to discuss some topic then their inter-action is similar to the section of one person by the other person. Our inter-action with a religious saint will produce some new ideas of religion in our brain. Our inter-action with a political leader will produce political ideas in our brain.

Remark 2. We already know from chapter 4 and Appendix after it, that plane sections of a conicoid are conics and sections by parallel planes are also similar conics. In this chapter we will study the nature of the plane sections of all conicoids as well as the equations to determine the magnitudes and directions of the axes.

Remark 3. We can also understand from common sense that orthogonal projection of a parabola is another parabola, of a hyperbola is another hyperbola and of an ellipse is another ellipse or in some cases circle.

2. A Note on Conics

If the lengths of the semi-axes of a conic are α, β and $\alpha > r > \beta$, the conic has two diameters of length $2r$, and they are equally inclined to the axes. If $r = \alpha$ or β, the two diameters of length $2r$ coincide with an axis. Hence to find the axes of the conic in which a given plane cuts a conicoid, we first form the equation of the cone whose vertex is the centre, C, of the conic and which has a generators the lines of length r which can be drawn in the plane from C to the conicoid. The lines of section of this cone and the given plane are the semi-diameters of length r of the conic. If $2r$ is the length of an axis, these are coincident, or the plane touches the cone, the generator of contact being the axis. The condition of tangency is expressed in an equation which determines r ; the comparison of the equations of the given plane and a tangent plane leads to the direction-cosines of the generator of contact.

3. To determine the Nature of plane sections of a central conicoid

Let the equation of the central conicoid be

$$ax^2 + by^2 + cz^2 = 1 \qquad ...(1)$$

Let the equation of the plane be

$$lx + my + nz = p \qquad ...(2)$$

We know from chapter 12, Art. 3 that the equation of the cylinder passing through the section given by (1) and (2) and having its generators parallel to z-axis is obtained by eliminating z from (1) and (2).

From eqn. (2) $\qquad z = \dfrac{p - lx - my}{n}.$

Putting this value of z in Eqn. (1), we have

$$ax^2 + by^2 + c\,\frac{(p - lx - my)^2}{n^2} = 1$$

Multiplying by L.C.M. $= n^2$, equation of the cylinder is

$$an^2x^2 + bn^2y^2 + c\,(p^2 + l^2x^2 + m^2y^2 - 2plx + 2lmxy - 2pmy) = n^2$$

or
$$(an^2 + cl^2)x^2 + 2clmxy + (bn^2 + cm^2)y^2 - 2cplx - 2cpmy + (cp^2 - n^2) = 0 \qquad ...(3)$$

We know that the plane $z = 0$ (*i.e.*, xy-plane) is perpendicular to the generating lines of the cylinder (3) which are parallel to z-axis.

The plane $z = 0$ cuts the cylinder (3) in the conic $z = 0$ and

$$(an^2 + cl^2)x^2 + 2clmxy + (bn^2 + cm^2)y^2 - 2cplx - 2cpmy + (cp^2 - n^2) = 0 \qquad ...(4)$$

Clearly the conic (4) is the orthogonal projection of the conic given by (1) and (2) on the plane $z = 0$.

This projection and therefore also the given section (By Remark 3, Art. 1) is a parabola, ellipse or Hyperbola according as

$$H^2 - AB = 0, < 0 \quad \text{or} \quad > 0$$

where A is Coeff. of x^2, B = Coeff. of y^2 and H $= \dfrac{1}{2}$ Coeff. of xy.

i.e., according as

$$c^2l^2m^2 - (an^2 + cl^2)(bn^2 + cm^2) = 0, < 0 \quad \text{or} \quad > 0$$

i.e., $$c^2l^2m^2 - abn^4 - acm^2n^2 - bcn^2l^2 - c^2l^2m^2 = 0, < 0 \quad \text{or} \quad > 0$$

i.e., $$-abn^4 - acm^2n^2 - bcn^2l^2 = 0, < 0 \quad \text{or} \quad > 0$$

Dividing every term by $-n^2$, (and hence reversing the sign of inequality)

$$abn^2 + acm^2 + bcl^2 = 0, > 0 \quad \text{or} \quad < 0$$

(Note) (Note)

Hence the given section is

(*i*) a parabola if $bcl^2 + cam^2 + abn^2 = 0$

(*ii*) an ellipse if $bcl^2 + cam^2 + abn^2 > 0$

(*iii*) a hyperbola if $bcl^2 + cam^2 + abn^2 < 0$.

4. Axes of a central plane section

To determine the lengths and direction cosines of the axes of the central section of the central conicoid

$$ax^2 + by^2 + cz^2 = 1 \qquad \text{...(1)}$$

by the central plane $$lx + my + nz = 0 \qquad \text{...(2)}$$

We know from chapter 4 that centre of conicoid (1) is also the centre of the central section given by Eqns. (1) and (2) and this centre here is origin (0, 0, 0) [obvious from Eqn. (1)].

Let λ, μ, γ be the direction cosines of a semi-diameter of conicoid (1) of length r, then its equations (through the centre (0, 0, 0) are

$$\frac{x-0}{\lambda} = \frac{y-0}{\mu} = \frac{z-0}{\gamma} = r \qquad \text{...(3)}$$

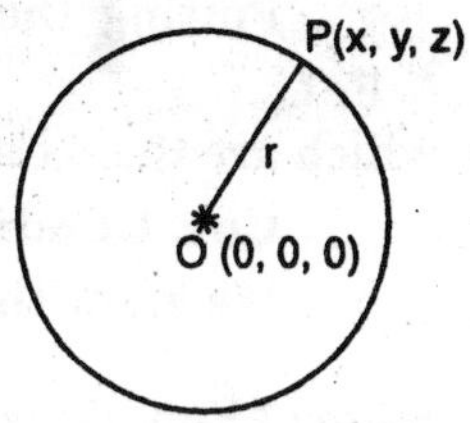

From (3), the co-ordinates of the extremity of this semi-diameter are $P(x, y, z) = (\lambda r, \mu r, \gamma r)$ and this point γ lies on conicoid (1).

$$\therefore \quad a\lambda^2r^2 + b\mu^2r^2 + c\gamma^2r^2 = 1 = \lambda^2 + \mu^2 + \gamma^2$$

$$[\because \ \lambda, \mu, \gamma \text{ are d.c's of the line} \Rightarrow \lambda^2 + \mu^2 + \gamma^2 = 1]$$

or $$(ar^2 - 1)\lambda^2 + (br^2 - 1)\mu^2 + (cr^2 - 1)\gamma^2 = 0 \qquad \text{...(4)}$$

Eliminating λ, μ, γ from Eqns. (3) and (4) ;

(Putting $\lambda = x, \mu = y, \gamma = z$ from (3) in (4))

We have $$(ar^2 - 1)x^2 + (br^2 - 1)y^2 + (cr^2 - 1)z^2 = 0 \qquad \text{...(5)}$$

Which being homogeneous equation of second degree in x, y, z represents a cone with centre (0, 0, 0) of conicoid as its vertex and the semi-diameters of the conicoid of length r as its generators.

We know that plane (2) passing through the origin, the vertex of cone (5), cuts the cone (5) in two generators which determine the directions of two equal diameters of length $2r$ of the section and which are therefore, equally inclined to the axes of the section. (By Art. 2).

In case $2r$ is the length of either axes of the section, the generators coincide and therefore the plane (2) touches the cone (5), the generator of contact being one of the axes and hence touches the cone at every point of the generator and hence at the extremity $(\lambda r, \mu r, \gamma r)$ also.

Now the condition for the plane (2) to be a tangent plane to cone (5) is

$$\frac{l^2}{ar^2-1}+\frac{m^2}{br^2-1}+\frac{n^2}{cr^2-1}=0 \quad ...(6)$$

[$\because$ By chapter 2 on cones, condition for the plane $lx + my + nz = 0$ to be a tangent plane to the cone $Ax^2 + By^2 + Cz^2 = 0$ is $\frac{l^2}{A}+\frac{m^2}{B}+\frac{n^2}{C}=0$]

Multiplying Eqn. (6) by L.C.M., we have

$$l^2(br^2-1)(cr^2-1)+m^2(ar^2-1)(cr^2-1)+n^2(ar^2-1)(br^2-1)=0$$

or

$$r^4(bcl^2+acm^2+abn^2)-r^2[(b+c)l^2+(c+a)m^2+(a+b)n^2]+(l^2+m^2+n^2)=0 \quad ...(7)$$

Eqn. (7) is a quadratic in r^2 and hence has two roots say r_1^2 and r_2^2 which are the squares of the semi-axes of the conic given by (1) and (2).

To find d.c's of axes of the conic

Now Equation of the tangent plane at the point $(\lambda r, \mu r, \gamma r)$ to cone (5) is

$$(ar^2-1)\,x\,.\,\lambda r+(br^2-1)\,y\,.\,\mu r+(cr^2-1)\,z\gamma r=0$$

[changing x^2 to xx_1 etc.]

Dividing every term by r,

$$(ar^2-1)\,\lambda x+(br^2-1)\,\mu y+(cr^2-1)\,\gamma z=0 \quad ...(8)$$

Comparing Eqns. (8) and (2) [$\because$ Both are the equations of the tangent plane to cone (5) along the same generator and hence at $(\lambda r, \mu r, \gamma r)$ also], we have

$$\frac{(ar^2-1)\lambda}{l}=\frac{(br^2-1)\mu}{m}=\frac{(cr^2-1)\gamma}{n} \quad ...(9)$$

Putting the values of $r^2 = r_1^2$ and r_2^2 from Eqn. (7) in Eqn. (9), we get the values of λ, μ, γ
which are the d.c's of the two axes of the conic.

Cor. Condition for the central section to be a rectangular hyperbola

We know that for a conic section to be a rectangular hyperbola $r_1^2 + r_2^2 = 0$ *i.e.*, sum of roots of Eqn. (7) is 0 *i.e.*, $\frac{-b}{a}=0$ or $b = 0$.

$\therefore$ From Eqn. (7), Coeff. of $r^2 = 0$

$$\therefore \quad (b+c)\,l^2+(c+a)\,m^2+(a+b)\,n^2=0.$$

5. ***To prove that area of a central plane section of a central conicoid, if it is an ellipse, in terms of p, the length of perpendicular from the centre of the central conicoid to the tangent plane parallel to plane section of conicoid.***

Reproduce Art. 4 upto Eqn. (7).

We know that area of ellipse (plane section) $= \pi r_1 r_2$ ...(8) | πab

From Eqn. (7), product of roots

$$= r_1^2 r_2^2 = \frac{c}{a} = \frac{l^2+m^2+n^2}{bcl^2+cam^2+abn^2}$$

$$\therefore \qquad r_1 r_2 = \frac{\sqrt{l^2 + m^2 + n^2}}{\sqrt{bcl^2 + cam^2 + abn^2}}.$$

Putting this value of $r_1 r_2$ in Eqn. (8), we have

$$\text{Area of ellipse plane section} = \frac{\pi\sqrt{l^2 + m^2 + n^2}}{\sqrt{bcl^2 + cam^2 + abn^2}} \qquad ...(9)$$

We know that equation of any plane parallel to plane (2) is

$$lx + my + nz = d \qquad ...(10)$$

If it is a tangent plane to conicoid (1), then

$$\frac{l^2}{a} + \frac{m^2}{b} + \frac{n^2}{c} = d^2 \qquad \left| \; \frac{l^2}{a} + \frac{m^2}{b} + \frac{n^2}{c} = p^2 \right.$$

$$\therefore \qquad d = \sqrt{\frac{l^2}{a} + \frac{m^2}{b} + \frac{n^2}{c}}.$$

Putting this value of d in Eqn. (10), equation of tangent plane to conicoid (1) parallel to given plane (2) is

$$lx + my + nz - \sqrt{\frac{l^2}{a} + \frac{m^2}{b} + \frac{n^2}{c}} = 0$$

$\therefore$ p = Length of $\perp$ from centre (0, 0, 0) of the conicoid (1) to this tangent plane

$$= \frac{\left| -\sqrt{\frac{l^2}{a} + \frac{m^2}{b} + \frac{n^2}{c}} \right|}{\sqrt{l^2 + m^2 + n^2}} \qquad \left| \; \frac{|ax_1 + by_1 + cz_1 + d|}{\sqrt{a^2 + b^2 + c^2}} \right.$$

or

$$p = \frac{\sqrt{bcl^2 + cam^2 + abn^2}}{\sqrt{abc}\sqrt{l^2 + m^2 + n^2}}$$

$$\therefore \qquad \sqrt{abc}\; p = \frac{\sqrt{bcl^2 + cam^2 + abn^2}}{\sqrt{l^2 + m^2 + n^2}}.$$

Putting this value in Eqn. (9), Area of ellipse is $\dfrac{\pi}{p\sqrt{abc}}$. ...(11)

Cor. If the central conicoid (1) is the ellipsoid $\frac{x^2}{a^2} + \frac{y^2}{b^2} + \frac{z^2}{c^2} = 1$; then replacing a, b, c by $\frac{1}{a^2}, \frac{1}{b^2}, \frac{1}{c^2}$ in Eqn. (11),

Area of ellipse (plane section) is

$$\frac{\pi}{p\sqrt{\frac{1}{a^2} \cdot \frac{1}{b^2} \cdot \frac{1}{c^2}}} = \frac{\pi abc}{p}.$$

Example 1. *Show that the section of the ellipsoid $9x^2 + 6y^2 + 14z^2 = 3$ by the plane $x + y + z = 0$ is an ellipse with semi-axes $\frac{1}{2}$ and $\frac{3}{\sqrt{22}}$. Also find direction cosines of the axes of this section. Also find equations of the axes.*

Sol. Equation of the ellipsoid is $9x^2 + 6y^2 + 14z^2 = 3$ ...(1)

Dividing every term by 3 to make R.H.S. as 1,

$$3x^2 + 2y^2 + \frac{14z^2}{3} = 1$$

Comparing with $ax^2 + by^2 + cz^2 = 1$, we have

$$a = 3, \quad b = 2 \quad \text{and} \quad c = \frac{14}{3}.$$

Equation of the plane is $x + y + z = 0$ ...(2)

Comparing with $lx + my + nz = 0$, we have

$$l = 1, \quad m = 1 \quad \text{and} \quad n = 1$$

Eqns. (1) and (2) together with the equations of the plane section.

Putting these values in Eqn. (7) of Art. 4, that lengths of the semi-axes of the conic section are given by

$$r^4(bcl^2 + acm^2 + abn^2) - r^2 ((b + c)\, l^2 + (c + a)\, m^2 + (a + b)n^2] + l^2 + m^2 + n^2 = 0 \quad ...(3)$$

i.e.,
$$r^4\left(2.\frac{14}{3}.1 + 3.\frac{14}{3}.1 + 3.2.1\right) - r^2\left[\left(2 + \frac{14}{3}\right).1 + \left(\frac{14}{3} + 3\right).1 + (2+3).1\right] + (1 + 1 + 1) = 0$$

i.e.,
$$r^4\left(\frac{88}{3}\right) - r^2\left(10 + \frac{28}{3}\right) + 3 = 0$$

i.e.,
$$88r^4 - 58r^2 + 9 = 0$$

or
$$88r^4 - 22r^2 - 36r^2 + 9 = 0$$

or
$$22r^2 (4r^2 - 1) - 9 (4r^2 - 1) = 0$$

or
$$(4r^2 - 1)(22r^2 - 9) = 0$$

$\therefore$
$$r^2 = \frac{1}{4}, \frac{9}{22} = r_1^2, r_2^2.$$

Here both these values of r^2 are positive, therefore section of the conic given by equations (1) and (2) is an ellipse whose semi-axes are given by

$$r_1^2 = \frac{1}{4} \quad \text{and} \quad r_2^2 = \frac{9}{22}$$

i.e.,
$$r_1 = \frac{1}{2} \quad \text{and} \quad r_2 = \frac{3}{\sqrt{22}}$$

$\therefore$ Lengths of the semi-axes of the given conic are $\frac{1}{2}$ and $\frac{3}{\sqrt{22}}$.

Also by Eqn. (9) of Art. 4, d.c's λ, μ, γ of the axes of section are given by

$$\frac{(ar^2 - 1)\lambda}{l} = \frac{(br^2 - 1)\mu}{m} = \frac{(cr^2 - 1)\gamma}{n} \quad ...(4)$$

Putting $r = \frac{1}{2}$, d.c's of one axes are given by

$$\frac{\left(3.\frac{1}{4}-1\right)\lambda}{1} = \frac{\left(2.\frac{1}{4}-1\right)\mu}{1} = \frac{\left(\frac{14}{3}.\frac{1}{4}-1\right)\gamma}{1}$$

or $$-\frac{1}{4}\lambda = -\frac{2\mu}{4} = \frac{2}{12}\gamma \quad \text{or} \quad -\frac{\lambda}{4} = -\frac{\mu}{2} = \frac{\gamma}{6}$$

Dividing denominators by – 2,

$$\frac{\lambda}{2} = \frac{\mu}{1} = \frac{\gamma}{-3}$$

∴ d.r's of one axis of the plane section are 2, 1, – 3.

Dividing each by $\sqrt{(2)^2+(1)^2+(-3)^2} = \sqrt{14}$,

d.c's of one axis of the conic are $\frac{2}{\sqrt{14}}, \frac{1}{\sqrt{14}}, \frac{-3}{\sqrt{14}}$.

Putting $r = \frac{3}{\sqrt{22}}$ and values of a, b, c, l, m, n in Eqn. (2),

$$\frac{\left(3.\frac{9}{22}-1\right)\lambda}{1} = \frac{\left(2.\frac{9}{22}-1\right)\mu}{1} = \frac{\left(\frac{14}{3}.\frac{9}{22}-1\right)\gamma}{1}$$

or $$\left(\frac{27}{22}-1\right)\lambda = \left(\frac{18}{22}-1\right)\mu = \left(\frac{42}{22}-1\right)\gamma$$

Multiplying by 22,

$$5\lambda = -4\mu = 20\gamma \quad \text{or} \quad \frac{\lambda}{\frac{1}{5}} = \frac{\mu}{\frac{-1}{4}} = \frac{\gamma}{\frac{1}{20}}$$

Multiplying denominators by L.C.M. = 20,

$$\frac{\lambda}{4} = \frac{\mu}{-5} = \frac{\gamma}{1}$$

∴ D.R's of second axis of the conic are 4, – 5, 1.

Dividing each by $\sqrt{(4)^2+(-5)^2+(1)^2} = \sqrt{42}$,

d.c's of second axis are $\frac{4}{\sqrt{42}}, \frac{-5}{\sqrt{42}}, \frac{1}{\sqrt{42}}$

We know that centre of ellipsoid (1) is origin (0, 0, 0)

∴ Equations of the axes (passing through the centre (0, 0, 0)) of the given conic section are

$$\frac{x-0}{2} = \frac{y-0}{1} = \frac{z-0}{-3} \quad \text{and} \quad \frac{x-0}{4} = \frac{y-0}{-5} = \frac{z-0}{1}$$

i.e., $$\frac{x}{2} = \frac{y}{1} = \frac{z}{-3} \quad \text{and} \quad \frac{x}{4} = \frac{y}{-5} = \frac{z}{1}.$$

Example 2. *Planes are drawn through the origin so as to cut the conicoid $ax^2 + by^2 + cz^2 = 1$ in rectangular hyperbolas. Prove that the normals to the planes through the origin lie on a quadric cone.*

Sol. We know that equation of any plane through the origin is $lx + my + nz = 0$ ...(1)

Because these planes cut the conicoid $ax^2 + by^2 + cz^2 = 1$ ...(2)

rectangular hyperbolas, (given), therefore

By Cor. Art. 4, $(b + c)l^2 + (c + a)\, m^2 + (a + b)\, n^2 = 0$ ...(3)

We know that d.r's of normal to plane (1) are

$$l, \; m, \; n \qquad | \text{ Coeffs. of } x, y, z$$

$\therefore$ Equations of the normal through the origin to plane (1) are

$$\frac{x-0}{l} = \frac{y-0}{m} = \frac{z-0}{n} \quad \text{or} \quad \frac{x}{l} = \frac{y}{m} = \frac{z}{n} \qquad ...(4)$$

Eliminating l, m, n from (3) and (4) *i.e.*, putting values of l, m, n from (4) in (3), the normals, in question lie on the surface

$$(b + c)\, x^2 + (c + a)y^2 + (a + b)\, z^2 = 0$$

which being a homogeneous equation of second degree in x, y, z represents a cone whose vertex is origin.

Example 3. *Prove that the axes of the section of the conicoid $ax^2 + by^2 + cz^2 = 1$ by the plane $lx + my + nz = 0$ lie on the cone $(b - c)\frac{l}{x} + (c - a)\frac{m}{y} + (a - b)\frac{n}{z} = 0$.*

Sol. We know by Eqn. (9), Art. 4 that d.c's λ, μ, γ of the axes of the conic $ax^2 + by^2 + cz^2 = 1$, $lx + my + nz = 0$ are given by

$$\frac{(ar^2 - 1)\lambda}{l} = \frac{(br^2 - 1)\mu}{m} = \frac{(cr^2 - 1)\gamma}{n} = k(\text{say})$$

$$\therefore \quad (ar^2 - 1)\, \lambda = lk, \; (br^2 - 1)\, \mu = mk \;\text{ and }\; (cr^2 - 1)\, \gamma = nk$$

$$\therefore \quad \frac{lk}{\lambda} = ar^2 - 1 \quad ...(1) \qquad \frac{mk}{\mu} = br^2 - 1 \quad ...(2)$$

$$\frac{nk}{\gamma} = cr^2 - 1. \qquad ...(3)$$

Multiplying Eqns. (1), (2) and (3) by $(b - c)$, $(c - a)$ and $(a - b)$ respectively (to eliminate r and k). (This step is also being suggested by the answer) and adding we have

$$\frac{lk}{\lambda}(b - c) + \frac{mk}{\mu}(c - a) + \frac{nk}{\gamma}(a - b) = 0$$

Dividing by k,
$$\frac{l(b - c)}{\lambda} + \frac{m(c - a)}{\mu} + \frac{n(a - b)}{\gamma} = 0 \qquad ...(4)$$

We know that equations of the axes of the conic (passing through centre (0, 0, 0)) are

$$\frac{x-0}{\lambda} = \frac{y-0}{\mu} = \frac{z-0}{\gamma} \quad i.e., \quad \frac{x}{\lambda} = \frac{y}{\mu} = \frac{z}{\gamma} \qquad ...(5)$$

Eliminating λ, μ, γ from (4) and (5) *i.e.*, putting values of λ, μ, γ from (5) in (4), the axes of the conic lie on

$$\frac{l(b - c)}{x} + \frac{m(c - a)}{y} + \frac{n(a - b)}{z} = 0$$

i.e., $l(b - c)\, yz + m\, (c - a)\, zx + n(a - b)\, xy = 0$

which being a homogeneous equation of second degree in x, y, z represents a cone whose vertex is origin.

6. *Find the conditions for the two lines*

$$\frac{x}{l_1}=\frac{y}{m_1}=\frac{z}{n_1}\ ; \qquad \frac{x}{l_2}=\frac{y}{m_2}=\frac{z}{n_2}$$

to be the axes of the section of the conicoid $ax^2 + by^2 + cz^2 = 1$ (by a plane through them).

(K.U. 2003)

Equation of the given conicoid is

$$ax^2 + by^2 + cz^2 = 1 \qquad \text{...(1)}$$

Equations of the given lines are

$$\frac{x}{l_1}=\frac{y}{m_1}=\frac{z}{n_1} \qquad \text{...(2)}$$

and

$$\frac{x}{l_2}=\frac{y}{m_2}=\frac{z}{n_2} \qquad \text{...(3)}$$

It is given that lines (2) and (3) are the axes of the section of the conicoid by a plane passing through them.

∴ By def. of axes of the conic section, the two axes given by (2) and (3) are perpendicular to each other.

∴ $$l_1l_2 + m_1m_2 + n_1n_2 = 0 \qquad \text{...(4)}$$

Because lines (2) and (3) are the axes of the section, therefore by def. of axes each of them will bisect chords parallel to the other. Hence each of the them (2) and (3) will lie in the diametral plane corresponding to the other. ...(5)

We know from chapter 4 that the equation of diametral plane corresponding to line (2) is

$$al_1x + bm_1y + cn_1z = 0$$

Now line (3) lies in this plane (By result (5)),

∴ $$al_1l_2 + bm_1m_2 + cn_1n_2 = 0 \qquad \text{...(6)} \ | \ \mathbf{al + bm + cn = 0}$$

Hence (4) and (6) are the required conditions.

Example 4. *One axis of a central section of the conicoid lies in the plane $ux + vy + wz = 0$. Show that the other lies on the cone $(b - c)\, uyz + (c - a)\, vzx + (a - b)\, wxy = 0$.*

Sol. Let the equations of the two axes of a central section be

$$\frac{x}{l_1}=\frac{y}{m_1}=\frac{z}{n_1} \qquad \text{...(1)}$$

and

$$\frac{x}{l_2}=\frac{y}{m_2}=\frac{z}{n_2} \qquad \text{...(2)}$$

Let axis (1) lie in the given plane $ux + vy + wz = 0$ (given)

∴ $$ul_1 + vm_1 + wn_1 = 0 \qquad \text{...(3)}$$

Again because lines (1) and (2) are axes of a central section, therefore by Art. 6, (Eqn. (4) and Eqn. (6) of Art. 6),

$$l_1l_2 + m_1m_2 + n_1n_2 = 0 \qquad \text{...(4)}$$

and

$$al_1l_2 + bm_1m_2 + cn_1n_2 = 0 \qquad \text{...(5)}$$

Eliminating l_1, m_1, n_1 from Eqns. (3), (4) and (5) ; we have

$$\begin{vmatrix} u & v & w \\ l_2 & m_2 & n_2 \\ al_2 & bm_2 & cn_2 \end{vmatrix} = 0$$

Expanding by first row

$$u(cm_2n_2 - bm_2n_2) - v(cl_2n_2 - al_2n_2) + w(bl_2m_2 - al_2m_2) = 0$$

or

$$-u(b-c)\, m_2n_2 - v(c-a)l_2n_2 - w(a-b)l_2m_2 = 0$$

Dividing every term by $-l_2m_2n_2$, we have

$$\frac{u(b-c)}{l_2} + \frac{v(c-a)}{m_2} + \frac{w(a-b)}{n_2} = 0 \qquad ...(6)$$

∴ The locus of the other axes (2) is obtained by eliminating l_2, m_2, n_2 from Eqns. (2) and (6).

[By putting values of l_2, m_2, n_2 from (2) in (6)],

and hence this locus is $\dfrac{u(b-c)}{x} + \dfrac{v(c-a)}{y} + \dfrac{w(a-b)}{z} = 0$

Multiplying by L.C.M = xyz,

$$u(b-c)yz + v(c-a)\, xz + w\, (a-b)xy = 0$$

which being a homogeneous of second degree in x, y, z represents a cone.

Example 5. ***If A_1, A_2, A_3 are the areas of three mutually perpendicular sections of an ellipsoid, then show that $A_1^{-2} + A_2^{-2} + A_3^{-2} = constant$.***

Sol. Let the equation of the ellipsoid be

$$\frac{x^2}{a^2} + \frac{y^2}{b^2} + \frac{z^2}{c^2} = 1 \qquad ...(1)$$

Let the equations of the three mutually perpendicular central planes (*i.e.*, planes passing through the centre (0, 0, 0) of ellipsoid (1)) be

$$l_1x + m_1y + n_1z = 0 \qquad ...(2)$$

$$l_2x + m_2y + n_2z = 0 \qquad ...(3)$$

$$l_3x + m_3y + n_3z = 0 \qquad ...(4)$$

Because planes (2), (3) and (4) are mutually perpendicular (given), therefore their normals having d.c's l_1, m_1, n_1 ; l_2, m_2, n_2 and l_3, m_3, n_3 (Coeffs. of x, y, z, in planes (2), (3) and (4), are also mutually perpendicular. Hence there exist twelve relations among these direction cosines (see Art. 4 Review IV) namely

$$l_1^2 + m_1^2 + n_1^2 = 1 \text{ etc.} \qquad l_1^2 + l_2^2 + l_3^2 = 1 \text{ etc.} \qquad ...(5)$$

$$l_1l_2 + m_1m_2 + n_1n_2 = 0 \text{ etc. and } l_1m_1 + l_2m_2 + l_3m_3 = 0 \qquad ...(6)$$

We know by eqn. (9) of Art. 5 that A_1 = Area of central section of ellipsoid (1) by central plane (2), (which must be an ellipse)

$$= \frac{\pi\sqrt{l_1^2 + m_1^2 + n_1^2}}{\sqrt{bcl_1^2 + cam_1^2 + abn_1^2}}$$

Here from (1) $a = \dfrac{1}{a^2}, b = \dfrac{1}{b^2}, c = \dfrac{1}{c^2}$.

$$\therefore \quad A_1 = \frac{\pi\sqrt{l_1^2 + m_1^2 + n_1^2}}{\sqrt{\dfrac{l_1^2}{b^2c^2} + \dfrac{m_1^2}{c^2a^2} + \dfrac{n_1^2}{a^2b^2}}} = \frac{\pi abc\sqrt{l_1^2 + m_1^2 + n_1^2}}{\sqrt{a^2l_1^2 + b^2m_1^2 + c^2n_1^2}}$$

$$\therefore \quad A_1^2 = \frac{\pi^2a^2b^2c^2(l_1^2 + m_1^2 + n_1^2)}{a^2l_1^2 + b^2m_1^2 + c^2n_1^2} = \frac{\pi^2a^2b^2c^2}{a^2l_1^2 + b^2m_1^2 + c^2n_1^2}$$

[Using Eqns. (5)]

$$\therefore \quad A_1^{-2} = \frac{1}{A_1^2} = \frac{a^2l_1^2 + b^2m_1^2 + c^2n_1^2}{\pi^2a^2b^2c^2}$$

Similarly, $A_2^{-2} = \dfrac{a^2l_2^2 + b^2m_2^2 + c^2n_2^2}{\pi^2a^2b^2c^2}$ and $A_3^{-2} = \dfrac{a^2l_3^2 + b^2m_3^2 + c^2n_3^2}{\pi^2a^2b^2c^2}$

Adding the above three equations

$$A_1^{-2} + A_2^{-2} + A_3^{-2} = \frac{a^2(l_1^2 + l_2^2 + l_3^2) + b^2(m_1^2 + m_2^2 + m_3^2) + c^2(n_1^2 + n_2^2 + n_3^2)}{\pi^2a^2b^2c^2}$$

$$= \frac{a^2 + b^2 + c^2}{\pi^2a^2b^2c^2} \qquad \text{[Using Eqns. (5)]}$$

which being independent of l_1, m_1, n_1 ; l_2, m_2, n_2 and l_3, m_3, n_3 is constant.

EXERCISE 1

1. Show that the curve $x^2 + 7y^2 - 10z^2 + 9 = 0$, $x + 2y + 3z = 0$ is a hyperbola whose transverse axis is 6 and the direction cosines of whose axes are proportional to 6, 3, – 4 and 17, – 22,9.
2. Any plane whose normal lies on the cone $bcx^2 + cay^2 + abz^2 = 0$ cuts the surface $ax^2 + by^2 + cz^2 = 1$ in a parabola.

   [**Hint.** Let the equation of the plane be $lx + my + nz = p$.

   $\therefore$ D.R's of normal to this plane are l, m, n.

   Because this normal lies on the cone $bcx^2 + cay^2 + abz^2 = 0$,

   therefore its d.r's satisfy the cone. $\therefore$ $bcl^2 + cam^2 + abn^2 = 0$

   Now reproduce Art. 3].
3. Show that the section of the surface $yz + zx + xy = a^2$ by the plane $lx + my + nz = p$ will be a parabola if

   $$\sqrt{l} + \sqrt{m} + \sqrt{n} = 0$$

   [**Hint.** Proceed as in Art. 3.]
4. Prove that the section of the conicoid $ax^2 + by^2 + cz^2 = 1$ by a tangent plane to the cone $\dfrac{x^2}{b+c} + \dfrac{y^2}{c+a} + \dfrac{z^2}{a+b} = 0$ is a rectangular hyperbola.
5. Prove that all plane sections of the conicoid $ax^2 + by^2 + cz^2 = 1$ which are rectangular hyperbolas and pass through the point (α, β, γ) touch the cone $\dfrac{(x-\alpha)^2}{b+c} + \dfrac{(y-\beta)^2}{c+a} + \dfrac{(z-\gamma)^2}{a+b} = 0$.
6. Prove that the axes of the section of the conicoid $ax^2 + by^2 + cz^2 = 1$ which pass through the line $\dfrac{x}{l} = \dfrac{y}{m} = \dfrac{z}{n}$ lie on the cone

$$\frac{(b-c)}{x}(mz-ny)+\left(\frac{c-a}{y}\right)(nx-lz)+\left(\frac{a-b}{z}\right)(ly-mx)=0$$

[**Hint.** Let $x/l_1 = y/m_1 = z/n_1$ and $x/l_2 = y/m_2 = z/n_2$ be the two axes.
Then from Art. 6, we have

$$al_1l_2 + bm_1m_2 + cn_1n_2 = 0 \quad \text{and} \quad l_1l_2 + m_1m_2 + n_1n_2 = 0.$$

Also since the three lines are coplanar, therefore

$$\begin{vmatrix} l & m & n \\ l_1 & m_1 & n_1 \\ l_2 & m_2 & n_2 \end{vmatrix} = 0,$$

or $l_2(mn_1 - m_1n) + m_2(nl_1 - n_1l) + n_2(lm_1 - l_1m) = 0.$
Eliminating l_2, m_2, n_2 from these three equations, we get

$$\begin{vmatrix} al_1 & bm_1 & cn_1 \\ l_1 & m_1 & n_1 \\ mn_1 - m_1n & nl_1 - n_1l & lm_1 - l_1m \end{vmatrix} = 0\cdot$$

Eliminating l_1, m_1, n_1 between this equation and the axis, we obtain

$$\begin{vmatrix} ax & by & cz \\ x & y & z \\ mz - ny & nx - lz & ly - mx \end{vmatrix} = 0,$$

or
$$\left.\frac{(b-c)}{x}(mz-ny)+\frac{(c-a)}{y}(nx-lz)+\frac{(a-b)}{z}(ly-mx)=0\right].$$

---

7. Axes of non-central plane sections

To determine the lengths and direction cosines of the section of the central conicoid

$$ax^2 + by^2 + cz^2 = 1 \qquad \text{...(1)}$$

by the plane
$$lx + my + nz = p \ (p \neq 0) \qquad \text{...(2)}$$

Here centre of plane section is **not** the origin because centre (0, 0, 0) of conicoid (1) **does** not satisfy plane (2).

Let (α, β, γ) be the centre of this section.

$\therefore$ Another equation of plane (2) of this section of conicoid (1) is given by $T = S_1$ (see chapter 4)

or
$$ax\alpha + by\beta + cz\gamma - 1 = a\alpha^2 + b\beta^2 + c\gamma^2 - 1$$

or
$$a\alpha x + b\beta y + c\gamma z = a\alpha^2 + b\beta^2 + c\gamma^2 \qquad \text{...(3)}$$

Comparing Eqns. (3) and (2), we have

$$\frac{a\alpha}{l} = \frac{b\beta}{m} = \frac{c\gamma}{n} = \frac{a\alpha^2 + b\beta^2 + c\gamma^2}{p} = k \text{ (say)}$$

$$\therefore \qquad a\alpha = lk \quad \text{or} \quad \alpha = \frac{lk}{a} \qquad \text{...(4)}$$

Similarly, $\beta = \dfrac{mk}{b}$ ...(5) and $\gamma = \dfrac{nk}{c}$ ...(6)

Also $\quad a\alpha^2 + b\beta^2 + c\gamma^2 = pk \qquad$ ...(7)

Putting values of α, β, γ from (4), (5), (6) in (7), we have

$$a \cdot \frac{l^2k^2}{a^2} + b\frac{m^2k^2}{b^2} + c\frac{n^2k^2}{c^2} = pk$$

Dividing by k, $$k\left(\frac{l^2}{a} + \frac{m^2}{b} + \frac{n^2}{c}\right) = p$$

For sake of brevity, let us write $\frac{l^2}{a} + \frac{m^2}{b} + \frac{n^2}{c}$ as p_0^2 ...(8)

$\therefore$ $$k\,p_0^2 = p \quad \text{or} \quad k = \frac{p}{p_0^2}$$

Putting $k = \frac{p}{p_0^2}$ in Eqns. (4), (5) and (6), we have

$$\alpha = \frac{lp}{a\,p_0^2}, \quad \beta = \frac{mp}{b\,p_0^2}, \quad \gamma = \frac{np}{c\,p_0^2} \quad \text{...(9)}$$

$\therefore$ Co-ordinates of the centre of this non-central conic section are (α, β, γ)

$$= \left(\frac{lp}{a\,p_0^2}, \frac{mp}{b\,p_0^2}, \frac{np}{c\,p_0^2}\right)$$

Now let us shift the origin to the centre

i.e., let us treat centre (α, β, γ) as origin.

Changing x to $x + \alpha$, y to $y + \beta$ and z to $z + \gamma$

(See chapter 0) in Eqn. (1), we have

$$a(x + \alpha)^2 + b(y + \beta)^2 + c(z + \gamma)^2 = 1$$

or $$ax^2 + by^2 + cz^2 + 2(ax\alpha + by\beta + cz\gamma) + (a\alpha^2 + b\beta^2 + c\gamma^2) = 1$$

Putting values of α, β, γ from (9), it becomes

$$ax^2 + by^2 + cz^2 + 2\left(\frac{ax\,lp}{ap_0^2} + \frac{by\,mp}{bp_0^2} + \frac{cznp}{cp_0^2}\right)$$
$$+\left(a\frac{l^2p^2}{a^2p_0^4} + b\frac{m^2p^2}{b^2p_0^4} + c\frac{n^2p^2}{c^2p_0^4}\right) = 1$$

or $$ax^2 + by^2 + cz^2 + 2\frac{p}{p_0^2}(lx + my + nz) + \frac{p^2}{p_0^4}\left(\frac{l^2}{a} + \frac{m^2}{b} + \frac{n^2}{c}\right) = 1$$

Putting $\frac{l^2}{a} + \frac{m^2}{b} + \frac{n^2}{c} = p_0^2$ from (8), it becomes

$$ax^2 + by^2 + cz^2 + \frac{2p}{p_0^2}(lx + my + nz) + \frac{p^2}{p_0^2} = 1 \quad \text{...(10)}$$

Alsc equation of plane (2) becomes $(x \to x + \alpha, y \to y + \beta, z \to z + \gamma)$

$$l(x + \alpha) + m(y + \beta) + n(z + \gamma) = p$$

i.e., $$lx + my + nz + l\alpha + m\beta + n\gamma = p$$

Putting values of α, β, γ from (9), it becomes

$$lx + my + nz + \frac{l^2 p}{a p_0^2} + \frac{m^2 p}{b p_0^2} + \frac{n^2 p}{c p_0^2} = p$$

or $$lx + my + nz + \frac{p}{p_0^2}\left(\frac{l^2}{a} + \frac{m^2}{b} + \frac{n^2}{c}\right) = p$$

or $$lx + my + nz + \frac{p}{p_0^2} \cdot p_0^2 = p \qquad \text{[By (9)]}$$

or $$lx + my + nz = 0 \qquad \text{...(11)}$$

Now the given conic (represented by (1) and (2)) is the same as represented by Eqns. (10) and (11) but is a central conic now [∵ plane (11) passes through centre (0, 0, 0) of conicoid (10)]

Putting $lx + my + nz = 0$ from (11) in (10),

equation of conicoid is $ax^2 + by^2 + cz^2 = 1 - \dfrac{p^2}{p_0^2} = d^2$ (say)

Dividing by d^2, $$\frac{ax^2}{d^2} + \frac{by^2}{d^2} + \frac{cz^2}{d^2} = 1 \qquad \text{...(12)}$$

where $$d^2 = 1 - \frac{p^2}{p_0^2} \qquad \text{...(13)}$$

Now replacing a by $\dfrac{a}{d^2}$, b by $\dfrac{b}{d^2}$ and c by $\dfrac{c}{d^2}$ in Eqn. (7) of Art. 4, we get the lengths of the semi-axes of the conic and from Eqn. (9) of Art. 4, we can find the d.c's of the two axes of the conic.

8. Area of non-central plane section

To find area of a non-central plane section of a central conicoid, if it is an ellipse.

We know by Eqn. (9) of Art. 5 that area of the plane section, if it is an ellipse

$$= \frac{\pi\sqrt{l^2 + m^2 + n^2}}{\sqrt{bcl^2 + cam^2 + abn^2}}$$

Now reproduce Art. 7 upto Eqn. (12). Replacing a by $\dfrac{a}{d^2}$, b by $\dfrac{b}{d^2}$ and c by $\dfrac{c}{d^2}$ in the above formula,

Area of ellipse $$= \frac{\pi\sqrt{l^2 + m^2 + n^2}}{\sqrt{\dfrac{bcl^2}{d^4} + \dfrac{cam^2}{d^4} + \dfrac{abn^2}{d^4}}} = \frac{\pi d^2\sqrt{l^2 + m^2 + n^2}}{\sqrt{bcl^2 + cam^2 + abn^2}}.$$

Example 1. *Prove that the area of the section of the ellipsoid* $\dfrac{x^2}{a^2} + \dfrac{y^2}{b^2} + \dfrac{z^2}{c^2} = 1$ *by the plane* $\dfrac{x}{a} + \dfrac{y}{b} + \dfrac{z}{c} = 1$ *is* $\dfrac{2\pi}{3\sqrt{3}}(b^2c^2 + c^2a^2 + a^2b^2)^{\frac{1}{2}}$. (M.D.U. 2003 S, 2007)

Sol. Equations of the given (conic) section are

$$\frac{x^2}{a^2}+\frac{y^2}{b^2}+\frac{z^2}{c^2}=1 \quad ...(1) \qquad \text{and} \qquad \frac{x}{a}+\frac{y}{b}+\frac{z}{c}=1 \quad ...(2)$$

Comparing (1) with $ax^2 + by^2 + cz^2 = 1$, we have

$$a=\frac{1}{a^2},\ b=\frac{1}{b^2} \text{ and } c=\frac{1}{c^2}$$

Comparing (2) with $lx + my + nz = p$, we have

$$l=\frac{1}{a},\ m=\frac{1}{b},\ n=\frac{1}{c} \text{ and } p=1.$$

(By Eqn. (8) of Art. 7) $p_0^2=\frac{l^2}{a}+\frac{m^2}{b}+\frac{n^2}{c}$

$$=\frac{\frac{1}{a^2}}{\frac{1}{a^2}}+\frac{\frac{1}{b^2}}{\frac{1}{b^2}}+\frac{\frac{1}{c^2}}{\frac{1}{c^2}}=1+1+1=3$$

$\therefore \qquad p_0=\sqrt{3}$

(Now by Eqn. (13) of Art. 7) $d^2=1-\frac{p^2}{p_0^2}=1-\frac{1}{3}=\frac{2}{3}$

Putting these values in the result of Art. 8,

Area of section of ellipsoid (1) by plane (2)

$$=\frac{\pi d^2\sqrt{l^2+m^2+n^2}}{\sqrt{bcl^2+cam^2+abn^2}}=\frac{\pi.\frac{2}{3}\sqrt{\frac{1}{a^2}+\frac{1}{b^2}+\frac{1}{c^2}}}{\sqrt{\frac{1}{b^2c^2}\frac{1}{a^2}+\frac{1}{c^2a^2}\frac{1}{b^2}+\frac{1}{a^2b^2}.\frac{1}{c^2}}}$$

$$=\pi.\frac{2}{3}\frac{\frac{\sqrt{b^2c^2+c^2a^2+a^2b^2}}{abc}}{\sqrt{\frac{3}{a^2b^2c^2}}}=\frac{2\pi}{3\sqrt{3}}\sqrt{b^2c^2+c^2a^2+a^2b^2}.$$

Example 2. *Find the co-ordinates of the centre and the lengths of the semi-axes of the section of the ellipsoid $3x^2 + 3y^2 + 6z^2 = 10$ by the plane $x + y + z = 1$.*

Sol. Equations of the given plane section are

$$3x^2 + 3y^2 + 6z^2 = 10$$

Dividing every term by 10 to make R.H.S. unity,

$$\frac{3}{10}x^2+\frac{3}{10}y^2+\frac{6}{10}z^2=1 \quad ...(1)$$

and

$$x+y+z=1 \quad ...(2)$$

Comparing (1) with $ax^2 + by^2 + cz^2 = 1$, we have

$$a=\frac{3}{10},\ b=\frac{3}{10},\ c=\frac{6}{10}.$$

Comparing (2) with $lx + my + nz = p,$ we have

$$l = 1, m = 1, n = 1, p = 1$$

$$p_0^2 = \frac{l^2}{a} + \frac{m^2}{b} + \frac{n^2}{c} = \frac{1}{\left(\frac{3}{10}\right)} + \frac{1}{\left(\frac{3}{10}\right)} + \frac{1}{\left(\frac{6}{10}\right)}$$

$$= \frac{10}{3} + \frac{10}{3} + \frac{10}{6} = \frac{20+20+10}{6} = \frac{50}{6}$$

$\therefore$
$$d^2 = 1 - \frac{p^2}{p_0^2} = 1 - \frac{1}{\left(\frac{50}{6}\right)} = 1 - \frac{6}{50} = 1 - \frac{3}{25} = \frac{22}{25}$$

We know (by Eqn. (9) of Art. 7) that centre of plane section is given by $\left(\frac{lp}{ap_0^2}, \frac{mp}{bp_0^2}, \frac{np}{cp_0^2}\right)$

$$= \left(\frac{1.1}{\frac{3}{10}\cdot\frac{50}{6}}, \frac{1.1}{\frac{3}{10}\cdot\frac{50}{6}}, \frac{1.1}{\frac{6}{10}\cdot\frac{50}{6}}\right) = \left(\frac{2}{5}, \frac{2}{5}, \frac{1}{5}\right)$$

We know (By Art. 7) that lengths of semi-axes of a non-central section are obtained by replacing a by $\frac{a}{d^2}$, b by $\frac{b}{d^2}$, c by $\frac{c}{d^2}$ in Eqn. (7) of Art. 4 namely in

$$r^4(bcl^2 + acm^2 + abn^2) - r^2[(b+c)\,l^2 + (c+a)m^2 + (a+b)n^2] + (l^2 + m^2 + n^2) = 0, \text{ we have}$$

$$r^4(bcl^2 + acm^2 + abn^2) - d^2r^2[(b+c)\,l^2 + (c+a)m^2 + (a+b)n^2] + d^4\,(l^2 + m^2 + n^2) = 0.$$

Putting values, we have

$$r^4\left[\frac{18}{100} + \frac{18}{100} + \frac{9}{100}\right] - \frac{22}{25}r^2\left[\frac{9}{10} + \frac{9}{10} + \frac{6}{10}\right] + \left(\frac{22}{25}\right)^2 (1+1+1) = 0$$

or
$$\frac{45}{100}r^4 - \frac{22\times 24}{250}r^2 + \frac{484}{625}\times 3 = 0$$

Dividing every term by $\frac{3}{25}$,

$$\frac{15r^4}{4} - \frac{176}{10}r^2 + \frac{484}{25} = 0$$

Multiplying by L.C.M. = 100,

$$15 \times 25r^4 - 1760r^2 + 484 \times 4 = 0$$

or
$$15 \times 25r^4 - 1100r^2 - 660r^2 + 1936 = 0$$

or
$$25r^2\,(15r^2 - 44) - 44(15r^2 - 44) = 0$$

or
$$(25r^2 - 44)(15r^2 - 44) = 0$$

$\therefore$
$$r^2 = \frac{44}{25}, \frac{44}{15}$$

$\therefore$
$$r = \sqrt{\frac{44}{25}} = \frac{\sqrt{44}}{5} \quad \text{and} \quad \sqrt{\frac{44}{15}}$$

$\therefore$ Lengths of the semi-axes of the conic are

$$\sqrt{\frac{44}{15}} \text{ and } \frac{\sqrt{44}}{5}.$$

EXERCISE 2

1. The ellipsoid $x^2 + 2y^2 + 3z^2 = 1$ is cut by parallel planes $2x + 3y + 4z = 2$ and $2x + 3y + 4z = 3$. Show that the areas of the sections of the conicoid made by these planes are in the ratio 59 : 29.
2. Find the lengths and directions of the axes of the section of the ellipsoid $9x^2 + 6y^2 + 14z^2 = 3$ by the plane $x + y + z = 1$.
3. Find the locus of the centres of the section of the ellipsoid $\frac{x^2}{a^2}+\frac{y^2}{b^2}+\frac{z^2}{c^2}=1$ which are of constant area πk^2.

[**Hint.** Let (α, β, γ) be the centre of the plane section of the ellipsoid $\frac{x^2}{a^2}+\frac{y^2}{b^2}+\frac{z^2}{c^2}=1$

$\therefore$ Equation of the plane of the conic section is $T = S_1$

i.e.,
$$\frac{x\alpha}{a^2}+\frac{y\beta}{b^2}+\frac{z\gamma}{c^2}=\frac{\alpha^2}{a^2}+\frac{\beta^2}{b^2}+\frac{\gamma^2}{c^2}$$

Compare it with $lx + my + nz = p$].

Answers

2. $\frac{3}{22}, \frac{\sqrt{22}}{44}$; 4, – 5, 1 and 2, 1, – 3

3. $a^2b^2c^2\left(\frac{x^2}{a^4}+\frac{y^2}{b^4}+\frac{z^2}{c^4}\right)\left(1-\frac{x^2}{a^2}-\frac{y^2}{b^2}-\frac{z^2}{c^2}\right)^2=k^2\left(\frac{x^2}{a^2}+\frac{y^2}{b^2}+\frac{z^2}{c^2}\right)$.

CIRCULAR SECTIONS

9. Introduction

If $F = 0$ is the equation of a coincoid and if it can be put in the form

$S + \lambda uv = 0$, where $S = 0$ is the equation of a sphere and $u = 0$, $v = 0$ represent planes ; then the common points of the conicoid $F = 0$ and the planes $uv = 0$ must also lie on $F - \lambda uv = 0$.

i.e., on sphere $S = 0$ $\qquad (\because \; F = S + \lambda uv \Rightarrow F - \lambda uv = S)$

But we know from chapter 10 that sections of spheres by planes are circles. Therefore sections of the conicoid by the planes are also circles.

Remark 1. Again we know from chapter 10 that section of a sphere by a plane passing through the centre of the sphere is called a **Great Circle.**

Also we know that Centre of **Great Circle** is centre of the sphere and radius of Great Circle is same as radius of the sphere.

Remark 2. Let us learn the locus of the equation $Ay^2 + Bz^2 = 0$ ...(1)

Case I. A and B are of opposite signs. $\therefore$ $\frac{B}{A}$ is negative.

From (1) $\quad Ay^2 = -Bz^2$

$$y^2 = -\frac{B}{A}z^2 \qquad \therefore \qquad y = \pm\sqrt{\frac{-B}{A}}\,z$$

which are the equations of two real planes passing through the origin.

$$\left[\because \frac{B}{A} \text{ is negative } \Rightarrow \frac{-B}{A} \text{ is positive}\right]$$

Case II. A and B are of same sign $\therefore \frac{B}{A}$ is positive.

From (1) $\quad Ay^2 = -Bz^2$

$$\therefore \qquad y = \pm\sqrt{\frac{-B}{A}}\,z$$

which are the equations of two imaginary planes. $\left[\because \frac{B}{A}\right.$ is positive $\Rightarrow \frac{-B}{A}$ is negative and square root of a negative number is imaginary]

10. The Central Circular Sections of an Ellipsoid

Let the equation of the ellipsoid be

$$\frac{x^2}{a^2} + \frac{y^2}{b^2} + \frac{z^2}{c^2} = 1 \qquad \text{...(1)}$$

Adding and subtracting $\frac{y^2}{a^2} + \frac{z^2}{a^2}$ in L.H.S. of Eqn. (1), we have

$$\frac{x^2}{a^2} + \frac{y^2}{a^2} + \frac{z^2}{a^2} + \frac{y^2}{b^2} - \frac{y^2}{a^2} + \frac{z^2}{c^2} - \frac{z^2}{a^2} = 1$$

or
$$\frac{x^2 + y^2 + z^2}{a^2} - 1 + y^2\left(\frac{1}{b^2} - \frac{1}{a^2}\right) + z^2\left(\frac{1}{c^2} - \frac{1}{a^2}\right) = 0$$

or
$$\frac{1}{a^2}(x^2 + y^2 + z^2 - a^2) + y^2\left(\frac{1}{b^2} - \frac{1}{a^2}\right) + z^2\left(\frac{1}{c^2} - \frac{1}{a^2}\right) = 0 \qquad \text{...(2)}$$

Similarly, adding and subtracting $\frac{z^2}{b^2} + \frac{x^2}{b^2}$ in L.H.S. of Eqn. (1), it can be written as

$$\frac{1}{b^2}(x^2 + y^2 + z^2 - b^2) + z^2\left(\frac{1}{c^2} - \frac{1}{b^2}\right) + x^2\left(\frac{1}{a^2} - \frac{1}{b^2}\right) = 0 \qquad \text{...(3)}$$

Similarly, adding and subtracting $\frac{x^2}{c^2} + \frac{y^2}{c^2}$ in L.H.S. of Eqn. (1), it can be written as

$$\frac{1}{c^2}(x^2 + y^2 + z^2 - c^2) + x^2\left(\frac{1}{a^2} - \frac{1}{c^2}\right) + y^2\left(\frac{1}{b^2} - \frac{1}{c^2}\right) = 0 \qquad \text{...(4)}$$

Equation (1) of ellipsoid has been rewritten in Equations (2), (3) and (4) in the form $S + \lambda uv = 0$.

For Eqn. (2), S, sphere is $x^2 + y^2 + z^2 - a^2 = 0$ which is a sphere of radius a

and
$$uv = y^2\left(\frac{1}{b^2} - \frac{1}{a^2}\right) + z^2\left(\frac{1}{c^2} - \frac{1}{a^2}\right) = 0$$

is the equation of a pair of planes passing through origin, the centre of sphere of radius a.

$\therefore$ By Remark 1 Art. 9, sections of sphere $S = 0$ by planes $uv = 0$ (*i.e.*, $u = 0$ and $v = 0$) are great circles of radius a, (= Radius of sphere). Similarly, we can conclude from Eqns. (3) and (4).

$\therefore$ From (2), (3) and (4), we can claim that the pair of planes given by

$$y^2\left(\frac{1}{b^2} - \frac{1}{a^2}\right) + z^2\left(\frac{1}{c^2} - \frac{1}{a^2}\right) = 0 \qquad \text{...(5)}$$

$$z^2\left(\frac{1}{c^2} - \frac{1}{b^2}\right) + x^2\left(\frac{1}{a^2} - \frac{1}{b^2}\right) = 0 \qquad \text{...(6)}$$

and
$$x^2\left(\frac{1}{a^2} - \frac{1}{c^2}\right) + y^2\left(\frac{1}{b^2} - \frac{1}{c^2}\right) = 0 \qquad \text{...(7)}$$

meet the ellipsoid (1) in circles of radii a, b, c, respectively.

Therefore sections of the ellipsoid by these planes are also circles.

Let us assume that a > b > c

$$\therefore \qquad \frac{1}{a} < \frac{1}{b} < \frac{1}{c} \text{ and therefore } \frac{1}{a^2} < \frac{1}{b^2} < \frac{1}{c^2}$$

$\therefore$ $\frac{1}{b^2} - \frac{1}{a^2}$ and $\frac{1}{c^2} - \frac{1}{a^2}$ are both positive (*i.e.*, of same sign) and hence by Remark 2, Art. 10 ; Equation (5) represents a pair of imaginary planes.

Again $\frac{1}{c^2} - \frac{1}{b^2}$ is positive and $\frac{1}{a^2} - \frac{1}{b^2}$ is negative

i.e., are of opposite signs and hence by Remark 2, Art. 9, Eqn. (6) represents a pair of real planes through the origin

Again $\frac{1}{a^2} - \frac{1}{c^2}$ and $\frac{1}{b^2} - \frac{1}{c^2}$ are both negative (*i.e.*, of same sign) and hence by Remark 2 Art. 9, Eqn. (7) represents a pair of imaginary planes.

Hence the only real pair of central circular ellipsoid (1) are given by real pair of planes (6)

i.e., by
$$z^2\,\frac{(b^2 - c^2)}{c^2b^2} = -\,x^2\,\frac{(b^2 - a^2)}{a^2b^2}$$

or
$$\frac{x^2(a^2 - b^2)}{a^2} = z^2\,\frac{(b^2 - c^2)}{c^2}$$

Taking square roots
$$\frac{x}{a}\sqrt{a^2 - b^2} = \pm\,\frac{z}{c}\sqrt{b^2 - c^2}$$

or
$$\frac{x}{a}\sqrt{a^2 - b^2} \pm \frac{z}{c}\sqrt{b^2 - c^2} = 0 \qquad \text{...(8)}$$

Also we know that parallel plane sections of a conicoid are similar and similarly situated conics.

Note : Therefore the equations

$$\frac{x}{a}\sqrt{a^2 - b^2} + \frac{z}{c}\sqrt{b^2 - c^2} = \lambda$$

and
$$\frac{x}{a}\sqrt{a^2 - b^2} - \frac{z}{c}\sqrt{b^2 - c^2} = \mu$$

give circular sections for all values of λ and μ.

11. To prove that any two circular sections of an ellipsoid of opposite systems lie on a sphere

We know (by Note Art. 10) that equations of two circular sections of an ellipsoid of opposite systems are

$$\frac{x}{a}\sqrt{a^2-b^2}+\frac{z}{c}\sqrt{b^2-c^2}=\lambda$$

and
$$\frac{x}{a}\sqrt{a^2-b^2}-\frac{z}{c}\sqrt{b^2-c^2}=\mu$$

Joint equation of these two planes is

(Making R.H.S. zero and multiplying the L.H. sides) is

$$\left(\frac{x}{a}\sqrt{a^2-b^2}+\frac{z}{c}\sqrt{b^2-c^2}-\lambda\right)\left(\frac{x}{a}\sqrt{a^2-b^2}-\frac{z}{c}\sqrt{b^2-c^2}-\mu\right)=0 \quad ...(1)$$

Equation of the ellipsoid whose circular sections are written above is

$$\frac{x^2}{a^2}+\frac{y^2}{b^2}+\frac{z^2}{c^2}-1=0 \quad ...(2)$$

$\therefore$ Equation of any conicoid which passes through the two circular sections given by (1) of ellipsoid (2) is $F+\lambda uv=0$

i.e.,
$$\frac{x^2}{a^2}+\frac{y^2}{b^2}+\frac{z^2}{c^2}-1+\lambda\left(\frac{x}{a}\sqrt{a^2-b^2}+\frac{z}{c}\sqrt{b^2-c^2}-\lambda\right)$$
$$\times\left(\frac{x}{a}\sqrt{a^2-b^2}-\frac{z}{c}\sqrt{b^2-c^2}-\mu\right)=0 \quad ...(3)$$

Let us try to know of Eqn. (3) represents a sphere. For this to happen, in equation (3), Coeff. of x^2 = Coeff. of y^2 = Coeff. of z^2

$$\therefore \quad \frac{1}{a^2}+\lambda\frac{(a^2-b^2)}{a^2}=\frac{1}{b^2}=\frac{1}{c^2}-\lambda\frac{(b^2-c^2)}{c^2}$$

From first two members

$$\frac{1}{a^2}+\lambda\frac{(a^2-b^2)}{a^2}=\frac{1}{b^2} \quad \text{or} \quad \lambda\frac{(a^2-b^2)}{a^2}=\frac{1}{b^2}-\frac{1}{a^2}=\frac{a^2-b^2}{a^2b^2}$$

$$\therefore \quad \lambda=\frac{1}{b^2}$$

Similarly from second and third members also $\lambda=\dfrac{1}{b^2}$.

$\therefore$ There exists a value of $\lambda=\dfrac{1}{b^2}$ for which conicoid (3) represents a sphere on which circular sections of an ellipsoid of opposite system lie.

Example 1. *Find the real central circular sections of the ellipsoid $x^2+2y^2+6z^2=8$.* (K.U. 2006)

Sol. Equation of the ellipsoid is

$$x^2+2y^2+6z^2=8 \quad \text{or} \quad x^2+2y^2+6z^2-8=0 \quad ...(1)$$

Rewriting Eqn. (1) as $(x^2+y^2+z^2-8)+(y^2+5z^2)=0$

It is of the form $S+\lambda uv=0$

but $uv=y^2+5z^2=0$ gives

$$y^2 = -5z^2 \quad \text{or} \quad y = \pm i\sqrt{5}\,z$$

which are imaginary planes.

Again rewriting Eqn. (1) as

$$2x^2 - x^2 + 2y^2 + 2z^2 + 4z^2 - 8 = 0$$

or $$2(x^2 + y^2 + z^2 - 4) - (x^2 - 4z^2) = 0$$

which is also of the form $$S + \lambda uv = 0$$

where $uv = x^2 - 4z^2 = 0$ gives $x^2 = 4z^2$ or $x = \pm 2z$

which are real planes passing through the centre (0, 0, 0) of ellipsoid (1).

Hence equation of (planes of) real central circular sections is $x^2 - 4z^2 = 0$ ...(2)

Eqns. (1) and (2) represent real central circular sections.

Again rewriting Eqn. (1) as

$$6x^2 - 5x^2 + 6y^2 - 4y^2 + 6z^2 - 8 = 0$$

or $$(6x^2 + 6y^2 + 6z^2 - 8) - (5x^2 + 4y^2) = 0$$

which is again of the form $$S + \lambda uv = 0$$

where $uv = 5x^2 + 4y^2 = 0$ gives $5x^2 = -4y^2$

or $$x^2 = -\frac{4}{5}y^2 \quad \text{or} \quad x = \pm i\sqrt{\frac{4}{5}}\,y$$

which are imaginary planes.

Example 2. *Prove that the radius of the circle in which the plane* $\frac{x}{a}\sqrt{a^2 - b^2} + \frac{z}{c}\sqrt{b^2 - c^2} = \lambda$ *cuts the ellipsoid* $\frac{x^2}{a^2} + \frac{y^2}{b^2} + \frac{z^2}{c^2} = 1$ *is* $b\,.\sqrt{1 - \frac{\lambda^2}{a^2 - c^2}}$.

Sol. We know by Art. 10 (Eqn. (6) and its simplied form (8)) that the plane $\frac{x}{a}\sqrt{a^2 - b^2} + \frac{z}{c}\sqrt{b^2 - c^2} = 0$

cuts the ellipsoid $$\frac{x^2}{a^2} + \frac{y^2}{b^2} + \frac{z^2}{c^2} = 1 \qquad ...(1)$$

in a circle of Radius b ...(2)

Therefore the parallel plane $\frac{x}{a}\sqrt{a^2 - b^2} + \frac{z}{c}\sqrt{b^2 - c^2} = \lambda$ ...(3)

will cut the ellipsoid in a circle of radius **b d. (Note)** ...(4)

Comparing (1) with $ax^2 + by^2 + cz^2 = 1$, we have

$$a = \frac{1}{a^2},\ b = \frac{1}{b^2},\ c = \frac{1}{c^2}$$

Comparing (3) with $lx + my + nz = p$, we have

$$l = \frac{1}{a}\sqrt{a^2 - b^2},\ m = 0\ ,\ n = \frac{1}{c}\sqrt{b^2 - c^2}\ ,\ p = \lambda$$

$$\therefore \qquad p_0^2 = \frac{l^2}{a} + \frac{m^2}{b} + \frac{n^2}{c} = \frac{\frac{1}{a^2}(a^2 - b^2)}{\frac{1}{a^2}} + 0 + \frac{\frac{1}{c^2}(b^2 - c^2)}{\frac{1}{c^2}}$$

or $$p_0^2 = a^2 - b^2 + b^2 - c^2 = a^2 - c^2$$

$\therefore$ $$d^2 = 1 - \frac{p^2}{p_0^2} = 1 - \frac{\lambda^2}{a^2 - c^2}$$

$\therefore$ $$d = \sqrt{1 - \frac{\lambda^2}{a^2 - c^2}}$$

Putting this value of d in (4),

Radius of required circle $= b\sqrt{1 - \dfrac{\lambda^2}{a^2 - c^2}}$.

EXERCISE 3

1. Show that the central circular sections of the hyperboloids $\frac{x^2}{a^2} + \frac{y^2}{b^2} - \frac{z^2}{c^2} = 1$ and $\frac{x^2}{a^2} - \frac{y^2}{b^2} - \frac{z^2}{c^2} = 1$ are given by the planes

$$\frac{y}{b}\sqrt{a^2 - b^2} \pm \frac{z}{c}\sqrt{a^2 + c^2} = 0$$

and $$\frac{x}{a}\sqrt{a^2 + b^2} \pm \frac{z}{c}\sqrt{b^2 - c^2} = 0.$$

[**Hint.** Proceed as in Art. 10]

2. Find the real circular sections of the following conicoids :

   (i) $2x^2 + 11y^2 + z^2 = 1$ (ii) $10x^2 - 2y^2 + z^2 + 2 = 0$.

   [**Hint.** Proceed as in Example 1 and then use the result of Note Art. 10]

3. Prove that the radius of a circular section of the ellipsoid at a distance p from the centre is $b \cdot \sqrt{1 - \dfrac{p^2 b^2}{a^2 c^2}}$.

4. Prove that the section of the hyperboloid $\frac{x^2}{a^2} - \frac{y^2}{b^2} - \frac{z^2}{c^2} = 1$ by the plane $\frac{x}{a}\sqrt{a^2 + b^2} + \frac{z}{c}\sqrt{b^2 - c^2} = \lambda$ is real if $\lambda^2 > a^2 + c^2$.

Answer

2. (i) $3y + z = \lambda,\quad 3y - z = \mu$ (ii) $\sqrt{3}\,x + y = \lambda,\ \sqrt{3}\,x - y = \mu$.

12. To determine the nature of plane sections of a paraboloid.

Let the equation of the paraboloid be

$$ax^2 + by^2 = 2cz \qquad \text{...(1)}$$

Let the equation of the plane be

$$lx + my + nz = p \qquad \text{...(2)}$$

We know from Chapter 3, Art. 3, that the equation of the cylinder passing through the section given by (1) and (2) and having its generators parallel to x-axis is obtained by eliminating x from (1) and (2).

From equation (2) $x = \dfrac{p - my - nz}{l}$

Putting this value of x in Eqn. (1), we have

$$a\,\frac{(p - my - nz)^2}{l^2} + by^2 = 2cz$$

Multiplying by L.C.M. = l^2, equation of the cylinder is

$$a(p^2 + m^2y^2 + n^2z^2 - 2pmy + 2mnyz - 2pnz) + bl^2y^2 = 2cl^2z$$

or $$(am^2 + bl^2)y^2 + 2amnyz + an^2z^2 - 2apmy - 2(apn + cl^2)\,z + ap^2 = 0 \qquad ...(3)$$

We know that the plane $x = 0$ (*i.e.*, yz-plane) is perpendicular to the generating lines of the cylinder (3) which are parallel to x-axis.

The plane $x = 0$ cuts the cylinder (3) in the conic $x = 0$

and $$(am^2 + bl^2)\,y^2 + 2amnyz + an^2z^2 - 2(apn + cl^2)\,z + ap^2 = 0 \qquad ...(4)$$

Clearly, the conic (4) is the orthogonal projection of the conic given by (1) and (2) on the plane $x = 0$. This projection and therefore also the given section (By Remark 3, Art. 1) is a parabola, ellipse or hyperbola according as

$$H^2 - AB = 0, < 0 \quad \text{or} \quad > 0$$

where A is Coeff. of y^2, B is Coeff. of z^2 and H = $\frac{1}{2}$ Coeff. of yz *i.e.*, according as

$$a^2m^2n^2 - (am^2 + bl^2)\,an^2 = 0, < 0 \quad \text{or} \quad > 0$$

or $$a^2m^2n^2 - a^2m^2n^2 - abn^2l^2 = 0, < 0 \quad \text{or} \quad > 0$$

or $$-\,abn^2l^2 = 0, < 0 \quad \text{or} \quad > 0$$

or $$abn^2l^2 = 0, > 0 \quad \text{or} \quad < 0 \qquad ...(5)$$

Let us assume that $l \neq 0$ so that plane (2) is not perpendicular to yz-plane ($x = 0$) which is a plane of symmetry of surfaces (1).

Therefore from (5) for a parabola, $abn^2 = 0$.

But $ab \neq 0 \quad \therefore \quad n = 0$

For $n \neq 0$ From (5), dividing by n^2l^2, the conic will be an ellipse or hyperbola according as $ab > 0$ or < 0.

13. Axes of plane sections of paraboloids

To determine the lengths and the direction ratios of the axes of the section of the paraboloid

$$ax^2 + by^2 = 2cz \qquad ...(1)$$

by the plane $$lx + my + nz = p \qquad ...(2)$$

Let (α, β, γ) be the centre of this section given by Equations (1) and (2).

$\therefore$ Another equation of plane (2) of this section of paraboloid (1) is given by T = S_1 (see Chapter 4)

or $$ax\alpha + by\beta - 2c\,\frac{(z + \gamma)}{2} = a\alpha^2 + b\beta^2 - 2c\gamma$$

or $$ax\alpha + by\beta - cz - c\gamma = a\alpha^2 + b\beta^2 - 2c\gamma$$

or $$a\alpha x + b\beta y - cz = a\alpha^2 + b\beta^2 - c\gamma \qquad ...(3)$$

Comparing equations (3) and (2), we have

$$\frac{a\alpha}{l} = \frac{b\beta}{m} = \frac{-c}{n} = \frac{a\alpha^2 + b\beta^2 - c\gamma}{p}$$

From first and third members $\qquad \alpha = -\frac{lc}{an}$

From second and third members $\qquad \beta = -\frac{mc}{bn}$

From fourth and third members $\qquad a\alpha^2 + b\beta^2 - c\gamma = -\frac{cp}{n}$

or
$$c\gamma = a\alpha^2 + b\beta^2 + \frac{cp}{n}$$

Putting values of α and β

$$c\gamma = \frac{al^2c^2}{a^2n^2} + \frac{bm^2c^2}{b^2n^2} + \frac{cp}{n} = \frac{c^2}{n^2}\left(\frac{l^2}{a} + \frac{m^2}{b} + \frac{np}{c}\right)$$

or
$$c\gamma = \frac{c^2k}{n^2}, \quad \text{where } k = \frac{l^2}{a} + \frac{m^2}{b} + \frac{np}{c}$$

∴
$$\gamma = \frac{ck}{n^2}$$

∴ Centre of plane section is $(\alpha, \beta, \gamma) = \left(-\frac{lc}{an}, -\frac{mc}{bn}, \frac{ck}{n^2}\right)$ ...(4)

where $k = \frac{l^2}{a} + \frac{m^2}{b} + \frac{np}{c}$ ...(5)

Now let us shift the origin to centre (α, β, γ) of the conic.

Changing x to $x + \alpha$, y to $y + \beta$ and z to $z + \gamma$ (see Chapter 0) in Eqn. (1), we have

$$a(x + \alpha)^2 + b(y + \beta)^2 = 2c(z + \gamma)$$

or
$$ax^2 + by^2 + 2(ax\alpha + by\beta - cz) + a\alpha^2 + b\beta^2 - 2c\gamma = 0$$

Putting values of α, β, γ from (4),

$$ax^2 + by^2 + 2\left(-\frac{lcx}{n} - \frac{mcy}{n} - cz\right) + \frac{al^2c^2}{a^2n^2} + \frac{bm^2c^2}{b^2n^2} - 2c \cdot \frac{ck}{n^2} = 0$$

or
$$ax^2 + by^2 - \frac{2c}{n}(lx + my + nz) + \frac{l^2c^2}{an^2} + \frac{m^2c^2}{bn^2} - \frac{2c^2k}{n^2} = 0$$

or
$$ax^2 + by^2 - \frac{2c}{n}(lx + my + nz) + \frac{c^2}{n^2}\left(\frac{l^2}{a} + \frac{m^2}{b} - 2k\right) = 0$$

Putting $\frac{l^2}{a} + \frac{m^2}{b} = k - \frac{np}{c}$ from (5), it becomes

$$ax^2 + by^2 - \frac{2c}{n}(lx + my + nz) + \frac{c^2}{n^2}\left(\frac{k - np - 2k}{c}\right) = 0$$

or
$$ax^2 + by^2 - \frac{2c}{n}(lx + my + nz) + \frac{c^2}{cn^2}(-ck - np) = 0$$

or $$ax^2 + by^2 - \frac{2c}{n}(lx + my + nz) = \frac{c(ck+np)}{n^2}$$

$$= \frac{p_0^2}{n^2}, \text{ where } p_0^2 = c(ck+np) \qquad ...(6)$$

Again equation (2) becomes [on changing x to $x + \alpha$, y to $y + \beta$, z to $z + \gamma$]

$$l(x+\alpha) + m(y+\beta) + n(z+\gamma) = p$$

or $$lx + my + nz = p - l\alpha - m\beta - n\gamma$$

Putting values of α, β, γ from (4),

$$lx + my + nz = p + \frac{l^2c}{an} + \frac{m^2c}{bn} - \frac{ck}{n}$$

$$= p + \frac{l^2c}{an} + \frac{m^2c}{bn} - \frac{c}{n}\left(\frac{l^2}{a} + \frac{m^2}{b} + \frac{np}{c}\right) \qquad \text{[By (5)]}$$

$$= 0$$

or $$lx + my + nz = 0 \qquad ...(7)$$

Putting $lx + my + nz = 0$ from (7) in (6), equation of the paraboloid referred to centre (α, β, γ) of conic as origin is $ax^2 + by^2 = \dfrac{p_0^2}{n^2}$ ...(8)

Eqns. (8) and (7) are equations of the conic referred to centre of conic as origin.

To find lengths of the axes

Let λ, μ, γ be the d.c's of a semi-diameter (a line through vertex) of paraboloid (1) of length r ; then its equations through the vertex (0, 0, 0) are

$$\frac{x-0}{\lambda} = \frac{y-0}{\mu} = \frac{z-0}{\gamma} = r \qquad ...(9)$$

From (9), the co-ordinates of the extremity of this semi-diameter are $(\lambda r, \mu r, \gamma r)$ and this point lies on paraboloid (8).

$$\therefore \quad a\lambda^2r^2 + b\mu^2r^2 = \frac{p_0^2}{n^2}$$

or $$a\lambda^2r^2 + b\mu^2r^2 = \frac{p_0^2}{n^2}(\lambda^2 + \mu^2 + \gamma^2)$$

$$| \because \lambda^2 + \mu^2 + \gamma^2 = 1 \text{ as } \lambda, \mu, \gamma \text{ are d.c's}$$

or $$a\,r^2n^2\lambda^2 + b\,r^2\mu^2n^2 = p_0^2\lambda^2 + p_0^2\mu^2 + p_0^2\gamma^2$$

or $$(ar^2n^2 - p_0^2)\lambda^2 + (br^2n^2 - p_0^2)\mu^2 - p_0^2\gamma^2 = 0 \qquad ...(10)$$

Eliminating λ, μ, γ from Eqns. (3) and (4)

(Putting $\lambda = x$, $\mu = y$, $\gamma = z$ from (9) in (10), we have

$$(ar^2n^2 - p_0^2)x^2 + (br^2n^2 - p_0^2)y^2 - p_0^2z^2 = 0 \qquad ...(11)$$

which being homogeneous equation of second degree in x, y, z represents a cone with vertex of paraboloid at its vertex and the semi-diameters of the conicoid of length r as its generators.

Plane (7) touches cone (11) (for detailed explanation see Art. 3 of this chapter)

$$\therefore \quad \frac{l^2}{ar^2n^2 - p_0^2} + \frac{m^2}{br^2n^2 - p_0^2} + \frac{n^2}{-p_0^2} = 0 \qquad \left| \frac{l^2}{A} + \frac{m^2}{B} + \frac{n^2}{C} = 0 \right.$$

Multiplying by L.C.M. $= p_0^2 (ar^2n^2 - p_0^2)(br^2n^2 - p_0^2)$ and simplifying, we have

$$abn^6r^4 - n^2r^2 p_0^2 [(a+b)n^2 + am^2 + bl^2] + p_0^4 (l^2 + m^2 + n^2) = 0 \qquad ...(12)$$

which is a quadratic in r^2 and hence has two roots say r_1^2 and r_2^2 which are the squares of the semi-axes of the conic given by (1) and (2).

To find d.c's of axes of the conic

Now equation of the tangent plane at the point $(\lambda r, \mu r, \gamma r)$ to cone (11) is

$$(ar^2n^2 - p_0^2)\, x\lambda r + (br^2n^2 - p_0^2)\, y\mu r - p_0^2\, z.\, \gamma r = 0$$

[Changing x^2 to xx_1 etc.]

Dividing every term by r

$$(ar^2n^2 - p_0^2)\, \lambda x + (br^2n^2 - p_0^2)\, \mu y - p_0^2 \gamma z = 0 \qquad ...(13)$$

Comparing Eqns. (13) and (7), we have

$$\frac{(ar^2n^2 - p_0^2)\lambda}{l} = \frac{(br^2n^2 - p_0^2)\mu}{m} = -\frac{p_0^2\gamma}{n} \qquad ...(14)$$

Putting the values of $r^2 = r_1^2$ and r_2^2 from Eqn. (12) in Eqn. (14), we get the values of λ, μ, γ which are the d.c's of the two axes of the conic.

Cor. This section will be a **rectangular hyperbola** if $r_1^2 + r_2^2 = 0$

i.e., if $\quad (a+b)\, n^2 + am^2 + bl^2 = 0 \qquad$ **(From Eqn. (12))**

14. To find Area of section of paraboloid if the section is an ellipse

Reproduce Art. 13 up to Eqn. (12)

We know that area of ellipse

$$= \pi r_1 r_2 = \pi\sqrt{r_1^2 r_2^2} = \pi\sqrt{\frac{p_0^4 (l^2 + m^2 + n^2)}{abn^6}} \qquad \text{(From Eqn. (12))}$$

$$= \frac{\pi p_0^2}{n^3}\sqrt{\frac{l^2 + m^2 + n^2}{ab}}$$

Putting the values of $p_0^2 = c(ck + np)$

$$= c\left(c\left(\frac{l^2}{a} + \frac{m^2}{b} + \frac{np}{c}\right) + np\right) = c\left(\frac{l^2c}{a} + \frac{m^2c}{b} + 2np\right),$$

$$\therefore \text{ Area of ellipse } = \frac{\pi c\left(\dfrac{l^2c}{a} + \dfrac{m^2c}{b} + 2np\right)}{n^3}\sqrt{\frac{l^2 + m^2 + n^2}{ab}}.$$

List of Important Results and Formulae

(Review IV and Chapters 10, 11, 12, 13, 14, 15, 16, 17)

REVIEW IV

1. Change the origin without changing the directions of axes

If the origin is shifted to the point (x_1, y_1, z_1), then x changes to $x + x_1$, y to $y + y_1$ and z to $z + z_1$.

2. Change of directions of axes without changing the origin

If l_1, m_1, n_1 ; l_2, m_2, n_2 ; l_3, m_3, n_3 are the d.c.'s of three mutually perpendicular lines taken as new axes OX′, OY′, OZ′.

Let the co-ordinates of any point P in space be (x, y, z) referred to axes OX, OY, OZ and (x', y', z') referred to new axes OX′, OY′, OZ′.

(*a*) **To express x, y, z in terms of x′, y′, z′**

$$x = l_1x' + l_2y' + l_3z'$$
$$y = m_1x' + m_2y' + m_3z'$$
$$z = n_1x' + n_2y' + n_3z'.$$

(*b*) **To express x′, y′, z′ in terms of x, y, z**

$$x' = l_1x + m_1y + n_1z$$
$$y' = l_2x + m_2y + n_2z$$
$$z' = l_3x + m_3y + n_3z.$$

3. Relations among the d.c's of three mutually perpendicular lines

If l_1, m_1, n_1 ; l_2, m_2, n_2 ; l_3, m_3, n_3 are the d.c's of three mutually ⊥ lines OX′, OY′, OZ′ ; then

(*i*)
$$\left.\begin{array}{l} l_1^2 + m_1^2 + n_1^2 = 1 \\ l_2^2 + m_2^2 + n_2^2 = 1 \\ l_3^2 + m_3^2 + n_3^2 = 1 \end{array}\right\} \quad \text{...(A)}$$

(*ii*)
$$\left.\begin{array}{l} l_1l_2 + m_1m_2 + n_1n_2 = 0 \\ l_1l_3 + m_1m_3 + n_1n_3 = 0 \\ l_2l_3 + m_2m_3 + n_2n_3 = 0 \end{array}\right\} \quad \text{...(B)}$$

[Also l_1, l_2, l_3 ; m_1, m_2, m_3 ; n_1, n_2, n_3 are the d.c.'s of three mutually ⊥ lines OX, OY, OZ referred to OX′, OY′, OZ′ as axes].

∴ (*iii*)
$$\left.\begin{array}{l} l_1^2 + l_2^2 + l_3^2 = 1 \\ m_1^2 + m_2^2 + m_3^2 = 1 \\ n_1^2 + n_2^2 + n_3^2 = 1 \end{array}\right\} \quad \text{...(C)}$$

$$(iv) \quad \left.\begin{aligned} l_1m_1 + l_2m_2 + l_3m_3 &= 0 \\ l_1n_1 + l_2n_2 + l_3n_3 &= 0 \\ m_1n_1 + m_2n_2 + m_3n_3 &= 0 \end{aligned}\right\} \quad ...(D)$$

$$(v) \quad \begin{vmatrix} l_1 & m_1 & n_1 \\ l_2 & m_2 & n_2 \\ l_3 & m_3 & n_3 \end{vmatrix} = \pm 1 \qquad (vi) \quad \begin{vmatrix} l_1 & l_2 & l_3 \\ m_1 & m_2 & m_3 \\ n_1 & n_2 & n_3 \end{vmatrix} = \pm 1$$

(vii) $l_1 = (m_2n_3 - m_3n_2)$ etc.

CHAPTER 10

1. Equations of the sphere :

(i) **Equation of the sphere whose centre is origin and radius a is**

$$x^2 + y^2 + z^2 = a^2 \qquad | \textit{ Standard form}$$

(ii) **The sphere whose centre is (a, b, c) and radius is**

$$(x - a)^2 + (y - b)^2 + (z - c)^2 = r^2 \qquad | \textit{ Central form}$$

(iii) The *general equation* of the sphere is

$$x^2 + y^2 + z^2 + 2ux + 2vy + 2wz + d = 0.$$

Its centre is $(-u, -v, -w)$

i.e., $\left[\frac{-1}{2} \text{co-eff. of } x, \frac{-1}{2} \text{co-eff. of } y, \frac{-1}{2} \text{co-eff. of } z\right]$

and radius $= \sqrt{u^2 + v^2 + w^2 - d}$.

(iv) The sphere on the join of (x_1, y_1, z_1) and (x_2, y_2, z_2) as diameter is $(x - x_1)(x - x_2) + (y - y_1)(y - y_2) + (z - z_1)(z - z_2) = 0$ | *Diameter form*

2. (a) **Any sphere through a given circle**

$$x^2 + y^2 + z^2 + 2ux + 2vy + 2wz + d = 0, \; Ax + By + Cz + D = 0$$

is $\quad x^2 + y^2 + z^2 + 2ux + 2vy + 2wz + d + K(Ax + By + Cz + D) = 0$

i.e., $\quad$ *sphere + K(plane) = 0.*

(b) **Any sphere through the circle of intersection of two spheres**

$S_1 = 0$ and $S_2 = 0$ is

$S_1 + KS_2 = 0$ or $S_1 + K(S_1 - S_2) = 0.$

3. Rule to write down the equation of tangent plane (or plane of contact or polar plane of) at (x_1, y_1, z_1) to the given sphere (or *conicoid*) is change :

x^2 to xx_1, $\quad y^2$ to yy_1, $\quad z^2$ to zz_1 ;

x to $\frac{1}{2}(x + x_1)$, $\quad y$ to $\frac{1}{2}(y + y_1)$, $\quad z$ to $\frac{1}{2}(z + z_1)$;

yz to $\frac{1}{2}(yz_1 + y_1z)$, $\quad zx$ to $\frac{1}{2}(zx_1 + z_1x)$, $\quad xy$ to $\frac{1}{2}(xy_1 + x_1y)$.

4. **Tangent plane (line) property.** *If a plane (line) touches the sphere, then ⊥ distance from the centre of the sphere on the plane (line) must be equal to its radius.*

5. If two spheres cut **orthogonally,** *then the square of the distance between the centres = sum of the squares of the radii.*

   The condition of orthogonality of the two spheres
   $$x^2 + y^2 + z^2 + 2u_1x + 2v_1y + 2w_1z + d_1 = 0$$
   and $$x^2 + y^2 + z^2 + 2u_2x + 2v_2y + 2w_2z + d_2 = 0$$
   is $$2u_1u_2 + 2v_1v_2 + 2w_1w_2 = d_1 + d_2.$$

6. **Angle of intersection of two spheres is given by**

   $\cos\theta = \dfrac{r_1^2 + r_2^2 - d^2}{2r_1r_2}$, where r_1, r_2 are the radii of the two spheres and d, the distance between the centres of the two spheres.

7. **Powers of a point** (x_1, y_1, z_1) w.r.t. the sphere $S = 0$ (equation in which co-effis. x^2, y^2, z^2 are unity) is given by S_1.

8. **Length of tangent of the point** (x_1, y_1, z_1) from the sphere $S = 0$ (the equation where co-effis. of x^2, y^2, z^2 are unity) is $\sqrt{S_1}$.

9. **Radical plane of two spheres :**
   $$S_1 = x^2 + y^2 + z^2 + 2u_1x + 2v_1y + 2w_1z + d_1 = 0$$
   $$S_2 = x^2 + y^2 + z^2 + 2u_2x + 2v_2y + 2w_2z + d_2 = 0$$
   is given by $2(u_1 - u_2)x + 2(v_1 - v_2)y + 2(w_1 - w_2)z + d_1 - d_2 = 0$

   i.e., by $S_1 - S_2 = 0$.

10. **Two spheres in the simplest form.** For problems relating to two given spheres, take their equations to be
    $$x^2 + y^2 + z^2 + 2u_1x + d = 0, \quad x^2 + y^2 + z^2 + 2u_2x + d = 0.$$

11. **Equation of co-axial system of spheres** in the simplest form is
    $$x^2 + y^2 + z^2 + 2ux + d = 0$$
    where u is the parameter

    The limiting point of this co-axial system are $(\pm\sqrt{d}, 0, 0)$.

CHAPTER 11

1. **General equation of the cone whose vertex is the origin is**
   $$ax^2 + by^2 + cz^2 + 2fyz + 2gzx + 2hxy = 0.$$
   Note. The d.c's of the generator of the cone (with vertex at origin) satisfy the equation of the cone.

2. **The equation of quadratic (second degree) cone through the axes is**
   $$fyz + gzx + hxy = 0.$$

3. **Equation of a right circular cone.** The equation of the right circular cone whose vertex is origin ; axis, the z-axis ; and the semi-vertical and α, is
   $$x^2 + y^2 = z^2 \tan^2\alpha \qquad | \textit{ Standard form}$$

4. **Equation of the enveloping cone from the point (x_1, y_1, z_1) to a sphere (or a conicoid)** is $SS_1 = T^2$

where $S = 0$ is the equation of the sphere (or conicoid)

S_1 = the result obtained on putting (x_1, y_1, z_1) in S

and $T = 0$ is the equation of tangent plane at (x_1, y_1, z_1).

5. **Rule to prove that a given numerical equation represents a cone and to find its vertex.**

(*i*) *Make the given equation homogeneous in x, y, z, t by introducing proper powers of t* and the R.H.S. = 0. Let it be denoted by $F(x, y, z, t) = 0$.

(*ii*) Find $\frac{\partial F}{\partial x}, \frac{\partial F}{\partial y}, \frac{\partial F}{\partial z}, \frac{\partial F}{\partial t}$ and put $t = 1$ and then put $\frac{\partial F}{\partial x} = 0$,

$$\frac{\partial F}{\partial y} = 0, \frac{\partial F}{\partial z} = 0, \frac{\partial F}{\partial t} = 0.$$

(*iii*) Solve for x, y, z any three of the above four equations.

(*iv*) Put these values of x, y, z in the fourth equation. If the equation is satisfied, then the given equation represents a cone whose vertex is given by values of x, y, z found in step (*iii*).

6. **Tangent plane at (x_1, y_1, z_1) to the cone**

$$ax^2 + by^2 + cz^2 + 2fyz + 2gzx + 2hxy = 0 \text{ is}$$

$$axx_1 + byy_1 + czz_1 + f(yz_1 + y_1z) + g(zx_1 + z_1x) + h(xy_1 + x_1y) = 0.$$

7. **The condition of tangency** of the plane $lx + my + nz = 0$ and the cone

$$ax^2 + by^2 + cz^2 + 2fyz + 2gzx + 2hxy = 0$$

is $$Al^2 + Bm^2 + Cn^2 + 2Fmn + 2Gnl + 2Hlm = 0.$$

8. **Equation of reciprocal cone of the cone**

$$ax^2 + by^2 + cz^2 + 2fyz + 2gzx + 2hxy = 0$$

is $$Ax^2 + By^2 + Cz^2 + 2Fyz + 2Gzx + 2Hxy = 0$$

where A, B, C, F, G, H are the co-factors of a, b, c, f, g, h respectively in the determinant

$$\begin{vmatrix} a & h & g \\ h & b & f \\ g & f & c \end{vmatrix}.$$

9. **Condition for the cone** $ax^2 + by^2 + cz^2 + 2fyz + 2gzx + 2hxy = 0$

(*i*) to have **three mutually ⊥ generators is** $a + b + c = 0$

[*i.e., sum of co-effis. of* $x^2, y^2, z^2 = 0$]

(*ii*) to have **three mutually ⊥ tangent planes is**

$$A + B + C = 0$$

i.e., $$ab + bc + ca - f^2 - g^2 - h^2 = 0.$$

CHAPTER 13. Section I—Conicoid

1. **Equations of conicoid in the simplest form** (*standard forms*)

(*a*) *Ellipsoid* $$\frac{x^2}{a^2} + \frac{y^2}{b^2} + \frac{z^2}{c^2} = 1$$

(b) *Hyperboloid of one sheet* $\quad \dfrac{x^2}{a^2}+\dfrac{y^2}{b^2}-\dfrac{z^2}{c^2}=1$

(c) *Hyperboloid of two sheets* $\quad \dfrac{x^2}{a^2}-\dfrac{y^2}{b^2}-\dfrac{z^2}{c^2}=1$

(d) *Central conicoid* $\quad ax^2+by^2+cz^2=1.$

Note. Equation of the central conicoid, when not stated, should be taken as $ax^2+by^2+cz^2=1$.

2. **Tangent plane at** (x_1, y_1, z_1) to the conicoid $ax^2+by^2+cz^2=1$ is $axx_1+byy_1+czz_1=1$.
3. **Condition of tangency of a plane and central conicoid.**
The condition that the plane $lx+my+nz=p$ should touch the conicoid $ax^2+by^2+cz^2=1$ is

$$\frac{l^2}{a}+\frac{m^2}{b}+\frac{n^2}{c}=p^2.$$

Cor. Any tangent plane to the central conicoid

$$ax^2+by^2+cz^2=1$$

is $\quad lx+my+nz=\sqrt{\dfrac{l^2}{a}+\dfrac{m^2}{b}+\dfrac{n^2}{c}}$.

4. **Condition** for the plane $lx+my+nz=p$ to be a tangent plane to the ellipsoid

$$\frac{x^2}{a^2}+\frac{y^2}{b^2}+\frac{z^2}{c^2}=1$$

is $\mathbf{a^2l^2+b^2m^2+c^2n^2=p^2}$.

5. **Equations of the normal at** (x_1, y_1, z_1) to the ellipsoid are

$$\frac{x-x_1}{x_1/a^2}=\frac{y-y_1}{y_1/b^2}=\frac{z-z_1}{z_1/c^2}$$

i.e., $$\frac{x-x_1}{\dfrac{\partial F}{\partial x_1}}=\frac{y-y_1}{\dfrac{\partial F}{\partial y_1}}=\frac{z-z_1}{\dfrac{\partial F}{\partial z_1}}$$

where $F(x, y, z)=0$ is the equation of the ellipsoid. *(or any conicoid)*

Equations of the normal at (x_1, y_1, z_1) of the ellipsoid, **in the actual direction cosines form, are**

$$\frac{x-x_1}{px_1/a^2}=\frac{y-y_1}{py_1/b^2}=\frac{z-z_1}{pz_1/c^2}$$

where p is the length of $\perp$ from the centre of the ellipsoid on the tangent plane at (x_1, y_1, z_1).

6. **Equation of polar plane and the plane of contact of** (x_1, y_1, z_1) w.r.t. the central conicoid $ax^2+by^2+cz^2=1$ is *the same as the equation of the tangent plane at* (x_1, y_1, z_1) *i.e.,* $axx_1+byy_1+czz_1=1$.
7. **Equation of enveloping cone** from the point (x_1, y_1, z_1) to the central conicoid $S=0$ is given by $\mathbf{SS_1=T^2}$.
8. **Equation of the plane of the section** of the central conicoid whose centre is the given point (x_1, y_1, z_1), is given by $\mathbf{T=S_1}$.

Section II—The Paraboloid

1. **Equation of the paraboloid** in the simplest form is

$$ax^2 + by^2 = 2z \qquad | \textit{ Standard form}$$

2. **Tangent plane at** (x_1, y_1, z_1) **is** $axx_1 + byy_1 = z + z_1$.
3. **Condition of tangency.** The condition for the plane

$lx + my + nz = p$ to touch the paraboloid $ax^2 + by^2 = 2z$, is

$$\frac{l^2}{a} + \frac{m^2}{b} = -2np.$$

4. **Equation of the polar plane or the plane of contact** of point (x_1, y_1, z_1) w.r.t. the paraboloid is $axx_1 + byy_1 = z + z_1$ [*This is same as the equation of tangent plane at* (x_1, y_1, z_1)].
5. **The enveloping cone** from the point (x_1, y_1, z_1) to paraboloid $S = 0$ is given by $SS_1 = T^2$.
6. **Equations of normal to an elliptic paraboloid**

$\frac{x^2}{a^2} + \frac{y^2}{b^2} = 2z$ at the point (x_1, y_1, z_1) are

$$\frac{x - x_1}{x_1/a^2} = \frac{y - y_1}{y_1/b^2} = \frac{z - z_1}{-1}.$$

CHAPTER 14. Generating Lines

1. Equations of the two systems of generating lines of hyperboloid of one sheet

$\frac{x^2}{a^2} + \frac{y^2}{b^2} - \frac{z^2}{c^2} = 1$ are

$$\text{(I)} \quad \frac{x}{a} - \frac{z}{c} = \lambda\left(1 - \frac{y}{b}\right),\ 1 + \frac{y}{b} = \lambda\left(\frac{x}{a} + \frac{z}{c}\right)$$

$$\text{(II)} \quad \frac{x}{a} - \frac{z}{c} = \mu\left(1 + \frac{y}{b}\right),\ 1 - \frac{y}{b} = \mu\left(\frac{x}{a} + \frac{z}{c}\right).$$

2. Equation of a generating line through any point of the hyperboloid of one sheet $\frac{x^2}{a^2} + \frac{y^2}{b^2} - \frac{z^2}{c^2} = 1$ can be written in the form

$$\frac{x - a\cos\alpha}{a\sin\alpha} = \frac{y - b\sin\alpha}{-b\cos\alpha} = \frac{z}{\pm c}.$$

3. The co-ordinates of the point of intersection of any two generating lines of two different systems of hyperboloid of one sheet (given by formula 1 above) are $x = a\,\frac{(1 + \lambda\mu)}{\lambda + \mu}$,

$y = b\,\frac{(\lambda - \mu)}{\lambda + \mu}$, $z = c\,\frac{(1 - \lambda\mu)}{\lambda + \mu}$.

These equations are also callded Parametric Equations of the hyperboloid ; λ, μ being the two parameters.

4. A point (θ, ϕ) on the hyperboloid $\frac{x^2}{a^2}+\frac{y^2}{b^2}-\frac{z^2}{c^2}=1$ is the point whose co-ordinates are $(a \cos \theta \sec \phi, b \sin \theta \sec \phi, c \tan \phi)$.

5. Equations of the two systems of generators of the paraboloid $\frac{x^2}{a^2}-\frac{y^2}{b^2}=2z$ are

(I) $\frac{x}{a}-\frac{y}{b}=\lambda z,\ 2=\lambda\left(\frac{x}{a}+\frac{y}{b}\right)$

(II) $\frac{x}{a}-\frac{y}{b}=2\mu,\ z=\mu\left(\frac{x}{a}+\frac{y}{b}\right)$.

CHAPTER 15. Confocal Conicoids

1. **Def. Confocal Conicoids**
   Conicoids whose principal sections *i.e.*, sections by principal planes are confocal conics are called confocal conicoids.
2. Equation of a system of confocal conicoids is

$$\frac{x^2}{a^2-\lambda}+\frac{y^2}{b^2-\lambda}+\frac{z^2}{c^2-\lambda}=1.$$

CHAPTER 16. General Equation of Second Degree in x, y, z

1. Equation of the tangent plane at any point (α, β, γ) on the conicoid $f(x, y, z) = ax^2 + by^2 + cz^2 + 2fyz + 2gzx + 2hxy + 2ux + 2vy + 2wz + d = 0$ is

$$T=0$$

i.e., $$x(a\alpha + h\beta + g\gamma + u) + y(h\alpha + b\beta + f\gamma + v) + z(g\alpha + f\beta + c\gamma + w) + u\alpha + v\beta + w\gamma + d = 0.$$

2. **Reduction of general equation of second degree in x, y, z.**
   The equation $f(x, y, z) = ax^2 + by^2 + cz^2 + 2fyz + 2gzx + 2hxy + 2ux + 2vy + 2wz + d = 0$ can be reduced to either of the following two standard forms by transformation of axes :
   (*i*) $\lambda_1 x^2 + \lambda_2 y^2 + \lambda_3 z^2 = \mu$
   (*ii*) $\lambda_1 x^2 + \lambda_2 y^2 = 2\mu z$
   Here below is the nature of the surfaces represented by (*i*) and (*ii*) in the form of two tables :
   (*i*) $\lambda_1 x^2 + \lambda_2 y^2 + \lambda_3 z^2 = \mu$

λ_1	λ_2	λ_3	μ	Surfaces
+	+	+	+	Ellipsoid
+	+	–	+	Hyperboloid of one sheet
+	–	–	+	Hyperboloid of two sheets

+	+	–	0	Cone
+	+	0	+	Elliptic cylinder
+	–	0	+	Hyperbolic cylinder
+	–	0	0	Pair of Intersecting planes
+	0	0	+	Pair of parallel planes

(ii) $\lambda_1 x^2 + \lambda_2 y^2 = 2\mu z$

λ_1	λ_2	μ	Surfaces
+	+	+	Elliptic paraboloid
+	–	+	Hyperbolic paraboloid
+	0	±	Parabolic cylinder

Remark. If λ_1, λ_2 are both negative and μ is positive, the equation still represents an elliptic paraboloid.

Note. Let $D = \begin{vmatrix} a & h & g \\ h & b & f \\ g & f & c \end{vmatrix}$

Expanding by first row $D = abc + 2fgh - af^2 - bg^2 - ch^2$. ...(1)

Let A, B, C, F, G, H denote the cofactors of a, b, c, f, g, h in D.

$$\therefore \quad A = bc - f^2, B = ca - g^2, C = ab - h^2$$

$$F = gh - af, G = hf - bg, H = fg - ch.$$

$$BC - F^2 = (ca - g^2)(ab - h^2) - (gh - af)^2$$

$$= a^2bc - ach^2 - abg^2 + g^2h^2 - g^2h^2 - a^2f^2 + 2afgh$$

or $$BC - F^2 = a(abc + 2fgh - af^2 - bg^2 - ch^2)$$

$$= aD$$

Similarly $CA - G^2 = bD$, $AB - H^2 = cD$.

Similarly $GH - AF = fD$, $HF - BG = gD$ etc.

3. Equation of the **Discriminating cubic** is

$$\begin{vmatrix} a-\lambda & h & g \\ h & b-\lambda & f \\ g & f & c-\lambda \end{vmatrix} = 0$$

or $\quad \lambda^3 - \lambda^2(a + b + c) + \lambda(A + B + C) - D = 0,$

where $\quad A = bc - f^2, B = ca - g^2, C = ab - h^2$

and $\quad D = abc + 2fgh - af^2 - bg^2 - ch^2.$

4. If the centre of the conicoid is origin, then coefficients of x, y, z are all zero in the equation of the conicoid.

5. Centre of the conicoid

Centre of the conicoid $f(x, y, z) = 0$ is obtained by solving the equations $\frac{\partial f}{\partial x} = 0, \frac{\partial f}{\partial y} = 0$ and $\frac{\partial f}{\partial z} = 0$ for x, y and z.

6. Working rule for reducing a general equation of second degree in x, y, z to standard form if the terms of second degree donot form a perfect square

Step I. Solve the discriminating cubic

$$\lambda^3 - \lambda^2 (a + b + c) + \lambda (A + B + C) - D = 0.$$

Sub-Case I

Step II. If all the three roots are different from zero say $\lambda_1, \lambda_2, \lambda_3$; then the centre of the quadric (conicoid) is obtained by solving the equations $\frac{\partial f}{\partial x} = 0, \frac{\partial f}{\partial y} = 0$ and $\frac{\partial f}{\partial z} = 0$ for x, y and z. Let the centre obtained be (x_1, y_1, z_1).

Step III. Shifting the origin to the centre and transforming the axes, the reduced equation will be

$$\lambda_1 x^2 + \lambda_2 y^2 + \lambda_3 z^2 = \mu$$

where $\mu = - d' = - (ux_1 + vy_1 + wz_1 + d)$.

Step IV. The direction cosines l, m, n of one of the new axes are given by

$$\frac{al + hm + gn}{l} = \frac{hl + bm + fn}{m} = \frac{gl + fm + cn}{n} = \lambda, \quad \text{...(1)}$$

where λ is the corresponding root of the discriminating cubic.

Note. If two roots of the discriminating cubic are equal, then it is a surface of revolution and axis of rotation is obtained by putting that value of λ which is different from the two equal values.

Sub-Case II. If one root say $\lambda_3 = 0$; then find l_3, m_3, n_3 for third new axis from the equations

$$al_3 + hm_3 + gn_3 = 0, \; hl_3 + bm_3 + fn_3 = 0$$

and

$$gl_3 + fm_3 + cn_3 = 0$$

(Putting $\lambda = 0$ in equation (1) and cross-multiplying)

Evaluate $k = ul_3 + vm_3 + wn_3$

If $k \neq 0$, reduced equation is the paraboloid $\lambda_1 x^2 + \lambda_2 y^2 + 2kz = 0$, and the **vertex** is obtained by solving following equations (2), (3), (4) and (5).

$$ax + hy + gz - l_3 k + u = 0 \quad \text{...(2)}$$

$$hx + by + fz - m_3 k + v = 0 \quad \text{...(3)}$$

$$gx + fy + cz - n_3 k + w = 0 \quad \text{...(4)}$$

and

$$k (l_3 x + m_3 y + n_3 z) + ux + vy + wz + d = 0 \quad \text{...(5)}$$

In all cases equation (5) must be used to find the **Vertex.**

If k = 0, then there is a line of centres given by any two of equations

$$\frac{\partial f}{\partial x} = 0, \frac{\partial f}{\partial y} = 0, \frac{\partial f}{\partial z} = 0.$$

Choose any point (α, β, γ) on this line as the centre and reduced equation is the cylinder $\lambda_1 x^2 + \lambda_2 y^2 + u\alpha + v\beta + w\gamma + d = 0$ because one variable here z is absent.

CHAPTER 17. Plane Sections of Conicoids

1. Nature of plane section of a conicoid

(Conic) section of the conicoid $ax^2 + by^2 + cz^2 = 1$ by the plane $lx + my + nz = p$ is a parabola or ellipse or hyperbola according as

$$bcl^2 + cam^2 + abn^2 = 0 \quad \text{or} \quad > 0 \quad \text{or} \quad < 0.$$

2. Axes of a central plane section

(*i*) Lengths of the semi-axes of the central section

$$ax^2 + by^2 + cz^2 = 1,\ lx + my + nz = 0 \quad \text{are given by}$$

$$r^4(bcl^2 + cam^2 + abn^2) - r^2[(b+c)l^2 + (c+a)m^2 + (a+b)n^2] + (l^2 + m^2 + n^2) = 0$$

(*ii*) D.C's λ, μ, γ of the axes of the central section $ax^2 + by^2 + cz^2 = 1$, $lx + my + nz = 0$ are given by

$$\frac{(ar^2-1)\lambda}{l} = \frac{(br^2-1)\mu}{m} = \frac{(cr^2-1)\gamma}{n}.$$

3. Condition for the above central section to be a rectangular hyperbola is

$$(b+c)l^2 + (c+a)m^2 + (a+b)n^2 = 0.$$

4. Area of central plane section ellipse

Area of central plane section (ellipse) is

$$\frac{\pi\sqrt{l^2 + m^2 + n^2}}{\sqrt{bcl^2 + cam^2 + abn^2}}.$$

5. Conditions for the two lines $\frac{x}{l_1} = \frac{y}{m_1} = \frac{z}{n_1}$ and $\frac{x}{l_2} = \frac{y}{m_2} = \frac{z}{n_2}$ to be the axes of the section of the conicoid $ax^2 + by^2 + cz^2 = 1$ are

$$l_1 l_2 + m_1 m_2 + n_1 n_2 = 0$$

and

$$al_1 l_2 + bm_1 m_2 + cn_1 n_2 = 0.$$

6. Axes of non-central plane sections

To find the lengths and direction cosines of the section of the central conicoid $ax^2 + by^2 + cz^2 = 1$ by the plane $lx + my + nz = p\ (p \neq 0)$;

Replace a by $\frac{a}{d^2}$, b by $\frac{b}{d^2}$, and c by $\frac{c}{d^2}$,

$$\left(\text{where } d^2 = 1 - \frac{p^2}{p_0^2} \text{ and } p_0^2 = \frac{l^2}{a} + \frac{m^2}{b} + \frac{n^2}{c}\right)$$

in the formulae (*i*) and (*ii*) of the result 2 above.

7. Area of non-central plane section

$$ax^2 + by^2 + cz^2 = 1,\ lx + my + nz = p\ (p \neq 0) \text{ is}$$

$$\frac{\pi d^2\sqrt{l^2 + m^2 + n^2}}{\sqrt{bcl^2 + cam^2 + abn^2}}$$

where $d^2 = 1 - \frac{p^2}{p_0^2}$ and $p_0^2 = \frac{l^2}{a} + \frac{m^2}{b} + \frac{n^2}{c}$.

8. Central circular sections of ellipsoid

Equations of central circular sections of ellipsoid $\frac{x^2}{a^2}+\frac{y^2}{b^2}+\frac{z^2}{c^2}=1$ are

$$\frac{x}{a}\sqrt{a^2-b^2}\pm\frac{z}{c}\sqrt{b^2-c^2}=0.$$

9. Equations of circular sections of the ellipsoid $\frac{x^2}{a^2}+\frac{y^2}{b^2}+\frac{z^2}{c^2}=1$ are

$$\frac{x}{a}\sqrt{a^2-b^2}+\frac{z}{c}\sqrt{b^2-c^2}=\lambda$$

and
$$\frac{x}{a}\sqrt{a^2-b^2}-\frac{z}{c}\sqrt{b^2-c^2}=\mu$$

for all values of λ and μ.

EXAMINATION PAPERS

M.D.U., ROHTAK

B.A., Part I, Examination, April 2007

Paper III (B.M. 103)

Vector Analysis and Geometry

Time allowed : 3 Hours] [*Maximum Marks : 30*

Note : Attempt **Five** questions in all, selecting at least **One** question from each unit. All questions carry equal marks.

UNIT I

1. (*a*) Given $\vec{a} = 2\hat{i} - \hat{j} + 3\hat{k}$, $\vec{b} = 2\hat{i} + \hat{j} - \hat{k}$ and $\vec{c} = \hat{i} + 3\hat{j} - \hat{k}$, find the reciprocal triad $\vec{a'}\ \vec{b'}\ \vec{c'}$ and verify $\left[\vec{a}\ \vec{b}\ \vec{c}\right]\left[\vec{a'}\ \vec{b'}\ \vec{c'}\right] = 1$. 3

(*b*) If $\vec{a} = \sin\theta\,\hat{i} + \cos\theta\,\hat{j} + \theta\,\hat{k}$, $\vec{b} = \cos\theta\,\hat{i} - \sin\theta\,\hat{j} - 3\,\hat{k}$, and $\vec{c} = 2\hat{i} + 3\hat{j} - \hat{k}$, find $\dfrac{d}{d\theta}\left[\vec{a} \times \left(\vec{b} \times \vec{c}\right)\right]$ at $\theta = \dfrac{\pi}{2}$. 3

2. (*a*) If $\vec{f} = \dfrac{x\hat{i} + y\hat{j} + z\hat{k}}{\sqrt{x^2 + y^2 + z^2}}$, show that $\nabla \cdot \vec{f} = \dfrac{2}{\sqrt{x^2 + y^2 + z^2}}$. 3

(*b*) Prove that $\nabla^2\left(r\,\vec{r}\right) = \left(\dfrac{4}{r}\right)\vec{r}$, where r and $\vec{r}$ have their usual meanings. 3

UNIT II

3. (*a*) If $\vec{f} = (2x + y)\,\hat{i} + (3y - x)\,\hat{j}$, evaluate $\int_C \vec{f} \cdot d\vec{r}$, where C is the curve in the xy–plane consisting of the straight lines from (0, 0) to (2, 0) and then to (3, 2). 3

(*b*) Evaluate $\iint_S \vec{f} \cdot \vec{n}\, dS$, where $\vec{f} = y\hat{i} + 2x\hat{j} - z\hat{k}$ and S is the surface of the plane $2x + y = 6$ in the first octant cut off by the plane $z = 4$. 3

4. (*a*) Prove that $\iiint_v \frac{1}{r^2}\, dV = \iint_S \frac{\vec{r}\,.\,\vec{n}}{r^2}\, dS.$ 3

(*b*) Verify Stoke's Theorem for $\vec{f} = y\hat{i} + 2\hat{j} + x\hat{k}$, where S is the upper half surface of the sphere $x^2 + y^2 + z^2 = 1$ and C is its boundary. 3

UNIT III

5. (*a*) Find the equation of the conic referred to axes through the centre parallel to the original axes. 3

(*b*) Find the equations of the tangents to the conic $x^2 + 4xy + 3y^2 - 5x - 6y + 3 = 0$ which are parallel to the straight line $x + 4y = 0$. 3

6. (*a*) Trace the conic

$$x^2 + xy + y^2 + x - 4y + 1 = 0.$$ 3

(*b*) Find the equation of the parabola which has contact of third order with the conic $ax^2 + 2hxy + by^2 + 2gx + 2fy + c = 0$ at origin. 3

7. (*a*) Show that the confocal hyperbola through the point on the ellipse $\frac{x^2}{a^2} + \frac{y^2}{b^2} = 1$, whose eccentric angle is θ, is $\frac{x^2}{\cos^2\theta} - \frac{y^2}{\sin^2\theta} = a^2 - b^2$. 3

(*b*) In any conic, prove that the sum of reciprocal of two perpendicular focal chords is constant. 3

UNIT IV

8. (*a*) Prove that there are five points on an elliptic paraboloid, the normals at which pass through a given point (α, β, γ). 3

(*b*) Prove that the area of the section of the ellipsoid $\frac{x^2}{a^2} + \frac{y^2}{b^2} + \frac{z^2}{c^2} = 1$ by the plane $\frac{x}{a} + \frac{y}{b} + \frac{z}{c} = 1$ is $\frac{2\pi}{3\sqrt{3}}(b^2c^2 + c^2a^2 + a^2b^2)^{1/2}$. 3

9. (*a*) Show that the two confocal paraboloids cut everywhere at right angle. 3

(*b*) Reduce the equation $9x^2 + 4y^2 + 4z^2 + 8yz + 12zx + 12xy + 4x + y + 10z + 1 = 0$ to the standard form. Show that it represents a parabolic cylinder. 3

10. (*a*) Obtain the equation of the sphere having the circle, $x^2 + y^2 + z^2 = 9$, $x - 2y + 2z - 5 = 0$ as great circle. Find its centre and radius also. 3

(*b*) Find the equation of the cone whose vertex is (α, β, γ) and whose generating lines pass through the conic $\frac{x^2}{a^2} + \frac{y^2}{b^2} = 1, z = 0$. 3

M.D.U., ROHTAK

B.Sc., Part I, Examination, April 2007

Paper III (B.M. 103)

Vector Analysis and Geometry

Time allowed : 3 Hours] [*Maximum Marks : 45*

Note : Attempt **Five** questions in all, selecting at least **One** question from each unit.

UNIT I

1. (*a*) If $\vec{a}, \vec{b}, \vec{c}$, are 3 unit vectors such that $\vec{b} \times \left(\vec{c} \times \vec{a}\right) = \frac{1}{2}\vec{c}$, find angles, which $\vec{b}$ makes with $\vec{c}$ and $\vec{a}$; $\vec{c}$ and $\vec{a}$ being non-parallel. 4

(*b*) If $\vec{a} = x^2yz\hat{i} - 2xz^3\hat{j} + xz^2\hat{k}$ and $\vec{b} = 2z\hat{i} + y\hat{j} - x^2\hat{k}$, find the value of $\frac{\partial^2}{\partial x\, \partial y}\left(\vec{a} \times \vec{b}\right)$ at the point $(1, 0, -2)$. 5

2. (*a*) Find the angle between the surfaces $x^2 + y^2 + z^2 = 9$ and $z = x^2 + y^2 - 3$ at the point $(2, -1, 2)$. 4

(*b*) Prove that $\nabla^2 (r\vec{r}) = \left(\frac{4}{r}\right)\vec{r}$. 5

UNIT II

3. (*a*) Evaluate $\iint_S \vec{f} \cdot \vec{n}\, dS$, where $\vec{f} = yz\hat{i} + zx\hat{j} + xy\hat{k}$ and S is that part of the surface of the sphere $x^2 + y^2 + z^2 = 1$, which lies in the first octant. 4

(*b*) Evaluate $\iint_S (\nabla \times \vec{f}) \cdot \vec{n}\, dS$, where S is the surface of the cone $z = 2 - \sqrt{x^2 + y^2}$ above xy plane and $\vec{f} = (x - 2)\,\hat{i} + (x^3 + yz)\,\hat{j} - 3xy^2\hat{k}$. 5

4. (*a*) Verify Green' theorem in the plane for $\oint_C (3x^2 - 8y^2)\, dx + (4y - 6xy)\, dy$, where C is the boundary of the region bounded by the parabolas : $y = \sqrt{x}$ and $y = x^2$. 4

(*b*) Verify Gauss's divergence theorem for the function $\vec{f} = 4\,xz\,\hat{i} - y^2\,\hat{j} + yz\,\hat{k}$ over the surface S of the cube bounded by $x = 0, x = 1, y = 0, y = 1, z = 0, z = 1$. 5

UNIT III

5. (*a*) What curves are represented by the equation $x^2 + 2xy + (1 + \lambda)y^2 + 2\lambda y - 1 = 0$, for different values of λ ? 4

(*b*) If $S = ax^2 + 2hxy + by^2 + 2gx + 2fy + c = 0$ is the equation of conic, then determine locus of the middle points of a system of parallel chords. 5

6. (*a*) Trace the conic $x^2 - 4xy - 2y^2 + 10x + 4y = 0$. 4

(*b*) Two conics have double contact. Prove that the polar of the centre of one conic with respect to the other is parallel to common chords. 5

7. (*a*) Prove that the normals to the confocals with an ellipse through any given point bisect the angles between the tangents drawn from that point to the ellipse. 4

(*b*) If the normals at the points whose vectorial angles are α, β, γ on the parabola $\frac{l}{r} = 1 + \cos\theta$ meet in the point (ρ, ϕ), then prove that $2\phi = \alpha + \beta + \gamma$. 5

UNIT IV

8. (*a*) Show that the equations of the polar of the line $\frac{x-1}{2} = \frac{y-2}{3} = \frac{z-3}{4}$ w.r.t. to the quadric $x^2 - 2y^2 + 3z^2 = 4$ are $\frac{x+6}{3} = \frac{y-2}{3} = z - 2$. 4

(*b*) Find the locus of the centres of sections of the ellipsoid $\frac{x^2}{a^2} + \frac{y^2}{b^2} + \frac{z^2}{c^2} = 1$, whose area is constant. 5

9. (*a*) Show that the perpendiculars from the origin on the generator of the hyperboloid $\frac{x^2}{a^2} + \frac{y^2}{b^2} - \frac{z^2}{c^2} = 1$, lie on the curve $\frac{a^2(b^2+c^2)^2}{x^2} + \frac{b^2(c^2+a^2)^2}{y^2} = \frac{c^2(a^2-b^2)^2}{z^2}$. 4

(*b*) Prove that the equation $x^2 + y^2 + z^2 + yz + zx + xy - 1 = 0$ represents an ellipsoid, the squares of whose semi-axes are 2, 2, $\frac{1}{2}$ and that the axis of revolution is $x = y = z$. 5

10. (*a*) Prove that the centre of spheres, which touches the lines, $y = mx, z = c$ and $y = -mx$, $z = -c$ lie upon the conicoid, $mxy + cz(1 + m^2) = 0$. 4

(*b*) Find the equation of the enveloping cylinder of the sphere, $x^2 + y^2 + z^2 - 2x + 4y = 1$ having its generators parallel to the line $x = y = z$. 5

K.U., KURUKSHETRA

B.A./B.Sc., Part I, Examination, April 2007

Paper III (B.M. 103)

Vector Analysis and Geometry

Time : Three Hours] [*Maximum Marks : B.A.–30*
B.Sc.–45

Note : Attempt **Five** questions in all, selecting at least **One** question from each unit marks for B.Sc. students are given in brackets.

UNIT I

1. (*a*) Define reciprocal trial of vectors $\vec{a}, \vec{b}, \vec{c}$. If a', b', c' are reciprocal set of vectors, then prove that

(*i*) $\mathbf{a}'.\,\mathbf{a} = \mathbf{b}'.\,\mathbf{b} = \mathbf{c}'.\,\mathbf{c} = 1$ (*ii*) $[\mathbf{a}'\ \mathbf{b}'\ \mathbf{c}'] = \dfrac{1}{[\mathbf{a}\ \mathbf{b}\ \mathbf{c}]}$. 3(4½)

(*b*) Show that $\text{div}\left[\dfrac{f(r)\,\vec{r}}{r}\right] = \dfrac{1}{r^2}\dfrac{d}{dr}(r^2 f(r))$, where r and $\vec{r}$ have their usual meanings.

3(4½)

2. (*a*) Prove that : $\text{Curl}\,\dfrac{\vec{a}\times\vec{r}}{r^3} = -\dfrac{\vec{a}}{r^3} + \dfrac{3\vec{r}}{r^5}\left(\vec{a}\,.\,\vec{r}\right)$, where $\vec{a}$ is a constant vector. 3(4½)

(*b*) Prove that $\nabla^2\left(r\,\vec{r}\right) = \left(\dfrac{4}{r}\right)\left(\vec{r}\right)$. 3(4½)

UNIT II

3. (*a*) If $\vec{f} = (3x^2 + 6y)\,\hat{i} - 14yz\,\hat{j} + 20xz^2\,\hat{k}$, evaluate $\int_C \vec{f}\,.\,dr$, where C is given by $x = t$, $y = t^2$, $z = t^3$ joining the points. (0, 0, 0) to (1, 1, 1). 3(4½)

(*b*) Verify Stoke's theorem for the function $\vec{f} = (x^2 + y^2)\hat{i} - 2xy\,\hat{j}$, taken around the rectangle bounded by $x = \pm\, a$, $y = 0$, $y = b$. 3(4½)

4. (*a*) If $\vec{f} = (2x^2 - 3z)\hat{i} - 2xy\,\hat{j} - 4x\hat{k}$, find $\iiint \nabla\,.\,\vec{f}\,dv$, where v is the region bounded by the co-ordinate planes and the plane $2x + 2y + z = 4$. 3(4½)

(b) Using Divergence theorem, prove that $\iiint_v \frac{1}{r^2}\, dv = \iint \frac{\vec{r}\cdot\vec{n}}{r^2}\, ds$, where $\vec{r}, \vec{n}$ have usual meanings. 3(4½)

UNIT III

5. (a) Trace the conic $9x^2 + 24xy + 16y^2 - 2x + 14y + 1 = 0$. 3(4½)
   (b) Find the equation of a sphere which passes through the circle $x^2 + y^2 + z^2 - 2x + 3y - 4z + 6 = 0$, $3x - 4y + 5z - 15 = 0$ and cuts orthogonally the sphere $x^2 + y^2 + z^2 + 2x + 4y - 6z + 11 = 0$. 3(4½)
6. (a) Find the equation of the right circular cylinder whose guiding circle is $x^2 + y^2 + z^2 = 9$, $x - y + z = 3$. 3(4½)
   (b) Find the equation of the right circular cone whose vertex is the origin, axis the z-axis and semi-vertical angle is α. 3(4½)
7. (a) Find the centres of the two spheres which touch the plane $x + 2y + 2z - 5 = 0$ at the point (1, 1, 1) and the sphere $x^2 + y^2 + z^2 + 2x + 4y + 6z - 11 = 0$. 3(4½)
   (b) Prove that the conics $x^2 - y^2 - 4x + 2y + 2 = 0$ and $x^2 + 3y^2 - 4x - 6y + 4 = 0$ are confocal. 3(4½)

UNIT IV

8. (a) Show that the plane $2x - 4y - z + 3 = 0$ touches the paraboloid $x^2 - 2y^2 = 3z$ and find the co-ordinates of the point of contact. 3(4½)
   (b) Find the condition that the line $\frac{x-\alpha}{l} = \frac{y-\beta}{m} = \frac{z-\gamma}{n}$ is a generator of the conicoid $ax^2 + by^2 + cz^2 = 1$. 3(4½)
9. (a) Find the equations of the tangent planes to $2x^2 - 6y^2 + 3z^2 = 5$ which pass through the line $x + 9y - 3z = 0$ $3x - 3y + 6z - 5$. 3(4½)
   (b) Show that two confocal paraboloids cut every where at right angle. 3(4½)
10. (a) Prove that the equation: $5x^2 + 5y^2 + 8z^2 + 8yz + 8zx - 2xy + 12x - 12y + 6 = 0$ represents a cylinder whose cross section is an ellipse of eccentricity $\frac{1}{\sqrt{2}}$ and find the equation of its axis. 3(4½)
    (b) Find the equations of the polar line of the line $\frac{x-x_1}{l} = \frac{y-y_1}{m} = \frac{z-z_1}{n}$ w.r.t. the conicoid $ax^2 + by^2 + cz^2 = 1$. 3(4½)